Methods in Cell Biology

VOLUME 102

Recent Advances in Cytometry,
Part A: Instrumentation, Methods

Series Editors

Leslie Wilson
Department of Molecular, Cellular and Developmental Biology
University of California
Santa Barbara, California

Paul Matsudaira
Department of Biological Sciences
National University of Singapore
Singapore

Methods in Cell Biology

VOLUME 102

Recent Advances in Cytometry,
Part A: Instrumentation, Methods

Edited by

Zbigniew Darzynkiewicz

Brander Cancer Research Institute, Department of Pathology,
New York Medical College, Valhalla, NY, USA

Elena Holden

CompuCyte Corporation, Westwood, MA, USA

Alberto Orfao

Cancer Research Center (CSIC/USAL),
University of Salamanca, Salamanca (Spain)

William Telford

National Cancer Institute, Bethesda, MD, USA

Donald Wlodkowic

The BioMEMS Research Group, Department of Chemistry,
University of Auckland, Auckland, New Zealand

AMSTERDAM • BOSTON • HEIDELBERG • LONDON
NEW YORK • OXFORD • PARIS • SAN DIEGO
SAN FRANCISCO • SINGAPORE • SYDNEY • TOKYO
Academic Press is an imprint of Elsevier

Academic Press is an imprint of Elsevier
225 Wyman Street, Waltham, MA 02451, USA
525 B Street, Suite 1900, San Diego, CA 92101-4495, USA
32, Jamestown Road, London NW1 7BY, UK
Linacre House, Jordan Hill, Oxford OX2 8DP, UK

Fifth edition 2011

ISBN: 978-0-12-374912-3
ISSN: 0091-679X

For information on all Academic Press publications
visit our website at elsevierdirect.com

Printed and bound in USA

11 12 13 14   10 9 8 7 6 5 4 3 2 1

Working together to grow
libraries in developing countries

www.elsevier.com | www.bookaid.org | www.sabre.org

ELSEVIER   BOOK AID International   Sabre Foundation

CONTENTS

IN MEMORIAM

I dedicate this book to the memory of my mentor Professor Kazimierz L. Ostrowski (1921–2010). He is remembered as a distinguished scientist with keen interest and eminent accomplishments in many fields of cell biology and medicine. As the Head of the Department of Histology and Embryology at the Medical University in Warsaw, Poland, he was a great educator and mentor of several generations of researchers and physicians. His passion and devotion to science as well as the virtues of integrity and ethics inspired dozens of his students who later became prominent researchers in Poland and abroad.

The evolutionary biologist Richard Dawkins coined the term "meme" (in the book: *The Selfish Gene*, 1976) that defines the creativity products of our brain, such as ideas or concepts, which propagate themselves in the meme pool by leaping from brain to brain, often through several generations. By means of lectures, publications, and collaborations the mental creativity of researchers is transmitted as ideas (memes) to students, colleagues, and collaborators. As a mentor of so many students who have become accomplished scientists, Professor Ostrowski was able to transmit a lot of his memes to their brains. These memes are alive and propagating. The realization of immortality through his memes makes his passing less sorrowful.

Zbigniew Darzynkiewicz

CONTRIBUTORS

Numbers in parentheses indicate the pages on which the author's contributions begin.

Yousef Al-Kofahi (161), CompuCyte Corporation, Westwood, Massachusetts, USA

Alison L. Allan (261), London Regional Cancer Program; London Health Sciences Centre, Lawson Health Research Institute; Departments of Anatomy & Cell Biology; Departments of Anatomy and Oncology, University of Western Ontario; London, Ontario, Canada

Anna C. Belkina (49), Cancer Center, Boston University Medical Center, Boston, Massachusetts, USA

Jacek Bielecki (411), Department of Applied Microbiology, Warsaw University, Warsaw, Poland

Matthew R. Chapman (127), Biophysics Graduate Group, University of California, Berkeley, California, USA

Pratip K. Chattopadhyay (463), Immuno Technology Section, Vaccine Research Center, NIAID, NIH, Bethesda, Maryland, USA

Man Ching Cheung (49), Departments of Biomedical Engineering/Electrical and Computer Engineering, Boston University, Boston, Massachusetts, USA

Jonathan M. Cooper (25), School of Engineering, University of Glasgow, UK

Zbigniew Darzynkiewicz (105), Brander Cancer Research Institute, Department of Pathology, NYMC, Valhalla, New York, USA

Gerald V. Denis (49), Cancer Center, Boston University Medical Center, Boston, Massachusetts, USA

Daniel J. Ehrlich (49), Departments of Biomedical Engineering/Electrical and Computer Engineering, Boston University, Boston, Massachusetts, USA

James G. Evans (49), Departments of Biomedical Engineering/Electrical and Computer Engineering, Boston University, Boston, Massachusetts, USA

Michael Fenech (321), CSIRO Food and Nutritional Sciences, Nutritional Genomics & Genome Health Diagnostics, Adelaide, SA, Australia

John Ferbas (309), Department of Clinical Immunology, Amgen, Inc., One Amgen Center Drive, Thousand Oaks, California, USA

Maxime François (321), CSIRO Food and Nutritional Sciences, Nutritional Genomics & Genome Health Diagnostics, Adelaide, SA; Edith Cowan University, Centre of Excellence for Alzheimer's Disease Research and Care, Joondalup, WA, Australia

Terry Gaige (77), CellASIC Corporation, Hayward, California, USA

David Goodale (261), London Regional Cancer Program, University of Western Ontario, London, Ontario, Canada

Ludomira Granicka (411), Institute of Biocybernetics and Biomedical Engineering PAS, Warsaw, Poland

Margaret M. Harnett (231), Institute of Infection, Immunity and Inflammation, College of Medical Veterinary & Life Sciences, Glasgow Biomedical Research Centre, University of Glasgow, Scotland, UK

Melvin Henriksen (161), CompuCyte Corporation, Westwood, Massachusetts, USA

Elena Holden (161, 321), CompuCyte Corporation, Westwood, Massachusetts, USA

Paul Hung (77), CellASIC Corporation, Hayward, California, USA

James W. Jacobberger (341), Case Comprehensive Cancer Center, Case Western Reserve University, Cleveland, Ohio, USA

Dayong Jin (479), Advanced Cytometry Labs, MQ Photonics Centre, Faculty of Science, Macquarie University, Sydney, Australia

Gloria Juan (309), Department of Clinical Immunology, Amgen, Inc., One Amgen Center Drive, Thousand Oaks, California, USA

Jerzy Kawiak (411), Department of Clinical Cytology, Medical Center Postgraduate Education; Institute of Biocybernetics and Biomedical Engineering PAS, Warsaw, Poland

Michael Keeney (261), Special Hematology/Flow Cytometry; London Health Sciences Centre, Lawson Health Research Institute, University of Western Ontario, London, Ontario, Canada

David L. Krull (291), GlaxoSmithKline, Safety Assessment, Investigative Pathology Laboratory, Research Triangle Park, North Carolina, USA

Philip Lee (77), CellASIC Corporation, Hayward, California, USA

Wayne R. Leifert (321), CSIRO Food and Nutritional Sciences, Nutritional Genomics & Genome Health Diagnostics, Adelaide, SA, Australia

Lori E. Lowes (261), London Regional Cancer Program; Departments of Anatomy & Cell Biology; Departments of Anatomy and Oncology, University of Western Ontario, London, Ontario, Canada

Ed Luther (321), Independent LSC Consultant, Wilmington, Massachusetts, USA

Marcin Lyżniak (411), Department of Clinical Cytology, Medical Center Postgraduate Education, Warsaw, Poland

Mairi A. McGrath (231), Institute of Infection, Immunity and Inflammation, College of Medical Veterinary & Life Sciences, Glasgow Biomedical Research Centre, University of Glasgow, Scotland, UK

Brian K. McKenna (49), Departments of Biomedical Engineering/Electrical and Computer Engineering, Boston University, Boston, Massachusetts, USA

Bruce Miller (161), CompuCyte Corporation, Westwood, Massachusetts, USA

Anja Mittag (1), Department of Pediatric Cardiology, Heart Centre; Translational Centre for Regenerative Medicine (TRM), University of Leipzig, Germany

Angela M. Morton (231), Institute of Infection, Immunity and Inflammation, College of Medical Veterinary & Life Sciences, Glasgow Biomedical Research Centre, University of Glasgow, Scotland, UK

Judith Newmark (161), CompuCyte Corporation, Westwood, Massachusetts, USA

John P. Nolan (515), La Jolla Bioengineering Institute, La Jolla; NanoComposix, Inc., San Diego, California, USA

Geoffrey W. Osborne (533), Queensland Brain Institute/Australian Institute for Bioengineering and Nanotechnology, The University of Queensland, Brisbane, Queensland, Australia

Richard A. Peterson (291), GlaxoSmithKline, Safety Assessment, Investigative Pathology Laboratory, Research Triangle Park, North Carolina, USA

Kiryl D. Piatkevich (431), Department of Anatomy and Structural Biology, and Gruss-Lipper Biophotonics Center, Albert Einstein College of Medicine, Bronx, New York, USA

Arkadiusz Pierzchalski (1), Department of Pediatric Cardiology, Heart Centre; Translational Centre for Regenerative Medicine (TRM), University of Leipzig, Germany

Mariusz Z. Ratajczak (207), Stem Cell Biology Institute, James Graham Brown Cancer Center, University of Louisville, Louisville, Kentucky, USA

David S. Sebba (515), La Jolla Bioengineering Institute, La Jolla; NanoComposix, Inc., San Diego, California, USA

David H. Sherr (49), Department of Environmental Health, Boston University School of Public Health, Boston, Massachusetts, USA

Lydia L. Sohn (127), Biophysics Graduate Group; Department of Mechanical Engineering, University of California, Berkeley, California, USA

Radoslaw Stachowiak (411), Department of Applied Microbiology, Warsaw University, Warsaw, Poland

Tammy Stefan (341), Case Comprehensive Cancer Center, Case Western Reserve University, Cleveland, Ohio, USA

Attila Tárnok (1), Department of Pediatric Cardiology, Heart Centre; Translational Centre for Regenerative Medicine (TRM), University of Leipzig, Germany

William G. Telford (375), Experimental Transplantation and Immunology Branch, National Cancer Institute, National Institutes of Health, Bethesda, Maryland, USA

Philip Thomas (321), CSIRO Food and Nutritional Sciences, Nutritional Genomics & Genome Health Diagnostics, Adelaide, SA, Australia

Vladislav V. Verkhusha (431), Department of Anatomy and Structural Biology, and Gruss-Lipper Biophotonics Center, Albert Einstein College of Medicine, Bronx, New York, USA

Donald Wlodkowic (105), The BioMEMS Research Group, Department of Chemistry, University of Auckland, Auckland, New Zealand

Michele Zagnoni (25), School of Engineering, University of Glasgow, UK

Stephen J. Zoog (309), Department of Clinical Immunology, Amgen, Inc., One Amgen Center Drive, Thousand Oaks, California, USA

Ewa K. Zuba-Surma (207), Department of Medical Biotechnology, Faculty of Biochemistry, Biophysics and Biotechnology, Jagiellonian University, Krakow, Poland

PREFACE TO FIFTH EDITION

Two hundred sixteen chapters presenting different cytometric methodologies and instrumentation consisting of six volumes (33, 41 & 42, 63 & 64, and 75) were published in the *four editions* (1990, 1994, 2001, and 2004) of the series of *Methods in Cell Biology* (MCB) dedicated to cytometry. The chapters presented the most widely used methods of flow- and quantitative image-cytometry, outlining their *principles, applications, advantages, alternative approaches, and potential pitfalls in their use.* These volumes received wide readership, high citation rates, and were valuable in promoting cytometric techniques across different fields of cell biology. Thirty-nine chapters from these volumes, selected based on high frequency of citations and relevance of methodology, were updated and recently published by Elsevier within the framework of the new series defined "Reliable Lab Solutions" as a special edition of the "Essential Cytometry Methods." Collectively, these volumes contain the *most inclusive assortment of articles on different cytometric methods and the associated instrumentation.*

The development in instrumentation and new methods as well as novel applications of cytometry continued at an accelerating pace since the last edition. This progress and the success of the earlier CYTOMETRY MCB editions, which become the proverbial "bible" for researchers utilizing these methods in a variety of fields of biology and medicine, prompted us to prepare the fifth edition. *The topics of all chapters in the present edition (Volumes A and B) are novel, covering the instrumentation, methods, and applications that were not included in the earlier editions.* The present volumes thus complement and not update the earlier editions.

There is an abundance of the methodology books presenting particular methods in a form of technical protocols such as "Current Protocols" by Wiley-Liss, "Practical Approach" series by Oxford Press, "Methods of Molecular Biology" series by Humana Press, and Springer or Nature Protocols. The commercially available reagent kits also provide protocols describing the use of these reagents. Because of the proprietary nature of some reagents the latter are often cryptic and do not inform about chemistry of the components or mechanistic principles of the kit. While the protocols provide the guidance to reproduce a particular assay their standard "cook-book" format is restrictive and does not allow one to explain in detail the principles of the methodology, discuss its limitations and possible pitfalls. Likewise the discussion on optimal choice of the assay for a particular task or cell system, or review of the method applications, is limited. Yet such knowledge is of importance for rational use of the methodology and for extraction of maximal relevant information from the experiment. *Compared to the protocol-format series the chapters in CYTOMETRY MCB volumes provide more comprehensive and often complementary to protocols description of particular methods.* The authors were invited to review and discuss the aspects of the methodology that cannot be included in the typical protocols, explain theoretical foundations of the methods, their applicability in experimental laboratory and clinical setting, outline common traps and pitfalls, discuss problems with data interpretation, and compare with alternative assays. While authors of some chapters did include specific protocols, a large number of chapters can be defined as *critical reviews of methodology and applications.*

The 35 chapters presented in CYTOMETRY Fifth Edition cover a wide range of diverse topics. Several chapters describe different approaches to *downsizing cytometry* instrumentation to the *microfluidic and lab-on-a-chip dimension.* Application of these miniaturized cytometric platforms in high-throughput analysis, as reported in these chapters, opens new possibilities in drug discovery studies. It also offers the means for real time, dynamic clinical assays that may be customized to individual patients, which could be a significant asset in targeted therapy. The microfluidic cytometry platforms are expected to play a major role in the era of the introduction of micro- and nanodimensional tools to modern biology and medicine, which we currently witness.

Imaging cytometry, by providing morphometric analytical capabilities, makes it possible to measure cellular attributes that cannot be assessed by flow cytometry. Different approaches and applications of imaging cytometry are addressed in several other chapters of this edition. Capturing intercellular interactions during the immune response *in situ,* quantifying, and imaging the blood-circulating tumor cells as well as measuring apoptosis in fine-needle biopsy aspirates are the chapters describing highly relevant *applications of imaging cytometry* with a potential for use in the *clinical setting.* Also of interest and of importance is the chapter addressing the *assessment of mutagenicity* by buccal micronucleus cytome assay. The use of imaging cytometry was also instrumental for dissecting consecutive *mitotic stages and states,* revealed by highly choreographed molecular and morphological changes, as presented in yet another chapter.

Further chapters describe advances in *development of flow cytometry instrumentation, new probes, and methods.* Among them are reviews on new lasers that are applicable to flow cytometry, applications of *quantum dots,* progress in development of *red fluorescent proteins* and biosensors, application of *lanthanide elements* to eliminate the autofluorescence background, *surface-enhanced Raman scattering cytometry* (SERC), and *recent advances in cell sorting.* The novel use of cytometry in analysis of bacteriological samples maintained on hollow fibers is also presented.

Reviews of new *applications of cytometry in cell biology* are presented in several other chapters. Two chapters of this genre are focused on the use of cytometry for *identification and isolation of stem cells.* Other chapters present the advances in use of cytometry in studies of *cell necrobiology,* in assessment of *oxidative DNA damage,* in *DNA damage response,* and in analysis of *cell senescence.*

Still another group of chapters present reviews on *preclinical and clinical applications of cytometry.* Of particular interest is the chapter addressing the use of cytometry in monitoring the intracellular signaling, which outlines the possibilities of assessing the effectiveness of the protein kinases-targeted therapies. The chapter describing advances in immunophenotyping of myeloid cell populations is very comprehensive, being illustrated by as many as 33 figures. Other chapters of interest for pathologists and clinicians describe the cytometry advances in *monitoring transplantation patients, progress in HLA antibody detection, in erythropoiesis and nonclonal red cell disorders, as well as in mast cells disorders.* The latter received recognition of the World Health Organization (WHO) as an example of the clinical utility of flow cytometry immunophenotyping in the diagnosis of mastocytosis.

Both volumes contain the introductory chapters from the laboratory of Dr. Attila Tarnok, the Editor-in-Chief of the Cytometry A, outlining in more general terms the advances in development in cytometry instrumentation, probes, and methods (Part A), as well as in applications of flow and image-assisted cytometry in different fields of biology and medicine (Part B).

In tradition with the earlier CYTOMETRY MCB editions, the chapters were prepared by the colleagues who either developed the described methods, contributed to their modification, or found new applications and have extensive experience in their use. The list of authors, thus, is a continuation of "Who's Who" directory in the field of cytometry. We are thankful to all contributing authors for the time they devoted to share their knowledge and experience.

Applications of cytometric methods have had a tremendous impact on research in various fields of cell and molecular biology, immunology, microbiology, and medicine. We hope that these volumes of MCB will be of help to many researchers who need these methods in their investigation, stimulate application of the methodology in new areas, and promote further progress in science.

Zbigniew Darzynkiewicz, Elena Holden,
Alberto Orfao, William G. Telford and
Donald Wlodkowic

Note to the readers:

For interpretation of the references to color in the figure legends, please refer to the web version of this book. Also, note that all the color figures will appear in color in online version.

CHAPTER 1

Introduction A: Recent Advances in Cytometry Instrumentation, Probes, and Methods—Review

Arkadiusz Pierzchalski,[*,†] **Anja Mittag**[*,†] **and Attila Tárnok**[*,†]

[*]Department of Pediatric Cardiology, Heart Centre, University of Leipzig, Germany

[†]Translational Centre for Regenerative Medicine (TRM), University of Leipzig, Germany

Abstract

Cytometric techniques are continually being improved, refined, and adapted to new applications. This chapter briefly outlines recent advances in the field of cytometry with the main focus on new instrumentations in flow and image cytometry as well as new probes suitable for multiparametric analyses. There is a remarkable

0091-679X/10 $35.00
DOI 10.1016/B978-0-12-374912-3.00001-8

trend for miniaturizing cytometers, developing label-free and fluorescence-free analytical approaches, and designing "intelligent" probes. Furthermore, new methods for analyzing complex data for extracting relevant information are reviewed.

I. Preface

Cytometry is the art and science of measuring phenotypical and functional characteristics of thousands to millions of cells in complex cell systems. Just a few decades ago, it became evident in cellular sciences that the scientific and diagnostic value of analyzing single-cell constituents that may be genes or gene products reached its limits. Cellular systems rely on a multitude of pathways reacting on external or internal stimuli and perturbations. This cognition gave rise to new disciplines in biomedical science with the "wholistic" approach of determining system-wide pattern alterations, termed "omics". The first omics approach was genomics soon followed by proteomics, cytomics, lipidomics, etc. Since the entire pattern of cell features changes in response to particular stimuli, the observation of the system in its totality (the "omics" approach), whether it is genome, proteome, etc., is closer to reality than the investigation of individual parameters alone.

Investigation of complex cell systems by the "bulk" techniques such as Western immunoblotting not allowing for the distinction between properties of their individual (cellular) members runs into the pitfall of overlaying specific signals of single highly relevant cells with that of an overbearing background (Szaniszlo *et al.*, 2006). Furthermore, the information on heterogeneity of cell populations, which is critical in many situations (e.g., to identify individual cells that are drug-resistant), is not available. This means that the system-wide determination also needs to recognize and analyze individual cells. Techniques that allow for obtaining information for cytomics or single-cell genomics and proteomics of hundreds to millions of individual cells would be advantageous. This perspective received particular attention by the progress in stem cell research, which opened new vistas to revolutionize in near future cellular therapy and regenerative medicine.

The potential of applications of stem cells in clinical medicine, in particular, distinctly exemplifies why there is a need for multiplexed and high-speed single-cell analysis. Each organ appears to have its own specialized stem cells type essential for its regeneration. However, these cells are extremely rare and can only be unequivocally identified by the characteristic expression pattern of a multitude of markers (Tárnok *et al.*, 2010b). Nowadays, stem cell characterization covers practically all possible progenitor cells from many tissues, for example, liver, cornea cells, hematopoietic cells, endothelial cells, very small embryonic stem cells, vascular progenitors from adipose tissue, and others (Adams *et al.*, 2009; Challen *et al.*, 2009; Möbius-Winkler *et al.*, 2009; Porretti *et al.*, 2010; Takács *et al.*, 2009; Zimmerlin *et al.*, 2010; Zuba-Surma *et al.*, 2008). Although presently not yet uniformly accepted in the whole scientific community, even tumors seem to have their own stem cells (Fábián *et al.*, 2009), which may evoke new therapeutic strategies for curing cancer.

Cytometry is the technology and science of choice for precisely identifying rare cells and describing the heterogeneity of cell populations in mixed systems. With all its different facets like flow cytometry (FCM), image cytometry, or chip-based technology, it quantitatively scrutinizes individual cells. This is based on binding of or reacting with a plethora of specific detecting molecules but is also realized by technologies that rely on physical properties such as electrical impedance or Raman light scattering. Although the foundations of cytometry date back to the mid-1960s, ongoing technological advances make a regular upgrade of the state-of-the-art technologies, new assays with all their advances, and consequently novel perspectives in cell analysis necessary.

Single-cell and multiplexed analyses are presently the shooting stars of biotechnology and they will alter our view on many mechanisms of biological processes, enforce completely innovative ways for diagnosis and treatment, and will improve the development of new drugs. This will be briefly outlined in the following and detailed in specific sections within this and the following chapters of this book.

II. Image Cytometry

Image cytometry, also termed slide-based cytometry or laser scanning cytometry (LSC) or image-assisted cytometry, is a high-content screening method. It is characterized by high reproducibility, capability of high-throughput analysis, and it can be standardized similar to FCM (Mittag and Tárnok, 2009). Image cytometry was used for many different applications and a wide range of biological, preclinical, and clinical materials (Gerstner *et al.*, 2009; Harnett, 2007; Pozarowski *et al.*, 2006; Rew *et al.*, 2006).

While FCM is unsurpassed in routine analysis of blood specimens, the analysis of solid tissue possesses unique challenges for which this technology is less suited. Most important in tissue analysis is to investigate cells in their spatial and topological context. Most often there is only limited amount of sample material available for the detailed functional and/or phenotypic analysis of specific cell subsets. In this context, image cytometry is a valuable tool for clinical analysis. It is feasible to perform diagnosis even from extremely small and/or hypocellular specimens such as body fluids and fine-needle aspiration biopsies (Gerstner *et al.*, 2002; Mocellin *et al.*, 2001, 2003; Pozarowski *et al.*, 2006). Cells or cell constituents of interest are generally tagged and identified by fluorescence labels. Measurement is comparable to FCM and fluorescence microscopy. This is making obtained data and its analysis familiar for users of these instruments. It is also possible to automatically image whole slides in multiple colors (Varga *et al.*, 2009). Also chromatically stained tissue, more familiar in pathology and immunohistochemistry (IHC), can be quantitatively analyzed by image cytometry. Advanced image analysis was also applied for automated classification of inflammation in histological sections (Ficsor *et al.*, 2008). LSC has been shown to be a reliable and efficient, relatively high-throughput, and high-content automated technology to quantify morphological endpoints in IHC labeled and nonfluorescent tissue samples (Peterson *et al.*, 2008).

A. Seeing Is Believing

Data analysis based on images allows for unambiguous identification of cells, cell aggregates, or biological constituents of interest based on morphology or fluorescence labeling. Data seem to be more reliable if one can verify results by eye as "a picture is worth a thousand dots" (Bisha and Brehm-Stecher, 2009). Morphometric image analysis allows for extracting a list of numerical parameters. Identified objects can be described in rates for shape, texture, size, intensity, etc.

It is possible to train classification algorithms to discriminate between cell phenotypes (Pepperkok and Ellenberg, 2006) with high accuracy. However, these algorithms are limited in recognizing new phenotypes. Suitable for that purpose are "intelligent" classification systems that automatically learn and define new classes with similar characteristics (Pepperkok and Ellenberg, 2006). It is a valuable tool in location proteomics, for quantitative classification of intracellular structures (Huh *et al.*, 2009; Newberg *et al.*, 2009; Shariff *et al.*, 2010). Also live cells can be imaged and monitored over time. Cell motility complicates direct retrieval of cell information from single captured images, but improved cell tracking algorithms allow for connecting objects in time, tracking of object splitting (cell division), or merging (cell fusion). Analysis of time-lapsed data sets provides information of individual cell cycle progression (Chen *et al.*, 2006), cell migration (Brown *et al.*, 2010; Degerman *et al.*, 2009), or cell motility behavior (Fotos *et al.*, 2006; Kamgoué *et al.*, 2009).

B. Image Cytometry Applications

Detection of apoptosis and cell proliferation by labeling DNA strand breaks was the first reported biological application of LSC (Li and Darzynkiewicz, 1995), demonstrating that simultaneously different information can be obtained by labeling intracellular DNA (nuclear and cytoplasmic DNA). Fluorescence labeling enables to determine DNA content, cell-cycle states, and cellular abnormalities. This represents the easiest way to identify abnormal, for example, tumor cells (Darzynkiewicz *et al.*, 2010; Tsujioka *et al.*, 2008; Zhao *et al.*, 2010b) and distinguish them from "normal" cells. Moreover, cell-cycle-specific markers highlight only cells in a certain development phase (Chakraborty and Tansey, 2009; Halicka *et al.*, 2005). Similarly, DNA condensation and chemical modification such as phosphorylation status of many proteins are also important parameters to study certain aspects of proliferation and death (Halicka *et al.*, 2005; Zhao *et al.*, 2008). Further examples of fluorescence-based LSC applications are spatial resolution of nuclear versus cytoplasmic fluorescence (Bedner *et al.*, 1998), cellular morphometry and cell-cycle analysis based on maximal pixel intensity (Haider *et al.*, 2003; Schwock *et al.*, 2005;

Pozarowski *et al.*, 2004), analysis of enzyme kinetics (Smolewski *et al.*, 2002), drug uptake (Rew *et al.*, 2006), ligand binding (Nagy and Szöllosi, 2009), evaluation of cytoplasmic/nuclear translocation (Peterson *et al.*, 2010; Usuku *et al.*, 2005), fluorescence *in-situ* hybridization (FISH) analysis (Ikemoto *et al.*, 2004; Smolewski *et al.*, 2001), and quantification of fluorescent IHC labeling in tissue sections (Peterson *et al.*, 2008).

Furthermore, LSC represents a powerful tool for qualitative and quantitative analysis of tissue sections in preclinical drug development (Peterson *et al.*, 2008). The high-throughput capability makes this instrument as well as other image cytometry systems suitable for single-cell analyses in drug-screening exercises (Esposito *et al.*, 2007; Galanzha *et al.*, 2007; Lövborg *et al.*, 2005). In drug discovery, high-throughput analyses are essential for excluding nonefficient or toxic and identify the (very rare) active agents (Tárnok *et al.*, 2010a). Therefore, a multitude of simple assays have to be run to test thousands of chemical compounds. Most often only one or two cellular parameters or functions are investigated at the same time. This may lead to neglect of potential drug candidates not able to induce the expected monitored biological effect but would pop-up with another more appropriate assay. The constructive approach, therefore, is to concurrently test for several cell functions (O'Brien *et al.*, 2006) using progressively more sensitive and specific probes (Tárnok *et al.*, 2010a).

III. New Instrumentations

A. Multiparametric Capabilities of Image Cytometry

In FCM, a multiparametric analysis has to rely on different labels, that is, different colors for different cellular properties, which have to be separated for unequivocal identification of the desired cell type or some functional aspect. There is a plethora of fluorescent dyes available, which are suitable for multicolor analysis, including "classical" and new organic dyes (Wessels *et al.*, 2010; Zhao *et al.*, 2009) with broad emission and low Stoke's shift as well as quantum dots that have a relatively narrow emission spectrum and higher Stokes' shift (Brown *et al.*, 2010; Mathur and Kelso, 2010; Smith and Giorgio, 2009). However, although up-to-date cytometers are capable of highly multiplexed multicolor analysis, limitations in hardware (excitation sources and detectors) and particularly spectral cross-talk between colors are often main hindrance in establishing multicolor panels in many laboratories. Only image cytometry is able to circumvent these limitations. As the same cells can be repeatedly analyzed, their restaining and sequential measurement enhance the depth of information manifold. With highly sophisticated techniques such as the MELC (multi-epitope-ligand cartography) technology, up to 100 different proteins have been investigated in (the identical) single cell enabling efficient target search for drug discovery (Schubert *et al.*, 2006).

Multiparametric analyses do not have to be multicolor. If the same cells can be interrogated a second time, different information can be obtained from the same fluorescence channel even if targets are labeled with the same color. The multiparameter single-cell analysis is of immense complexity but can be substantially simplified by the use of a single photobleachable fluorochrome (Mittag, 2008; Mittag *et al.*, 2006a, 2006b). Cell microwell arrays or regular microscope slide assays may be used for intracellular and surface antigen staining to a practically unlimited complexity (Hennig *et al.*, 2009; Tajiri *et al.*, 2009).

The emergence of powerful probes and dyes as well as fluorescence microscopy techniques, such as fluorescence recovery after photobleaching (FRAP) (Noda et al 2010, Mochizuki *et al* 2001), fluorescence resonance energy transfer (FRET) (Roszik *et al.*, 2009), total internal reflection fluorescence (TIRF) (Angres *et al.*, 2009; Weber *et al.*, 2006), fluorescence correlation spectroscopy (FCS) (Allen and Thompson, 2006; Gombos *et al.*, 2008), or fluorescence uncaging (Warther *et al.*, 2010), has made fluorescence microscopy an indispensable tool for cell biology. They particularly have opened opportunities for quantitative measurement of molecules *in vivo*.

Although most of the above technologies are presently still low-throughput, large efforts are being made to increase sample analysis speed for large-scale screening (Bruns *et al.*, 2009). For high-content and high-throughput cytometric analysis, new tools like automatic stations (robots) are being introduced, which are the part and parcel of modern and future cytometry development (Naumann and Wand, 2009).

B. The Merge of Systems

Basically, there are two different cytometry systems: flow- and microscope-based. Both have advantages and disadvantages. So, why not combining their virtues? Image cytometry and also FCM are capable of high-content analyses by multiplexed assays. The link between image cytometry and FCM represents the image stream cytometer (Zuba-Surma *et al.*, 2007; see also Chapter . . . in this issue). It combines conventional FCM with single-cell image acquisition and analysis. Thereby, the advantages of image analysis, mainly the fluorescence localization in the cell, are added to the high-throughput capability of cell suspension analysis of FCM for quantitative analysis of receptor internalization, phagocytosis, or nuclear translocation (Elliott, 2009). Imaging FCM incorporates certainly some very useful features of image analysis, but, nevertheless, continuous cell monitoring with high structural resolution can only be done with microscope-based imaging systems.

C. Modifications of the Well-Known – The Microcytometers

Tracking and understanding cell-to-cell variability is fundamental for systems biology, cytomics, and computational modeling. The rapid augmentation of instrument complexity allows an increased number of parameters to be analyzed simultaneously. Increasing velocity for multiparameter measurements is of key importance for time-efficient data acquisition and subsequent meaningful data analysis (Roederer, 2008). Reduction of sample volume for analysis leads to cost reduction of reagents and reduces the time needed for analysis (Zagnoni and Cooper, 2009).

Measurement at the bedside (point-of-care testing) is the goal of today's clinical diagnosis approaches. Limitations of conventional cell-based techniques, such as FCM and single-cell imaging, however, make the high-throughput dynamic analysis of cellular and subcellular processes tedious and exceedingly expensive. Hence, downsizing of high-tech instruments for their broad availability is the key goal of modern diagnostics. The concept of sample downsizing is realized by lab-on-a-chip, an approach which requires new developments of microchips including microfluidics, signal creation, and detection microdevices (Zagnoni and Cooper, 2009). The development of microfluidic lab-on-a-chips is one of the most innovative and cost-effective approaches toward integrated cytomics. These devices promise greatly reduced costs, increased sensitivity, and ultrahigh throughput by implementing parallel sample processing (Wlodkowic and Cooper, 2010).

It is largely anticipated that advances in microfluidic technologies should aid in tailoring investigational therapies and support the current computational efforts in systems biology. Microfluidics is an emerging technology with a multitude of applications in high-throughput drug-screening routines, high-content personalized clinical diagnostics, and diagnostics in resource-poor areas (Wlodkowic and Cooper, 2010). Chip-based devices enable precise cell phenotype identification. With such systems, it is possible to analyze a virtually unlimited number of intracellular and surface markers even on living immune cells (Hennig *et al.*, 2009).

D. Better – Easier – Affordable

FCM has become essential for CD4 cell count monitoring in HIV patients and leukemia diagnosis. Challenging are the relatively high instrument costs, which make FCM unaffordable for those regions of the world that need it most. One factor for high costs is the hydrodynamic focusing of cells in flow. The introduction of a novel flow cell that uses ultrasonic acoustic energy to focus small particles to the center of a flow stream has clearly increased sensitivity and speed of analysis (Goddard *et al.*, 2006). Such features offer the possibility of a truly versatile low-cost portable flow cytometer for field applications (Goddard *et al.*, 2007). An alternative method for particle positioning in FCM was presented recently (Swalwell *et al.*, 2009). Three position-sensitive photodetectors can be used to create a virtual core in the sample stream eliminating the need for sheath fluid. Furthermore, costs for preparation of blood samples should not be neglected and with no-lyse, no-wash flow-cytometric methods it is possible to significantly reduce costs per sample (Cassens *et al.*, 2004; Greve *et al.*, 2003).

Beside FCM, image cytometry with simplified optics, low-cost detectors, and data analysis tools may also lead to affordable cytometers and therewith appropriate diagnosis and health care in resource-limited countries (Shapiro and Perlmutter, 2006). An example for such an affordable HIV diagnostics device utilizes immobilized anti-CD4 antibodies, a CCD sensor, and an automatic cell-counting software (Moon *et al.*, 2009). Image cytometry as technique may even be more appropriate for affordable cytometers than FCM as it is normally of low-maintenance and easier to use.

E. Off the Beaten Track – Non–fluorescent Analyses

FCM at its beginning provided only information on unlabeled cells (before fluorescence dyes were developed and linked to antibodies). Nowadays it is almost forgotten that also "untouched" (label-free) cells can provide relevant information on cells' quality and condition. Label-free approaches have the main advantage that cells are less affected by sample preparation (mainly labeling procedures). Such assays may be important for preparative stem cell applications in cell therapy as medicinal products. Technologies on the horizon include impedance cytometry, Raman spectroscopy, near-infrared spectroscopy, multiple angles light scatter, and photoacoustic cytometry (Cheung *et al.*, 2005; Galanzha *et al.*, 2008; Lee *et al.*, 2006; Rajwa *et al.*, 2008; Rappaz *et al.*, 2008; Steiner *et al.*, 2008).

1. Electrical Impedance Cytometry

Flow system measurements of cell impedance properties have been performed for many decades (Coulter, 1956; Hoffman and Britt, 1979). In impedance measurement, the electric field in the detection volume is perturbed by each individual cell while the cells are passing through a capillary. This perturbation results in the creation of positive and negative signals, which are processed to provide the impedance (Cheung *et al.*, 2005). Also impedance-based cytometric systems exhibit the potential to become point-of-care blood analysis systems (Holmes *et al.*, 2009). Microfabricated impedance analysis devices offer high sensitivity combined with reduction in sample size. Impedance cytometry has been widely used to measure the dielectric properties of cells, determining membrane capacitance, membrane resistance, cytoplasmic conductivity, and permittivity (Cheung *et al.*, 2010; Holmes *et al.*, 2009; Holmes and Morgan, 2010).

Differential leukocyte identification based on dielectric properties of cells is one application of impedance cytometry (Holmes *et al.*, 2009). The dielectric properties of cells in impedance analyses are sensitive to stimuli arising from exposure to drug molecules and a variety of mitogens derived from bacterial and viral products. Hence, the technology may also find applications in cell-cycle analysis, apoptosis, and toxicity/viability assays. Impedance analysis may be further refined through the development of dielectric labels to identify cells with similar impedance properties (e.g., for determination of $CD4^+$ T-cell counts for HIV diagnostics). To this end, a new approach for impedance-based antibody identification was proposed by Holmes and Morgan (2010) using small antibodies conjugated to beads for $CD4^+$ cell identification and enumeration. Furthermore, DNA content can be estimated label-free based on the linear relationship between the DNA content of eukaryotic cells and the change in capacitance that is evoked by the passage of individual cells across a 1-kHz electric field (Sohn *et al.*, 2000). This technique is termed "capacitance cytometry."

Nowadays, it is possible to analyze dynamic mechanisms involving cells in real time and label free by microelectromechanical systems (BioMEMS) (Debuisson *et al.*, 2008). The concept of nanoscale devices has developed over the last decade with successful applications for monitoring cell-membrane conductivity, cell monolayer permeability, morphology, migration, and cellular micromotion. In addition to these efforts, some researchers have worked on the monitoring of cellular consequences of ligand–receptor interactions and ion channel activities (Debuisson *et al.*, 2008).

Another highly sensitive and label-free method for characterizing cells is aimed at cell-surface receptors and is called protein-functionalized pore. It measures cell retardation while the cell is passing a pore. The retardation of the cell is caused by interaction with a pore-coating protein and indicates the presence of a specific marker on the cell surface (Carbonaro *et al.*, 2008).

2. Raman Scatter Cytometry

There is an increasing interest in alternate, nonfluorescent probes since spectral overlap of various fluorochromes limits simultaneous measurement of multiple parameters. New methods for multiplex analysis are at the reach. One such alternative involves Raman-based probes (Goddard *et al.*, 2010). Intrinsic Raman scattering from molecules is orders of magnitude less intense than fluorescence from commonly used fluorochromes. Surface-enhanced Raman scattering provides a partial solution of this problem. Raman scattering can be enhanced by many Raman-active compounds in the presence of a metal surface such as gold or silver (Watson *et al.*, 2008). Raman vibrations based optical probes are inherently suitable for advanced multiplexed analysis. However, there remain significant challenges realizing Raman-based multiplexing in flow (Goddard *et al.*, 2010). Instruments have been developed for full Raman fingerprint region signal acquisition (Goddard *et al.*, 2010; Watson *et al.*, 2008). These instruments are modified in a way that the Raman spectrum from cells labeled with nanoparticles can be acquired and used as additional parameter (Watson *et al.*, 2008). Raman FCM opens up new possibilities for multiplexing using a simple optical configuration with a single detector and light source (Watson *et al.*, 2008) and can be applied even for whole organisms and large particles (Watson *et al.*, 2009).

3. Mass–Spectrometry Cytometry

With the advent of multimodular systems combining advantages of well-established modules, the capability of simultaneously measured parameters increased. The introduction of inductively coupled plasma mass spectrometry (ICP-MS) fulfills the expectations for nonambiguous antigen identification. If many different metal-isotope-tagged antibodies are used for simultaneous staining of antigens, complex immunophenotyping is possible (Ornatsky *et al.*, 2008). ICP-MS possesses several advantages that can enhance the performance of immunoassays. It exhibits high precision, low detection limits, and a large dynamic range, both for each antigen and between antigens. There are lower matrix effects from other components of the biological sample, that is, contaminating proteins in the sample have no effect on elemental analysis. Moreover, there is a lower background since plastic containers do not cause interference on elemental detection as they can with fluorescence. Another advantage is the absence of "unspecific" background, that is, there is no autofluorescence. Likewise, an analytical response from incubation or storage times is irrelevant as protein degradation does not affect analysis of an elemental tag. Problems with changing signal intensities such as bleaching of fluorochromes cannot be observed in ICP-MS. Furthermore, ICP-MS exhibits a large multiplexing capability (potentially up to 167 isotopes, realistically around 100 distinguishable tags) and there is a better spectral resolution (abundance sensitivity) (Ornatsky *et al.*, 2008). Since signals from element tags are essentially nonoverlapping, there is no need for compensation.

Recently, the introduction of flow system with MS detection unit (FL-MS) has brought the technology closer to common use (Ornatsky *et al.*, 2008). More than 20 antigens in the same sample have been successfully measured by FL-MS technology (Bandura *et al.*, 2009), and still there is a high potential to increase the amount of simultaneously measurable antigens (with different elemental tags) to 30–50, which allow for complex analysis of the cellular status. It is believed that the determination of the cellular status of patients suffering from different diseases will enable fast and accurate diagnosis and new therapy. It may even guarantee therapy success, as proposed by the cytomics approach used for individualized therapy (Tárnok *et al.*, 2010a). Also drug discovery will be much more effective once dozens of parameters are estimated on the single-cell level. Alternatively, the ability to highly multiplex cell authentication by image cytometry can be combined with the high molecular resolution of MS to detect specific cellular products in single cells as shown by Brown *et al.* (2010). This method combines single-cell capillary electrophoresis for quantitation and separation of analytes with MS for analyte identification.

IV. New Probes, Components, and Methods

Over the last decade, many improvements have been implemented to increase sensitivity, refine sorting, miniaturization, and many others. Cytometric techniques are being adapted to new applications and concepts such as cytomics. Complex multiparametric analyses are developed as well. New lasers (or even diodes nowadays) and filters are implemented or an assortment of different scatter angles – not to mention new fluorescence dyes, "intelligent" probes, or the increasing capabilities of software.

A. Let There Be Light

Appropriate laser selection for accurate dye excitation is crucial in multiparameter analysis. There is a bunch of lasers tailored for numerous applications. New developments like fiber optics technology, improved green lasers (550 nm) (Telford *et al.*,

2009a), or a super-continuum white light laser (Telford *et al.*, 2009b) practically extend the range of usable excitation wavelengths. The advantages of flexible laser selection are reduction in cellular autofluorescence and improvements in signal-to-noise ratio and detection sensitivity of fluorochromes. By selective filtering the wavelength range of interest of a white laser, almost any laser wavelength can be separated and used for cytometric analysis. This means, if almost any wavelength range can be made available for excitation, virtually any fluorescent probe can be analyzed (Telford *et al.*, 2009b).

B. More Colorful World

The portfolio of accessible dyes is still growing. With an appropriate combination of detecting molecules labeled with different colors as well as site-specific structural and functional targeting, it is possible to quantify different functional aspects of cellular response in a single experiment. Fluorescent tags such as the already mentioned quantum dots (Chattopadhyay *et al.*, 2006, 2007, 2010; Michalet *et al.*, 2005), a plethora of fluorescent proteins (Shaner *et al.*, 2005), and switchable molecular colors (PS-CFP, PA-GFP) (Ando *et al.*, 2004) are beneficial for imaging selectively labeled cells and their interaction *in vitro* and *in situ* with an excellent signal-to-noise ratio. If molecular targets are stained with a multitude of fluorescent molecules, single-cell-based analyses will be more specific and sensitive (Giuliano and Taylor, 1998).

Another group of dyes named NorthernLights has been introduced recently to the market. These dyes are excitable at different wavelengths, very stable, almost unbleachable, and importantly exhibit a very interesting feature: under red light excitation, the NorthernLight NL637 increase fluorescence intensity (excitation max) over excitation time (Wessels *et al.*, 2010). As this is in contrary to photobleaching, it can be combined with bleachable dyes. The combination of Alexa dyes (known to be stable, e.g., Alexa633), bleachable dyes (e.g., APC), and NL637 is suitable for triple differential fluorochrome identification in the red channel adding new parameters to hyperchromatic image cytometry (Mittag *et al.*, 2006b).

C. Revealing Cell Fates

The best way to investigate cellular behavior is to do that in their natural environment, that is, *in vivo*. However, a main challenge in fluorescence *in-vivo* imaging is tissue penetration and subsequent signal detection of fluorescent dyes. New solutions are now available for improving *in-vivo* single-cell signal detection for a wide range of applications comprising of red and far red emitting fluorescence proteins (Morozova *et al.*, 2010; Piatkevich *et al.*, 2010; Subach *et al.*, 2010, 2009). With the possibility to track and trace cells *in vivo*, not only information on biodistribution of administered cells (e.g., in stem cell therapy) can be obtained but also the investigation of the interaction of different cells is possible. Functional analysis with a specific metabolic insight has much developed, thanks to new enzyme-specific fluorogenic substrates. Together with extensive phenotyping, it enables precise estimation of the activity of cells *in vitro* or *in vivo* (Packard *et al.*, 2007; Packard and Komoriya, 2008; Telford *et al.*, 2002).

Development of fluorescent, organelle-targeted probes has been driven by discovering new dyes that excite and emit in the visible spectrum. These dyes possess specific subcellular localization features so that they can be used as organelle markers or physiological biosensors (Giuliano and Taylor, 1998; Merzlyak *et al.*, 2007; Subach *et al.*, 2010). One of the outstanding examples of fluorescent proteins was presented recently by the group of Allan Waggoner. They developed protein reporters that generate fluorescence from otherwise dark molecules (fluorogens) (Szent-Gyorgyi *et al.*, 2008). Eight unique fluorogen-activating proteins (FAPs) have been isolated by screening a library of human single-chain antibodies using derivatives of thiazole orange and malachite green. These FAPs bind fluorogens with nanomolar affinity, resulting in a thousand-fold increase in green or red fluorescence, up to brightness levels typically achieved by fluorescent proteins. Visualization of FAPs on the cell surface or within the secretory apparatus of mammalian cells can be achieved by membrane-permeant or impermeant fluorogens, respectively. This enables live cell imaging and the analysis of subcellular locations of interest as well as surface proteins (Holleran *et al.*, 2010).

Still another feature of fluorescent bioimaging probes is based on chemical address tags namely styryl compounds derivatives (Shedden and Rosania, 2010). Upon chemical modification, they tend to luminesce at different wavelength and provide therewith cell- and compartment-specific information. These probes seem to possess internal sensitivity for cellular states and cell types enabling accurate cell identification in heterogeneous cell populations (Shedden and Rosania, 2010). Yet more permeable probes are being introduced enabling control of RNA and DNA synthesis for life cell imaging. The approach is based on "click" chemistry, which relies on efficient nucleotide analog (EdU) incorporation in activated or proliferating cells, respectively, and then subsequent detection by a fluorescent azide (Zhao *et al.*, 2010a). The small size of azides allows the staining of whole-mount preparations of large tissues and organs (Jao and Salic, 2008; Salic and Mitchison, 2008).

V. New Strategies for Data Analysis

Multiparametric analyses produce a vast quantity of data. If the data are analyzed in terms of cytomics by a hypothesis-free approach (which is preferable to gain insights into heterogeneous systems over purely hypothesis driven approach), powerful data analysis software and algorithms are needed. Multicolor analysis leads to creation of huge databases. Multidimensional view of data allows to determine and understand cellular complexity, but it requires new tools for data analysis (Lugli *et al.*, 2010; Novo and Wood, 2008). Supervised or unsupervised data-mining algorithms allow for an effective analysis of multiparametric datasets (Pyne *et al.*, 2009). One step in this direction is the analysis of FCM data analogous to gene expression studies. This approach represents cytometric profiling and enables identification of significant parameters for classification of several groups (Steinbrich-Zöllner *et al.*, 2008). Clustering helps to arrange multidimensional datasets based on differences and similarities between analyzed objects (Lugli *et al.*, 2010; Steinbrich-Zöllner *et al.*, 2008; Zeng *et al.*, 2007). Application of cluster and principal component analysis to FCM data may promote the human cytome project (Kitsos *et al.*, 2007; Steinbrich-Zöllner *et al.*, 2008) and will lead to more efficient panel development and detection of suitable biomarkers for diagnosis and predictive medicine (Pierzchalski *et al.*, 2008).

The data need to be properly organized according to international standards and be comprehensible for a wider audience. To this end, much effort has been done by introducing improved cytometric data standards (FCS 3.1) (Spidlen *et al.*, 2010), gating descriptors (Spidlen *et al.*, 2008), and minimal experimental requirements for cytometric data publication called MIFlowCyt (Lee *et al.*, 2008). The latter has been for the first time implemented into a study for B-cell identification (Blimkie *et al.*, 2010). Growing multidimensionality requires new display tools, which have been proposed and are being used by many cytometry leaders (Appay *et al.*, 2008; Apweiler *et al.*, 2009; Pedreira *et al.*, 2008; Roederer and Moody, 2008; Steinbrich-Zöllner *et al.*, 2008). Such display tools are polychromatic plots and a "super" multicolor staining display for a virtually infinite number of colors. Further analysis tools are under development and of high importance for understanding and interpretation of complex multiparametric analyses. Automation in complex data analysis, that is, implementation of automatic processing tools, makes it easier to tease out the requested data from a vast amount of information collected (Jeffries *et al.*, 2008).

VI. Perspective

Cytometry is by nature a multidisciplinary field of science aimed at quantitative cell analysis. Over the last half century, cytometry has been maturing and is catching the attention of diverse scientific fields. Nowadays, instruments are capable for truly multiparametric analyses and the creation of very complex data. For the interpretation of these data and the understanding of the complexity of cell subsets and their interaction, new data analysis tools are mandatory. A few software tools for handling analysis of complex data have been released or are under development. Nevertheless, development of analysis tools for the illustration of multiparametric data sets and automatic or at least semiautomatic gating and analysis tools will be a trend in the upcoming years.

Unlike the progressive increase in complexity of cytometric analyses, the last years have also introduced simplification of instruments for the use in resource-poor areas. Approaches for instrument simplification are being introduced to the market (Cossarizza, 2010; Greve *et al.*, 2009). This goes hand-in-hand with the increasing demand for cheap, reliable instruments in HIV high-incidence areas for accurate diagnosis and therapy control. This progress is still going on and is hopefully making cytometric technologies available for those who desperately need it.

Another trend points toward label-free approaches for cell analyses. Presently available label-free technologies are regaining attention for on-site cellular sample quality control. Taking into account the pace of development, these technologies are expected to reach the market within next 5 years (Cheung *et al.*, 2010). Also multiparametric but non-fluorescent analyses (e.g., FL-MS) may gain importance as data interpretation should be easier without the bothersome spillover problems of fluorescence dyes.

There are not only developments and refinements in cytometric technologies and instrumentation but also the bunch of applications is steadily growing. More and more biomedical questions are addressed by cytometry, for example, in the field of nanotoxicology (Tárnok, 2010). Hence, the next years will provide a lot of new applications for FCM and image cytometry.

References

Adams, V., Challen, G. A., Zuba-Surma, E., Ulrich, H., Vereb, G., Tárnok, A. (2009). Where new approaches can stem from: focus on stem cell identification. *Cytometry A* **75**, 1–3.

Allen, N. W., and Thompson, N. L. (2006). Ligand binding by estrogen receptor beta attached to nanospheres measured by fluorescence correlation spectroscopy. *Cytometry A* **69**, 524–532.

Ando, R., Mizuno, H., and Miyawaki, A. (2004). Regulated fast nucleocytoplasmic shuttling observed by reversible protein highlighting. *Science* **306**, 1370–1373.

Angres, B., Steuer, H., Weber, P., Wagner, M., and Schneckenburger, H. (2009). A membrane-bound FRET-based caspase sensor for detection of apoptosis using fluorescence lifetime and total internal reflection microscopy. *Cytometry A* **75**, 420–427.

Appay, V., van Lier, R. A. W., Sallusto, F., and Roederer, M. (2008). Phenotype and function of human T lymphocyte subsets: consensus and issues. *Cytometry A* **73**, 975–983.

Apweiler, R., Aslanidis, C., Deufel, T., Gerstner, A., Hansen, J., Hochstrasser, D., Kellner, R., Kubicek, M., Lottspeich, F., Maser, E., *et al.* (2009). Approaching clinical proteomics: current state and future fields of application in cellular proteomics. *Cytometry A* **75**, 816–832.

Bandura, D. R., Baranov, V. I., Ornatsky, O. I., Antonov, A., Kinach, R., Lou, X., Pavlov, S., Vorobiev, S., Dick, J. E., Tanner, S. D. (2009). Mass cytometry: technique for real time single cell multitarget immunoassay based on inductively coupled plasma time-of-flight mass spectrometry. *Anal. Chem.* **81**, 6813–6822.

Bedner, E., Melamed, M. R., and Darzynkiewicz, Z. (1998). Enzyme kinetic reactions and fluorochrome uptake rates measured in individual cells by laser scanning cytometry. *Cytometry* **33**, 1–9.

Bisha, B., and Brehm-Stecher, B. F. (2009). Flow-through imaging cytomet2ry for characterization of Salmonella subpopulations in alfalfa sprouts, a complex food system. *Biotechnol. J.* **4**, 880–887.

Blimkie, D., Fortuno, E. S., Thommai, F., Xu, L., Fernandes, E., Crabtree, J., Rein-Weston, A., Jansen, K., Brinkman, R. R., Kollmann, T. R. (2010). Identification of B cells through negative gating – an example of the MIFlowCyt standard applied. *Cytometry A* **77**, 546–551.

Brown, M. R., Summers, H. D., Rees, P., Chappell, S. C., Silvestre, O. F., Khan, I. A., Smith, P. J., and Errington, R. J. (2010). Long-term time series analysis of quantum dot encoded cells by deconvolution of the autofluorescence signal. *Cytometry A* **77**, 925–932.

Bruns, T., Angres, B., Steuer, H., Weber, P., Wagner, M., Schneckenburger, H. (2009). Forster resonance energy transfer-based total internal reflection fluorescence reader for apoptosis. *J. Biomed. Opt.* **14**, 021003.

Carbonaro, A., Mohanty, S. K., Huang, H., Godley, L. A., and Sohn, L. L. (2008). Cell characterization using a protein-functionalized pore. *Lab. Chip* **8**, 1478–1485.

Cassens, U., Göhde, W., Kuling, G., Gröning, A., Schlenke, P., Lehman, L. G., Traoré, Y., Servais, J., Henin, Y., Reichelt, D., *et al.* (2004). Simplified volumetric flow cytometry allows feasible and accurate determination of CD4 T lymphocytes in immunodeficient patients worldwide. *Antivir. Ther. (Lond.)* **9**, 395–405.

Chakraborty, A. A., and Tansey, W. P. (2009). Inference of cell cycle-dependent proteolysis by laser scanning cytometry. *Exp. Cell Res.* **315**, 1772–1778.

Challen, G. A., Boles, N., Lin, K. K., and Goodell, M. A. (2009). Mouse hematopoietic stem cell identification and analysis. *Cytometry A* **75**, 14–24.

Chattopadhyay, P. K., Price, D. A., Harper, T. F., Betts, M. R., Yu, J., Gostick, E., Perfetto, S. P., Goepfert, P., Koup, R. A., De Rosa, S. C., *et al.* (2006). Quantum dot semiconductor nanocrystals for immuno-phenotyping by polychromatic flow cytometry. *Nat. Med.* **12**, 972–977.

Chattopadhyay, P. K., Yu, J., and Roederer, M. (2007). Application of quantum dots to multicolor flow cytometry. *Methods Mol. Biol.* **374**, 175–184.

Chattopadhyay, P. K., Perfetto, S. P., Yu, J., and Roederer, M. (2010). The use of quantum dot nanocrystals in multicolor flow cytometry. *Wiley Interdiscip. Rev. Nanomed. Nanobiotechnol.* **2**, 334–348.

Chen, X., Zhou, X., and Wong, S. T. C. (2006). Automated segmentation, classification, and tracking of cancer cell nuclei in time-lapse microscopy. *IEEE Trans. Biomed. Eng.* **53**, 762–766.

Cheung, K., Gawad, S., and Renaud, P. (2005). Impedance spectroscopy flow cytometry: on-chip label-free cell differentiation. *Cytometry A* **65**, 124–132.

Cheung, K. C., Di Berardino, M., Schade-Kampmann, G., Hebeisen, M., Pierzchalski, A., Bocsi, J., Mittag, A., Tárnok, A. (2010). Microfluidic impedance-based flow cytometry. *Cytometry A* **77**, 648–666.

Cossarizza, A. (2010). HIV infection in the era of cytomics. *Cytometry A* **77**, 721–724.

Coulter, W. H. (1956). High speed automatic blood cell counter and cell size analyzer. *Proc. Natl. Electron. Conf.* **12**, 1034–1040.

Darzynkiewicz, Z., Halicka, H. D., and Zhao, H. (2010). Analysis of cellular DNA content by flow and laser scanning cytometry. *Adv. Exp. Med. Biol.* **676**, 137–147.

Debuisson, D., Treizebré, A., Houssin, T., Leclerc, E., Bartès-Biesel, D., Legrand, D., Mazurier, J., Arscott, S., Bocquet, B., Senez, V. (2008). Nanoscale devices for online dielectric spectroscopy of biological cells. *Physiol. Meas.* **29**, S213–225.

Degerman, J., Thorlin, T., Faijerson, J., Althoff, K., Eriksson, P. S., Put, R. V. D., Gustavsson, T. (2009). An automatic system for *in vitro* cell migration studies. *J. Microsc.* **233**, 178–191.

Elliott, G. S. (2009). Moving pictures: imaging flow cytometry for drug development. *Comb. Chem. High Throughput Screen* **12**, 849–859.

Esposito, A., Dohm, C. P., Bähr, M., and Wouters, F. S. (2007). Unsupervised fluorescence lifetime imaging microscopy for high content and high throughput screening. *Mol. Cell. Proteomics* **6**, 1446–1454.

Fábián, A., Barok, M., Vereb, G., and Szöllosi, J. (2009). Die hard: are cancer stem cells the Bruce Willises of tumor biology? *Cytometry A* **75**, 67–74.

Ficsor, L., Varga, V. S., Tagscherer, A., Tulassay, Z., and Molnar, B. (2008). Automated classification of inflammation in colon histological sections based on digital microscopy and advanced image analysis. *Cytometry A* **73**, 230–237.

Fotos, J. S., Patel, V. P., Karin, N. J., Temburni, M. K., Koh, J. T., Galileo, D. S. (2006). Automated time-lapse microscopy and high-resolution tracking of cell migration. *Cytotechnology* **51**, 7–19.

Galanzha, E. I., Tuchin, V. V., and Zharov, V. P. (2007). Advances in small animal mesentery models for *in vivo* flow cytometry, dynamic microscopy, and drug screening. *World J. Gastroenterol.* **13**, 192–218.

Galanzha, E. I., Shashkov, E. V., Tuchin, V. V., and Zharov, V. P. (2008). In vivo multispectral, multiparameter, photoacoustic lymph flow cytometry with natural cell focusing, label-free detection and multicolor nanoparticle probes. *Cytometry A* **73**, 884–894.

Gerstner, A. O. H., Machlitt, J., Laffers, W., Tárnok, A., and Bootz, F. (2002). Analysis of minimal sample volumes from head and neck cancer by laser scanning cytometry. *Onkologie* **25**, 40–46.

Gerstner, A. O. H., Laffers, W., and Tárnok, A. (2009). Clinical applications of slide-based cytometry – an update. *J. Biophotonics* **2**, 463–469.

Giuliano, K. A., and Taylor, D. L. (1998). Fluorescent-protein biosensors: new tools for drug discovery. *Trends Biotechnol.* **16**, 135–140.

Goddard, G., Martin, J. C., Graves, S. W., and Kaduchak, G. (2006). Ultrasonic particle-concentration for sheathless focusing of particles for analysis in a flow cytometer. *Cytometry A* **69**, 66–74.

Goddard, G. R., Sanders, C. K., Martin, J. C., Kaduchak, G., and Graves, S. W. (2007). Analytical performance of an ultrasonic particle focusing flow cytometer. *Anal. Chem.* **79**, 8740–8746.

Goddard, G., Brown, L. O., Habbersett, R., Brady, C. I., Martin, J. C., Graves, S. W., Freyer, J. P., Doorn, S. K. (2010). High-resolution spectral analysis of individual SERS-active nanoparticles in flow. *J. Am. Chem. Soc.* **132**, 6081–6090.

Gombos, I., Steinbach, G., Pomozi, I., Balogh, A., Vámosi, G., Gansen, A., László, G., Garab, G., Matkó, J. (2008). Some new faces of membrane microdomains: a complex confocal fluorescence, differential polarization, and FCS imaging study on live immune cells. *Cytometry A* **73**, 220–229.

Greve, B., Cassens, U., Westerberg, C., Jun, W., Sibrowski, W., Reichelt, D., Gohde, W. (2003). A new no-lyse, no-wash flow-cytometric method for the determination of CD4 T cells in blood samples. *Transfus. Med. Hemother.* **30**, 8–13.

Greve, B., Weidner, J., Cassens, U., Odaibo, G., Olaleye, D., Sibrowski, W., Reichelt, D., Nasdala, I., Göhde, W. (2009). A new affordable flow cytometry based method to measure HIV-1 viral load. *Cytometry A* **75**, 199–206.

Haider, A. S., Grabarek, J., Eng, B., Pedraza, P., Ferreri, N. R., Balazs, E. A., Darzynkiewicz, Z. (2003). In vitro model of "wound healing" analyzed by laser scanning cytometry: accelerated healing of epithelial cell monolayers in the presence of hyaluronate. *Cytometry A* **53**, 1–8.

Halicka, H. D., Huang, X., Traganos, F., King, M. A., Dai, W., Darzynkiewicz, Z. (2005). Histone H2AX phosphorylation after cell irradiation with UV-B: relationship to cell cycle phase and induction of apoptosis. *Cell Cycle* **4**, 339–345.

Harnett, M. M. (2007). Laser scanning cytometry: understanding the immune system in situ. *Nat. Rev. Immunol.* **7**, 897–904.

Hennig, C., Adams, N., and Hansen, G. (2009). A versatile platform for comprehensive chip-based explorative cytometry. *Cytometry A* **75**, 362–370.

Hoffman, R. A., and Britt, W. B. (1979). Flow-system measurement of cell impedance properties. *J. Histochem. Cytochem.* **27**, 234–240.

Holleran, J., Brown, D., Fuhrman, M. H., Adler, S. A., Fisher, G. W., Jarvik, J. W. (2010). Fluorogen-activating proteins as biosensors of cell-surface proteins in living cells. *Cytometry A* **77**, 776–782.

Holmes, D., and Morgan, H. (2010). Single cell impedance cytometry for identification and counting of CD4 T-cells in human blood using impedance labels. *Anal. Chem.* **82**, 1455–1461.

Holmes, D., Pettigrew, D., Reccius, C. H., Gwyer, J. D., van Berkel, C., Holloway, J., Davies, D. E., Morgan, H. (2009). Leukocyte analysis and differentiation using high speed microfluidic single cell impedance cytometry. *Lab. Chip* **9**, 2881–2889.

Huh, S., Lee, D., and Murphy, R. F. (2009). Efficient framework for automated classification of subcellular patterns in budding yeast. *Cytometry A* **75**, 934–940.

Ikemoto, K., Furuya, T., Matsuda, K., Kawauchi, S., Oga, A., Ikeda, J., Liu, X. P., Sasaki, K. (2004). Multicolor FISH and cytometric analyses allow classification of urothelial carcinomas into two subtypes, low- and high-grade tumors. *Int. J. Oncol.* **25**, 893–898.

Jao, C. Y., and Salic, A. (2008). Exploring RNA transcription and turnover *in vivo* by using click chemistry. *Proc. Natl. Acad. Sci. U.S.A* **105**, 15779–15784.

Jeffries, D., Zaidi, I., de Jong, B., Holland, M. J., and Miles, D. J. C. (2008). Analysis of flow cytometry data using an automatic processing tool. *Cytometry A* **73**, 857–867.

Kamgoué, A., Ohayon, J., Usson, Y., Riou, L., and Tracqui, P. (2009). Quantification of cardiomyocyte contraction based on image correlation analysis. *Cytometry A* **75**, 298–308.

Kitsos, C. M., Bhamidipati, P., Melnikova, I., Cash, E. P., McNulty, C., Furman, J., Cima, M. J., Levinson, D. (2007). Combination of automated high throughput platforms, flow cytometry, and hierarchical clustering to detect cell state. *Cytometry A* **71**, 16–27.

Lee, E., Kidder, L. H., Kalasinsky, V. F., Schoppelrei, J. W., and Lewis, E. N. (2006). Forensic visualization of foreign matter in human tissue by near-infrared spectral imaging: methodology and data mining strategies. *Cytometry A* **69**, 888–896.

Lee, J. A., Spidlen, J., Boyce, K., Cai, J., Crosbie, N., Dalphin, M., Furlong, J., Gasparetto, M., Goldberg, M., Goralczyk, E. M., *et al.* (2008). MIFlowCyt: the minimum information about a flow cytometry experiment. *Cytometry A* **73**, 926–930.

Li, X., and Darzynkiewicz, Z. (1995). Labelling DNA strand breaks with BrdUTP. Detection of apoptosis and cell proliferation. *Cell. Prolif.* **28**, 571–579.

Lövborg, H., Gullbo, J., and Larsson, R. (2005). Screening for apoptosis – classical and emerging techniques. *Anticancer Drugs* **16**, 593–599.

Lugli, E., Roederer, M., and Cossarizza, A. (2010). Data analysis in flow cytometry: the future just started. *Cytometry A* **77**, 705–713.

Mathur, A., and Kelso, D. M. (2010). Multispectral image analysis of binary encoded microspheres for highly multiplexed suspension arrays. *Cytometry A* **77**, 356–365.

Merzlyak, E. M., Goedhart, J., Shcherbo, D., Bulina, M. E., Shcheglov, A. S., Fradkov, A. F., Gaintzeva, A., Lukyanov, K. A., Lukyanov, S., Gadella, T. W. J., *et al.* (2007). Bright monomeric red fluorescent protein with an extended fluorescence lifetime. *Nat. Methods* **4**, 555–557.

Michalet, X., Pinaud, F. F., Bentolila, L. A., Tsay, J. M., Doose, S., Li, J. J., Sundaresan, G., Wu, A. M., Gambhir, S. S., Weiss, S. (2005). Quantum dots for live cells, *in vivo* imaging, and diagnostics. *Science* **307**, 538–544.

Mittag, A. (2008). Merging of data files in laser scanning cytometry-seeing is believing? *Cytometry A* **73**, 880–883.

Mittag, A., Lenz, D., Bocsi, J., Sack, U., Gerstner, A. O. H., Tárnok, A. (2006a). Sequential photobleaching of fluorochromes for polychromatic slide-based cytometry. *Cytometry A* **69**, 139–141.

Mittag, A., Lenz, D., Gerstner, A. O. H., and Tárnok, A. (2006b). Hyperchromatic cytometry principles for cytomics using slide based cytometry. *Cytometry A* **69**, 691–703.

Mittag, A., and Tárnok, A. (2009). Basics of standardization and calibration in cytometry – a review. *J. Biophotonics* **2**, 470–481.

Möbius-Winkler, S., Höllriegel, R., Schuler, G., and Adams, V. (2009). Endothelial progenitor cells: implications for cardiovascular disease. *Cytometry A* **75**, 25–37.

Mocellin, S., Fetsch, P., Abati, A., Phan, G. Q., Wang, E., Provenzano, M., Stroncek, D., Rosenberg, S. A., Marincola, F. M. (2001). Laser scanning cytometry evaluation of MART-1, gp100, and HLA-A2 expression in melanoma metastases. *J. Immunother.* **24**, 447–458.

Mocellin, S., Wang, E., Panelli, M., Rossi, C. R., and Marincola, F. M. (2003). Use of laser scanning cytometry to study tumor microenvironment. *Histol. Histopathol.* **18**, 609–615.

Mochizuki, N., Yamashita, S., Kurokawa, K., Ohba, Y., Nagai, T., Miyawaki, A., Matsuda, M. (2001). Spatio-temporal images of growth-factor-induced activation of Ras and Rap1. *Nature* **411**, 1065–1068.

Moon, S., Keles, H. O., Ozcan, A., Khademhosseini, A., Haeggstrom, E., Kuritzkes, D., Demirci, U. (2009). Integrating microfluidics and lensless imaging for point-of-care testing. *Biosens. Bioelectron.* **24**, 3208–3214.

Morozova, K. S., Piatkevich, K. D., Gould, T. J., Zhang, J., Bewersdorf, J., Verkhusha, V. V. (2010). Far-red fluorescent protein excitable with red lasers for flow cytometry and superresolution STED nanoscopy. *Biophys. J.* **99**, L13–15.

Nagy, P., and Szöllosi, J. (2009). Proximity or no proximity: that is the question – but the answer is more complex. *Cytometry A* **75**, 813–815.

Naumann, U., and Wand, M. P. (2009). Automation in high-content flow cytometry screening. *Cytometry A* **75**, 789–797.

Newberg, J., Hua, J., and Murphy, R. F. (2009). Location proteomics: systematic determination of protein subcellular location. *Methods Mol. Biol.* **500**, 313–332.

Noda, K., Zhang, J., Fukuhara, S., Kunimoto, S., Yoshimura, M., Mochizuki, N. (2010). Vascular endothelial-cadherin stabilizes at cell-cell junctions by anchoring to circumferential actin bundles through alpha- and beta-catenins in cyclic AMP-Epac-Rap1 signal-activated endothelial cells. *Mol. Biol. Cell* **21**, 584–596.

Novo, D., and Wood, J. (2008). Flow cytometry histograms: transformations, resolution, and display. *Cytometry A* **73**, 685–692.

O'Brien, P. J., Irwin, W., Diaz, D., Howard-Cofield, E., Krejsa, C. M., Slaughter, M. R., Gao, B., Kaludercic, N., Angeline, A., Bernardi, P., *et al.* (2006). High concordance of drug-induced human hepatotoxicity with *in vitro* cytotoxicity measured in a novel cell-based model using high content screening. *Arch. Toxicol* **80**, 580–604.

Ornatsky, O. I., Lou, X., Nitz, M., Schäfer, S., Sheldrick, W. S., Baranov, V. I., Bandura, D. R., and Tanner, S. D. (2008). Study of cell antigens and intracellular DNA by identification of element-containing labels and metallointercalators using inductively coupled plasma mass spectrometry. *Anal. Chem.* **80**, 2539–2547.

Packard, B. Z., Telford, W. G., Komoriya, A., and Henkart, P. A. (2007). Granzyme B activity in target cells detects attack by cytotoxic lymphocytes. *J. Immunol.* **179**, 3812–3820.

Packard, B. Z., and Komoriya, A. (2008). Intracellular protease activation in apoptosis and cell-mediated cytotoxicity characterized by cell-permeable fluorogenic protease substrates. *Cell Res.* **18**, 238–247.

Pedreira, C. E., Costa, E. S., Barrena, S., Lecrevisse, Q., Almeida, J., van Dongen, J. J. M., Orfao, A. (2008). Generation of flow cytometry data files with a potentially infinite number of dimensions. *Cytometry A* **73**, 834–846.

Pepperkok, R., and Ellenberg, J. (2006). High-throughput fluorescence microscopy for systems biology. *Nat. Rev. Mol. Cell Biol.* **7**, 690–696.

Peterson, R. A., Krull, D. L., and Butler, L. (2008). Applications of laser scanning cytometry in immunohistochemistry and routine histopathology. *Toxicol. Pathol.* **36**, 117–132.

Peterson, A. W., Halter, M., Tona, A., Bhadriraju, K., and Plant, A. L. (2010). Using surface plasmon resonance imaging to probe dynamic interactions between cells and extracellular matrix. *Cytometry A* **77**, 895–903.

Piatkevich, K. D., Hulit, J., Subach, O. M., Wu, B., Abdulla, A., Segall, J. E., Verkhusha, V. V. (2010). Monomeric red fluorescent proteins with a large Stokes shift. *Proc. Natl. Acad. Sci. U.S.A* **107**, 5369–5374.

Pierzchalski, A., Robitzki, A., Mittag, A., Emmrich, F., Sack, U., O'Connor, J., Bocsi, J., Tárnok, A. (2008). Cytomics and nanobioengineering. *Cytometry B Clin. Cytom.* **74**, 416–426.

Porretti, L., Cattaneo, A., Colombo, F., Lopa, R., Rossi, G., Mazzaferro, V., Battiston, C., Svegliati-Baroni, G., Bertolini, F., Rebulla, P., *et al.* (2010). Simultaneous characterization of progenitor cell compartments in adult human liver. *Cytometry A* **77**, 31–40.

Pozarowski, P., Huang, X., Gong, R. W., Priebe, W., and Darzynkiewicz, Z. (2004). Simple, semiautomatic assay of cytostatic and cytotoxic effects of antitumor drugs by laser scanning cytometry: effects of the bis-intercalator WP631 on growth and cell cycle of T-24 cells. *Cytometry A* **57**, 113–119.

Pozarowski, P., Holden, E., and Darzynkiewicz, Z. (2006). Laser scanning cytometry: principles and applications. *Methods Mol. Biol.* **319**, 165–192.

Pyne, S., Hu, X., Wang, K., Rossin, E., Lin, T., Maier, L. M., Baecher-Allan, C., McLachlan, G. J., Tamayo, P., Hafler, D. A., *et al.* (2009). Automated high-dimensional flow cytometric data analysis. *Proc. Natl. Acad. Sci. U.S.A* **106**, 8519–8524.

Rajwa, B., Venkatapathi, M., Ragheb, K., Banada, P. P., Hirleman, E. D., Lary, T., Robinson, J. P. (2008). Automated classification of bacterial particles in flow by multiangle scatter measurement and support vector machine classifier. *Cytometry A* **73**, 369–379.

Rappaz, B., Barbul, A., Emery, Y., Korenstein, R., Depeursinge, C., Magistretti, P. J., Marquet, P. (2008). Comparative study of human erythrocytes by digital holographic microscopy, confocal microscopy, and impedance volume analyzer. *Cytometry A* **73**, 895–903.

Rew, D. A., Woltmann, G., and Kaur, D. (2006). Near-clinical applications of laser scanning cytometry. *Methods Mol. Biol.* **319**, 193–212.

Roederer, M. (2008). How many events is enough? Are you positive? *Cytometry A* **73**, 384–385.

Roederer, M., and Moody, M. A. (2008). Polychromatic plots: graphical display of multidimensional data. *Cytometry A* **73**, 868–874.

Roszik, J., Lisboa, D., Szöllosi, J., and Vereb, G. (2009). Evaluation of intensity-based ratiometric FRET in image cytometry – approaches and a software solution. *Cytometry A* **75**, 761–767.

Salic, A., and Mitchison, T. J. (2008). A chemical method for fast and sensitive detection of DNA synthesis *in vivo*. *Proc. Natl. Acad. Sci. U.S.A* **105**, 2415–2420.

Schubert, W., Bonnekoh, B., Pommer, A. J., Philipsen, L., Böckelmann, R., Malykh, Y., Gollnick, H., Friedenberger, M., Bode, M., Dress, A. W. M. (2006). Analyzing proteome topology and function by automated multidimensional fluorescence microscopy. *Nat. Biotechnol.* **24**, 1270–1278.

Schwock, J., Geddie, W. R., and Hedley, D. W. (2005). Analysis of hypoxia-inducible factor-1alpha accumulation and cell cycle in geldanamycin-treated human cervical carcinoma cells by laser scanning cytometry. *Cytometry A* **68**, 59–70.

Shaner, N. C., Steinbach, P. A., and Tsien, R. Y. (2005). A guide to choosing fluorescent proteins. *Nat. Methods* **2**, 905–909.

Shapiro, H. M., and Perlmutter, N. G. (2006). Personal cytometers: slow flow or no flow? *Cytometry A* **69**, 620–630.

Shariff, A., Murphy, R. F., and Rohde, G. K. (2010). A generative model of microtubule distributions, and indirect estimation of its parameters from fluorescence microscopy images. *Cytometry A* **77**, 457–466.

Shedden, K., and Rosania, G. R. (2010). Chemical address tags of fluorescent bioimaging probes. *Cytometry A* **77**, 429–438.

Smith, R. A., and Giorgio, T. D. (2009). Quantitative measurement of multifunctional quantum dot binding to cellular targets using flow cytometry. *Cytometry A* **75**, 465–474.

Smolewski, P., Ruan, Q., Vellon, L., and Darzynkiewicz, Z. (2001). Micronuclei assay by laser scanning cytometry. *Cytometry* **45**, 19–26.

Smolewski, P., Grabarek, J., Halicka, H. D., and Darzynkiewicz, Z. (2002). Assay of caspase activation in situ combined with probing plasma membrane integrity to detect three distinct stages of apoptosis. *J. Immunol. Methods* **265**, 111–121.

Sohn, L. L., Saleh, O. A., Facer, G. R., Beavis, A. J., Allan, R. S., Notterman, D. A. (2000). Capacitance cytometry: measuring biological cells one by one. *Proc. Natl. Acad. Sci. U.S.A* **97**, 10687–10690.

Spidlen, J., Leif, R. C., Moore, W., Roederer, M., and Brinkman, R. R. (2008). Gating-ML: XML-based gating descriptions in flow cytometry. *Cytometry A* **73A**, 1151–1157.

Spidlen, J., Moore, W., Parks, D., Goldberg, M., Bray, C., Bierre, P., Gorombey, P., Hyun, B., Hubbard, M., Lange, S., *et al.* (2010). Data file standard for flow cytometry, version FCS 3.1. *Cytometry A* **77**, 97–100.

Steinbrich-Zöllner, M., Grün, J. R., Kaiser, T., Biesen, R., Raba, K., Wu, P., Thiel, A., Rudwaleit, M., Sieper, J., Burmester, G., *et al.* (2008). From transcriptome to cytome: integrating cytometric profiling, multivariate cluster, and prediction analyses for a phenotypical classification of inflammatory diseases. *Cytometry A* **73**, 333–340.

Steiner, G., Küchler, S., Hermann, A., Koch, E., Salzer, R., Schackert, G., Kirsch, M. (2008). Rapid and label-free classification of human glioma cells by infrared spectroscopic imaging. *Cytometry A* **73A**, 1158–1164.

Subach, F. V., Subach, O. M., Gundorov, I. S., Morozova, K. S., Piatkevich, K. D., Cuervo, A. M., Verkhusha, V. V. (2009). Monomeric fluorescent timers that change color from blue to red report on cellular trafficking. *Nat. Chem. Biol.* **5**, 118–126.

Subach, F. V., Zhang, L., Gadella, T. W. J., Gurskaya, N. G., Lukyanov, K. A., Verkhusha, V. V. (2010). Red fluorescent protein with reversibly photoswitchable absorbance for photochromic FRET. *Chem. Biol* **17**, 745–755.

Swalwell, J. E., Petersen, T. W., and van den Engh, G. (2009). Virtual-core flow cytometry. *Cytometry A* **75**, 960–965.

Szaniszlo, P., Rose, W. A., Wang, N., Reece, L. M., Tsulaia, T. V., Hanania, E. G., Elferink, C. J., Leary, J. F. (2006). Scanning cytometry with a LEAP: laser-enabled analysis and processing of live cells in situ. *Cytometry A* **69**, 641–651.

Szent-Gyorgyi, C., Schmidt, B. F., Schmidt, B. A., Creeger, Y., Fisher, G. W., Zakel, K. L., Adler, S., Fitzpatrick, J. A. J., Woolford, C. A., Yan, Q., *et al.* (2008). Fluorogen-activating single-chain antibodies for imaging cell surface proteins. *Nat. Biotechnol.* **26**, 235–240.

Tajiri, K., Kishi, H., Ozawa, T., Sugiyama, T., and Muraguchi, A. (2009). SFMAC: a novel method for analyzing multiple parameters on lymphocytes with a single fluorophore in cell-microarray system. *Cytometry A* **75**, 282–288.

Takács, L., Tóth, E., Berta, A., and Vereb, G. (2009). Stem cells of the adult cornea: from cytometric markers to therapeutic applications. *Cytometry A* **75**, 54–66.

Tárnok, A. (2010). Cytometry and single cell analysis: 30 years of coevolution. *Cytometry A* **77**, 589–590.

Tárnok, A., Pierzchalski, A., and Valet, G. (2010a). Potential of a cytomics top-down strategy for drug discovery. *Curr. Med. Chem.* **17**, 1719–1729.

Tárnok, A., Ulrich, H., and Bocsi, J. (2010b). Phenotypes of stem cells from diverse origin. *Cytometry A* **77**, 6–10.

Telford, W. G., Komoriya, A., and Packard, B. Z. (2002). Detection of localized caspase activity in early apoptotic cells by laser scanning cytometry. *Cytometry* **47**, 81–88.

Telford, W. G., Babin, S. A., Khorev, S. V., and Rowe, S. H. (2009a). Green fiber lasers: an alternative to traditional DPSS green lasers for flow cytometry. *Cytometry A* **75**, 1031–1039.

Telford, W. G., Subach, F. V., and Verkhusha, V. V. (2009b). Supercontinuum white light lasers for flow cytometry. *Cytometry A* **75**, 450–459.

Tsujioka, T., Tochigi, A., Kishimoto, M., Kondo, T., Tasaka, T., Wada, H., Sugihara, T., Yoshida, Y., Tohyama, K. (2008). DNA ploidy and cell cycle analyses in the bone marrow cells of patients with megaloblastic anemia using laser scanning cytometry. *Cytometry B Clin. Cytom.* **74**, 104–109.

Usuku, T., Nishi, M., Morimoto, M., Brewer, J. A., Muglia, L. J., Sugimoto, T., Kawata, M. (2005). Visualization of glucocorticoid receptor in the brain of green fluorescent protein-glucocorticoid receptor knockin mice. *Neuroscience* **135**, 1119–1128.

Varga, V. S., Ficsor, L., Kamarás, V., Jónás, V., Virág, T., Tulassay, Z., and Molnár, B. (2009). Automated multichannel fluorescent whole slide imaging and its application for cytometry. Cytometry A **75**, 1020-1030.

Warther, D., Bolze, F., Léonard, J., Gug, S., Specht, A., Puliti, D., Sun, X., Kessler, P., Lutz, Y., Vonesch, J., *et al.* (2010). Live-cell one- and two-photon uncaging of a far-red emitting acridinone fluorophore. *J. Am. Chem. Soc.* **132**, 2585–2590.

Watson, D. A., Brown, L. O., Gaskill, D. F., Naivar, M., Graves, S. W., Doorn, S. K., Nolan, J. P. (2008). A flow cytometer for the measurement of Raman spectra. *Cytometry A* **73**, 119–128.

Watson, D. A., Gaskill, D. F., Brown, L. O., Doorn, S. K., and Nolan, J. P. (2009). Spectral measurements of large particles by flow cytometry. *Cytometry A* **75**, 460–464.

Weber, P., Wagner, M., and Schneckenburger, H. (2006). Microfluorometry of cell membrane dynamics. *Cytometry A* **69**, 185–188.

Wessels, J. T., Busse, A. C., Mahrt, J., Hoffschulte, B., Mueller, G. A., Tárnok, A., Mittag, A. (2010). NorthernLights in slide-based cytometry and microscopy. *Cytometry A* **77**, 420–428.

Wlodkowic, D., and Cooper, J. M. (2010). Microfabricated analytical systems for integrated cancer cytomics. *Anal. Bioanal. Chem.* **398**, 193–209.

Zagnoni, M., and Cooper, J. M. (2009). On-chip electrocoalescence of microdroplets as a function of voltage, frequency and droplet size. *Lab. Chip* **9**, 2652–2658.

Zeng, Q. T., Pratt, J. P., Pak, J., Ravnic, D., Huss, H., Mentzer, S. J. (2007). Feature-guided clustering of multi-dimensional flow cytometry datasets. *J. Biomed. Inform.* **40**, 325–331.

Zhao, H., Traganos, F., and Darzynkiewicz, Z. (2008). Kinetics of histone H2AX phosphorylation and Chk2 activation in A549 cells treated with topotecan and mitoxantrone in relation to the cell cycle phase. *Cytometry A* **73**, 480–489.

Zhao, H., Traganos, F., Dobrucki, J., Wlodkowic, D., and Darzynkiewicz, Z. (2009). Induction of DNA damage response by the supravital probes of nucleic acids. *Cytometry A* **75**, 510–519.

Zhao, H., Oczos, J., Janowski, P., Trembecka, D., Dobrucki, J., Darzynkiewicz, Z., Wlodkowic, D. (2010a). Rationale for the real-time and dynamic cell death assays using propidium iodide. *Cytometry A* **77**, 399–405.

Zhao, H., Traganos, F., and Darzynkiewicz, Z. (2010b). Kinetics of the UV-induced DNA damage response in relation to cell cycle phase. Correlation with DNA replication. *Cytometry A* **77**, 285–293.

Zimmerlin, L., Donnenberg, V. S., Pfeifer, M. E., Meyer, E. M., Péault, B., Rubin, J. P., Donnenberg, A. D. (2010). Stromal vascular progenitors in adult human adipose tissue. *Cytometry A* **77**, 22–30.

Zuba-Surma, E. K., Kucia, M., Abdel-Latif, A., Lillard, J. W., and Ratajczak, M. Z. (2007). The ImageStream System: a key step to a new era in imaging. *Folia Histochem. Cytobiol.* **45**, 279–290.

Zuba-Surma, E. K., Kucia, M., Wu, W., Klich, I., Lillard, J. W., Ratajczak, J., Ratajczak, M. Z. (2008). Very small embryonic-like stem cells are present in adult murine organs: ImageStream-based morphological analysis and distribution studies. *Cytometry A* **73A**, 1116–1127.

Down-sizing cytometry to "micro" dimension

CHAPTER 2

Droplet Microfluidics for High-throughput Analysis of Cells and Particles

Michele Zagnoni and Jonathan M. Cooper

Centre for Microsystems and Photonics, Dept. Electron. Electric. Eng., University of Strathclyde, Glasgow, G1 1XW, UK. Email:michele.zagnoni@eee.strath.ac.uk

Abstract

Droplet microfluidics (DM) is an area of research which combines lab-on-a-chip (LOC) techniques with emulsion compartmentalization to perform high-throughput, chemical and biological assays. The key issue of this approach lies in the generation, over tens of milliseconds, of thousands of liquid vessels which can be used either as a carrier, to transport encapsulated particles and cells, or as microreactors, to perform parallel analysis of a vast number of samples. Each compartment comprises a liquid droplet containing the sample, surrounded by an immiscible fluid. This microfluidic technique is capable of generating subnanoliter and highly monodispersed liquid droplets, which offer many opportunities for developing novel single-cell and single-molecule studies, as well as high-throughput methodologies for the detection and sorting of encapsulated species in droplets. The aim of this chapter is to give an

0091-679X/10 $35.00
DOI 10.1016/B978-0-12-374912-3.00002-X

overview of the features of DM in a broad microfluidic context, as well as to show the advantages and limitations of the technology in the field of LOC analytical research. Examples are reported and discussed to show how DM can provide novel systems with applications in high-throughput, quantitative cell and particle analysis.

I. Introduction

Over the last 20 years, the development of microfluidics has steadily increased toward the implementation of high-throughput analytical techniques at the microscale, providing novel lab-on-a-chip (LOC) systems to be used for biological and chemical applications (Atencia and Beebe, 2005; Mark *et al.*, 2010; Squires and Quake, 2005; Weibel *et al.*, 2007; Weibel and Whitesides, 2006). Proposed as an alternative to standard laboratory procedures, these systems are characterized by faster analysis time and reduced sample volumes, minimizing the need of expensive and rare biological reagents. Microfluidic systems typically operate at low Reynolds number regimes, which dictate that the fluid flow is laminar (Beebe *et al.*, 2002; Squires and Quake, 2005). This characteristic of the flow, combined with active LOC techniques, enables the transport of particles suspended in the fluids to be controlled with a high degree of precision within the microchannels.

In this particular context, a major distinction between two different systems has to be made which concerns the type of fluids used in microfluidic devices: these are continuous-flow systems and multiphase systems. In the first case, continuous-flow systems operate with fluids that can be mixed together by molecular diffusion, resulting in a homogeneous flow with a single velocity field. In the second case, multiphase systems (also known as segmented flow) are characterized by using two or more immiscible fluids (e.g., oil and water), where each of the phases is considered to have a separately defined volume fraction and a distinct velocity field. In this chapter, we will focus our attention onto high-throughput applications addressed by multiphase microfluidic systems for single-cell and particle analysis.

If we consider droplet technology, we can seek to define two distinct microfluidic approaches that have been developed in the last decade, known as "digital microfluidics" (DMF) and "droplet microfluidics" (DM) (Fair, 2007; Teh *et al.*, 2008). DMF concerns with the formation and transport of discrete liquid droplets (i.e., water-in-air droplets (W/A)) across the surface of an array of electrodes, where drops can be controlled individually by means of electromechanical actions exerted on the drops using electric fields. DM concerns, instead, with the formation and transport of micro- and nano-sized emulsions in diameter, mainly obtained by hydrodynamic means in microfluidic devices. The most elementary emulsion is a mixture of two immiscible fluids, comprising a liquid core suspended in a second immiscible liquid, as water-in-oil droplets (W/O) or in oil-in-water droplets (O/W).

The two approaches have been widely used to address both biological and chemical applications, generating small liquid volumes that can be transported, mixed, and analyzed within LOC devices. Unlike in continuous-flow systems, droplets serve as discrete microcompartments, in which biological and chemical reactions can be carried out without cross-contamination between different drops. The main difference between DMF and DM techniques resides in the timescale, size dispersion, and number of droplets that can be formed and handled in a microfluidic device. Whilst DMF enables droplets to be formed and individually manipulated in the 1–10 Hz range, DM is characterized by drop formation up to kHz frequencies. Using DM, highly monodisperse emulsions can be produced in the nanometer to micrometer diameter range, enabling parallel processing of reactions to be performed in shorter times and in higher numbers than with DMF. As an example of DM capability, the reader can consider that if W/O drops having a diameter of 50 μm are produced, each of which constitutes a reactor, approximately 40,000 reactions can be performed simultaneously in a few minutes within a microfluidic chamber having an area of 1 cm^2 and a depth of 50 μm. These characteristics offer greater potential for high-throughput and scalability than other microfluidic approaches. DM technology has been successfully used in a variety of microfluidic applications, both in chemistry and in biology, including the following: for janus particle, colloidosomes, microcapsules, and sol–gel bead formation; for enzymatic reactions, PCR and cell screening; for biomolecules synthesis, drug delivery, and diagnostic testing (Teh *et al.*, 2008; Theberge *et al.*, 2010).

In the following sections, we focus our attention only to those applications dedicated to the analysis of cells and particles achieved using droplet microfluidic technology. We first describe the properties needed by microfluidic architectures in order to controllably form, store, and handle on-chip microemulsions, providing a summary of: the most common device geometries utilized for drop formation. Subsequently, we provide a summary of the key issues in hydrodynamic theory for drop generation, droplet stabilization, particle encapsulation techniques, and droplet biocompatibility. After a brief introduction about the detection techniques used in DM, we present more in details relevant protocols used in cell- and particle-based analytical applications. Finally, we discuss specific advantages and limitations of DM approaches, together with perspectives and advances that may provide novel applications in the future. The reader will also be referred to the most relevant papers and reviews (both classic and recent) on droplet microfluidic methodologies.

II. Droplet Microfluidics

Microfluidic systems need to satisfy some critical requirements to address emulsion formation, storage and handling. The most important factors to be taken into account are the surface properties of the channel walls, the microfluidic geometry, and the effects produced by the microfluidic system onto the emulsions.

A. Microchannel Characteristics

When moving from single-phase systems to multiphase systems, surface tension phenomena, both between the phases and between each phase and a solid surface, must be taken into account, as these strongly influence the behavior of the liquids in the microfluidic channels. In LOC microenvironments, the high surface area to volume ratio enhances the interfacial effects, which typically become dominant over inertial and viscous effects in multiphase systems. Generally speaking, in order to stably obtain a dispersed phase in another one (i.e., an emulsion) within confined geometries (i.e., a microchannel), the inner phase must be completely surrounded by the outer phase. The implication is that the outer phase must be "more willing" to wet the solid surface of the channel wall than the inner one. In addition, the presence of surfactant molecules also alters the surface tension of the phases. Therefore, the hydrophobic and hydrophilic properties of a surface and the type of surfactants used will play a fundamental role in determining the orientation and the stability of the emulsions. Excellent reviews on emulsion stability in bulk and in microfluidic devices can be found in (Baroud *et al.*, 2010; Becher, 2001; Boyd *et al.*, 1972; Christopher and Anna, 2007; Gelbart *et al.*, 1994; Gunther and Jensen, 2006; Kabalnov and Weers, 1996; Leal-Calderon *et al.*, 2007).

The materials used in the fabrication of microfluidic devices include moldable elastomeric polymers (i.e., poly(dymethil)siloxane (PDMS)), hard polymers (i.e., poly(methyl methacrylate) (PMMA)), photocurable polymers, glass, and silicon (Becker and Locascio, 2002). These materials present very different surface wettabilities. Typically, in order to improve the emulsion stability and manipulation in LOC devices, surface treatments (such as silanization, oxygen plasma treatment, and film coating) are often required. These can be used either to change the hydrophobic/hydrophilic properties of the channel surfaces or to create specific hydrophobic/hydrophilic patterns within the microfluidic devices. Examples of such treatments can be found in (Abate *et al.*, 2008a, 2008b; Bauer *et al.*, 2010; Chae *et al.*, 2009; Darhuber and Troian, 2005; Lee *et al.*, 2005; Li *et al.*, 2007; Seo *et al.*, 2007).

B. Droplet Formation

One characteristic of DM is the generation of highly monodisperse emulsions in microchannels, achieving precise control over the drop size, shape, and composition in a high-throughput fashion (Gunther and Jensen, 2006; Teh *et al.*, 2008). The operation principle is based on passive microfluidic techniques that enable a droplet stream to be produced at a prescribed rate. By exploiting the flow field to deform the interface between two immiscible fluids, the dispersion of one phase into another is achieved. Three different microfluidic approaches have been developed for droplet generation: co-flowing streams, cross-flowing streams, and T-shaped junctions. However, T-junction and cross-flowing junctions are predominantly used in LOC devices, as illustrated in Figs. 1a and 1b, respectively.

The phase flow rates are usually controlled independently by syringe pumps and a local flow field, depending on the particular geometry and on the fluid properties, leads to droplet formation and transport in a microchannel. A review of the current understanding of the drop formation mechanisms occurring using these three geometries has been recently given in (Baroud *et al.*, 2010; Christopher and Anna, 2007).

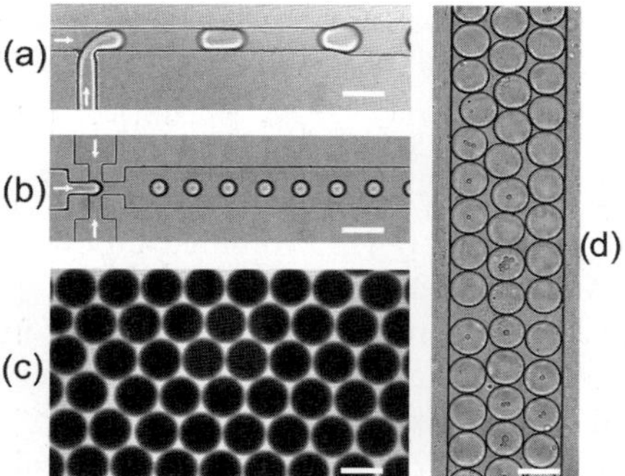

Fig. 1   Microemulsion formation and storage, obtained using droplet microfluidic techniques: (a) and (b) Examples of a T-junction and a cross-flowing junction geometry, respectively. Depending on the surface properties of the channels, emulsion orientation is controlled: W/O droplets are obtained when the phases are injected in hydrophobic channel walls and O/W droplets are obtained when the phases are injected in hydrophilic channel walls. (c) Examples of O/W droplet storage in hydrophilic microchannels, where fluorescein has been added only to the water phase. (d) Example of W/O droplet storage in hydrophobic microchannels. Cells have been encapsulated within the microemulsions. Scale bars are 100 μm.

Briefly, droplet formation involves pinch off at a junction, due to the competition between hydrodynamic pressure, viscous shear stresses, and the pressure arising due to interfacial tension between the phases. Whereas hydrodynamic forces tend to deform the interface between the phases, interfacial forces oppose the drop deformation produced by the flow field. In DM, the dimensionless capillary number Ca is typically used to compare the relative importance of viscous stresses with respect to interfacial tension phenomena. This is expressed as

$$Ca = \frac{\eta U}{\gamma} \tag{1}$$

where η is the larger dynamic viscosity in the system, U represents a characteristic velocity scale, and γ is the interfacial tension between the phases (with typical values of Ca ranging from 10^{-3} to 10^{1}). Low values of Ca indicate that the surface tension forces dominate over the viscous forces, leading to flowing droplets characterized by spherical ends (due to surface area minimization). High values of Ca indicate the opposite trend, leading to flowing droplets that can be easily deformed by the flow, characterized by asymmetric shapes. The influence of the capillary number on the behavior of multiphase flow has been described more in details in the literature (Bretherton, 1961; De Menech *et al.*, 2008; Stone and Leal, 1990; Zagnoni *et al.*, 2010a).

Apart from hydrodynamic conditions (arising due to the specific channel geometry and to physical properties of the fluid) and interfacial tension, droplet size can also be influenced by the ratio of the phase flow rates (Garstecki *et al.*, 2004, 2006; Nie *et al.*, 2008). Generally, when fluids are driven by constant volumetric flow rates and the volume fraction of the phases is fixed, the drop generation frequency f can be approximated as

$$f = \frac{Q_D}{w \cdot d \cdot R_D} \qquad (2)$$

where f is in Hz, Q_D the volumetric flow rate of the dispersed phase at the junction, R_D is the representative length of the droplet after formation, and w and d are the width and the depth of the channel at the junction, respectively. As an alternative to passive techniques, droplet generation has also been investigated using either integrated microvalves (Churski *et al.*, 2010; Galas *et al.*, 2009; Lin and Su, 2008; Oh and Ahn, 2006; Zeng *et al.*, 2009), optical means (Baroud *et al.*, 2007a), electrical means (He *et al.*, 2005b, 2006; Kim *et al.*, 2007; Link *et al.*, 2006) or microheaters (Baroud *et al.*, 2007b; Nguyen *et al.*, 2007; Ting *et al.*, 2006).

C. Particle Encapsulation

One of the main features offered by droplet microfluidic technology is that each drop can serve as a compartment within which individual reactions can be performed. The combination of speed, containment, and small drop volumes, from few femtoliter (10^{-15} l) to hundreds of picoliter (10^{-12} l), is highly valuable for the encapsulation in droplets of cells, organisms, beads, and other discrete reagents. This property has resulted in the development of a new class of microfluidic, high-throughput applications for the detection and analysis of particles, examples of which will be reported in the next section.

Typically, particle encapsulation in droplets has been accomplished by diluting a suspension of particles into the inner phase. This results in an encapsulation process that follows a Poisson statistics. The Poisson distribution for particle insertion into droplets is given by

$$p(M; n) = \frac{M^n e^{-M}}{n!} \qquad (3)$$

where n is the number of particles in a drop and M is the average number of particles per drop (M is usually adjusted by controlling the cell suspension concentration). Therefore, if single-particle encapsulation is required, the methods are inefficient, leading to a large number of empty drops with a much smaller number of drops containing a single particle (Clausell-Tormos *et al.*, 2008; Koster *et al.*, 2008).

This reason has led to the development of new techniques to improve the efficiency of particle encapsulation in microemulsions (Abate *et al.*, 2009a; Chabert and Viovy, 2008; Edd *et al.*, 2008; He *et al.*, 2005a). In particular, to guarantee high throughput, inertial ordering has been proposed as an efficient method to passively encapsulate particles within droplets (Edd *et al.*, 2008). This can be achieved under appropriate flow conditions and channel geometries, which lead to the generation of regular spacing between flowing particles prior to encapsulation, as shown in Fig. 2.

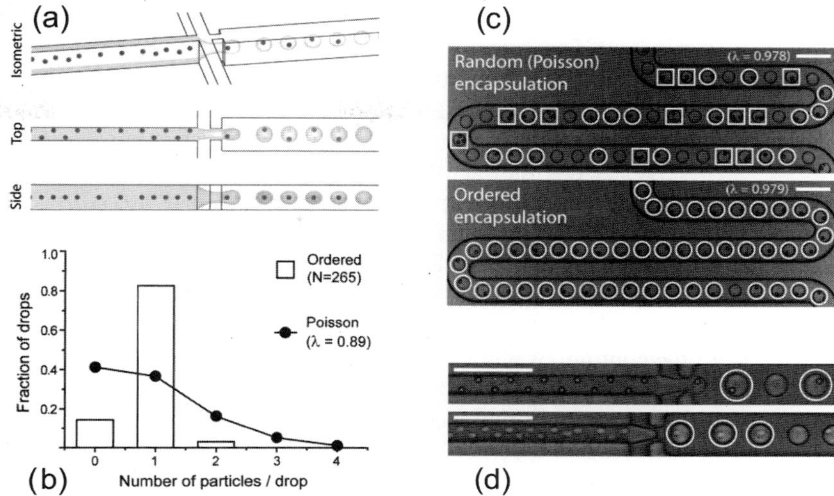

Fig. 2   Example of ordered cell and particle encapsulation in droplets, using a cross-flowing junction in a microfluidic device: (a) Schematic representation of hydrodynamic interactions that cause particles to self-organize along one side of the microchannel or into a diagonal/alternating pattern. (b) Comparison between particle encapsulation results obtained from inertial ordering and stochastic Poisson loading. (c) Ordered encapsulation of particles generates more single-particle drops (circles) and fewer empty (not marked) or multiple-particle drops (boxes) than stochastic Poisson loading. (d) Examples of self-organization during particle (top) and cell (bottom) encapsulation in droplets. Scale bars are 100 μm. Images reproduced with permission from (Edd *et al.*, 2008).

As a result, by matching the periodicity of the drop generation with that obtained for the particles, encapsulation efficiencies greater than 80% have been achieved. This technique, producing closely packed particles, offers advantages over Poisson statistics not only for its encapsulation efficiency, but also because particle periodicity can be controlled independently of drop formation, thus allowing controlled multiple particle encapsulation (Abate *et al.*, 2009a). However, a serious drawback of the inertial ordering technique is that undesired particle clogging in microchannel constrictions can also be obtained.

D. Biocompatibility and Emulsion Lifetime

To fully exploit the high-throughput characteristics of DM for particle- and cell-based applications, three important requirements must be fulfilled: 1) droplets must be stored either on- or off-chip for long period of times (i.e. from hours to days), retaining their initial character; 2) droplets must be resistant to coalescence; 3) the system must provide a biocompatible environment when encapsulating living particles (i.e. cells and organisms). Typically, due to the nature of the experiments when using cells and other organisms, W/O droplets are used and two factors play an important role in determining the above-mentioned conditions: the choice of surfactant molecules and the gas permeability of the material with which the device has been fabricated.

Surface active agents, or surfactants, are amphiphilic molecules whose function is to lower the interfacial tension between two immiscible phases. These compounds, containing both a water-soluble and an oil-soluble component, self-orient themselves at the interface between immiscible phases by energy minimization (Rosen and National Science Foundation (U.S.), 1987; Tadros, 1984). Surfactants are employed both to improve the stability of the emulsions, resistance to coalescence, and also to regulate the amount of small molecules that can diffuse through the surfactant layer between the inner and outer phase (Bai *et al.*, 2010). Therefore, the choice of surfactants in DM is of foremost importance and depends on the nature of the continuous phase (i.e., hydrocarbon or fluorocarbon oil) and on the experimental requirements. The nature of the hydrophilic head of the surfactant has an effect on the viability of the encapsulated cell and on the rate of absorption of encapsulated molecules at the interface of the emulsion. Recent reports have demonstrated that the choice of surfactants is essential to these aims, achieving emulsion stability up to 14 days on-chip, maintaining cell viability and enabling also the emulsions to be handled off- and on-chip, providing excellent condition of stability and biocompatibility (Clausell-Tormos *et al.*, 2008; Holtze *et al.*, 2008). For further and more detailed information on surfactant characteristics and effects in microfluidic devices, we point the reader to the literature (Baret *et al.*, 2009a; Kreutz *et al.*, 2009; Lee and Pozrikidis, 2006; Liu *et al.*, 2009; Roach *et al.*, 2005; Stone and Leal, 1990; Theberge *et al.*, 2010; Wang *et al.*, 2009b).

Finally, gas permeability (i.e., oxygen and carbon dioxide are required for cells to stay viable) is another important parameter to be considered when encapsulating living cells or organisms within emulsions. In this respect, both the continuous phase (i.e., fluorocarbon oils improve gas permeability compared to hydrocarbon oils) and the material with which the microfluidic device is fabricated (i.e., PDMS allows gas permeation through its porous structure whilst glass does not) are important as they must allow for the desired gas exchange between the inside and the outside of the device channels (Huebner *et al.*, 2009; Lee *et al.*, 2003; Shim *et al.*, 2007).

III. Detection Techniques and Methodologies in Droplet Microfluidics

Chemical and biological assays in DM systems are usually achieved by using a set of microfluidic library of operations (see Fig. 3) that can be sequentially implemented in a device to accomplish the desired droplet-based function (Mazutis *et al.*, 2009a). These include: formation (Abate *et al.*, 2009b, 2009c; Anna *et al.*, 2003; Baroud *et al.*, 2010; Bauer *et al.*, 2010; Christopher and Anna, 2007; Cramer *et al.*, 2004; Gupta *et al.*, 2009; Hsiung *et al.*, 2006; Lin *et al.*, 2008; Ota *et al.*, 2009; Stone, 1994; Tice *et al.*, 2003; Wang *et al.*, 2009c; Zhang and Stone, 1997; Zheng *et al.*, 2004), storage (Boukellal *et al.*, 2009; Clausell-Tormos *et al.*, 2008; Huebner *et al.*,

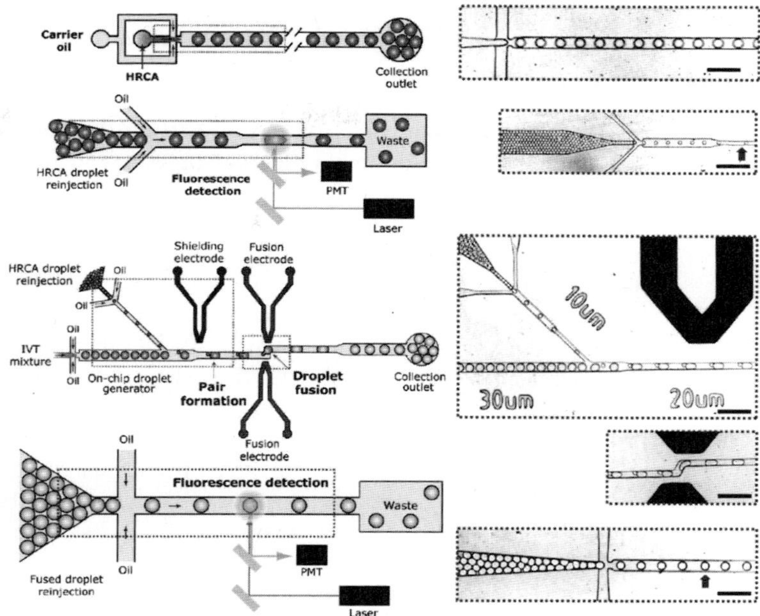

Fig. 3 Schematic representation and experimental results of droplet-based libraries implemented in microfluidic devices. Droplets can be produced and extracted from the microfluidic device for further processing and subsequently reinjected into the device. By combining droplet electrocoalescence with fluorescence spectroscopy, biological and chemical assays can be performed on a chip. Scale bars are 100 μm. Images reproduced with permission from (Mazutis *et al.*, 2009a).

2009; Koster *et al.*, 2008; Schmitz *et al.*, 2009; Trivedi *et al.*, 2010), splitting (Christopher *et al.*, 2009; Link *et al.*, 2004), sorting (Ahn *et al.*, 2006b; Baret *et al.*, 2009b; Chabert and Viovy, 2008; Niu *et al.*, 2007), passive (Fidalgo *et al.*, 2007; Hung *et al.*, 2006; Mazutis *et al.*, 2009b; Niu *et al.*, 2008; Tan *et al.*, 2007; Wang *et al.*, 2009d) and active drop coalescence (Ahn *et al.*, 2006a; Bremond *et al.*, 2008; Christopher *et al.*, 2009; Priest *et al.*, 2006; Zagnoni *et al.*, 2009, 2010b; Zagnoni and Cooper, 2009), trapping and strategic emulsion positioning (Bai *et al.*, 2010; Shi *et al.*, 2008; Tan and Takeuchi, 2007b; Wang *et al.*, 2009d; Zagnoni and Cooper, 2010). The engineering of these functionalities in a microfluidic device enables several parameters to be controlled passively. As examples, the residence time of a droplet inside a channel, as well as their velocity, packing, and position can be controlled by the accurate design of the channel geometries and by the choice of the physical parameters of the phases. In addition, active components can be implemented in the microfluidic architectures to further improve functionality. As examples, microheaters can be used to control the temperature of the phases (Yap *et al.*, 2009); electric fields can be employed either to coalesce droplets or to sort them in bifurcating channels (Ahn *et al.*, 2006b); both surface acoustic waves (Franke *et al.*,

2009, 2010), magnetic fields (Zhang *et al.*, 2009), lasers (Baroud *et al.*, 2007a, 2007b) and optical tweezers (Dixit *et al.*, 2010; He *et al.*, 2005a; Jeffries *et al.*, 2007) can be used to manipulate droplets.

Different approaches have been used to perform analysis of cells and particles in microdroplets. In one case, the species to be detected is maintained within the emulsion. In this condition, droplets can be stored on a chip, performing the analysis in static conditions.

As an example, a microfluidic platform suitable to store thousands of individual micron-sized droplets encapsulating cells has been reported to monitor β-galatosidase activity (Schmitz *et al.*, 2009). Alternatively, a detection procedure of encapsulated species, similar to fluorescent activated sorting systems (FACS), has been carried out, exploiting the full potential of high-throughput offered by DM technology. Baret *et al.* (2009b) have used electric fields to sort droplets based on the fluorescent readout obtained by enzymatic reactions from encapsulated bacteria in drops. In a further case, emulsions have been stably extracted from the device for additional analytical steps that require off-chip handling. Previously encapsulated cells in drops have been extracted and recultured to build single-cell statistics (Koster *et al.*, 2008). Finally, phase separation has also been induced to recover the encapsulated samples from the dispersed phase. This has been achieved by destabilizing emulsions containing cells or precipitates or by inline fusion of droplets with a phase streams. As an example, phase and contained particle separation has been shown by combining fluorescence intensity detection with selective emulsion fusion into a continuous aqueous stream using electric fields (Fidalgo *et al.*, 2008).

The ability to reliably integrate several functionalities in a microfluidic platform renders these systems valuable for use as powerful tools for biological and chemical research. Several detection techniques have been integrated and employed in LOC architectures, obtaining both parallel multidrop measurements and single-drop measurements for chemical and biological assays.

Notwithstanding brightfield microscopy, fluorescence is the most common and successful technique used to analyze the content of a droplet. However, a tradeoff in sensitivity arises due to the transient time of a drop under the excitation beam and the exposure time required for detection. Therefore, fluorescence microscopy has been mostly used for generating statistics and analysis for population studies in static conditions (Courtois *et al.*, 2009; Huebner *et al.*, 2009; Schmitz *et al.*, 2009) or when detecting processes characterized by slow kinetics (Damean *et al.*, 2009; Liau *et al.*, 2005). Alternatively, to enable high-throughput screening to be achieved using DM, laser-induced fluorescence spectroscopy has been utilized. This has been shown to provide higher sensitivity and shorter detection times. Examples of this technique have been reported for cell-based assays (Huebner *et al.*, 2007) and binding assays either using fluorescence lifetime imaging (FLIM) (Solvas *et al.*, 2010; Srisa-Art *et al.*, 2008a, 2009) or fluorescence energy transfer (Srisa-Art *et al.*, 2008b), resolving events at kHz frequencies.

Both Raman spectroscopy and surface-enhanced Raman spectroscopy (SERS) have also been proven successful in providing information on the detection of chemical structures and concentration of substances in droplets (Barnes *et al.*, 2006; Sarrazin *et al.*, 2008; Wang *et al.*, 2009a). Similarly to the case of fluorescence microscopy, averaging techniques for sample analysis has been required when using Raman spectroscopy, due to the same tradeoff between drop speed and acquisition time. In contrast, with SERS, the use of colloids amplifies the Raman signal, providing increased sensitivity and reduced time measurements from encapsulated species in drops.

Finally, another method has been employed for encapsulated species detection. This involves the implementation in LOC devices of electrochemical methodologies to obtain information about the physical and chemical properties of the phases. In this context, amperometric techniques have been integrated onto a chip to detect solute concentrations in droplets (Liu *et al.*, 2008) and to study enzyme kinetics (Han *et al.*, 2009).

IV. High–Throughput Cell and Particle Analysis in Droplet Microfluidics

The main advantage offered by DM for cell and particle analysis derives from the encapsulation properties of the technique and the characteristic monodispersity of the generated emulsions. In fact, not only encapsulated quantities can be transported within the drops accordingly to the geometry of the microchannels, but also the amount of substances enclosed within the drop can be controlled with a high degree of accuracy, providing reagent delivery with fL precision. These features are highly desirable especially for single-cell studies. In addition, the droplet content can also be adjusted for different substances and concentrations, after encapsulation, by fusing two or more droplets together. This choice of operations constitutes a unique way to build single-cell statistics and to develop new analytical tools in a controlled, cell-sized environment.

A. Cell–Based Analysis

Aqueous microcompartments obtained by DM techniques have been recently used as miniaturized vessels within which one can perform novel cell-based applications. These approach has been demonstrated using bacteria (Boedicker *et al.*, 2008, 2009; Koster *et al.*, 2009), yeast cells (Choi *et al.*, 2007; Luo *et al.*, 2006), mammalian cells (Clausell-Tormos *et al.*, 2008; Tan and Takeuchi, 2007a) and vermiform organisms, such as *Caenorhabditis elegans* (Clausell-Tormos *et al.*, 2008; Shi *et al.*, 2008). Examples of these applications are shown in Fig. 4.

All these reports have offered a good indication of the biocompatible nature of the on-chip emulsions, maintaining encapsulated cells and multicellular organisms viable within the drops for several days. Apart from cell growth and high-throughput viability tests of cells in drops, achieved in static conditions (Clausell-Tormos *et al.*,

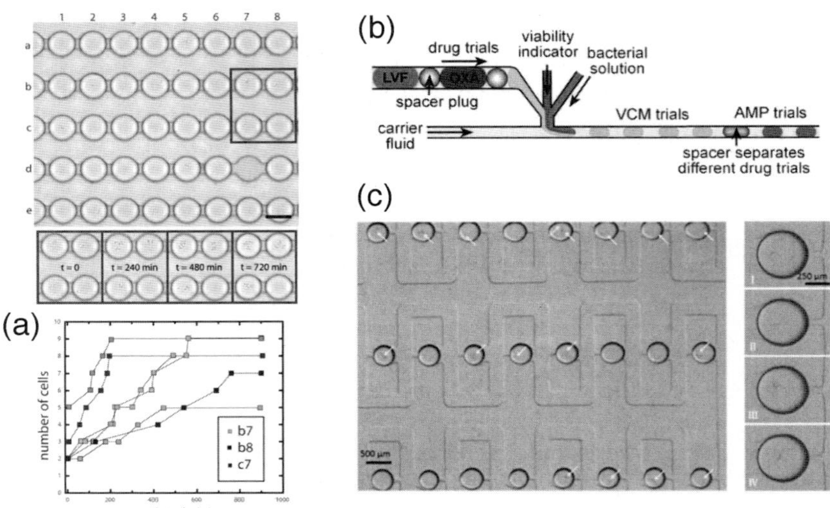

Fig. 4 (a) Microdroplets containing yeast cells are stored in an array of chambers in a microfluidic device to monitor growth rates of single cells (Top). Graph showing the number of cells grown in individual drops over 15 h incubation period (Bottom). Scale bar is 40 µm. Images reproduced with permission from (Schmitz *et al.*, 2009). (b) Schematic drawing illustrating the formation of oblong droplets containing bacteria, viability indicator, and antibiotic from a preformed array of drops of different antibiotics. Images reproduced with permission from (Boedicker *et al.*, 2008). (c) Image showing 24 array droplets encapsulated with worms mechanically trapped within the microchannels. The white arrows indicate the positions of *C. elegans* worms (Left). Representative images of the mobility shapes of a single worm in response to drugs (Right). Images reproduced with permission from (Shi *et al.*, 2008).

2008; Schmitz *et al.*, 2009), different analytical assays have also been developed, including, for example, the rapid laser photolysis of single cell in droplets (He *et al.*, 2005a). This procedure enables the cell lysate to be confined within the small volume of the droplet, providing analytical tools for detection of enzymatic activity at the single-cell level. Cell-based enzymatic assays are often used in cell biology for drug screening and droplet compartmentalization provides a highly valuable tool which enables reliable measurements of low substance concentrations to be performed avoiding diffusion of the product outside of the discrete drop volume.

Other enzymatic assays and particle analysis have also been reported, such as: the investigation of phosphatise activity produced by *Escherichia coli* cells, providing time-resolved kinetic measurements of wild type and mutant enzymes in picoliter droplets (Huebner *et al.*, 2008); the detection and analysis of human cell surface protein biomarkers using enzymatic amplification inside microdroplets (Joensson *et al.*, 2009). This last method has also provided parallel analysis of several cell samples by incorporating optical labels (i.e., quantum dots) within the droplets, combining higher optical sensitivity than standard FACS-like techniques together with drop-based high throughput. In a further example, a microfluidic enzymatic assay, using bacteria encapsulated in droplets, has been developed in static conditions by simultaneously measuring the fluorescent readout obtained by time-dependent protein expression and cellular enzymatic activity (Shim *et al.*, 2009).

Cell electroporation in droplets has also been demonstrated in a high-throughput manner (Luo *et al.*, 2006; Zhan *et al.*, 2009). This was achieved by flowing cell containing W/O droplets through a pair of microelectrodes to which a constant voltage was applied. By carefully selecting the electric potential applied to the electrodes and the drop velocity (obtained by setting opportune volumetric flow rates), encapsulated cells in droplets were exposed to the electric field for periods of a few milliseconds, whilst flowing past the electrodes. Cell electroporation has been demonstrated by delivering enhanced green fluorescent protein plasmid into Chinese hamster ovary cells, obtaining cell viability levels up to 80% after electroporation. This technique has not been used extensively, but has great potential to be implemented into DM architectures for high-throughput functional genomics studies.

Reducing droplet volumes also reduces the diffusion lengths within that volume which, when combined with particle encapsulation, provides faster and functional tools for analytical processes that depend on volumetric particle concentration. Examples of these have been given by (Boedicker *et al.*, 2008, 2009; Kim *et al.*, 2008), investigating the response to antibiotics from bacteria in human blood plasma. By confining single cells into microdroplets of nanoliter volumes, the detection time is dramatically reduced with respect to standard laboratory procedures. Confinement also increases cell density and allows released molecules to accumulate around the cell in shorter times, eliminating preincubation steps. These results have outlined the potential of DM to develop new and faster functional assays on different research areas, such as in the detection of contaminated food or water, in clinical diagnostics, and in monitoring industrial bioprocesses.

Such experiments have been carried out using both adherent and nonadherent cells within droplets, maintaining cells viability for up to 9 days and offering the option of recovering cells from drops for recultivation. However, some open questions remain and further investigation is required to clarify particular biological questions. For example, it has not yet been convincingly demonstrated that the response obtained from adherent cells in a droplet environment (thus in nonadherent conditions) is representative of the natural situation. Moreover, whilst the advantages of single organisms or blood cells studies in drops are unquestionable, cell studies using microdroplets that target applications other than for detection and sorting (i.e., FACS-like applications) still have to be further improved to extract biologically relevant information. One important aspect to be taken into account when developing cell-based procedures using DM is to reproduce the complex cell-to-cell interactions and environmental stimuli, to mimic the "real" biological environment. These are essential factors that must be addressed in the future to develop high-throughput cell-based assays in confined drop volumes to unravel important cellular signal pathways mechanisms.

A different approach to cell encapsulation in droplets that has been reported provides a possible solution to the nonadherable nature of a liquid emulsion. By forming biocompatible, hydrogel particles or capsules through on-chip gelation (Shah *et al.*, 2008, 2010), cell encapsulation conditions can be improved. For instance, this microfluidic procedure can be used for the generation of monodisperse spherical alginate beads (either using photocurable or chemically curable gels), otherwise not readily formed using conventional external gelation procedures. Gel microbeads offer a solid matrix that acts as a support for the encapsulated cell, providing an environment for growth and diffusion of fuels and metabolites. In addition, the gellification process facilitates the extraction of the beads from the oil phase, providing also the possibility to immerse the gellified drops in different aqueous solutions. Several examples of microfluidic techniques have been shown using embryonic carcinoma cells (Kim *et al.*, 2009), yeast cells (Choi *et al.*, 2007), and Jurkat cells in hydrogel alginate beads (Workman *et al.*, 2007, 2008).

B. Polymerase Chain Reaction, Particle Synthesis and Analysis

DM also offers outstanding potential for emulsion-based polymerase chain reaction (Williams *et al.*, 2006). This is a technique used to amplify single or few copies of DNA molecules, generating thousands to millions of copies of a particular DNA sequence. Performing this technique within miniaturized emulsions enables fast and high-throughput results to be obtained, preventing inactivation of polymerase and cross-contamination between samples. Examples of the techniques in DM have been used for quantification of rare events in large populations using encapsulated beads in drops to capture the amplified sequence for the detection of mutated cancer cells (Kumaresan *et al.*, 2008) and for high-throughput screening of transcription factor targets (Kojima *et al.*, 2005). Microfluidic approaches to continuous-flow PCR in W/O droplets of nanoliter volumes have also been reported by Schaerli *et al.* (2009), using a circular device design which allows droplets to pass through alternating temperature zones, completing tens of cycles of PCR in less than half an hour (Fig. 5a). The architecture allows the temperatures to be adjusted according to requirements, by measuring the temperature inside the droplets using FLIM. Results showed amplification from a single molecule of DNA per droplet.

Because of the compartmentalization features of DM, the technique can be used to develop useful tools in mimicking artificial cell environments. In this respect, protein transcription and translation processes can be performed *in vitro* within microdroplets, providing advantageous platform for evolutionary experiments. This microfluidic alternative to commercially available bacteria or cell-based techniques enables proteins that can be harmful and toxic for the host cell to be expressed without causing host death. Moreover, nonnatural molecules can also be artificially synthesized in these processes. DM has been recently used for *in vitro* high-throughput expression of GFP (Dittrich *et al.*, 2005) (see Fig. 5b) and for expression and detection of enzymes (Holtze *et al.*, 2008). A future challenge in this field will be to combine the high-throughput efficiency of microdroplet technology with the *in vitro*

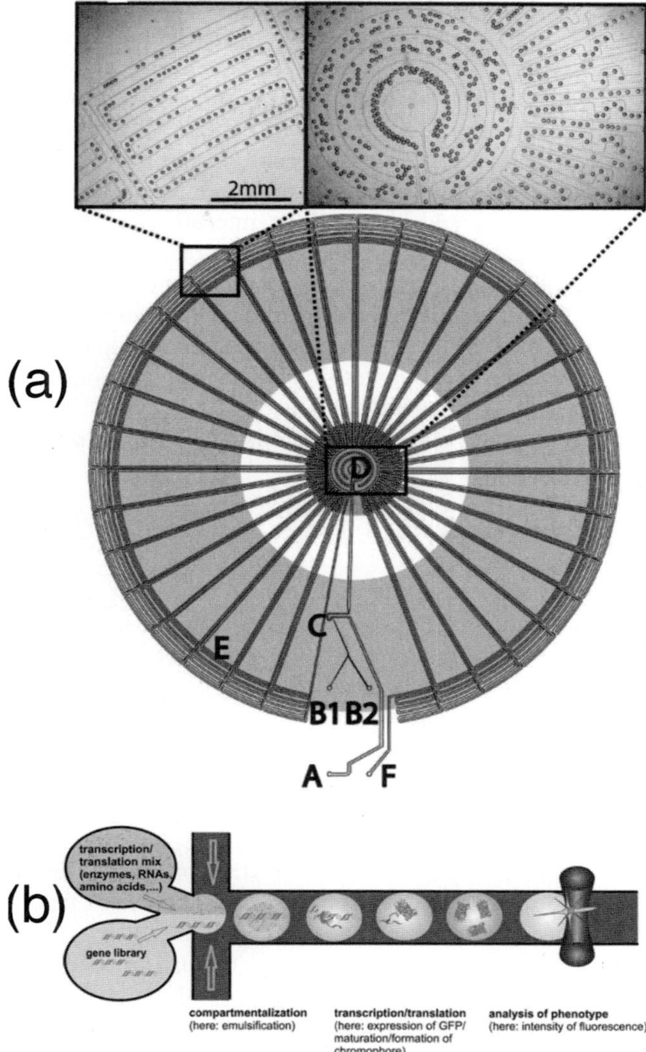

Fig. 5 (a) Images and design of a radial PCR microfluidic chip. The device contains an oil inlet (A) that joins two aqueous inlet channels (B1 and B2) to form droplets at a T-junction (C). The droplets pass through the inner circles in the hot zone (D) to ensure initial denaturation of the template and travel on to the periphery were primer annealing and template extension occur (E). The droplets then flow back to the center, where the DNA is denatured and a new cycle begins. The droplets exit the device after 34 cycles (F). Images reproduced with permission from (Schaerli *et al.*, 2009). (b) Schematic illustrating the principle of operation of *in vitro* evolution of proteins in microfluidic channels. During continuous formation of a W/O droplets, the encapsulated compounds for cell-free expression of proteins are mixed with templates from a gene library. *In vitro* expression takes place in biomimicking artificial cells during transport through the channel. Images reproduced with permission from (Dittrich *et al.*, 2005).

expression of transmembrane proteins, using the emulsion as a cellular chassis, toward the development of total artificial cells.

Finally, DM has also been shown to be a powerful platform for high-throughput synthesis and screening of micro- and nanoparticles. Multiple reactions can be performed in droplets by varying the reaction conditions, involving the controlled addition of reagents to a mixture, the mixing of reagents and the control of the reaction time. The main advantage in using DM, with respect to standard laboratory procedures and single-phase microfluidic approaches, is the ability to perform precipitate-forming reactions without clogging the microfluidic channels. Applications include the high-throughput studies of protein crystallization through different methods, such as gradient screening, protein diffusion, and X-ray diffraction; the synthesis of monodispersed nanoparticles; and the synthesis of organic molecules and synthesis of functional reaction networks. For detailed information regarding the methodologies used in DM for particle synthesis and screening, an excellent review has been published by Ismagilov and co-workers (Song *et al.*, 2006).

V. Perspectives

The area of research of DM has grown exponentially since what can be considered the first report in 2001 (Thorsen *et al.*, 2001), adding considerable value to emulsion-based science in bulk in terms of reproducibility, reliability, and high throughput. This progress has been driven by the constant development of new microfluidic techniques, ranging from the fabrication of new platforms and the improvement of surface treatments to the synthesis of new surfactants. Recent reports have proven DM to be highly suitable to address the requirements needed for both biological and chemical experiments, exploiting the compartmentalization and high-throughput characteristics of the technology.

DM has great potential to develop highly sensitive LOC tools to be used for laboratory-based analysis and diagnostics. In fact, as for many other microfluidic approaches, droplet-based microfluidics has not been demonstrated to be suitable for point-of-care applications and for use in industry. Other challenges involve the improvement of the capabilities of DM to provide new solutions for fluid actuation and for the fabrication of novel devices (i.e., architectures, materials and surface treatments).

Furthermore general challenges, which are common to many areas of microfluidics, concern the multidisciplinary approach needed to address biological problems from a technological point of view. More efforts must be engaged toward the development of LOC devices and procedures that use real biological samples, without requiring trained personal to actuate the architectures and without introducing artificial environments that are not representative of the natural cellular conditions. Generally speaking, the highly interdisciplinary nature required in LOC applications using DM (involving fluidic phenomena, electronic detection/control, chemistry/biochemistry, and biology) makes it attractive but also highlights the challenging nature of the field.

Finally, specific problems related solely to DM, concern the development of automatable control of the movement of thousands of droplets within a device. In this context, the use of clever engineered microfluidic geometries would provide enhanced functionalities to be obtained in droplet-based systems. This will enable not only automatic control and positioning of the droplets within the devices, but also of the encapsulated particles, as recently described by Bai *et al.* (2010), Stanley *et al.* (2010), and Zagnoni and Cooper (2010). In addition, these characteristics will also prove extremely valuable for the creation of interfaces between droplets allowing high-throughput analysis of membrane-based processes, toward the realization of artificial cell environments for drug screening.

VI. Conclusions

Droplet microfluidics provides novel and attractive procedures for high-throughput cell and particle analysis. Retaining all the well-known advantages offered by microfluidic techniques (i.e., reduced sample volumes and faster analysis times), DM provides means to form and control a large numbers of highly monodisperse and miniaturized compartments that can be used to perform thousands of reactions in parallel. This recent branch of microfluidics is increasingly attracting the attention of diverse groups of researchers due to the range of multidisciplinary applications that can be addressed, from physics and chemistry to biology and diagnostics. In the last 10 years, a set of droplet-based functions have been developed, including the control of droplet generation, droplet fission/fusion, mixing, and sorting. This allows a broad range of applications to be addressed using this technology. In chemistry, reactions will be controlled most precisely in droplets using smaller sample concentrations and allowing the study of kinetic conditions in reduced times. In biology, biomolecular and cellular events in cell-like environments will be reproduced most faithfully leading to real applications.

DM technology is certainly mature enough to be used to develop microsystems characterized by improved robustness and reproducibility, enabling new applications to be addressed, such as at the interface between biomedicine and engineering. As a result, recently, microfluidic companies have started to market droplet-based microfluidic products, primarily addressed at the scientific community and to biomedical and biopharmaceutical industries.

References

Abate, A. R., Chen, C. H., Agresti, J. J., and Weitz, D. A. (2009a). Beating Poisson encapsulation statistics using close-packed ordering. *Lab on a Chip* **9**, 2628–2631.

Abate, A. R., Krummel, A. T., Lee, D., Marquez, M., Holtze, C., Weitz, D. A. (2008a). Photoreactive coating for high-contrast spatial patterning of microfluidic device wettability. *Lab on a Chip* **8**, 2157–2160.

Abate, A. R., Lee, D., Do, T., Holtze, C., and Weitz, D. A. (2008b). Glass coating for PDMS microfluidic channels by sol–gel methods. *Lab on a Chip* **8**, 516–518.

Abate, A. R., Poitzsch, A., Hwang, Y., Lee, J., Czerwinska, J., Weitz, D. A. (2009b). Impact of inlet channel geometry on microfluidic drop formation. *Phys. Rev. E* 80.

Abate, A. R., Romanowsky, M. B., Agresti, J. J., and Weitz, D. A. (2009c). Valve-based flow focusing for drop formation. *Appl. Phys. Lett.* 94.

Ahn, K., Agresti, J., Chong, H., Marquez, M., and Weitz, D. A. (2006a). Electrocoalescence of drops synchronized by size-dependent flow in microfluidic channels. *Appl. Phys. Lett.* 88.

Ahn, K., Kerbage, C., Hunt, T. P., Westervelt, R. M., Link, D. R., Weitz, D. A. (2006b). Dielectrophoretic manipulation of drops for high-speed microfluidic sorting devices. *Appl. Phys. Lett.* 88.

Anna, S. L., Bontoux, N., and Stone, H. A. (2003). Formation of dispersions using "flow focusing" in microchannels. *Appl. Phys. Lett.* **82**, 364–366.

Atencia, J., and Beebe, D. J. (2005). Controlled microfluidic interfaces. *Nature* **437**, 648–655.

Bai, Y. P., He, X. M., Liu, D. S., Patil, S. N., Bratton, D., Huebner, A., Hollfelder, F., Abell, C., Huck, W. T. S. (2010). A double droplet trap system for studying mass transport across a droplet–droplet interface. *Lab on a Chip* **10**, 1281–1285.

Baret, J. C., Kleinschmidt, F., El Harrak, A., and Griffiths, A. D. (2009a). Kinetic aspects of emulsion stabilization by surfactants: a microfluidic analysis. *Langmuir* **25**, 6088–6093.

Baret, J. C., Miller, O. J., Taly, V., Ryckelynck, M., El-Harrak, A., Frenz, L., Rick, C., Samuels, M. L., Hutchison, J. B., Agresti, J. J., Link, D. R., Weitz, D. A., Griffiths, A. D. (2009b). Fluorescence-activated droplet sorting (FADS): efficient microfluidic cell sorting based on enzymatic activity. *Lab on a Chip* **9**, 1850–1858.

Barnes, S. E., Cygan, Z. T., Yates, J. K., Beers, K. L., and Amis, E. J. (2006). Raman spectroscopic monitoring of droplet polymerization in a microfluidic device. *Analyst* **131**, 1027–1033.

Baroud, C. N., de Saint Vincent, M. R., and Delville, J. P. (2007a). An optical toolbox for total control of droplet microfluidics. *Lab on a Chip* **7**, 1029–1033.

Baroud, C. N., Delville, J. P., Gallaire, F., and Wunenburger, R. (2007b). Thermocapillary valve for droplet production and sorting. *Phys. Rev. E* 75.

Baroud, C. N., Gallaire, F., and Dangla, R. (2010). Dynamics of microfluidic droplets. *Lab on a Chip* **10**, 2032–2045.

Bauer, W. A. C., Fischlechner, M., Abell, C., and Huck, W. T. S. (2010). Hydrophilic PDMS microchannels for high-throughput formation of oil-in-water microdroplets and water-in-oil-in-water double emulsions. *Lab on a Chip* **10**, 1814–1819.

Becher, P. (2001). *Emulsions: Theory and Practice.* Oxford University Press, American Chemical Society, New York, Oxford, Washington D.C.

Becker, H., and Locascio, L. E. (2002). Polymer microfluidic devices. *Talanta* **56**, 267–287.

Beebe, D. J., Mensing, G. A., and Walker, G. M. (2002). Physics and applications of microfluidics in biology. *Ann. Rev. Biomed. Eng.* **4**, 261–286.

Boedicker, J. Q., Li, L., Kline, T. R., and Ismagilov, R. F. (2008). Detecting bacteria and determining their susceptibility to antibiotics by stochastic confinement in nanoliter droplets using plug-based microfluidics. *Lab on a Chip* **8**, 1265–1272.

Boedicker, J. Q., Vincent, M. E., and Ismagilov, R. F. (2009). Microfluidic confinement of single cells of bacteria in small volumes initiates high-density behavior of quorum sensing and growth and reveals its variability. *Angew. Chem. – Int. Edition* **48**, 5908–5911.

Boukellal, H., Selimovic, S., Jia, Y. W., Cristobal, G., and Fraden, S. (2009). Simple, robust storage of drops and fluids in a microfluidic device. *Lab on a Chip* **9**, 331–338.

Boyd, J., Sherman, P., and Parkinso, C. (1972). Factors affecting emulsion stability, and Hlb concept. *J. Colloid Interface Sci.* **41**, 359.

Bremond, N., Thiam, A. R., and Bibette, J. (2008). Decompressing emulsion droplets favors coalescence. *Phys. Rev. Lett.* 100.

Bretherton, F. P. (1961). The motion of long bubbles in tubes. *J. Fluid Mech.* **10**, 166–188.

Chabert, M., and Viovy, J. L. (2008). Microfluidic high-throughput encapsulation and hydrodynamic self-sorting of single cells. *Proc. Nat. Acad. Sci. U.S.A.* **105**, 3191–3196.

Chae, S. K., Lee, C. H., Lee, S. H., Kim, T. S., and Kang, J. Y. (2009). Oil droplet generation in PDMS microchannel using an amphiphilic continuous phase. *Lab on a Chip* **9**, 1957–1961.

Choi, C. H., Jung, J. H., Rhee, Y. W., Kim, D. P., Shim, S. E., Lee, C. S. (2007). Generation of monodisperse alginate microbeads and *in situ* encapsulation of cell in microfluidic device. *Biomed. Microdevices* **9**, 855–862.

Christopher, G. F., and Anna, S. L. (2007). Microfluidic methods for generating continuous droplet streams. *J. Phys. D – Appl. Phys.* **40**, R319–R336.

Christopher, G. F., Bergstein, J., End, N. B., Poon, M., Nguyen, C., Anna, S. L. (2009). Coalescence and splitting of confined droplets at microfluidic junctions. *Lab on a Chip* **9**, 1102–1109.

Churski, K., Michalski, J., and Garstecki, P. (2010). Droplet on demand system utilizing a computer controlled microvalve integrated into a stiff polymeric microfluidic device. *Lab on a Chip* **10**, 512–518.

Clausell-Tormos, J., Lieber, D., Baret, J. C., El-Harrak, A., Miller, O. J., Frenz, L., Blouwolff, J., Humphry, K. J., Koster, S., Duan, H., Holtze, C., Weitz, D. A., Griffiths, A. D., Merten, C. A. (2008). Droplet-based microfluidic platforms for the encapsulation and screening of mammalian cells and multicellular organisms. *Chem. Biol.* **15**, 427–437.

Courtois, F., Olguin, L. F., Whyte, G., Theberge, A. B., Huck, W. T. S., Hollfelder, F., Abell, C. (2009). Controlling the retention of small molecules in emulsion microdroplets for use in cell-based assays. *Anal. Chem.* **81**, 3008–3016.

Cramer, C., Fischer, P., and Windhab, E. J. (2004). Drop formation in a co-flowing ambient fluid. *Chem. Eng. Sci.* **59**, 3045–3058.

Damean, N., Olguin, L. F., Hollfelder, F., Abell, C., and Huck, W. T. S. (2009). Simultaneous measurement of reactions in microdroplets filled by concentration gradients. *Lab on a Chip* **9**, 1707–1713.

Darhuber, A. A., and Troian, S. M. (2005). Principles of microfluidic actuation by modulation of surface stresses. *Ann. Rev. Fluid Mech.* **37**, 425–455.

De Menech, M., Garstecki, P., Jousse, F., and Stone, H. A. (2008). Transition from squeezing to dripping in a microfluidic T-shaped junction. *J. Fluid Mech.* **595**, 141–161.

Dittrich, P. S., Jahnz, M., and Schwille, P. (2005). A new embedded process for compartmentalized cell-free protein expression and on-line detection in microfluidic devices. *Chembiochem* **6**, 811-+.

Dixit, S. S., Kim, H., Vasilyev, A., Eid, A., and Faris, G. W. (2010). Light-driven formation and rupture of droplet bilayers. *Langmuir* **26**, 6193–6200.

Edd, J. F., Di Carlo, D., Humphry, K. J., Koster, S., Irimia, D., Weitz, D. A., Toner, M. (2008). Controlled encapsulation of single-cells into monodisperse picolitre drops. *Lab on a Chip* **8**, 1262–1264.

Fair, R. B. (2007). Digital microfluidics: is a true lab-on-a-chip possible? *Microfluid. Nanofluid.* **3**, 245–281.

Fidalgo, L. M., Abell, C., and Huck, W. T. S. (2007). Surface-induced droplet fusion in microfluidic devices. *Lab on a Chip* **7**, 984–986.

Fidalgo, L. M., Whyte, G., Bratton, D., Kaminski, C. F., Abell, C., Huck, W. T. S. (2008). From micro-droplets to microfluidics: selective emulsion separation in microfluidic devices. *Angew. Chem. – Int. Ed.* **47**, 2042–2045.

Franke, T., Abate, A. R., Weitz, D. A., and Wixforth, A. (2009). Surface acoustic wave (SAW) directed droplet flow in microfluidics for PDMS devices. *Lab on a Chip* **9**, 2625–2627.

Franke, T., Braunmuller, S., Schmid, L., Wixforth, A., and Weitz, D. A. (2010). Surface acoustic wave actuated cell sorting (SAWACS). *Lab on a Chip* **10**, 789–794.

Galas, J. C., Bartolo, D., and Studer, V. (2009). Active connectors for microfluidic drops on demand. *New J. Phys.* **11**.

Garstecki, P., Fuerstman, M. J., Stone, H. A., and Whitesides, G. M. (2006). Formation of droplets and bubbles in a microfluidic T-junction – scaling and mechanism of break-up. *Lab on a Chip* **6**, 437–446.

Garstecki, P., Gitlin, I., DiLuzio, W., Whitesides, G. M., Kumacheva, E., Stone, H. A. (2004). Formation of monodisperse bubbles in a microfluidic flow-focusing device. *Appl. Phys. Lett.* **85**, 2649–2651.

Gelbart, W. M., Ben-Shaul, A., and Roux, D. (1994). *Micelles, membranes, microemulsions, and mono-layers.* Springer, New York, London.

Gunther, A., and Jensen, K. F. (2006). Multiphase microfluidics: from flow characteristics to chemical and materials synthesis. *Lab on a Chip* **6**, 1487–1503.

Gupta, A., Murshed, S. M. S., and Kumar, R. (2009). Droplet formation and stability of flows in a microfluidic T-junction. *Appl. Phys. Lett.* 94.

Han, Z. Y., Li, W. T., Huang, Y. Y., and Zheng, B. (2009). Measuring rapid enzymatic kinetics by electrochemical method in droplet-based microfluidic devices with pneumatic valves. *Anal. Chem.* **81**, 5840–5845.

He, M., Kuo, J. S., and Chiu, D. T. (2006). Effects of ultrasmall orifices on the electrogeneration of femtoliter-volume aqueous droplets. *Langmuir* **22**, 6408–6413.

He, M. Y., Edgar, J. S., Jeffries, G. D. M., Lorenz, R. M., Shelby, J. P., Chiu, D. T. (2005a). Selective encapsulation of single cells and subcellular organelles into picoliter- and femtoliter-volume droplets. *Anal. Chem.* **77**, 1539–1544.

He, M. Y., Kuo, J. S., and Chiu, D. T. (2005b). Electro-generation of single femtoliter- and picoliter-volume aqueous droplets in microfluidic systems. *Appl. Phys. Lett.* 87.

Holtze, C., Rowat, A. C., Agresti, J. J., Hutchison, J. B., Angile, F. E., Schmitz, C. H. J., Koster, S., Duan, H., Humphry, K. J., Scanga, R. A., Johnson, J. S., Pisignano, D., Weitz, D. A. (2008). Biocompatible surfactants for water-in-fluorocarbon emulsions. *Lab on a Chip* **8**, 1632–1639.

Hsiung, S. K., Chen, C. T., and Lee, G. B. (2006). Micro-droplet formation utilizing microfluidic flow focusing and controllable moving-wall chopping techniques. *J. Micromech. Microeng.* **16**, 2403–2410.

Huebner, A., Bratton, D., Whyte, G., Yang, M., deMello, A. J., Abell, C., Hollfelder, F. (2009). Static microdroplet arrays: a microfluidic device for droplet trapping, incubation and release for enzymatic and cell-based assays. *Lab on a Chip* **9**, 692–698.

Huebner, A., Olguin, L. F., Bratton, D., Whyte, G., Huck, W. T. S., de Mello, A. J., Edel, J. B., Abell, C., Hollfelder, F. (2008). Development of quantitative cell-based enzyme assays in microdroplets. *Anal. Chem.* **80**, 3890–3896.

Huebner, A., Srisa-Art, M., Holt, D., Abell, C., Hollfelder, F., Demello, A. J., Edel, J. B. (2007). Quantitative detection of protein expression in single cells using droplet microfluidics. *Chem. Commun.* 1218–1220.

Hung, L. H., Choi, K. M., Tseng, W. Y., Tan, Y. C., Shea, K. J., Lee, A. P. (2006). Alternating droplet generation and controlled dynamic droplet fusion in microfluidic device for CdS nanoparticle synthesis. *Lab on a Chip* **6**, 174–178.

Jeffries, G. D. M., Kuo, J. S., and Chiu, D. T. (2007). Dynamic modulation of chemical concentration in an aqueous droplet. *Angew. Chem. – Int. Ed.* **46**, 1326–1328.

Joensson, H. N., Samuels, M. L., Brouzes, E. R., Medkova, M., Uhlen, M., Link, D. R., Andersson-Svahn, H. (2009). Detection and analysis of low-abundance cell-surface biomarkers using enzymatic amplification in microfluidic droplets. *Angew. Chem. – Int. Ed* **48**, 2518–2521.

Kabalnov, A., and Weers, J. (1996). Macroemulsion stability within the Winsor III region: theory versus experiment. *Langmuir* **12**, 1931–1935.

Kim, C., Lee, K. S., Kim, Y. E., Lee, K. J., Lee, S. H., Kim, T. S., Kang, J. Y. (2009). Rapid exchange of oil-phase in microencapsulation chip to enhance cell viability. *Lab on a Chip* **9**, 1294–1297.

Kim, H., Luo, D. W., Link, D., Weitz, D. A., Marquez, M., Cheng, Z. D. (2007). Controlled production of emulsion drops using an electric field in a flow-focusing microfluidic device. *Appl. Phys. Lett.* 91.

Kim, H. J., Boedicker, J. Q., Choi, J. W., and Ismagilov, R. F. (2008). Defined spatial structure stabilizes a synthetic multispecies bacterial community. *Proc. Nat. Acad. Sci. U.S.A.* **105**, 18188–18193.

Kojima, T., Takei, Y., Ohtsuka, M., Kawarasaki, Y., Yamane, T., Nakano, H. (2005). PCR amplification from single DNA molecules on magnetic beads in emulsion: application for high-throughput screening of transcription factor targets. *Nucleic Acids Res* 33.

Koster, S., Angile, F. E., Duan, H., Agresti, J. J., Wintner, A., Schmitz, C., Rowat, A. C., Merten, C. A., Pisignano, D., Griffiths, A. D., Weitz, D. A. (2008). Drop-based microfluidic devices for encapsulation of single cells. *Lab on a Chip* **8**, 1110–1115.

Koster, S., Evilevitch, A., Jeembaeva, M., and Weitz, D. A. (2009). Influence of internal capsid pressure on viral infection by phage lambda. *Biophys. J.* **97**, 1525–1529.

Kreutz, J. E., Li, L., Roach, L. S., Hatakeyama, T., and Ismagilov, R. F. (2009). Laterally mobile, functionalized self-assembled monolayers at the fluorous-aqueous interface in a plug-based micro-fluidic system: characterization and testing with membrane protein crystallization. *J. Am. Chem. Soc.* **131**, 6042-+.

Kumaresan, P., Yang, C. J., Cronier, S. A., Blazei, R. G., and Mathies, R. A. (2008). High-throughput single copy DNA amplification and cell analysis in engineered nanoliter droplets. *Anal. Chem.* **80**, 3522–3529.

Leal-Calderon, F., Schmitt, V., Bibette, J., and SpringerLink (Online service). (2007). Emulsion science basic principles, pp. xi, 227 p. Springer, New York.

Lee, G. B., Lin, C. H., Lee, K. H., and Lin, Y. F. (2005). On the surface modification of microchannels for microcapillary electrophoresis chips. *Electrophoresis* **26**, 4616–4624.

Lee, J., and Pozrikidis, C. (2006). Effect of surfactants on the deformation of drops and bubbles in Navier–Stokes flow. *Comput. Fluids* **35**, 43–60.

Lee, J. N., Park, C., and Whitesides, G. M. (2003). Solvent compatibility of poly(dimethylsiloxane)-based microfluidic devices. *Anal. Chem.* **75**, 6544–6554.

Li, W., Nie, Z. H., Zhang, H., Paquet, C., Seo, M., Garstecki, P., Kumacheva, E. (2007). Screening of the effect of surface energy of microchannels on microfluidic emulsification. *Langmuir* **23**, 8010–8014.

Liau, A., Karnik, R., Majumdar, A., and Cate, J. H. D. (2005). Mixing crowded biological solutions in milliseconds. *Anal. Chem.* **77**, 7618–7625.

Lin, B. C., and Su, Y. C. (2008). On-demand liquid-in-liquid droplet metering and fusion utilizing pneumatically actuated membrane valves. *J. Micromech. Microeng.* 18.

Lin, Y. H., Lee, C. H., and Lee, G. B. (2008). Droplet formation utilizing controllable moving-wall structures for double-emulsion applications. *J. Microelectromech. Syst.* **17**, 573–581.

Link, D. R., Anna, S. L., Weitz, D. A., and Stone, H. A. (2004). Geometrically mediated breakup of drops in microfluidic devices. *Phys. Rev. Lett.* 92.

Link, D. R., Grasland-Mongrain, E., Duri, A., Sarrazin, F., Cheng, Z. D., Cristobal, G., Marquez, M., Weitz, D. A. (2006). Electric control of droplets in microfluidic devices. *Angew. Chem. – Int. Ed.* **45**, 2556–2560.

Liu, S. J., Gu, Y. F., Le Roux, R. B., Matthews, S. M., Bratton, D., Yunus, K., Fisher, A. C., Huck, W. T. S. (2008). The electrochemical detection of droplets in microfluidic devices. *Lab on a Chip* **8**, 1937–1942.

Liu, Y., Jung, S. Y., and Collier, C. P. (2009). Shear-driven redistribution of surfactant affects enzyme activity in well-mixed femtoliter droplets. *Anal. Chem.* **81**, 4922–4928.

Luo, C. X., Yang, X. J., Fu, O., Sun, M. H., Ouyang, Q., Chen, Y., Ji, H. (2006). Picoliter-volume aqueous droplets in oil: electrochemical detection and yeast electroporation. *Electrophoresis* **27**, 1977–1983.

Mark, D., Haeberle, S., Roth, G., von Stetten, F., and Zengerle, R. (2010). Microfluidic lab-on-a-chip platforms: requirements, characteristics and applications. *Chem. Soc. Rev.* **39**, 1153–1182.

Mazutis, L., Araghi, A. F., Miller, O. J., Baret, J. C., Frenz, L., Janoshazi, A., Taly, V., Miller, B. J., Hutchison, J. B., Link, D., Griffiths, A. D., Ryckelynck, M. (2009a). Droplet-based microfluidic systems for high-throughput single DNA molecule isothermal amplification and analysis. *Anal. Chem.* **81**, 4813–4821.

Mazutis, L., Baret, J. C., and Griffiths, A. D. (2009b). A fast and efficient microfluidic system for highly selective one-to-one droplet fusion. *Lab on a Chip* **9**, 2665–2672.

Nguyen, N. T., Ting, T. H., Yap, Y. F., Wong, T. N., Chai, J. C. K., Ong, W. L., Zhou, J., Tan, S. H., Yobas, L. (2007). Thermally mediated droplet formation in microchannels. *Appl. Phys. Lett.* 91.

Nie, Z. H., Seo, M. S., Xu, S. Q., Lewis, P. C., Mok, M., Kumacheva, E., Whitesides, G. M., Garstecki, P., Stone, H. A. (2008). Emulsification in a microfluidic flow-focusing device: effect of the viscosities of the liquids. *Microfluid. Nanofluid.* **5**, 585–594.

Niu, X., Gulati, S., Edel, J. B., and deMello, A. J. (2008). Pillar-induced droplet merging in microfluidic circuits. *Lab on a Chip* **8**, 1837–1841.

Niu, X. Z., Zhang, M. Y., Peng, S. L., Wen, W. J., and Sheng, P. (2007). Real-time detection, control, and sorting of microfluidic droplets. *Biomicrofluidics* 1.

Oh, K. W., and Ahn, C. H. (2006). A review of microvalves. *J Micromech. Microeng.* **16**, R13–R39.

Ota, S., Yoshizawa, S., and Takeuchi, S. (2009). Microfluidic formation of monodisperse, cell-sized, and unilamellar vesicles. *Angew. Chem. – Int. Ed.* **48**, 6533–6537.

Priest, C., Herminghaus, S., and Seemann, R. (2006). Controlled electrocoalescence in microfluidics: targeting a single lamella. *Appl. Phys. Lett.* 89.

Roach, L. S., Song, H., and Ismagilov, R. F. (2005). Controlling nonspecific protein adsorption in a plug-based microfluidic system by controlling interfacial chemistry using fluorous-phase surfactants. *Anal. Chem.* **77**, 785–796.

Rosen, M. J., and National Science Foundation (U.S.). (1987). Surfactants in Emerging Technologies. Dekker, New York.

Sarrazin, F., Salmon, J. B., Talaga, D., and Servant, L. (2008). Chemical reaction imaging within microfluidic devices using confocal Raman spectroscopy: the case of water and deuterium oxide as a model system. *Anal. Chem.* **80**, 1689–1695.

Schaerli, Y., Wootton, R. C., Robinson, T., Stein, V., Dunsby, C., Neil, M. A. A., French, P. M. W., deMello, A. J., Abell, C., Hollfelder, F. (2009). Continuous-flow polymerase chain reaction of single-copy DNA in microfluidic microdroplets. *Anal. Chem.* **81**, 302–306.

Schmitz, C. H. J., Rowat, A. C., Koster, S., and Weitz, D. A. (2009). Dropspots: a picoliter array in a microfluidic device. *Lab on a Chip* **9**, 44–49.

Seo, M., Paquet, C., Nie, Z. H., Xu, S. Q., and Kumacheva, E. (2007). Microfluidic consecutive flow-focusing droplet generators. *Soft Matter* **3**, 986–992.

Shah, R. K., Kim, J. W., Agresti, J. J., Weitz, D. A., and Chu, L. Y. (2008). Fabrication of monodisperse thermosensitive microgels and gel capsules in microfluidic devices. *Soft Matter* **4**, 2303–2309.

Shah, R. K., Kim, J. W., and Weitz, D. A. (2010). Monodisperse stimuli-responsive colloidosomes by self-assembly of microgels in droplets. *Langmuir* **26**, 1561–1565.

Shi, W. W., Qin, J. H., Ye, N. N., and Lin, B. C. (2008). Droplet-based microfluidic system for individual *Caenorhabditis elegans* assay. *Lab on a Chip* **8**, 1432–1435.

Shim, J. U., Cristobal, G., Link, D. R., Thorsen, T., Jia, Y. W., Piattelli, K., Fraden, S. (2007). Control and measurement of the phase behavior of aqueous solutions using microfluidics. *J. Am. Chem. Soc.* **129**, 8825–8835.

Shim, J. U., Olguin, L. F., Whyte, G., Scott, D., Babtie, A., Abell, C., Huck, W. T. S., Hollfelder, F. (2009). Simultaneous determination of gene expression and enzymatic activity in individual bacterial cells in microdroplet compartments. *J. Am. Chem. Soc.* **131**, 15251–15256.

Solvas, X. C. I., Srisa-Art, M., Demello, A. J., and Edel, J. B. (2010). Mapping of fluidic mixing in microdroplets with 1 mu s time resolution using fluorescence lifetime imaging. *Anal. Chem.* **82**, 3950–3956.

Song, H., Chen, D. L., and Ismagilov, R. F. (2006). Reactions in droplets in microflulidic channels. *Angew. Chem. – Int. Ed.* **45**, 7336–7356.

Squires, T. M., and Quake, S. R. (2005). Microfluidics: fluid physics at the nanoliter scale. *Rev. Mod. Physics* **77**, 977–1026.

Srisa-Art, M., deMello, A. J., and Edel, J. B. (2008a). Fluorescence lifetime imaging of mixing dynamics in continuous-flow microdroplet reactors. *Phys. Rev. Lett.* 101.

Srisa-Art, M., Dyson, E. C., Demello, A. J., and Edel, J. B. (2008b). Monitoring of real-time streptavidin-biotin binding kinetics using droplet microfluidics. *Anal. Chem.* **80**, 7063–7067.

Srisa-Art, M., Kang, D. K., Hong, J., Park, H., Leatherbarrow, R. J., Edel, J. B., Chang, S. I., deMello, A. J. (2009). Analysis of protein–protein interactions by using droplet-based microfluidics. *Chembiochem* **10**, 1605–1611.

Stanley, C. E., Elvira, K. S., Niu, X. Z., Gee, A. D., Ces, O., Edel, J. B., deMello, A. J. (2010). A microfluidic approach for high-throughput droplet interface bilayer (DIB) formation. *Chem. Commun.* **46**, 1620–1622.

Stone, H. A. (1994). Dynamics of drop deformation and breakup in viscous fluids. *Ann. Rev. Fluid Mech.* **26**, 65–102.

Stone, H. A., and Leal, L. G. (1990). The effects of surfactants on drop deformation and breakup. *J. Fluid Mech.* **220**, 161–186.

Tadros, T. F. (1984). *Surfactants.* Academic Press, London, Orlando.

Tan, W. H., and Takeuchi, S. (2007a). Monodisperse alginate hydrogel microbeads for cell encapsulation. *Adv. Mater.* **19**, 2696-+.

Tan, W. H., and Takeuchi, S. (2007b). A trap-and-release integrated microfluidic system for dynamic microarray applications. *Proc. Nat. Acad. Sci. U.S.A.* **104**, 1146–1151.

Tan, Y. C., Ho, Y. L., and Lee, A. P. (2007). Droplet coalescence by geometrically mediated flow in microfluidic channels. *Microfluid. Nanofluid.* **3**, 495–499.

Teh, S. Y., Lin, R., Hung, L. H., and Lee, A. P. (2008). Droplet microfluidics. *Lab on a Chip* **8**, 198–220.

Theberge, A. B., Courtois, F., Schaerli, Y., Fischlechner, M., Abell, C., Hollfelder, F., Huck, W. T. S. (2010). Microdroplets in microfluidics: an evolving platform for discoveries in chemistry and biology. *Angew. Chem. – Int. Ed.* **49**, 5846–5868.

Thorsen, T., Roberts, R. W., Arnold, F. H., and Quake, S. R. (2001). Dynamic pattern formation in a vesicle-generating microfluidic device. *Phys. Rev. Lett.* **86**, 4163–4166.

Tice, J. D., Song, H., Lyon, A. D., and Ismagilov, R. F. (2003). Formation of droplets and mixing in multiphase microfluidics at low values of the Reynolds and the capillary numbers. *Langmuir* **19**, 9127–9133.

Ting, T. H., Yap, Y. F., Nguyen, N. T., Wong, T. N., Chai, J. C. K., Yobas, L. (2006). Thermally mediated breakup of drops in microchannels. *Appl. Phys. Lett.* **89**.

Trivedi, V., Doshi, A., Kurup, G. K., Ereifej, E., Vandevord, P. J., Basu, A. S. (2010). A modular approach for the generation, storage, mixing, and detection of droplet libraries for high throughput screening. *Lab on a Chip* **10**, 2433–2442.

Wang, G., Lim, C., Chen, L., Chon, H., Choo, J., Hong, J., deMello, A. J. (2009a). Surface-enhanced Raman scattering in nanoliter droplets: towards high-sensitivity detection of mercury(II) ions. *Anal. Bioanal. Chem.* **394**, 1827–1832.

Wang, K., Lu, Y. C., Xu, J. H., and Luo, G. S. (2009b). Determination of dynamic interfacial tension and its effect on droplet formation in the T-shaped microdispersion process. *Langmuir* **25**, 2153–2158.

Wang, W., Yang, C., and Li, C. M. (2009c). Efficient on-demand compound droplet formation: from microfluidics to microdroplets as miniaturized laboratories. *Small* **5**, 1149–1152.

Wang, W., Yang, C., and Li, C. M. (2009d). On-demand microfluidic droplet trapping and fusion for on-chip static droplet assays. *Lab on a Chip* **9**, 1504–1506.

Weibel, D. B., DiLuzio, W. R., and Whitesides, G. M. (2007). Microfabrication meets microbiology. *Nature Rev. Microbiol.* **5**, 209–218.

Weibel, D. B., and Whitesides, G. M. (2006). Applications of microfluidics in chemical biology. *Curr. Opin. Chem. Biol.* **10**, 584–591.

Williams, R., Peisajovich, S. G., Miller, O. J., Magdassi, S., Tawfik, D. S., Griffiths, A. D. (2006). Amplification of complex gene libraries by emulsion PCR. *Nature Methods* **3**, 545–550.

Workman, V. L., Dunnett, S. B., Kille, P., and Palmer, D. D. (2007). Microfluidic chip-based synthesis of alginate microspheres for encapsulation of immortalized human cells. *Biomicrofluidics* 1.

Workman, V. L., Dunnett, S. B., Kille, P., and Palmer, D. D. (2008). On-chip alginate microencapsulation of functional cells. *Macromol. Rapid Commun.* **29**, 165–170.

Yap, Y. F., Tan, S. H., Nguyen, N. T., Murshed, S. M. S., Wong, T. N., Yobas, L. (2009). Thermally mediated control of liquid microdroplets at a bifurcation. *J. Phys. D – Appl. Phys.* 42.

Zagnoni, M., Anderson, J., and Cooper, J. M. (2010a). Hysteresis in multiphase microfluidics at a T-junction. *Langmuir* **26**, 9416–9422.

Zagnoni, M., Baroud, C. N., and Cooper, J. M. (2009). Electrically initiated upstream coalescence cascade of droplets in a microfluidic flow. *Phys. Rev. E* 80.

Zagnoni, M., and Cooper, J. M. (2009). On-chip electrocoalescence of microdroplets as a function of voltage, frequency and droplet size. *Lab on a Chip* **9**, 2652–2658.

Zagnoni, M., and Cooper, J. M. (2010). A Microdroplet-based Shift Register. *Lab on a Chip* **10**, 3069–3073.

Zagnoni, M., Le Lain, G., and Cooper, J. M. (2010b). Electrocoalescence mechanisms of microdroplets using localized electric fields in microfluidic channels. *Langmuir* **26**, 14443–14449.

Zeng, S. J., Li, B. W., Su, X. O., Qin, J. H., and Lin, B. C. (2009). Microvalve-actuated precise control of individual droplets in microfluidic devices. *Lab on a Chip* **9**, 1340–1343.

Zhan, Y. H., Wang, J., Bao, N., and Lu, C. (2009). Electroporation of cells in microfluidic droplets. *Anal. Chem.* **81**, 2027–2031.

Zhang, D. F., and Stone, H. A. (1997). Drop formation in viscous flows at a vertical capillary tube. *Phys. Fluids* **9**, 2234–2242.

Zhang, K., Liang, Q. L., Ma, S., Mu, X. A., Hu, P., Wang, Y. M., Luo, G. A. (2009). On-chip manipulation of continuous picoliter-volume superparamagnetic droplets using a magnetic force. *Lab on a Chip* **9**, 2992–2999.

Zheng, B., Tice, J. D., and Ismagilov, R. F. (2004). Formation of arrayed droplets of soft lithography and two-phase fluid flow, and application in protein crystallization. *Adv. Mater.* **16**, 1365–1368.

CHAPTER 3

Parallel Imaging Microfluidic Cytometer

Daniel J. Ehrlich,[*] **Brian K. McKenna,**[*] **James G. Evans,**[*]
Anna C. Belkina,[†] **Gerald V. Denis,**[†] **David H. Sherr**[‡]
and Man Ching Cheung[*]

[*]Departments of Biomedical Engineering/Electrical and Computer Engineering, Boston University, Boston, Massachusetts, USA

[†]Cancer Center, Boston University Medical Center, Boston, Massachusetts, USA

[‡]Department of Environmental Health, Boston University School of Public Health, Boston, Massachusetts, USA

METHODS IN CELL BIOLOGY, VOL 102

0091-679X/10 $35.00
DOI 10.1016/B978-0-12-374912-3.00003-1

Abstract

By adding an additional degree of freedom from multichannel flow, the parallel microfluidic cytometer (PMC) combines some of the best features of fluorescence-activated flow cytometry (FCM) and microscope-based high-content screening (HCS). The PMC (i) lends itself to fast processing of large numbers of samples, (ii) adds a 1D imaging capability for intracellular localization assays (HCS), (iii) has a high rare-cell sensitivity, and (iv) has an unusual capability for time-synchronized sampling. An inability to practically handle large sample numbers has restricted applications of conventional flow cytometers and microscopes in combinatorial cell assays, network biology, and drug discovery. The PMC promises to relieve a bottleneck in these previously constrained applications. The PMC may also be a powerful tool for finding rare primary cells in the clinic.

The multichannel architecture of current PMC prototypes allows 384 *unique* samples for a cell-based screen to be read out in ~6–10 min, about 30 times the speed of most current FCM systems. In 1D intracellular imaging, the PMC can obtain protein localization using HCS marker strategies at many times for the sample throughput of charge-coupled device (CCD)-based microscopes or CCD-based single-channel flow cytometers. The PMC also permits the signal integration time to be varied over a larger range than is practical in conventional flow cytometers. The signal-to-noise advantages are useful, for example, in counting rare positive cells in the most difficult early stages of genome-wide screening. We review the status of parallel microfluidic cytometry and discuss some of the directions the new technology may take.

I. Introduction

Relatively narrow sets of methods define eras like genomics and proteomics. The instruments used to practice these methods are often badly mismatched to the biological agenda. We argue that such a bottleneck now exists in cell-resolved measurement. The various "omics" have increased the encyclopedia of molecules and interactions to the point where we can practice broad combinatorial experiments in cells. The primary tools for the readout of these experiments remain microscopy, cytometry, arrays, fluorimeters, and a handful of biochemical assays.

Because it can quickly produce a statistically significant reading, one of the most important of these tools is the fluorescence-activated flow cytometer (FCM)

(Givan, 2001; McCoy, 2007; Shapiro, 2003). However, in several dimensions, FCM is inadequate to the agenda. It is only practical to make measurements on a few variables at a time and at a compromised sample throughput. In contrast, HCS (i.e., automated microscopy) (Bullen *et al.*, 2008; Eggert *et al.*, 2006; Gough *et al.*, 2007; Haney *et al.*, 2008; Lee *et al.*, 2006; Pepperkok *et al.*, 2006; Taylor *et al.*, 2007) is an attempt to add more information content to cell cytometry. Throughput of both FCM and HCS is an issue for readout of combinatorial biology in general, but particularly with live cells. For example, nuclear transcription kinetics often have a half-time response of 5–10 min (Ding *et al.*, 1998). In a live-cell kinetic study, it is usually not possible to read a single 96-well HCS plate in this time. Furthermore, for either flow cytometry or HCS, fixing cells causes protein reorganization, and many cytokine modifiers can show alternatively agonism *or* antagonism in a dose-dependent fashion. Therefore, the biology of combinatorial biology such as large RNAi screens or small molecule studies calls out for dose–response curves taken over many concentrations, on live cells, and with time response on the order of several minutes. The current methods remain orders of magnitude mismatched in speed for the real needs of network biology. Furthermore, as an entirely separable point, the 1D imaging ability of the PMC is *new* to high-speed flow cytometry. The movement of FCM toward higher content has been expressed in recent years by adding lasers and more color channels. The addition of 1D imaging can be thought of continuing this trend. Adding 1D imaging to high-speed FCM is equivalent to adding many color channels, however.

Limitations implicit in the architecture of single-channel flow cytometers restrict applications for studying rare-cell types and for massively parallel screening. These are, principally, (i) serial sample processing, which is bounded by sample changeover and (ii) a short (usually microsecond) data acquisition time, which in turn limits signal averaging. Commercial flow cytometers have been demonstrated with positive abundances as low as parts per million. However, depending on available sample and background noise, single-channel machines are generally not seen as practical for screening when the abundance of "positives" is lower than about 1:10,000 or when the total sample is less than 10–50 thousand cells (Shapiro, 2003). In many cases, autofluorescence and nonspecific markers limit minimum abundances to higher ratios (1:1000 or 1:100). Recent developments in cytometers have explored automated sample loaders to minimize the disadvantage of serial analysis (Edwards *et al.*, 2004); however, sample changeover times still remain on the order of a minute for most commercial FCM machines that are in the field.

In this chapter, we review considerations in adding a high degree of microfluidic parallelism to flow cytometry. Specifically, we review results from a prototype PMC, which was designed with particular attention to the needs of rare-cell counting (McKenna *et al.*, 2009). Rare-cell capability (detection of rare positives within a high background of negatives) is the priority for detection in cancer and also, quite generally, at the early stages of genome-wide screening.

II. Background

A. Flow Cytometry

Flow cytometry is an impressive technology that has been optimized to extraordinary refinement (Givan, 2001; McCoy, 2007; Shapiro, 2003). There is also a large body of more recent work on elegant microfluidic manipulations of cells, including sorting and switching of biological cells in single channels and in dispensing of cells into arrayed well devices. Some examples are cited here (Cheung *et al.*, 2010; Dittrich *et al.*, 2003; Emmelkamp *et al.*, 2004; Fu *et al.*, 2002; Gawd, *et al.*, 2004; McClain *et al.*, 2001; Wang *et al.*, 2005; Wolff *et al.*, 2003; Yi *et al.*, 2006) but the full microfluidics literature is far too extensive to review in this chapter.

B. High–Content Screening

HCS is frequently done with CCD-based microscopes in open wells (Bullen *et al.*, 2008; Ding *et al.*, 1998; Gonzales and Woods, 2008; Gough *et al.*, 2007; Haney *et al.*, 2008; Lee *et al.*, 2006; Taylor *et al.*, 2007), on spotted slides (Carpenter *et al.*, 2006; Wheeler *et al.*, 2005), or in flow (George *et al.*, 2006, http://www.amnis.com). Even on high-density slides, the state of the art is largely determined by the performance of low-signal scientific CCD cameras. At 1024×1024-pixel image size, the frame rate due to buffering restrictions is either 15 or (conditionally) 30 frames a second. However, even much slower rates are often mandated by low signal. Analysis of a single high-density spotted slide may take many hours (Carpenter *et al.*, 2006; Wheeler *et al.*, 2005). Autofocusing and mechanical motions further limit throughput (accounting for the majority of the time budget on wide-field imaging systems (Taylor *et al.*, 2007). CCD-based imaging flow cytometers are more limited in throughput. Users typically report raw data acquisition (unclassified cells) from such a machine at 100–1000 objects/s (http://www.amnis.com/applications. asp#link2). The bottom line is that high content microscope-based systems for HCS are frequently too slow for scaled-up applications. A second drawback can be that, with full 2D imaging, data storage rapidly requires terabytes and overflows even large data-storage resources.

Several of the most common high-content assays implemented on microscopes (in 2D) are (Bullen *et al.*, 2008; Ding *et al.*, 1998; Gonzales and Woods, 2008; Gough *et al.*, 2007; Haney *et al.*, 2008; Lee *et al.*, 2006; Taylor *et al.*, 2007):

(a) *Nuclear translocation (NT).* The most common NT assay is NF-κB translocation. NF-κB is a transcription factor that is critical to cellular stress response. The p65 subunit is a sensitive to several known stimulants, for example, by altered interleukin ILa1 or tumor necrosis factor. The translocation to the nucleus is required to induce gene expression. (b) *Apoptosis.* Image-based assays for apoptosis can provide more information than FCM. For examplen, by determining nucleus size, it is possible to ascertain necrotic or late apoptotic cells. The nucleus is stained and the image algorithm determines shape and size relative to the cell dimensions. *(c) Target activation.* A very wide class of assays measure localization and total intensity from GFP fusions or other fluorescent markers. Cell cycle, receptor internalization, or drug resistance are commonly measured. (d) *Colocalization of markers.* Colocalization is highly informative about biological mechanism. This is enormous area of active research particularly in the field of biological development. Imaging information is highly useful. (e) *Intracellular trafficking.* Several microscope-based assays track the intracellular migration of molecules by programmed endocytosis. Amnis, Inc., has introduced an assay where the antibody CD20 is monitored and correlated with markers for endosomes and lysosomes. (f) *Morphology.* The most obvious markers for phenotype are cell shape and area; however, more subtle rearrangements of the cytoskeleton and location of organelles are also often used in microscope assays. (g) *Cell cycle.* The progression of cell cycle is widely used in screening cancer therapies. The phase of individual cells is correlated with markers for specific proteins. Measurements are often also made on the dimensions or total DNA of the nucleus.

C. HCS Instruments

Several commercial 2D HCS instruments are (i) CCD/automated microscopes (Thermo Scientific – Cellomics ArrayScan^{TM}, GE Healthcare – inCell^{TM}, PerkinElmer – EvoTech Opera^{TM}, Molecular Devices IsoCyte^{TM}); (ii) TDI CCD/ flow cytometer (Amnis ImageStream^{TM}); and (iii) low-resolution laser scanners (CompuCyte iColor^{TM}, Acumen – Explorer^{TM} and Cyntellect, LEAP). These systems generally achieve assay rates of about 2–6 wells/min for real HCS assays. The Amnis ImageStream is a CCD-based *flow* imaging system. However, it is a single-channel instrument. The laser scanning instruments (CompuCyte iColor^{TM}, Acumen – Explorer^{TM}, and Cyntellect, LEAP) are not flow-based.

III. Instrument Design

The design of a PMC differs from that of a FCM in (1) its need for a wide field of view detector (rather than focused point detector), (2) its need for automation to support parallel sample transfer, (3) its differing needs for data processing, and (4) the design of the microfluidic itself. The microfluidic, when all fabrication and flow considerations are taken into account, becomes a big opportunity for broadly novel design. One specific consideration is how to rethink flow focusing in order to make best use of the small-sample capability of microfabricated devices. The detector becomes more complex than a FCM since the wide-field requirement more or less mandates a scanner (arguments below). However, once the additional mechanical complexity of scanning is accepted, there is a large and important freedom in signal-collection strategies. This is also what permits high-speed imaging. We discuss these design aspects below.

A. PMC System Architecture

A prototype automated PMC is shown in Figs. 1 and 2. The microfluidic flow devices are mounted on a top plate and are serviced with a gantry robot combined with a sample elevator that handles 384-well microtiter plates. The fluid handling is via an automated 96-tip pipettor. The sample deck includes positions for nutrient/wash trays that can also be accessed by the pipettor. As a result, live-cell cultures can be sustained for several days on the system or can be loaded from off-system culture apparatus. All 384 channels can be loaded from a microtiter plate in <30 s. Flow is actuated by suction using syringe pumps. The optical detector is a photomultiplier-based rotary scanner located under the microfluidics (see below). The system is operated from a graphical user interface that displays data during real time. However, data reduction is done on exported files off line (see below).

B. Robotics

In the prototype of a PMC, it was thought to be important to include competent robotics. The stages of the X-Y-Z axes (GL16S with 1250 mm of travel for the X-axis, a KR46 with 540 mm of travel for the Y-axis, and KR26 with 65 mm of travel for the Z-axis T.H.K. Ltd, Tokyo, Japan) are driven by a set of three servo drivers (SGDH-01AE, Yaskawa Electric, Japan) and are coordinated by a six-axis motion controller (Model 6k6 Compumotor, Rohnert Park, CA). The pipettor head has programmable suction and injection capability for volumes between 2 and 20 µl, and is driven by a DC brushless servo amplifier (Model 503, Copley Controls Corp., Westwood, MA). The 96-tip robot head accesses water and buffer reservoirs, an ultrasound washing station, a microtiter plate elevator with up to 32 sample plates for continuous operations (Packard Instrument Co, Meriden, CT).

C. PMC Detector

The requirements for a PMC detector are mesoscale sensor field, high time response, and variable integration time. It is also highly desirable to have out-of-plane (confocal) background-light rejection. The combined factors led us to strongly favor photomultiplier or avalanche photodiode detection (rather than a CCD or CMOS imager). We therefore modified a PMT fluorescent scanner that we had developed for DNA sequencing (El-Difrawy *et al.*, 2005) and applied it to the new application.

A 100-mW multiline argon-ion laser beam (Melles Griot #532-MA-A04, Melles Griot, Carlsbad, CA) is passed through a rotating head that moves a 0.5-numerical-aperture (NA) aspheric lens (# 350240, Thor Labs, Corp., Newton NJ) and is driven by a DC brushless motor (BEI# DIP20-17-0027, BEI Kimco, Vista, CA). The laser focus is adjusted between NA 0.01 and NA 0.50 and excites fluorescence as the rotating head moves under the detection window. The fluorescence is collected at NA 0.5 through the rotating head (Fig. 2), is reimaged through a pinhole, then is separated into four wavelength bands using dichroics and bandpass filters, and is distributed onto four PMTs (H957-8 Hamamatsu, Bridgewater, NJ).

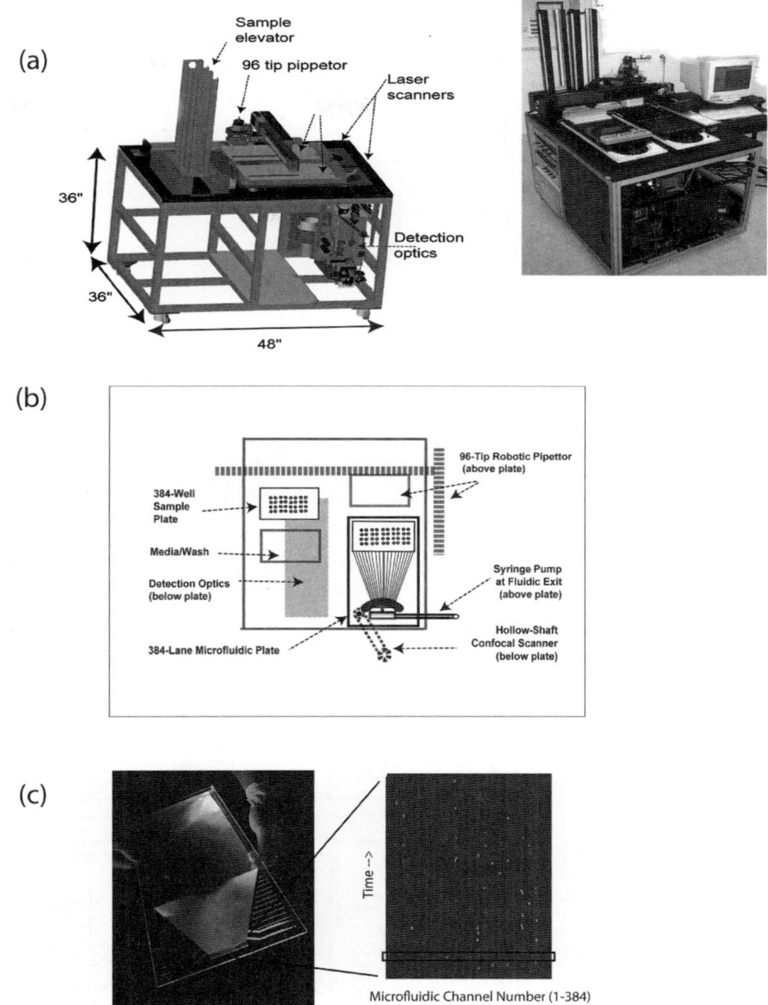

Fig. 1 **(a)** Parallel microfluidic cytometer (PMC) for cell-based assays. The system is designed for automated fluorescence measurements on 384-channel microfluidic plates and comprises up to two temperature-controlled microfluidic "chips" (16–384 channels each), a scanning detector, and automated pipettor/sample elevator for automated maintenance of cell suspensions/cultures. Cell suspensions are pulled by vacuum suction from injection wells using a positive-displacement syringe pump. Multicolor detection is via a scanned confocal detector that oscillates below the microfluidics. **(b)** Plan-view detail of the PMC showing microfluidics, sample, and wash plates and the optical scanner located beneath the fluidics. **(c)** A 384-channel microfluidic plate and a segment of data collected from several channels. A short time sequence from one of four photomultipliers is shown with each pixel representing 35 µm in the (horizontal) scan direction. Data is collected at a rate of 3 scans/s, (0.33 s vertical displacement of each row of pixels in the data image); 240 s of data shown. Signal amplitude is shown in RGB color scale with blue representing low signal and red high signal. [(Modified from El-Difrawy *et al.* (2005) with permission of American Institute of Physics) and from McKenna *et al.* (2009) with permission of the Royal Society of Chemistry.] (See plate no. 1 in the color plate section.)

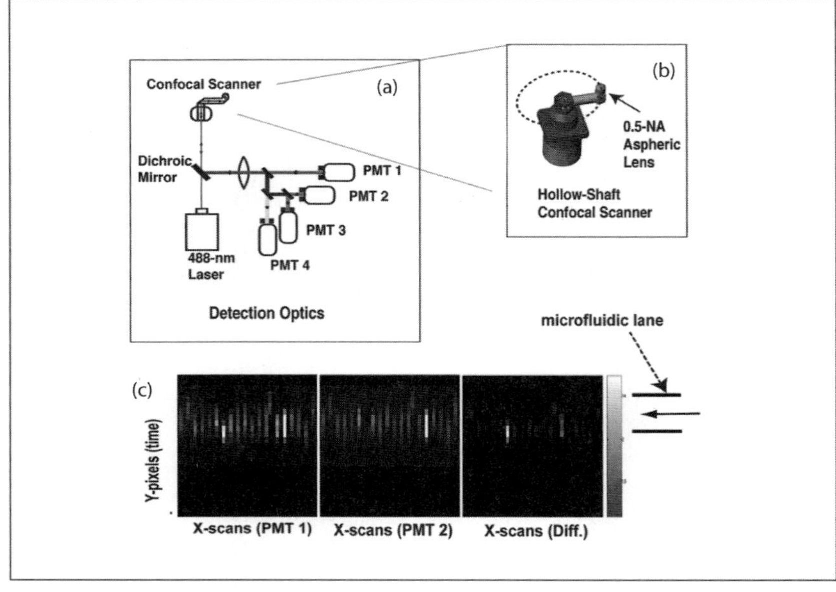

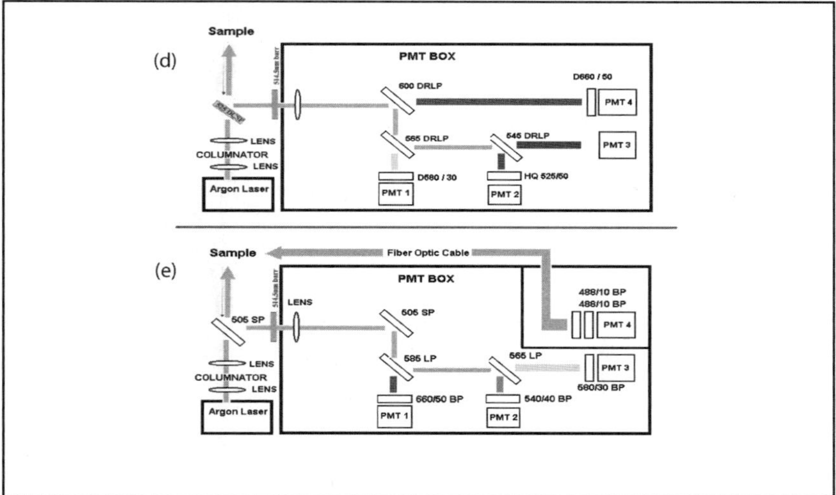

Fig. 2 **(a/b)** Optical diagram of the laser-induced fluorescence (LIF) detector including detail of rotary scanner that introduces the 488-nm laser beam and returns fluorescence onto four photomultipliers (PMTs, other configurations shown below in Fig. 4). The scanner uses a 3-in. radius of rotation, a DC rotary motor, and an optical encoder. **(c)** Detail of a typical condensed-time data scans for a single microfluidic channel [c, left and center] and reduced-difference scan to identify positives [c, right]. See text. **(d)** Two configurations of the optical detector to match cell assays used for a genome-wide cloning screen (Fig. 7) and for the dilution study on primary leukemia cells (Fig. 6c, d). The scattered forward light sensor **(e)** is a fiber optic (910-μm diameter, 0.22 NA) on a rotatable mount that can be adjusted in the range from 20° to 70° off the forward direction. [Modified from McKenna *et al.* (2009). Reproduced with permission of the Royal Society of Chemistry.]

Nonuniform velocity and cycle-to-cycle fluctuations cause variations in the level of the collected signal and results in added noise. Therefore, the position of the rotating head is measured using an optical encoder (HT30P156 D14 N4096, Dynamic Research Corp., Wilmington, MA), and a proportional-integral-derivative (PID) controller maintains the speed profile and is tuned to minimize scan-to-scan speed fluctuations while maximizing accelerations outside the collection arc.

The rotating head is programmed to a sawtooth velocity profile at 12 or 3 Hz recording 300–2800 data points across a window of ~3.5–80 mm. For each encoder point, the values of the four PMTs are recorded and the 16-bit digital value is saved to the PC hard drive 12(3) times a second.

The constancy and reproducibility of the speed profile have been measured and show a standard deviation of less than 1% from the target velocity (10,000 scans, all flow channels). The sensitivity of the system was evaluated with fluorescence standards and shown to have a 10 pM (fluorescein) detection limit in a 60-μm-deep channel, which is near the state of the art for on-column laser-induced fluorescence detectors.

D. Data Processing

The data acquisition and PID control are synchronized using a digital signal processor (DSP, ADSP-2181, Analog Devices, Norwood, MA) running at a clock rate of 33 MHz. The PID controller runs at a servo rate of 1 kHz, whereas the data acquisition is performed at a faster rate of 200 kHz. The collected sequencing data is uploaded from the DSP memory to the PC through the PC parallel port, which runs in the enhanced parallel port mode. A simple control circuit is used to provide the PC with direct memory access to the program and data memory of the DSP processor.

Raw PMT data (Fig. 3(a)) is saved in four files representing 16-bit data at each location separated by 10 μm in the scan window. Each file contains the data for one PMT. The data is first processed to eliminate scan areas without cells. This data is then reformatted into $512 \times 512 \times 16$ bit gray scale images and saved as TIFF images for each PMT channel. A calculated image as shown in Fig. 3(b) is then created by subtracting the red channel from green (negative values are set to zero).

E. Microfluidics

Microdevices with 16, 32, and 384 channels (Fig. 4) were fabricated in alumina-silicate glass (Corning, Eagle™). Unaligned single-mask contact lithography was followed by high-temperature fusion bonding of 0.7-mm-thick plates of 25×50 and 25×25-cm size. (Aborn *et al.*, 2005; Goedecke *et al.*, 2004). The microchannels had a hemispherical cross section with a radius of 60 μm and converged to a density of 5 channels/mm in the scan zone. In the network layouts, channels (20–40 mm length, 1–3 μl internal volume) were matched to a few percent in flow resistance. Access for introduction of the cell suspension was through laser-drilled ports, which were conical in shape and terminated at the flow channel with an exit of ~80–100 μm

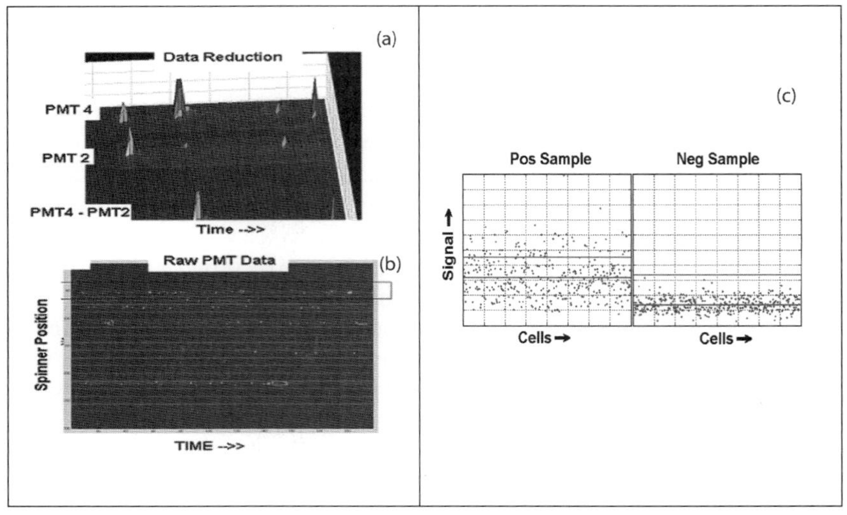

Fig. 3 **(a/b)** Partially reduced image data: Raw data is plotted as an image of the microfluidic channel cross section (vertical axis labeled spinner position) versus time, raw data is collected at 300 pixels per 16-channel (or 384-channel) scan and 12 scans/s. **(c)** Distinguishing positives from autofluorescence. A scatter-plot comparison of the ratio of the GFP channel to yellow channel for objects with a sufficient maximum GFP value (threshold) in positive and negative samples. The low ratio in the negative sample shows how autofluorescence cells can be rejected as negatives. By determining the mean and standard deviation for cells in the negative sample, it is possible to calculate an outlier threshold ($>$ mean $+ 4$ SD). (For interpretation of the references to color in this figure legend, the reader is referred to the Web version of the chapter.)

diameter. Composite G-10 fiberglass boards were mechanically machined with 2-mm-diameter sample wells distributed on 4.5- or 9.0-mm centers, and were glued with thermally curing epoxy on top of the bonded glass devices.

F. Flow and Flow Focusing

Microfluidic systems, created by lithographic methods, are generally constrained as 2D (X,Y) flow networks. One-dimensional squeezing, *in the plane of the flow network*, is relatively easy to accomplish simply by using T-junctions. However, "vertical" hydrodynamic focusing (in the plane perpendicular to the network) is more germane for narrow-depth-of-field optical detection of the PMC. A good discussion, albeit for a slightly different application, can be found elsewhere (Cheung *et al.*, 2010).

In order to focus microfluidic flows vertically, it is necessary to utilize a torque (out of the plane of the network) or to merge flows as vertically distinct layers. From a fabrication standpoint, the geometry in which layers are introduced by intersecting two vertically displaced channels is easiest; this approach requires only a simple unaligned (or weakly aligned) two-level network structure, with no significant microfabrication changes from our normal unaligned procedure. From a modeling point of view, the geometry is slightly more complicated since the normal isotropic

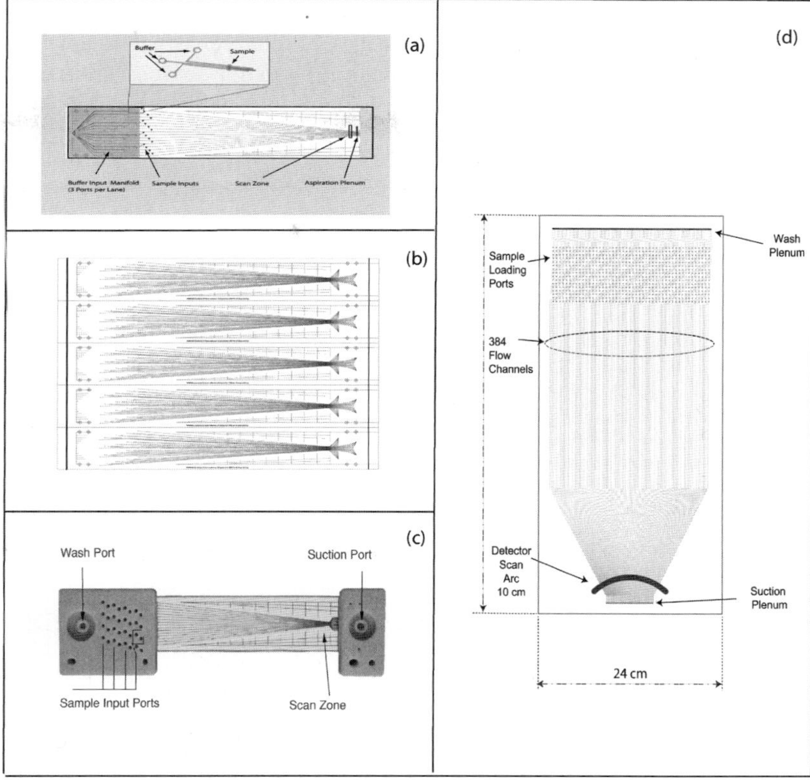

Fig. 4 **(a)** A 16-channel PMC microdevice with 3-sided hydrodynamic focusing. This design can be fabricated with one microlithographically defined fluidic level and captures the three (out of four) directions for flow focusing. The glue-on fiberglass block (see, e.g., Fig. 8 above) is machined to combine the three "blue" buffer flows into a single manifold and reservoir. Other flow focusing designs are provided in Section 3.4. **(b)** A plate of 32-channel PMC microdevices at the lithography stage of fabrication. Five devices are fabricated simultaneously on a 250 × 250 mm alumina silicate glass plate. There are economies of scale from batch fabrication – particularly yield improvements at bonding stage. As a last step, individual devices are separated by diamond sawing. **(c)** A finished PMC microdevice similar to (b) (slightly different design) but after attachment of G-10 fiberglass pumping block and fluid reservoirs. The suction port and wash port are threaded to receive standard 10–32 HPLC fittings. The 32 open sample ports are 2-mm diameter and 10-mm deep, on 9 mm centers (other designs use 4.5 mm centers), and are compatible with a standard multitip pipettors. **(d)** A 384-channel PMC microdevice plate at mask stage, finished device shown in Figure 1c. The flow channels fan out on the "loading" (top) end to allow room for the sample-well array that must match the 4.5-mm spacing of the robotic pipettor. At the "scan" end, the flow channels converge to a maximum density allowed by the bonding process, five channels per millimeter. The channel cross section is hemispherical, 60-μm radius. This channel structure is etched into the glass plate (flat-panel display glass), the access holes are laser drilled, conical shape is terminating with a 80-μm diameter at the etched channel, then the plate is sealed by high-temperature fusion bonding. (For interpretation of the references to color in this figure legend, the reader is referred to the Web version of the chapter.)

                                                                    Daniel J. Ehrlich *et al.*

wet etching procedure produces a nearly hemispherical channel cross section, and flow profiles are highly sensitive to relatively small changes in channel cross section.

To understand how to design focusing devices for the PMC, we explored low-Reynolds-number, fully reversible, pressure-driven Stokes flow, in the geometry of Fig. 5(a/b) through two CFD simulation packages (Lin *et al.*, 2009). Based on

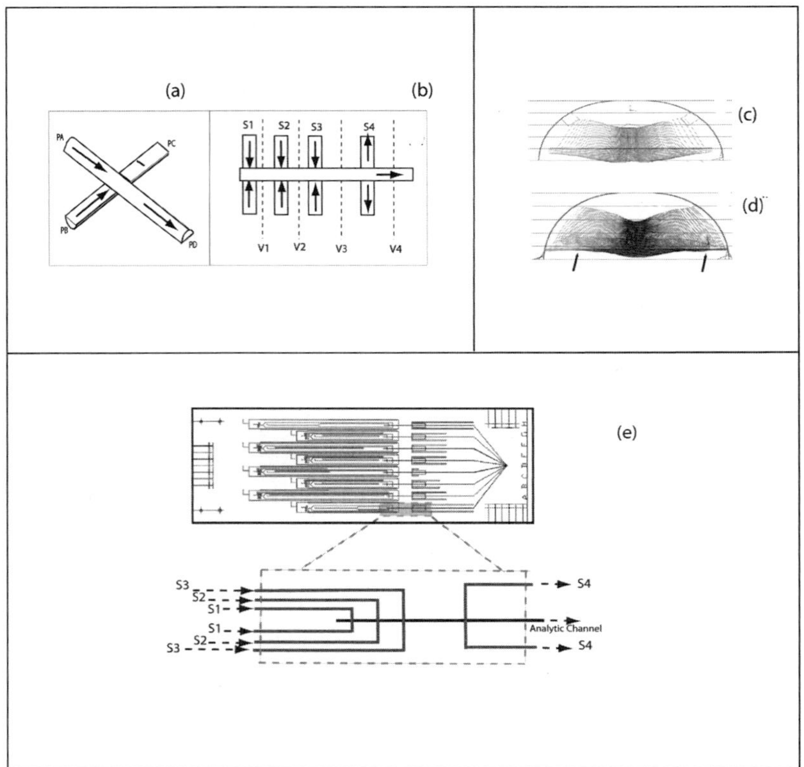

Fig. 5   Microfluidics for hydrodynamic focusing. **(a)** A simple crossing junction used as a design element in software and imaging calibrations of flow focusing; two inlet flows from PA and PB and single outlet flow from PD. No flow allowed through PC (wall boundary condition). The analysis channel is on top. The sheath channel is on bottom. Percentages of flow from PA and PB are in reference to PD, the total flow after the junction. **(b)** Illustrating a four-level compensated vertical focusing device. Additive sheath (symmetric sheath inputs S1 and S3) and additive analysis (symmetric S2) are combined upstream of a subtractive correction flow (symmetric S4). The device is driven by suction from a port at the right end. Adjustable flow resistances on the channels S1–S4 are used to tune the device. Simulations of four-layer focused flow before **(c)** and after **(d)** the channel S4 junction and subtractive correction flow (plane V4). As the traces pass beyond the channel S4 junction, they are preferentially pulled downward and outward. The flow interface indicated by the arrows is most strongly altered by the subtractive flow. **(e)** Plan view layout of the device designed to test vertical flow focusing and subtractive compensation. Eight variations are included on the single test die, labeled A–H' (right side of die). A single laser-drilled hole is provided for each input or output (S1–S4) and for a common suction port (common to configurations A–H, right side of die). The full die size is 3 × 7 cm. (Reprinted with permission from Lin *et al.* (2009). Copyright 2009, American Institute of Physics.)

the resulting models, we constructed simple three-level and four-level vertical focusing devices and tested their performance via 3D optical imaging in a confocal microscope (Lin *et al.*, 2009). The models show that the profile created by combining two flows in isotropically etched channels progresses nonlinearly as a function of the flow-rate ratio of the several fluid streams. That is, an addition of 50% fluid B to A does not give the same result as two sequential 25% additions of fluid B to A. However, through comparison with experimental data, we found that the models are highly accurate in predicting flow profiles (Fig. 5(c/d)).

G. Sorting

A number of innovative microfluidic cell-sorting devices have been designed and implemented on single-channel microfluidic cytometers (Dittrich *et al.*, 2003; Emmelkamp *et al.*, 2004; Fu *et al.*, 2002; Gawd, *et al.*, 2004; McClain *et al.*, 2001; Wang *et al.*, 2005; Wolff *et al.*, 2003; Yi *et al.*, 2006). However, many of these single-channel switches are difficult to multiplex, or lack the switching speed needed for a PMC. A truly impressive parallel switch has been designed and implemented on a PMC by Bohm *et al.* (2007). This system uses 144 parallel channels and a flow switch capable of a 0.5-ms activation cycle. These researchers have announced ambitious applications in the purification of therapeutic quantities of human blood (http://www.cytonome.com).

IV. Operating Methods

For the most part, the operating methods and the sample preparation for PMC applications are identical to the well-established protocols of flow cytometry and microscopy. A few aspects are summarized in the paragraphs below. We also provide specific protocols used to prepare the samples used in the demonstrations described in Section V.

A. Microdevice Maintenance

All flow cytometers require certain routine operating procedures and maintenance. The PMC is no exception. A 1% concentration of bovine serum albumin in phosphate buffered saline (PBS) buffer is periodically pumped through the microdevice to reduce protein adhesion (not more than once a week even with heavy use). As with single-channel cytometers, cell suspensions are treated with established cytometry prefiltration methods (Shapiro, 2003). An iodixanol (OptiPrep®, Sigma Aldrich) gradient-medium buoyancy agent is typically added to the samples to assist buoyancy of the suspended cells. After about 100 h of use, the microdevices are usually cleaned with chlorine bleach; however, there are no extraordinary difficulties with channel fouling or clogging. With careful handling, devices appear to be reusable for an indefinite number of cycles.

Sample loading onto the microdevice is with the automated pipettor out of 96-well or 384-well plates. To counteract settling, the pipettor is also used to periodically mix the sample suspensions by returning at an interval of ~10 min to each well, aspirating, then reloading a portion of each well volume on the microfluidic device.

B. Sample Preparation

Cytometry samples were prepared by standard protocols. Several details relevant to Section V are given in the next several paragraphs.

C. Samples for the Primary-Cell (Lymphoma Model) Dilution Studies

For the sensitivity trials (Section 5.A), Eμ-*BRD2*-/GFP large B-cell lymphoma cells were obtained from the spleens of female 20-week-old FVB mice (Greenwald *et al.*, 2004). Unstained splenocytes (negatives) were obtained from female 16-week-old FVB mice. Fresh cells were frozen in freezing media (50% complete – 10% RPMI-medium (developed at Roswell Park Memorial Institute), 40% fetal bovine serum (FBS), 10% dimethyl sulfoxide (DMSO)), then thawed in small batches as needed, diluted to calibrated ratios in PBS buffer and scanned on the PMC.

D. Cell Line for CPTHR Screen

For this large-scale screen (Section V.C), clonal osteocytic cells, expressing a high level of the C-terminal region of parathyroid hormone receptor (CPTHR), were derived from fetuses in which the majority of exons encoding PTH1R had been ablated by gene targeting. These clonal osteocytic cell lines expressed 1,900,000–3,400,000 CPTHR binding sites per cell, a level 6- to 10-fold higher than observed on osteoblastic cells obtained from the same fetal calvarial bones and at least 5-fold higher than in ROS 17/2.8 cells. Biotinylated [Tyr 34] human PTH (24–84) was synthesized at the Massachusetts General Hospital Peptide and Oligonucleotide Core Laboratory (Boston, MA).

E. The cDNA Library for CPTHR Screen

The cDNA library (Section V.C) was constructed using both random and oligo dT primers to synthesize the first strand DNA. This approach enriches the library with the 5′ portions of large cDNAs compared with cDNA libraries prepared using oligo dT primers only. Inserts were cloned in Lambda Zap pCMV-script expression vector (Stratagene). Since insert size represented in the library is crucial for the successful expression cloning, we examined the insert size in single colonies from different pools of the library. For this purpose, we used PCR analysis approach using T3 and T7 primers and cDNA preps from the single colonies. An average size of 2 kb was obtained. The library was divided into 100 pools of 10,000 PFUs/each and single pools were transiently transfected into COS-7 cells using Fugene 6 (Roche) according to the manufacture's protocol.

The cDNA library, average insert size 2 kb, was divided into 100 pools of 10,000 PFUs/each and single pools were transiently transfected into COS-7. We calculated that a 200-µL sample (1000 cells/µL) would produce 20–40 positive events in a positive pool. Osteocyte cells without fluorescently labeled ligand were used as a negative control.

V. Results

A. Sensitivity Trials on Primary B-cell Lymphoma Cells

From work to date we know that two of the strengths of the PMC are (1) rare-cell measurements and (2) measurements on primary cells or on cultures where available sample is limited. Below we show results for a simple dilution study using murine B-cell lymphoma cells (Fig. 6). The study was undertaken to prepare for larger studies that will use, in one case, human clinical samples and, in a second case, murine blood samples for active monitoring of cancer treatment and regression in mouse models. We used splenocytes from the fresh spleen of an existing transgenic mouse model that constitutively expresses a double bromodomain-containing 2 (BRD2) GFP fusion (Greenwald, 2004).

Samples were prepared by quantitative dilution from cell stocks, then presented to the PMC at a flow rate of $\sim$200 µm/s using the detection arrangement of Fig. 2(e). Frozen extracts were used; hence, the preliminary study represents a more difficult case in terms of S/N (weaker GFP marker) relative to fresh clinical samples. However, we expect additional sources of variability in the clinical samples. The high discrimination and, in particular, the high S/N ratio of the PMC allowed statistically significant quantification of the weak markers even down to 1 part in 10,000 (Fig 6(d)).

B. Dilution Study on Clonal Osteocytes

A second dilution-curve study was performed in clonal osteocytic cells (Fig. 6(e)), in preparation for a large-scale screen (Section V.C). Positive dsRed-expressing cells were serially diluted in a background of GFP-expressing cells. Figure 6(e) plots all microfluidic channels for a 384-lane microdevice, but uses eight channels redundantly to collect data for each dilution. This procedure makes use of one of the inherent attributes of the PMC, namely high channel count, to average out flow nonuniformities. The results are same as for the primary cell study above, but with different scan settings.

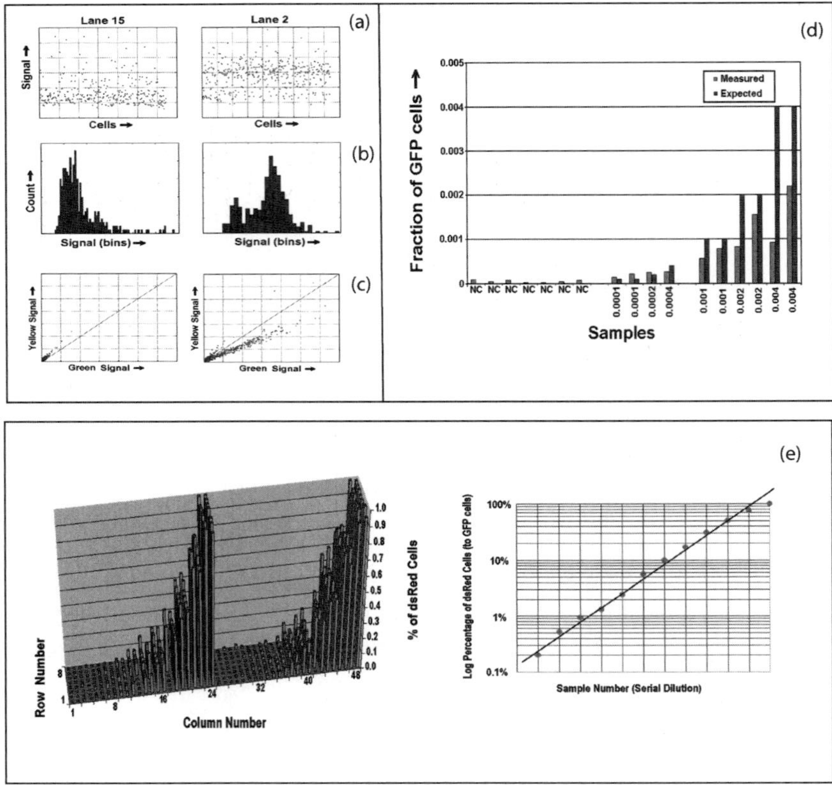

Fig. 6   **(a/b)** Calibration of dilution study on primary leukocytes. For all the objects identified by the scatter detector, we plot the maximum GFP channel value versus the yellow channel value. Note that most objects in the negative sample (Lane 15) have lower fluorescence than the positive sample (Lane 2). **(c)** A more sensitive measure is made by using the ratio of the two PMTs. **(d)** Results for dilution study on primary splenocytes. Measured percentage and expected percentage of GFP-labeled cells for all samples (ordered by expected percentage) shows a clear distinction between negative samples and positive samples down to dilutions of 0.01%. **(e)** Histogram (*left*) of counts for a second cell dilution curve (dsRed-expressing osteocytes diluted serially with GFP-expressing osteocytes). The histogram is organized by well placement on the PMC fluidics. Counts for all 384 microfluidic channels are shown. Sample dilutions are run redundantly in 2-ea. columns of 8-well rows (layout on the microfluidic device), that is, 24 channels for each dilution. Total counts are summed for each sample and used to generate the serial dilution curve (*right*, log vertical scale) that shows slight saturation at the highest concentration of positives (100% positives, right side of the figure). [From McKenna *et al.* (2009). Reproduced with permission of the Royal Society of Chemistry.] (For interpretation of the references to color in this figure legend, the reader is referred to the Web version of the chapter.)

C. Genome-Wide cDNA Screen

The longer integration times of a PMC should increase rare-cell selectivity and thereby allow increased pool sizes for early stages of large screens. This has major implications for a genome-wide screen where the target must be found in an initial pool of a many negatives and where the number of positive cells may number in the single digits per microliter. As a test (McKenna *et al.*, 2009), we chose an ongoing genome-wide cDNA screen for the CPTHR. The classical way to approach a screen of this kind is to (a) separate the several million potential target sequences into a manageable number of initial pools (usually about 10–100 pools), (b) to identify the pool containing the positive sequence, and then to (c) subdivide this pool. This process is repeated until the positive pool is enriched to the level of a single candidate. The most demanding part of the screen occurs in the initial stage, since it requires finding as few as several-dozen positives (antibody-stained clonal oes-teocytes) in a background of a million negatives.

Clonal oesteocytes were incubated with 0.5 mM EGTA for 20 min at 4 °C. Cells were then centrifuged for 3 min at 3000 rpm at room temperature and resuspended in binding buffer. Cell suspensions were incubated with 10^{-6} M biotinylated hPTH (24–84) and streptavidin Texas red for 1 h at 4 °C. Cells were then washed by centrifugation for 3 min at 3000 rpm at 4 °C, then were resuspended in binding buffer.

Cells in a 200-µl buffer volume were loaded into multiple sample wells and pulled through the detection zone of the PMC at a flow rate of 10–20 µL/h per channel. This corresponds to a flow velocity of several hundred micrometer per second. The laser spot (nominal diameter $\sim$30 µm) was adjusted to traverse the biological cell at a much faster scan rate of 10–40 mm/s (0.8–3 ms nominal dwell). Each sample was sampled in 4–10 duplicate channels in our experiments.

To partially automate data reduction, we developed a post-scan data process using Matlab. First, the raw data signal of the red PMT (4) is subtracted from the green PMT (2) (see Fig. 2) to compensate for autofluorescence. The channel locations are then overlaid to segment the data into individual-channel time sequences – about 15 pixels wide by the total number of scans ($\sim$50,000 pixels) long. Each channel segment is searched for scans that contain signal above a noise threshold. These scans are then automatically "cut-and-pasted" to a new image that represents the objects in one channel (accumulated for the run), and the number of events are determined by a software counting algorithm.

Final bright cell counts were entered into a spreadsheet and compared across samples in order to determine run-group statistics median, average, and standard deviation. These values were used to determine the probability that a given pool was negative. Those pools that were above the median plus two standard deviations were retested, and if they still contained outliers were designated for further expansion.

The workflow of the screen is shown in Fig. 7. The initial stage included nine sample pools and one control, all of which were run in redundant microfluidic channels. All samples showed a few positive cells with a median count of 4 and a standard deviation of 12.58. We calculated the boundary for outliers, median plus two sigma, to be 28. One pool was an outlier with 39 positives, and when tested again produced 35 positives. The outlier was subdivided into 20 subpools and each was tested twice. A count of positives produced a median of 5.5, a sample deviation of 12.47, and an outlier boundary of 47. One subpool showed 95, then on recount 98,

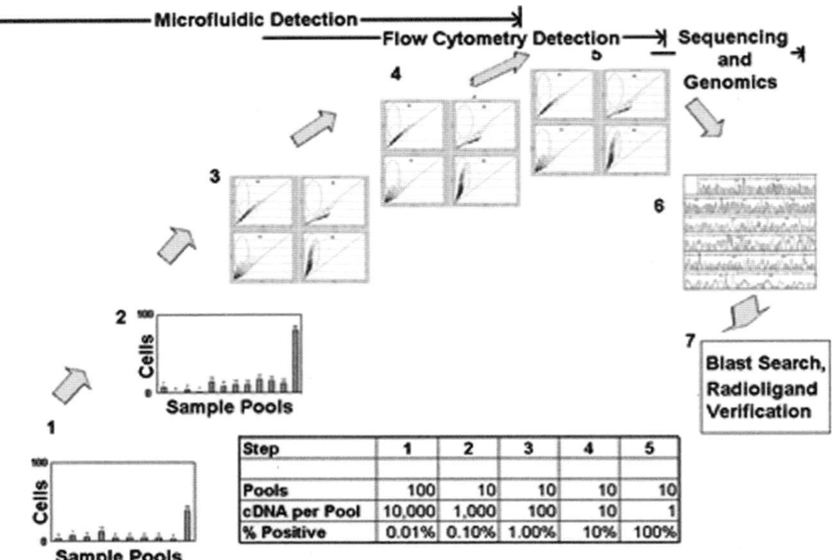

Step	1	2	3	4	5
Pools	100	10	10	10	10
cDNA per Pool	10,000	1,000	100	10	1
% Positive	0.01%	0.10%	1.00%	10%	100%

Fig. 7 Schematic representation of a cDNA expression cloning study that identified a new target for the CPTHR receptor. The most difficult first two stages were completed on the PMC using the rare-cell detection advantages of the variable integration detector.

positive cells. This process was repeated for two more subdivisions until a sample was produced that was overwhelmingly positive ($>$10,000 on the PMC). Levels 3–5, which had much higher abundances of positives, were conducted in parallel on the PMC and on a conventional single-channel cytometer (FACS-Caliber[TM], BD Biosciences). Finally, we isolated a candidate cDNA, which was sequenced by capillary electrophoresis and found to include a seven-transmembrane domain belonging to a family of G-protein-coupled receptors. The sequence was run against the BLAST database and found to be a novel candidate. The end result is that the PMC was able to rapidly perform a full genome-wide cDNA-screening assay with statistically significant results on positive counts of only several dozen cells in background of several million negatives and with sample pools of 200 μl.

D. Adding 1D Imaging to the PMC

The PMC offers a way to increase the throughput of *image-based* HCS into the domain of FACS through a flow architecture rather than static imaging. Specifically, our approach circumvents the rate limitations of the CCD (microscopes and CCD-based flow cytometers) by using a 1D scanner and photomultiplier detection. The principal PMC instrument adjustment is to increase the spatial resolution of the scanner in Fig. 2, and thereby collect multiple intracellular pixels on each cell that is detected in the flow (Fig. 8). The scanner then collects a multicolor image, from each microfluidic channel.

E. Classification of Phenotypes by 1D Images

The economy of 1D images (when compared with CCD images) is a computational advantage (Gonzales and Woods, 2008). However, less image information means more ambiguity. The question becomes: "Can a 1D image provide sufficient information for a high-content screen?" A key aspect for fluorescence localization assays will be a fast analysis algorithm for the binning of image events.

The classification ambiguity typical from 1D imaging as it relates to a protein-localization assay is illustrated for two-colors in Fig. 8. With 1D images, the feature

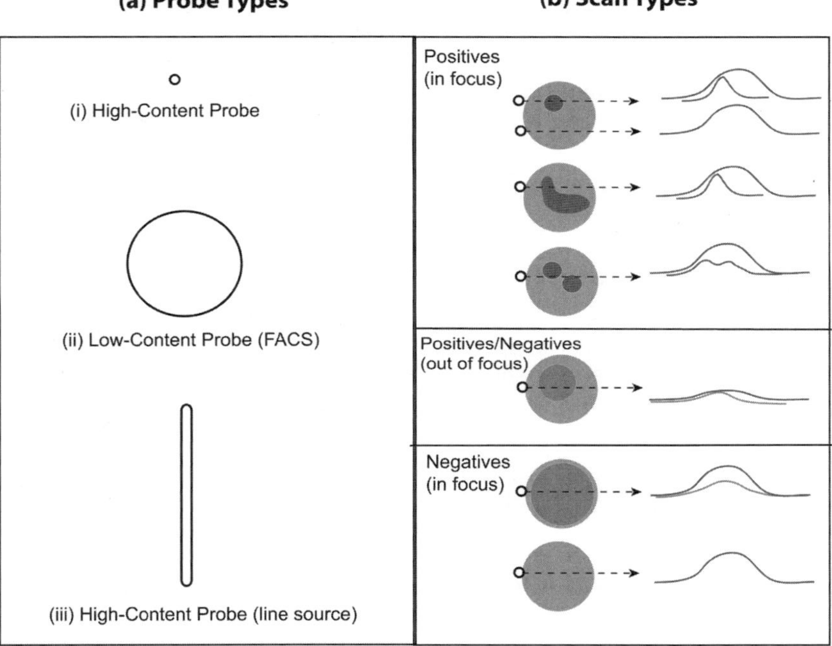

Fig. 8 Cartoon of typical 1D images that are encountered in a protein localization assay. The left column (a) shows several choices of laser probes, both for (HCS) *imaging* (i, iii) and for unresolved total fluorescence (FACS) (ii). The right column (b) shows models of both 2D (microscope) images and 1D scan types, with the marker (green) and cytoplasm (pink). Three positives are shown on top; three negatives on bottom. The confocal slit in our detector discriminates strongly against out-of-focus images. The right side of the right column shows the several principal 1D image types using the probe spot in the upper left (a, i) that are generated depending on how the laser scanner traverses the cell. The dashed arrow shows the location of the single-line scan that is taken per cell. Some of the most diagnostic signatures are surprising. (See plate no. 2 in the color plate section.)

set is greatly reduced relative to 2D. Distinguishing features become asymmetries, profile shape factors, and relative curve heights. However, there is a great deal of information available in 1D; furthermore, even in our initial system, there are four different 1D color images for each cell. Figure 8 also illustrates an interesting problem that was unknown at the onset; it is not clear if the problem of phenotype classification in 1D will become *easier* or *harder* with higher resolution in the scan (smaller laser spot). More detail does not necessarily add to the efficiency of phenotype classification.

Our task is to acquire/model various forms of multicolor 1D images from typical cells and to then partition them into the "positives" and "negatives" typical of a cell assay. The problem is complicated by the several trade-off choices in the optical system and the illumination. In addition, a real assay sample will contain both positives and negatives. The cell types are on a continuum of size/shape/cell-cycle factors, which causes a heterogeneous distribution of 1D images. The exact position of cells in the Z-focus is a complicating factor for all imaging methods (although it is minimized for our confocal detector). The traditional way to approach these problems, all of which are also encountered in CCD imaging, is to set up data filters and thresholds that eliminate ambiguous data. We used the same approach; however, the algorithms and filters are unique to 1D imaging. The metrics of success are partitioning confidence factor (e.g., the Students T test) and the sampling efficiency (as measured in time per assay). For a simple binary (yes/no) assay, the number of discriminating (qualified) objects is as few as 50–100 objects (Taylor *et al.*, 2007). Therefore, since many thousand events per second can often be processed, it is possible to "throw away" a large number of the events and still end up with a fast high-confidence assay.

The problem was addressed with a combination of empirical modeling and data reduction from our data libraries of 2D images. Obviously, the actual 1D data as acquired on real live and fixed-cell samples must ultimately be used to refine the models.

F. Confirmation of 1D Imaging on a PMC

For a feasibility study, we began by modifying one of our prototype PMC systems to reduce the spot size of the scanner to the extent possible (from 30 to 3.5 μm). Next we programmed the signal processing hardware to collect 100 points at 1-μm spacing across the channel. However, our current hardware had the limitation of processing pixels at a maximum 8000 per second. To get around this restraint we limited the range of the scanner to 320 data points at 1-μm resolution and 20 Hz. (However, the optical resolution remained at 3.5 μm.)

We utilized *Saccharomyces cerevisiae* mutants engineered to overexpress the amyloid protein αSynuclein (αSyn-GFP) (Shorter and Lindquist, 2005). In the native state, cells show a uniform distribution of the fusion protein along the membrane and in the cytoplasm. Under induction, the protein condenses to one or

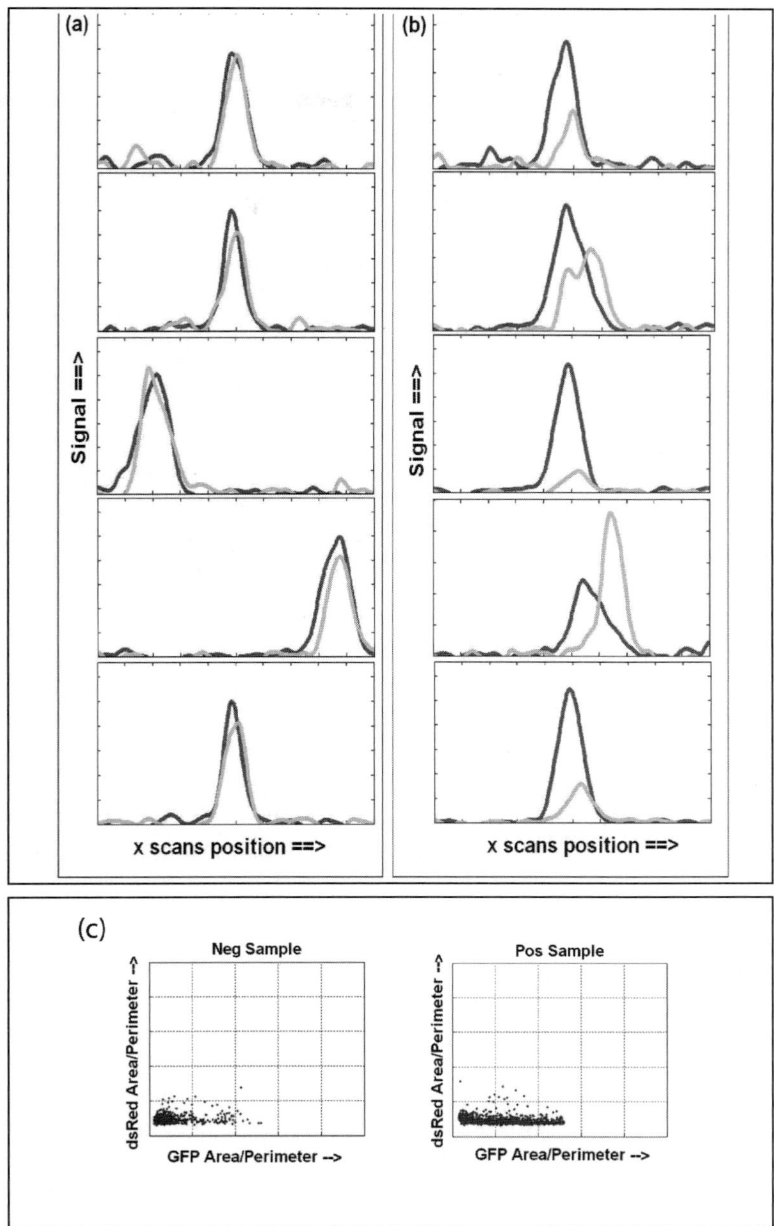

Fig. 9   **(a,b)** Results showing 1D HCS data using a 3.5-μm laser spot to scan αSyn-GFP expression patterns in *S. cerevisiae*. The first version of the detector is (just) able to distinguish the localization patterns. Raw scans for whole-cell (red) and αSyn-GFP (green) in negative cells (a) and positives cells (b), showing αSyn aggregates. Below: Filtered data using a modified "roundness" parameter distinguishes positive (induced) sample from a negative with baseline αSyn expression. (See plate no. 3 in the color plate section.)

several focal conjugates per cell of 1–2 μm diameter. Cells were fixed and fluorescently labeled with a red whole cell dye. Our samples contained a negative control with normally expressed αSyn and positive sample with $\sim$50% of cells overexpressing. Cells were fixed, suspended in PBS at a density of 1000 cells/μl, and then run on the PMC.

A post-scan algorithm identified cells, created a Gaussian-smoothed image for each color channel, and used various comparative color-channel algorithms to categorize images and identify cell metadata. These data were filtered to select a target diameter (red FWHM) of 4–6 μm, and an algorithm modeled after (2D) "roundness" was applied to the red and green channels. As shown in Fig. 9(c), the two populations are clearly distinguished. This was repeated for filters set to various signal X-widths. When we analyzed some subgroups we were surprised to find that we could separate the positive and negative samples using some less-obvious signatures. For example, for small-width thresholding (red FWHM $\sim$ 2 or 3 μm after deconvolution of the laser spot), we found that green signal would occur over threshold in 5–20% of negative samples, but less then 1% for positive samples. Our explanation is that this group represents scans that skirt the center of the cell, and that such scans often entirely miss the αSyn-GFP focal conjugates. This is a novel indirect way to infer the condensed-state positive.

G. Proof of Principle for NT Assay by 1D Imaging

Next we simulated the NT assay. We used mouse fibroblast cells (Swiss-3T3) that were treated with Trypsin EDTA (Cellgro) to make them nonadherent, and then fixed (3.7% formaldehyde) and labeled these cells with Sytox Orange nuclear stain (2.5 μM, Invitrogen). Half of this sample was dyed with a second nuclear stain, (0.5 μM Sytox Green, Invitrogen) and the other half with carboxyfluorescein diacetate succinimidyl ester (CFSE) whole cell stain (5 μM, Invitrogen). Three singly stained, control samples were also scanned in order to obtain PMT color correction information. As above, the cells were scanned in the PMC with a laser spot size of $\sim$3.5 μm and an image digital capture resolution of 1-μm per time point under the lane. A post-scan algorithm identified the cells, smoothed, digitally zoomed the images, color corrected, and normalized the fluorescence levels. We found two methods to separate the samples. The first was by comparing the object width (FWHM) of the orange and green channels (Fig. 10). A more powerful method appears to be to use the orange channel as a (1D) mask of the nucleus, and quantifying the green signal outside that mask (Fig. 10(b)). *Therefore*, even with 3.5-μm spot resolution, 1D line scans can resolve nuclear versus cytoplasmic location of the green marker. A next-generation optical scanner with 1-μm resolution (rather than 3.5-μm) and updated digitizing electronics will greatly increase the number of color channels, allow 1D and 2D line scanning, and enable data collection at increased speed.

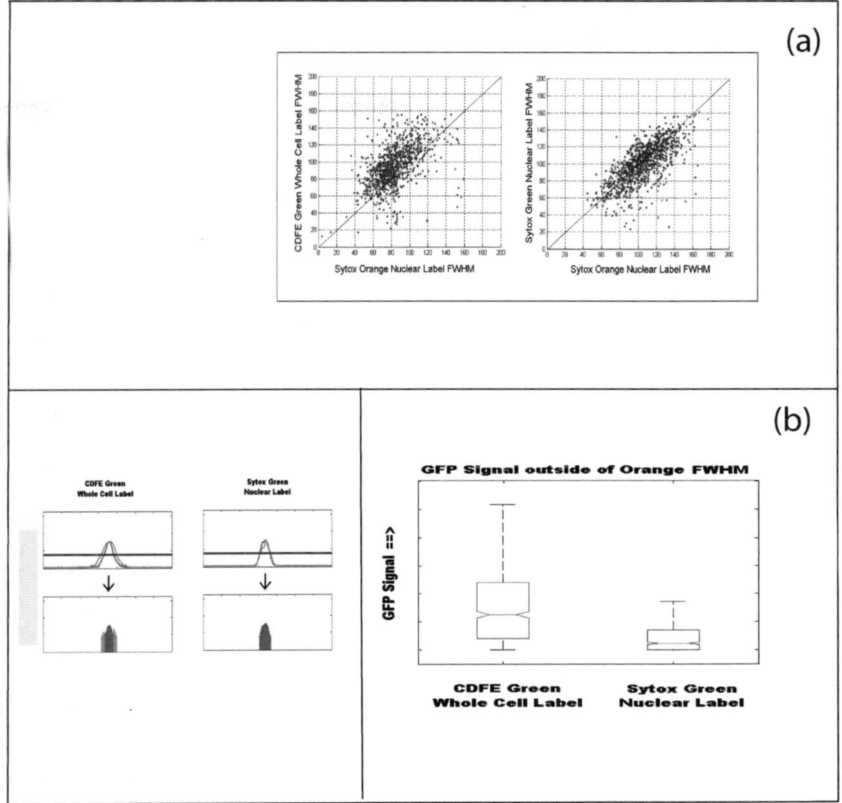

Fig. 10   Proof of principle for a nuclear translocation HCS normally done with 2D microscope images, but done here with 1D images from the PMC. (**a**) The first data reduction uses a ratio of the FWHM diameters of the nuclear and cytoplasmic green markers, respectively, relative to the orange cytoplasmic marker. The results show 1D HCS data using a 3.5-μm laser with 3T3 cells in suspension. The detector is able to distinguish the localization patterns based purely on size (**b,** *left*). (**b,** *right*) The same data evaluated using KS means statistics and an image feature derived as a modified "roundness" parameter. The marker in the nucleus (CFSE) "positive" sample is clearly distinguished from the marker in the cytoplasm (Sytox Green) "negative". (For interpretation of the references to color in this figure legend, the reader is referred to the Web version of the chapter.).

VI. Conclusions

Although parallel microfluidic cytometry is at early stages of development, nonetheless, some of the predicted features have been proven. Four key aspects of the architecture are (i) *parallelism* from the microfluidics, (ii) *high sensitivity* from an optical scanner with variable integration time, (iii) *Parallel flow imaging* with a high-speed analog detector (rather than CCD), and (iv) a *small-sample capability* from the microfluidics.

The 384-channel parallelism, most importantly, improves sample-throughput, but also sidesteps the *time biasing* between samples due to sample changeover in a single channel. The elimination of time biasing addresses issues with unstable samples or degrading markers and, importantly, allows rigorously time-synchronized comparative assays, for example, for biological process with fast kinetics. The scanner permits practical adjustment of integration time, including lengthened signal averaging, which greatly improves performance in rare-cell analyses. The microfluidic flow allows efficient handling of very small and rare-cell samples, for example, a few microliters of primary cells. Single-channel cytometers continue to be improved in *some* of these features (e.g., Goddard *et al.*, 2007; Haynes *et al.*, 2009); however, none combines these features.

An increased detection sensitivity relative to conventional flow cytometers, as seen in our dilution studies, is reasonable, given simple S/N arguments. The PMC and single-channel cytometers utilize nearly the same spectral separation and PMT-based photodetection, both operate in the high-signal (rather than photon-counting) regime, both have a dominant noise contribution from the shot noise, and both systems can be operated near photobleaching. This implies a comparable number of signal photons for the two detectors. In the experiments above, we have varied the integration time between 0.8 and 60 ms, up to 3–4 orders of magnitude longer than is typically used in a single-channel cytometer. This permits a 3–4 order-of-magnitude smaller amplification bandwidth and, for a Poisson statistical distribution of noise, an advantage of 1.5–2 orders of magnitude in the S/N for our detector.

Moreover the integration time of the PMC is an elective setting in the system; it is set by the scanner velocity and is independent of minimum flow requirements. On the high-count-rate end both PMC and single-channel cytometers (FCM) are ultimately limited by essentially the same digitizing electronics; therefore, the PMC, when it is run at high flow velocity, can achieve approximately the same total count rates as a high-end FCM. In some initial trials, we have adjusted the PMC for rare-cell capability and high sample-number throughput. This is the optimization for early stages of a genome-wide screen. We have been able to confirm the improved performance in this domain of optimization.

For a binary assay, closer to a classical flow-cytometer assay, that is, abundance of "positives" $\sim$0.1% or higher, we can operate the PMC at an integration time closer to that used in FCM. A realistic sample throughput for a binary assay on the PMC with this tuning is 384 unique samples in 6 min (384-ea. 1 μL samples, 10^3 cells/μL). This might compare with a maximum of approximately 10 unique samples in 6 min for a typical commercial single-channel FCM. However, the 384-well-plate automation that permits the PMC to be integrated with existing high-throughput cell culture is important in order to realize these advantages.

In the imaging application, the PMC has demonstrated an, perhaps, unexpected proficiency in separating samples via highly economical 1D images. Even with a 3.5-μm resolution on a relatively small (5–6 μm) yeast cell, we are able to see condensation of a GFP marker. On mammalian cells, the classical NT assay was simulated, also using a relaxed laser resolution. We definitely expect that 1D imaging on the PMC will be further developed as a means to add "high-content" to flow cytometry. While it is difficult to project the ultimate speed, it is clear that the useful HCS assays will certainly be possible on the PMC at the speed of current (nonimaging) FCM.

There are several intriguing directions that will be developed in the near future. First, there remains a space to be explored at ultrahigh count rate on a PMC. This has only recently become possible with improvements in digitizing electronics. Digitization rates now exceed the maximum rates that can be used on as single-channel flow cytometer. By expanding the flow stream over parallel channels, the throughput of the latest digitization electronics can be used to full effect. The S/N trade-offs need to be explored and low integration time can introduce a trade-off in single-cathode PMTs. However, multicathode PMTs can be used, so there is little real question that substantial count-rate improvements can be achieved over the first PMC prototypes with a further large advance over single-channel cytometers. The *sample* throughput of the PMC already exceeds FCM (via parallel sample loading); in the future, the PMC will also exceed FACS in absolute (single-sample) count rate. We can anticipate an improvement of 10 or more over the current state. This will push flow cytometry into the domain near two 384-plates per minute for a binary assay, that is, well into a space useful for drug discovery.

A second area that needs expanded engineering is the integration of cell sorting onto the PMC (e.g., Bohm *et al.*, 2007), with further addition of good independent logic controllers on each channel and with isolated-well fraction collection. The added value of fraction sorting on a PMC is enormous. It will allow downstream analysis, for example, qPCR or mass spectrometry, on sorted fractions in a massively combinatorial way.

A third unexplored direction for the PMC is into high-time-response kinetics. This is a domain where the comments in the introduction about "tools limiting science" apply. Since it has not been practically possible, without heroic measures, to do cell-resolved studies of kinetics in a massively parallel way, it has not been possible to do statistically significant studies of many aspects of biological kinetics with high time response. We know that the majority of signaling pathways are dynamic on time-scales of minutes. But since there has been no efficient way to measure them, this fundamental aspect of systems biology has remained outside the realm of practical investigation.

Acknowledgements

This work was supported by National Institutes of Health under grant HG-01389. We thank Hafez Salim, F. Richard Bringhurst of the Endocrine Unit, Massachusetts General Hospital for their collaboration in the CTPHR screen, and Brooke Bevis and Susan Lindquist of the Whitehead Institute for providing the *S. cerevisiae* mutants.

References

Aborn, J. H., El-Difrawy, S. A., Novotny, M., Gismondi, E. A., Lam, R., Matsudaira, P., McKenna, B. K., O'Neil, T., Streechon, P., Ehrlich, D. J. (2005). A 768-lane microfabricated system for high-throughput DNA sequencing. *Lab Chip* **6**, 669–674.

Bohm, S., Gilbert, J., and Deshpande, M. (2007). Method and apparatus for sorting particles. US Patent No. 7,157,274.

Bullen, A. (2008). Microscopic imaging techniques for drug discovery. *Nature Rev.* **7**, 54–93.

Carpenter, A. E., Jones, T. R., Lamprecht, M. R., Clarke, C., Kang, I. H., Friman, O., Guertin, D. A., Chang, J. H., Lindquist, R. A., Moffat, J., Golland, P., Sabatini, D. M. (2006). CellProfiler: image analysis software for identifying and quantifying cell phenotypes. *Genome Biol.* **7**, R100 Epub 2006 Oct 31.

Cheung, K., Berardino, C., Di, M., Schade-Kampmann, G., Hebeisen, M., Pierzchalski, A., Bocsi, J., Mittag, A., Tarnok, A. (2010). Microfluidic impedance-based flow cytometry. *Cytometry A* **77A**, 648–666.

Ding, G. J. F., Fischer, P. A., Boltz, R. C., Schmidt, J. A., Colaianne, J. J., Gough, A., Rubin, R. A., Miller, D. K. (1998). Characterization and quantitation of NF-κB nuclear translocation induced by interleukin-1 and tumor necrosis factor-α: development and use of a high capacity fluorescence cytometric system. *J. Biol. Chem.* **273**, 28897–28905.

Dittrich, P. S., and Schwille, P. (2003). An integrated microfluidic system for reaction, high-sensitivity detection, and sorting of fluorescent cells and particles. *Anal. Chem.* **75**, 5767–5774.

Edwards, B. S., Oprea, T., Psossnitz, E. R., and Sklar, L. A. (2004). Flow cytometry for high-throughput, high content screening. *Curr. Opin. Chem. Biol.* **8**, 392–398.

Eggert, U. S., and Mitchinson, T. J. (2006). Small molecule screening by imaging. *Curr. Opin. Chem. Biol.* **10**, 232–237.

El-Difrawy, S. A., Lam, R., Aborn, J. H., Novotny, M., Gismondi, E. A., Matsudaira, P., McKenna, B. K., O'Neil, T., Streechon, P., Ehrlich, D. J. (2005). High throughput system for DNA sequencing. *Rev. Sci. Instrum.* **76**, 074301–074301-7.

Emmelkamp, J., Wolbers, F., Andersson, H., DaCosta, R. S., Wilson, B. C., Vermes, I., Van den Berg, A. (2004). The potential of autofluorescence for the detection of single living cells for label-free cell sorting in microfluidic systems. *Electrophoresis* **25**, 3740–3745.

Fu, A. Y., Chou, H. P., Spence, C., Arnold, F. H., and Quake, S. R. (2002). An integrated microfabricated cell sorter. *Anal. Chem.* **74**, 2451–2457.

Gawad, S., Cheung, K., Seger, U., Bertsch, A., and Renaud, P. (2004). Dielectric spectroscopy in a micromachined flow cytometer: theoretical and practical considerations. *Lab Chip* **4**, 241–251.

George, T. C., Fanning, S. L., Fitzgerald-Bocarsly, P., Medeiros, R. B., Highfill, S., Shimizu, Y., Hall, B. E., Frost, K., Basiji, D., Ortyn, W. E., Morrissey, P. J., Lynch, D. H. (2006). Quantitative measurement of nuclear translocation events using similarity analysis of multispectral cellular images obtained in flow. *J. Immunol. Methods* **311**, 117–129 Epub 2006 Mar 10.

Givan, A. L. (2001). The sorting of cells. Wiley-Liss, Inc, Wilmington, DE 159-L 174.

Goddard, G. R., Sanders, C. K., Martin, J. C., Kaduchak, G., and Graves, S. W. (2007). Analytical performance of an ultrasonic particle focusing flow cytometer. *Anal. Chem.* **79**, 8740–8746.

Goedecke, N., McKenna, B., El-Difrawy, S., Carey, L., Matsudaira, P., Ehrlich, D. (2004). A high-performance multilane microdevice system designed for the DNA forensics laboratory. *Electrophoresis* **25**, 1678–1686.

Gonzales, R. C., and Woods, R. E. (2008). Chapter 10: Image segmentation. pp. 689–794. Prentice-Hall, Upper Saddle, NJ.

Gough, A. H., and Johnston, P. A. (2007). Requirements, features, and performance of high content screening platforms. *Methods Mol. Biol.* **356**, 41–61.

Greenwald, R. J., Tumang, J. R., Sinha, A., Currier, N., Cardiff, R. D., Rothstein, T. L., Faller, D. V., Denis, G. V. (2004). E mu-BRD2 transgenic mice develop B-cell lymphoma and leukemia. *Blood* **103**, 1475–1484.

Haney, S. A., LaPan, P., Pan, J., and Shang, J. (2008). High-content screening moves to the front of the line. *Drug Discovery Today* **11**, 889–894.

Haynes, M. K., Strouse, J. J., Walter, A., Leitao, A., Curpan, R. F., Bologa, C., Oprea, T. I., Prossnitz, E. R., Edwards, B. S., Sklar, L. A., Thompson, T. A. (2009). Detection of intracellular granularity induction in prostate cancer cell lines by small molecules using the HyperCyt® high-throughput flow cytometry system. *J. Biomol. Screen* **14**, 596–609.

Lee, S., and Howel, B. J. (2006). High-content screening: emerging hardware and software technologies. *Methods Enzymol.* **414**, 468–483.

Lin, A., Hosoi, A., and Ehrlich, D. J. (2009). Vertical hydrodynamic focusing in microchannels. *Biomicrofluidics* **3**, 014101–014112.

McClain, M., Culbertson, C., Jacobson, S., and Ramsey, M. (2001). Flow cytometry of *Escherichia coli* on microfluidic devices. *Anal. Chem.* **73**, 5334–5338.

McCoy, J. P. (2007). Basic principles in clinical flow cytometry. *In* "Flow Cytometry in Clinical Diagnostics," (D. F. Keren, J. P. McCoy, and J. L. Carey, eds.), pp. 15–34. American Society for Clinical Pathology Press, Chicago.

tMcKenna, B. K., Salim, H., Bringhurst, F. R., and Ehrlich, D. J. (2009). 384-Channel parallel microfluidic cytometer for rare-cell screening. *Lab Chip* **9**, 305–310.

Pepperkok, R., and Ellenberg, J. (2006). High-throughput fluorescence microscopy for systems biology. *Nat. Rev. Cell Biol.* **7**, 690–696.

Shapiro, H. M. (2003). Chapter 6: Flow sorting. pp. 257–272. Wiley-Liss, Inc, Wilmington, DE.

Taylor, D. L., Haskins, J. R., and Giuliano, K. A. (2007). Assays and applications of high content screening. pp. 353–434. Humana Press, Totowa, NJ.

Wang, M. M., Tu, E., Raymond, D. E., Yang, J. M., Zhang, H., Hagen, N., Dees, B., Mercer, E. M., Forester, A. H., Kariv, I., Marchand, P. J., Butler, W. F. (2005). Microfluidic sorting of mammalian cells by optical force switching. *Nature Biotech.* **23**, 83–87.

Wheeler, D. B., Carpenter, A. E., and Sabatini, D. M. (2005). Cell microarrays and RNA interference chip away at gene function. *Nature Genetics* **37**, s25–s30.

Wolff, A., Perch-Nielsen, I. R., Larsen, U. D., Friis, P., Goranovic, G., Poulsen, C. R., Kutter, J. P., Telleman, P. (2003). Integrating advanced functionality in a microfabricated high-throughput fluorescent-activated cell sorter. *Lab Chip* **3**, 22–27.

Yi, C., Li, C. W., Ji, S., and Yang, M. (2006). Microfluidics technology for manipulation and analysis of biological cells. *Anal. Chim. Acta* **560**, 1–23.

CHAPTER 4

Microfluidic Systems for Live Cell Imaging

Philip Lee, Terry Gaige and Paul Hung

CellASIC Corporation, Hayward, California, USA

METHODS IN CELL BIOLOGY, VOL 102

0091-679X/10 $35.00
DOI 10.1016/B978-0-12-374912-3.00004-3

Abstract

Microfluidic systems provide many advantages for live cell imaging, including improved cell culture micro-environments, control of flows and dynamic exposure profiles, and compatibility with existing high resolution microscopes. Here, we will discuss our approach for design and engineering of microfluidic cell culture environments as well as interfacing with standard laboratory tools and protocols. We focus on an application specific design concept, whereby a shared fabrication process is used to deliver multiple products for different biological applications. As adoption of advanced *in vitro* models increases, we envision the use of microfluidic cell culture technology to become commonplace.

I. Introduction

The ability to observe live cells *in vitro* is critical for cell biology research. Advances in microscopy technology (Gerlich and Ellenberg, 2003) and fluorescent intracellular probes (Giepmans *et al.*, 2006) provide researchers with unprecedented access to the inner dynamics of living cells. However, the culture environments used for such studies still rely on static monolayer culture on plastic or glass dishes. There is currently a need for improved culture systems that can maintain live cells in more physiologically relevant environments to give the researcher the ability to perform experiments not possible with existing methods. Microfluidics technology offers a promising solution to this challenge (Kim *et al.*, 2007; Wu *et al.*, in press). Using technologies originally borrowed from the semiconductor industry, microfluidics enables bioengineers to create microscale cell culture devices with properties similar to those found in living tissues.

In this chapter, we will discuss the use of microfluidics technology to create systems for live cell experimentation. We will cover the key physical properties of microfluidic environments as they pertain to cell culture, current microfabrication techniques, control systems, design aspects, and example applications. The intention of this work is to provide a resource for biologists interested in understanding the basic concepts, engineering methodologies, and applications of microfluidic systems for live cell imaging.

II. Physical Properties of Microfluidic Cell Culture

The typical microfluidic channel has a minimum dimension on the scale of 1–1,000 μm. A standard microfluidic network may consist of a set of 10 interconnected channels, each 100 μm in height and width, and 10 mm in length, giving a total fluid volume of 1 μl. On this scale, it is important to consider the fluid mechanics and mass transport differences in comparison with a standard culture dish (a 60 mm culture dish has no flow, and roughly 4,000 μl of volume). The key physical properties in a microfluidic environment are discussed in this section, with emphasis on cell culture. The core concept is that since human cells *in vivo* survive in a microfluidic (tissue) environment, the benefits of artificially engineered microfluidic culture environments will be advantageous for *in vitro* experimentation (Fig. 1).

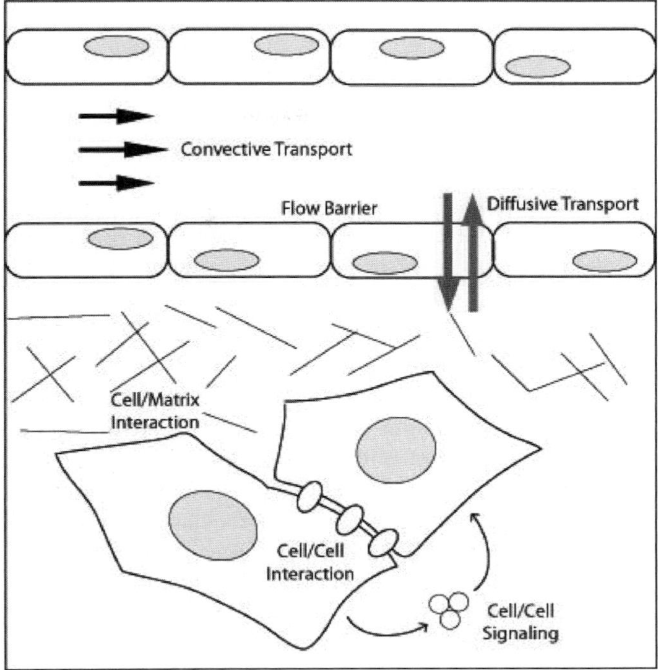

Fig. 1 Physical aspects of the cell culture microenvironment.

A. Volume, Surface Area, and Cell/Medium Ratio

Anyone who has cultured cells *in vitro* and also worked with isolated tissues realizes the drastic difference in the volumetric cell density between the two. For example, a typical cell culture dish grown to confluence provides roughly 1×10^6 cells/ml. The density of cells in tissue is approximately 1×10^8 cells/ml, 100 times higher than that in a culture dish. The average cell monolayer is 1–4 μm in thickness, and occupies 0.2% of the culture volume (Fig. 2). In a microfluidic chamber, a ceiling height of 50–100 μm is typical, allowing the cells to occupy 4% of the culture volume. More advanced microfluidic designs can increase this ratio to 50% or higher.

B. Batch versus Continuous Reactors

A fundamental problem of the culture dish is that it operates as a static batch reactor. This means that the cells sit in a bath of medium until it is emptied and refilled. The concentration of nutrients depletes, and waste products accumulate. A sufficient volume of medium is necessary to buffer against starvation, leading to the large medium/cell volume discussed above. Most troubling is the fact that the soluble

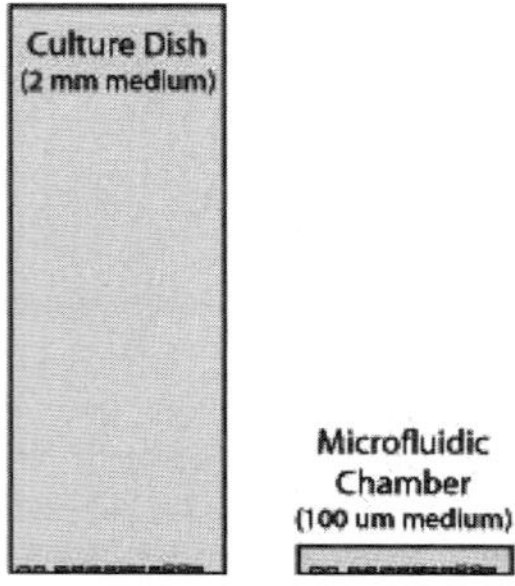

Fig. 2 Cell/medium volume in a typical culture dish and microfluidic chamber. Cell monolayer represented at the bottom of the culture vessel.

factor environment is constantly changing over time, making it impossible to maintain a steady concentration of solutes. In the engineering industry (and *in vivo*) these problems are avoided by using a continuous flow reactor where the cells are fixed in a culture chamber with continuous perfusion of medium in and out of the chamber. Almost all microfluidic cell culture chambers operate on this principle. This reactor type allows much smaller vessels while providing the same exposure of medium per cell per day. From a biological standpoint, the greatest benefit of a perfusion culture environment versus the static dish is the ability to preserve steady-state environments where fresh medium flows in and waste products are removed (Fig. 3). This also provides a more physiologically accurate model of drug exposure and mass transport to cultured cells.

Property	Dish	Microfluidic
Reactor Type	Static Batch	Continuous Flow
Kinetics	Time Dependent	Flow Rate Dependent
Steady State?	No	Yes
Volume	4-10 ml	10-100 nl
Residence Time	3 days	5-15 min
Medium/Cell/Day	2 nl/cell/day	5 nl/cell/day

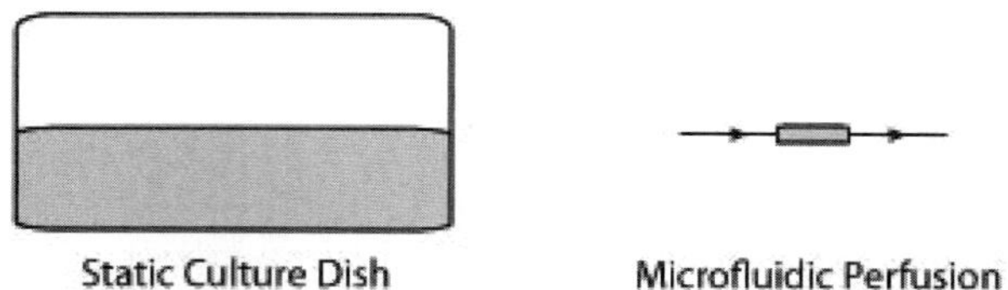

Fig. 3 Comparison of static batch culture with continuous perfusion. Typical values for dish and microfluidic cultures given in the table.

C. Laminar Flow

For a fluidic system, the flow characteristics can be determined with the Navier–Stokes equation, which describes a momentum balance on fluid within the channel (Janakiraman *et al.*, 2008). The general form used for pressure-driven flow is

$$\rho(d_t v + (v \cdot \Delta)v) = -\Delta P + \eta \Delta^2 v$$

where ρ is the fluid density, d_t is the material time derivative, v is velocity, P is the pressure, and η is the viscosity. The left-hand side of the equation describes the momentum of the fluid. The right-hand side of the equation describes the body forces acting on the fluid, and consists (in this case) of a pressure gradient and a viscous drag. Due to the small-length scales of microfluidic flows, the Reynolds number is very small (typically < 1) and fluid motion is described as laminar. This can be construed to mean that fluid particles move in generally straight, predictable paths with no translation or time variance. The invariance of the laminar flow at low Reynolds number means that the left-hand side of the equation can be neglected (the derivatives go to zero), and the simplified relation is thus

$$\Delta P = \eta \Delta^2 v$$

This relation can be integrated to obtain the velocity profile in the form

$$v = f(y, z, L, h, w, \Delta P, \eta)$$

where the velocity is a function of position within the channel, channel geometry, pressure gradient, and viscosity. Integrating the velocity over the cross-sectional area gives the flow rate through the channel Q:

$$Q = f(L, h, w, \Delta P, \eta) = \frac{\Delta P}{R}$$

which turns out to be a linear relationship between the pressure drop and a term R called the hydraulic resistance. Note the resemblance of this formula to Ohm's law for electrical circuits. For microfluidic systems, it is then possible to describe fluid velocity in terms of resistive networks. For a cylindrical pipe, the exact solution to this equation is given by the Hagen–Poiseuille equation. Since the typical microfluidic channel is rectangular in the cross section, there is no finite analytical solution, but a reasonable approximation is

$$R_{rect} \cong \frac{12 \eta L}{1 - 0.63(h/w)} \frac{1}{h^3 w}$$

Laminar flow provides a number of physical properties that are beneficial for live cell studies. The lack of turbulence in the flow ensures that the cells are exposed to a uniform velocity profile. The defined flow lines in the laminar regime also provide crisp transitions between solutions. This is commonly evidenced in two scenarios: solution switching and parallel flow. During solution switching (when cells are

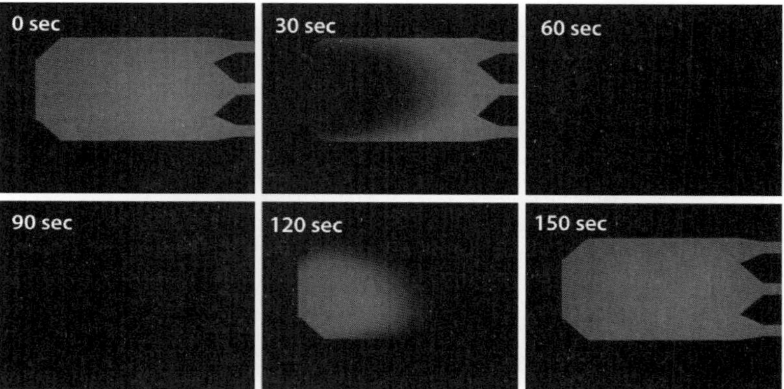

Fig. 4 Laminar flow switching through a microfluidic cell culture chamber. A $2 \times 1 \times 0.1$ mm chamber is switched between a fluorescent dye (Texas Red Dextran-10 kDa) and buffer solution. Flow is from left to right. Note the clear boundary between the two solutions, and the complete washout as the new solution flows in. (For interpretation of the references to color in this figure legend, the reader is referred to the Web version of this chapter.)

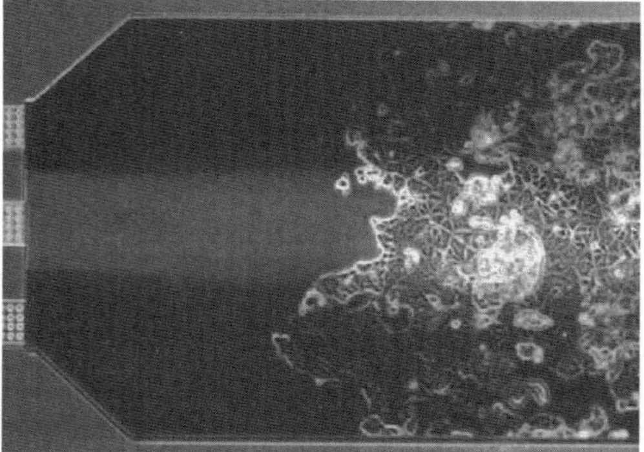

Fig. 5 Parallel laminar flows. Here, three solutions (red, clear, blue) are flowing in parallel from left to right across cells cultured in a $2 \times 1 \times 0.1$ mm microfluidic chamber. Due to the laminar dynamics, the three solutions form three distinct regions in the chamber with minimal diffusion across the interface. (See plate no. 4 in the color plate section.)

exposed to different mediums over time), the laminar flow profile creates a sharp transition between the solutions with minimal mixing at the interface (Fig. 4). During solution change, the new solution rapidly and completely "sweeps" away the old solution, enabling more precise monitoring of cell response.

When more than one laminar stream is flowed in parallel, the solutions create a distinct boundary between the flows, with only diffusive mixing at the interface. This allows creation of unique flow profiles as depicted in Fig. 5.

D. Nutrient Exchange

Mass transport on the microscale is significantly different than that on the macroscale. The most notable difference is the increased contribution of diffusion at the microscale. Diffusion is the random motion of molecules from regions of high concentration to regions of lower concentration. The typical simplified form of the diffusion equation is

$$t = \frac{x^2}{2D}$$

where t is the mean diffusion time, x is distance, and D is the molecular diffusivity. Note that since t scales to x squared, the impact of diffusion is negligible at larger length scales. The typical diffusivity of biological molecules in solution is between 2×10^{-5} cm/s (ions) and 7×10^{-7} cm/s (proteins). Plugging in a value of $D = 10^{-6}$ cm/s gives a diffusion distance of 0.8 mm over 1 h. For a typical microchannel with a 100 μm width, the diffusion time is 50 s. To diffuse 3 cm (across a 60 mm culture dish), it will take 52 days. This means that diffusive transport of nutrients and wastes to/from cells is only reasonable when the length scale is below 1 mm. In living tissues, an extensive blood capillary network delivers nutrients to within a few hundred microns of all cells.

The second component to a diffusive transport method for cell culture is the convective flow. Diffusion requires a constant concentration gradient, meaning there needs to be a continuous supply of fresh medium to the channels. In a microfluidic system (as in your body) this is achieved through rapid convective flow through transport channels coupled with diffusion out of the channels to the cells. A useful engineering term that relates convection with diffusion is the Peclet number, defined as $Pe = Lv/D$, where L is the length scale, v is the velocity, and D is the diffusivity. The value of Pe gives the ratio of convective to diffusive transport, also an estimate of how quickly nutrients are being replenished by flow compared to the amount diffusing out. For microfluidic cell culture, a Pe between 10 and 100 is preferable.

For maintaining healthy cell cultures, it is important to maintain favorable mass transport conditions. If convective transport is insufficient, the cells will starve. If it is too fast, important signaling factors may be washed out. Similarly, if the diffusion distance is too far, there will be a concentration gradient based on the cell location. If the diffusion distance is too short, it becomes difficult to fit sufficient cells in the limited space.

In order to separate convective and diffusive transport, it is necessary to utilize a flow barrier that does not limit molecular diffusion. The human body achieves this goal with the use of endothelial cell membranes. The cells shield interstitial cells from the blood flow, but freely transport nutrients and waste across their membranes. In a similar fashion, microfluidic barriers can be fabricated with similar properties (see Fig. 6). The principle is to create a high fluidic resistance barrier that blocks

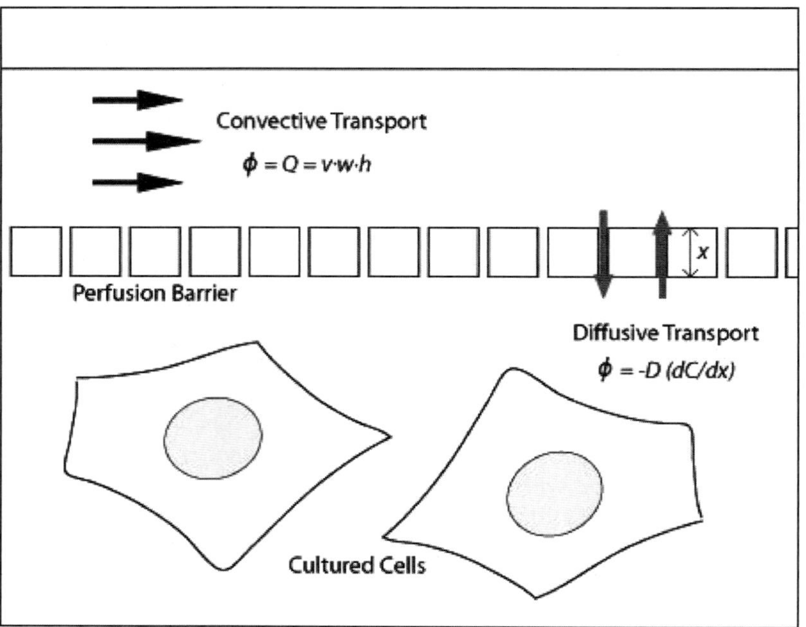

Fig. 6 Convective and diffusive transport in a cell culture system. Convective flux is given by the flow rate, equal to velocity times the channel cross-sectional area. Diffusive flux is given by the diffusivity times the concentration gradient.

convective flow, but allows free diffusion. This is most commonly achieved with a porous wall running parallel to the flow channel such that nutrient diffusion can occur along the length of the barrier. The practical implementation of this design will be discussed in Section IV.D.

E. Shear Stress

Liquid flowing past a surface (or cell) will create a shear force. The shear is proportional to the flow velocity at the surface of the cell. While it is difficult to calculate the shear stress on a dynamic 3D object, like a cell, a useful approximation for shear at the surface of a pipe under laminar flow is:

$$\gamma = \frac{4Q}{\pi R^3}$$

where γ is the shear rate (1/s), Q is the volumetric flow rate, and R is the pipe radius. The shear stress is γ multiplied by the liquid viscosity. For a microfluidic channel, the flow rate Q is proportional to R^4, which means the shear rate is proportional to

$R^4/R^3 = R$. Therefore, the smaller the channel dimension, the lower the shear stress (for a fixed pressure drop). A table of example shear stresses is given below:

Case	Channel radius (μm)	Flow rate (μl/h)	Velocity (μm/s)	Shear stress (dyn/cm^2)
Arteriole	15	10	3900	9
Microfluidic channel (slow)	25	4	570	0.8
Microfluidic channel (fast)	25	40	5700	8
Cell culture chamber (slow)	100	4	35	0.01
Cell culture chamber (fast)	100	40	350	0.1

While shear is physiologically important for many types of cells (mainly endothelial cells), a large number of cell types are not normally exposed to shear. For most cell culture applications, it is beneficial to reduce the shear stress on cells. The important variables to keep in mind here are to avoid high flow rates through narrow channels. For endothelial cells, it is generally observed that shear stresses around 10 dyn/cm^2 will elicit cellular responses (Janakiraman *et al.*, 2008). This is typical of shear in physiologic blood flow (Renard *et al.*, 2003). For a normal microfluidic cell culture channel, the range of shear stresses is between 0.5 and 10 dyn/ cm^2. In the cell culture chamber, this is reduced to below 0.1 dyn/cm^2, approximating interstitial flow (Rutkowski and Swartz, 2007).

1. Material Interactions

Cells are sensitive to the materials they are in contact with. While there are numerous materials used for cell culture, this chapter will deal with polydimethyl siloxane (PDMS) on glass microfluidic channels and chambers. This is a well-studied combination of materials for cell culture in microfluidics, and has the benefits of cell compatibility, optical transparency, and ease of fabrication (Chiu *et al.*, 2000; Hung *et al.*, 2005b; Regehr *et al.*, 2009). It should also be noted that PDMS (in the absence of surface modification) absorbs organic solutions and low molecular weight hydrophobic compounds. Therefore, this type of system may not be suitable for experiments requiring very low concentrations (nM) and volumes (<10 μl) of highly hydrophobic compounds (Regehr *et al.*, 2009). For our microfluidic cell culture applications, the flow volumes tend to be 100–300 μl/day, minimizing the absorption problem.

III. Microfabrication Methods

Microfabrication technology, which paved the way for the established semiconductor industry in the past decades, enables the fabrication of microsized features to approximate the physiologic cellular time and length scale. This section will introduce the associated methods to fabricate our microfluidic cell culture array with a modified version of PDMS replicate molding for more robust device manufacturing and an easy chip-to-world interface.

A. PDMS Molding

PDMS is a soft polymer with backbones consisting of repeating monomer [SiO$(CH_3)_2$] units. By mixing the liquid monomers with cross-linking molecules to connect the backbones, PDMS exhibits a wide range of elastic properties depending on the ratio between the monomer and the cross-linker. PDMS is optically transparent in the visible spectrum, considered inert and nontoxic to cell cultures. With the low cost of the material itself, PDMS replicate molding processes have become a popular tool in academia for rapid prototyping of microfluidic devices for various biological applications (5). However, the conventional thick slab PDMS devices require labor intensive hole-punching processes for the fluidic connections. In addition, the tedious tubing and syringe pump connections also increase the difficulty and reliability of setting up experiments in a timely fashion. In order to solve these issues, we have developed a modified PDMS replicate molding process to reduce the thickness of the PDMS layer to around 250 μm under a 1 mm thick acrylic sheet. The resulting device can be machined by a laser for fluidic connections. The acrylic sheet backing further provides easy integration with fluidic reservoirs such as standard well plates using adhesives.

Figure 7 shows the schematics of the modified PDMS replicate molding processes. The processes begin with a master template having microfluidic structures made with negatively toned epoxy SU-8 (Microchem) on a 6″ silicon wafer. A PMMA sheet 6″ in diameter and 1 mm in thickness is chemically modified to attach to PDMS, forming a sandwich configuration where the PDMS replicate molding process takes place between the PMMA sheet and the master template. After detaching the microfluidic device from the master template, a CO_2 laser is used to create fluidic connections. Adhesives are then applied to the bottom of the fluidic reservoirs for attachment of the microfluidic device. Finally, the device is enclosed by a coverglass to complete the microfluidic plate.

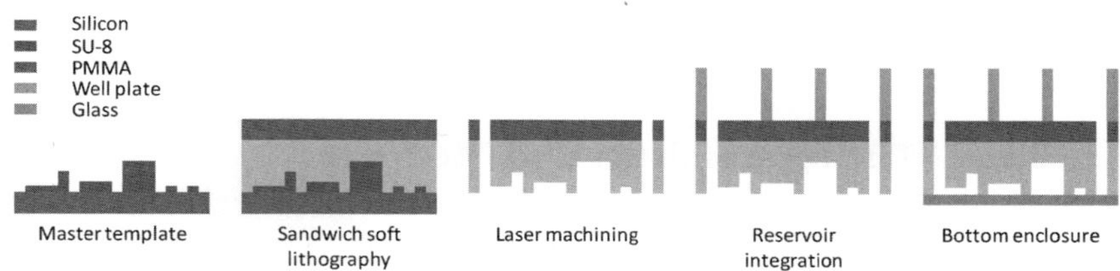

Fig. 7   Process flow of the modified PDMS replicate molding.

The overall processes could further be broken down into the following sectors: master template fabrication, replicate molding, and packaging, which will be described in details in the following sections.

B. Master Template Fabrication

The design of the microfluidic architecture is accomplished using computer assisted design (CAD) software. Most of our designs require various geometries of microfluidic structures, each with a different height for different design considerations. For example, we use 4 μm tall channels to create high-fluidic resistance regions to modulate the flow rate and minimize the back-flow between channels, while we use a 130 μm layer to construct the cell culture chamber. Each layer needs its own photomask because the thickness of the layer is different. After photomasks are generated, the SU-8 epoxy photoresist is used to create the microfluidic structures on 6″ silicon substrates. After each layer of the epoxy is spin coated to its desired thickness, a mask aligner is used to photolithographically transfer the patterns from the mask onto the epoxy layer. For multi-layer SU-8 lithography, the preceding layers of the unexposed SU-8 are left on the surface to preserve the surface flatness for spin coating of the succeeding layers. After the completion of the SU-8 3D microfluidic architecture on the silicon substrate, to improve the robustness of the master template, the master template is coated with a thin layer of fluoropolymer to increase the surface tension. The master templates made using this process have reliably provided more than 200 replications of PDMS molding without noticeable master template degradation.

C. Replicate Molding

As described earlier, the basic principle behind replicate molding is to cast the PDMS over a master template. After the polymer is cured, it will replicate the geometry of the master template with opposite polarity. For example, a post on the master template will result in a hole on the PDMS replicate. The manufacturing processes are very similar to that in the mature injection molding industry, where metal master templates are machined to accommodate melted thermoplastics at temperatures above the glass transition temperature of the plastics. When the temperature cools, the thermoplastic solidifies, and can be separated from the template. PDMS Sylgard 184 is purchased from Dow Corning Corporation. Depending on applications, different tones of PDMS can be prepared. The recommended mixture is 10:1 in weight, with 10 parts of monomers and one part of the crosslink chemical. A PTFE disposable beaker and a paint mixer can be used to mix the PDMS. After mixing, the beaker is placed under a vacuum (25 inHg for 30 min) to remove the bubbles generated during the mixing. After dispensing about 3 mL of the PDMS in the center of the master template, a sheet of chemically modified PMMA is pressed against the mold to sandwich the PDMS between the sheet and the mold with the final PDMS thickness determined by Teflon spacers. The assembly is then cured in a 70 °C oven for 4 h to replicate the microfluidic structures in PDMS. The cured PDMS microfluidic structures adhere preferentially to the PMMA than the fluoropolymer coated master template. After detaching the microfluidic structures from the mold, a CO_2 laser cutter is used to create through-holes at certain locations on the PMMA/PDMS composite sheet as the fluidic connections. To attach the microfluidic device to the well plate, a biocompatible UV adhesive is applied to the bottom of the plate and the microfluidic device is manually attached to the plate. The assembly is then exposed under flood UV lights to completely cure the adhesives.

D. Packaging

The surface of the PDMS can be oxidized temporarily to provide hydroxyl-rich groups for covalent bonding to glass-based materials. To enclose the bottom of the microfluidic structures, the bottom of the plate is treated with oxygen plasma and a 103 mm × 97 mm #1.5 coverglass is bonded. Sterile PBS solution is then pipette into each well and a manifold is used to apply pressure from the top of all wells to remove any bubbles in the microfluidic channels through the out-gassing of the gas-permeable PDMS. After confirmation of the removal of bubbles through visual inspection, the plate is sterilized in a UV-ozone chamber. Finally, a vacuum sealing system is used to package the plate in a sterile medical-grade pouch.

In summary, unlike the conventional PDMS soft lithography where microfluidic channels are formed at the bottom of a bulk slab of elastomer, we have developed a manufacturing process (Fig. 8) to mold the microfluidic features with PDMS on a thin plastic backing out of a fluoropolymer-coated 6″ silicon wafer with SU-8 microarchitectures. The thickness of the PDMS layer is reduced to 250 μm, making it compatible with laser cutting and integration with other plastic parts such as a standard well plate and subsequently sealed with a #1.5 coverglass after oxygen plasma activation. The advantages over "slab" PDMS devices include the

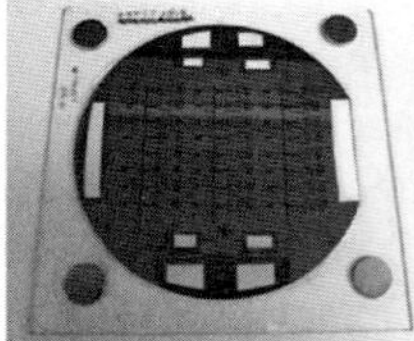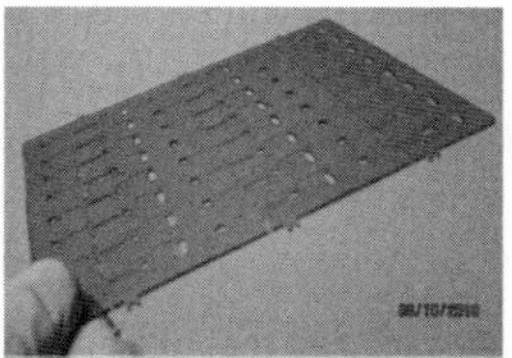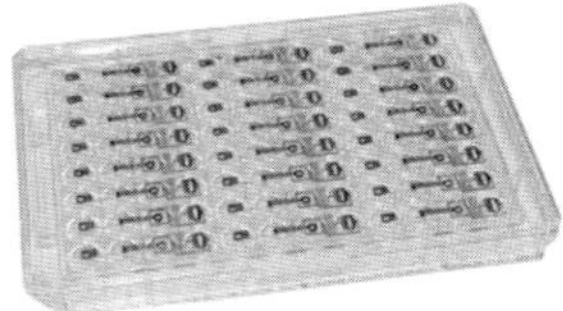

Fig. 8   Fabrication steps. (Left) the microfluidic architecture is patterned on 6″ silicon wafer using SU8 photoresist. (Middle) the novel molding process attached with a thin PDMS microfluidic layer to a 1 mm acrylic sheet. The fluidic/air ports are laser drilled before glass bonding. The blue tape is used for surface protection. (Right) the molded array is glued to an open bottom 96-well plate and sealed with a #1.5 coverglass. (For interpretation of the references to color in this figure legend, the reader is referred to the Web version of this chapter.)

significant reduction of the PDMS per device, no mechanical hole punching through the slabs, the easy assembly with fluidic reservoirs, and the standardized format which is compatible with state-of-the-art analytical instruments.

IV. Flow Control

An ability to control fluid flow is essential for most microfluidic applications. At the most basic level, all fluid flow is caused by pressure differentials. The methods for controlling flow through channels can be categorized into two general ways: volumetrically or pressure driven. Volumetrically controlled flow pushes a specific volume of fluid in a set duration of time; methods to achieve this include syringe pumps and peristaltic pumps. Pressure-driven flows employ an ability to set the pressure at specific locations in a fluid path, often achieved at air/liquid interfaces. For microfluidic systems, pressure-driven flow control tends to have several advantages compared to volumetric control. This chapter covers three major methods of generating pressure differentials: air pressure, gravity, and surface tension. Another method of controlling pressure, electro-osmotic, is not covered because of the material characteristics, manufacturing cost, and operation requirements are not generally compatible with cell culture and microscopy. The flow rate is directly proportional to the pressure drop between two points. The pressure differentials created by air pressure are on the order of 10,000 Pa, surface tension about 1000 Pa, while gravity differentials are around 100 Pa.

Advantages of pressure-driven flow control are numerous. In particular, the flow rate response time and stability in a pressure-driven system are independent of the tubing and microsystem characteristics. Elastic downstream tubing will

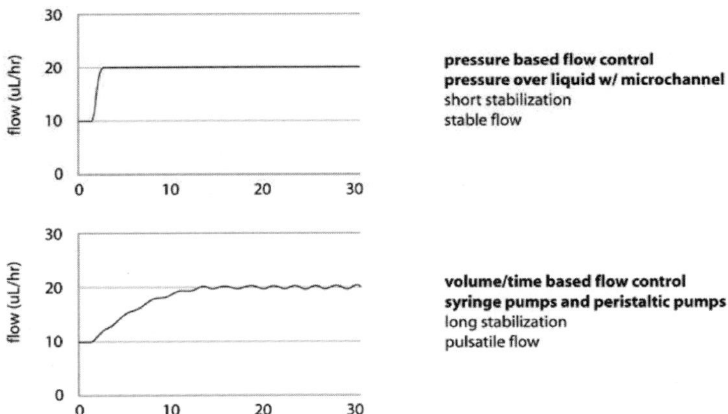

Fig. 9 Comparison of stability of typical pressure-driven versus volume-driven pumps.

reach a force and pressure equilibrium rapidly. Volumetrically driven alternatives will exhibit extremely long stabilization time if downstream tubing is elastic. If the system is very inelastic it becomes highly sensitive to the pulsations caused by the mechanical drives used in syringe and peristaltic pumps (Fig. 9). It is valuable to consider potential failure modes and outcomes as well. In the case of a blocked channel, volumetrically driven flow has the potential to create damagingly high pressures, while a pressure-driven system is naturally limited to the drive pressure.

A. Air Pressure–Driven Flow

Air pressure can modulate flows by directly applying, adjusting or removing pressure above an air–liquid interface. Pressures in the range of 7–70 kPa are useful because they are an order of magnitude larger than pressures caused by surface tension and gravity, but small enough so that damage to the microfluidic device is unlikely and leak-tight seals are easier to construct.

Control of air pressure can be achieved in many ways; a few of the most common are illustrated in Fig. 10. Pulse-width modulated valves as well as variable orifice electronic pressure controllers are generally available for purchase, while piston type pressure controllers can be custom developed. All three methods actively control an output pressure using feedback from a pressure transducer and some physical way to modify the pressure. Each method has certain advantages and disadvantages. For example, the variable orifice controller can be stable, quiet and have a fast response time, but it will consume a lot of supply air even in the steady state when the downstream volume is closed. Accuracies of pressure transducers are typically about 0.2% full scale, no matter what the scale is. If a pressure controller is selected to be accurate in the range of 7–70 kPa, flow rates can still be accurately controlled over several orders of magnitude by using different fluidic resistances (see section V.C).

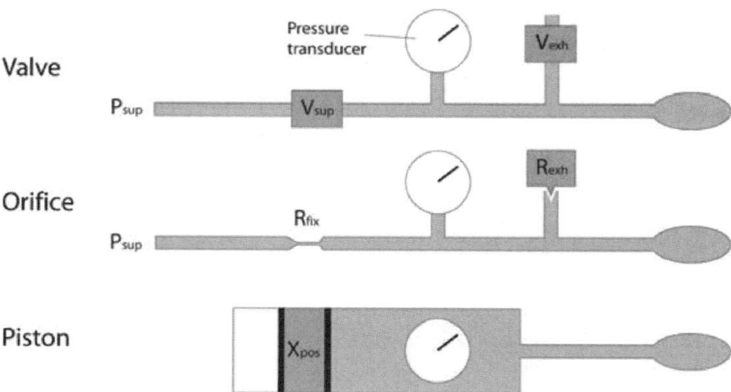

Fig. 10   Common mechanical methods to control fluid pressure.

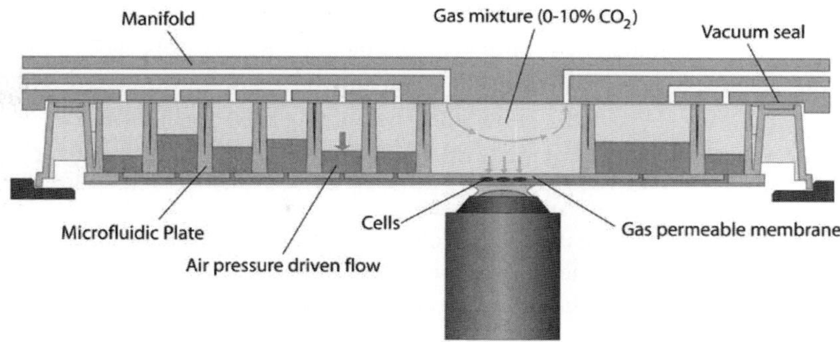

Fig. 11 Schematic of a manifold interface. The manifold seals to the microfluidic plate, allowing control of pneumatic pressures to individual wells.

B. Manifold Interface

With a higher number of inlet and outlet streams per fluidic device, the complexity of tubing and connections becomes extremely difficult to manage. Many reasons drive the development of a manifold style interface as shown in Fig. 11. One, there is a need to create an enclosed pressurized volume for each fluid, while facilitating the addition or subtraction of media. Two, multiple identical parallel flow units can be easily created by splitting a single pneumatic input. Three, a reusable manifold and disposable plate optimally reduce the number of contaminated surfaces. Instead of using a physical clamping method to seal the manifold to the microfludic plate, applying a vacuum creates an evenly distributed mating force which ensures a positive seal on all wells. Additionally, the presence of a negative pressure in the voids between wells completely eliminates the possibility of accidental pressurization of a neighboring well. A simple micro-to-macro interface has the additional advantage of being repeatable and consistent between users thereby making manufacturing, marketing, and user operation more efficient.

C. Microincubation and Microscopy

The manifold and microfluidic plate should be compatible with bright field, phase contrast, and fluorescence microscopy. Most live cell microscopy is performed on inverted microscopes. In order to acquire quality images using the high numerical aperture objectives necessary for higher magnification, the surface on which the cells are cultured should be a thin glass slide. Typically, objectives are optimized for #1.5 glass coverslips which are 170 μm thick. Most plastics are not transparent enough for fluorescence and are often autofluorescent so they should be eliminated from the fluorescent light path. Plastics also disrupt the polarization of light so if differential interference contrast microscopy will be used, plastic in the visible light path should be avoided.

Microfluidic devices have culture volumes on the order of tens on nanoliters and the geometry and construction materials have a low thermal mass and low thermal conductivity compared to typical culture vessels. These characteristics make it possible to use novel microincubation techniques to maintain the desired temperature and pH of the cell environment. High numerical aperture lenses typically have very short working distances and will act as a heat sink drawing heat away from the sample and therefore must be heated to the desired temperature. If the entire cell culture region is within a 5 mm diameter circle and an oil immersion objective is used, it is quite feasible to use the objective heater as the sole heat source. Perfusion of the gas mixture through the gas permeable PDMS to the cell culture region provides the desired dissolved gas concentrations locally without issues caused by evaporation.

D. Gravity-Driven Flow

The ideal method of flow control depends on the particular application. In some situations, such as high throughput screening, features such as constant flow rate and

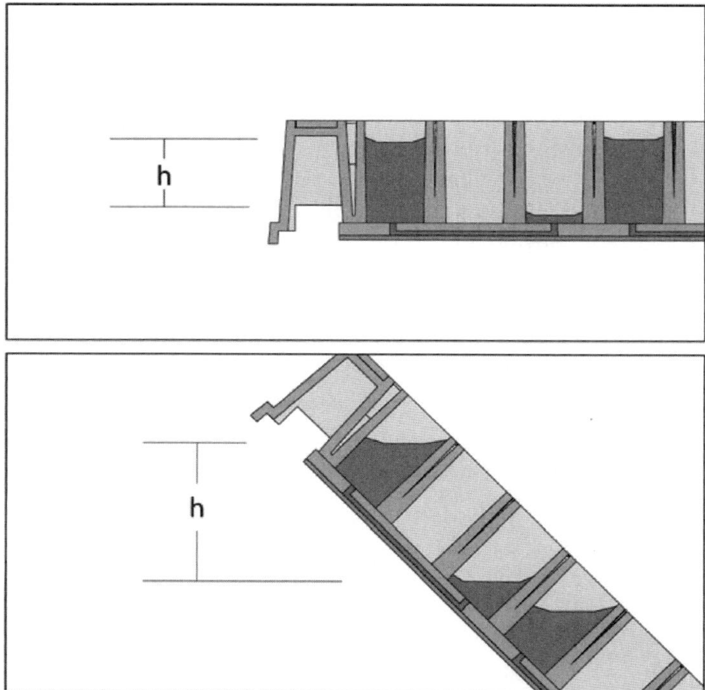

Fig. 12   Schematic of gravity-driven flow as a result of hydrostatic pressure between in the inlet and outlet wells of a microfluidic plate. When laid flat, the liquid level difference is largely determined by well geometry. Tilting the plate enables modulation of the hydrostatic pressure.

the ability to switch between solutions while imaging on a microscope are less important than a flow control method that is highly parallelizable and inexpensive. Gravity-driven flow is a convenient solution. The pressure drop caused by a difference in height between the inlet and outlet will drive flow until the inlet and outlet are leveled. The dimensions of a standard microplate limit the largest pressure differentials that can be created using gravity to the order of 100 Pa.

$$P_{gravity} = \rho g h$$

Microfluidic devices operated with gravity-driven flow can be used with standard automation equipment for repeated feeding operations during the long-term culture. If faster flow rates are desired for certain assays, the device can be tilted, increasing the height differential for faster flow (Fig. 12).

E. Surface Tension Flow

On the microfluidic scale surface tension is significant and the resulting pressures dominate gravitational and even low pneumatic pressures. Surface tension is caused by the cohesive forces between molecules. The molecules at the interfaces between different substances behave differently than those fully embedded because of different attractive forces to neighboring molecules. Where a liquid, solid and gas meet, they form a contact angle, θ, which depends on the relative strengths of these attractive forces. In Fig. 13, tension forces are shown for the liquid–air interface, the liquid–solid interface, and the solid–air interface. Cell culture plastics are typically hydrophobic, having a contact angle greater than 90°, but can be modified to become hydrophilic through certain processes. One such process, exposure to oxygen plasma, is a commonly used technique during fabrication of PDMS microfluidic devices.

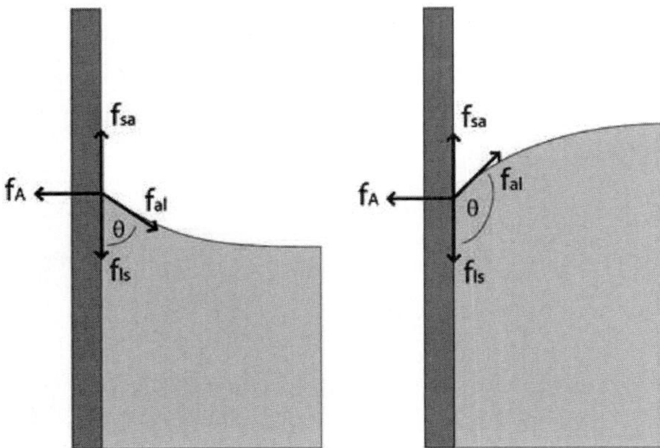

Fig. 13   Surface tension at air/liquid/solid interfaces results in an angle of incidence at the contact point. This angle is useful in describing surface tension forces.

A force balance in the vertical direction gives

$$f_{ls} - f_{sa} = -f_{la}\cos\theta$$

Since the forces are in direct proportion to their respective surface tensions, we also have

$$\gamma_{ls} - \gamma_{sa} = -\gamma_{la}\cos\theta$$

where γ_{ls} the liquid–solid surface tension, γ_{la} is the liquid–air surface tension, and γ_{sa} is the solid–air surface tension.

If the geometry is similar to Fig. 14 such that two wells of different diameters are fluidically connected, the final resting state of the fluid will exhibit a difference in height, h, as a function of the liquid–air surface tension, γ_{la}, the density of the fluid, ρ, the radius of the smaller well, r_s, the radius of the large well, r_l, gravity, g, and the angle of contact described above.

$$h = \frac{2\gamma_{la}\cos\theta}{\rho g}\left(\frac{1}{r_s} - \frac{1}{r_l}\right)$$

The act of removing fluid from the smaller well, effectively reducing h to 0, initiates a pressure to restore the equilibrium of

$$P_{surfacetension} = 2\gamma_{la}\cos\theta\left(\frac{1}{r_s} - \frac{1}{r_l}\right)$$

Surface tension pressure differentials are approximately 1000 Pa in wells with radii between 1 and 3 mm.

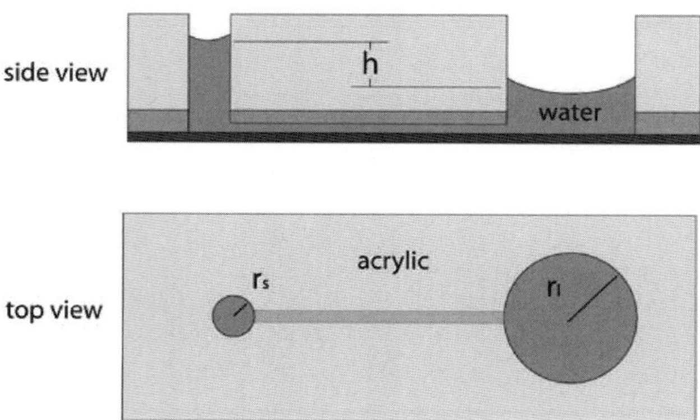

Fig. 14 Surface tension-driven flow depends on the dimensions of the air/liquid interface. If wells of different diameters connect a microfluidic channel, the surface tension will drive liquid flow until the liquid height difference balances the surface force.

V. Design Aspects

In this section, we introduce the practical design issues for creating microfluidic networks for cell culture studies. The process of designing microfluidic devices should follow an "application-specific" path, with careful consideration of the intended features and end users. We also advocate that the job of the microfluidics engineer is to deliver finished products that can be put to widespread use. Therefore, the focus should not always be on novelty, but also robustness, cost/benefit, and ease-of-use. While the actual design process varies by application, the following concepts outline important features of the microfluidic cell culture design.

A. Plate Format

The easiest to use microfluidic format is the "well-plate" design. In this configuration, the microfluidic plate resembles a standard microtiter plate, such as a 96-well plate. The culture media and sample solutions are dispensed using a pipette into the wells, and a network of microfluidic channels are integrated underneath the wells of the plate. The major benefit of this design is to have all the wetted components in a self contained, disposable plate. This eliminates the use of liquid tubing, syringe pumps, and device assembly common in early microfluidic setups. In addition, the standard format of the plate allows multiple microfluidic designs to be used with the same process and control systems.

The most common microfluidic array is designed on a 96-well plate. In this format, the microfluidic units are designed to span the wells of a standard 96-well plate. For example, in one type of design, the plate is divided into 32 identical units, with each flow unit connecting three wells of the plate. The independent units are

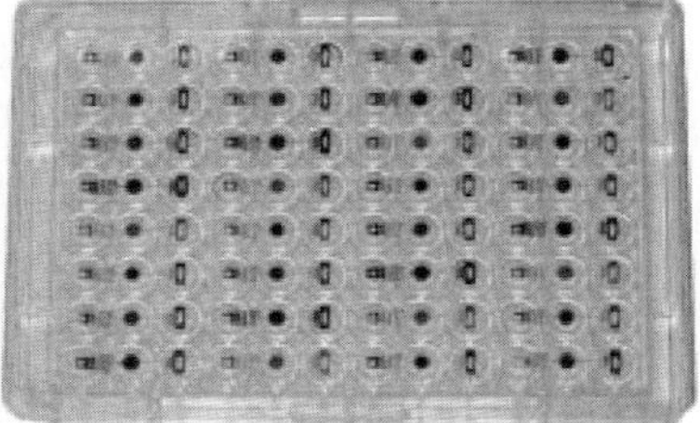
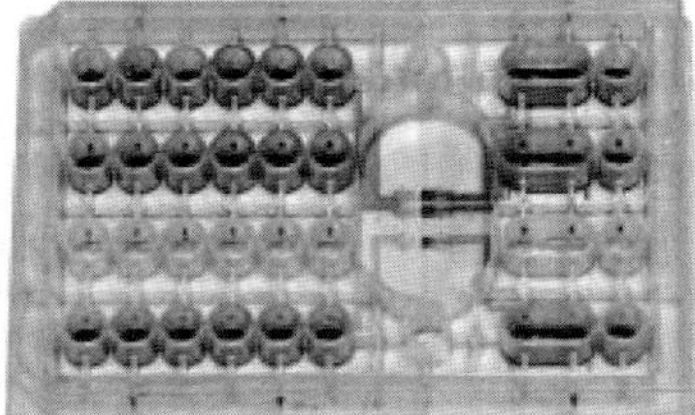

Fig. 15   Microfluidic plate formats. (Left) a photograph of the bottom of a 96-well microfluidic plate. In this example, three wells make up one unit, with 32 independent units per plate. (Right) a modified well-plate format designed for microfluidic live cell imaging. The cell culture chambers are centralized under a large imaging window. Each of the four units is addressed by the wells in a single row. The microfluidic channels in both plates are filled with colored dyes for visualization.

tiled across the 96-well plate. This type of design is favorable for screening applications due to compatibility with existing bioscience tools (liquid dispensers, automation, plate readers, etc.).

For live cell imaging applications, we utilize a customized well plate depicted in Fig. 15. This plate design contains a large imaging window where the cell culture chambers are localized. This is ideal for microscopy applications as it provides a wider light path (for phase contrast and brightfield imaging), reduces stage travel distance between units, and enables microincubation control on the microscope stage (see Section IV.C). Typically, the plate is routed such that each row addresses one flow unit (four units per plate). The six wells to the left of the imaging window provide the inlet solutions, with the wells to the right as the outlet. The dimensions of the wells and the plate footprint are the same as in a 96-well plate. Example microfluidic plates using this format are described in Section VI.

B. Unit Layout

The design of the flow unit determines the function of the microfluidic cell culture plate. A flow unit is defined as a network of fluidically interconnected channels. Separate flow units have no fluidic connections with each other. The first step in designing a flow unit is typically to determine how many wells are needed to accommodate the cell culture chamber(s), inlet well(s), and outlet well(s). In the following example, we will consider a four-well layout (Fig. 16). In this case, we wanted a flow inlet well, a flow outlet well, a cell culture/imaging well, and a cell inlet well. The next step is to determine the routing of the channels, and the arrangements of the wells. While many considerations come into play here, the following reasons led us to the layout depicted in Fig. 16: (1) There should be minimal distance between the inlet and cell culture chamber, (2) the cells would be loaded using capillary force, and the cell loading distance should be minimized, (3) the outlet would collect flows from the cell loading as well as from the inlet.

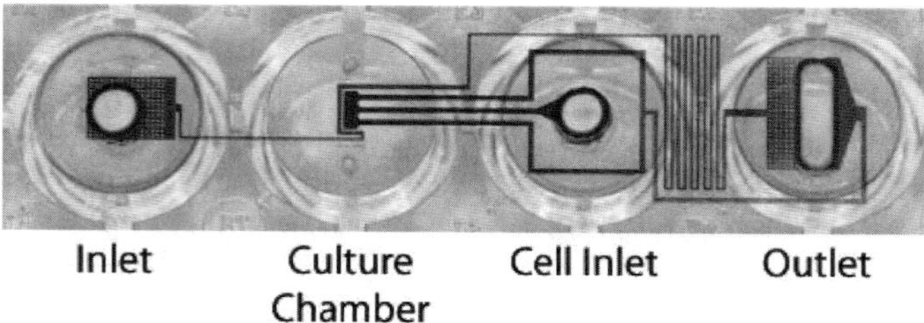

Fig. 16 Example fluidic unit layout. This photo shows the bottom side of a four-well microfluidic culture unit. The channels are filled with red dye for visualization. (For interpretation of the references to color in this figure legend, the reader is referred to the Web version of this chapter.)

C. Resistance Calculations

One of the most useful design tools for microfluidic networks is the fluidic resistance calculation (Fig. 17). Using the relations from Section II.C, it is possible to get a very good approximation of flow distribution in the channel network. Continuing with the above example, the goal was to get $\sim$50 µl/day gravity-driven flow from inlet to outlet, and a cell loading resistance of $\sim$10^{12} Ns/m^5. From previous experiments, we had determined that for a 1.5×0.5 mm area culture chamber, a flow rate of 50 µl/day of culture medium was suitable for the cell growth. Additionally, for effective surface tension based loading, the resistance needed to be approximately 10^{12} Ns/m^5. The gravity flow is from inlet to outlet, and the cell-loading path is from cell inlet to outlet. A perfusion barrier separates the two, which will be discussed in the next section.

Starting with the gravity flow path, we know that $Q = \frac{P}{R}$ for laminar flow, where Q is the volumetric flow rate (m^3/s), P is the pressure drop (Pa), and R is the resistance (Ns/m^5). For a 96-well plate, the typical pressure head between the inlet (full) well and the outlet (empty) well is 6.5 mm. This gives a pressure differential of 64 Pa. Plugging in the flow rate and pressure drop, we get a desired resistance

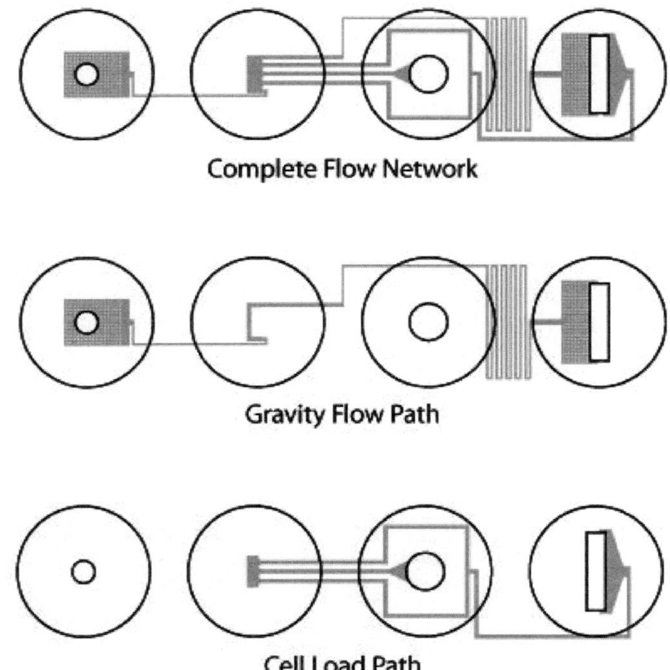

Fig. 17 Resistance calculations. The flow unit has two paths: a gravity flow path, and a cell load path, separated by a perfusion barrier. For the purpose of this calculation, the perfusion barrier is assumed to be of infinite resistance.

of 1.1E14 Ns/m^5. If we set the channel cross section as 100 μm wide by 40 μm tall, we can then solve for the total channel length from the flow resistance equation (section I.C.) to get 44 mm. Now we know the total length between the inlet and outlet wells. Since the direct line path between the wells is 27 mm, it is necessary to add channel length, which was achieved via the serpentine channels near the outlet.

For the cell-loading path, we used a capillary loading geometry, which requires a very low resistance. Here we used a 200 μm wide channel with 120 μm height. Using a minimal distance path from the cell inlet to chamber and two paths from chamber to outlet, we get a total resistance of 8E11 Ns/m^5, which is within our 10^{12} Ns/m^5 criteria.

A second example is for designing resistances to accommodate pressure-driven flow from a pneumatic source (see Section IV.A). Here, we typically have pressure control from 1–10 pounds/in^2, or 7–70 kPa. Typical experiment criteria call for perfusion for up to 72 h. With a 300 μl inlet well, this comes to 100 μl/day or 4.2 μl/h. Note for a typical chamber size of 2 × 2 × 0.1 mm (0.4 μl), this gives a full turnover every 6 min. In order to achieve flow rates on this scale, the resistance for pressure-driven flow is typically set around 10^{16} Ns/m^5, which translates to a range of 2.5-25 μl/h over the pressure control range.

When multiple paths and branching paths are to be modeled, it is useful to apply methods from Ohm's law for electronic circuits. For example, the resistances of channels in series are additive, and channels in parallel sum via inverses. Pressure drops occur linearly to resistance along channels, and flows always move from high to low pressure.

D. Perfusion Barriers

A key concept of our microfluidic cell culture chambers is the use of perfusion barriers to separate cell culture regions from flow channels (Hung *et al.*, 2005a; Lee *et al.*, 2007a, 2007b). This creates a biomimetic environment that is favorable for cell health, as well as to prevent fouling of channels by cells. The perfusion barrier consists of a region where the microchannels are 2–8 μm in size (width and height), and generally 75–150 μm in length (Fig. 18). Due to the small cross section of the

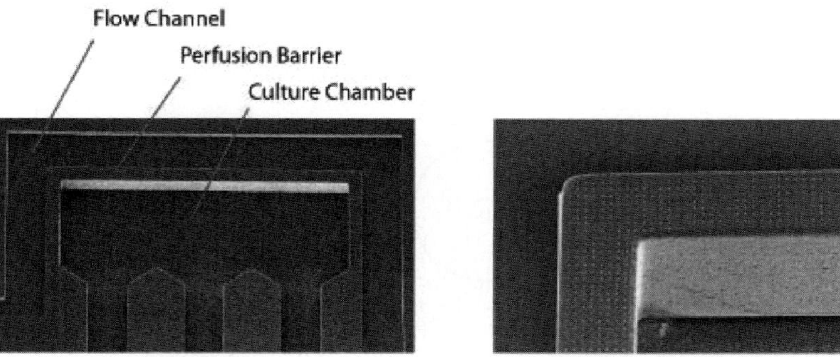

Fig. 18   Perfusion barrier. (Left) SEM image showing a perfusion barrier separating the culture chamber from a flow channel; (Right) close-up of the perfusion barrier showing a grid-like network of 4 μm channels.

channels, the fluidic resistance through the perfusion barrier is very high (a 2 μm diameter channel has 6E6 more resistance than a 100 μm channel). Because of this high resistance, there is limited pressure-driven flow across the perfusion barrier. Instead, nutrients diffuse through the pores, simulating biological transport.

E. Cell Barriers

A corollary to the perfusion barrier design is that it also serves as a cell/particle barrier similar to a size exclusion filter (Fig. 19). In most cases, it is desirable to localize cells to a specified region for observation, culture, and manipulation. In addition, if growing cells are allowed to populate the flow channels, they may alter the transport kinetics or block flow. In our experience, the microfabricated perfusion barriers effectively localize cells of all types as well as beads and other microparticles.

VI. Example Applications

A key benefit of establishing engineering principles for microfluidic cell culture is to be able to design devices to address important biological applications. Here, we will describe microfluidic designs that have been implemented for a variety of purposes. The goal of this section is to provide real world examples of how the engineering and design principles described in this chapter can be applied.

A. Microfluidic Culture Array

There is a fundamental difference between perfusion cell culture and static culture. For that reason, it is logical to develop a microfluidic analog to the standard 96-well plate. The microfluidic culture array design serves this purpose (Fig. 20). It consists of 32 perfusion culture units on a 96-well plate (three wells per unit). Each unit consists of three well positions: a flow inlet, a cell chamber, and a flow outlet. The culture chamber has a 2 mm diameter open top, enabling direct dispensing of cells into the chamber. Medium perfuses from the inlet well to the outlet well via gravity at a rate of ~100 μl/day. The chamber is surrounded by a 4 μm perfusion barrier that separates the cells from the flow channels. As the medium flows past the chamber, diffusive

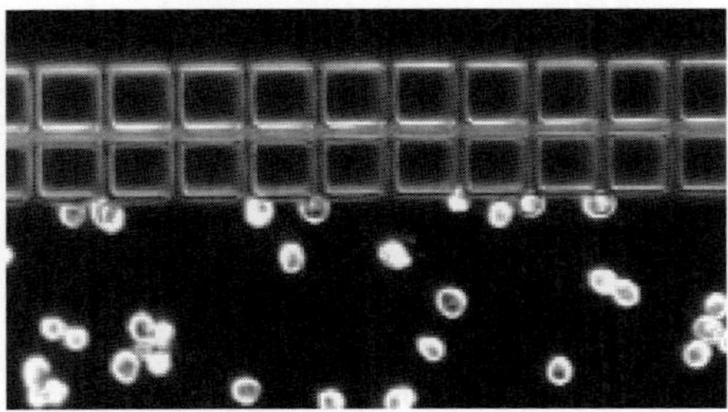

Fig. 19   Cell/particle barrier. A 4 μm pore perfusion barrier effectively prevents cells from crossing into the flow channel.

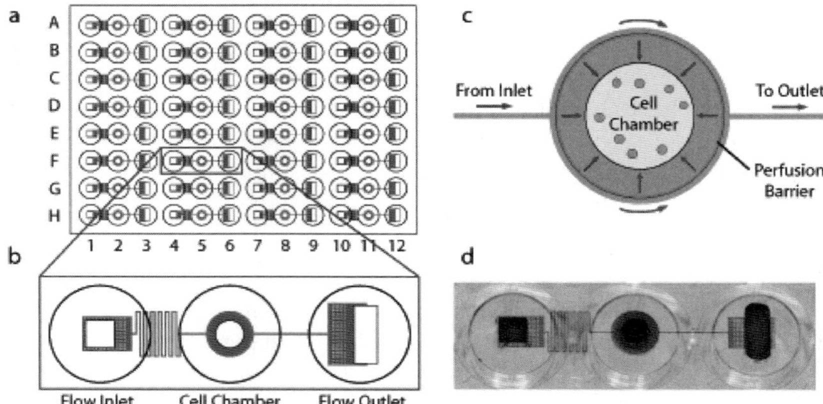

Fig. 20 Microfluidic perfusion culture array. (a) 32 flow units are tiled on a 96-well plate; (b) each unit has three-well positions; (c) the culture chamber has a 2 mm open top for direct dispensing of cells, surrounded by a perfusion barrier (green); flow moves from the inlet, around the chamber, to the outlet, with diffusive exchange with the chamber; (d) photograph of the single flow unit filled with blue dye. (See plate no. 5 in the color plate section.)

transport feeds the cells. This microfluidic culture array is suitable for 2D and 3D perfusion culture. In 2D culture, the surface tension of the liquid above the chamber prevents medium leaking to the middle well. In 3D culture, the gel is unable to cross the perfusion barrier, enabling free diffusion into and out of the gel using gravity flow.

B. Solution Switching

A key advantage of a fluidic cell culture system is the ability to introduce dynamic stimuli. In a common example, it is desirable to culture cells in a neutral state, expose them to a drug for a set amount of time, and wash out the drug (Lee *et al.*, 2009). This type of experiment is perfectly suited to microfluidic cell culture systems. In this example, a chamber with six different inlet solutions is designed for live cell imaging experiments (Fig. 21). Four independent culture chambers are put on a single plate.

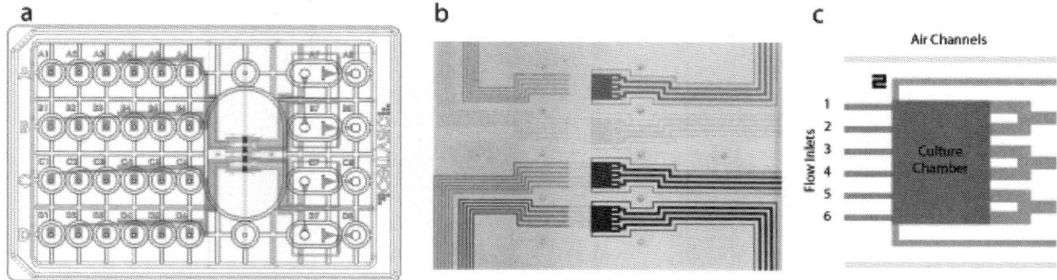

Fig. 21 Solution switching design. (a) The plate contains four separate flow units, each with six upstream inlet wells; (b) photo of the microchannels and chambers for the four culture units filled with dye; (c) drawing of the culture chamber, showing the flow inlets, air channels, and outlet paths. (See plate no. 6 in the color plate section.)

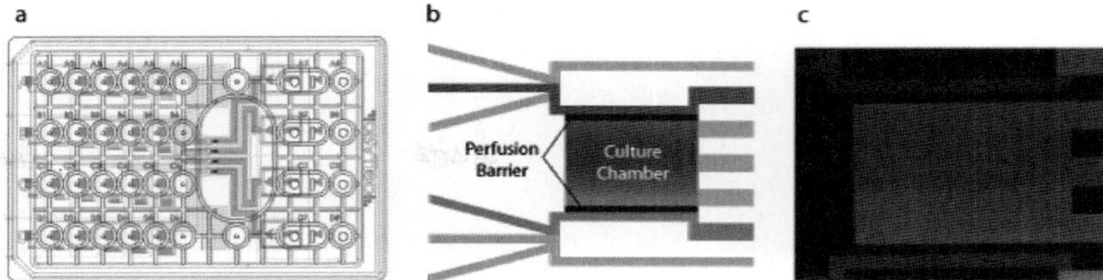

Fig. 22   Diffusion gradient design. (a) Four flow units are put on each plate, with three inlets for the top channel and three inlets for the bottom channel; (b) the culture chamber has two perfusion barriers such that different solutions will diffuse from the opposite sides of the chamber, forming a stable concentration gradient; (c) micrograph of a diffusion gradient between two fluorescent dyes.

Cells are loaded into the chamber using capillary action from the channels to the right. Active pressure-driven flow is used to control the six inlet flows during live cell microscopy. The laminar flow mechanics ensures a uniform and rapid solution switch with minimal mixing. Since the chamber is only 0.3 µl in volume, long-term experiments can be operated without replacing the medium.

C. Spatial Gradient Control

A unique feature of microscale laminar flow is the ability to create stable spatial gradients (Keenan and Folch, 2008; Wu *et al.*, in press). In nature, spatial gradients exist due to diffusion of molecules from "sources" (high concentration) to "sinks" (low concentration). In this microfluidic design, we utilized two perfusion barriers on opposite sides of the culture chamber fed by separate inlet streams. If the two streams have different concentrations of solutes, there will be diffusion across the chamber from the source to the sink. Since both streams are continuously replenished (forming infinite source and sink conditions), the spatial gradient in the chamber will be stable for long periods of time. In addition, the ability to switch the solutions in the flow channels allows dynamic gradient switching (Fig. 22).

D. 3D Extracellular Matrix Culture

The perfusion barrier is also effective at localizing 3D gels, preventing gel from crossing from one side to the other. This enables a chamber to be filled with 3D gel without blocking the flow channel. Perfusion of medium from the flow channel provides continuous diffusion mass transport into the gel, sustaining long-term 3D culture. We performed 3D culture of breast epithelial cells in laminin-rich extracellular matrix in the microfluidic cell array (see Section VI.A). The cells/gel were loaded at 4 °C, and allowed to polymerize in the chamber at 37 °C. The cells were then fed with medium perfusion for multiple days, demonstrating 3D culture (Fig. 23).

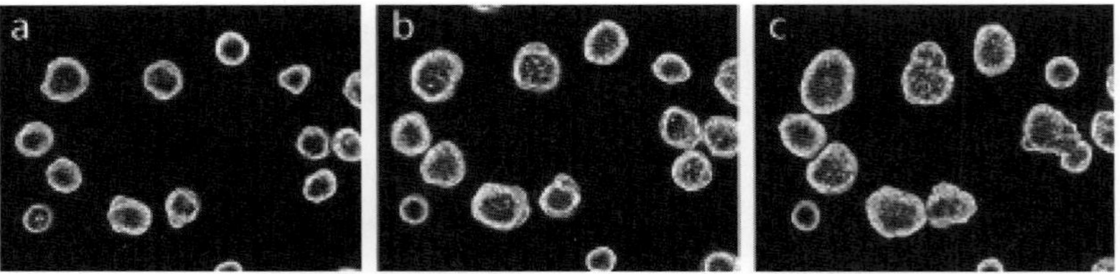

Fig. 23 3D Cell culture in a microfluidic chamber. Images of MCF-10A breast epithelial cells cultured in Matrigel after 3, 4, and 5 days perfusion in the microfluidic chamber.

E. Nonadherent Cell Imaging

Using traditional culture vessels, it is impossible to follow individual nonadherent cells over long periods of time, as they will move, drift, and experience Brownian motion (Fig. 24). In this example, we developed a microfluidic chamber for imaging yeast cells. Yeast cells are a common model organism in cell biology, but are traditionally difficult to image due to their nonadherent nature. The design utilizes a ceiling height roughly the same height of the yeast cells. Since the ceiling is elastic (made of silicone) and the cells are relatively rigid, pressure-driven loading will enable the ceiling to expand and trap the cells against the glass floor when the pressure is released.

VII. Conclusion

Recent advances in microfluidics technology offer improved methods for live cell analysis experimentation. The major benefits compared with nonmicrofluidic approaches include: (1) an improved cell microenvironment, (2) more accurate mass transport kinetics, (3) reduced volumes, and (4) ease-of-use. The intention of this chapter was to present the key engineering principles and design processes to enable microfluidic live cell imaging systems. As the use of microfluidic cell culture systems continues to expand in the life sciences, we believe it is important for

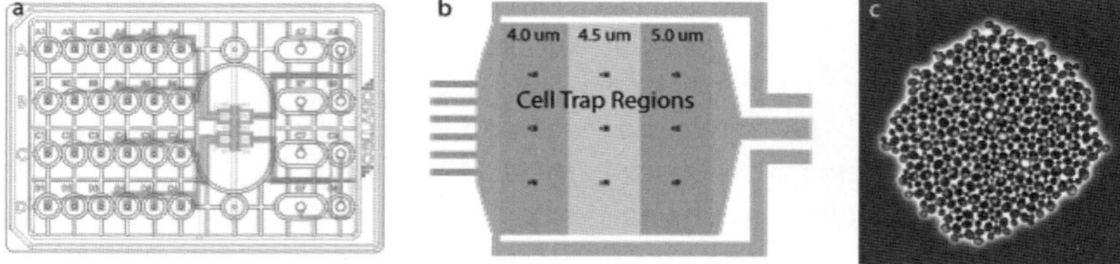

Fig. 24 Nonadherent cell trapping. (a) The four-unit yeast cell imaging plate has four units and six inlets per unit for solution switching; (b) the chamber contains three elastic trap regions with ceiling heights of 4.0, 4.5, and 5.0 μm (spanning the average yeast cell diameter); (c) once trapped, nonadherent cells are fixed in x, y, z for long-term growth monitoring. Here a single colony of *S. cerevisiae* is imaged after 12 h.

end-users to understand the features and limitations of the technology. Future growth of this industry will likely be in an "application specific" manner, whereby the fluidic design, layout, and instrumentation are engineered to suit the needs to specific biological applications. This maturation of technology will occur alongside with improvements in optical imaging systems, cell-based assays, and better live cell models. All of these areas are currently undergoing rapid advances, helping spur the demand for microfluidic cell culture systems for the next generation of biomedical research.

References

Chiu, D. T., Jeon, N. L., Huang, S., Kane, R. S., Wargo, C. J., Choi, I. S., Ingber, D. E., Whitesides, G. M. (2000). Patterned deposition of cells and proteins onto surfaces by using three-dimensional microfluidic systems. *Proc. Natl. Acad. Sci. U.S.A.* **97**, 2408–2413.

Gerlich, D., and Ellenberg, J. (2003). 4D imaging to assay complex dynamics in live specimens. *Nat. Cell Biol.* S14–S19.

Giepmans, B. N., Adams, S. R., Ellisman, M. H., and Tsien, R. Y. (2006). The fluorescent toolbox for assessing protein location and function. *Science* **312**, 217–224.

Hung, P. J., Lee, P. J., Sabounchi, P., Aghdam, N., Lin, R., Lee, L. P. (2005a). A novel high aspect ratio microfluidic design to provide a stable and uniform microenvironment for cell growth in a high throughput mammalian cell culture array. *Lab Chip* **5**, 44–48.

Hung, P. J., Lee, P. J., Sabounchi, P., Lin, R., and Lee, L. P. (2005b). Continuous perfusion microfluidic cell culture array for high-throughput cell-based assays. *Biotechnol. Bioeng.* **89**, 1–8.

Janakiraman, V., Sastry, S., Kadambi, J. R., and Baskaran, H. (2008). Experimental investigation and computational modeling of hydrodynamics in bifurcating microchannels. *Biomed. Microdevices* **10**, 355–365.

Keenan, T. M., and Folch, A. (2008). Biomolecular gradients in cell culture systems. *Lab Chip* **8**, 34–57.

Kim, L., Toh, Y. C., Voldman, J., and Yu, H. (2007). A practical guide to microfluidic perfusion culture of adherent mammalian cells. *Lab Chip* **7**, 681–694.

Lee, P. J., Gaige, T. A., Ghorashian, N., and Hung, P. J. (2007a). Microfluidic tissue model for live cell screening. *Biotechnol. Prog.* **23**, 946–951.

Lee, P. J., Gaige, T. A., and Hung, P. J. (2009). Dynamic cell culture: a microfluidic function generator for live cell microscopy. *Lab Chip* **9**, 164–166.

Lee, P. J., Hung, P. J., and Lee, L. P. (2007b). An artificial liver sinusoid with a microfluidic endothelial-like barrier for primary hepatocyte culture. *Biotechnol. Bioeng.* **97**, 1340–1346.

McDonald, J. C., Duffy, D. C., Anderson, J. R., Chiu, D. T., Wu, H., Schueller, D. J., Whitesides, G. M. (2000). Fabrication of microfulidic systems in poly(dimethylsiloxane). *Electrophoresis* **21**, 27.

Regehr, K. J., Domenech, M., Koepsel, J. T., Carver, K. C., Ellison-Zelski, S. J., Murphy, W. L., Schuler, L. A., Alarid, E. T., Beebe, D. J. (2009). Biological implications of polydimethylsiloxane-based microfluidic cell culture. *Lab Chip* **9**, 2132–2139.

Renard, M., Heutte, F., Boutherin-Falson, O., Finet, M., and Boisseau, M. R. (2003). Induced changes of leukocyte slow rolling in an in flow pharmacological model of adhesion to endothelial cells. *Biorheology* **40**, 173–178.

Rutkowski, J. M., and Swartz, M. A. (2007). A driving force for change: interstitial flow as a morphoregulator. *Trends Cell Biol.* **17**, 44–50.

Wu, M. H., Huang, S. B., and Lee, G. B. (2010). Microfluidic cell culture systems for drug research. *Lab Chip.* **10**(8), 939–956.

CHAPTER 5

Rise of the Micromachines: Microfluidics and the Future of Cytometry

Donald Wlodkowic[*] and Zbigniew Darzynkiewicz[†]

[*]The BioMEMS Research Group, Department of Chemistry, University of Auckland, Auckland, New Zealand

[†]Brander Cancer Research Institute and Department of Pathology, New York Medical College, Valhalla, New York, USA

Abstract

The past decade has brought many innovations to the field of flow and image-based cytometry. These advancements can be seen in the current miniaturization trends and simplification of analytical components found in the conventional flow cytometers. On the other hand, the maturation of multispectral imaging cytometry in flow imaging and the slide-based laser scanning cytometers offers great hopes for improved data quality and throughput while proving new vistas for the multiparameter, real-time analysis of cells and tissues. Importantly, however, cytometry remains a viable and very dynamic field of modern engineering. Technological milestones and innovations made over the last couple of years are bringing the next generation of cytometers out of centralized core facilities while making it much more affordable and user friendly. In this context, the development of microfluidic, lab-on-a-chip (LOC) technologies is one of the most innovative and cost-effective approaches

105

0091-679X/10 $35.00
DOI 10.1016/B978-0-12-374912-3.00005-5

toward the advancement of cytometry. LOC devices promise new functionalities that can overcome current limitations while at the same time promise greatly reduced costs, increased sensitivity, and ultra high throughputs. We can expect that the current pace in the development of novel microfabricated cytometric systems will open up groundbreaking vistas for the field of cytometry, lead to the renaissance of cytometric techniques and most importantly greatly support the wider availability of these enabling bioanalytical technologies.

I. Introduction

Advances in conventional flow and image-assisted cytometry provide the instrumentation of choice for studies requiring quantitative and multiparameter analysis with a single cell resolution (Darzynkiewicz *et al.*, 1999; Wlodkowic *et al.*, 2010). The advancements in conventional cytometric technologies can be easily seen in the current miniaturization trends and substantial simplification of analytical components (Shapiro, 2004). Technological milestones have already brought flow cytometry (FCM) out of centralized core facilities while making it much more affordable and user friendly. These trends will surely drive the renaissance of cytometry and growing interest in cytometric techniques worldwide (Shapiro, 2004; Wlodkowic *et al.*, 2010).

Although the present array of conventional technologies has delivered many options and innovative solutions, there are still numerous areas for technological improvements (Eisenstein, 2006; Melamed, 2001). Not surprisingly, many new enabling strategies that can reduce expenditures while at the same time increase portability, throughput, and content of collected information are attracting growing interest among both academic and industrial markets (Hill, 1998). The improvements in such capabilities are of particular importance for the development of personalized therapeutic approaches (point-of-care diagnostics) and the increasing role of cost and time savings in drug discovery pipelines (Hill, 1998; Kiechle and Holland, 2009; Mayr and Bojanic, 2009; Myers and Lee, 2008).

In this context, the transfer of traditional bioanalytical methods to a microfabricated format can greatly facilitate the reduction of drug screening expenditures (Andersson and van den Berg, 2004; Chung and Kim, 2007; Kiechle and Holland, 2009). Microfabricated technologies can also vastly increase throughput and content of information from a given sample (Svahn and van den Berg, 2007; Whitesides, 2006). Advances in physics, electronics as well as material sciences have recently facilitated the development of miniaturized bioanalytical systems collectively known as lab-on-a-chip (LOC) (Andersson and van den Berg, 2004; Dittrich and Manz, 2006; Sims and Allbritton, 2007). LOC represents the next generation of analytical laboratories that have been miniaturized to the size of a matchbox, and represent one of the most groundbreaking offshoots of nanotechnology, delivering, among others, platforms for high-throughput drug discovery and content-rich personalized diagnostics (Mauk *et al.*, 2007; Myers and Lee, 2008; Wang *et al.*, 2009;

Weigl *et al.*, 2008; Whitesides, 2006). By providing an alternative to expensive instrumentation such as flow or laser scanning cytometers and sorters, while offering new functionalities, user-friendly LOC technologies can prospectively enable development of bench-top or even highly portable cytometers of the size of the modern notebooks or even mobile phones (Chung and Kim, 2007; Huh *et al.*, 2005; Myers and Lee, 2008; Wlodkowic and Cooper, 2010c). Based on the progress in the field of LOC, we envisage that in a couple of years from now the microflow cytometers may well be as ubiquitous as standard PCR thermocyclers.

Attempts providing such groundbreaking capabilities have already been made and microfluidic chip-based cytometry is slowly entering a commercial stage with appearance of portable devices capable of multiparameter fluorescent interrogation. In this chapter we will discuss some of the selected and innovative microflow technologies nearing the commercial stage. Some of them are particularly attractive for the basic research, clinical and diagnostic laboratories as they allow rapid analysis of only small numbers of patient-derived cells. As such these can be particularly suitable for point-of-care diagnostics and distributed telemedicine (Kiechle and Holland, 2009; Myers and Lee, 2008; Weigl *et al.*, 2008).

II. The Smaller the Better: Microfluidics and Enabling Prospects for Single Cytomics

Microfluidics is an emerging field of engineering aimed at manipulating liquids in networks of microchannels with dimensions between 1 and 1000 µm (Fig. 1). The dimensionless parameter called the Reynolds number (*Re*) describes unique physical principles of the fluid in micrometer-sized channels as a function of the geometry, fluid viscosity, and average flow rates (Sia and Whitesides, 2003; Whitesides, 2006). As described by the *Re*, the fluids at the microscale exhibit dissimilar physicochemical properties, when compared to their behavior at the macroscale (Holmes and Gawad, 2010; Pfohl *et al.*, 2003; Polson and Hayes, 2001). For example, fluids at the microscale are largely dominated by viscous rather than inertial forces. This means that fluid flow in microchannels is laminar. Under laminar conditions, the fluid flow has no inertia and mass transport will be dominated by local diffusion rates (Beebe *et al.*, 2002; Franke and Wixforth, 2008; Kastrup *et al.*, 2008). As such convective contributions to mixing can be negated, and the supply of nutrients, gases, and drugs to cells can be spatiotemporally controlled that is aliquots of fluid can be delivered to particular positions at controlled timings (Takayama *et al.*, 2001, 2003). Most importantly, however, the dimensions of microfluidic environment are comparable to the intrinsic dimensions of cells and blood vessels (Fig. 1) (Irimia and Toner, 2009; Liu *et al.*, 2009; Stroock and Fischbach, 2010; Wlodkowic and Cooper, 2010c). Therefore, gas and drug diffusion rates, shear stress, and even microscale cellular niches can be artificially recreated on chip, mimicking the physiological microenvironment encountered in the human body (Warrick *et al.*, 2008; Wlodkowic and Cooper, 2010c).

A)

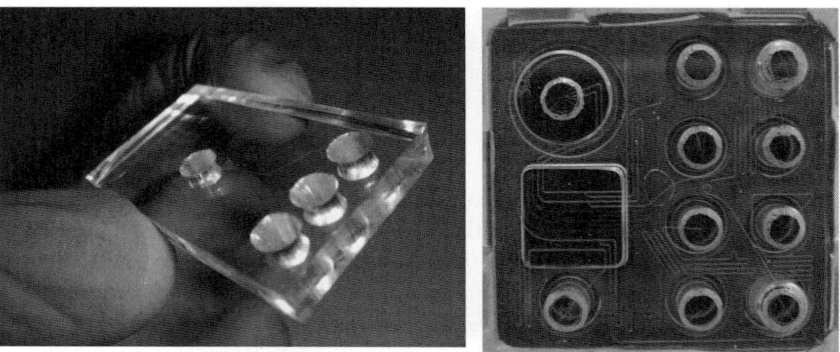

B)

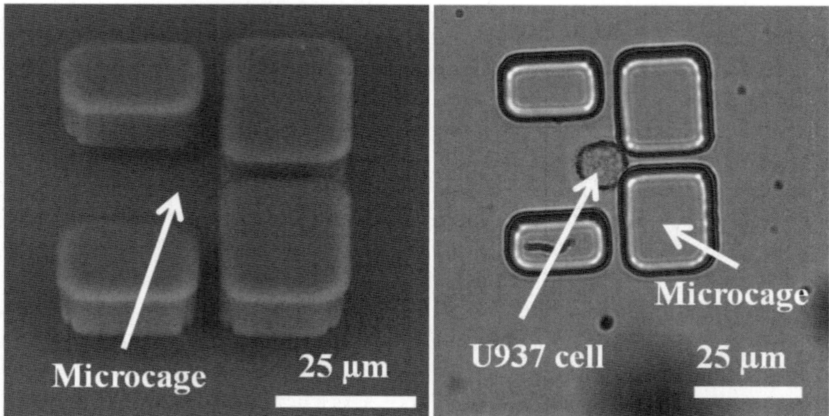

Fig. 1   The design of miniaturized LOC technologies is a promising avenue to address the inherent complexity of cellular systems with massive experimental parallelization, throughput, and analysis on a single cell level. (A) Microfluidics is an emerging field of engineering aimed at manipulating ultralow volumes of liquids in networks of microchannels with dimensions between 1 and 1000 µm. Fluid flow in microfluidic channels is dominated by viscous rather than inertial forces. Laminar flow describes the conditions where all fluid particles move in parallel to the flow direction. Note that during microfluidic circuitry is dominated only by a limited and local diffusion. Miniaturization of LOC promises greatly reduced equipment costs, simplified operation, increased sensitivity, and throughput by implementing many innovative and integrated on-chip analytical modules. (B) Microfabrication allows for design of new analytical functionalities. These for instance enable immobilization, manipulation, treatment, and analysis of single cells. An example of microcage (microjail) structure is shown that allows for micro-mechanical trapping and immobilization of single human cells.

The enclosed and sterile format of microfluidic circuitry eliminates the evaporative water loss from microsized channels (Fig. 1) (Beebe *et al.*, 2002; Berthier *et al.*, 2008a, b; Wlodkowic *et al.*, 2009a). Inexpensive and biocompatible polymers such as poly(dimethylsiloxane) (PDMS) are often materials of choice for rapid prototyping of microfluidic devices (Sia and Whitesides, 2003; Weibel and Whitesides, 2006; Zaouk *et al.*, 2006; Zhang *et al.*, 2009). Glass, silica, or thermoplastics such as, for example, acrylic (poly(methyl methacrylate; PMMA) are on the other hand common substrates found in commercial chip-based designs (Fig. 1) (Pugmire *et al.*, 2002; Qi *et al.*, 2002; Zhang *et al.*, 2009). The latter technologies enhance the durability and application range to reusable and/or high-pressure applications. The disposable format of many LOC devices is particularly suitable for the point-of-care diagnostics and future personalized therapy. LOC devices also promise greatly reduced costs, increased sensitivity, and ultrahigh throughput by implementing parallel sample processing and a vast miniaturization of integrated on-chip components (Fig. 1) (Holmes and Gawad, 2010; Mauk *et al.*, 2007; Michelsen, 2007; Myers and Lee, 2008; Weigl *et al.*, 2008; Wlodkowic and Cooper, 2010d).

III. Microflow Cytometry (μFCM)

Conventional FCM is a powerful analytical and diagnostic tool that provides the multiparameter and high-speed measurements of cells 3D focused into a single file and integrated fluorescence from every cell collected by photomultiplier tubes (PMTs) (Mach *et al.*, 2010; Melamed, 2001; Shapiro, 2004). FCM overcomes a frequent problem of traditional bulk techniques. Its advantages include the correlation of different cellular events at a time, single cell analysis (the avoidance of bulk analysis), and measuring thousand of cells per second. It suffers, however, from high cost, complex operation, and very limited portability (Mach *et al.*, 2010; Melamed, 2001). Inherent limitations of traditional FCM have recently stimulated the fast development of chip-based microfluidic cytometers (Bhagat *et al.*, 2010b; Cheung *et al.*, 2010; Chung and Kim, 2007; Oakey *et al.*, 2010; Skommer *et al.*, 2010; Wlodkowic and Cooper, 2010d; Wlodkowic *et al.*, 2010).

Application of microfluidic technologies can supplant many of the conventional disadvantages through the development of on-chip μFCM (Cheung *et al.*, 2010; Chung and Kim, 2007; Huh *et al.*, 2005). The major advantage of μFCM is that they sample a greatly reduced number of cells when compared with conventional FCM (Wlodkowic and Cooper, 2010a; Wlodkowic *et al.*, 2009c, 2010). This is of a particular value when studying, for example, rare, patient-derived cells and monitoring their reactions with therapeutic compounds (Wlodkowic *et al.*, 2009c). Recent progress has opened many innovative opportunities to obtain confinement and regulation of laminar streams of cells in two or even in three dimensions thus miniaturizing the bulky 3D flow cells associated with conventional FCM (Kummrow *et al.*, 2009; Lee *et al.*, 2009; Mao *et al.*, 2009; Scott *et al.*, 2008). Some progress in sheath-free microcytometers that create ordered streams of cells

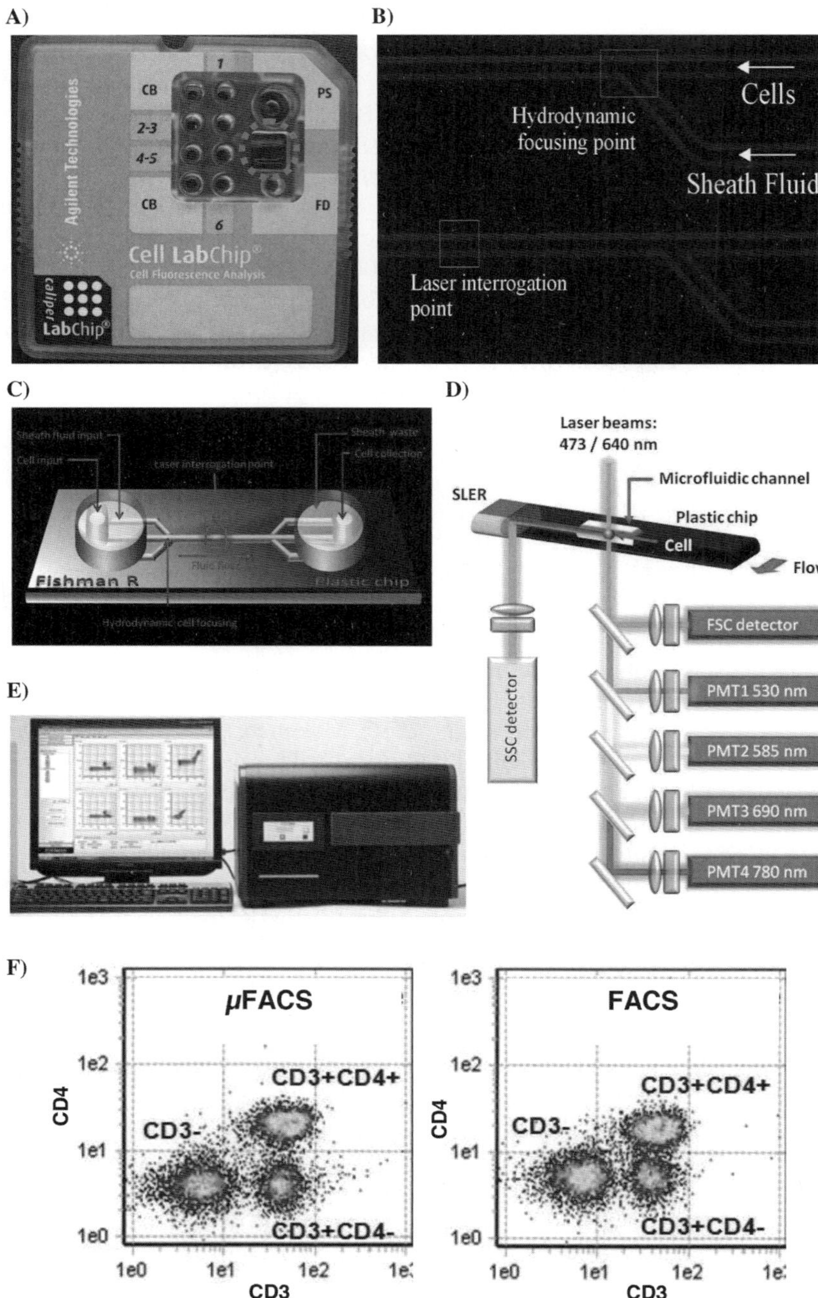

Fig. 2   Innovative microfluidic flow cytometers (μFCM). (A) (B) Overview of the Agilent CellLab Chip technology. A glass chip with an etched network of microfluidic channels is mounted into the plastic

has also been reported. Similarly to a conventional system, single file of cells on μFCMs can be sequentially interrogated by independent lasers providing substantial multiplexing opportunities. Moreover, the enclosed, sterile, and disposable nature of microfluidic circuitry makes μFCMs particularly suitable for the analysis of highly infective samples. This may be of particular importance during monitoring of HIV infections in resource poor areas.

Interestingly, apart from a mushrooming number of academic publications, microfluidic chip-based cytometry is slowly entering a commercial stage. The arrival of user-friendly, reasonably priced, and portable devices capable of multiparameter fluorescent analysis based on microfluidic principles heralds a sudden change in the field (Wlodkowic and Cooper, 2010a, c). Examples include a multitude of experimental prototypes with flow rates controlled either by step motor-driven syringe pumps, positive air pressure applied to input reservoirs, or vacuum applied to output reservoirs of the microfluidic chips fabricated in polymers or glass (Chan *et al.*, 2003; Palkova *et al.*, 2004). While in the conventional flow cytometers the transfer rates through the flow chamber can be as high as 10^4–10^6 cells/s, most microfluidic planar chips maintain much lower transfer rates of 10–300 cells/s (Wlodkowic and Cooper, 2010a, d; Wolff *et al.*, 2003; Zhu *et al.*, 2010). The reduced transfer rates can be, in turn, effectively compensated for by parallel processing and simultaneous analysis of multiple and parallel streams of cells on each chip (Hur *et al.*, 2010; Wlodkowic and Cooper, 2010a, c). Low transfer rates and pressures are also advantageous for preservation of living cells. In this regard, some of the most recent innovations in the μFCM chip design facilitate also the collection of an undiluted cell population for many functional studies to be performed following the microcytometric analysis. This feature is not available in any of the conventional FCM systems apart from the complex fluorescently activated cell sorters.

The most notable examples of commercial stage μFCM involve the CellLab Chip (Agilent Technologies, Santa Clara, CA, USA) and Fishman-R (On-chip Biotechnologies Co., Tokyo, Japan) (Fig. 2 A, B). CellLab chip was the first

---

cartridge providing the interface for the modified Agilent 2100 Bioanalyzer. Note the principles of microcytometry such as microhydrodynamic cell focusing using isobuoyant sheath buffer and double-point laser interrogation point (right panel). Systems permit for an automated, clog-free operation. No laser alignment or operator engagement is necessary for a sequential analysis of up to six samples on one chip. (C) Disposable microfluidic cartridge of the Fishman-R microfluidic flow cytometer with a two-way 2D hydrodynamic focusing of cells. (D) Optical configuration of Fishman-R microfluidic flow cytometers. Multiparameter detection capabilities are comparable to the conventional flow cytometers. Note forward scatter (FSC), side scatter (SSC), and four fluorescence detectors used in combination with spatially separated solid state 473 and 640 nm solid-state diode lasers. Side scatter detection is performed using innovative SLER technology (side scattered light detection using edge reflection) recently developed by On-chip Biotechnologies Co. (Tokyo, Japan). (E) Off-chip interface of the Fishman-R microfluidic flow cytometer housing the pneumatic and optical modules (data courtesy of Dr Kazuo Takeda, On-chip biotechnologies Co., Tokyo, Japan). (F) Multicolor immunophenotyping performed on the microfluidic Fishman-R flow cytometer as compared to conventional flow cytometers. Note that μFACS analysis requires merely 20 μl of blood and yields comparative multiparameter data to FACS (data courtesy of Dr Kazuo Takeda, On-chip Biotechnologies Co., Tokyo, Japan).

commercially available microcytometry system utilizing disposable glass chips (cartridges) based on patented Caliper LabChip® technology (Fig 2 A, B). The microfluidic cartridges are run by a modified Agilent 2100 Bioanalyzer (Agilent Technologies) (Chan *et al.*, 2003; Palkova *et al.*, 2004). This technology allows for simultaneous analysis of up to six independent samples with a throughput of about 2.5 cells/s (Chan *et al.*, 2003; Palkova *et al.*, 2004). It provides a very low cell consumption and complete automation that alleviates laser aligning and operator engagement. The optical configuration of the Agilent 2100 Bioanalyzer features two excitation light sources (LED 525 nm and solid state laser 685 nm) (Chan *et al.*, 2003). Substantial fluorescence signal separation between emission channels (FL1 *Em* 525 nm vs. FL2 *Em* 685 nm) alleviates need for any spectral compensation. In our recent work we have proven that sensitivity of Caliper LabChip® technology is adequate to detect subtle changes in fluorescence intensity during two color microcytometric analysis of drug-induced apoptosis (Wlodkowic *et al.*, 2009c).

Fishman-R is on the other hand an example of a quantum leap in the development of μFCM (Fig 2C-F) (Takeda and Jimma, 2009; Takao *et al.*, 2009). The system developed by the group of Takeda at On-chip Biotechnologies (Tokyo, Japan) introduces for the first time a fully fledged and user friendly microcytometer with (i) 2D hydrodynamic sheath focusing, (ii) multi-laser excitation, (ii) up to four color detection, and (iv) detection with fluorescence area, height, and width parameters for all fluorescent channels (Fig 2C-F) (Takeda and Jimma, 2009; Takao *et al.*, 2009). Sensitivity of this microfluidic cytometer is well within specifications of conventional analyzers (i.e., FITC <600 MESF, PE <200 MESF). Fishman-R is also the first microflow cytometer that incorporates true Forward Site Scatter and Side Site Scatter detection (Fig 2C-F) (Takeda and Jimma, 2009; Takao *et al.*, 2009).

Current progress in on-chip cytometry leverages advances in microfluidic technologies with the ultimate outcome to produce user friendly, reasonably priced and portable devices (Takeda and Jimma, 2009; Takao *et al.*, 2009). They are particularly attractive for the clinical and diagnostic laboratories. Based on the progress in the field of LOC, one can already envisage that in a couple of years from now the microflow cytometers may well be as ubiquitous as standard PCR thermocyclers (Fig 2E).

IV. Microfluidic Cell Sorting (μFACS)

The accurate detection, analysis, and sorting of defined cell subpopulations is of importance not only for basic and clinical research but also for biotechnology and agriculture (Eisenstein, 2006; Melamed, 2001). In this respect, conventional FCM still remains the technology of choice with an array of high-speed fluorescently activated cell sorters (FACS) available on the market. Indeed, modern FACS have an impressive capabilities that support algorithms based on up to 16 optical parameters measured from a single cell and enormous acquisition rates that exceed 2.5×10^4

events/s (Eisenstein, 2006; Melamed, 2001). The deployment of FACS is, however, currently limited to only centralized core facilities. This is mostly due to their high complexity, power consumption, and resulting intrinsic cost of the equipment. The need for specialized and dedicated personnel capable of operating these machines is also yet another limiting factor that restricts widespread access to this technology (Eisenstein, 2006).

Not surprisingly, there is an increasing interest and demand for cost-effective cell sorting systems that could supplement conventional FCM. In this context microfluidics has an immense potential to meet these demands, due to the inherent ease of rapid prototyping, potential of flexible and scalable designs, enhanced analytical performance and economical fabrication. Numerous emerging microfluidic cell sorting devices have been designed, and are based on various cell sorting principles such as piezoelectrical actuation, dielectrophoresis (DEP), magnetophoresis, optical gradient forces, and hydrodynamic flow switching (Adams *et al.*, 2008; Adams and Tom Soh, 2009; Arndt-Jovin and Jovin, 1974; Baret *et al.*, 2009; Bhagat *et al.*, 2010a; Gluckstad, 2004; Kuntaegowdanahalli *et al.*, 2009; Revzin *et al.*, 2005; Wang *et al.*, 2005; Wlodkowic and Cooper, 2010a; Wolff *et al.*, 2003).

Only recently, several microfluidic FACS (μFACS) have been proposed. They have mostly suffered, however, from low throughput, lack of integrated mechanisms for actuation, poor sensitivity of fluorescence detection, and low sorting accuracy as compared to the conventional FACS. Recent innovations made in the field seem, however, to alleviate most of these dilemma and pave the way for the development of fully integrated and automated microsorters. In this context, recent work of Cho and colleagues introduced a new design with microfluidic, as well as optical, acoustical, and piezoelectronic components, all embedded into a single chip (Cho *et al.*, 2010). This superior design features also a new automated sorting control system and a signal processing algorithm ("space-time coding technology") that permit high-speed sorting of particles combined with a superior purification enrichment factors (Cho *et al.*, 2010).

The multiple sample processing procedures are often required during conventional FACS that decrease the efficiency and make separation of clinically relevant rare subpopulations challenging. High pressures and electrostatic charges used to deflect sorted cells can further affect the recovery of fragile cell subpopulations such as apoptotic cells and cancer stem cells. At the same time large-size conventional equipment coupled with complex fluidic technologies are major impediments for the development of clinical grade sorting. Also undesired aerosol formation usually associated with conventional FACS often preclude safe sorting of highly infectious specimens without dedicated containment rooms necessary for biohazard FACS sorting. Most recently, an innovative μFACS system has been commercialized that provides a new solution to the above-mentioned problems. The Gigasort™ Clinical Grade Cell Sorter developed by the CytonomeST LLC (Boston, MA, USA) is the fastest fully sterile optical cell sorter ever produced and can be named as the most revolutionary advancement in FACS since its invention in early 1960s. It employs fully enclosed, disposable chip sorting cartridges that

     Donald Wlodkowic and Zbigniew Darzynkiewicz

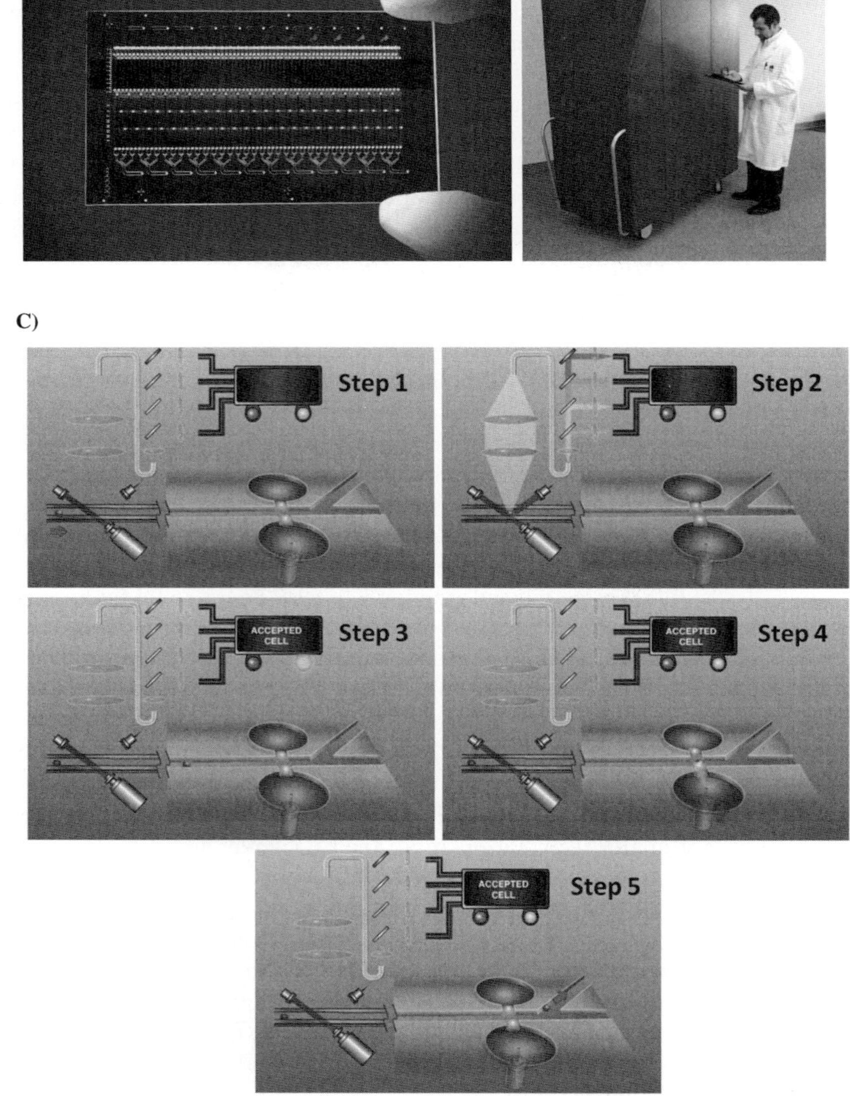

Fig. 3 Gigasort[TM] Clinical Grade Cell Sorter (CytonomeST LLC, Boston, MA, USA) is the fastest fully sterile optical cell sorter ever produced and can be named as the most revolutionary advancement in FACS since its invention in the early 1960s. (A) A microfabricated glass die measuring 2×3 inches and containing 72 parallel microfluidic switches (microsorters). Using a

enable gentle, clinical grade, sterile sorting without any undesired side-effects of the procedure (Fig. 3). As such clinical-grade cell sorting can be performed that was not possible with other solid or flow-based cell sorting methods. Importantly, the Gigasort's enclosed and disposable cartridges support all staining protocols developed for conventional FCM with up to two optical (extinction and side scatter) and seven simultaneous fluorescent parameters from each cell. In the heart of this groundbreaking technology lies a system of innovative microfluidic particle switches used to deflect and sort cells in the microchannels (Fig. 3). The switch (microsorter) operates at a rate of 2000 cell sorting operations per second. By leveraging the potential for modular design, Gigasort^{TM} increases the throughput by using a parallel sample processing paradigm well known in the microprocessor industry (Fig. 3). Using a massively parallel array of up to 72 concurrently operating microfluidic sorter cores, the technology achieves ultra-high sorting speeds of up to 1 billion events/h. The 72 microfluidic switches have been fabricated onto a single glass chip that measures 2×3 in. (Fig. 3). Reportedly, the single 72 channel chip can sort up to 1×10^5 cells/s (0.5×10^9 cells/h) and can operate at over seven times the throughput of any conventional FACS. According to CytonomeST data, the new generation of the Gigasort sorters will boast the configuration of up to 144 sorter cores reaching the combined sorting speed of up to 2.8×10^5 cells/s (1.0×10^9 cells/h). To support such ultra-high throughput, the Cytonome's cartridges exceed 2 billion cells capacity that is optimized for a sample concentration of 5×10^6/ml.

We expect many new microfabricated cell sorters coming up to the market during next couple of years. Considering the advantages and simplicity of the on-chip cell sorting protocols, these platforms have a wide potential to be used for automated diagnostic and laboratory routines. Similarly to the μFCM, we anticipate that rapid advances in microfluidics, electronics, miniaturized optical components, micron tolerance mechanics and control algorithms will bring fluorescently activated cell sorting out of the centralized core facilities and onto the benches of regular biomedical laboratories.

---

massively parallel array of simultaneously operating microfluidic sorter cores, the technology achieves ultra-high sorting speeds of up to 1 billion events per hour. As such it can operate at over seven times the throughput of any conventional FACS. (B) Early prototype of Gigasort^{TM} Clinical Grade Cell Sorter. (C) Principles of Gigasort operation. Each switch (microsorter) operates in sequential steps at a rate of 2000 cell sorting operations per second. In Step 1, cells pass through the microfluidic channel. In Step 2, laser beams illuminate the cells and optical emission occurs. In Step 3, emission is analyzed and accept/reject sort decisions are made by the off-chip hardware and software modules. In Step 4, microactuator pushes the diaphragm that moves the sorted cell into the upper portion of the laminar fluid stream. In Step 5, the desired cell (green) is moved into the keep area, whereas non-sorted cells (red) are directed to waste compartment. Data courtesy of Dr John C Sharpe, CytonomeST LLC (Boston, MA, USA). (See plate no. 7 in the color plate section.)

V. Real–Time Cell Analysis: Living Cell Microarrays and a Real–Time Physiometry on a Chip

A common drawback of both conventional FCM and μFCM is that the cells flow suspended in a laminar stream of fluid and fluorescence is collected by PMTs or photodiodes without collecting the valuable information on subcellular and morphological features (Skommer *et al.*, 2010; Wlodkowic and Cooper, 2010b, c; Wlodkowic *et al.*, 2009d, 2010). Moreover, flow cytometric analysis suffers from a lack of capabilities to monitor single living cells in real time (dynamic analysis) and as such still represent a binary system that averages the results by capturing only a snapshot of the intermittent cellular reaction at a particular point in time (Skommer *et al.*, 2010; Wlodkowic and Cooper, 2010c). Cellular processes, however, are dynamic and feature very large cell-to-cell variations. Moreover, the majority of cellular events have highly ordered interactions not only between different intracellular compartments but also between the neighboring cells (Skommer *et al.*, 2010; Wlodkowic and Cooper, 2010a, b). Monitoring of these biological phenomena at a single cell level requires real-time and spatial detection systems with a micrometer scale resolution (Faley *et al.*, 2009; Skommer *et al.*, 2010; Wlodkowic and Cooper, 2010b, d; Wlodkowic *et al.*, 2009b).

These drawbacks of microflow-based methods have recently led to the development of new LOC technologies with an attempt to transfer cell culture and real-time cell assays to a format commonly known as living cell microarrays (Di Carlo *et al.*, 2006; Situma *et al.*, 2006; Wang *et al.*, 2007; Wheeler *et al.*, 2005; Wlodkowic and Cooper, 2010b; Yarmush and King, 2009). Cell microarrays in general allow creating positioned arrays composed of highly miniaturized cell culture. These unique LOC systems can be easily integrated with imaging or laser-scanning cytometers. Unlike FCM, measurements are made at multiple time points, and in contrast to conventional time-lapse microscopy, image analysis is greatly simplified by arranging the cells in a spatially defined pattern and/or by their physical separation (Di Carlo *et al.*, 2006; Wang *et al.*, 2007; Wlodkowic and Cooper, 2010b, d; Wlodkowic *et al.*, 2009b; Yarmush and King, 2009). Cell microarrays are scalable for massively parallel processing and as such they are ideal for drug screening routines. They also have the ability to perform kinetic and multivariate analysis of signaling events on a single cell level. Thus, cell microarray technology seems to be particularly suitable to uncover intricacies in cell-to-cell variability (Skommer *et al.*, 2010; Wlodkowic and Cooper, 2010b, c; Wlodkowic *et al.*, 2009b, 2010).

Successful attempts have been made to provide high-density living cell arrays in both static and microperfusion formats and several interesting commercial designs have already reached the market. Picovitro plates (Picovitro AB, Stockholm, Sweden) are one of the examples of static cell microarrays where cells can be seeded, cultivated, and treated in microwells with sizes of the order of 600 μm and the net volume of 500 nl (Fig. 4A) (Lindstrom and Andersson-Svahn, 2010; Lindstrom *et al.*, 2009c). This cell microarray design utilizes glass bottom with a silicon

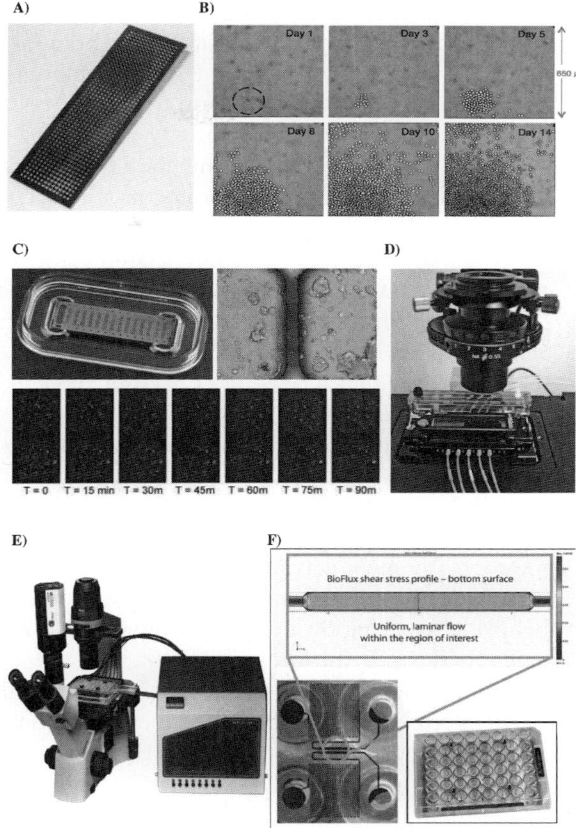

Fig. 4 High-throughput drug screening on innovative microfabricated chips. (A) Picovitro microarray plates and slides as an example of a proprietary, microfabricated high-density cell array. Microwells 650×650 µm are fabricated by anodically bonding the silicon grid wafer to a 500 µm borofloat glass substrate (left panel). (B) Example of a long-term cell proliferation analysis on a Picovitro array. Long-term clone formation was started with a single K-562 cell FACS sorted to one well and cultured for up to 2 weeks. Data courtesy of Dr Sara Lindström, Picovitro AB (Stockholm, Sweden). (C) (D) CellTRAY®– a novel micro-etched live cell screening technology. Independently addressable regions of glass or plastic microwells allow for a multiplexed and time-resolved experimentation at a single cell level. Fully integrated and automated CellTRAY® system mounted on a microscope stage. On-microscope incubator and integrated microfluidics system allow for long-term experiments with automated, precise time-lapse imaging of live cells over the course of several days. Data courtesy of Dr Cathy Owen, Nanopoint Inc. (Honolulu, HI, USA). (E) (F) The BioFlux System (Fluxion Biosciences, San Francisco, CA, USA) that provides the state-of-the-art ability to emulate the physiological shear flow in an *in vitro* model. Shear force, flow rates, temperature, and compound addition, however, can be independently controlled and automated through the dedicated software. The BioFlux system leverages the advantages of microfluidics to create a network of laminar flow cells integrated into standard microtiterwell plates to ensure compatibility with common microscope stages making it thus compatible with bright field, fluorescence, confocal microscopy, and possibly also laser scanning cytometry. Data courtesy of Dr Mike Schwartz, Fluxion Biosciences (San Francisco, CA, USA).

microgrid that forms the walls of the wells (Lindstrom *et al.*, 2009a, 2009b, 2009c). A top level semi-permeable PDMS membrane is added after cell seeding that seals the array from external world. PDMS membrane facilitates efficient gas diffusion but profoundly limits the evaporative water loss (Lindstrom *et al.*, 2009a). These high-density microplates reportedly allow long-term and simultaneous culture in 3243 microwells vastly increasing the assay throughput while being compatible with conventional liquid handling protocols (Fig 4B).

The precise delivery and exchange of reagents using static cell microarrays still requires macroscale liquid handling equipment. Most important, it does not allow for a precise spatiotemporal control over the artificial microenvironment on a chip (Wlodkowic and Cooper, 2010a, b; Yarmush and King, 2009). Microfluidic cell arrays provide here a considerable technological improvement over static microarrays that allow for fabrication of parallelized and fully addressable arrays (Di Carlo *et al.*, 2006; Wang *et al.*, 2007; Yarmush and King, 2009). In contrast to static cell microarrays, the constant microperfusion under low Reynold numbers facilities physiologically relevant exchange of stimulants and metabolites. Furthermore, the straightforward computer-aided design (CAD) of microfluidic chips and computational modeling of their fluid mechanics supports the precise control over the microenvironment surrounding growing cells. As a result, microfluidic cell arrays are ideally positioned for studies on real time and spatiotemporal control of cell behavior to changing microenvironment, a feature not attainable with any static technologies (Di Carlo *et al.*, 2006; Wang *et al.*, 2007; Wlodkowic and Cooper, 2010b, c; Wlodkowic *et al.*, 2009b; Yarmush and King, 2009).

An interesting proprietary design of perfusion array technology has recently been introduced by the Nanopoint Inc. (Honolulu, HI, USA) under the trademark of CellTRAY® (Fig 4C, D). CellTRAY® consists of independent subregions of glass or plastic microwells that allow for a multiplexed, time-resolved imaging of single cells (Fig 4C). Apart from the innovative chip design, Nanopoint Inc. provides also fully integrated and automated control hardware and software that is user friendly and reportedly can be installed within minutes (Fig 4D). Microchip and integrated microfluidics control system can be mounted on an inverted microscope stage and allow for long-term experiments with time-lapse imaging of live cells over the course of several days (Fig 4C, D).

Many conventional assays do not incorporate physiological processes that are normally encountered by cells/tissues in the human body, such as microperfusion, gas/drug diffusion rates and shear stress (Hallow *et al.*, 2008; Schaff *et al.*, 2007). These design limitations of macroscale analytical systems have led to biased understanding of many transient and intermittent physiological processes. This is often reflected by failure of many therapeutic leads, selected after *in vitro* screening, to perform *in vivo* in animal models. In this context, another noteworthy commercial technology for automated real-time analysis under shear flow has recently been proposed by Fluxion Biosciences (San Francisco, CA, USA) (Fig 4E). The BioFlux System is a versatile platform for conducting cellular interaction assays which overcomes the limitations of static well plates and conventional laminar flow

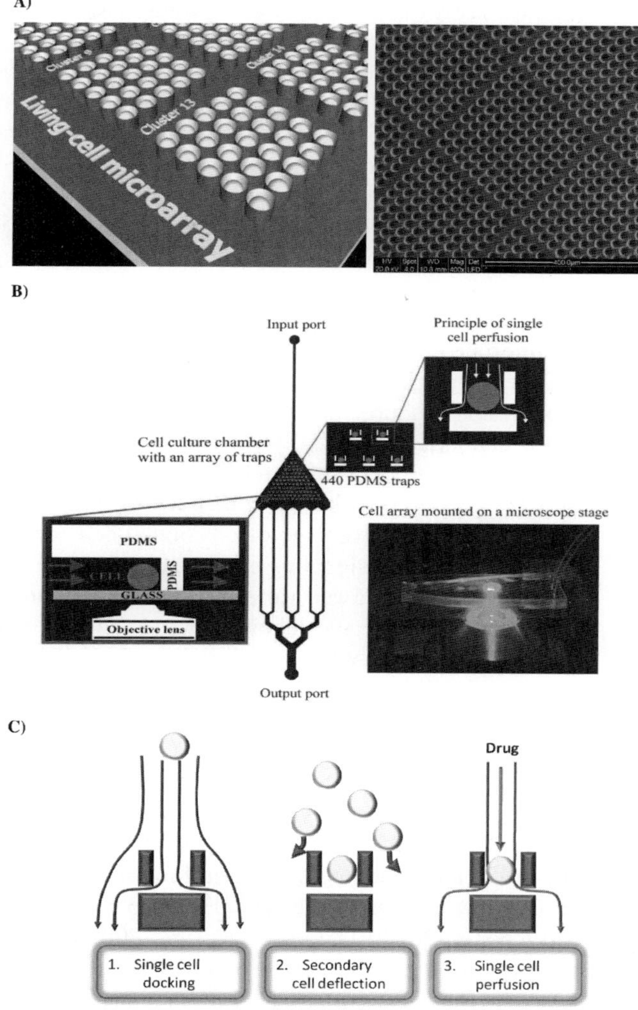

Fig. 5 Living cell microarrays. (A) Single cell microarray platform fabricated in a glass substratum using femtosecond pulse laser μ-machining. Static cell microarray allows for a single cell docking into an array of microfabricated wells. Single wells are 25 μm in diameter. Cells passively sediment into a predefined pattern and are inaccessible to neighboring cells. This minimizes the influence of extrinsic factors, such as physical cell-to-cell contacts and paracrine signaling. This design facilitates real-time analysis at both single cell and population level. (B) Design of the microfluidic cell array (microfluidic array cytometer) with an active (hydrodynamic) cell docking into an array of microfabricated cell traps (microjails). Note the triangular chamber that contains a low-density cell positioning array. Array of microjails was fabricated in a biocompatible elastomer, polydimethylsiloxane (PDMS), and bonded to a glass substrate. The microfluidic array cytometer allows for a gentle trapping of single live cells for prolonged periods of time. (C) Principles of hydrodynamic cell docking and continuous microperfusion on a microfluidic array cytometer.

chambers (Fig 4E, F). The system provides the state-of-the-art ability to emulate the physiological shear flow in an *in vitro* model (Fig 4F). It leverages the advantages of microfluidics to create a network of laminar flow cells integrated into standard microtiterwell plates to ensure compatibility with common microscope stages making it thus compatible with bright field, fluorescence, confocal microscopy, and possibly also laser scanning cytometry (Fig 4E, F). Shear force, flow rates, temperature, and compound addition, however, can be independently controlled and automated through the dedicated software. The microfluidic environment closely mimics the physiological microenvironment, including gas and drug diffusion rates, shear stress, and cell confinement. In the context of, that is, tumor, vascular, developmental, and stem cell biology, the BioFlux system warrants a major "quantum leap" for the improved drug discovery pipelines.

Other emerging microfluidic technologies provide innovative ways to simultaneously analyze large population of living cells whereby the position of every single cell is encoded and spatially maintained over extended periods of time (Fig. 5) (Di Carlo *et al.*, 2006; Rettig and Folch, 2005; Revzin *et al.*, 2005; Wlodkowic and Cooper, 2010b; Wlodkowic *et al.*, 2009b; Wood *et al.*, 2010; Yarmush and King, 2009). The main advantage of positioned cell arrays lies in the ability to study the kinetic multivariate signaling events on a single cell level, which is particularly useful for analysis of cell-to-cell variability and its relevance to cancer therapy (Wlodkowic and Cooper, 2010b; Wlodkowic *et al.*, 2009b). Many single cell immobilization designs have recently been explored that include static microwells, DEP, as well as micromechanical, chemical, and hydrodynamic cell trapping (Di Carlo *et al.*, 2006; Jang *et al.*, 2009; Lan and Jang, 2010; Rettig and Folch, 2005; Revzin *et al.*, 2005; Thomas *et al.*, 2009; Wlodkowic and Cooper, 2010b; Wlodkowic *et al.*, 2009b; Wood *et al.*, 2010). Most recently, the noteworthy reports have proposed hydrodynamic positioning and immobilization of single cells in arrays of micromechanical traps, designed to passively cage individual non-adherent cells (Fig. 5). They reportedly allow for rapid trapping of cells in low shear stress zones while being continuously perfused with drugs and sensors (Di Carlo *et al.*, 2006; Faley *et al.*, 2009; Wlodkowic and Cooper, 2010b; Wlodkowic *et al.*, 2009b). The density and cell trapping efficiency can be easily tailored by changing the number and geometry of microjails (Fig. 5). These emerging techniques create a living single cell and dynamic arrays, ideal for modeling cellular microenvironment and inherently scalable for high-throughput screening (Faley *et al.*, 2009; Wlodkowic and Cooper, 2010b, c; Wlodkowic *et al.*, 2009b; Yarmush and King, 2009).

VI. Conclusions

Microfluidic LOC technologies emerge as relatively uncomplicated and effective solutions for modern cytometry and single cytomics (Wlodkowic and Cooper, 2010a, b). Most importantly, they often provide functionalities and sensitivity that often cannot be easily achieved with any conventional analytical platforms. There is a reason to believe that progress in novel LOC technologies is just a prelude to the major transformation of the field of cytometry. There are still key challenges that lie ahead and include on-chip integration and simplification of many functional components. Robust incorporation within the clinical and research laboratory infrastructure and standardization of innovative LOC cytometers will be an important step toward the rapid expansion of these enabling technologies. As discussed, the recent emergence on the market of user-friendly microscale equipment heralds arrival of the new era in the field of cytometry. We envisage that technologies described here will soon have direct implications for cost and time savings that play an ever increasing role in industrial drug screening pipelines. We anticipate that maturation of such microfluidic technologies in the form of user friendly and integrated enabling systems will attract a mounting interest from biopharmaceutical and academic communities, and will help to reduce drug development expenditures while increasing throughput and content of information.

Expectedly, we are also soon to witness the rise of even more innovative microtechnologies with vast potential in clinical and point-of-care diagnostics (Kiechle and Holland, 2009; Lee *et al.*, 2010; Linder, 2007; Mansfield *et al.*, 2006; Mauk *et al.*, 2007; Ong *et al.*, 2010; Sia and Kricka, 2008). Finally, based on the incredible progress in the field and looking beyond the next 10 years, we anticipate that a new generation of cytometers will prospectively employ microfluidic systems possibly with integrated on-chip cell culture, drug delivery, cytometric analysis and sorting modules. As such they will provide an all-in-one workstations for the future generations of cytometrists.

Acknowledgments

This work was supported by a grant from the Foundation for Research Science and Technology NZ (FRST). Authors thank: Drs Kazuo Takeda from On-chip Biotechnologies Co. (Tokyo, Japan), John C Sharpe from CytonomeST LLC (Boston, MA, USA), Sara Lindström from Picovitro AB (Stockholm, Sweden), Cathy Owen from Nanopoint Inc. (Honolulu, HI, USA), and Mike Schwartz from Fluxion Biosciences (San Francisco, CA, USA) for sharing data and providing materials on proprietary microfabricated analytical systems.

The authors declare no conflicting financial interest.

References

Adams, J. D., Kim, U., and Soh, H. T. (2008). Multitarget magnetic activated cell sorter. *Proc. Natl. Acad. Sci. U.S.A.* **105**, 18165–18170.

Adams, J. D., and Tom Soh, H. (2009). Perspectives on utilizing unique features of microfluidics technology for particle and cell sorting. *JALA Charlottesv Va* **14**, 331–340.

Andersson, H., and van den Berg, A. (2004). Microtechnologies and nanotechnologies for single-cell analysis. *Curr. Opin. Biotechnol.* **15**, 44–49.

Arndt-Jovin, D. J., and Jovin, T. M. (1974). Computer-controlled multiparameter analysis and sorting of cells and particles. *J. Histochem. Cytochem.* **22**, 622–625.

Baret, J. C., Miller, O. J., Taly, V., Ryckelynck, M., El-Harrak, A., Frenz, L., Rick, C., Samuels, M. L., Hutchison, J. B., Agresti, J. J., Link, D. R., Weitz, D. A., Griffiths, A. D. (2009). Fluorescence-activated droplet sorting (FADS): efficient microfluidic cell sorting based on enzymatic activity. *Lab Chip* **9**, 1850–1858.

Beebe, D. J., Mensing, G. A., and Walker, G. M. (2002). Physics and applications of microfluidics in biology. *Annu. Rev. Biomed. Eng.* **4**, 261–286.

Berthier, E., Warrick, J., Yu, H., and Beebe, D. J. (2008a). Managing evaporation for more robust microscale assays. Part 1. Volume loss in high throughput assays. *Lab Chip* **8**, 852–859.

Berthier, E., Warrick, J., Yu, H., and Beebe, D. J. (2008b). Managing evaporation for more robust microscale assays. Part 2. Characterization of convection and diffusion for cell biology. *Lab Chip* **8**, 860–864.

Bhagat, A. A., Bow, H., Hou, H. W., Tan, S. J., Han, J., Lim, C. T. (2010a). Microfluidics for cell separation. *Med. Biol. Eng. Comput.* **48**, 999–1014.

Bhagat, A. A., Kuntaegowdanahalli, S. S., Kaval, N., Seliskar, C. J., and Papautsky, I. (2010b). Inertial microfluidics for sheath-less high-throughput flow cytometry. *Biomed. Microdevices* **12**, 187–195.

Chan, S. D., Luedke, G., Valer, M., Buhlmann, C., and Preckel, T. (2003). Cytometric analysis of protein expression and apoptosis in human primary cells with a novel microfluidic chip-based system. *Cytometry A* **55**, 119–125.

Cheung, K. C., Di Berardino, M., Schade-Kampmann, G., Hebeisen, M., Pierzchalski, A., Bocsi, J., Mittag, A., Tarnok, A. (2010). Microfluidic impedance-based flow cytometry. *Cytometry A* **77**, 648–666.

Cho, S. H., Chen, C. H., Tsai, F. S., Godin, J. M., and Lo, Y. H. (2010). Human mammalian cell sorting using a highly integrated micro-fabricated fluorescence-activated cell sorter (microFACS). *Lab Chip* **10**, 1567–1573.

Chung, T. D., and Kim, H. C. (2007). Recent advances in miniaturized microfluidic flow cytometry for clinical use. *Electrophoresis* **28**, 4511–4520.

Darzynkiewicz, Z., Bedner, E., Li, X., Gorczyca, W., and Melamed, M. R. (1999). Laser-scanning cytometry: A new instrumentation with many applications. *Exp. Cell Res.* **249**, 1–12.

Di Carlo, D., Wu, L. Y., and Lee, L. P. (2006). Dynamic single cell culture array. *Lab Chip* **6**, 1445–1449.

Dittrich, P. S., and Manz, A. (2006). Lab-on-a-chip: microfluidics in drug discovery. *Nat. Rev. Drug Discov.* **5**, 210–218.

Eisenstein, M. (2006). Cell sorting: divide and conquer. *Nature* **441**, 1179–1185.

Faley, S. L., Copland, M., Wlodkowic, D., Kolch, W., Seale, K. T., Wikswo, J. P., Cooper, J. M. (2009). Microfluidic single cell arrays to interrogate signalling dynamics of individual, patient-derived hematopoietic stem cells. *Lab Chip* **9**, 2659–2664.

Franke, T. A., and Wixforth, A. (2008). Microfluidics for miniaturized laboratories on a chip. *Chemphyschem* **9**, 2140–2156.

Gluckstad, J. (2004). Microfluidics: sorting particles with light. *Nat Mater.* **3**, 9–10.

Hallow, D. M., Seeger, R. A., Kamaev, P. P., Prado, G. R., LaPlaca, M. C., Prausnitz, M. R. (2008). Shear-induced intracellular loading of cells with molecules by controlled microfluidics. *Biotechnol. Bioeng.* **99**, 846–854.

Hill, D. C. (1998). Trends in development of high-throughput screening technologies for rapid discovery of novel drugs. *Curr. Opin. Drug Discov. Devel.* **1**, 92–97.

Holmes, D., and Gawad, S. (2010). The application of microfluidics in biology. *Methods Mol. Biol.* **583**, 55–80.

Huh, D., Gu, W., Kamotani, Y., Grotberg, J. B., and Takayama, S. (2005). Microfluidics for flow cytometric analysis of cells and particles. *Physiol. Meas.* **26**, R73–98.

Hur, S. C., Tse, H. T., and Di Carlo, D. (2010). Sheathless inertial cell ordering for extreme throughput flow cytometry. *Lab Chip* **10**, 274–280.

Irimia, D., and Toner, M. (2009). Spontaneous migration of cancer cells under conditions of mechanical confinement. *Integr. Biol. (Camb)* **1**, 506–512.

Jang, L. S., Huang, P. H., and Lan, K. C. (2009). Single-cell trapping utilizing negative dielectrophoretic quadrupole and microwell electrodes. *Biosens. Bioelectron.* **24**, 3637–3644.

Kastrup, C. J., Runyon, M. K., Lucchetta, E. M., Price, J. M., and Ismagilov, R. F. (2008). Using chemistry and microfluidics to understand the spatial dynamics of complex biological networks. *Acc. Chem. Res.* **41**, 549–558.

Kiechle, F. L., and Holland, C. A. (2009). Point-of-care testing and molecular diagnostics: miniaturization required. *Clin. Lab. Med.* **29**, 555–560.

Kummrow, A., Theisen, J., Frankowski, M., Tuchscheerer, A., Yildirim, H., Brattke, K., Schmidt, M., Neukammer, J. (2009). Microfluidic structures for flow cytometric analysis of hydrodynamically focussed blood cells fabricated by ultraprecision micromachining. *Lab Chip* **9**, 972–981.

Kuntaegowdanahalli, S. S., Bhagat, A. A., Kumar, G., and Papautsky, I. (2009). Inertial microfluidics for continuous particle separation in spiral microchannels. *Lab Chip* **9**, 2973–2980.

Lan, K. C., and Jang, L. S. (2010). Integration of single-cell trapping and impedance measurement utilizing microwell electrodes. *Biosens Bioelectron.* **26**, 2025–2031.

Lee, M. G., Choi, S., and Park, J. K. (2009). Three-dimensional hydrodynamic focusing with a single sheath flow in a single-layer microfluidic device. *Lab Chip* **9**, 3155–3160.

Lee, W. G., Kim, Y. G., Chung, B. G., Demirci, U., and Khademhosseini, A. (2010). Nano/microfluidics for diagnosis of infectious diseases in developing countries. *Adv. Drug Deliv. Rev.* **62**, 449–457.

Linder, V. (2007). Microfluidics at the crossroad with point-of-care diagnostics. *Analyst* **132**, 1186–1192.

Lindstrom, S., and Andersson-Svahn, H. (2010). Miniaturization of biological assays – Overview on microwell devices for single-cell analyses. *Biochim. Biophys. Acta.* **1810**, 308–316.

Lindstrom, S., Eriksson, M., Vazin, T., Sandberg, J., Lundeberg, J., Frisen, J., Andersson-Svahn, H. (2009a). High-density microwell chip for culture and analysis of stem cells. *PLoS One* **4**, e6997.

Lindstrom, S., Hammond, M., Brismar, H., Andersson-Svahn, H., and Ahmadian, A. (2009b). PCR amplification and genetic analysis in a microwell cell culturing chip. *Lab Chip* **9**, 3465–3471.

Lindstrom, S., Mori, K., Ohashi, T., and Andersson-Svahn, H. (2009c). A microwell array device with integrated microfluidic components for enhanced single-cell analysis. *Electrophoresis* **30**, 4166–4171.

Liu, T., Li, C., Li, H., Zeng, S., Qin, J., Lin, B. (2009). A microfluidic device for characterizing the invasion of cancer cells in 3-D matrix. *Electrophoresis* **30**, 4285–4291.

Mach, W. J., Thimmesch, A. R., Orr, J. A., Slusser, J. G., and Pierce, J. D. (2010). Flow cytometry and laser scanning cytometry, a comparison of techniques. *J. Clin. Monit. Comput.* **24**, 251–259.

Mansfield, C. D., Man, A., and Shaw, R. A. (2006). Integration of microfluidics with biomedical infrared spectroscopy for analytical and diagnostic metabolic profiling. *IEE Proc. Nanobiotechnol.* **153**, 74–80.

Mao, X., Lin, S. C., Dong, C., and Huang, T. J. (2009). Single-layer planar on-chip flow cytometer using microfluidic drifting based three-dimensional (3D) hydrodynamic focusing. *Lab Chip* **9**, 1583–1589.

Mauk, M. G., Ziober, B. L., Chen, Z., Thompson, J. A., and Bau, H. H. (2007). Lab-on-a-chip technologies for oral-based cancer screening and diagnostics: capabilities, issues, and prospects. *Ann. N.Y. Acad. Sci.* **1098**, 467–475.

Mayr, L. M., and Bojanic, D. (2009). Novel trends in high-throughput screening. *Curr. Opin. Pharmacol.* **9**, 580–588.

Melamed, M. R. (2001). A brief history of flow cytometry and sorting. *Methods Cell Biol.* **63**, 3–17.

Michelsen, U. (2007). Microfluidics in medical applications. *Med. Device Technol.* **18**, 12–14.

Myers, F. B., and Lee, L. P. (2008). Innovations in optical microfluidic technologies for point-of-care diagnostics. *Lab Chip* **8**, 2015–2031.

Oakey, J., Applegate Jr., R. W., Arellano, E., Di Carlo, D., Graves, S. W., Toner, M. (2010). Particle focusing in staged inertial microfluidic devices for flow cytometry. *Anal. Chem.* **82**, 3862–3867.

Ong, P. K., Lim, D., and Kim, S. (2010). Are microfluidics-based blood viscometers ready for point-of-care applications? A review. *Crit. Rev. Biomed. Eng.* **38**, 189–200.

Palkova, Z., Vachova, L., Valer, M., and Preckel, T. (2004). Single-cell analysis of yeast, mammalian cells, and fungal spores with a microfluidic pressure-driven chip-based system. *Cytometry A* **59**, 246–253.

Pfohl, T., Mugele, F., Seemann, R., and Herminghaus, S. (2003). Trends in microfluidics with complex fluids. *Chemphyschem* **4**, 1291–1298.

Polson, N. A., and Hayes, M. A. (2001). Microfluidics: controlling fluids in small places. *Anal. Chem.* **73**, 312A–319A.

Pugmire, D. L., Waddell, E. A., Haasch, R., Tarlov, M. J., and Locascio, L. E. (2002). Surface characterization of laser-ablated polymers used for microfluidics. *Anal. Chem.* **74**, 871–878.

Qi, S., Liu, X., Ford, S., Barrows, J., Thomas, G., Kelly, K., McCandless, A., Lian, K., Goettert, J., Soper, S. A. (2002). Microfluidic devices fabricated in poly(methyl methacrylate) using hot-embossing with integrated sampling capillary and fiber optics for fluorescence detection. *Lab Chip* **2**, 88–95.

Rettig, J. R., and Folch, A. (2005). Large-scale single-cell trapping and imaging using microwell arrays. *Anal. Chem.* **77**, 5628–5634.

Revzin, A., Sekine, K., Sin, A., Tompkins, R. G., and Toner, M. (2005). Development of a microfabricated cytometry platform for characterization and sorting of individual leukocytes. *Lab Chip* **5**, 30–37.

Schaff, U. Y., Xing, M. M., Lin, K. K., Pan, N., Jeon, N. L., Simon, S. I. (2007). Vascular mimetics based on microfluidics for imaging the leukocyte–endothelial inflammatory response. *Lab Chip* **7**, 448–456.

Scott, R., Sethu, P., and Harnett, C. K. (2008). Three-dimensional hydrodynamic focusing in a microfluidic Coulter counter. *Rev. Sci. Instrum.* **79**, 046104.

Shapiro, H. M. (2004). The evolution of cytometers. *Cytometry A* **58**, 13–20.

Sia, S. K., and Kricka, L. J. (2008). Microfluidics and point-of-care testing. *Lab Chip* **8**, 1982–1983.

Sia, S. K., and Whitesides, G. M. (2003). Microfluidic devices fabricated in poly(dimethylsiloxane) for biological studies. *Electrophoresis* **24**, 3563–3576.

Sims, C. E., and Allbritton, N. L. (2007). Analysis of single mammalian cells on-chip. *Lab Chip* **7**, 423–440.

Situma, C., Hashimoto, M., and Soper, S. A. (2006). Merging microfluidics with microarray-based bioassays. *Biomol. Eng.* **23**, 213–231.

Skommer, J., Darzynkiewicz, Z., and Wlodkowic, D. (2010). Cell death goes LIVE: technological advances in real-time tracking of cell death. *Cell Cycle* 9.

Stroock, A. D., and Fischbach, C. (2010). Microfluidic culture models of tumor angiogenesis. *Tissue Eng. Part A* **16**, 2143–2146.

Svahn, H. A., and van den Berg, A. (2007). Single cells or large populations? *Lab Chip* **7**, 544–546.

Takayama, S., Ostuni, E., LeDuc, P., Naruse, K., Ingber, D. E., Whitesides, G. M. (2001). Subcellular positioning of small molecules. *Nature* **411**, 1016.

Takayama, S., Ostuni, E., LeDuc, P., Naruse, K., Ingber, D. E., Whitesides, G. M. (2003). Selective chemical treatment of cellular microdomains using multiple laminar streams. *Chem. Biol.* **10**, 123–130.

Takeda, K., and Jimma, F. (2009). Maintenance free biosafety flowcytometer using disposable microfluidic chip (FISHMAN-R). *Cytometry B* **76B**, 405–406.

Takao, M., Jimma, F., and Takeda, K. (2009). Expanded applications of new-designed microfluidic flow cytometer (FISHMAN-R). *Cytometry B* **76B**, 405.

Thomas, R. S., Morgan, H., and Green, N. G. (2009). Negative DEP traps for single cell immobilisation. *Lab Chip* **9**, 1534–1540.

Wang, J., Ren, L., Li, L., Liu, W., Zhou, J., Yu, W., Tong, D., Chen, S. (2009). Microfluidics: a new cosset for neurobiology. *Lab Chip* **9**, 644–652.

Wang, M. M., Tu, E., Raymond, D. E., Yang, J. M., Zhang, H., Hagen, N., Dees, B., Mercer, E. M., Forster, A. H., Kariv, I., Marchand, P. J., Butler, W. F. (2005). Microfluidic sorting of mammalian cells by optical force switching. *Nat. Biotechnol.* **23**, 83–87.

Wang, Z., Kim, M. C., Marquez, M., and Thorsen, T. (2007). High-density microfluidic arrays for cell cytotoxicity analysis. *Lab Chip* **7**, 740–745.

Warrick, J. W., Murphy, W. L., and Beebe, D. J. (2008). Screening the cellular microenvironment: a role for microfluidics. *IEEE Rev. Biomed. Eng.* **1**, 75–93.

Weibel, D. B., and Whitesides, G. M. (2006). Applications of microfluidics in chemical biology. *Curr. Opin. Chem. Biol.* **10**, 584–591.

Weigl, B., Domingo, G., Labarre, P., and Gerlach, J. (2008). Towards non- and minimally instrumented, microfluidics-based diagnostic devices. *Lab Chip* **8**, 1999–2014.

Wheeler, D. B., Carpenter, A. E., and Sabatini, D. M. (2005). Cell microarrays and RNA interference chip away at gene function. *Nat. Genet.* **37**(Suppl), S25–30.

Whitesides, G. M. (2006). The origins and the future of microfluidics. *Nature* **442**, 368–373.

Wlodkowic, D., and Cooper, J. M. (2010a). Microfabricated analytical systems for integrated cancer cytomics. *Anal. Bioanal. Chem.* **398**, 193–209.

Wlodkowic, D., and Cooper, J. M. (2010b). Microfluidic cell arrays in tumor analysis: new prospects for integrated cytomics. *Expert Rev. Mol. Diagn.* **10**, 521–530.

Wlodkowic, D., and Cooper, J. M. (2010c). Tumors on chips: oncology meets microfluidics. *Curr. Opin. Chem. Biol.* **14**, 556–567.

Wlodkowic, D., Faley, S., Skommer, J., McGuinness, D., and Cooper, J. M. (2009a). Biological implications of polymeric microdevices for live cell assays. *Anal. Chem.* **81**, 9828–9833.

Wlodkowic, D., Faley, S., Zagnoni, M., Wikswo, J. P., and Cooper, J. M. (2009b). Microfluidic single-cell array cytometry for the analysis of tumor apoptosis. *Anal. Chem.* **81**, 5517–5523.

Wlodkowic, D., Skommer, J., and Darzynkiewicz, Z. (2010). Cytometry in cell necrobiology revisited. Recent advances and new vistas. *Cytometry A* **77**, 591–606.

Wlodkowic, D., Skommer, J., Faley, S., Darzynkiewicz, Z., and Cooper, J. M. (2009c). Dynamic analysis of apoptosis using cyanine SYTO probes: from classical to microfluidic cytometry. *Exp. Cell Res.* **315**, 1706–1714.

Wlodkowic, D., Skommer, J., McGuinness, D., Faley, S., Kolch, W., Darzynkiewicz, Z., Cooper, J. M. (2009d). Chip-based dynamic real-time quantification of drug-induced cytotoxicity in human tumor cells. *Anal. Chem.* **81**, 6952–6959.

Wolff, A., Perch-Nielsen, I. R., Larsen, U. D., Friis, P., Goranovic, G., Poulsen, C. R., Kutter, J. P., Telleman, P. (2003). Integrating advanced functionality in a microfabricated high-throughput fluorescent-activated cell sorter. *Lab Chip* **3**, 22–27.

Wood, D. K., Weingeist, D. M., Bhatia, S. N., and Engelward, B. P. (2010). Single cell trapping and DNA damage analysis using microwell arrays. *Proc. Natl. Acad. Sci. U.S.A.* **107**, 10008–10013.

Yarmush, M. L., and King, K. R. (2009). Living-cell microarrays. *Annu. Rev. Biomed. Eng.* **11**, 235–257.

Zaouk, R., Park, B. Y., and Madou, M. J. (2006). Fabrication of polydimethylsiloxane microfluidics using SU-8 molds. *Methods Mol. Biol.* **321**, 17–21.

Zhang, W., Lin, S., Wang, C., Hu, J., Li, C., Zhuang, Z., Zhou, Y., Mathies, R. A., Yang, C. J. (2009). PMMA/PDMS valves and pumps for disposable microfluidics. *Lab Chip* **9**, 3088–3094.

Zhu, T., Luo, C., Huang, J., Xiong, C., Ouyang, Q., Fang, J. (2010). Electroporation based on hydrodynamic focusing of microfluidics with low dc voltage. *Biomed. Microdevices* **12**, 35–40.

CHAPTER 6

Label-Free Resistive-Pulse Cytometry

M.R. Chapman[*] **and L.L. Sohn**[*,†]

[*]Biophysics Graduate Group, University of California, Berkeley, California, USA

[†]Department of Mechanical Engineering, University of California, Berkeley, California, USA

Abstract

Numerous methods have recently been developed to characterize cells for size, shape, and specific cell-surface markers. Most of these methods rely upon exogenous labeling of the cells and are better suited for large cell populations ($>$10,000). Here, we review a label-free method of characterizing and screening cells based on the Coulter-counter technique of particle sizing: an individual cell transiting a microchannel (or "pore") causes a downward pulse in the measured DC current

METHODS IN CELL BIOLOGY, VOL 102

127

0091-679X/10 $35.00
DOI 10.1016/B978-0-12-374912-3.00006-7

across that "pore". Pulse magnitude corresponds to the cell size, pulse width to the transit time needed for the cell to pass through the pore, and pulse shape to how the cell traverses across the pore (i.e., rolling or tumbling). When the pore is functionalized with an antibody that is specific to a surface-epitope of interest, label-free screening of a specific marker is possible, as transient binding between the two results in longer time duration than when the pore is unfunctionalized or functionalized with a nonspecific antibody. While this method cannot currently compete with traditional technology in terms of throughput, there are a number of applications for which this technology is better suited than current commercial cytometry systems. Applications include the rapid and nondestructive analysis of small cell populations (<100), which is not possible with current technology, and a platform for providing true point-of-care clinical diagnostics, due to the simplicity of the device, low manufacturing costs, and ease of use.

I. Introduction

Cell characterization through identification of membrane components is an essential element in cell biology (Kemper and Atkinson, 2007; Lane *et al.*, 2005), disease diagnosis and monitoring (Hrusak and Porwit-MacDonald, 2002; Lau *et al.*, 2006), and drug discovery (Hruby, 2002). Although current methods for cell analysis, such as flow cytometry (Shapiro, 2003) and magnetic-bead column selection (Lang *et al.*, 2002), play an invaluable role in both research laboratories and clinical settings, they do present limitations. For example, traditional approaches often require advanced preparation, including exogenous labeling of cells. Such labeling leads to added incubation time, additional costs, and the possibility of modifying cell physiology and function. As another example, data analysis can be challenging when the available number of cells to be screened is on the order of just a few hundred or less. Finally, traditional approaches do not lend themselves to portability, which can be desirable in certain clinical situations. Given these limitations and constraints, there is still a clear need for new, or at the very least, complementary methods for cell characterization.

In this chapter, we describe a highly sensitive, accurate, and *label-free* method for characterizing individual cells. The method is based on employing resistive-pulse sensing (RPS) with a microchip pore. This electronic-sensing technique, as will be described, measures the current pulse caused by a cell traversing a micron-sized channel or pore. Analysis of the current pulse leads to multiparametric information regarding size, shape, and surface markers of the cell. We will discuss RPS in depth and examine its strengths and weaknesses for cell analysis as compared to the flow cytometry.

II. Resistive–Pulse Sensing

RPS, or Coulter technique of particle sizing (Coulter, 1953), is based on measuring a DC current (or equivalently, resistance) pulse as a particle passes through a

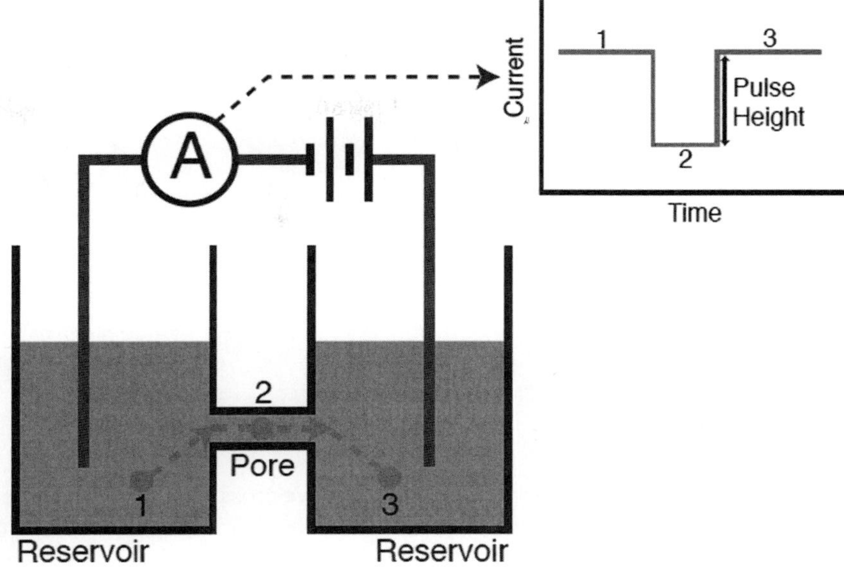

Fig. 1 Schematic of a basic resistive-pulse sensor, which consists of two reservoirs connected to a pore. A constant baseline current is measured (1) until a particle passes through the pore; (2) the particle displaces, by volume, conducting fluid, which results in a drop in current. When the particle successfully leaves the pore; (3) the pore once again fills with fluid and the current rises to the baseline. The magnitude of the current pulse corresponds to the size of the particle, and the pulse width to the transit time the particle needs to pass through the pore.

micron-sized channel (hereafter referred to as a "pore") that is connected to two fluid-filled reservoirs (Fig. 1). The particle displaces by volume conducting fluid within the pore, resulting in a drop in current across the channel, itself. RPS can be used to measure the size of particles whose dimensions are on the order of the pore dimensions. While this method has long been used to characterize cells several microns in diameter (Gregg and Steidley, 1965) its relative simplicity has led to many efforts to detect nanoscale particles (Deblois and Bean, 1970; Kobayashi and Martin, 1997; Koch $et\ al.$, 1999; Sun and Crooks, 1999), including viruses (Deblois $et\ al.$, 1977), with it. More recently, it has been used to detect single molecules (Bayley and Cremer, 2001; Bezrukov $et\ al.$, 1994; Gu $et\ al.$, 2001) and their interactions (Gu $et\ al.$, 2001). In our own group, we have used RPS to detect the nanometer increase in diameter of a submicron latex colloid upon binding to an unlabeled specific antibody (Carbonaro and Sohn, 2005; Saleh and Sohn, 2003b) and to sense single molecules of unlabeled lambda (λ) phage DNA (Saleh and Sohn, 2003a).

The sensitivity of RPS relies upon the relative size of the pore and the particle to be measured (Fig. 2). The resistance of a pore, R_p, increases by δR_p when a particle enters since the particle displaces conducting fluid. δR_p can be estimated for a pore

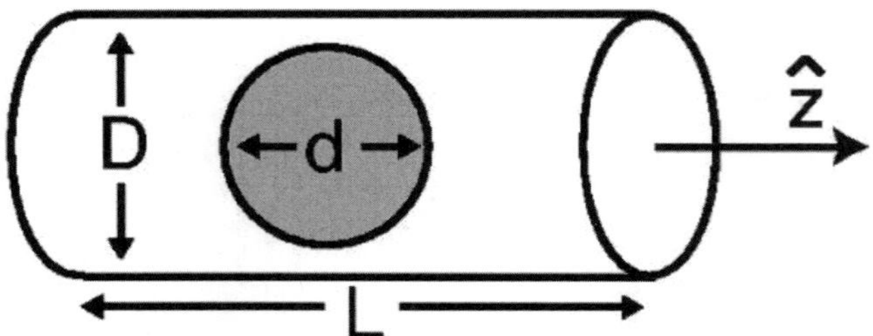

Fig. 2   Schematic of a spherical particle of diameter d in a pore of diameter D and length L. (Saleh and Sohn, 2001).

aligned along the z-axis by

$$\delta R_p = \rho \int \frac{dz}{A(z)} - R_p \tag{1}$$

where $A(z)$ represents the successive cross sections of the pore containing a particle, and ρ is the resistivity of the solution. For a spherical particle of diameter d in a pore of diameter D and length L, the relative change in resistance is

$$\frac{\delta R_p}{R_p} = \frac{D}{L} \left| \frac{\arcsin(d/D)}{(1 - (d/D)^2)^{1/2}} - \frac{d}{D} \right| \tag{2}$$

Eqs. (1) and (2) assume that the current density is uniform across the channel, and thus is not applicable for cases where the cross section $A(z)$ varies quickly, that is, when $d \ll D$. For that particular case, Deblois and Bean (1970) formulated an equation for δR_p based on an approximate solution to the Laplace equation

$$\frac{\delta R_p}{R_p} = \frac{d^3}{LD^2} \left[\frac{D^2}{2L^2} + \frac{1}{\sqrt{1 + (D/L)^2}} \right] F\left(\frac{d^3}{D^3}\right) \tag{3}$$

where $F(d^3/D^3)$ is a numerical factor that accounts for the bulging of the electric field lines into the pore wall. When employing Eq. (3) to predict resistance changes, one finds an effective value for d by equating the cross-sectional area of a rectangular channel with that of a cylindrical one.

If R_p is the dominant resistance of the measurement circuit, then relative changes in the current I are equal in the magnitude to the relative changes in the resistance, that is, $|\delta I/I| = |\delta R_p/R_p|$, and Eqs. (2) and (3) can be both directly compared to measure current changes. This comparison is disallowed if R_p is similar in magnitude to other series resistances, such as the electrode/fluid interfacial resistance, $R_{e/f}$. One can completely remove $R_{e/f}$ from the electrical circuit by performing a *four-point* measurement of the current (described in further detail below).

III. Coulter Counter on a Chip

The Coulter counter (Coulter, 1953), which employs RPS, was specifically developed during World War II as a means to count plankton particles for the Navy. Coulter's original device consisted of two reservoirs connected by a small aperture (i.e., a pore, as discussed in the previous section). A single electrode in each

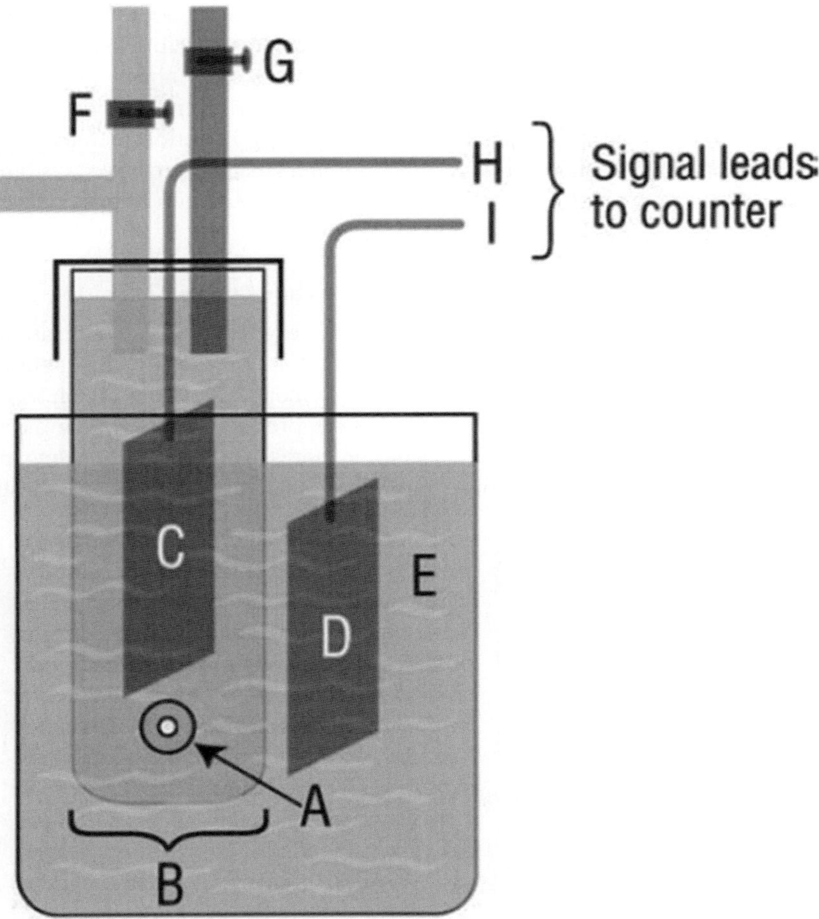

Fig. 3 Basic schematic of a commercial Coulter counter, where (A) is the sensing aperture, (B) is the aperture tube, (C) and (D) are the sensing electrodes, (E) is the input reservoir, (F) and (G) are the valves to control metering of the sample and aperture tube flushing, and (H) and (I) are the leads which attach the sensing electrodes to the detector. (Graham, 2003)

reservoir measured the current across the aperture when single particles were driven across using pressure. Current commercial Coulter counters use a glass tube, which has a small "sensing" aperture at one of its end, suspended in a large reservoir (Fig. 3) (Graham, 2003). Electrodes are placed in both the glass tube and the reservoir to monitor current across the sensing aperture. Typical aperture diameters range from 50 to 200 μm, thereby allowing for the counting of particles that range from 1 to 120 μm in size, depending upon which aperture tube is installed.

Microfabricated Coulter counters offer a number of advantages over current commercial systems, including low cost, small sample volumes, portability, and most importantly, increased sensitivity. Commercial Coulter counters are only sensitive down to 1 μm-sized particles. It is possible to measure smaller-sized particles; however, high salt concentrations – well above physiological conditions – are needed. Because the sensitivity of RPS is directly related to the diameter of the sensing aperture, modern microfabrication technology provides the means to make microfluidic Coulter counters that are sensitive to submicron particles, without the need for the aforementioned high-salt conditions.

A. Achieving nm Precision and High Throughput using Colloids

In 2001, our group created the very first working realization of a Coulter counter on a microchip (Saleh and Sohn, 2001). As shown in Fig. 4, our device was fabricated on top of a quartz substrate using standard microfabrication techniques. (We have much simplified the device fabrication process, which will be discussed in the subsequent sections.) The reservoirs and pores were patterned using photolithography and/or electron-beam lithography and then etched into the substrate using reactive-ion

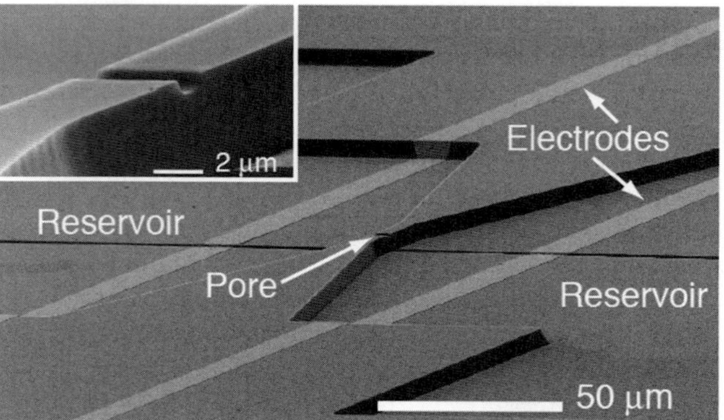

Fig. 4 Scanning electron micrograph of our microchip Coulter counter. 3.5 μm deep reservoirs and the inner Ti/Pt electrodes, which control the voltage applied to the pore but pass no current, are only partially shown. The outer electrodes, which inject current into the solution, are not visible in this image. The four electrodes enable the four-point measurement described in the text. The inset shows a magnified view of the device's pore, which has dimensions $5.1 \times 1.5 \times 1.0 \ \mu m^3$. (Saleh and Sohn, 2001)

etching. Four platinum electrodes, used to perform a four-point measurement of the current across the pore, were subsequently created on the substrate before the device was sealed with a polydimethylsiloxane (PDMS)-coated glass coverslip. Using pores with lateral dimensions ranging from 1 to 10.5 µm and cross-sectional areas ranging from 160 nm^2 to 1.5 µm^2, we were able to measure latex colloidal particles as small as 87 nm in diameter (Fig. 5a). Measurements of polydisperse submicron colloidal suspensions showed that our device had a resolution of $\pm$ 10 nm in diameter of the measured colloids. This precision approached the intrinsic variations in colloid diameter of 2–4%, as given by the manufacturer (Fig. 5b and c).

While Coulter counters with submicron sensitivity are not necessary for analyzing most eukaryotic cells, there is an increasing interest in apoptotic microparticles and exosomes, both of which are submicron in size, as possible targets for clinical diagnostics. Analysis of RNA and DNA fragments found in apoptotic microparticles and exosomes (Halicka *et al.*, 2000) have shown great promise as noninvasive alternatives for genetic profiling of cancer tumors (Andre *et al.*, 2002; Koga *et al.*, 2005; Ratajczak *et al.*, 2006) and prenatal diagnostics (Orozco *et al.*, 2006, 2008). Furthermore, rapid and label-free detection and determination of the concentration of various pathogens such as viral capsids (Berg *et al.*, 2003) could help physicians monitor disease progression and treatment efficacy without lengthy and costly traditional assays.

While they provide increasing sensitivity to smaller particles, as well as increased resolution of particle size, smaller apertures do have a tendency to clog. Lower sample concentration, and integrated filters designed to prevent larger particles or aggregates from entering the pore can help alleviate this issue. In 2004, Nieuwenhuis *et al.* used a "liquid aperture" to overcome the issue of channel clogging, all the while achieving the sensitivity of smaller apertures (Nieuwenhuis *et al.*, 2004). The "liquid aperture" was created using hydrodynamic focusing to achieve a noncoaxial sheath of conductive fluid surrounded by nonconductive fluid on three sides (Fig. 6). Sample, suspended in conductive fluid, was introduced through a port in the bottom of the main input channel. While the height of the liquid aperture was controlled by the relative flow rates of the sample injection and flow of the nonconductive fluid in the main input channel, the horizontal dimensions were controlled by the injection of additional nonconductive fluid from a channel on either side of the main channel.

Use of a liquid aperture for RPS helps prevent channel blockage without compromising sensitivity, as the physical channel is much larger than the sensing aperture. Furthermore, dynamic modulation of the aperture is possible. This, in turn, enables the study of a much wider range of particle sizes since the aperture can spontaneously be adapted to the appropriate size by varying relative flow rates of conductive and nonconductive fluid. While there are many advantages to using a sensing aperture that can be modulated, the use of a reference bead of known size in one's sample is critical to providing an accurate real-time measurement of the "liquid aperture", that is, for calibration and final data analysis.

In addition to a high probability of clogging, throughput does decrease with decreasing pore cross section. In order to improve throughput, one can either

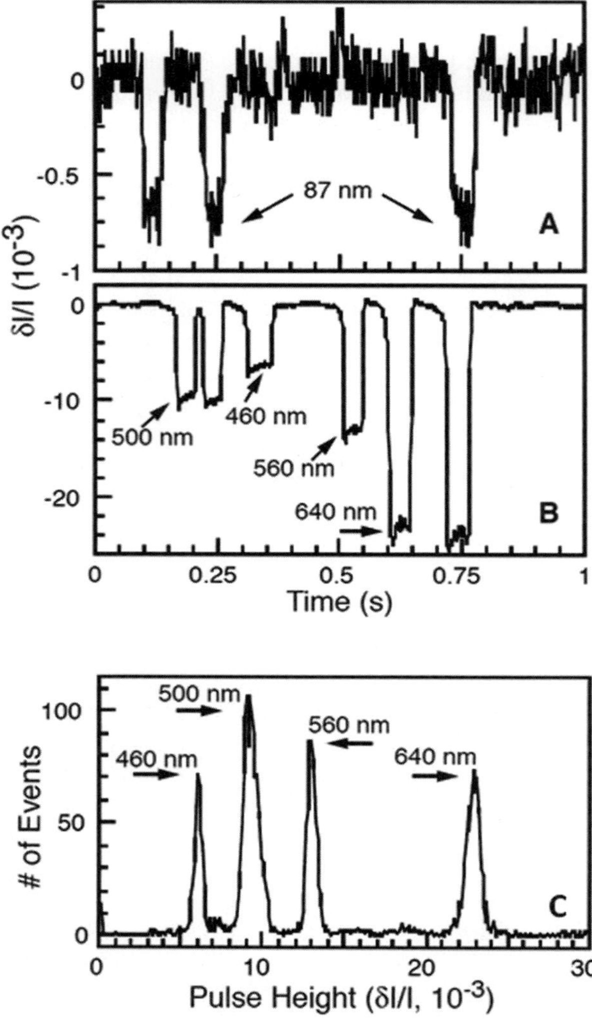

Fig. 5 Relative changes in baseline current $\delta I/I$ versus time for (A) a monodisperse solution of 87 nm diameter latex colloids measured with a pore of length 8.3 μm and cross section 0.16 μm^2, and (B) a polydisperse solution of latex colloids with diameters 460, 500, 560, and 640 nm measured with a pore of length 9.5 μm and cross section 1.2 μm^2. Each downward current pulse represents an individual particle entering the pore. The four distinct pulse heights in (B) correspond to the four different colloid diameters. (C) A histogram of pulse heights resulting from measuring the polydisperse solution shown in (B). The resolution for this particular device is ±10 nm in diameter for the particles measured. (Saleh and Sohn, 2001)

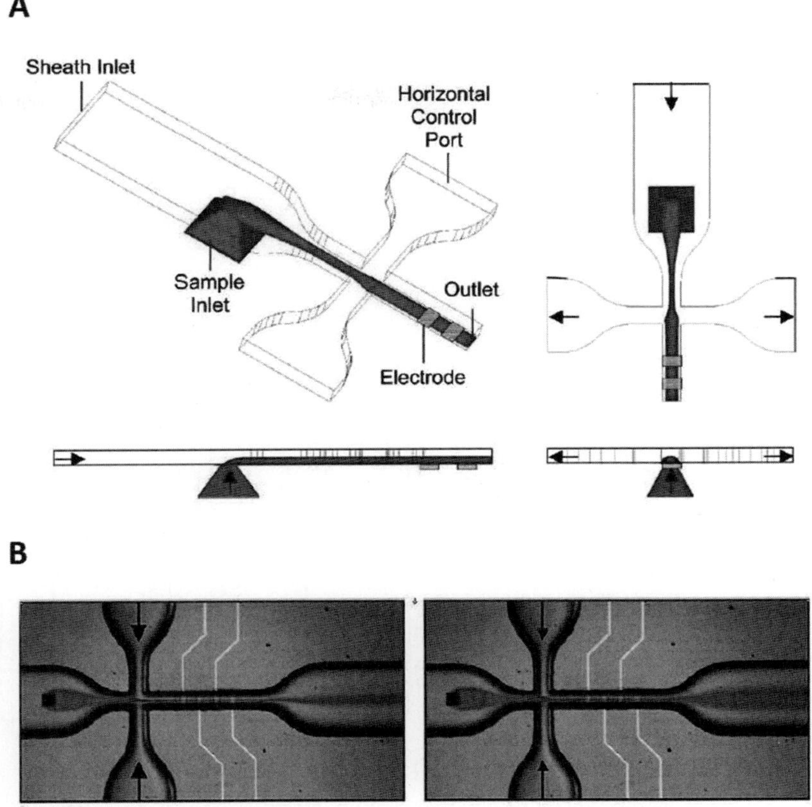

Fig. 6 (A) is a schematic of the chip and all the input ports used to control the sheath flow. (B) is an image of the device with the sample liquid colored red to demonstrate stable sheath flow and control of the diameter. (Nieuwenhuis *et al.*, 2004) (For interpretation of the references to color in this figure legend, the reader is referred to the Web version of this chapter.)

fabricate a device that can detect smaller particles in a larger channel, or integrate additional channels in parallel into the device. In 2008, Wu *et al.* used two parallel channels with a two-stage differential amplification detection scheme (Fig. 7) to achieve higher throughput, increased sensitivity, and greater dynamic range (Wu *et al.*, 2008). In general, RPS is subject to two primary sources of noise: (i) fluctuations in the system power supply which typically has a characteristic frequency of 60 Hz; (ii) the intrinsic noise from thermal fluctuations. Because a perfectly fabricated symmetric dual channel device should have identical noise in each channel, connection of the two outputs to a differential amplifier with a high common-mode rejection ratio reduces the noise of the measurement. As Wu *et al.*'s system was only a single output from the differential amplifier, the readout contained

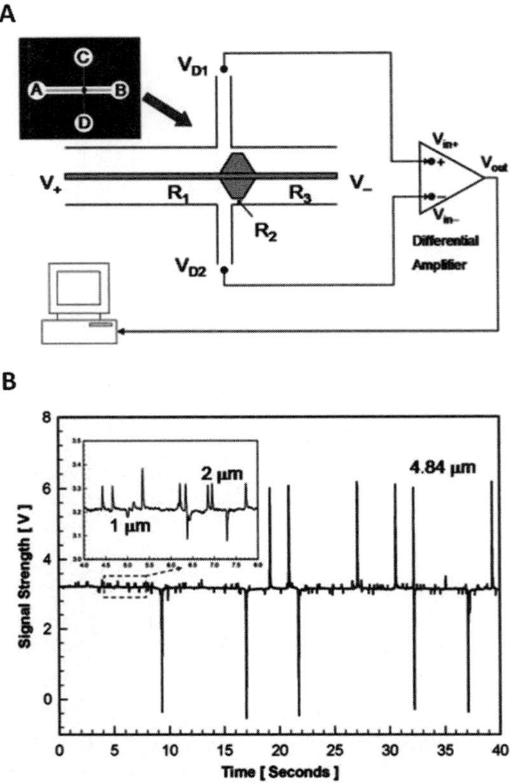

Fig. 7 (A) is a schematic for a dual pore device with one-stage differential amplification. A DC voltage is applied to drive particles from reservoir "A" to "B". The trans-aperture voltage modulation is detected by the electrodes in reservoirs "C" and "D". (B) is a plot of the resistive pulses for 1, 2, and 4.84 μm particles. (Wu *et al.*, 2008), http://dx.doi.org/10.1002/elps.200700912.

both upward and downward pulses, where a positive pulse corresponded to a translocation event in one channel, and a downward pulse corresponded to one in the other channel. The differential amplification of the two channels provides a very high signal-to-noise ratio, thereby allowing the authors to detect 520 nm sized colloids with a 20 μm × 50 μm × 16 μm (L × H × W) sized channel, which is an order of magnitude more sensitive than that reported in the literature and in commercial systems. Use of a larger Coulter aperture not only increased volumetric throughput, but also increased the sensing range of the device. However, it is important to note that due to the differential amplification, dilute samples must be used, as simultaneous translocation of particles in both channels would result in the cancellation of one pulse by the other. Furthermore, this design is particularly sensitive to microfabrication errors as both channels must be identical in all regards.

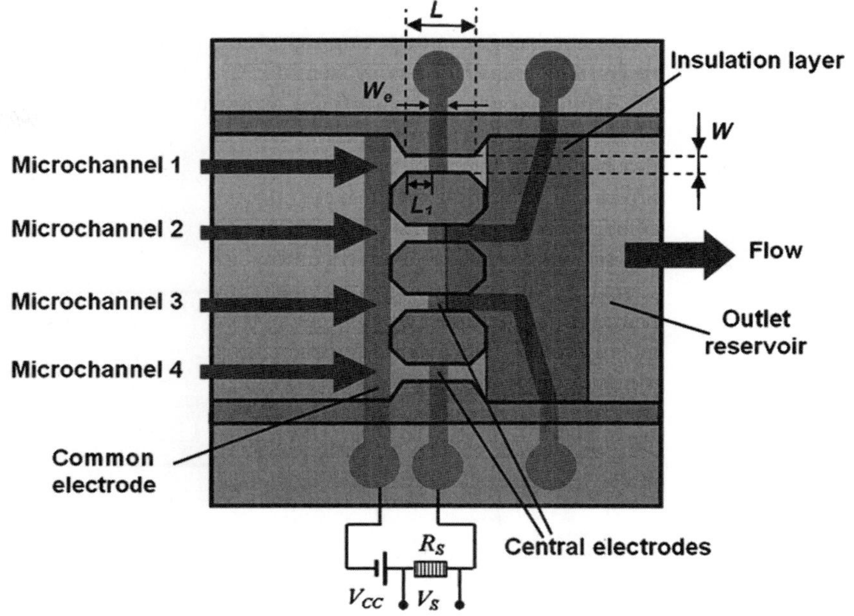

Fig. 8   Schematic of a multichannel microfluidic Coulter counter. (Zhe *et al.*, 2007), http://dx.doi.org/10.1088/0960-1317/17/2/017.

While microfluidic RPS-based devices have exceptional sensitivity when compared to commercial Coulter counters, they lag in throughput. Although throughput on a single chip can be increased by fabricating multiple distinct devices on a single substrate (Carbonaro and Sohn, 2005), or by using a single input with multiple sensing apertures branching off into separate output reservoirs (Jagtiani *et al.*, 2006), commercial viability of high-throughput RPS devices requires the simplicity of a single input and output reservoir connected by multiple channels in parallel. The major challenge toward this realization has been isolating the resistance of adjacent channels to minimize crosstalk between channels. In 2007, Zhe *et al.* demonstrated that by positioning the electrodes in the center of each channel, they could decouple four parallel channels sufficiently, and count colloids and juniper pollen spores with all four, simultaneously (Fig. 8) (Zhe *et al.*, 2007).

B. Counting Cells with Microfluidic Coulter Counters

While most work with microfluidic RPS to date has been with colloids or microparticles, there is significant interest in using microfluidic cytometry to study live cells. The vast majority of microfluidic-based cytometry has focused on optical detection of labeled cells, that is, a true miniaturization of flow cytometry (Kummrow *et al.*, 2009; Mckenna *et al.*, 2009). Very little work, however, has been done using microfluidic RPS to screen cells. This is surprising given the numerous advantages: label-free detection, which eliminates the need for sample preparation; standard microfabrication techniques that enable scaling up of design for high throughput; and simple electronics that could lead to a highly compact, hand-held

A

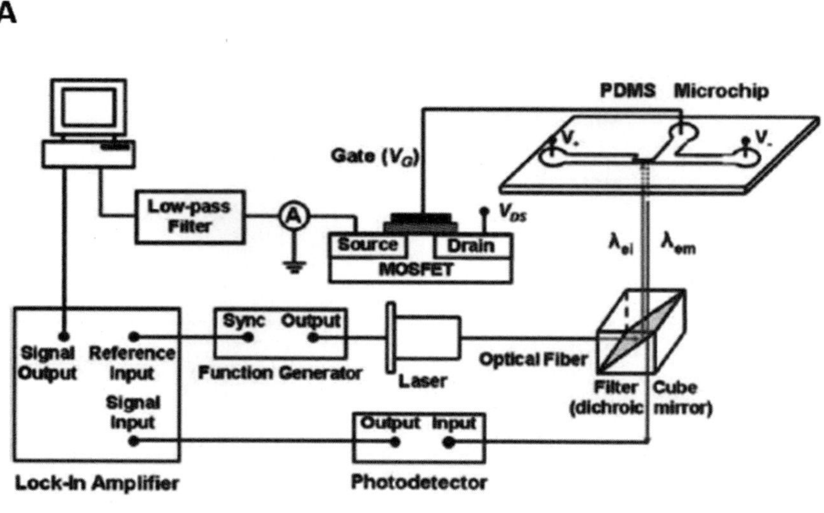

B

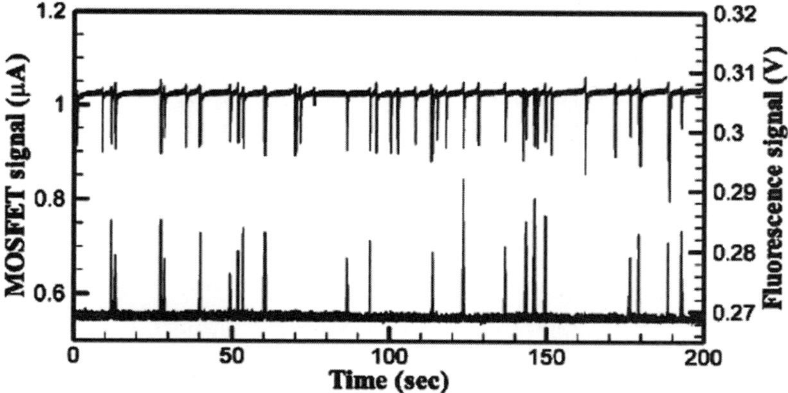

Fig. 9 (A) Schematic of the experimental setup for a resistive pulse device positioned above a fiber optic cable for simultaneous fluorescence detection. (B) A plot of both the fluorescence and resistive-pulse measurements of 50% stained CD4 cells. The upper plot and the left axis indicate the electronic measurement; the lower plot and the right axis indicate the fluorescent signal. (Wang *et al.*, 2008), http://dx.doi.org/10.1039/B713932B.

system. Because of these advantages, microfluidic RPS has the potential to provide true point-of-care (POC) medical diagnostics – from routine cellular blood counts to monitoring disease progression – and to screen samples that are "precious" and could be lost during the preparation. In 2002, Satake *et al.* used a microfabricated RPS-based device to count red blood cells (RBCs) (Satake *et al.*, 2002). While their device was able to distinguish between RBCs and white blood cells (WBCs), it was only able to count 65% of RBCs in a given sample. However, due to the consistency of this value, the authors assumed that this was not due to inherent problems with their device, but rather the size cut-off value used for data analysis.

One of the strengths of fluorescence-activated cell sorting (FACS) versus the Coulter counter is its ability to detect specific cell-surface markers. This detection enables the differentiation of similarly sized cells. In 2008, Wang *et al.* used optical fibers attached to a microfluidic RPS-based device to count the number and percentage of labeled CD4+ lymphocytes (Wang *et al.*, 2008). A 150 μm × 16 μm × 30 μm (L × W × H) channel was fabricated using standard soft lithography (Xia and Whitesides, 1998), sensing electrodes were inserted into the reservoirs on either side of the aperture, and the entire device was positioned above a fiber optic cable for fluorescence detection (Fig. 9a). To prepare a sample for measurement, peripheral blood mononuclear cells were separated from whole blood using Ficoll-Hypaque (Pharmacia), and CD4+ T cells were purified using Robosep CD4+ T cell enrichment kit (StemCell Technologies) before being washed in PBS and stained with SYTO-62 (molecular probes). The sample was then driven through the pore by electroosmotic flow induced by an applied DC voltage, and the current and fluorescence were recorded simultaneously (Fig. 9b). While the authors' device was capable of simultaneous sizing and receptor detection, the fluorescent labeling negated the very label-free nature of the Coulter counter that has always been so attractive.

IV. Multiparametric RPS for Cell Cytometry

In contrast to the work described in the previous section, our group has recently developed the first microchip Coulter counter that, in a single label-free measurement, accurately determines not only the size and shape of a cell, but also specific cell-surface markers and their levels of expression (Carbonaro *et al.*, 2008). As individual cells pass through our device's pore, the magnitude of the current pulse generated corresponds to cell size; as discussed in the previous sections, this is the basis of the Coulter technique of particle sizing. The current pulse width corresponds to the time needed for the cell to pass through the pore, and the current pulse shape indicates how the cell transited through the pore. When the pore is functionalized with proteins that have a high affinity to a particular surface epitope of the cell, specific interactions between a cell-surface marker and those proteins retard the cell passing through the pore. This correspondingly leads to an increase in the pulse width, thereby indicating the presence of that specific surface marker. Cells do not have to be labeled with an exogenous label, as the label, itself, is incorporated in the device. Our device is fabricated with great ease and control using common micro-fabrication and soft-lithography techniques (Xia and Whitesides, 1998). The chip-based fabrication outlined below provides flexibility in the device design, thus greatly extending the capabilities and cell-based applications of the microchip Coulter counter.

V. Device Fabrication and Experimental Methods

Fig. 10 shows a photograph of a typical pore device we now fabricate in our laboratory: a PDMS mold sealed to a glass substrate with predefined platinum (Pt) electrodes. The PDMS mold consists of reservoirs, a set of filters, and a pore whose diameter is comparable to that of the cells to be interrogated and whose length is typically several hundred microns. We lithographically create a negative master of the reservoirs, filters, and pore, which is subsequently cast into a PDMS slab. We create the master in two steps, each involving patterning SU-8 resist (MicroChem Corp.) on a polished silicon substrate to form the negatives of the pore, filters, and reservoirs. SU-8 resist is exceptionally durable once cross-linked, thereby allowing us to reuse a master indefinitely. Following standard micromolding techniques (Xia and Whitesides, 1998), we pour PDMS (Sylgard 184, Dow Corning) over the master and cure it. We subsequently cut a slab surrounding the PDMS mold, core the reservoirs to provide inlet and outlet ports, align the mold with the glass substrate under a microscope, and permanently seal it to a glass substrate.

A. Functionalization of the Pore

Prior to sealing the PDMS slab onto the glass substrate, we functionalize the glass substrate, itself. After exposing the glass substrate with the predefined Pt electrodes with an oxygen plasma, we coat, via microcontact printing (Wilbur *et al.*, 1994), the region between the electrodes with amino-silane groups using a solution of amino-propyltriethoxysilane (APTES) (Sigma Aldrich) in anhydrous toluene (10% w/w) (Sigma Aldrich) at room temperature. To ensure a stable APTES layer, we bake the substrate in an oven at 80 °C for 4–5 h, thereby cross-linking the APTES. Following baking, we soak the cured substrate in toluene (10 min) and then soak and rinse in 18 MΩ deionized (DI) water (twice, 10 min each) to remove any unbound APTES. We have used infrared spectroscopy to characterize the treated substrate to confirm that we have successfully coated the amino-silane groups (data not shown). Once the APTES is patterned onto the glass substrate, we use a micropipette to deliver a droplet (50 mL) of a 5 mM solution of N-5-azido-nitrobenzoyloxysuccinimide (ANB-NOS) (Pierce Biotechnology) dissolved in dimethylsulphoxide (Sigma Aldrich) (10% of the final reaction volume) in 20 mM HEPES (Invitrogen) (pH 7.3) to the area between the electrodes (Karrasch *et al.*, 1993). Use of this flexible cross-linker ensures that the functionalized proteins extend beyond the substrate

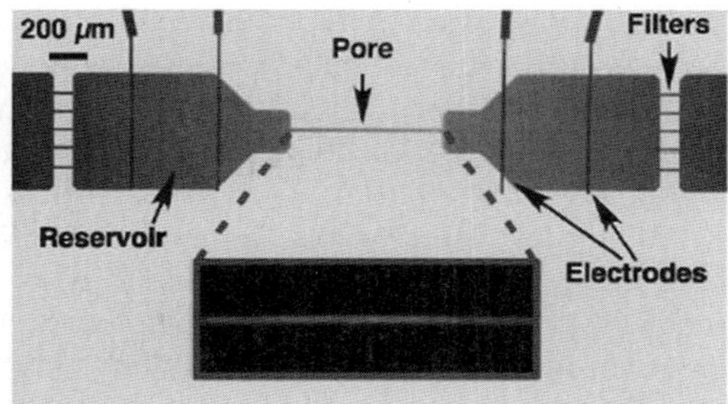

Fig. 10   Optical image of one of our pore devices, consisting of 30 μm deep reservoirs connected by an 800 × 15 × 15 μm (L × W × H) pore. The reservoirs and pore are embedded in a PDMS slab that is sealed to a glass substrate. Filters, of the same cross-sectional area as the pore, are embedded in the PDMS and prevent the pore from clogging. A nonpulsatile pressure drives cells through the pore, and a four-point technique, where the inner (outer) electrodes control the voltage (current), measures the electrical current across the pore. Inset: epifluorescent image of a pore functionalized with 1500 μg/mL of FITC-conjugated anti-CD34 antibody. (Carbonaro *et al.*, 2008), http://dx.doi.org/10.1039/b801929k

surface (Patel *et al.*, 1995). After incubating the substrate in the dark, we remove the excess ANB-NOS by first washing the substrates with HEPES (10 min) and then rinsing with 18 MΩ DI water. We subsequently thermally bond the PDMS mold, embedded with the pore, filters, and reservoir, to the prepared substrate.

We next inject a protein solution (5–1500 μg/mL) into the pore and incubate for 3–4 h. For the covalent binding of proteins with the ANB-NOS-treated glass substrate, we place the sealed device directly under a UV-light source. UV light activates the ANB-NOS aryl-azide groups and subsequently binds the protein covalently to the substrate. We remove any excess protein by flushing the device thoroughly with buffer solution (1× PBS at 25 °C). In general, antibodies functionalized via this procedure may have reduced activity; however, previous work has shown that they are still able to interact specifically with protein receptors (Cozens-Roberts *et al.*, 1990). Fig. 10 (inset) is an epifluorescent image of a pore functionalized with 1500 μg/mL of FITC-conjugated anti-CD34 antibody (eBioscience).

B. Estimate of the Density of Functionalized Antibody

Following Cozens-Roberts *et al.* (1990) who investigated the kinetics of receptor-mediated cell adhesion to a ligand-coated surface, we estimate the functionalized-antibody density, N_{Ab}, within our pores as

$$N_{AB} = 0.7[Ab]V_p \left(\frac{A_V}{A_p M_W} \right) \tag{4}$$

Here, [Ab] is the antibody concentration used to incubate the pore, V_p is the pore volume, A_v is Avogadro's number, A_p is the area of the functionalized glass, and M_w is the molecular weight of the antibody. Like Cozens-Roberts *et al.* (1990) and Clausen (1981), we assume that only 70% of the antibodies bind to the surface. Thus, for the pores described here, the estimated functionalized-antibody densities are $N_{Ab} \sim 2.6 \times 10^2$ Ab/μm^2, when using an antibody concentration of [Ab] = 5 μg/mL, and $N_{Ab} \sim 7.8 \times 10^4$ Ab/μm^2, when using a saturating antibody concentration of [Ab] = 1500 μg/mL. If only steric limits in a single layer of antibodies [$\sim$80 nm^2/antibody] (Saleh and Sohn, 2003b) were to be considered, then the upper limit to the functionalized antibody density would be $\sim 1.2 \times 10^4$ antibodies/μm^2. The quantitative discrepancy between the antibody density derived from Eq. (4) using [Ab] = 1500 μg/mL and that from steric considerations is due to the aforementioned assumption that 70% of the antibodies bind to the surface. This assumption is highly dependent on the particular functionalization protocol, its efficiency, and substrate type, to name only a few. Nonetheless, we do have agreement regarding the order of magnitude of antibody concentration, that is, $\sim 10^4$ Ab/μm^2.

C. Fluid Handling, Data Acquisition, and Analysis

Upon device completion, we are able to inject a sample of cells ($\sim 10^5$ to 10^6 cells/mL) into one of the outer reservoirs and apply a nonpulsatile pressure (typically 10.5 kPa, although pressures as low as 10.0 kPa and as high as 34.0 kPa have been employed) using a commercial microfluidic pump (Fluidigm Inc.) to drive individual cells ($\sim$100 cells/min) through the filters, the inner reservoir, and finally through the pore. Cellular debris and clumps are excluded from the pore by the filters (Fig. 10).

We use RPS to measure the current pulse caused by a single cell transiting the pore. As discussed earlier, a four-point measurement is ideal, as it allows one to measure solely the resistance of the pore by removing both the resistance of the electrodes and the interfacial resistance ($R_{e/f}$) between the electrodes and the buffer solution (Saleh and Sohn, 2001). In all performed measurements, the current is low-pass filtered below 0.3 ms in the rise time and is sampled at 50 kHz. This sampling

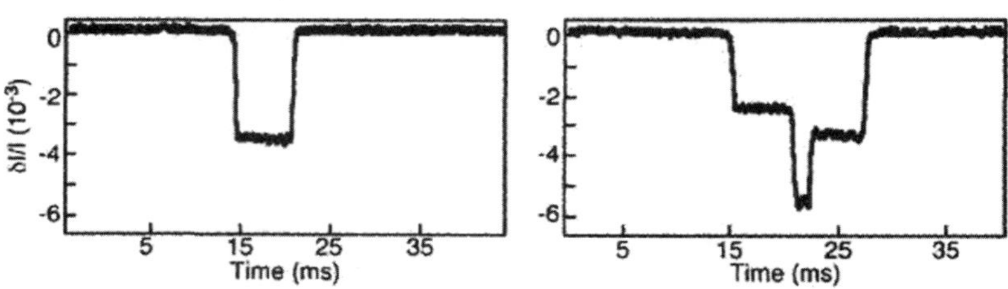

Fig. 11 Comparison of current pulses. (Left) Pulse generated by a single cell passing through a pore; (right) pulse generated when two cells occupy the pore at the same time. (Carbonaro and Sohn, 2008), http://dx.doi.org/10.1039/b801929k.

frequency corresponds to a time resolution of 20 μs and ensures a high-resolution pulse width. Data are recorded and analyzed using custom-written software. Pulses generated by the rare occurrence of two cells flowing simultaneously across the pore have a characteristic shape (Fig. 11) that is easily recognized and discarded.

VI. Cell Size

As we have stated previously, RPS has been used to determine accurately the size of cells [9]. The pulse magnitude resulting from a cell transiting a pore is directly related to the volume ratio of the cell to pore, V_{cell}/V_{pore} (Carbonaro and

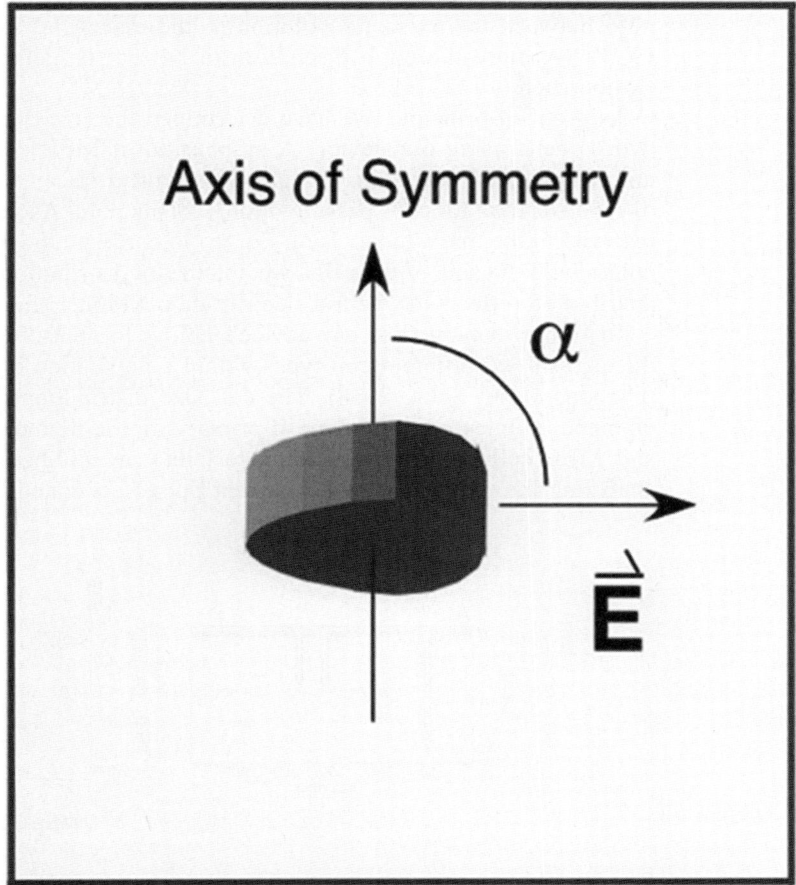

Fig. 12   Schematic defining an aspherical particle's axis of symmetry and α, the orientation of the axis of revolution with respect to the applied electric field $\vec{E}$.

Sohn, 2005; Deblois and Bean, 1970; Deblois *et al.*, 1977; Saleh and Sohn, 2001, 2003a) and to the cell orientation with respect to the pore's central axis, thus enabling the accurate and direct determination of cell size. In more detail, for an aspherical particle with an axis of symmetry, the orientation of its axis of revolution with respect to the applied electric field affects the electrical resistance (Fig. 12). If we assume $|\delta I/I| = |\delta R_{\mathrm{p}}/R_{\mathrm{p}}|$, the relationship between the normalized current $\delta I/I$ and the volume ratio $V_{\mathrm{cell}}/V_{\mathrm{pore}}$ is

$$\frac{\delta I}{I} = [f_{\perp} + (f_{\parallel} - f_{\perp})\cos^2\alpha][V_{\mathrm{cell}}/V_{\mathrm{pore}}] \tag{5}$$

where α is the angle between the electric field, E, and the axis of revolution. In Eq. (5), $f_{\perp}$ and $f_{\parallel}$ are the shape factors perpendicular and parallel components, respectively. Further, both are a function of the particle shape, which is defined by γ, the ratio between the axis of revolution to the equatorial axis (Golibersuch, 1972, 1973). (A more thorough discussion of γ, that is, cell shape, can be found in Golibersuch (1972, 1973).

As proof-of-principle, we have determined the size of murine erythroleukemia (MEL) cells using our device. A suspension of MEL cells was injected into an unfunctionalized pore. Fig. 13A shows a typical trace of the normalized current, $\delta I/I$, versus time for cells passing through blank pore. As shown, the pulses are well resolved in the magnitude. Using Eq. (5) to correlate the magnitude of the current pulses with the size of the cells, we obtained a distribution of size (Fig. 13B). The distribution reflects the natural size variation within a single population of cells.

To further demonstrate our device's ability to measure cell size, we used it to distinguish two different cell types within a population of varying concentrations (Carbonaro and Sohn, 2008). The cell size distributions shown in Fig. 14 were obtained by injecting a mixture of primary mouse thymocytes (6 μm in diameter) and MEL cells (8–15 μm in diameter) into an unfunctionalized pore and then analyzing the magnitude of the current pulses generated by each single cell as it

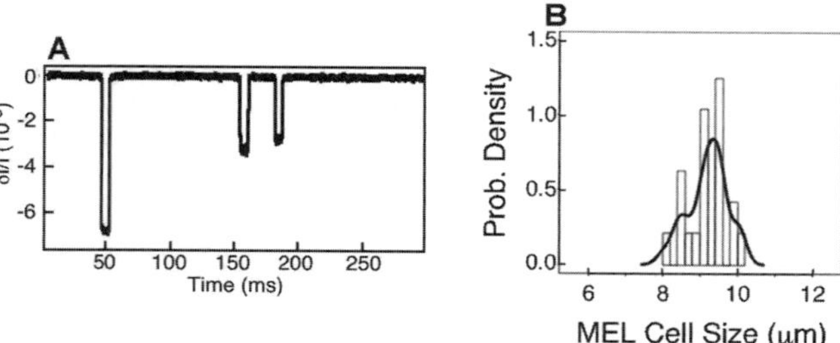

Fig. 13   (A) A typical measurement of the normalized current across a nonfunctionalized pore when MEL cells pass through. Each downward pulse corresponds to a single cell transiting the pore. (B) Distribution of cell size obtained by analyzing the magnitudes of the resistive pulses that resulted when 66 MEL cells flowed across a blank pore. (Carbonaro *et al.*, 2008), http://dx.doi.org/10.1039/b801929k.

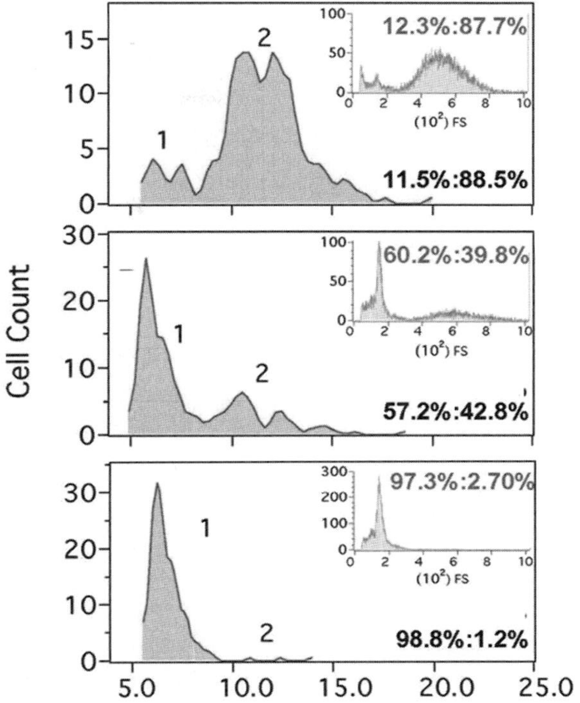

Fig. 14 Size measurement of primary mouse thymocytes and MEL cells mixed in different proportions and injected through a pore. Labels "1" and "2" correspond to primary mouse thymocytes and MEL cells, respectively. The ratio of primary mouse thymocytes to MEL cells as derived from the histograms is provided. FACS data obtained with the same sample of cells are shown in the inset. (Carbonaro and Sohn, 2008), http://dx.doi.org/10.1039/b801929k.

moved through the pore. Fig. 14 shows the resulting cell-size distributions. The cell population designated as "2" in the figure corresponds to the thymocytes, whereas the cell (Carbonaro *et al.*, 2007) population designated as "2" corresponds to the MEL cells. For a comparison, we also derived a cell-size distribution of the same population of cells using forward scattering in traditional flow cytometry (Beckman Coulter) (inset). A comparison between the two different cell distributions shows that there is excellent agreement between our pore device and flow cytometry. Our ability to measure cell size accurately, and with high precision, offers another advantage of our pore device, since there is no exact model to correlate the forward scatter to cell size when performing FACS.

To test our pore's ability to differentiate the cellular components of peripheral blood and to test whether the pore clogs easily, we injected clotted murine peripheral blood through an unfunctionalized pore (400 μm × 25 μm × 16 μm (L × W × H)) (Carbonaro *et al.*, 2007). Fig. 15 shows the distribution of cell sizes that we obtained.

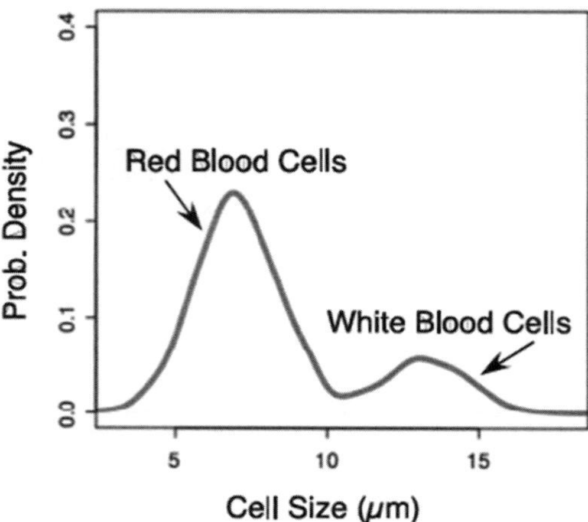

Fig. 15 Cell size distribution for clotted murine peripheral blood measured with an unfunctionalized pore (16 µm high, 25 µm wide and 400 µm long) at 3 psi. Blood was first diluted (1:5000) in 1× PBS. Analyzed data were first binned and then fitted with two Gaussian peaks. Peak corresponding to RBCs is centered at 7.01 ± 1.09, while the WBC (lymphoblasts) peak is centered at 12.88 ± 0.59. (Carbonaro *et al.*, 2007)

The pore can distinguish clearly between RBCs (5–10 µm) and WBCs (<10 um). Using this pore size, we were unable to distinguish platelets (<5 µm), which is most likely due to insufficient signal-to-noise ratio of the pore, that is, the employed pore was simply too large. We observed no clogging during numerous experiments.

VII. Cell–Surface Marker Screening

In RPS, the resistive-pulse *width* corresponds to the transit time, τ, of a cell passing through a pore (Fig. 16). A cell's transit time through a functionalized pore (τ_2 or τ_3) is greater than that through an unfunctionalized (i.e., "blank") pore (τ_1), due to interactions between the cell-surface marker of interest and the functionalized proteins. If the interactions are nonspecific, then $\tau_2 \geq \tau_1$ (II in Fig. 16). If the interactions are specific, then $\tau_3 \gg \tau_1$ (III in Fig. 16). Thus, by comparing τ of a cell passing through a functionalized or a blank pore, we are able to screen for a specific cell-surface marker (Carbonaro *et al.*, 2008). Of great importance is that at no time are cells strongly bound to the functionalized proteins, as they may be with cell-affinity chromatography or even magnetic activated cell sorting (MACS). Cells are flowing at a high enough shear force such that they do not have time to bind to the proteins; yet, the shear force is orders of magnitude less than that founds in MACS or FACS, thereby leaving the cells undamaged. Finally, while transient binding does occur, we have performed numerous tests to show that the cells are not activated (Mohanty *et al.*, 2010).

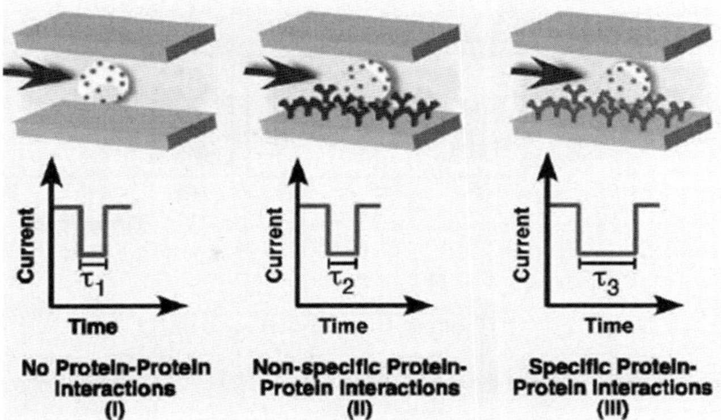

Fig. 16   As each cell passes through the pore, a current pulse is detected (I–III). The magnitude reflects cell size and the width indicates cell transit time, τ. A cell travels fastest (τ_1) through an unfunctionalized or "blank" pore (I). Interactions between a specific cell-surface marker and a functionalized pore retard the cell: $\tau_2 \geq \tau_1$ for nonspecific interactions (II) and $\tau_3 \gg \tau_1$ for specific interactions (III). (Carbonaro *et al.*, 2008), http://dx.doi.org/10.1039/b801929k.

A. Detection of CD34 in a Single Population of Cells

As proof-of-principle, we detected the CD34 receptors expressed on the surface of MEL cells. A suspension of MEL cells ($\sim 10^5$ cells/mL) was injected into (1) a blank pore; (2) a pore functionalized with rat IgG2a isotype control antibody (eBioscience); (3) a pore functionalized with anti-CD34 monoclonal antibody (eBioscience) (clone: RAM34; isotype IgG2a, k). Fig. 17A shows typical traces of the normalized current, $\delta I/I$, versus time for cells passing through the three different pores: the pulses are well resolved in width. A comparison among the different t-distributions shows that MEL cells traveled slowest through the anti-CD34 pore. Further, as shown in Figs. 17B–D, τ significantly depends on the antibody concentration used to functionalize the pore. When the pores were functionalized with 100 µg/mL of antibody, the average transit time of MEL cells passing through an anti-CD34 pore was $\tau_{avg} = 7.35$ ms. This transit time was 1.21 and 0.67 ms longer than through the blank and the isotype-control pores, respectively. More importantly, when a 15-fold higher concentration of antibody was used, τ_{avg} significantly increased only for the anti-CD34 pore. Equally important, the variance of the τ-distribution increased with antibody concentration only for the anti-CD34 pore (Figs. 17B and C). We attribute this increase to the number of binding events that occur between cells with varying CD34 densities and the anti-CD34 pore.

In Fig. 17D, we show ($\tau_{avg} - \tau_{avg}^*$), the difference in the mean τ between the functionalized pore (τ_{avg}) and that of the blank pore (τ_{avg}^*), versus the antibody concentration, used to functionalize the pore. Only the interaction between MEL cells and the functionalized anti-CD34 antibodies shows saturation. As the concentration increases, the functionalized anti-CD34 antibody reaches maximum density at ~ 500 µg/mL, thus leading to a maximum number of protein–protein binding

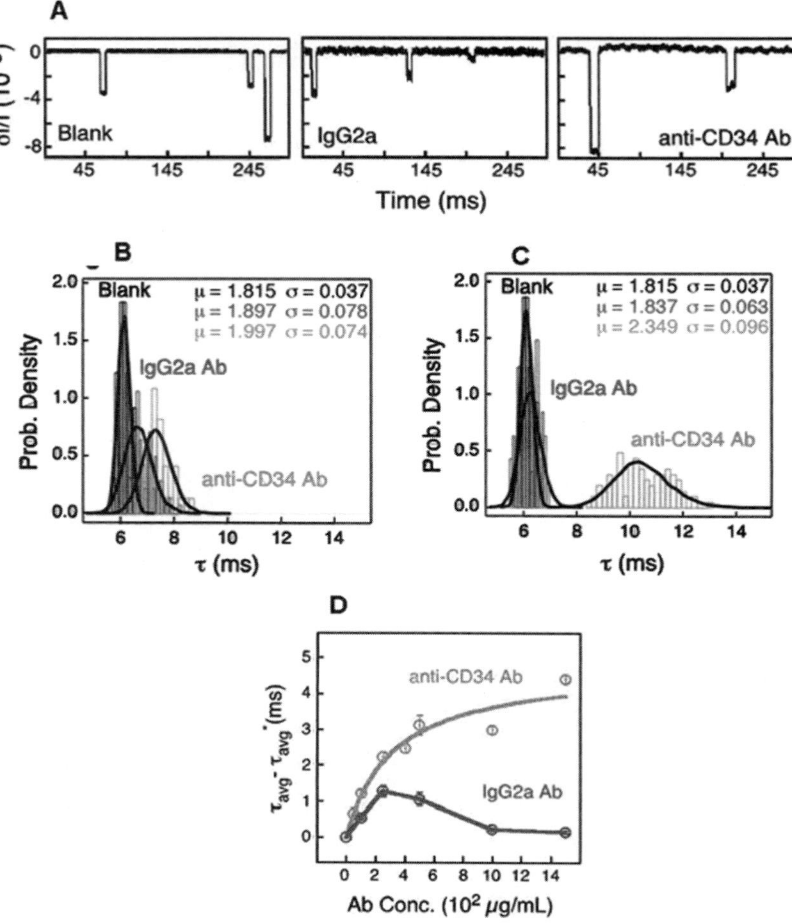

Fig. 17　Detection of CD34 receptors on MEL cells. (A) A typical measurement of the normalized current across a blank pore (top), one functionalized with IgG2a antibody (middle), and one functionalized with anti-CD34 antibody (bottom), as MEL cells ($\sim$3 $\times$ 10^5 cells/mL) flow (10.5 kPa) one by one. Each downward pulse corresponds to a single cell transiting the pore. (B) and (C) τ-distributions of MEL cells flowing through a blank pore and a pore functionalized with either IgG2a or anti-CD34 antibody. In (B), the antibody concentration used was 100 μg/mL. The number of cells measured was 66 cells in the blank pore, 139 cells in the IgG2a pore, and 37 cells in the anti-CD34 pore. In (C), the antibody concentration was 1500 μg/mL. The number of cells measured was 48 cells in the IgG2a pore, and 103 cells in the anti-CD34 pore. (D) $(\tau_{avg} - \tau^{*}_{avg})$ versus antibody concentration, where τ_{avg} and τ^{*}_{avg} is average cell transit time for a functionalized pore and blank pore, respectively. Saturation is only observed with increasing anti-CD34 antibody concentrations. (Carbonaro *et al.*, 2008), http://dx.doi.org/10.1039/b801929k. (See plate no. 8 in the color plate section.)

events possible. A slight increase in average transit time is observed at low antibody concentrations observed at the lower antibody concentrations for the isotype control. This surprising observation is most likely due to the fact that, at higher antibody densities, the surface charges collectively play a role in shielding the antibodies, thereby reducing nonspecific interactions. Nonetheless, despite the slight increase in nonspecific interaction at low concentrations of isotype-control antibody, the specific interaction between MEL cells and the functionalized anti-CD34 antibodies is clearly demonstrated.

B. Screening Cells in a Mixed Cell Population Based on the CD34 Receptor

For proof-of-principle, we characterized a *single* population of cells that were all positive for one particular cell-surface marker; however, a more powerful demonstration would be to screen a *mixed* population in which only a subpopulation of cells are positive for a specific marker of interest. Therefore, we mixed two different types of cells – MEL and U937 – in different proportions (1:1, 1:3, and 1:9) and subsequently screened the resulting mixtures with a pore functionalized with 500 µg/mL of anti-CD34 antibody. For control purposes, we also screened the same mixtures of cells with a blank pore and a pore functionalized with 500 µg/mL IgG2a isotype control antibody. If we compare the τ-distributions of a *single* population of MEL cells and U937 cells traveling in a blank pore (Fig. 18), we observe that the two distributions are shifted apart by $\sim$1 ms. This shift is most likely due to different cell properties (i.e., size, density, and stiffness) (Rosenbluth *et al.*, 2008), although further experiments are needed to confirm this. A–C show the τ-distributions (black histograms) obtained for the different mixtures, 1:1, 1:3, and 1:9 of MEL and U937 cells, respectively, when screened with the blank pore (left), the IgG2a antibody pore (middle), and the anti-CD34 antibody pore (right). From the figures, we first see that the left column (corresponding to cells passing through a blank pore) presents a distribution shift in increasing time from the top to the bottom, consistent with the fact that U937 cells travel slower than MEL cells in a blank pore (Fig. 18) and the fact

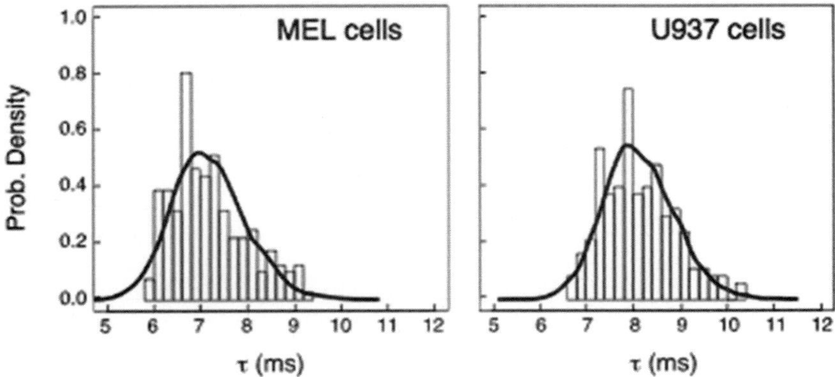

Fig. 18 τ-distribution for a single population of MEL cells (left) and U937 cells (right) flowing through a blank pore. Experimental data were fitted using a log-normal distribution (black line). The number of MEL and U937 cells measured was 206 and 189 cells, respectively. (Carbonaro *et al.*, 2008), http://dx.doi.org/10.1039/b801929k.

that the percentage of U937 cells increases from the top to the bottom in Fig. 19. More importantly, the left column figures also show that we are unable to distinguish between MEL and U937 cells when they transit the blank pore as a mixture. This agrees well with the observed large overlap of τ-distributions for single populations of MEL and U937 cells in Fig. 18. In the middle column of Fig. 19, where a mixed population of MEL and U937 cells transits a pore functionalized with an isotype control antibody, we make similar observations that the distributions for different types of cells look nonseparable. This is expected since only nonspecific interactions are present. In contrast, when a mixture of cells transit through an anti-CD34 antibody pore, only the MEL cells interact specifically with the functionalized anti-CD34 antibodies, thus increasing their transit time through the pore; the U937 cells are not affected, as these cells are human and therefore do not interact with the *murine-specific* monoclonal anti-CD34 antibody. The net result is that the overall τ-distribution is significantly broader than that obtained when the same mixture of cells passed through a blank or an IgG2a pore. This higher data spreadness, together with the observed separate distribution peaks, increases the promise of separating the distributions for different cell types.

C. Cell Shape

The specific protein–protein interactions that occur in our protein-functionalized pores are observable in not only the cell transit time but also the *shape* of individual resistive pulses. When α in Eq. (5) is constant, the pulse is flat (Fig. 20a, top). In contrast, when α periodically changes, as it does for a cell tumbling across the pore with period T, the pulse exhibits oscillations (Fig. 20a, bottom). We observed both types of pulses (Fig. 20b, top) when MEL cells passed through either a blank (Fig. 20b, middle) or an anti-CD34 (Fig. 20b, bottom) pore. As shown in Fig. 20c, the observed period of oscillation in a blank pore (red) is always less than that in an anti-CD34 pore (green), reflecting the fact that CD34 receptors transiently bind to the functionalized anti-CD34 antibodies and subsequently retard the cell's motion. Although further studies are needed, a cell's transit time combined with an analysis of its pulse shape could ultimately provide quantitative information regarding the density of a specific receptor on the cell's surface.

VIII. Applications

While microfluidic cytometry may not be able to achieve high throughput of flow cytometry, there are a number of applications in research and clinical environments for which current RPS is already well suited. At the research level, there has been significant interest in studying biological phenomena at the single-cell level in diverse areas ranging from circulating tumor cells and metastatic potential to stem cell biology. Often times when studying rare cells, sample size can be limited to as few as 5–10 cells. While traditional cytometry systems typically require samples in

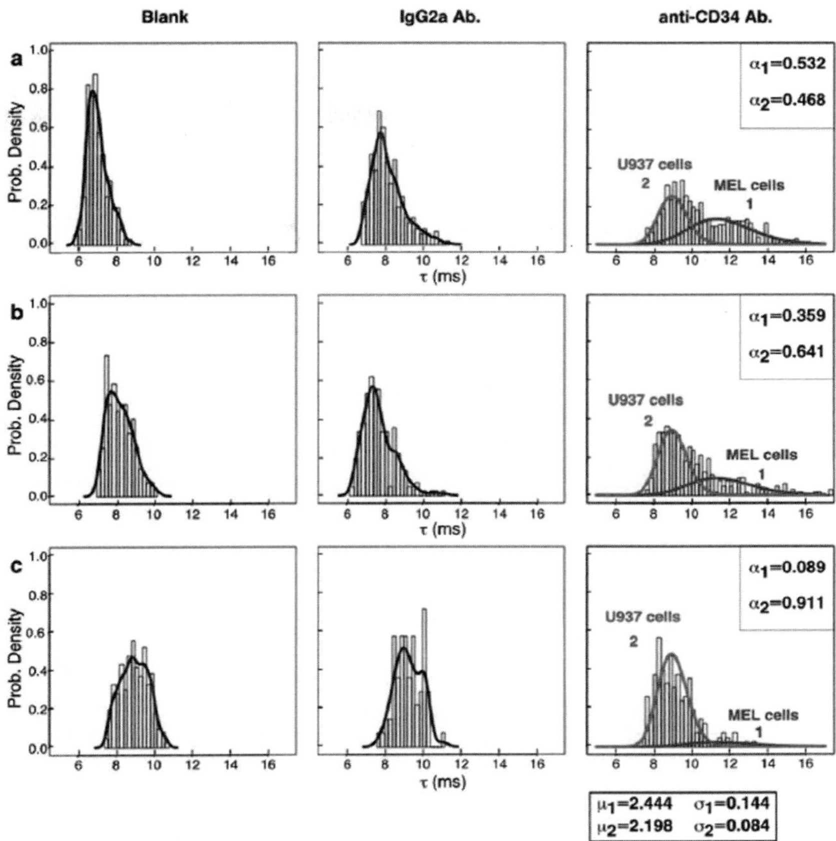

Fig. 19 Detection of MEL cells within a mixed cell population. τ-distributions and their fitting curves for mixtures with different proportions of MEL cells and U937 cells ($\sim$5 $\times$ 10^5 cells/mL) flowing (10.0 kPa) through a blank pore (left column), a pore functionalized with either IgG2a (middle column) or anti-CD34 antibody (right column). The antibody concentration for pores functionalized with IgG2a and anti-CD34 antibody was 500 μg/mL. Each distribution obtained (black histogram) when cells traveled through an anti-CD34 antibody pore was modeled with a mixture of two log-normal distributions (right column), identified as 1 and 2 in the figure. The distribution parameters were estimated through a unified constrained model. The parameters of the fittings (μ = mean and σ = standard deviation of the distribution calculated after taking the logarithm of the data) as well as the mixture ratios (a) are provided in the figure. MEL cells (blue distribution) traveled significantly slower than U937 cells (red distribution) in the anti-CD34 antibody pore due to specific interactions (right-most column). (a) The ratio of MEL to U937 cells was 1:1. The number of cells measured in the blank pore was 182 cells; in the IgG2a pore, 366 cells; and in the anti-CD34 pore, 541 cells. (b) The ratio of MEL to U937 cells was 1:3. The number of cells measured in the blank pore was 137 cells; in the IgG2a pore, 335 cells; and in the anti-CD34 pore, 336 cells. (c) The ratio of MEL to U937 cells was 1:9. The number of cells measured in the blank pore was 431 cells; in the IgG2a pore, 69 cells; and in the anti-CD34 pore, 213 cells. (Carbonaro *et al.*, 2008), http://dx.doi.org/10.1039/b801929k.

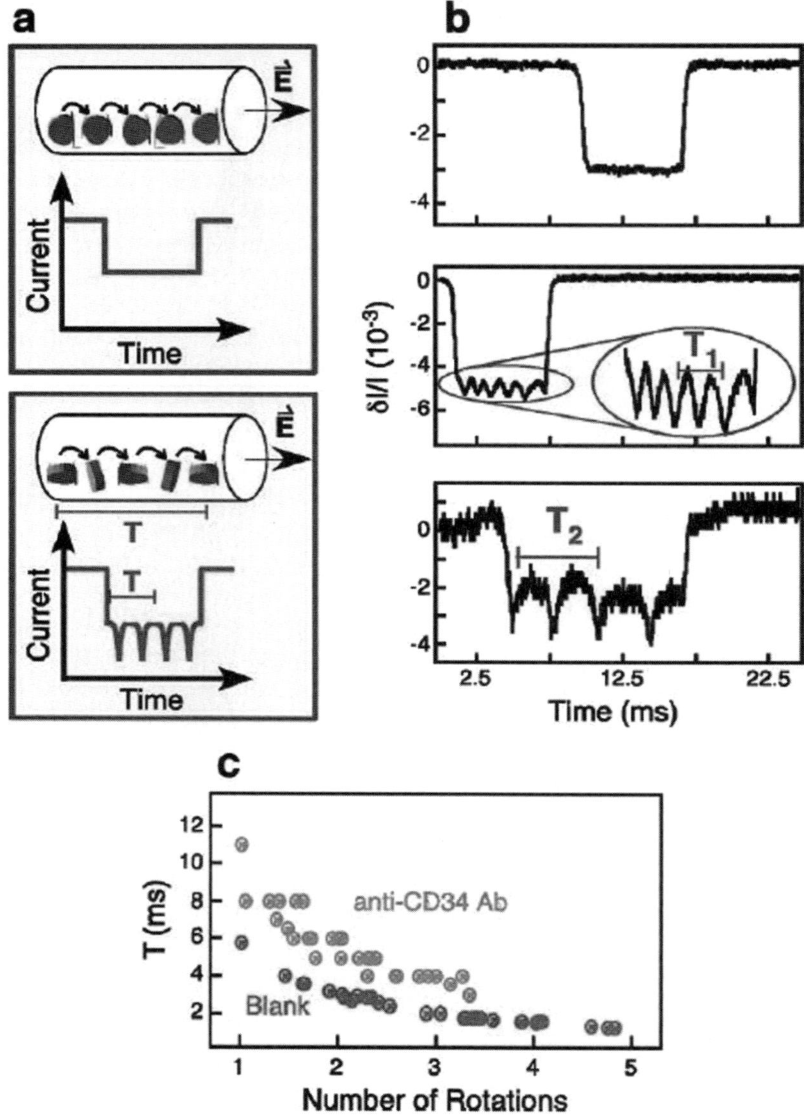

Fig. 20 Pulse-shape analysis. (a) Effect of cell orientation on the pulse shape. Flat resistive pulses occur when the orientation of the cell remains constant with respect to the electric field parallel to the pore's central axis (top). In contrast, pulses show oscillations due to the different orientations of the cell with respect to as it flows across the pore (bottom). T corresponds to the period of one complete cell revolution. (b) Examples of different pulse shapes obtained when MEL cells flowed through a blank and a functionalized anti-CD34 pore: a flat pulse (top); a pulse that oscillates with period T_1 in a blank pore (middle); and a pulse that oscillates with period T_2 in a functionalized pore (bottom). (c) T versus number of rotations in a blank pore (red) and in a pore functionalized with anti-CD34 antibody (green). (Carbonaro *et al.*, 2008), http://dx.doi.org/10.1039/b801929k. (See plate no. 9 in the color plate section.)

excess of 100,000 cells, our group has begun to use our microfluidic Coulter counter to perform quantitative analysis of single muscle satellite cells that have been directly, and unambiguously, isolated from their original microanatomical niche (Mohanty *et al.*, 2010). By leveraging the capabilities of our microfluidic device, we have been able to analyze satellite cells taken from individual murine muscle fibers and immediately quantify surface marker expression with no sample processing. Early results have been extremely exciting and indicate that bulk sample processing that has been used to study this and many other similar systems fail to identify a number cellular phenotypes. Furthermore, the label free nature of our multiparametric measurement enables further propagation and secondary analysis of the screened cells, since they are not damaged and there are no residual reagents left to interfere. We believe that the ability to handle small samples in a label-free manner is one of the greatest strengths of microfluidic cytometry and are very excited with the possibility of using this technology to access biological phenomena that have been elusive using traditional bulk cytometry.

Besides the potential use of microfluidic cytometry in research, clinical diagnostics represent a multibillion-dollar industry. Most current assays used in diagnostic labs require expensive reagents, large equipment that is expensive to purchase and maintain, highly trained technicians, and long turn-around times due to the fact that sample collection, transportation, and analysis are usual done by many individuals. Furthermore, these requirements typically prevent access to medical diagnostics in developing countries. We believe that our resistive-pulse cytometry technology could be integrated into a hand-held device, similar in size and scope to an iPhone, with disposable cartridges or chips designed for any number of diagnostic tests from complete blood counts to diagnosing life-threatening acute promyelocytic leukemia to detecting minimal residual disease. In contrast to current flow cytometry methods of screening leukemic cells, our method is *label free*, requiring little to no sample preparation, and can produce much-needed accurate results within five minutes at the bedside. Furthermore, our method is simple to use – any doctor, nurse, family member, or even patient, can perform the screening. Because of its compactness, low-power requirements, and simplicity, the device upon which our method is based can be *used in low-resource settings*, without the need for a larger or central instrument for read-out. Furthermore, our device is inexpensive to produce potentially leading to a significant reduction in the cost of clinical diagnostics. For instance, acute leukemias are currently diagnosed via immunophenotyping, which entails labeling the leukemic cells with fluorescently-tagged antibodies and using a flow cytometer for signal detection. The cost of immunophenotyping for a new acute leukemic sample is on the order of \$2,000–5,000, depending on how many antigens are tested, and excluding the cost of the flow cytometer, itself. Label-free RPS utilizes low-cost microfluidic technology and relies on measuring the transit times of cells passing through a pore functionalized with antibodies corresponding to specific cellular epitopes, as discussed above. The entire device can be manufactured for less than a dollar, including the cost of the antibodies. Overall, the extreme cost savings is more than 1000-fold due to the small amount of antibody needed for device functionalization, the lack of requirement for a fluorescent label, and the low cost of chip production. Besides potential savings associated with use of this technology, we anticipate that development of RPS for identify types of cells based on cell-surface markers and subsequently to perform cell separations, could provide an important new technology applicable to cancer, immunology, and other clinical situations in which monitoring at the POC provides patient benefit.

IX. Conclusion

The exquisite sensitivity and simplicity of RPS-based device imparts a number of general advantages over other current cell characterization methods, such as flow cytometry (Shapiro, 2003), fluorescence microscopy, MACS (Lang *et al.*, 2002), cell-affinity chromatography (Hage, 1999; Katoh, 1987; Killion and Kollmorgen, 1976), and impedance spectroscopy (Holmes and Morgan, 2010; Sun and Morgan, 2010). First, exogenous labeling is not required. This significantly reduces both cost and sample preparation time and is especially attractive in cases where it is not known whether labeling affects cell physiology and function. Second, accuracy is not lost as a result of the small number of cells interrogated. Thus, our method is suitable for cases in which only a few hundred cells may be available for interrogation. Third, separation (using microfluidic valves) and recovery of cells for further propagation are possible. In this initial work, we have demonstrated only proof-of–principle of our ability to use a functionalized pore to characterize cells. Although we have chosen to perform all our measurements using functionalized pores whose size is 800 μm × 15 μm × 15 μm (L × W × H), our signal-to-noise ratio (based largely on $\delta I/I \sim V_{\mathrm{cell}}/V_{\mathrm{pore}}$) is such that we could employ pores with much larger cross-sectional area (e.g., greater than 25 μm × 25 μm) and, correspondingly, shorter length to accommodate a mixture of cells whose components range in size from 5 μm to beyond 25 μm. The nonpulsatile flow rate of 10.0 kPa is the lower limit of the microfluidic pump employed. As mentioned, a flow rate as high as 34.0 kPa has been used with no detectable loss in pulse-width (or height) resolution. Further studies, however, are needed to identify the upper-limit flow rate to detect specific interactions between ligand and receptor with a specific affinity. That our data acquisition provides a time resolution of 20 μs strongly suggests that there is sufficient flexibility in the pore design.

We envision that, with minor modifications to our microfluidic device, that is, placing a blank, a control, and a specifically functionalized pore in series, single-cell analysis based on surface receptors can be realized. Similarly, full-phenotypic characterization of cells can be achieved, as a "decision tree" of multiple pores, each linked sequentially to one another and each functionalized with a different protein corresponding to a difference cell-surface marker, can easily be created. Although the throughput of a single pore is not comparable to that of flow cytometry (100 cells/min versus 10,000 cells/s), our microfluidic platform enables the construction of arrays of pores (Carbonaro and Sohn, 2005; Jagtiani *et al.*, 2006; Zhe *et al.*, 2007) for performing many measurements or assays in parallel. Finally, our measurements are sensitive to the density of receptors on the cell surface. For those cells that express low levels of receptors on the cell surface, the flow rate in our pore could be readily reduced to increase the sensitivity.

Acknowledgments

The authors would like to thank the numerous people who have, over the years, contributed much to the work described: O. A. Saleh, L. A. Carbonaro, A. Godley, H. Huang, I. Conboy, M. J. Conboy, S. K. Mohanty, K. Balakrishnan, E. Kelp-Stebbins, and G. Anwar.

This work has been partially funded by the W. M Keck Foundation Medical Research Program and NSF grant no. CBET-0651799.

References

Andre, F., Schartz, N. E., Movassagh, M., Flament, C., Pautier, P., Morice, P., Pomel, C., Lhomme, C., Escudier, B., Le Chevalier, T., Tursz, T., Amigorena, S., Raposo, G., Angevin, E., Zitvogel, L. (2002). Malignant effusions and immunogenic tumour-derived exosomes. *Lancet* **360**, 295–305.

Bayley, H., and Cremer, P. S. (2001). Stochastic sensors inspired by biology. *Nature* **413**, 226–230.

Berg, T., Sarrazin, C., Herrmann, E., Hinrichsen, H., Gerlach, T., Zachoval, R., Wiedenmann, B., Hopf, U., Zeuzem, S. (2003). Prediction of treatment outcome in patients with chronic hepatitis C: significance of baseline parameters and viral dynamics during therapy. *Hepatology* **37**, 600–609.

Bezrukov, S. M., Vodyanoy, I., and Parsegian, V. A. (1994). Counting polymers moving through a single ion channel. *Nature* **370**, 279–281.

Carbonaro, A., Godley, L. A., and Sohn, L. L. (2007). University of California, Berkeley.

Carbonaro, A., Mohanty, S. K., Huang, H., Godley, L. A., and Sohn, L. L. (2008). Cell characterization using a protein-functionalized pore. *Lab Chip* **8**, 1478–1485.

Carbonaro, A., and Sohn, L. L. (2008). University of California, Berkeley.

Carbonaro, A., and Sohn, L. L. (2005). A resistive-pulse sensor chip for multianalyte immunoassays. *Lab Chip* **5**, 1155–1160. -Reproduced by permission of the Royal Society of Chemistry.

Clausen, J. (1981). *Immunochemical Techniques for the Identification and Estimation of Macromolecules.* North-Holland, Amsterdam.

Coulter, W. H. (1953). Means for Counting Particles Suspended in a Fluid. In: U. S. P. Office, (Ed.), Vol. 2,656,508.

Cozens-Roberts, C., Lauffenburger, D. A., and Quinn, J. A. (1990). Receptor-mediated cell attachment and detachment kinetics. I. Probabilistic model and analysis. *Biophys J.* **58**, 841–856.

Deblois, R. W., and Bean, C. P. (1970). Counting and sizing of submicron particles by the resistive pulse technique. *Rev. Sci. Instrum.* 41.

Deblois, R. W., Bean, C. P., and Wesley, R. K. A. (1977). Electrokinetic measurements with submicron particles and pores by the resistive pulse technique. *J. Colloid Interface Sci.* **61**, 323–335.

Golibersuch, D. C. (1972). Observation of aspherical particle rotation in Poiseuille flow via the resistance pulse technique: I. application to human erythrocytes. *Biophys J.* **13**, 265–280.

Golibersuch, D. C. (1973). Observation of aspherical particle rotation in Poiseuille flow via resistance pulse technique II. application to fused sphere "Dumbells". *J. Appl. Phys.* **44**, 2580–2584.

Graham, M. D. (2003). The Coulter Principle: Foundation of an Industry. JALA.

Gregg, E. C., and Steidley, K. D. (1965). Electrical counting and sizing of mammalian cells in suspension. *Biophys. J.* **5**, 393–405.

Gu, L. Q., Cheley, S., and Bayley, H. (2001). Capture of a single molecule in a nanocavity. *Science* **291**, 636–640.

Halicka, H. D., Bedner, E., and Darzynkiewicz, Z. (2000). Segregation of RNA and separate packaging of DNA and RNA in apoptotic bodies during apoptosis. *Exp. Cell Res.* **260**, 248–255.

Hage, D. S. (1999). Affinity chromatography: a review of clinical applications. *Clin. Chem.* **45**, 593–615.

Holmes, D., and Morgan, H. (2010). Single cell impedance cytometry for identification and counting of CD4 T-cells in human blood using impedance labels. *Anal. Chem.* **82**, 1455–1461.

Hruby, V. J. (2002). Designing peptide receptor agonists and antagonists. *Nat. Rev. Drug Discov.* **1**, 847–858.

Hrusak, O., and Porwit-MacDonald, A. (2002). Antigen expression patterns reflecting genotype of acute leukemias. *Leukemia* **16**, 1233–1258.

Jagtiani, A. V., Sawant, R., and Zhe, J. (2006). A label-free high throughput resistive-pulse sensor for simultaneous differentiation and measurement of multiple particle-laden analytes. *J. Micromech. Microeng.* **16**, 1530–1539.

Karrasch, S., Dolder, M., Schabert, F., Ramsden, J., and Engel, A. (1993). Covalent binding of biological samples to solid supports for scanning probe microscopy in buffer solution. *Biophys. J.* **65**, 2437–2446.

Katoh, S. (1987). Scaling-up affinity chromatography. *Trends in Biotechnol.* **5**, 328–331.

Kemper, C., and Atkinson, J. P. (2007). T-cell regulation: with complements from innate immunity. *Nat. Rev. Immunol.* **7**, 9–18.

Killion, J. J., and Kollmorgen, G. M. (1976). Isolation of immunogenic tumor cells by cell-affinity chromatography. *Nature* **259**, 674–676.

Kobayashi, Y., and Martin, C. R. (1997). Toward a molecular Coulter counter type device. *J. Electroanal. Chem.* **431**, 29–33.

Koch, M., Evans, G. R., and Brunnschweiler, A. (1999). Design and fabrication of a micromachined Coulter counter. *J. Micromech. Microeng.* **9**, 159–161.

Koga, K., Matsumoto, K., Akiyoshi, T., Kubo, M., Yamanaka, N., Tasaki, A., Nakashima, H., Nakamura, M., Kuroki, S., Tanaka, M., Katano, M. (2005). Purification, characterization and biological significance of tumor-derived exosomes. *Anticancer Res.* **25**, 3703–3707.

Kummrow, A., Theisen, J., Frankowski, W. A. T., Yildirim, H., Brattke, K., Schmidt, M., Neukammera, J. (2009). Microfluidic structures for flow cytometric analysis of hydrodynamically focused blood cells fabricated by ultraprecision micromachining. *Lab Chip* **9**, 72–981.

Lane, P. J., Gaspal, F. M., and Kim, M. Y. (2005). Two sides of a cellular coin: CD4(+)CD3- cells regulate memory responses and lymph-node organization. *Nat. Rev. Immunol.* **5**, 655–660.

Lang, P., Pfeiffer, M., Handgretinger, R., Schumm, M., Demirdelen, B., Stanojevic, S., Klingebiel, T., Kohl, U., Kuci, S., Niethammer, D. (2002). Clinical scale isolation of T cell-depleted CD56+ donor lymphocytes in children. *Bone Marrow Transplant.* **29**, 497–502.

Lau, D., Guo, L., Liu, R., Marik, J., and Lam, K. (2006). Peptide ligands targeting integrin alpha3beta1 in non-small cell lung cancer. *Lung Cancer* **52**, 291–297.

Mckenna, B. K., Selim, A. A., Bringhurst, F. R., and Ehrlich, D. J. (2009). 384-Channel parallel microfluidic cytometer for rare-cell screening. *Lab Chip* **9**, 305–310.

Mohanty, S. K., Jabart, E., Conboy, M. J., Huang, H., Conboy, I.M., and Sohn, L. L. (2010). Isolation and Characterization of Organi Stem Cells from Their Microanatomical Niche.

Nieuwenhuis, J. H., Kohl, F., Bastemeijer, J., Sarro, P. M., and Vellekoop, M. J. (2004). Integrated Coulter counter based on 2-dimensional liquid aperture control. *Sens. Actuators B.* **102**, 44–50.

Orozco, A. F., Bischoff, F. Z., Horne, C., Popek, E., Simpson, J. L., Lewis, D. E. (2006). Hypoxia-induced membrane-bound apoptotic DNA particles: potential mechanism of fetal DNA in maternal plasma. *Ann. N.Y. Acad. Sci.* **1075**, 57–62.

Orozco, A. F., Jorgez, C. J., Horne, C., Marquez-Do, D. A., Chapman, M. R., Rodgers, J. R., Bischoff, F. Z., Lewis, D. E. (2008). Membrane protected apoptotic trophoblast microparticles contain nucleic acids: relevance to preeclampsia. *Am. J. Pathol.* **173**, 1595–1608.

Patel, K. D., Nollert, M. U., and McEver, R. P. (1995). P-selectin must extend a sufficient length from the plasma membrane to mediate rolling of neutrophils. *J. Cell Biol.* **131**, 1893–1902.

Ratajczak, J., Miekus, K., Kucia, M., Zhang, J., Reca, R., Dvorak, P., Ratajczak, M. Z. (2006). Embryonic stem cell-derived microvesicles reprogram hematopoietic progenitors: evidence for horizontal transfer of mRNA and protein delivery. *Leukemia* **20**, 847–856.

Rosenbluth, M. J., Lam, W. A., and Fletcher, D. A. (2008). Analyzing cell mechanics in hematologic diseases with microfluidic biophysical flow cytometry. *Lab Chip* **8**, 1062–1070.

Saleh, O. A., and Sohn, L. L. (2001). Quantitative sensing of nanoscale colloids using a microchip Coulter counter. *Rev. Sci. Inst.* **72**, 4449–4451.

Saleh, O. A., and Sohn, L. L. (2003a). An artificial nanopore for molecular sensing. *Nano Lett.* **3**, 37–38.

Saleh, O. A., and Sohn, L. L. (2003b). Direct detection of antibody-antigen binding using an on-chip artificial pore. *Proc. Natl. Acad. Sci. U.S.A.* **100**, 820–824.

Satake, D., Ebi, H., Oku, N., Matsuda, K., Takao, H., Ashiki, M., Ishida, M. (2002). A sensor for blood cell counter using MEMS technology. *Sens. Actuators B.* **83**, 77–81.

Shapiro, H. M. (2003). Practical Flow Cytometry. *Wiley*.

Sun, L., and Crooks, R. M. (1999). Fabrication and characterization of single pores for modeling mass transport across porous membranes. *Langmuir* **15**, 738–741.

Sun, T., and Morgan, H. (2010). Single-cell microfluidic impedance cytometry: a review. *Microfluid Nanofluid* **8**, 423–443.

Wang, Y. N., Kang, Y., Xu, D., Chon, C. H., Barnett, L., Kalams, S. A., Li, D., Li, D. (2008). On-chip counting of the number and the percentage of CD4+ T lymphocytes. *Lab Chip.* **8**, 309–315. -Reproduced by permission of the Royal Society of Chemistry.

Wilbur, J. L., Kumar, A., Kim, E., and Whitesides, G. M. (1994). Microfabrication by microcontact printing of self-assembled monolayers. *Adv. Mater.* **6**, 600–604.

Wu, X., Kang, Y., Wang, Y. N., Xu, D., Li, D., Li, D. (2008). Microfluidic differential resistive pulse sensors. *Electrophoresis* **29**, 2754–2759. -Copyright Wiley-VCH Verlag GmBH & Co. KGaA. Reproduced with permission.

Xia, Y. N., and Whitesides, G. M. (1998). Soft Lithography. *Angew. Chem. Int. Ed* **37**, 551–575.

Zhe, J., Jagtiani, A. V., Dutta, P. J. H., and Carletta, J. (2007). A micromachined high throughput Coulter counter for bioparticle detection and counting. *J. Micromech. Microeng.* **17**, 304–313.

Imaging cytometry

Laser Scanning Cytometry and Its Applications: A Pioneering Technology in the Field of Quantitative Imaging Cytometry

Melvin Henriksen, Bruce Miller, Judith Newmark, Yousef Al-Kofahi and Elena Holden

CompuCyte Corporation, Westwood, Massachusetts, USA

METHODS IN CELL BIOLOGY, VOL 102

161

0091-679X/10 $35.00
DOI 10.1016/B978-0-12-374912-3.00007-9

Abstract

Imaging cytometry plays an increasingly important role in all fields of biological and medical sciences. It has evolved into a complex and powerful discipline amalgamating image acquisition technologies and quantitative digital image analysis. This chapter presents an overview of the complex and ever-developing landscape of imaging cytometry, highlighting the imaging and quantitative performance of a wide range of available instruments based on their methods of sample illumination and the detection technologies they employ. Each of these technologies has inherent advantages and shortcomings stemming from its design. It is therefore paramount to assess the appropriateness of all of the imaging cytometry options available to determine the optimal choice for specific types of studies.

Laser scanning cytometry (LSC), the original imaging cytometry technology, is an attractive choice for analysis of both cellular and tissue specimens. Quantitative performance, flexibility, and the benefits of preserving native sample architecture and avoiding the introduction of artificial signals, particularly in cell-signaling studies and multicolor tissue analysis, are speeding the adoption of LSC and opening up new possibilities for developing sophisticated applications.

I. Introduction

Unquestionably, revolutions in genomics and proteomics have transformed academic life science research and the drug discovery process, and have brought the concept of personalized medicine to life. Identifying key proteins and understanding how, where, and when they interact have become the standard of practice in both academia and the pharmaceutical industry. The need for this information now drives development and dissemination of powerful instrumentation that provides an understanding of intercellular communications and complex biological processes at the subcellular level. Two previously disparate disciplines – imaging and cytometry – have converged with unprecedented speed, creating a cluster of new tools that we refer to as "imaging cytometry."

Imaging cytometry is the culmination of a series of important developments in two primary disciplines, microscopy and flow cytometry. Although early microscopy goes as far back as the 16th century, the key developments that have enabled modern-era imaging date from the second half of the 20th century (Coons and Kaplan, 1950; Minsky, 1988; Ploem, 1967, 1993). In parallel, the fundamentally new concept of quantitatively measuring the characteristics of large numbers of cells gave rise to flow cytometry, recognized today as the gold standard in quantitative cellular analysis (Coulter, 1956; Kamentsky *et al.*, 1963; Fulwyler, 1965; Kamentsky and Melamed, 1967; Van Dilla *et al.*, 1969). While classic flow cytometers represented the first truly high-content and high-throughput cellular analysis technology, they could not provide images of analyzed cells, nor could they work with adherent cells and tissue specimens without disturbing their native states and architecture. Microscopy instrumentation, while setting standards for visualization of cellular morphology and spatial localization, neither allowed easy, unbiased, and quantitative analysis of statistically significant numbers of cells, nor lent itself to quantification of cellular constituents. Multidisciplinary breakthroughs in the early 1990s, made possible by advances in optics, electronics, computing power, laser technologies, digital imaging, fluorescent-based reagents, and fluorescent proteins, have fostered pairing of imaging and cytometry (Kamentsky and Kamentsky, 1991).

II. Definition of Quantitative Imaging Cytometry (QIC) and Key Features Distinguishing Imaging Cytometry Platforms

A. Definition of QIC

We define quantitative imaging cytometry as the image-based quantification of the amount, size and shape of any selected marker in cells and tissues, in conjunction with their associated localization information at a subcellular, cellular, or intercellular level. From the perspective of the instrumentation involved, it encompasses both data acquisition platforms and digital image analysis.

The panorama of imaging cytometry platforms is diverse and continues to develop rapidly in response to classic demands: better (in terms of quality, efficiency, and degree of multiplexing), faster, and cheaper. There is no single characterization of the wide variety of platforms that are currently available, but it may be reasonable to begin with the type of samples that can be analyzed: cells in suspension, adherent cells, and tissue specimens. It is reasonable, therefore, to define two categories: flow-based and solid-phase imaging cytometry. The former category is fairly young and is currently represented by one commercially available imaging flow cytometer, ImageStream (Basiji *et al.*, 2007). The first solid-phase imaging cytometry system, the laser scanning cytometer (LSC), was developed by Louis Kamentsky and commercialized by CompuCyte Corporation, a company he founded in the early 1990s for the purpose of advancing the science and practice of imaging cytometry (Kamentsky, 2001; Kamentsky and Kamentsky, 1991). To enable precise quantification capabilities comparable to flow cytometry, LSC technology was built with laser excitation and PMT detection optics similar to those in flow cytometers, but applied to the analysis of samples positioned on a microscope slide. Various developments in the field of solid-phase imaging cytometry have followed, creating a

Table I
Imaging cytometry technology landscape

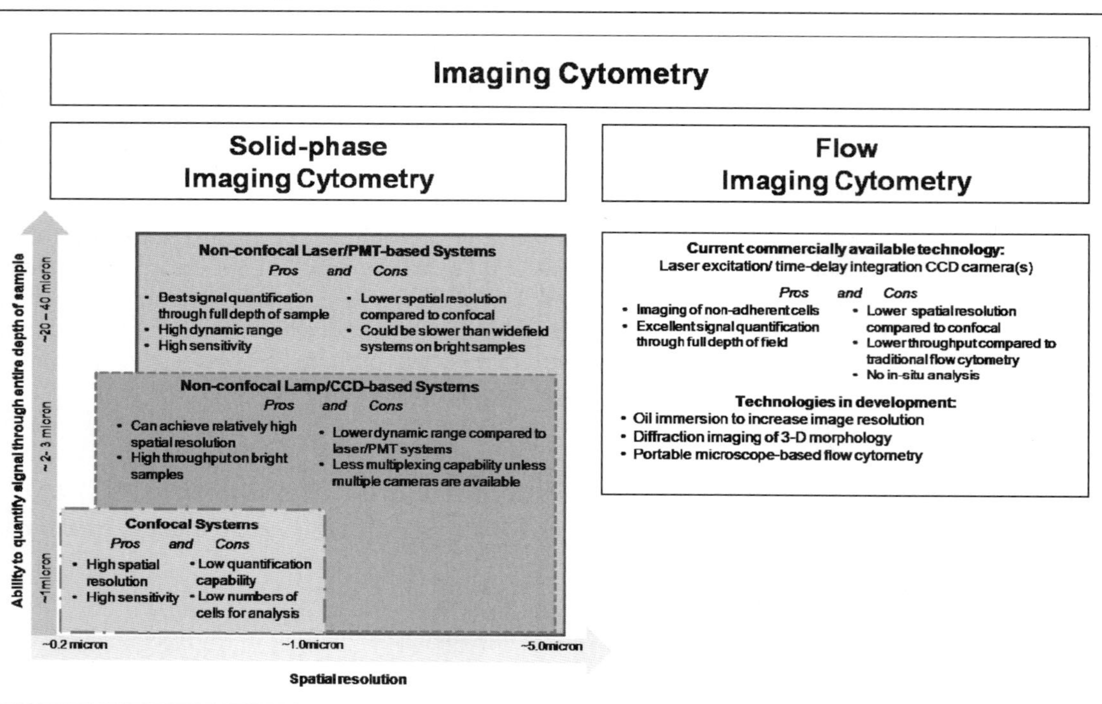

diverse instrumentation scene that can be further classified based on five main qualities: depth of field, light sources and illumination, signal detection, imaging modes, and type of sample labeling required for analysis. These modalities determine where in the continuum of imaging cytometry and quantitative performance these systems reside (Table I).

B. Key Features Distinguishing Imaging Cytometry Platforms

1. Depth of Field

The quantitative nature of an imaging system is directly related to the axial range of signal collection, referred to as "depth of field." Depth of field is a measure of the distance above and below the sample's focal plane from which meaningful, measurable signal is collected. Systems designed with high depths of field, such as LSCs and flow cytometers, have excellent quantitative capability by measuring the amount of signal of interest through the entire depth of biological specimen. Confocal systems have, by design, a very narrow depth of field (Pawley and James, 2006), typically less than 1 μm, and therefore are not able to capture signal through the full thickness of most samples in a single measurement. Images obtained from confocal systems exhibit a very high degree of spatial resolution but, due to the purposeful lack of information from outside the focal plane, will be less quantitative than a comparable image obtained on a system with a high depth of field. To accommodate this limitation, images from a confocal system can be obtained at different sample heights (optical sectioning) (Lodish, 2008), and then these images can be assembled (stacked) to create a volume image of the sample. Stacking confocal images improves the signal quantification accuracy; however, this process is time-consuming.

2. Sample Illumination

Robust cellular and tissue analysis requires a high-intensity light source, especially for fluorescent samples where optimal fluorophore emission signal is dependent on saturating the fluorophore with excitation energy. In imaging cytometry systems, two main types of illumination sources are used: (i) mercury (or xenon mercury) arc lamps and (ii) lasers. Light-emitting diodes (LEDs) are emerging as an additional option for illumination.

Arc lamps are the typical illumination sources for epi-fluorescence microscopy systems. Their illumination is typically "widefield," meaning that the entire field of view is illuminated at once. The output produced by the ionized gas employed by arc lamps is relatively intense. Although they emit light over a broad spectrum, their output includes some very strong, narrow-wavelength bands or "emission lines" (Shapiro, 2003). Coupled with bandpass or shortpass filters centered on the dominant spectral emission lines, these light sources can be used to excite fluorophores over a range of excitation wavelengths across the visible spectrum. The broad-spectral characteristics of these lamps also provide relatively true color imaging of chromatically stained specimens. Relative to laser illumination, systems utilizing arc lamp illumination typically produce lower excitation power densities at the sample plane, resulting in lower sensitivity or requiring longer exposure time. Longer exposure times improve signal-to-noise performance, but may result in "photobleaching" (degradation of fluorochromes) or susceptibility to system vibration. Widefield lamp illumination can provide faster image capture, at least for relatively bright sample signals.

Lasers typically produce monochromatic light so that multiple lasers are required to excite dyes of differing excitation wavelengths. In recent years, a broader range of laser wavelengths has become available, expanding the portfolio of dyes that may be utilized with laser-based systems.

Lasers produce very intense, very stable beams with low spatial divergence. The high intensity of the laser beams allows for short exposure times. Additionally, the high power density of laser light provides better sensitivity, even though exposure times are fixed. Caution must be exercised, however, in order to minimize photobleaching. Most new laser devices have user-controllable power outputs, allowing adjustment of laser power to accommodate samples with different staining levels.

Laser spots are typically used with an oscillating mirror or beam deflector to create a "scan line" in the sample plane. With this illumination method, a very small portion of the sample (the size of the laser spot) is illuminated at any given time, reducing the background signal from neighboring areas of the sample. The high stability over time also results in a very low contribution to signal noise.

3. Signal Detection

There are two main types of detectors used in imaging cytometry platforms: charge-coupled devices (CCDs) and photomultiplier tubes (PMTs). CCDs are structured as a dense array of sensors (typically millions of individual sensors – or pixels – per device) that, when placed at the image plane of a microscope or other optical imaging device, produce a digital image of the sample. Imaging platforms employing CCD detectors typically employ widefield illumination light sources. In these systems, one or two CCDs are usually used. As a result, for highly multiplexed samples with large numbers of dyes, each channel (emission wavelength) must be imaged sequentially by placing the appropriate filter in front of the CCD camera. CCDs have an inherently limited dynamic range, but can produce images over a wide range of signal levels by controlling the exposure time (the length of time over which the signal is integrated). Additionally, the signal-to-noise ratio is increased as exposure time lengthens. Dynamic range can also be increased by "binning" CCD pixels, a technique that combines the signals produced by a number of contiguous pixels. Binning increases the signal amplitude without increased exposure time, but reduces the spatial resolution of the image because there are fewer pixels contributing unique responses to the output image.

Flow cytometry, LSC, and laser-scanning confocal systems use PMTs for signal detection. PMTs are typically combined with laser illumination in spot-scanning systems and are arranged to collect the entire field of view of the illuminated sample, creating one output pixel at a time. Images are assembled pixel by pixel at a very rapid rate as the laser spot is moved across the sample. Spot scanning allows the data acquisition resolution to be varied to match the requirements of the scan. PMTs have significantly higher dynamic range (greater than four decades) (Pawley and James, 2006) than CCDs and the range can be increased by adjusting the input voltage to the device. While signal integration time is typically fixed so that dim and bright signals are collected in the same predetermined timeframe, the signal-to-noise ratio can be improved by increasing the voltage. PMT-based systems typically employ multiple PMTs for simultaneous signal acquisition, with each PMT coupled with an optical filter that isolates a unique section of the spectrum to be captured by the PMT.

4. Imaging Mode: Spot Scanning, Widefield Illumination, or Both

Widefield, spot, and Nipkow scanning are all utilized in fluorescent imaging. Because widefield imaging utilizes a white-light illumination source, many widefield imaging systems provide brightfield as well as fluorescent images, although most require these to be acquired sequentially. Spot-scanning systems are typically limited to fluorescent imaging, but some (e.g., LSC, which will be discussed in detail later in this chapter) do provide the ability to image both chromatic and fluorescent dyes. The combination of both chromatic and fluorescent dyes on a single sample allows for traditional brightfield reference images as well as quantification and high levels of multiplexing.

Both lasers and arc lamps can be used in spot scanning, although the high intensity and highly focused nature of lasers make them more attractive for this type of scanning. Widefield illumination can provide faster image acquisition on bright signals because large areas of the sample (the "field of view") are illuminated at once.

Nipkow disk scanning combines spot scanning with quasi-widefield illumination. Systems utilizing this technology employ either a laser or an arc lamp to create a large spot on a spinning disk containing an array of pinholes. The illumination is split into many parallel paths by the spinning disk pinholes to produce multiple illumination spots on the sample. The spots move across the sample as the disk spins, creating the effect of widefield illumination over time. Signal is collected back through the spinning disk, and the optically filtered output is imaged onto a CCD camera.

5. Type of Sample Labeling Required for Analysis: Fluorescent or Chromatic

Most QIC systems are designed to image and analyze samples stained either with fluorescent or chromatic dyes, but not both. As most histopathology specimens continue to be analyzed with traditional chromatic dyes, a system's ability to manage both modes of analysis becomes an attractive feature.

C. Imaging Cytometry Instrumentation Landscape

Table II summarizes the most common commercial systems presently available for imaging cytometry, categorized by the types of sample illumination, detection technology, and other technology features they employ. Each of these technologies has advantages and shortcomings stemming from its inherent design.

As a rule of thumb, there is an inverse correlation between image resolution and the degree of quantification various systems provide. Even within one technology type, a manufacturer's focus on a specific market or markets can result in its technology being geared more toward one end of the resolution/quantification spectrum. For example, a particular system may sacrifice imaging performance in order to achieve faster throughput. It is highly advisable to perform a well-designed

Table II

Summary of common[a] commercial imaging cytometry systems

Platform/provider	Type of imaging: confocal, Nonconfocal (NC), widefield (WF), Line scan Depth of field (Low+ to high+++)	Illumination	Detectors	Image resolution Low to moderate + Moderate to high++ High to ultrahigh +++	Sample types (cellular, tissue) Dye types
			Solid-phase imaging cytometry systems		
iGeneration LSC (iCyte, iCys, iColor) CompuCyte	Laser scanning NC +++	1–4 lasers	3–4 PMTs 1–3 scatter/ absorbance detectors	++	Various types of cellular and tissue samples Fluorescent, chromatic, unstained scatter imaging
Acumen EX3 TTP LabTech	Laser scanning NC +++	1–3 lasers	4 PMTs	+	Cellular Fluorescent
ArrayScan VTI Cellomics	WF NC Confocal option +/++	1 arc lamp	CCD	++	Cellular Fluorescent
Operetta PerkinElmer	WF NC or Confocal +	1 arc lamp LED	CCD	++ +++	Cellular Fluorescent or chromatic
IN Cell 2000 GE Healthcare	WF NC ++	1 arc lamp LED	Standard or large-chip CCD	++	Cellular and tissue Fluorescent and chromatic, DIC, phase contrast
Image-Xpress Micro MDS/ Danaher	WF NC ++	1 arc lamp	CCD	++	Cellular Fluorescent, phase contrast
Pathway 435 BD Biosciences	WF NC or Confocal +/++	1 arc lamp LED	CCD	++ +++	Fluorescent and chromatic

Scan^R Olympus	WF NC ++	1 arc lamp	CCD	++	Cellular and tissue Fluorescent, chromatic, DIC & phase contrast
Opera PerkinElmer	Laser scanning Confocal UV NC +	Up to 5 lasers UV lamp (NC)	3 CCD 1 CCD (NC)	+++	Cellular Fluorescent
Aperio ScanScope	Line scan NC ++	1 arc lamp	TDI CCD	++	Tissue Chromatic
Aperio ScanScope FL	Line scan NC ++	1 arc lamp	TDI CCD	++	Tissue Fluorescent
BioImagene iScan Coreo Au	Line scan NC ++	LED	CCD	++	Tissue Chromatic
BioImagene iScan Concerto	Line scan NC ++	1 arc lamp LED	CCD	++	Tissue Fluorescent and chromatic
Leica/Genetix Ariol	WF NC ++	1 arc lamp	CCD	++	Tissue Fluorescent and chromatic

Automated image analysis software coupled with microscopy systems

Various automated microscopes and image analysis software packages (proprietary and open source):

- *Noncommercial/open source.* CellProfiler (Broad Institute), Image J (NIH), Farsight (Rensselaer), OMERO (Open Microscopy Environment)
- *Commercial.* Metamorph (Molecular Devices), Image Pro (MediaCybernetics), Imaris (Bitplane), LSM Image Browser (Zeiss)
- *Widefield Microscopy Systems.* Eclipse Ti (Nikon), IX2 series (Olympus), Axiovert 40 (Zeiss), DM series (Leica)
- *Confocal Microscopy Systems.* A1R-A1 (Nikon), FluoView FV1000 (Olympus), LSM 700 series (Zeiss), TCS series (Leica)

Flow imaging cytometry systems					
ImageStream Amnis	NC +++	Up to 5 lasers	2 TDI CCDs	++	Cells in suspension Fluorescent

[a] This table is representative of commercially available systems and is not all-inclusive.

feasibility study to evaluate various platforms in order to decide which technology may be optimal for an institution's specific needs.

As the performance of solid-phase imaging cytometry systems can match the analytical performance of flow cytometry, it is highly likely that the "traditional" approach of disaggregating adherent cell lines and tissue sections before analyzing them via flow will become obsolete. The benefits of preserving native sample architecture and avoiding introduction of artificial signals, particularly in cell-signaling studies, are speeding adoption of solid-phase imaging cytometry systems into routine practice. In reality, we are already witnessing traditional flow cytometry core facilities complementing the "flow" portion of their services with "solid-phase" imaging cytometry techniques.

III. Technical and Analytical Features of iGeneration Laser Scanning Cytometry

One of the technologies offering the benefits of cytometry for adherent cells and tissues is LSC, the original imaging cytometry technology first developed by Louis Kamentsky. The iGeneration LSC instruments are nonconfocal systems that utilize laser-spot scanning illumination. The detection system, which uniquely and elegantly combines PMTs for fluorescence detection and photodiodes for absorbance and scatter detection, offers excellent quantitative cytometric performance and imaging capabilities. The systems provide fully automated or interactive operation for the analysis of cellular and tissue samples stained with fluorescent or chromatic dyes. Fluorescent excitation, laser light absorbance by chromatic dyes, and scatter signals may all be acquired simultaneously. Data analysis and image processing are performed either in real time as the scan is progressing or at the post-data acquisition stage so that the data analysis strategies may be modified to explore alternative scenarios. Numerical data are displayed as traditional cytometric histograms or bivariate scattergram plots and are available for analysis in third-party software applications. These data can be related in a variety of ways to the image data from which they are derived.

A. Illumination Method and Depth of Field

iGeneration LSCs utilize up to four lasers of different wavelengths, which currently can be chosen from six available wavelengths: 405, 488, 532, 561, 594, and 633 nm. (See Fig. 1 for a schematic diagram of the iCys® Research Imaging Cytometer.) Light from the four lasers is combined into a single coincident beam and directed to a scanning galvanometer mirror. The beam is then directed through a scan lens and the microscope objective onto the sample plane. The iGeneration LSC instruments are built on an inverted platform; the laser illumination is therefore presented to the underside of the sample. As the galvanometer oscillates, the beam

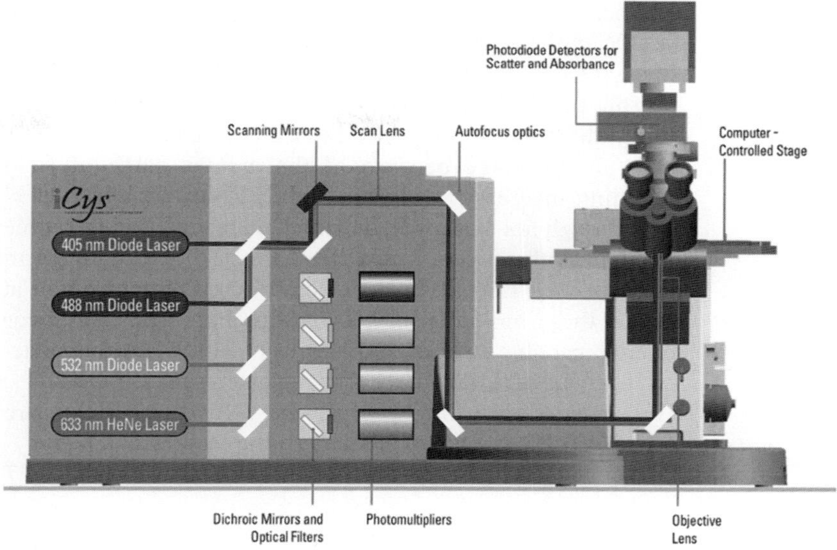

Fig. 1 iCys® Research Imaging Cytometer with the 405, 488, 532, and 633 nm laser option.

sweeps in a line across the sample plane. The scan lens ensures that the beam size and focus are constant at the sample plane, throughout the sweep of the beam.

LSC systems are designed so that the laser beams diverge slowly through the sample plane, creating the broad depth of field that gives LSC its high quantitative precision. This nonconfocal design allows signal capture through the full thickness of most samples. Fig. 2 shows confocal and LSC images of the same

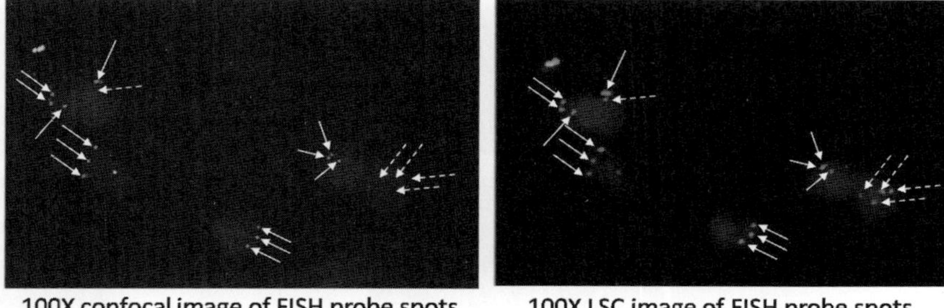

Fig. 2 100X confocal and LSC images of FISH probe spots. Solid arrows indicate spots present in the focal plane of the confocal image. Dashed arrows indicate spots present through the full depth of the cells as captured by LSC imaging.

cellular sample. In the *confocal* image on the left, the solid arrows point to spots visible in the current focal plane. The dashed arrows point to probes at different focal planes that are not visible in the focal plane of the image because of the narrow depth of focus of the confocal system. In the *laser scan* image on the right, all of the fluorescence *in-situ* hybridization (FISH) probe spots are visible in a single image because of the greater depth of field inherent with laser-scanning images.

Although not as severe as with confocal microscopy, camera-based systems have similar depth-of-field limitations (Fig. 3). Side-by-side images of dermal tissue generated by LSC and fluorescent microscopy, obtained at various focal planes, reflect these limitations. At 2 microns from optimum focus, some degradation in quality can be observed in the microscope images. The degradation becomes increasingly pronounced until at 5 microns, nuclear details are no longer discernible. By contrast, nuclear details in the LSC scan images are still visible, even at 10 microns from optimum focus.

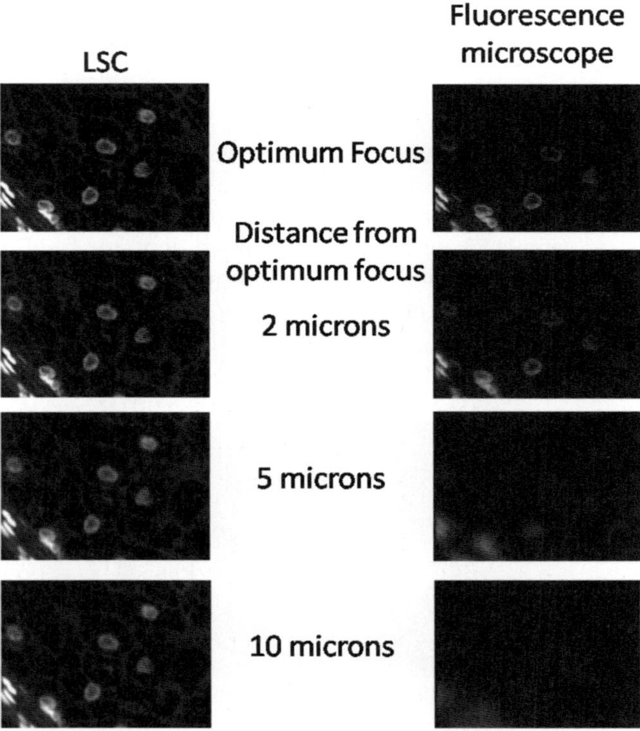

Fig. 3   40X images of dermal tissue at increasing distances from optimal focus, acquired by LSC (left) and fluorescent microscopy (right).

B. Signal Detection Method

Three distinct detection modalities work in concert on iGeneration LSC platforms: (a) quantification of excited fluorescent light, (b) laser light absorbed by the specimen, and (c) scattered laser light. This unique design enables analysis of samples stained with fluorescent and chromatic dyes, either sequentially or simultaneously, when the dyes are combined on the same sample.

Fluorescent light from the sample is collected by the objective lens and directed back along the laser illumination path until it reaches a dichroic filter that separates the excitation light from the fluorescence emission light based on their wavelengths, and sends the fluorescent emission light to the PMTs (Fig. 4). A series of dichroic mirrors and bandpass filters in front of each PMT separate the light into four wavelength bands, one for each PMT. These fluorescence detection filters are tailored to the specific laser wavelength combination of the individual system, with additional optional filter sets available. PMT exposure parameters are automatically set on user-specified areas of the sample in order to ensure proper exposure and background levels.

Laser light-loss (for use with chromatic dyes) is measured by collecting laser light transmitted through the sample and directing it to two photodiode detectors. Transmitted light from lasers of two different wavelengths may be measured simultaneously. Selectable filter wheels in front of each photodiode control the laser wavelengths allowed to reach that diode. In Fig. 5A, laser light of two wavelengths (405 and 633 nm) is transmitted through the sample (a Her2/neu-positive human breast tissue section stained with the chromatic dyes hematoxylin and anti-Her2 antibody developed with DAB). The two wavelengths are separated by the optical

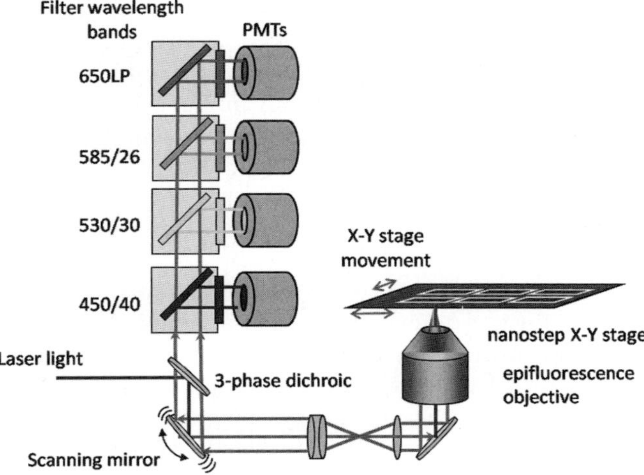

Fig. 4 iCys® Research Imaging Cytometer – Fluorescence detection path.

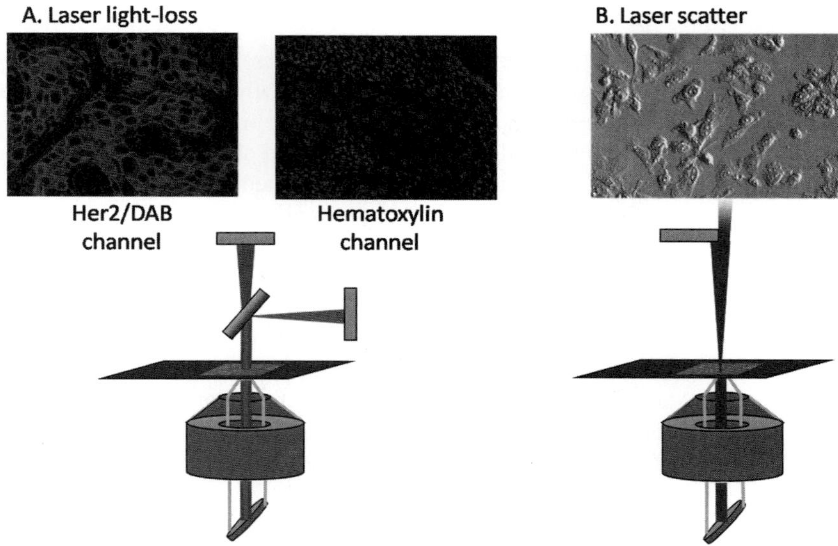

Fig. 5   Optical path and images for (A) laser light-loss measured on Her2-positive breast tissue stained with hematoxylin and anti-Her2 Ab developed with DAB, and (B) laser scatter measured on unstained CHO cells.

filters so that the photodiode detector at the top of the diagram collects only the transmitted 405 nm (violet) laser light, while the one to the right collects only the transmitted 633 nm (red) laser light.

For scatter imaging, the top photodiode detector is repositioned so that only a portion of the transmitted light *scattered away from the beam axis* is measured. Specific cellular structures scatter light differently; measuring the light scattered by the sample, therefore, reveals details of its structure and morphology. (See the light-scatter images of CHO cells in Fig. 5B.) This simple, patented configuration produces forward scatter images very similar to those produced using differential interference contrast imaging.

C. Autofocus

Various robust autofocus routines are available on iGeneration LSC systems.

Basic autofocus (Fig. 6) is accomplished when the system moves the selected objective lens to a nominal focus position, as defined by the sample carrier descriptor. (All standard carriers have a predefined software descriptor containing carrier parameters to be used in system operation. A system utility allows the user to define new descriptors for any nonstandard carrier.) From this nominal focus position, the system evaluates the reflected scan line at various focal planes through a predefined range, selecting as optimum the focal plane where the scan line is of highest contrast.

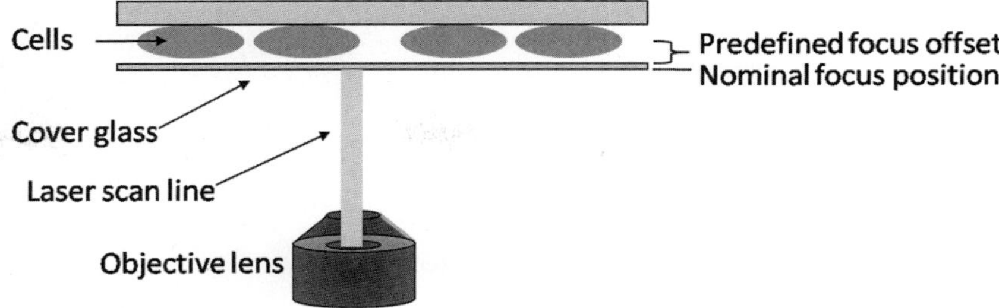

Fig. 6 Basic autofocus function. The microscope objective lens moves through a predetermined range, centered on the expected position of the cover slip, while the autofocus controller samples the contrast of the autofocus input signal. Upon completion, the system moves to the position that generated the highest contrast, and then moves by a predefined offset to the focal plane of the sample.

Focus is performed on the sample substrate, typically the bottom surface of the carrier (e.g., a microscope slide cover slip). A predefined offset is then applied to move the objective to the sample plane. Because focus is not performed directly on the sample, it is independent of the presence of cells or tissue in the area where the focus measurement is being made. Focus parameters allow for variations in this basic method in order to optimize focus for specific carriers or samples.

An alternative focus method, in which focus is performed directly on the sample, may be employed where there is significant variation in the sample depth (as is often the case with tissue sections). Additionally, either of these methods may be used in a "three-point" focus where a local plane is defined by three focus measurements. This technique is employed when the surface of the carrier is not fully orthogonal to the axis of incident laser light.

D. Image and Data Acquisition

As the laser spot sweeps across the sample, PMT signals are digitized 768 times along the length of the laser scan line, producing 768 image pixels for each of the six detector channels (four PMTs and two photodiode channels on iCyte® and iCys® systems, and three PMTs and three photodiodes on iColor® systems). After each scan line is completed, the sample is advanced by a high-precision motorized stage at user-defined stage steps (from 0.05 to 20 microns) and the next scan line is initiated. Field images are assembled and stored for every 1000 scan lines. Fig. 7 illustrates the image acquisition process. Field image sizes vary depending on the objective magnifications and scan resolution, as indicated in Table III.

PMTs have an inherently high dynamic range. In LSC technology, this translates into the ability to measure approximately three decades of fluorescence intensity for a given PMT setting, a significantly higher range than in most CCD imaging systems. As a result, measurements in a range from a few hundred fluorescence intensity units (very dim) to several hundred thousand units (very bright) can be

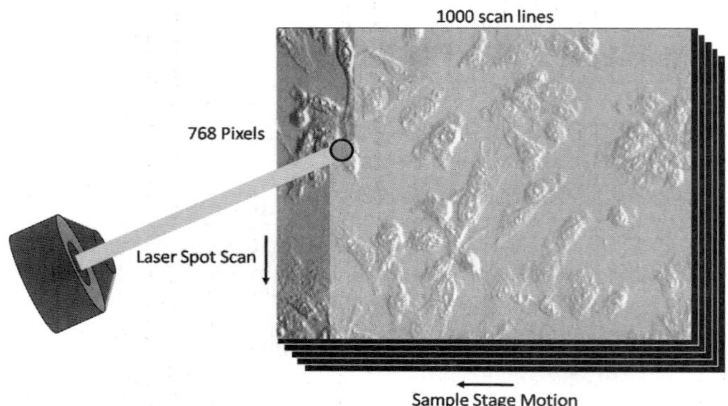

Fig. 7   Creating a field image from the scanning process. As the laser spot scans the sample, signal is digitized at 768 discrete points. The sample then advances for the next scan until 1000 scan lines of data have been acquired. Up to six channels are acquired simultaneously.

Table III

Nominal pixel resolution for different objective lenses

Objective lens	Nominal scan resolution (μm) (square pixels)	Pixel size (μm)	Image size (μm)
60X	0.15	0.15×0.167	128×150
40X	0.25	0.25×0.25	192×250
20X	0.5	0.5×0.5	384×500
10X	1.0	1.0×1.0	768×1000
4X	2.5	2.5×2.5	1920×2500

measured at the same setting. The center of this three-decade range may be moved higher or lower by adjusting the applied voltage of the PMT. The range of acquired signals is divided into a number of gray-scale values, each representing a different signal increment within the range. In order to resolve small signal increments within the range of acquired signals, a digitization depth (bit depth) of 14 bits is used, ensuring that subtle signal level changes will be resolved.

Overview scans are performed at lower scan resolution, typically to identify areas for high-resolution quantitative scanning. High-resolution field scan areas are assembled into mosaic region images. (See Fig. 8. Sample is courtesy of David Krull, GSK Safety Assessment, Research Triangle Park, NC. Scanning was performed at CompuCyte Corporation, Westwood, MA.)

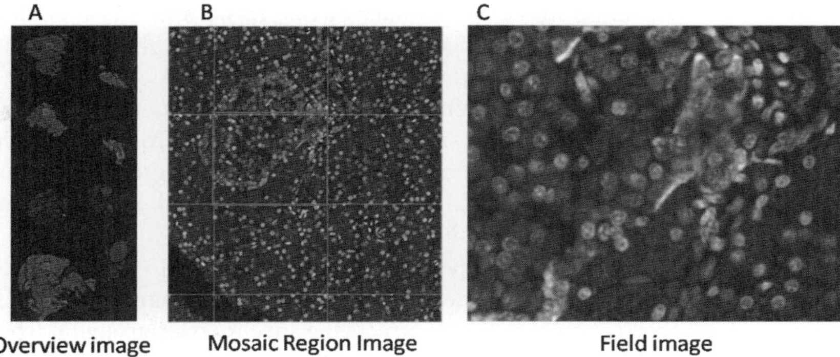

Fig. 8 (A) Overview image of multiple rat pancreas tissue samples. Areas designated for high-resolution scanning are defined around islets. (B) High-resolution mosaic region images are assembled from (C) field images.

E. Compensation

Fluorescent dyes typically have broad emission spectra, and the emitted light can often be observed in multiple PMT channels. The example in Fig. 9A displays the emission spectra for two commonly used fluorescent dyes – FITC (fluorescein) and PE (R-phycoerythrin). Also shown in Fig. 9A are the filtered spectra observed by two of the PMTs used in the iGeneration LSC system: green (530 nm center wavelength/30 nm bandwidth) and orange (580 nm center wavelength/ 30 nm bandwidth). In the example below, the green PMT is used as the primary PMT channel for FITC and the orange PMT is primary for PE. It can be seen, however, that some amount of the FITC emission spectrum falls within the orange PMT spectral band. This is termed "spectral overlap" and will be observed as an

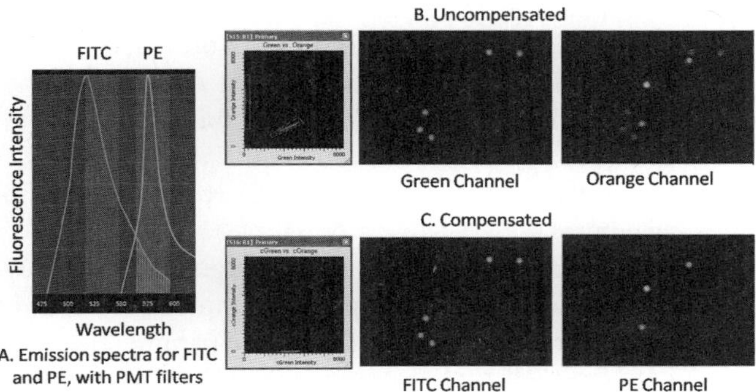

Fig. 9 Applying compensation removes the FITC fluorescence emission present in the neighboring orange channel, leaving only the PE beads in the compensated green (PE) channel. (See plate no. 10 in the color plate section.)

elevation in the PE primary channel wherever FITC is located within the sample. To produce an accurate representation of the PE signal in the sample, it is necessary to remove the FITC-generated signal from the orange channel. Removal of spectral overlap signals is referred to as "compensation." (Chromatic stains similarly absorb over a broad spectrum, and typically require compensation.) Uncompensated (acquired) green and orange fluorescence channel images of FITC- and PE-labeled beads are shown in Fig. 9B. The compensated images are shown in Fig. 9C.

Automated compensation allows the user to select one of two methods of compensation: least squares fit (LSF) or linear unmixing. The LSF method calculates a matrix equation that subtracts the spectral overlap signal from each relevant channel, creating compensated channels. Linear unmixing also removes spectral overlap signals but then reassigns these signals to the compensated channels for each dye. In this way, the spectral overlap signals are not lost. Both methods require the operator to identify single-positive dye areas (or events) within the sample. Each method has its own mode for making this selection. In LSF, event populations stained with a single dye are selected in scattergrams by gating regions in a manner similar to that used in flow cytometry. In linear unmixing, areas of isolated dye signal are identified on the laser scan images by the operator.

F. Multiplexing

As mentioned earlier, data are acquired simultaneously from up to six channels per pass. For multiple-pass scanning, the stage repositions automatically at the start of each scan field and additional sets of field images are acquired using the laser and detector settings for each pass. Multiple-pass scans may be employed to extend the number of acquired data channels, as a means of avoiding the need to compensate for spectral overlap or to increase the signal-to-noise ratio for very faint signals.

In the assay shown in Fig. 10, rat pancreas tissue was stained with five dyes – four fluorescent dyes (DAPI to measure DNA, AlexaFluor 488 for mitochondria, AlexaFluor 532 for glucagon, and AlexaFluor 647 for insulin) and the chromatic stain DAB to measure Ki67. The ability to use both fluorescent and chromatic dyes provides added flexibility in multiplexing assays. Autofluorescence can also be a useful parameter in evaluating necrotic areas, sample artifacts, and specific sample structures.

Data from two runs may be merged using the Merge Runs feature. Merging runs can be useful when comparing samples before and after treatment, at multiple time points, or to increase the number of parameters measured. When runs are merged, the data from all runs become available as though they were acquired in a single run, allowing images from separate acquisitions to be processed together. For example, a segmentation strategy may be developed that utilizes an image channel generated from each of the several merged runs.

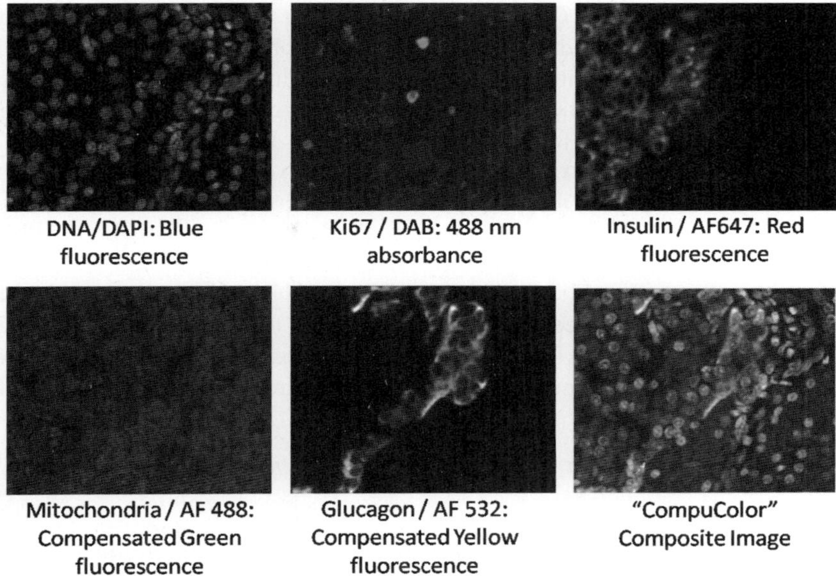

Fig. 10 High-content assay of rat pancreas tissue using a combination of four fluorescent dyes and one chromatic dye. (For interpretation of the references to color in this figure legend, the reader is referred to the Web version of this chapter.)

G. Throughput and Variable–Resolution Scanning

Variable-resolution scanning enables the scan resolution to be tailored to the requirements of a given application, resulting in image resolution and throughput optimization. Resolution is dependent on both the objective lens magnification and the rate of stage motion (referred to as "stage resolution"). Objective lenses with 4X, 10X, 20X, 40X, or 60X magnification are available. Objective lens selection defines the image resolution in the vertical dimension, whereas the horizontal resolution may be defined independently by selecting a stage resolution from 0.05- to 20-micron steps. In Fig. 11 below, bovine pulmonary artery endothelial (BPAE) cells (Invitrogen,

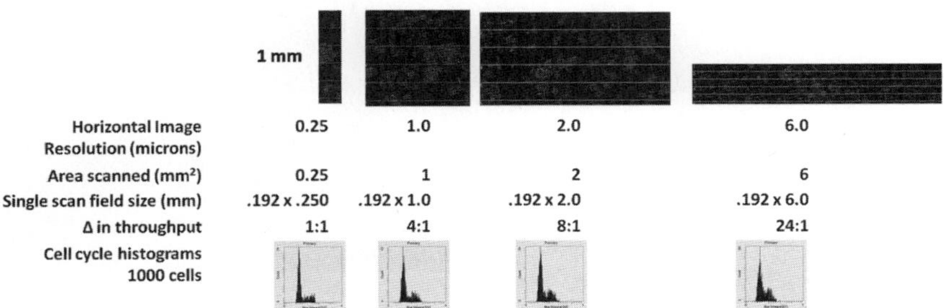

Fig. 11 Resolution, scanned area, change in throughput, and DNA histogram of BPAE cells for four different resolutions using a 40X objective.

Carlsbad, CA, FluoCell slide #1, Cat. no. F36924) labeled with MitoTracker[®] red for mitochondria, Alexa Fluor[®] 488 phalloidin for F-actin, and DAPI for nuclei are scanned at four different stage resolutions using the 40X objective.

Note the increase in scanned area as the resolution is decreased. The rate of scanning (and thus the throughput) increases eightfold when moving from 0.25- to 2-micron resolution. Assay precision (measured as the precision of the G_1 population) remains relatively constant over this range. Decreasing resolution further to 6 microns results in an additional threefold increase in throughput without significant decrease in G_1 precision.

Variable-resolution modality enables both high-throughput applications (e.g., cell-cycle screens on a 96-well plate in less than 10 minutes) and high-content analysis (HCA) (where scanning and analysis can take longer, often hours, depending on the level of multiplexing required, desired resolution, and size of the scan area).

H. Basic Thresholding, Event Contouring, and Analytical Features

1. Thresholding

Quantitative data analysis starts with identifying events, a process similar to thresholding in flow cytometry. Signal levels are derived from the pixel values that comprise each image. Values rising above the threshold value are used to identify discrete events for quantification.

2. Event Contouring

Once the threshold level is established, the iGeneration cytometric software draws a contour around the "events" (Fig. 12). Four types of contours may be drawn:

1. The *threshold* contour (solid) defines the edge of the event at the threshold limit. In Fig. 12, this is based on the threshold value of the nuclear signal.

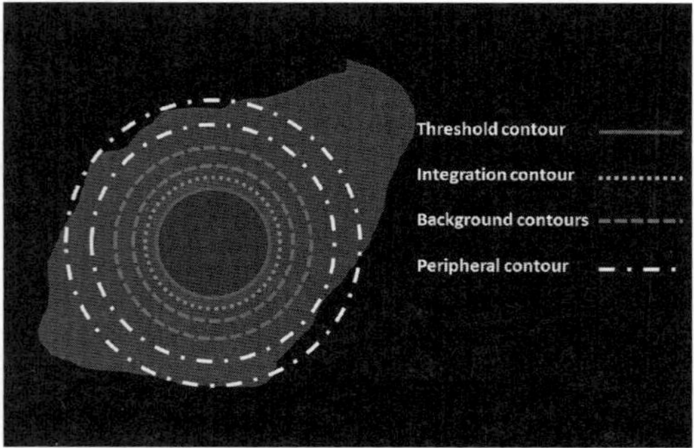

Fig. 12 Diagram of cell showing threshold, integration, background, and peripheral contours. Background contour is used to correct measurement within the integration contour. Peripheral contours sample cytoplasmic staining.

2. The *integration* contour (dotted) is a user-defined number of pixels outside of the threshold contour, ensuring that the full signal is measured. The feature data are generally based on this contour.
3. The *background* contours (dashed) define an annular area around the event and are used to "background-correct" those features based on the *integration* contour by subtracting the mean background value from the *integration* contour feature value.
4. The *peripheral* contour (dashed and dotted) is used to sample and quantify signal from an area external to the *integration* contour. In Fig. 12, the peripheral contour is used to sample the signal of interest in the cytoplasm.

Morphological, signal-based, and relational features are generated from the sets of pixels within these contours. For example, area is generated from the threshold contour, while integral (the sum of all the pixel values), intensity (the average of all the pixel values), and max pixel (the brightest pixel value) are all generated based on the integral contour and the peripheral contours. Note that these features are generated for each channel. A more detailed review of the range of available features is presented in Table IV.

Table IV
Numerical features generated from segmented or random sampling events

	Morphological features
Area	The area of the event in μm^2 based on the integration contour.
Perimeter	The perimeter of the event in μm based on the integration contour.
Circularity	A measure of the "roundness" of an event, calculated as the ratio *perimeter^2:area*. A perfectly round event will have a circularity of 4π.

	Signal-level related features (generated for each channel)
Integral	The sum of the pixel values within the integration contour. Integral is typically used to assess the total amount of target of interest in the event. In cell-cycle analysis, the integral of the DNA-intercalating dye channel is a measure of the total DNA content of the cell.
Max pixel	The value of the brightest pixel within the integration contour. In cell-cycle analysis, the max pixel of the DNA-intercalating dye channel provides an indication of the state of chromatin condensation and is used to distinguish mitotic from G2 cells.
Intensity	The average pixel values for a given channel within the integration contour. Intensity is the ratio *integral:area*, essentially normalizing the signal level to the area.
Peripheral integral, max pixel, and intensity	These features are analogous to their counterparts above, for the area enclosed by the peripheral contour.
Subcontour integral, max pixel, and intensity	These features are analogous to their counterparts above, for the sum of the associated subcontour events.

	Location and relational features
XY position	The coordinate positions of the events on the sample carrier.
Parent ID	The identification number of an event to which a subevent belongs. This gives visibility to the association of sub- and parent events and allows plotting and segregating subevents by their parents.
Scan position	The position along the scan line at which an event lies. This is most frequently utilized in QC procedures or when trouble-shooting the system's response along the scan.
Time	The time at which data for an event was scanned. This is utilized in multipass or repeat scans.

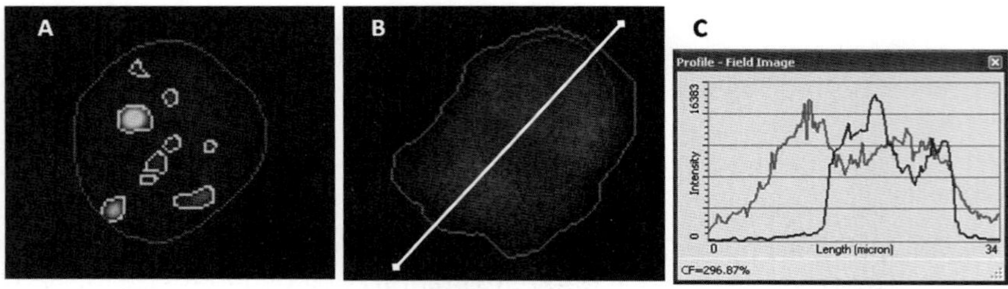

Fig. 13   CHO cells stained with DAPI for DNA, AlexaFluor 488 for phosphorylated histone H3, and AlexaFluor 633 for phalloidin. (A) Nuclear (green) and pH3 foci (cyan) contours. (B) Cytoplasmic AlexaFluor 633 phalloidin (magenta) subcontour. Note that "subcontour" does not imply relative size. White "profile line" allows plotting the relative signal level of multiple channels as a check of exposure settings and threshold values. (C) Plot of relative intensity across the "profile line" in panel B for nuclear and cytoplasmic signal. (See plate no. 11 in the color plate section.)

3. Multicomponent Event Contouring

Multiple segmentation routines may be generated for different types of events or cellular structures (called event components), each one using different segmentation criteria. In the example illustrated in Fig. 13, three independent event components are defined:

1. Nuclear – DNA, based on DAPI signal
2. Subnuclear – based on the phosphorylated histone H3 (pH3)/AF488 signal
3. Cytoplasmic – based on the phalloidin/AF633 signal

Any number of event components may be logically related to one another to represent their physical association. Here, the histone H3 foci event components are found within nuclear event components. The nuclear event components are found within the cytoplasm event components. Therefore, all three components can be associated with one another within the analysis protocol. This association allows the iGeneration software to "know" which and how many histone foci are within a given nucleus and which nucleus lies within which cytoplasm. Where an association exists between two or more event components, one is referred to as the parent and the others the children (or subcontours). The term "subcontour" is generic, referring to any event component with a parent event component. Multiple event components may also be defined without any association.

4. Random Sampling

In cases where the sample is not conducive to establishing boundaries between individual cellular events – such as with confluent adherent cellular samples or tumorous tissue samples – random sampling may be applied as an alternative strategy to generate quantitative expression data. Circular sampling elements of user-defined size and frequency are overlaid on the images in either a grid or random

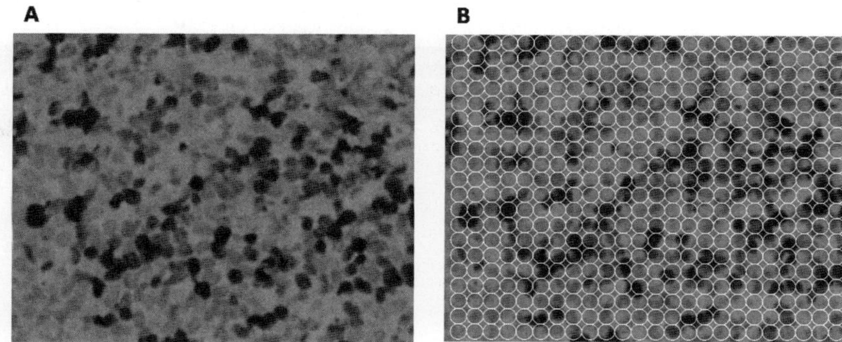

Fig. 14   (A) Composite image of colon tissue stained with Ki67-DAB and hematoxylin. (B) Random sampling elements overlaid on tissue image.

orientation (Fig. 14). These sampling elements act like segmentation contours from which analytical features are generated. This method allows sample analysis which would otherwise be impossible by conventional segmentation. Data are reported on a per-area basis rather than a per-cell basis; per-cell data are not possible because the sampling elements will frequently overlay portions of multiple cells. Random sampling can be performed simultaneously with event segmentation so that a comparison may be made. These methods typically yield very similar results on samples conducive to event segmentation. Dissimilar results are usually an indication that one or both of the analysis strategies are flawed in some way; this feature, therefore, can be used as an internal quality control (QC) measure for an application.

5. Analytical Features

As a result of applying segmentation, each event can be represented by a set of numerical features described in Table IV.

Additional advanced features continue to be developed. These advanced features fall into two main categories. The first includes features that describe geometric or shape properties of the events, such as elongation, eccentricity, and the major-to-minor axis ratio. The second category captures the textural properties of events, i.e., the patterns of pixel intensity level within the event. Since texture features are a function of signal level, they may be computed for different channels.

I. Data Display, Gating, and Relating Image and Cytometric Data

The typical mode of viewing feature data is through histogram and scattergram graphs. This approach is familiar to flow cytometrists but is not as frequently employed by the imaging community. It can be demonstrated with data from a

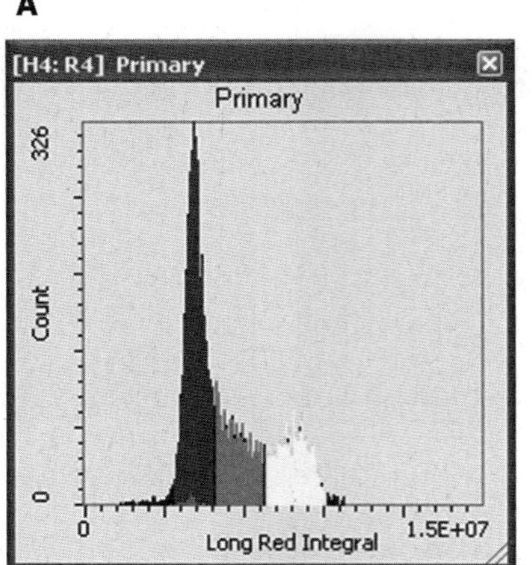
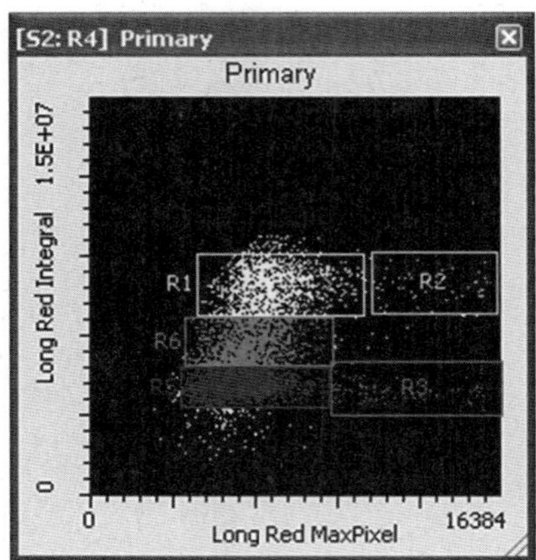

Fig. 15   Data display for a simple cell-cycle analysis: (A) DNA content histogram. (B) A scattergram of DNA content vs. chromatin condensation (as measured by the max pixel feature) showing gating of cell-cycle phases: G1 (blue) = R5, S (magenta) = R6, G2 (yellow) = R1, mitotic (cyan) = R2, and postmitotic cells (red) = R3. (See plate no. 12 in the color plate section.)

simple cell-cycle analysis of asynchronously cycling CHO cells stained with pro-pidium iodide (PI), an intercalating DNA dye.

A histogram is a frequency plot in which fluorescence intensity (or some other parameter) is plotted on the X-axis and the number of cells at each intensity level is plotted on the Y-axis. The histogram in Fig. 15A plots long red fluorescence (repre-senting DNA content) on the X-axis and cell count on the Y-axis. The blue peak represents the subpopulation of cells in G_1 of the cell cycle and the yellow area represents the subpopulation of cells in G_2. The long red fluorescence integral mean of this G_2 population is twice that of the G_1 population, reflecting the doubling of DNA content in these cells vs. those in G_1. The scattergram in Fig. 15B also displays DNA content, but here it is shown on the Y-axis. The X-axis is PI max pixel, a measure of the degree of condensation of the chromatin. When condensed, the chromatin presents much higher fluorescence intensity than in its uncondensed state. Thus the G_2 (yellow) population and the mitotic cell population (cyan) may be separated using the max pixel feature even though the total DNA content is the same for these two phases. Similarly, cells having just divided but whose DNA is still in the condensed state (red population) may be separated from the main G_1 cell population (shown in blue).

The regions providing coloration in these graphs are termed *gating* regions or simply *gates*, and may be defined to segregate any specific population of events for further analysis or review.

The power of LSC imaging cytometry technology stems from its ability to reliably, accurately, and reproducibly quantitate markers of interest and relate them to

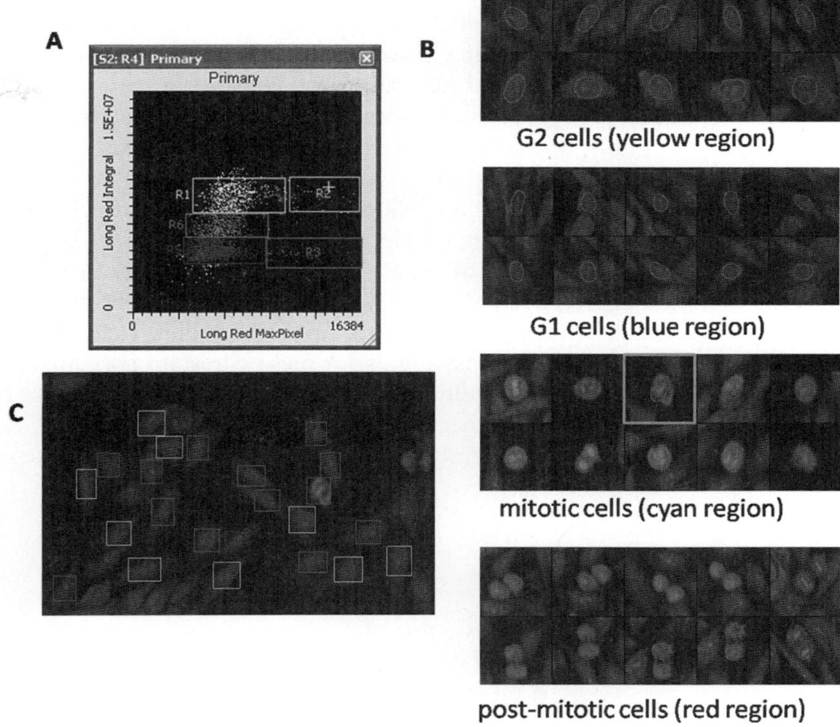

Fig. 16 Relating quantitative and image data. (A) Subpopulations corresponding to cell-cycle phases are gated and colored in the scattergram. (B) Image galleries are generated for those cells that lie within a specific gating region. A specific cell in the mitotic cell gallery (highlighted with blue rectangle) is located within the scattergram in 16A (white cross in gating region R2). (C) Individual cells in the field image are identified based on the gating region in which they lie. (See plate no. 13 in the color plate section.)

corresponding images. Using gating regions to identify event populations according to specified criteria allows the cytometric and image data for those populations to be displayed and related in the following ways (shown in Fig. 16):

1. Display of *galleries* of cells from particular gating regions
2. Colored overlays on cells in field and region images, reflecting the color of the gating region in which data from those cells lies
3. The location within scattergrams and histograms of specific cells identified in galleries or field images

J. iBrowser® Data Integration Software and iNovator Advanced Application Analysis Toolkit

Statistics and other parameters associated with the feature data can be exported to the iBrowser® Data Integration Software or to third-party applications for further analysis. These parameters include the count, mean, median, standard deviation, coefficient of variation, minimum, maximum, and sum selected for any feature, for the entire population or for any defined subpopulations. The iBrowser allows data display in a format resembling the physical format of the assayed sample, such as microtiter plate, tissue microarray (TMA), multiwell immunophenotyping slide, etc.

Fig. 17 presents data from the analysis of a breast TMA stained with anti-Her2/DAB and counterstained with hematoxylin. Panel A is the iBrowser display of the region images for the cores, arranged in the same array format as the TMA itself. The red areas are those of high Her2 expression. Panel B shows a similar iBrowser display, this time of the Kolmogorov–Smirnov (K-S) test. The K-S test measures the difference between the integrals of two histograms. Negative control cores are selected (in this case, cores 13,7 and 13,8, highlighted with rectangles in Fig. 17), the Her2 expression histograms for these cores are normalized and averaged, and the integral of the resulting curve is calculated. Similarly, the Her2 expression histograms of all the other cores are normalized and integrated. K-S curves (and values) are calculated by subtracting the averaged control integral from each of the

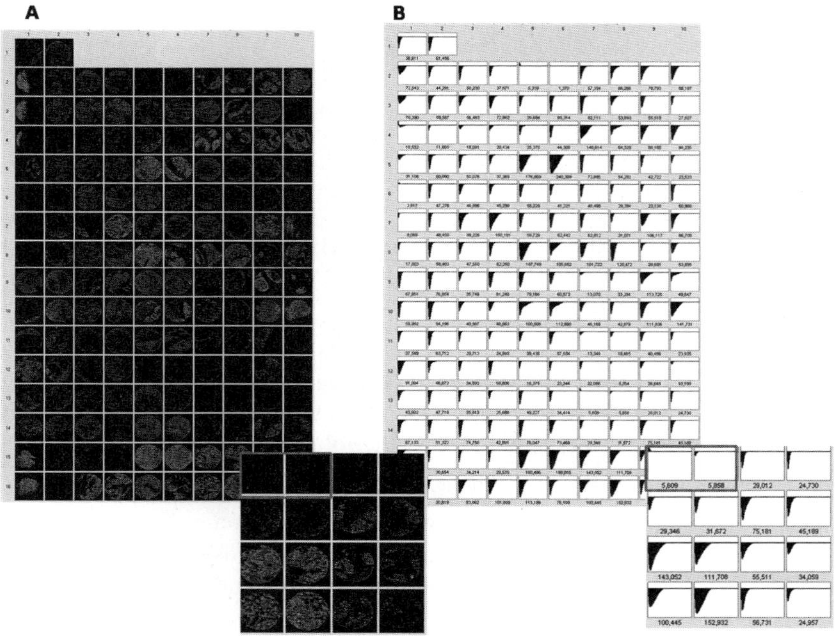

Fig. 17   iBrowser display of breast TMA stained with anti-Her2/DAB and counterstained with hematoxylin. (A) Region images of each core presented in the same array pattern as the core elements. Her2-positive areas appear as gray. (B) Kolmogorov–Smirnov test comparing the Her2 expression to the Her2-negative control cores (highlighted with rectangles in both panels). The magnitude and breadth of the downward peak indicates the relative level of Her2 expression.

individual core integrals. The result provides an excellent visual (as well as quantitative) measure of the degree to which the Her2 expression in each core differs from that of the control cores. The larger the downward sloping curve, the more the Her2 expression for that core differs from that of the control cores.

The iNovator Application Development Toolkit, an advanced software option, provides a variety of image analysis and processing tools for image enhancement, noise removal, and segmentation. These tools fall into the broad categories of image filtering, morphological operators, image operators, image feature enhancement tools, segmentation tools, and distance transforms.

The watershed segmentation method (Vincent and Soille, 1991) interprets a gray-scale image as a kind of topographic map for drawing segmentation boundaries. It is used widely for segmenting cells and nuclei; however, in some cases it may result in oversegmentation. To overcome this problem, a seeded watershed tool utilizes a set of markers (seeds) to aid in the segmentation process. In this way, only those potential events containing seeds are contoured as events. In the example in Fig. 18, breast tumor tissue stained with anti-Her2/DAB and counterstained with hematoxylin is segmented using the seeded watershed method in conjunction with the Frangi

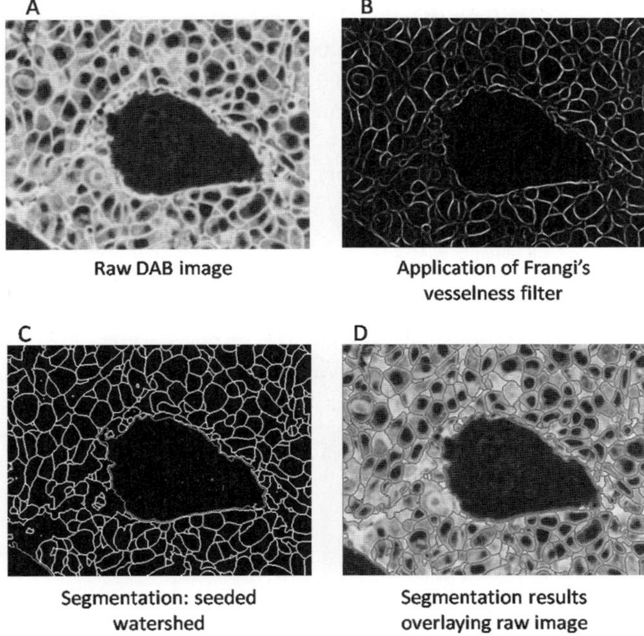

Fig. 18 Use of a ridge-enhancement filter and seeded watershed in segmenting cell membranes. (A) Raw image of a DAB-stained cell membrane. (B) The outcome of applying the Frangi vesselness filter. (C) Segmentation results using the image in (B) and the seeded watershed technique. (D) Membrane contours overlaid on the raw image in (A).

vesselness filter, which enhances the cell membrane signal and suppresses signals away from the membrane. The LoG filter is used to identify nucleus seeds that are then utilized by the seeded watershed segmentation tool.

K. Analytical Performance Characteristics

Monitoring analytical and imaging performance is essential in cellular and tissue biomarker development. iGeneration LSC systems have a proven performance track record and well-established procedures for daily and study-specific QC. System performance is assessed by measuring standard cytometric calibration microbeads (Beckman Coulter flow-check beads, Spherotech 8-peak rainbow beads, and BD Biosciences CaliBrite beads), along with some well-qualified biological samples (e.g., breast tissue chromatically stained with hematoxylin and antibody to Her2 protein developed with DAB).

Flow-check beads (Beckman Coulter, Fullerton, CA, # 9434167) are 10-μm diameter, broad-spectrum fluorescent beads produced in consistent size and dye loading within each large population. They are useful for determining both within-run precision and day-to-day precision of system fluorescence quantification. Fluorescence integral coefficients of variance should be less than 3% (Fig. 19A).

CaliBrite beads (BD Biosciences, Mountain View, CA, #340486, #340487) are labeled with FITC (blue-excited green fluorescence), phycoerythrin (blue-excited orange), and allophycocyanin (red-excited long red). The fluorescence levels of these beads are in a range typical of many samples, making these beads useful for monitoring day-to-day system fluorescence measurement. Mean of fluorescence intensity measurements should be within ±2 standard deviations (SDs) of the mean (Fig. 19B).

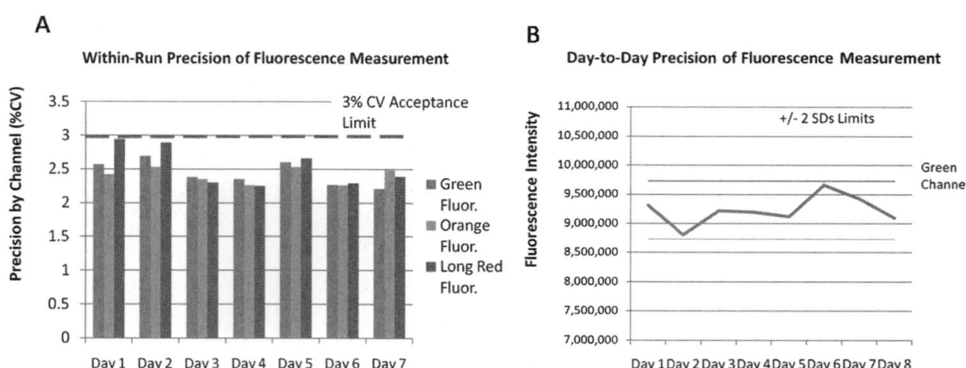

Fig. 19 (A) Within-run fluorescence precision for a period of 7 days, as assessed using Beckman Coulter flow-check beads. (B) Day-to-day fluorescence integral control chart showing performance of green-fluorescent BD CaliBrite beads over an 8-day period. Values are within the ±2 SD limits. (Data for orange and red channels show similar performance.) (For interpretation of the references to color in this figure legend, the reader is referred to the Web version of this chapter.)

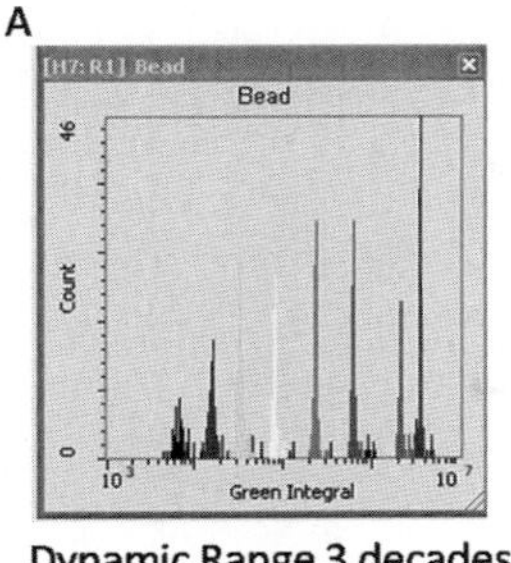

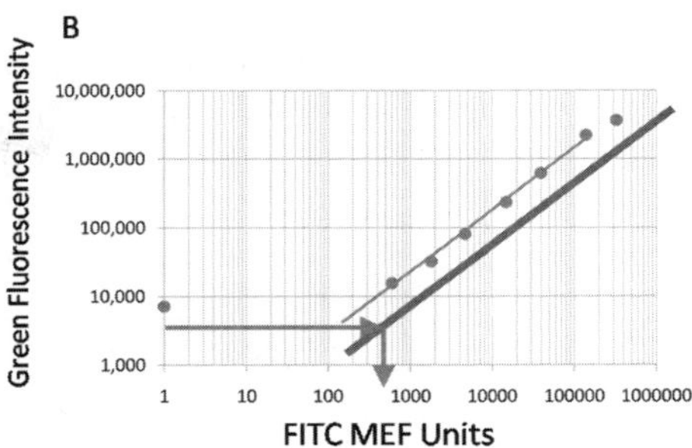

Fig. 20   LSC dynamic range and sensitivity assessment with 8-peak Spherotech rainbow beads. (A) Eight bead populations are measured at one PMT setting. (B) Seven of the eight peaks have fluorescence values established by their manufacturer. The eighth "negative" bead is assigned a value based on a calibration curve generated from the measurement of the other seven beads. Extrapolating the MEF value for this negative bead provides a measure of the lowest detectable signal.

Eight-peak rainbow calibration particles(Spherotech, Lake Forest, IL, #RCP-30-5A) contain a mixture of several similarly sized particles with different fluorescent intensities. The particles are 3-micron diameter beads, each containing a mixture of fluorophores. The Spherotech rainbow bead sample contains seven different populations of beads, each population containing different levels of fluorochrome. Additionally, a blank population is included in the sample. Although not a part of the regular QC procedure, these beads are used to determine system sensitivity and dynamic range. Sensitivity is assessed by ensuring that the fluorescence of the blank bead population is quantified at below 1000 molecules of equivalent fluorochrome (MEF) units. Dynamic range is assessed by ensuring that all of the bead populations can be distinctly resolved (separated), indicating a dynamic range of three decades (Fig. 20).

Chromatic tissue sections for the QC process. The TMA cores, labeled with antibodies to Her2 protein developed with DAB and counterstained with hematoxylin, were used in this example to determine day-to-day precision of laser light-loss measurements. Expected day-to-day precision on the Her2-positive and -negative tissue cores should be $< \pm 2$ SD (Fig. 21).

L. LSC Analysis Workflow

The iGeneration LSC technology employs graphical modular analysis protocols. These modules come "preassembled" into a variety of protocol templates that may be applied to different assays sharing similar analysis strategies. This modularized workflow allows for efficient, fully automated analysis. All protocol parameters can be easily adjusted by the operator if the application development process so requires.

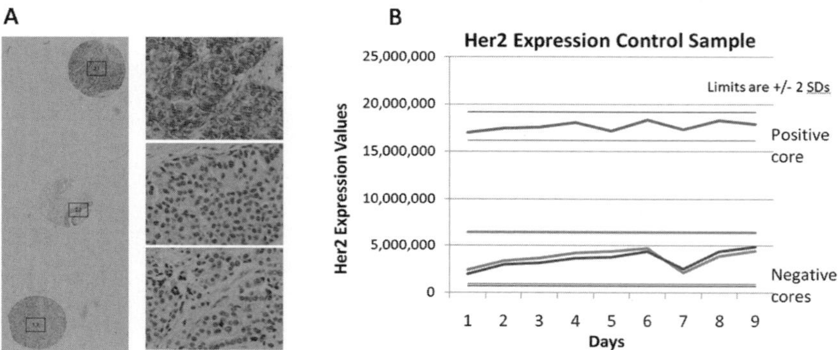

Fig. 21   (A) Three punch-cores of human breast tissue (Her2-negative and -positive) are labeled with Her2 antibody developed with DAB and counterstained with hematoxylin, for laser light-loss precision assessment. (B) Within-run chromatic precision for a period of 9 days.

Once an analysis strategy is completed, it can be saved for repeated use. By utilizing an optional robotic sample handler, multiple analyses may be run on up to 45 microtiter plates or 180 microscope slides in a walk-away fashion.

IV. Selected Application Areas of LSC

iGeneration LSC, pioneered by Dr. Louis Kamentsky, created the discipline of imaging cytometry and now is well integrated into it. The technology has become an essential tool for characterization and measurement of cells and cellular constituents when a specimen resides in a carrier or is attached to a horizontal surface. The array of LSC applications is vast, ranging from accurate quantification of the amount of biochemical cellular constituents (DNA, proteins, lipids, etc.), to spatial location and distribution of various markers, to morphological characteristics of cellular, subcellular, and super-cellular features (length, shape, texture, etc.). Several key distinguishing characteristics of the iGeneration LSC technology explain the speed of its adoption throughout life- and medical science disciplines:

1. Total signal quantification, permitting highly quantitative, biochemistry-like analysis of cellular and tissue specimens.
2. Ability to analyze label-free chromatically and fluorescently stained samples either simultaneously or sequentially. This feature alone allows a very high degree of multiplexing, often defined as high content analysis (HCA).
3. Highly developed cytometric analysis routines and advanced image analysis modalities.
4. Operator-adjustable, variable-image-resolution scanning, allowing optimization of assay throughput.

5. Flexibility in sample carriers: traditional sample carriers, such as microtiter plates, chamber slides, microscope slides, and Petri dishes, along with proprietary devices (e.g., rare-cell enrichment cartridges, immunophenotyping slides, and others) can be utilized.

Recent reviews have already provided thorough coverage of a range of LSC applications across a number of disciplinary fields (Harnett, 2007; Luther *et al.*, 2004; Pozarowski *et al.*, 2006). Some were specific to preclinical drug safety (Peterson *et al.*, 2008), while others to research cytology and pathology (Geddie, 2007; Luther and Geddie, 2008). The discussion that follows addresses both fundamental application categories and recent developments.

A. Cell Cycle

Cell-cycle analysis using direct, highly precise measurement of DNA content and chromatin condensation is one of the best examples to highlight the imaging, quantification, and multiplexing capabilities of the LSC technology (Chakraborty and Tansey, 2009; Jul-Larsen *et al.*, 2009; Kawamura *et al.*, 2004; Luther and Kamentsky, 1996; Zheng *et al.*, 2007). For simple cell-cycle analysis, LSC utilizes a single DNA dye. PI, DAPI, Feulgen, or any number of other stoichiometric nuclear dyes (both chromatic and fluorescent) can be used. Laser scan images of nuclear fluorescence or laser light-loss are generated and segmentation contours are drawn around the nuclei to determine the total nuclear fluorescence (fluorescence integral) or laser light absorption integral, which is proportional to the DNA content. (See the fluorescent example in Figs. 15 and 16.) LSC generates DNA histograms with sharp G_1 peaks, a definite S-phase, and a clearly separated G_2 population – all classic and robust measures of the quality of DNA content analysis. The max pixel feature provides a direct measurement of chromatin condensation, allowing mitotic cells to be separated from G_2 cells independent of a specific mitotic marker. Because cell-cycle data may be determined directly from DNA content using a single dye, other data acquisition channels can be used to explore cell-cycle related expression of other proteins of interest.

A representative example of a highly multiplexed LSC analysis of M-phase and mitotic cell states based on DNA, phospho-S10-histone H3, and either cyclin A2 or cyclin B1 expression is thoroughly described by Stefan and Jacobberger in Chapter 14 of this volume and elsewhere (Jacobberger *et al.*, 2008). A comprehensive review covering DNA content analysis methodology in individual cells and enumeration of cells in particular phases of the cell cycle by flow cytometry and LSC was published by Darzynkiewicz's group (Darzynkiewicz, Halicka, and Zhao, 2010). DNA content analysis by LSC is often used for DNA ploidy determinations in cellular specimens (Tsujioka *et al.*, 2008) or chromatin texture, DNA index, and S-phase fraction in tumor tissues (Kuliffay *et al.*, 2010; Mora *et al.*, 2007). In a recent study (Kawauchi *et al.*, 2010), LSC measurement of nuclear DNA content in touch-smear samples of surgically removed breast tumors revealed DNA copy number aberrations associated with aneuploidy and chromosomal instability.

B. DNA Damage, Apoptosis, and Senescence

Various forms of DNA damage, activation of repair pathways, and apoptosis or senescence are successfully assessed by LSC technology (Albino *et al.*, 2009; Ayllon and O'Connor, 2006; Dmitrieva, 2008; Holme *et al.*, 2007; Jorgensen *et al.*, 2010; Kato *et al.*, 2009; Rieber *et al.*, 2007a, 2007b; Sakaguchi and Steward, 2007; Tanaka *et al.*, 2009; Zhao *et al.*, 2007, 2009, 2008a, 2008b, 2010b.). Analyses typically take advantage of stoichiometric DNA stains, the availability of antibodies detecting specific molecular biomarkers, and LSC's ability to quantitatively assess changes in nuclear morphology. An extensive body of publications addresses direct assessment of DNA fragmentation, activation of checkpoint kinases, phosphorylation of H2AX, ATM, and ATR (Darzynkiewicz, Halicka and Zhao, 2010; Darzynkiewicz and Zhao, 2010; Kunos *et al.*, 2010; Tanaka *et al.*, 2007a, 2007b), phosphorylation and transcriptional activation of p53, caspase activation (Darzynkiewicz, Pozarowski, *et al.*, 2010; Stegh *et al.*, 2008; Ghavami *et al.*, 2010), and other related topics (Miranda-Carboni *et al.*, 2008).

One of the simplest and most effective methods for evaluating DNA damage remains the comet assay. Fig. 22 illustrates an LSC analysis identifying comet heads and tails, and calculating the amount of DNA present. Comet sizes increase proportionately with the increase in concentration of the DNA-damaging agent (etoposide). (Sample is courtesy of Trevigen Corporation.) An LSC-generated comet assay index measured the total DNA content of the entire cell (head and tail) and that of the tail alone, and then calculated the percentage of overall DNA content found in the tail. The higher the percentage of DNA content in the tail, the more DNA damage has occurred. Results showed excellent correlation with Trevigen reference data calculated in an identical manner.

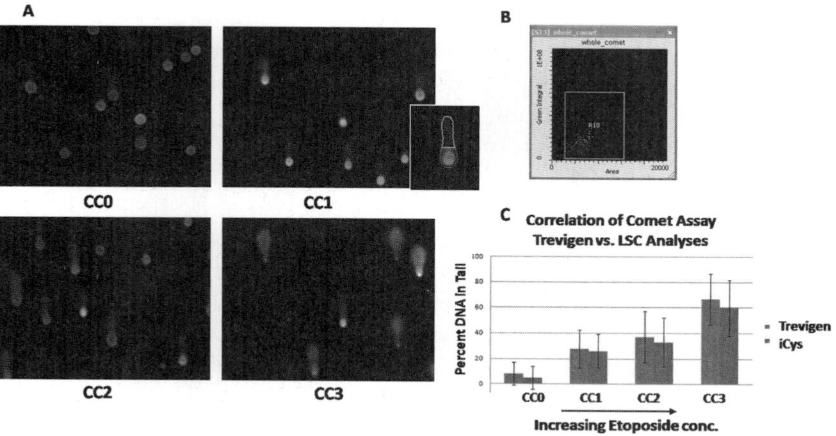

Fig. 22 Comet assay: (A) field images of control cells (top left) and cells treated with increasing concentrations of etoposide showing comet "tail" indicative of DNA damage. Inset shows segmentation of the comet head and tail. (B) Valid cells are segregated based on total cell DNA content and area. (C) Comparison of LSC analysis with Trevigen reference data.

Because LSC technology is particularly well-suited for multiplexed quantitative analysis, it is commonly used for high-content assays incorporating multiple markers (e.g., detection of DNA content and DNA double-strand breaks (DSBs) by labeling cells with antibody to phosphorylated histone H2AX and p53 Binding Protein 1 (53BP1). Fig. 23 shows an analysis of A549 cells treated with gamma radiation, compared to untreated controls. Individual cellular events are identified based on DNA (DAPI) staining. Scattergrams of DAPI vs. H2AX or 53BP1 fluorescence measure the intensity and number of individual foci. Individual foci as well as pan-nuclear staining can be quantified and related to their cell-cycle phase.

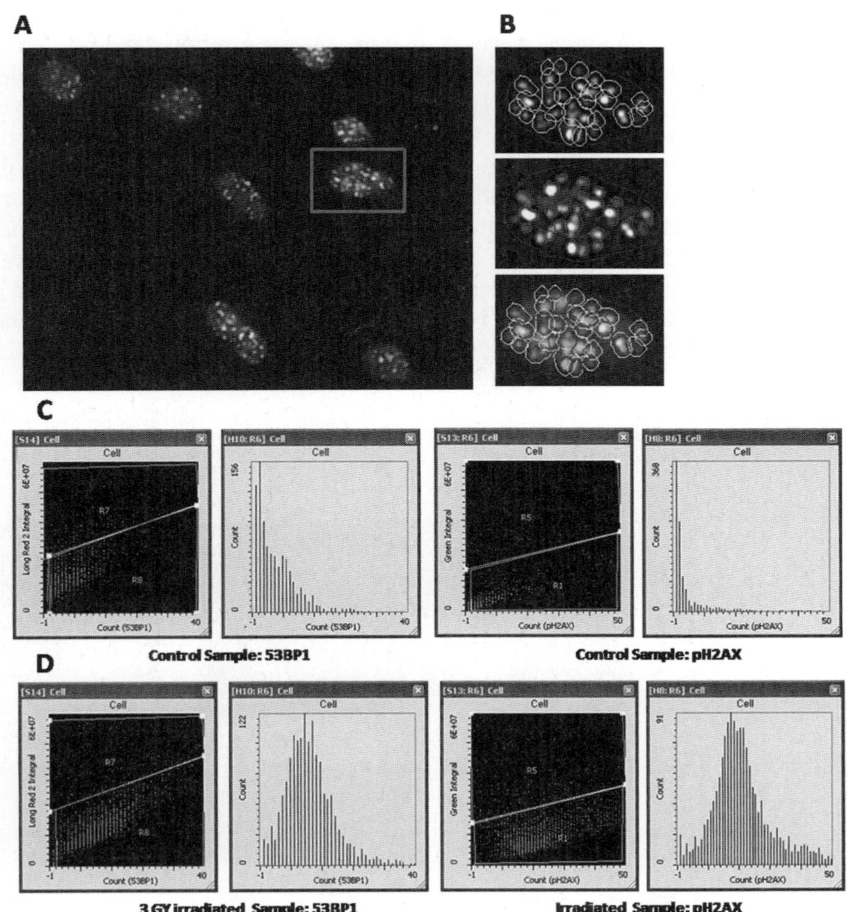

Fig. 23   (A) Laser scan image of A549 cells showing pH2AX and 53BP1 expression foci. (B) Contours on DNA of entire cell as well as pH2AX foci (top), 53BP1 foci (middle), and both foci (bottom). (C) Scattergram of total expression vs. foci count and histogram of foci count for control sample: 53BP1 left, pH2AX right, and (D) 3GY irradiated sample.

LSC technology also provides a novel approach for automated scoring of micronuclei (MN) in different types of mammalian cells, serving as a biomarker of genotoxicity and mutagenicity (Darzynkiewicz, Smolewski *et al.*, 2011). The technology's
flexibility fosters a diversity of MN applications: simultaneous measurement of multiple features such as DNA content and cytoplasmic and nuclear area; mono- and
multinucleated cells; nuclear texture and intensity of nuclear and MN staining; protein
content and cytoplasm density; and additional features using relevant molecular
probes. These investigations can be carried out in cell lines, buccal cells, erythrocytes,
dermal model systems, and a number of other specimen types. (See the detailed review
of buccal cell analysis by Leifert *et al.* in Chapter 13 of this volume.)

LSC technology successfully overcomes challenges presented by the complexity
of biological specimens. Fig. 24, for example, shows a buccal mucosa specimen.
Note the cellular heterogeneity typical of this specimen type, with numerous cells
not completely separated. Advanced imaging techniques available in the iNovator
software accurately define the cytoplasmic boundaries of individual cellular events
within a challenging clump of cells. Importantly, this feature does not rely on the cell
nucleus to provide a "seed" for cytoplasmic segmentation. Instead, the seed is
developed from the cytoplasmic staining itself, allowing this method to be used to
define karyolitic cells and removing any ambiguity that would otherwise be associated with nuclear seeds in binucleated cells. Well-defined single cells are identified for analysis, while the remaining cell clumps can be "gated out" based on larger
cytoplasmic area and noncircularity. MN assays by LSC are increasingly used in
routine toxicology screening and genome health studies. New developments focused
on MN analysis in exfoliated cells and dermal cell models are being introduced.

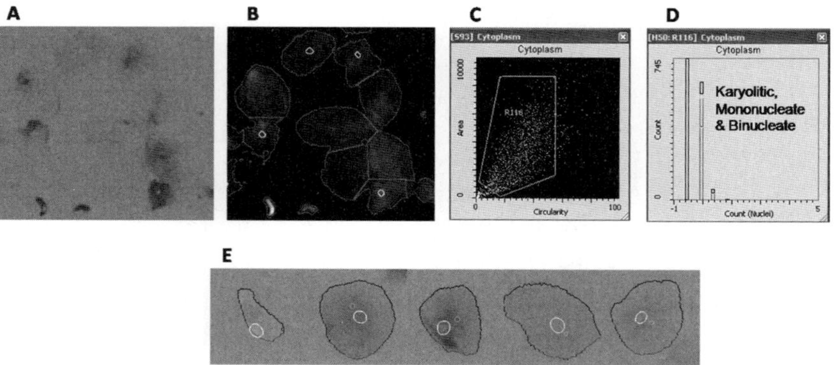

Fig. 24   (A) Colored composite image showing buccal cell heterogeneity: basal cells (dark cytoplasmic
staining), differentiated cells (lighter cytoplasmic staining), karyolitic cells (without nuclei). (B) Seeded
watershed segmentation using the cytoplasmic staining to generate a seed. (C) Valid cells are gated based
on area and circularity and (D) karyolitic, mono- and binucleated cells are identified. (E) Cells with MN
are identified by a separate part of the protocol (not shown) and displayed. (For interpretation of the
references to color in this figure legend, the reader is referred to the Web version of this chapter.)

The strength of LSC technology for nuclear morphometry has generated a novel approach to quantifying senescence-associated alterations in cell morphology and measuring β-galactosidase activity and cyclin-dependent kinase inhibitors p21[WAF1] and p27[KIP1] (Zhao *et al.*, 2010a). Taking advantage of LSC's multiparameter analysis capabilities, the ratio of max pixel to nuclear area was calculated, morphometric nuclear changes and CDK inhibitor expression were measured on a cell-by-cell basis, and the results were correlated with the senescence morphology phenotype. Quantification of senescence in cells treated with variable concentrations of inducers or inhibitors of cell senescence was successfully achieved. Fig. 25 illustrates mitoxantrone-induced, senescence-associated β-galactosidase activity in A549 cells. (Sample is courtesy of Drs. Zhao and Darzynkiewicz.)

LSC studies by researchers at the Beckman Research Institute of City of Hope (Duarte, CA) resulted in the development of a novel fluorescence-based multiparameter assay for assessing islet cell composition and β-cell apoptosis by TUNEL assay in undissociated whole islet preparations prior to islet transplantation (Todorov *et al.*, 2010). As the number of apoptotic β-cells was found to correlate inversely with success in reversing diabetes in mice, the need for a robust multiparameter assay combining apoptosis analysis with quantification of insulin, amylase,

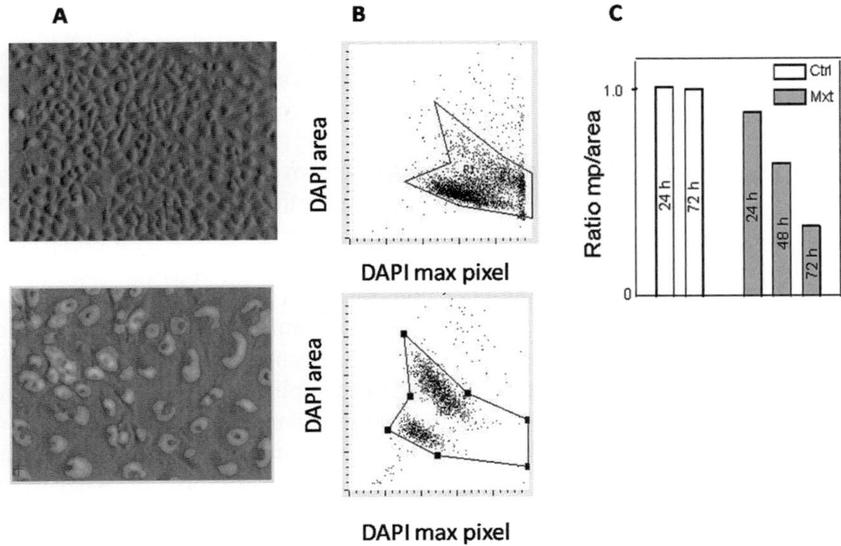

Fig. 25   Cell senescence analysis by LSC. (A) Senescence-associated b-galactosidase activity in A549 cells: 72-h cell growth in control (top) and in the presence of 2nM mitoxantrone (bottom). Note that b-galactosidase-positive cells are larger and have lower saturation density. (B) Scattergrams of nuclear (DNA/DAPI) area vs. max pixel for control and 72-h 2 nM mitoxantrone treatment showing increase in area and decrease in max pixel for the treated cells. (C) Ratio of max pixel and area for control and treated cells at three time points. A significant effect is seen in the treated cells. (Sample courtesy of Z. Darzynkiewicz, Scattergrams and bar chart, Ref. Zhao *et al.*, 2010a)

glucagon, somatostatin, and cytokeratin markers became critical. The result was development of a robust, highly reproducible quantitative assay. The total time required for the assay, from islet sectioning and sample preparation until results are available, is less than 24 h, allowing results to be available in real time before the start of transplantation procedures. Characterization of human pancreatic cells (Ichii *et al.*, 2007, 2008; Iglesias *et al.*, 2008; Omori *et al.*, 2007, 2010; Sweet *et al.*, 2008) and canine islets (Ito *et al.*, 2008) have also been reported.

C. Automated Tissue and TMA Analysis

A wide range of tissue analysis applications are performed on the iGeneration LSC instruments, from routine enumeration of cellular events expressing marker(s) of interest to cutting-edge quantitative *in-situ* protein expression analysis and multi-color colocalization studies (Faiola *et al.*, 2008; Friedman *et al.*, 2009; Harnett, 2007; Hjelmeland *et al.*, 2010; MacDonald *et al.*, 2010; Olive *et al.*, 2009; Schwock *et al.*, 2007; Swanberg *et al.*, 2009; Tellez *et al.*, 2007; Williams *et al.*, 2009; Zoog *et al.*, 2008, 2010; Takahashi *et al.*, 2009).

Automated "scoring" of protein expression in tissue specimens is a typical application for tissue analysis. The essence of a simple analysis routine is described in Fig. 26, depicting an assessment of SMAD4 expression in pancreatic tissue. This figure shows a typical two-scale automated workflow in which samples are rapidly scanned for a quick overview of morphology and any staining artifacts. Regions for high-resolution scanning are then selected either manually or in automated fashion. Compensation is applied to isolate the dye signals. The linear unmixing method of compensation reassigns spectral overlap signals to the compensated channels for each dye, retrieving what would otherwise be "lost" signal. (See the discussion of compensation, above.) Data from these high-resolution scans is plotted and exported for further analysis.

LSC's flexibility enables its use in truly high-content tissue applications where the simultaneous analysis of a large number of targets at high analytical precision is required. The ability to assay simultaneously in both fluorescent and chromatic modes further extends its broad applicability and is essential in tissue analysis, since chromatic dyes remain a mainstream approach in pathology while fluorescent dyes are used often in research and biomarker discovery applications. An example of such high-content multiplexed analysis was described earlier in the five-color pancreas tissue assay (Fig. 10) and is also presented by David Krull in Chapter 11 of this volume. The example in Fig. 26, which utilizes three chromatic dyes, also makes use of the fluorescent properties of permanent red to allow data for all three targets to be acquired simultaneously, increasing the assay throughput. Automated acquisition and analysis allows the flexibility to employ LSC technology in a walk-away fashion for established assays while still providing for hands-on custom assay development. The easy transition from custom development to validated assay makes LSC readily adaptable to regulated environments.

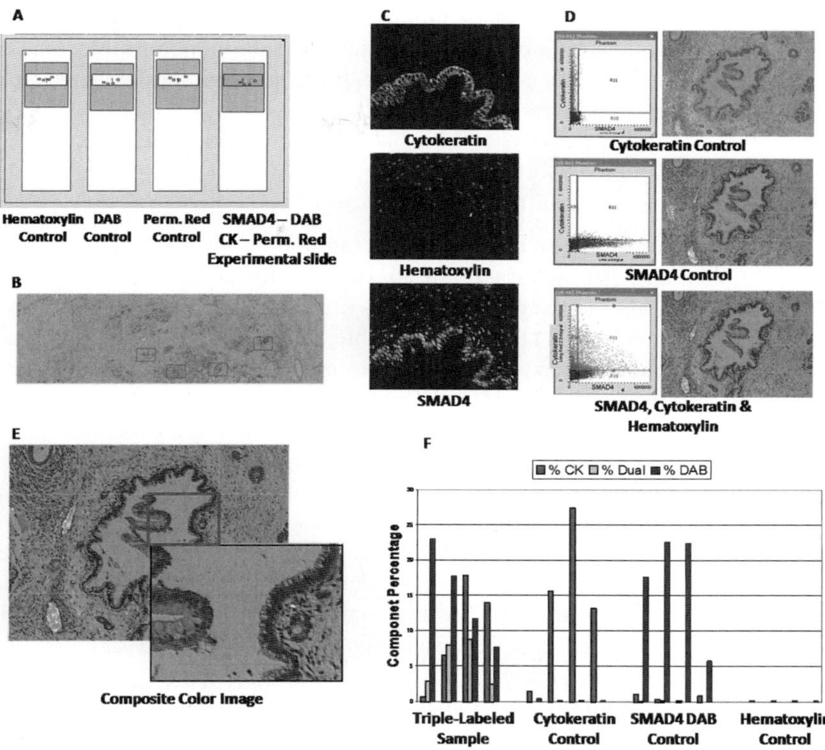

Fig. 26 Automated analysis of SMAD4 expression in epithelial pancreatic tissue sections. (A) The experimental sample stained with cytokeratin/permanent red, SMAD4/DAB, and counterstained with hematoxylin is analyzed along with three control samples, each singly stained and counterstained. (B) A fast overview scan is performed on all the slides, shown here for the experimental sample. Areas are selected for high-resolution scanning. (C) Data are acquired at high resolution for all three markers simultaneously. Permanent red is measured as a fluorescent dye. Acquired channels are compensated to isolate the three dye signals. (D) Gating boundaries are set based on the single-stain controls. (E) The individual compensated images are assembled into a composite color image. The inset shows the 3D-like appearance created using a process unique to LSC. (F) Data are exported and the percentage of area that is positive for SMAD4, for cytokeratin, and for both is plotted for each of the four slides. (Sample courtesy of John Mahoney and Alexei Protopopov, Dana Farber Cancer Institute, Boston, MA.) (For interpretation of the references to color in this figure legend, the reader is referred to the Web version of this chapter.)

Automated analysis of tissue specimens is the subject of extensive reviews (Harnett, 2007; Peterson, 2008). This volume contains several chapters dedicated to this subject, specifically by Gloria Juan (Chapter 12), David Krull and Richard Peterson (Chapter 11), and Margaret Harnett's group (Chapter 9).

Among notable new developments is the program initiated by Leslie Silberstein at Children's Hospital, Boston, MA (Nombela-Arrieta *et al.*, 2008). LSC is employed to objectively quantitate the spatial distribution of the anatomical localization of hematopoietic stem and progenitor cells (HSPCs) in bone marrow niches. The entire cellular content of nondecalcified longitudinal femoral bone sections is mapped with single-cell-level precision, making it possible to objectively quantify and statistically compare the localization of cellular bone marrow populations, including extremely rare populations.

D. Cell Surface and Tissue Immunophenotyping

Immunophenotyping, a common application in flow cytometry, allows multiple cell surface markers to be simultaneously characterized on a per-cell basis. Immunophenotyping can be difficult by flow cytometry, however, when only a small number of cells are available. In these cases, LSC analysis is a methodology of choice because of its low sample requirements. Immunophenotyping by LSC, originally developed by Richard Clatch (Clatch, 2001), is based on a custom-made carrier, the Clatch slide (Fig. 27A). Specimens derived from fine-needle aspirates, body fluid, tissue, or blood are introduced into 12 "lanes" or wells and stained with different antibody-conjugate combinations. The lymphocytes are gated from other leukocytes based on cell area and CD45 expression. Data for each "lane" of the Clatch slide are plotted in its own scattergram, dividing the cells into four subpopulations. Cell immunophenotype can be linked to cell morphology, based on the images generated during analysis (Fig. 27). Because the cells are left *in situ* during LSC analysis, they can be reprocessed afterward for cytoplasmic, immunological, or genetic markers.

While flow cytometry remains the standard for peripheral blood immunophenotyping, several independently performed correlation studies have shown that the results produced by LSC are equivalent to those from flow cytometry. The most

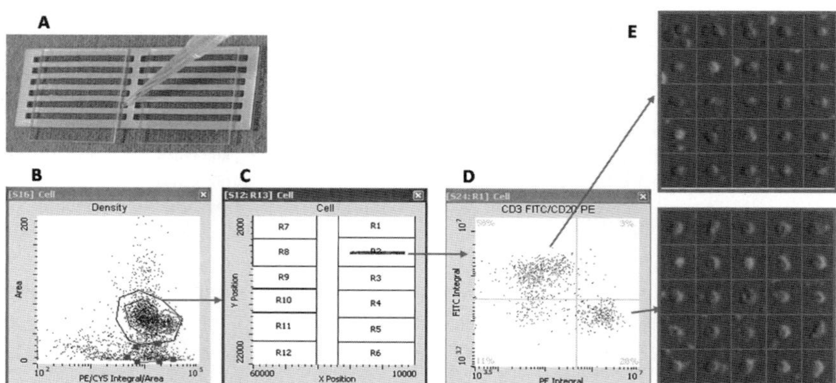

Fig. 27 (A) The Clatch slide. (B) Lymphocytes are isolated from leukocytes based on cell area and CD45 expression. (C) Each "lane" of the Clatch slide is plotted in its own scattergram (D), where the quadrants are defined for FITC+/PE+, FITC-/PE+, FITC+/PE-, and FITC-/PE-. (E) Cell galleries allow an examination of cell morphology for any of the subpopulations.

recent and comprehensive reviews (Al-Za'abi *et al.*, 2008; Geddie, 2007) revealed agreement between flow cytometry and LSC in classification of immunophenotype as reactive vs. abnormal and presented specific clinical cases.

Because LSC allows *in-situ* analysis, immunophenotyping of tissue sections or TMAs is one of the most frequently performed applications. An example of such analysis is shown in Fig. 28. A human multitumor TMA (Invitrogen 75-4053) was analyzed for kappa and lambda light chains. The slide was chromatically stained with antibodies to kappa and lambda chains, developed with DAB (kappa) and permanent red (lambda) and counterstained with hematoxylin. Nuclei were segmented based on hematoxylin staining and kappa and lambda expression per cell was measured. Gating boundaries were set to define elevated expression levels, and the number of positive cells per core was reported for each marker.

Rare-cell detection and immunophenotype determination, for example, in circulating tumor cells, is performed using LSC in both research applications and near-clinical settings (Camara *et al.*, 2007; Goodale *et al.*, 2008; Hamamoto *et al.*, 2007; Pachmann *et al.*, 2008; Sanoslo *et al.*, 2010). LSC's ability for multiparameter and

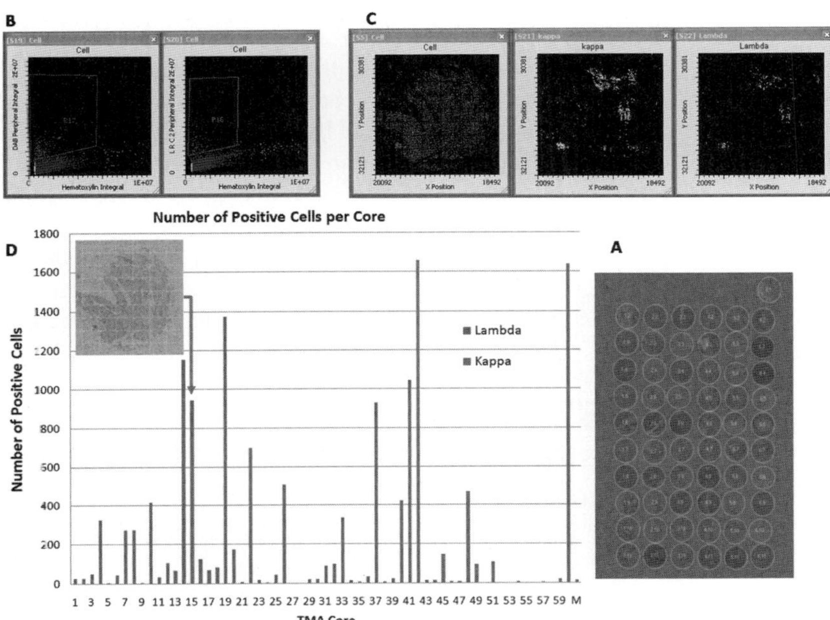

Fig. 28   (A) Overview scan of 61-core TMA. Each core is scanned at high resolution. (B) Scattergrams showing kappa (left) and lambda expression within gating regions. (C) XY map showing location of kappa- and lambda-positive cells; isolated kappa- and lambda-positive cells in XY maps. (D) Histogram of number of kappa- and lambda-positive cells allows easy correlation to each core of the TMA. A high-resolution mosaic of Core 15 is shown in the inset.

morphological analysis and cellular localization of antigens, as well as its high sensitivity and relatively small-volume sample requirements, recommend it for these types of analyses. Lowes *et al.* address this subject in detail in this volume in Chapter 10.

E. Live–Cell Analysis

To enable extended live-cell analysis, a stage-mounted environmental chamber (Tokai Hit INU series) has been recently validated on the LSC. Similar environmentally controlled stage inserts from other manufacturers should be easily adaptable as well. Included in the Tokai Hit validation were assays such as cell growth, phagocytosis, and apoptosis using caspase-3 FRET sensors. Fig. 29A shows initial and 24-h scans of unstained HeLa cells. Figs. 29B and 29C show phagocytosis of beads that become fluorescent within the environment of the phagocyte. Coupled with previously established LSC features such as repetitive scanning, file merging, and event positioning, along with LSC's quantitative precision, the introduction of on-board environmental controls opens up an entirely new area of HCA for the technology.

By making use of the Merge Runs feature described in Section III.F, live cells can be initially analyzed, then restained or fixed, followed by repeated live- or fixed-cell scanning. The acquired data from several scans performed on the same cells can then be combined into one file so that multiple parameters from different scans can be analyzed. For example, by combining measurement of mitochondrial membrane potential in apoptosis-induced live cells with DNA content measurement in the same cells after fixation, a reduction in mitochondrial membrane potential can be correlated with S-phase of the cell cycle. In a representative example (Li and Darzynkiewicz, 1999), mitochondrial membrane potential measured before fixation was correlated with both DNA content and DNA strand-break measurements made on the same cells after fixation. Thus, file merging provides yet an additional level of multiplexing: live- and fixed-cell parameter measurement on the same sample.

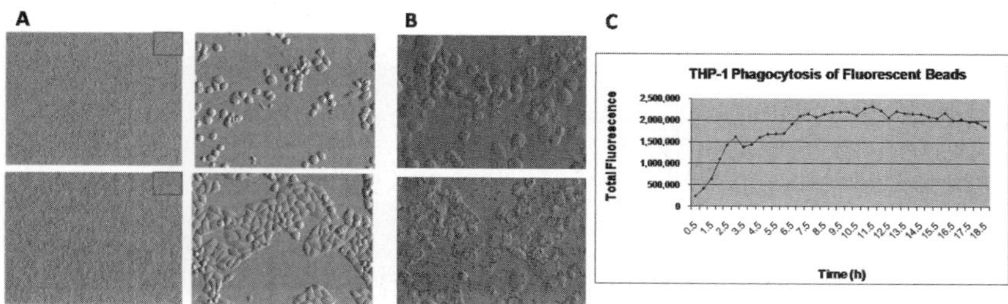

Fig. 29 (A) Live unstained HeLa cells: region image and selected field image from initial scan (top) and after 24 h. (B) Phagocytosis of pH-dependent fluorescent beads by THP-1 phagocytes. Beads in media show little if any fluorescence (top). Phagocytosed beads exhibit fluorescence (bottom). (C) Total fluorescence of sample increases with time. (Samples courtesy of Natasha Barteneva, Immune Disease Institute, Boston, MA.)

V. Concluding Remarks

LSC represents the first entrant into the field of QIC. Since its introduction, it has evolved into the iGeneration LSC product line of iCyte®, iCys®, and iColor® imaging cytometers, offering a powerful combination of performance features. Its analytical capabilities are characterized by high sensitivity, precision, and dynamic range. On the imaging side, it offers variable-resolution scanning with a high-resolution option.

LSC's ability to analyze various sample types, representing practically any cellular or tissue specimen attached to a horizontal surface, makes it the platform of choice for research environments where flexibility is important and applications are constantly changing. The technology works well as a complement to flow cytometry, which is limited to the analysis of cells in suspension, and to confocal microscopy, with its very high resolution imaging. As a result, it is recognized as an essential tool for basic life science research in any flow cytometry or imaging core facility. Although it is not FDA-approved for clinical use, LSC technology has been validated for regulated studies performed under good laboratory practice (GLP) guidelines, resulting in its utilization in drug safety and efficacy studies and in biomarker discovery programs.

Active developments in the field of QIC by LSC have led to the establishment of the Quantitative Imaging Cytometry Centers of Excellence program (www.imagingcytometrycenter.com), a network of educational and scientific exchange resources focusing on training programs and symposia for scientists employing quantitative imaging cytometry in their research. The culmination of the Centers' activities is the annual QIC Symposium, fostering cross-disciplinary collaboration in areas of research requiring cell- and tissue-based analysis, most recently focusing on cancer and stem cell research programs (Mei *et al.*, 2009, 2010).

References

Al-Za'abi, A. M., and Geddie, W. B., *et al*. (2008). Equivalence of laser scanning cytometric and flow cytometric immunophenotyping of lymphoid lesions in cytologic samples. *Am. J. Clin. Pathol.* **129**(5), 780–785.

Albino, A. P., Jorgenson, E. D., and Rainey, P., *et al*. (2009). H2AX: A potential DNA damage response biomarker for assessing toxicological risk of tobacco products. *Mutat. Res.* **678**, 43–52.

Ayllon, V., and O'Connor, R. (2006). PBK//TOPK promotes tumour cell proliferation through p38 MAPK activity and regulation of the DNA damage response. *Oncogene* **26**(24), 3451–3461.

Basiji, D.A., Ortyn, W.E., Liang, L., et al. (2007). Cellular image analysis and imaging by flow cytometry. *In* "Clin Lab Med." Vol. 27(3), pp. 653–viii. Amnis Corporation, Seattle, Washington.

Camara, O., Rengsberger, M., and Egbe, A., *et al*. (2007). The relevance of circulating epithelial tumor cells (CETC) for therapy monitoring during neoadjuvant (primary systemic) chemotherapy in breast cancer. *Ann. Oncol.* **18**, 1484–1492.

Chakraborty, A. A., and Tansey, W. P. (2009). Inference of cell cycle-dependent proteolysis by laser scanning cytometry. *Exp. Cell Res.* **315**(10), 1772–1778.

Clatch, R. J. (2001). Immunophenotyping of hematological malignancies by laser scanning cytometry. Vol. 64, pp. 313–342. Elsevier Inc, San Diego, CA.

Coons, A. H., and Kaplan, M. M. (1950). Localization of antigen in tissue cells. II. Improvements in a method for the detection of antigen by means of fluorescent antibody. *J. Exper. Med.* **91**, 1–13.

Coulter, W. H. (1956). High speed automatic blood cell counter and cell size analyzer. *Proc. Natl. Electron. Conf.* **12**, 1034.

Darzynkiewicz, Z., Halicka, D., and Zhao, H. (2010). Analysis of cellular DNA content by flow and laser scanning cytometry. *In* "Polyploidization in Cancer," (and R. Poon, ed.), Landes Bioscience, Austin, TX.

Darzynkiewicz, Z., and Pozarowski, P., *et al.* (2010). Fluorochrome-labeled inhibitors of caspases: convenient in vitro and in vivo markers of apoptotic cells for cytometric analysis. *In* "DNA Damage Detection *In Situ, Ex Vivo*, and *In Vivo*: Methods and Protocols, Methods in Molecular Biology," (and V. Didenko, ed.), Vol. 682, Springer Science, Berlin.

Darzynkiewicz, Z., and Zhao, H. (2010). Detection of DNA strand breaks in apoptotic cells by flow- and image-cytometry. *In* "DNA Damage Detection *In Situ*, *Ex Vivo*, and *In Vivo*: Methods and Protocols, Methods in Molecular Biology," (and V. Didenko, ed.), Vol. 682, Springer Science, Berlin.

Darzynkiewicz, Z., and Smolewski, P., *et al.* (2011). Laser scanning cytometry for automation of the micronucleus assay. *Mutagenesis* **26**(1), 153–161.

Dmitrieva, N. I. (2008). Analysis of DNA breaks, DNA damage response, and apoptosis produced by high NaCl. *Am. J. Physiol. Renal Physiol.* **295**, F1678–F1688.

Faiola, B., Falls, J. G., and Peterson, R. A., *et al.* (2008). PPAR alpha more than PPAR delta, mediates the hepatic and skeletal muscle alterations induced by the PPAR agonist GW0742. *Toxicol. Sci.* **105**, 384–394.

Friedman, B., Shachtrup, C., and Tsai, P. S., *et al.* (2009). Acute vascular disruption and aquaporin 4 loss after stroke. *Stroke* **40**(6), 2182–2190.

Fulwyler, M. J. (1965). Electronic separation of biological cells by volume. *Science* **150**, 910.

Geddie, William R. (2007). Cytology and laser scanning cytometry. *In* "Wintrobe's Atlas of Clinical Hematology," (D. Tkachuk, and J. V. Hirschmann, eds.), Lippincott Williams & Wilkins, Baltimore, MD.

Ghavami, S., *et al.* (2010). Statin-triggered cell death in primary human lung mesenchymal cells involves p53-PUMA and release of Smac and Omi but not cytochrome c. *Biochem. Biophys. Acta* **1803**(4), 452–467.

Goodale, D., Phay, C., and Postenka, C., *et al.* (2008). Characterization of tumor cell dissemination patterns in preclinical models of cancer metastasis using flow cytometry and laser scanning cytometry. *Cytometry A* **75**(4), 344–355.

Hamamoto, T., Suzuki, K., and Kodama, H., *et al.* (2007). Correlation of malignant phenotypes of human tumour cell lines with augmented expression of Hsp72 protein measured by laser scanning cytometry. *Int. J. Hyperthermia* **23**(4), 363–370.

Harnett, M. (2007). Laser scanning cytometry: Understanding the immune system in situ. *Nature Rev. Immunol.* **7**, 897–904.

Hjelmeland, L. M., and Fujikawa, A., *et al.* (2010). Quantification of retinal pigment epithelial phenotypic variation using laser scanning cytometry. *Mol. Vision* **16**, 1108–1121.

Holme, A. L., Yadav, S., and Pervaiz, S. (2007). Automated laser scanning cytometry: a powerful tool for multi-parameter analysis of drug-induced apoptosis. *Cytometry A* **71**(2), 80–86.

Iglesias, I., Bentsi-Barnes, K., and Umeadi, C., *et al.* (2008). Comprehensive analysis of human pancreatic islets using flow and laser scanning cytometry. *Transplant. Proc.* **40**(2), 351–354.

Ichii, H., Miki, A., and Yamamoto, T., *et al.* (2008). Characterization of pancreatic ductal cells in human islet preparations. *Lab Invest.* **88**(11), 1167–1177.

Ichii, H., Sakuma, Y., and Pileggi, A., *et al.* (2007). Shipment of human islets for transplantation. *Am. J. Transplant.* **7**(4), 1010–1020.

Ito, T., Omoro, K., and Rawson, J., *et al.* (2008). Improvement of canine islet yield by donor pancreas infusion with a p38MAPK inhibitor. *Transplantation* **86**, 321–329.

Jacobberger, J. W., Frisa, P., and Sramkoski, R., *et al.* (2008). A new biomarker for mitotic cells. *Cytometry A* **73**(1), 5–15.

Jorgensen, E. D., and Zhao, H., *et al.* (2010). DNA damage response induced by exposure of human lung adenocarcinoma cells to smoke from tobacco- and nicotine-free cigarettes. *Cell Cycle* **9**(11), 2170–2176.

Jul-Larsen, A., Grudic, A., Bjerkvig, R., and Ove Boe, R. (2009). Cell-cycle regulation and dynamics of cytoplasmic compartments containing the promyelocytic leukemia protein and nucleoporins. *J. Cell Sci.* **122**, 1201–1210.

Kamentsky, L. A. (2001). Laser scanning cytometry. *In* "Methods in Cell Biology" **63**, 51-87.

Kamentsky, L. A., Derman, H., and Melamed, M. R. (1963). Ultraviolet Absorption in Epidermoid Cancer Cells. *Science* **142**, 1580–1583.

Kamentsky, L. A., and Kamentsky, L. D. (1991). Microscope-based multiparameter laser scanning cytometer yielding data comparable to flow cytometry data. *Cytometry* **12**, 381–387.

Kamentsky, L. A., and Melamed, M. R. (1967). Spectrophotometric cell sorter. *Science* **156**, 1364–1365.

Kato, M., Putta, S., and Wang, M., *et al.* (2009). TGF-B activates Akt kinase through a micro RNA-dependent amplifying circuit targeting PTEN. *Nature Cell Biol.* **11**(7), 881–889.

Kawamura, K., *et al.* (2004). Induction of Centrosome Amplification and Chromosome Instability in Human Bladder Cancer Cells by p53 Mutation and Cyclin E Overexpression. *Cancer Res.* **64**, 4800–4809.

Kawauchi, S., and Furiya, T., *et al.* (2010). DNA copy number aberrations associated with aneuploidy and chromosomal instability in breast cancers. *Oncol. Rep.* **24**(4), 875–883.

Kuliffay, P., Sanislo, L., and Galbavy, S. (2010). Chromatin texture, DNA index, and S-phase fraction in primary breast carcinoma cells analysed by laser scanning cytometry. *Bratis Lek Listy* **111**(1), 4–8.

Kunos, C., Radivoyevitch, T., and Pink, J., *et al.* (2010). Ribonucleotide reductase inhibition enhances chemoradiosensitivity of human cervical cancers. *Radiat. Res.* **174**(5), 574–581.

Li, X., and Darzynkiewicz, Z. (1999). The Schrödinger's cat quandary in cell biology: integration of live cell functional assays with measurements of fixed cells in analysis of apoptosis. *Exp. Cell Res.* **249**(2), 404–412.

Lodish, Harvey F. (2008). *Molecular Cell Biology.* St. Martin's Press, New York, p. 387.

Luther, E., and Geddie, W. (2008). Laser scatter in clinical applications. *Proc. SPIE* **6864**, 686412; DOI:10.1117/12.763996, Online Publication Date: 15 February 2008.

Luther, E., and Kamentsky, L. A. (1996). Resolution of mitotic cells using laser scanning cytometry. *Cytometry* **23**, 272–278.

Luther, E., and Kamentsky, L., *et al.* (2004). Next-generation laser scanning cytometry. 75, pp. 185–218. Academic Press, New York.

MacDonald, K., Palmer, J., et al. (2010). An antibody against the colony-stimulating factor 1 receptor (CSF1R). depletes the resident subset of monocytes and tissue and tumor-associated macrophages but does not inhibit inflammation. *Blood* prepublished online August 3, 2010; DOI:10.1182/blood 2010-02-266296.

Mei, Y., and Saha, K., *et al.* (2010). Combinatorial development of biomaterials for clonal growth of human pluripotent stem cells. *Nature Mater.* **9**, 768–778.

Mei, Ying., *et al.* (2009). Mapping the interactions among biomaterials, adsorbed proteins, and human embryonic stem cells. *Adv. Mater.* **21**, 1–6.

Minsky, M. (1988). Memoir on inventing the confocal scanning microscope. *Scanning* **10**, 128–138.

Miranda-Carboni, G. A., Krum, S., and Yee, K., *et al.* (2008). A functional link between Wnt signaling and SKP2-independent p27 turnover in mammary tumors. *Genes Dev.* **22**, 31221–31234.

Mora, J., and Lavarino, C., *et al.* (2007). Comprehensive analysis of tumoral DNA content reveals clonal ploidy heterogeneity as a marker with prognostic significance in locoregional neuroblastoma. *Genes Chromosomes Cancer* **46**(4), 385–396.

Nombela-Arrieta, C., *et al.* (2008). Spatial analysis of hematopoietic stem and progenitor cells in the bone marrow. *Blood (ASH Annual Meeting Abstract)* **112**, 3570.

Olive, K. P., *et al.* (2009). Inhibition of hedgehog signaling enhances delivery of chemotherapy in a mouse model of pancreatic cancer. *Science* **10**, 1126.

Omori, K., and Valiente, K., *et al.* (2007). Improvement of Human Islet Cryopreservation by a p38 MAPK Inhibitor. *Am. J. Transplantation* **7**(5), 1224–1232.

Omori, K., and Mistuhashi, M., *et al.* (2010). Microassay for glucose-induced preproinsulin mRNA expression to assess islet transplantation. *Transplantation* **89**(2), 146–154.

Pachmann, K., and Camara, O., *et al.* (2008). Monitoring the response of circulating epithelial tumor cells to adjuvant chemotherapy in breast cancer allows detection of patients at risk of early relapse. *J. Clin. Oncol.* **26**(8), 1208–1215.

Pawley., &., and James, B. (2006). *Handbook of Biological Confocal Microscopy.* Springer, New York, pp. 4, 223ff.

Peterson, R. A., and Krull, D., *et al.* (2008). Applications of laser scanning cytometry in immunohistochemistry and routine histopathology. *Toxicol. Pathol.* **36**(1), 117–132.

Ploem, J. S. (1967). The use of a vertical illuminator with interchangeable dielectric mirrors for fluorescence microscopy with incident light. *Z. Wiss. Mikrosk.* **68**, 129–142.

Ploem, J. S. (1993). Fluorescence microscopy. *In* "Fluorescent and Luminescent Probes for Biological Activity," (and W. T. Mason, ed.), pp. 1–11. Academic Press, London.

Pozarowski, P., Holden, E., and Darzynkiewicz, Z. (2006). Laser scanning cytometry: principles and applications. Vol. 319, pp. 165–192. Humana Press, Clifton, NJ.

Rieber, M. S., Viola-Rhenals, M., and Rieber, M. (2007a). Attenuation of genotoxicity under adhesion-restrictive conditions through modulation of p53, γ H2AX and nuclear DNA organization. *Apoptosis* **12**, 449–458.

Rieber, M. S., Viola-Rhenals, M., and Rieber, M. (2007b). Role of peroxidases, thiols and Bak/Bax in tumor cell susceptibility to Cu[DEDTC]2. *Biochem. Pharmacol.* **74**, 841–850.

Sakaguchi, A., and Steward, R. (2007). Aberrant monomethylation of histone H4 lysine 20 activates the DNA damage checkpoint in *Drosophila melanogaster. J. Cell Biol.* **176**, 155–162.

Sanoslo, L., and Kuliffay, P., *et al.* (2010). Advanced detection and measurement of cells on membrane from peripheral blood by laser scanning cytometry (LSC) in early stage breast cancer patients. *Bratisl Lek Listy* **111**(1), 13–19.

Schwock, J., Ho, J. C., and Luther, E., *et al.* (2007). Measurement of signaling pathway activities in solid tumor fine-needle biopsies by slide-based cytometry. *Diagn. Molec. Pathol.* **16**(3), 130–140.

Shapiro, H. M. (2003). *Practical Flow Cytometry, Fourth ed. Wiley-Liss, New York.*

Stegh, A. H., and Kesari, S., *et al.* (2008). Bcl2L12-mediated inhibition of effector caspase 3 and 7 via distinct mechanisms in gliobastoma. *Proc. Natl. Acad. Sci. U.S.A.* **105**(31), 10703–10708.

Swanberg, S. E., Nagarajan, R. P., and Peddada, S., *et al.* (2009). Reciprocal co-regulation of EGR2 and MECP2 is disrupted in Rett syndrome and autism. *Human Mol. Gen.* **18**(3), 525–534.

Sweet, I. R., and Gilbert, M., *et al.* (2008). Glucose-stimulated increment in oxygen consumption rate as a standardized test of human islet quality. *Am. J. Transplant.* **8**, 183–192.

Takahashi, H., Ruiz, P., Ricordi, C., and Miki, A., *et al.* (2009). In situ quantitative immunoprofiling of regulatory T cells using laser scanning cytometry. *Transplant. Proc.* **41**(1), 238–239.

Tanaka, T., Halicka, D., Traganos, F., and Darzynkiewicz, Z. (2009). Cytometric analysis of DNA damage: phosphorylation of histone H2AX as a marker of DNA double-strand breaks (DSBs). *Methods in Molecular Biology* **523**, 161–168.

Tanaka, T., and Huang, X., *et al.* (2007a). Cytometry of ATM activation and histone H2AX phosphorylation to estimate extent of DNA damage induced by exogenous agents. *Cytometry A* **71**(9), 648–661.

Tanaka, T., and Huang, X., *et al.* (2007b). ATM activation accompanies histone H2AX phosphorylation in A529 cells upon exposure to tobacco smoke. *BMC Cell Biol.* **8**, 1–11 26.

Tellez, C. S., and Davis, D. W., *et al.* (2007). Quantitative analysis of melanocytic tissue array reveals inverse correlation between activator protein-2alpha and protease-activated receptor-1 expression during melanoma progression. *J. Invest. Dermatol.* **127**(2), 387–393.

Todorov, I., *et al.* (2010). Quantitative assessment of β-cell apoptosis and cell composition of isolated, undisrupted human islets by laser scanning cytometry. *Transplantation* **90**(8), 836–842.

Tsujioka, T., and Tochigi, A., *et al.* (2008). DNA ploidy and cell cycle analyses in the bone marrow cells of patients with megaloblastic anemia using laser scanning cytometry. *Cytometry B Clin. Cytom.* **74**(2), 104–109.

Van Dilla, M. A., Trujillo, T. T., Mullaney, P. F., and Coulter, J. R. (1969). Cell microfluorometry: a method for rapid fluorescence measurement. *Science* **163**, 1213–1214.

Vincent, L., and Soille, P. (1991). Watersheds in digital spaces: an efficient algorithm based on immersion simulations. *IEEE Trans. Pattern Anal. Mach. Intell.* **13**(6), 583–598.

Williams, Kirsten M., *et al.* (2009). Single cell analysis of complex thymus stromal cell populations: rapid thymic epithelia preparation characterizes radiation injury. *Clin. Transl. Sci.* **2**(4), 279–285.

Zhao, Hong., *et al.* (2009). Induction of DNA damage response by the supravital probes of nucleic acids. *Cytometry A* **75**(6), 510–519.

Zhao, H., Halicka, D., Traganos, F., Jorgensen, E., and Darzynkiewicz, Z. (2010a). New biomarkers probing depth of cell senescence assessed by laser scanning cytometry. *Cytometry A* **77**(11), 999–1007.

Zhao, H., *et al.* (2008a). The cytotoxic ribonuclease onconase targets RNA interference (siRNA). *Cell Cycle* **7**(20), 3258–3261.

Zhao, H., Traganos, F., and Darzynkiewicz, Z. (2008b). Kinetics of histone H2AX phosphorylation and Chk2aActivation in A549 cells treated with topotecan and mitozantrone in relation to cell cycle phase. *Cytometry A* **73**, 480–489.

Zhao, H., Traganos, F., and Darzynkiewicz, Z. (2010b). Kinetics of the UV-induced DNA damage response in relation to cell cycle phase. Correlation with DNA replication. *Cytometry A* **77**(3), 285–293.

Zhao, H., and Tanaka, T., *et al.* (2007). Cytometric assessment of DNA damage by exogenous and endogenous oxidants reports aging-related processes. *Cytometry A* **71A**, 905–914.

Zheng, X. L., and Yuan, S. G., *et al.* (2007). Phenotype-specific inhibition of the vascular smooth muscle cell cycle by high glucose treatment. *Diabetologia* **50**(4), 881–890.

Zoog, S. J., *et al.* (2010). Measurement of conatumumab-induced apoptotic activity in tumors by fine needle aspirate sampling. *Cytometry A* **77A**, 849–860.

Zoog, S. J., *et al.* (2008). Antagonists of CD117 (cKit). Signaling inhibit mast cell accumulation in healing skin wounds. *Cytometry A* **75**(3), 189–198.

CHAPTER 8

Analytical Capabilities of the ImageStream Cytometer

Ewa K. Zuba-Surma[*] **and Mariusz Z. Ratajczak**[†]

[*]Department of Medical Biotechnology, Faculty of Biochemistry, Biophysics and Biotechnology, Jagiellonian University, Krakow, Poland

[†]Stem Cell Biology Institute, James Graham Brown Cancer Center, University of Louisville, Louisville, Kentucky, USA

Abstract

Imaging cytometry has recently become an important achievement in development of flow cytometric technologies. The ImageStream cytometer combines the vast features of classical flow cytometry including an impartial analysis of great number of cells in short period of time which results in strong statistical data output, with essential features of fluorescence microscopy such us collecting of real multiparameter images of analyzed objects.

0091-679X/10 $35.00
DOI 10.1016/B978-0-12-374912-3.00008-0

In this chapter, we would like to introduce an overview of imaging cytometry platform and emphasize the potential advantages of using this system for several experimental purposes. Moreover, both well established as well as potential applications of imaging cytometry will be described. Eventually, we would like to illustrate the unique use of ImageStream cytometer for identification and characterization of subpopulations of stem/ progenitor cells present in different biological specimens.

I. Introduction

The aim of this chapter is to provide a comprehensive overview of characteristics and potential applications of imaging flow cytometry (IFC) which is currently represented by the ImageStream system (ISS) introduced as the first commercially available instrument developed in this technology. It should be noted that another type of image cytometry instrumentation is represented by laser scanning cytometry (LSC), also defined as "slide-based cytometry," which interrogates individual cells not in flow but positioned on microscope slides or on microwells (Pozarowski *et al.*, 2006). Several chapters in this volume (Chapters 7, 9-14) describe different applications of LSC.

The IFC fully integrates the features of classical flow cytometry (FC) and fluorescence microscopy combined with extended software for digital image analysis. Similarly to classical FC, the IFC allows for acquisition and analysis of a large number of cells according to their fluorescent parameters as well as provides statistically strong information about the collected objects. Moreover, the collected multicolor images of the acquired cells may be further utilized for multiparameter analysis resulting in the quantitative characterization of their multiple morphological features. Finally the photometric and morphometric features calculated by the system may be correlated and provide comprehensive information about morphology and functional status of acquired cells. Such unique capabilities of the ISS have introduced an entirely novel quality into flow cytometric technology and cell analysis.

The ISS technology was primarily developed as a novel flow cytometric method for multiparameter analysis of cells supporting the classical FC in its widely established applications. However, the innovative features of the instrument created multitude of possible new applications, especially for experimental approaches where classical FC may not provide enough information about analyzed processes.

Thus, the current chapter would provide an overview of the main characteristics and applications of the ISS when compared to classical FC and LSC.

A. Origin of Imaging Flow Cytometry – Historical Perspectives and Place Among Other Imaging Platforms

Unquestionably, the invention of the first microscope has opened the absolutely new and challenging opportunities for imperative discoveries on the cellular, subcellular and molecular levels which had not been possible before. Since the 17th century, the technical aspects of *microscopy*, the first cell imaging technology, have developed over the time resulting in the innovation of sophisticated systems including modern confocal laser scanning microscopy, transmission electron microscopy and scanning electron microscopy (TEM and SEM, respectively), scanning tunneling microscopy, and atomic force microscopy (Conchello and Lichtman, 2005; Hansma *et al.*, 1988; Haupt *et al.*, 2006; Henderson, 2004; Jones *et al.*, 2005; Lucitti and Dickinson, 2006; Subramaniam and Milne, 2004). Today, the advanced microscopic systems allow for multidimensional visualization of not only cells and organelle, but also molecular structures in various samples that must be fixed to specific carriers required for each microscopic platform.

FC has been successfully introduced to the field of cellular imaging as a supplemental method to microscopy and allowed analyzing objects in suspension. FC acquires a great number of objects in a relatively short time and provides multiparameter analysis of all acquired objects as well as selected cellular populations (Baumgarth and Roederer, 2000; Bonetta, 2005; Jaroszeski and Radcliff, 1999; Radcliff and Jaroszeski, 1998; Roederer, 2001; Shapiro, 1983; Shapiro, 2005). Advanced flow cytometric platforms employ various combinations of optical elements including multicolor lasers, filters and splitters resulting in a variety of fluorochromes and probes which may be used for cell labeling and detection. Thus, the classical FC provides multitude research and clinical applications vigorously developing since early 1900s when technology was initiated. Based on the great technological progress achieved during the past few decades, modern FC instruments use up to 17 color detectors and are armed with software capable of compensating the collected signals (De Rosa *et al.*, 2003; De Rosa and Roederer, 2001). The strength of the flow cytometric optical and analytical platforms has been used to build the first imaging cytometry – ISS.

FC provides great quantitative information about collected cells predominantly based on their fluorescent features. Unfortunately, neither the morphology of cells nor localization of the fluorescence signals may be analyzed in details by classical FC. The initial attempts to analyze morphometric features by FC were based on the use of "slit-scanning" of the cells in flow illuminated by a very narrow laser beam and analysis of the shape of the fluorescence signal (Wheeless *et al.*, 1991). However, because of relatively low resolution this approach was used only to eliminate the nonspecific DNA signals such as nuclear fragments from the analysis and to distinguish bi-nucleated cells from the tetraploid mononuclear cells.

The first technology which attempted to incorporate some features of FC into fluorescence microscopy to enlarge its analytical capacity was LSC. This advanced microscopic technology successfully provides quantitative multiparameter analysis of multiple objects based on their imaginary and combines the phenotypic analysis of cells based on the expression of cellular markers with their extended morphological analysis (Darzynkiewicz *et al.*, 1999; Deptala *et al.*, 2001; Kamentsky, 2001; Kamentsky *et al.*, 1997; Tarnok and Gerstner, 2002; Wlodkowic *et al.*, 2010).

However, the LSC type of acquisition requires cells to be fixed on carriers such as microscopic slides or plates and may not be used for analysis of cells in suspension

(Darzynkiewicz *et al.*, 1999; Deptala *et al.*, 2001; Kamentsky, 2001; Kamentsky *et al.*, 1997; Tarnok and Gerstner, 2002; Wlodkowic *et al.*, 2010).

The methodology of multiparameter cellular analysis, especially objects acquired in suspension, has taken a significant jump with IFC and its first instrument – ISS. The first generation of the ISS was launched in 2005 and recently the second advanced model of the platform has been introduced as ImageStream X (Basiji *et al.*, 2007; Ortyn *et al.*, 2007).

IFC combines the majority of the advanced capabilities of FC with features of fluorescence microscopy (Basiji *et al.*, 2007; George *et al.*, 2004b; Ortyn *et al.*, 2006; Ortyn *et al.*, 2007; Zuba-Surma *et al.*, 2007, 2008b). Similarly to advanced microscopy including LSC, IFC acquires and analyzes fluorescent signals emitted by cells as well as computes several parameters related to cellular morphology. On the other hand, similarly to classical FC, IFC acquires and analyzes large number of cells in suspension (Basiji *et al.*, 2006, 2007; George *et al.*, 2004a; Morrissey *et al.*, 2004; Zuba-Surma *et al.*, 2007, 2008b). The advantage of using this system in comparison to fluorescent microscopy is its ability to analyze imagery from a large number of cells acquired in suspension. This provides a unique opportunity for statistical analysis of large populations. The analysis of collected data is also more quantitative with the ISS when compared with conventional microscopy (Basiji *et al.*, 2007; Parsons *et al.*, 2006). On the other hand, IFC possesses also great advantages when compared to classical FC since it identifies the collected objects by their real images. Such capacity of IFC allows avoiding artifacts and falsely positive objects during direct analysis and inspection of collected image galleries, which made this method an important supporting tool for both classical FC and microscopy. Direct comparison of some features of ImageStream X with FC and LSC has been shown in Table I.

The IFC has become the first flow cytometric method that "decodes the dot" – the cell displayed as a dot on the flow cytometric plot – providing a real multicolor image of this cell. Such capacity occurs to be especially appreciated in applications detecting rare untypical cellular objects or analyzing difficult types of biological samples, which we will emphasize in later part of this chapter.

II. Background

A. Overview of the ImageStream Platform

The ISS has been developed as the first commercially available imaging flow cytometer. As is shown in Table II the instrument combines the advanced features of both fluorescence microscopy and FC.

Similar to FC, the ISS analyzes a large number of cells in suspension collecting various fluorescent signals in a short period of time. The idea of analysis is similar to classical FC (Basiji *et al.*, 2007; George *et al.*, 2004a, 2004b; Ortyn *et al.*, 2006; Zuba-Surma *et al.*, 2007). Briefly, the cells are loaded into a fluidic system in suspension and they are hydrodynamically focused into a core stream, which is

Table I

Comparison of selected features of the imaging flow cytometer: ImageStream X (ISS) with FC and LSC (Basiji *et al.*, 2007; Darzynkiewicz *et al.*, 1999; Deptala *et al.*, 2001; Perfetto *et al.*, 2004; Pozarowski *et al.*, 2006; Shapiro, 2005; Wlodkowic *et al.*, 2010; Zuba-Surma *et al.*, 2007; Zuba-Surma *et al.*, 2008b)

Feature:	ImageStream X system (ISS)	Classical flow cytometer (FC)	Laser scanning cytometer (LSC)
Material for acquisition	Cells in suspension	Cells in suspension	Cells bound to a microscopic slide, glass or plastic culture plate/dish, etc.
Speed of acquisition	Up to 1000 cells per second	Up to 70,000 cells per second	Up to 100 cells per second
Correlation of fluorescent signals with morphological features of cell	Available	Not available	Available
estimation of translocation fluorochromes between cellular compartments (exp. from cytoplasm to nucleus)	Available	Not available	Available
Localization of fluorescence in different cellular compartments	Available	Not available	Available
Option of cell sorting	Not available	Available	Not available
Analysis of the same object multiple time	Not available	Not available	Available
Identification of intracellular sites with various intensities of fluorescence	Available	Not available	Available
Lasers	488 nm, 658 nm[a] 405 nm[a] 561 nm[*] 592 nm[a]	488 nm 635 nm 405 nm[a] 375 nm[a] Others[a]	488 nm, 633 nm[a] 405 nm[a]
Objectives	Multiple exchangeable microscopic objectives (20×, 40×, 60×)	Nonexchangable collection lenses (distinct physical features than microscopic objectives – longer focal length, high-NA aspheric lenses)	Multiple exchangeable microscopic objectives (10×, 20×, 40×)
Other optical elements	Up to nine detectors of fluorescence	Up to 17 detectors of fluorescence.	Up to four detectors of fluorescence for laser (maximum 12 with exchangeable filters)

[a] Lasers available optionally with appropriate optical elements.

Table II

Selected features of ImageStream X system, outcomes and advantages for users (Basiji *et al.*, 2007; George *et al.*, 2004b; Ortyn *et al.*, 2007; Spibey *et al.*, 2001)

Features	Significances
488 nm standard laser; up to four additional lasers (405 nm, 561 nm, 592 nm, 658 nm); 12 channels – 2 brightfield (one for each camera); side scatter and up to nine fluorescence channels;	Allows us to employ the most common fluorescent dyes and probes with optical features similar to: FITC, PE, PE-Texas Red, PE-Cy5, PE-Cy7, Pacific Blue, Pacific Orange, APC, APC-Cy7[a]
MultiMag: three different objectives: 20X (0.5NA; 1 (m per pixel resolution) 40X (0.75NA; 0.5 (m per pixel resolution) 60X (0.9NA; 0.33 ((m per pixel resolution)	Allows different resolution of imaging and field of view (120, 60, and 40 μm, respectively) applicable for objects with vastly distinct sizes;
Two CCD cameras working in the time delay integration (TDI) mode;	Provides high sensitivity of instrument and allow collection images of cells in flow;
Extended depth of field (EDF) for each camera;	Generates "confocal-like" maximum projection image with entire cell in focus; useful for small spot counting;
Acquisition of 12,000 cellular images per second which corresponds to 1000 cells/s including 12 images/cell (with 40X standard objective);	Collects great number of cells in relatively short time;
Collection of up to 12 independent images of the cell in 12 separate channels (depending on laser option)	Allows for comparative analysis of images collected in different channels;
Pixel-by-pixel spectral overlap compensation between collected fluorescent images;	Allows us to exclude spectral contamination between images collected by fluorescence detecting channels;
Data files containing the information from 10,000 cells require about 500 MB file storage (for raw and processed files);	Data can be stored and analyzed using standard personal computers;
calculation of more than 200 various morphometric and photometric features for each cell;	Fluorescent signals can be co-localized as well as localized in exact compartments of the cell;
IDEAS software employing standard statistical calculations and dot-plots and histograms for visualization of analyzed cells;	Allows for statistical computing of various features of single cells and selected populations as well as their graphical visualization combined with image galleries;
IDEAS software identifying a single object (dot) on plots or gated population by loading their image/images.	Identification and verification of cellular objects; distinguishing of cellular objects from debris, positive objects from false "positive", etc..

[a] Detail information for possible fluorochromes and dyes depending on laser excitation and detection channels are summarized in Table III.

illuminated by a brightfield light source and a 488 nm solid-state laser. The signals that are collected by the system include transmitted (brightfield) and scattered (darkfield) lights as well as fluorescence signals emitted by acquired objects. Fluorescence emitted by cells is decomposed into separate predefined wavelength ranges that are collected in separate detection channels. The photodiodes and photo-multiplier tubes that are commonly used in classical flow cytometers have been replaced by two multichannel CCD cameras functioning as the detectors of the system (Basiji *et al.*, 2007; George *et al.*, 2004a, 2004b; Ortyn *et al.*, 2006). The standard collection system of the second generation ImageStream X system includes one CCD camera collecting signals excited by standard 488 nm blue laser. In such system, by the technique of spectral decomposition, the combined cellular image is optically split into a set of six subimages including four fluorescence images (corresponding to 4 fluorescence signals collected in multicolor FC) as well as brightfield and darkfield images. Optionally, ImageStream X system may be equipped in second CCD camera and additional up to four lasers which expand the capacity of the system to twelve detection channels including nine fluorescence colors (Table II). Typical three-laser system encloses blue (488 nm), red (658 nm), and violet (405 nm) lasers and allows for collecting the signals from the most commonly used fluorochromes and dyes such as, e.g., FITC, R-PE, PI, 7-AAD, PE-Cy7, APC, APC-Cy7, Pacific Blue, Pacific Orange, and DAPI as shown in Table III. Currently, the ISS may be equipped in the carousel holding three exchangeable objectives of $20\times$, $40\times$ and $60\times$ magnification. Such advancement allows for analysis of cells with different size and provides more detailed and magnified images.

The CCD cameras convert photons of collected light signals into photocharges in an array of pixels that may be analyzed by image analyzing software, called IDEAS. The high sensitivity of the instrument may be achieved by operating the CCD cameras in the time delay integration (TDI) mode. TDI imaging is a method of electronically panning the detector to track object motion that increases signal collection time and preserves sensitivity and image quality even with fast relative movement between the detector and the objects (cells) (George *et al.*, 2004b; Spibey *et al.*, 2001).

Similarly to FC, in postacquisition image processing, the images of each object derived from all channels are corrected because of spectral overlap by compensating pixel by pixel using single-fluorochrome stained samples. More than 200 features may be subsequently computed based on each collected cellular image. Importantly, both photometric and morphometric cellular features may be further correlated which eventually leads to localization of fluorescence signals in exact cellular compartments. Distinct features may be computed and compared not only between single cells but also among different selected populations (Basiji *et al.*, 2007; George *et al.*, 2004a, 2004b; Ortyn *et al.*, 2006; Zuba-Surma *et al.*, 2007). The analysis and display methods, together with the well-developed compensation feature allow not only co-localization of fluorescence and morphological signals in cells but also measurement of signals from various fluorescent probes (Table III).

III. Methods

A. Sample Preparation for Analysis by Imaging Cytometry

The IFC has been designed for collecting and analyzing of a large number of predominantly fluorescent images of cells that are acquired in suspension. Importantly, the samples for ImageStream analysis may be prepared and stained in a fashion similar to standard FC methodologies (Shapiro, 2005). Both freshly isolated as well as fixed cells can be analyzed by IC.

Table III
Fluorochromes and dyes applicable in ImageStream X system depending on excitation lasers and range of emitted light collected by specific channels

Channel 1 (420-480nm)	Channel 2 (480-560nm)	Channel 3 (560-595nm)	Channel 4 (595-660nm)
Brightfield	FITC ✳	PE ✳ ⬤	7-AAD ✳
	AF 488 ✳	Cy3 ✳ ⬤	Propidium Iodide ✳ ⬤
	eGFP, eYFP ✳	Calcium Orange ⬤	PE-Texas Red ✳ ⬤
	Spectrum Green ✳	AF 546 ⬤	PE-AF 610 ✳ ⬤
	SYTO 16, SYTOX Green ✳	AF 555 ⬤	AF 568 ⬤

Channel 5 (660-740nm)	Channel 6 (740-800nm)	*Channel 7 (420-505nm)*	*Channel 8 (505-570nm)*
QD-705 ✳	Darkfield (SSC)	DAPI, Hoechst ◆	QD-525, QD-545 ◆
PerCP, PerCP-Cy5.5 ✳	PE-Cy7 ✳ ⬤	AF 405 ◆	Pacific Orange ◆
PE-Cy5, PE-Cy5.5 ✳ ⬤	PE-AF 750 ✳ ⬤	Marina Blue, Pacific Blue, Cascade Blue ◆	Cascade Yellow, Lucifer Yellow, ◆
DRAQ5 ✳ ⬤		Live/Dead violet ◆	Alexa Fluor 430 ◆
PE-AF 680 ✳ ⬤			
PE-AF 647 ✳ ⬤			

Channel 9 (570-595nm)	*Channel 10 (595-660nm)*	*Channel 11 (660-740nm)*	*Channel 12 (740-800nm)*
QD-565, QD-585 ◆	QD-605, QD-625 ◆	QD-705 ◆	QD-800 ◆
	eFlour 605 ◆	eFlour 650 ◆	APC-Cy7 ⬤▲
	Texas Red ⬤	APC, APC-Cy5.5 ⬤▲	APC-AF 750 ⬤▲
	mCherry ⬤	Mito Tracker Deep Red ⬤▲	APC-eFlour 780 ⬤▲
	AF 568 ⬤	Cy5, Cy5.5 ▲	
	AF 594 ⬤	DRAQ5 ▲	
	AF 610 ⬤	AF 647 ▲	
		AF 660 ▲	
		AF 680 ▲	

Excitation:	⬤ 561nm laser (green)	⬤ 592nm laser (orange)
✳ 488nm laser (blue)	◆ 405nm laser (violet)	▲ 658nm laser (red)

Channels 1–6: for camera 1.
Channels 7–12: for camera 2 (*italic*).

Appropriate fixation might be effectively performed with 2–4% of paraformalde-hyde solution or other commercially available fixating buffers recommended for classical flow cytometric protocols (Arechiga *et al.*, 2005; George *et al.*, 2006b; Ortyn *et al.*, 2006; Parsons *et al.*, 2006; Zuba-Surma *et al.*, 2008a, 2008c, 2009). However, surface antigen staining is typically recommended before fixation and may be performed according to standard flow cytometric procedures (Shapiro, 2005).

In the case of staining for intracellular proteins, a permeabilization step will be required following the cell fixation. Widely accepted permeabilization agents, which may also be employed for IFC, include TritonX-100, saponin and methanol (George *et al.*, 2006b; Ortyn *et al.*, 2006; Parsons *et al.*, 2006; Zuba-Surma *et al.*, 2008a, 2008c, 2009).

The antibodies and other reagents used for cell labeling and detection in IC should be selected from those recommended for FC. Fortunately, most of widely used and well-validated fluorochromes and dyes employed in classical FC and excited by typical blue, red and violet laser may be efficiently used for IFC analysis. Other fluorochromes are also possible to be used depending on available ISS laser options. Standard DNA binding dyes compatible with ImageStream imaging and analysis include propidium iodide (PI), 7-aminoactinomycin D (7-AAD), dihydroxyanthraqui-none (DRAQ5), 4′-6-diamidino-2-phenylindole (DAPI) and Hoechst 33342 and they can be used for staining according to standard protocols recommended by vendors for classical FC (George *et al.*, 2006b; Gillard and Farr, 2006; Morrissey *et al.*, 2004; Ortyn *et al.*, 2006; Zuba-Surma *et al.*, 2008a, 2008d, 2009). Both 7-AAD, PI and DAPI are viable cell impermeant and thus can be used to determine cellular viability. If all of the cells need to be stained by one of the mentioned nuclear dyes, the fixation step would be required before staining (Glisic-Milosavljevic *et al.*, 2005; Klein *et al.*, 2006; Totino *et al.*, 2006). On the other hand DRAQ5 and Hoechst 33342 effectively intercalate into DNA of living as well as fixed cells (Martin *et al.*, 2005; Njoh *et al.*, 2006; Yuan *et al.*, 2004). The detailed list of fluorochromes and dyes recom-mended for particular lasers applicable in the ISS platform is given in Table III.

Experimental samples to be analyzed by IFC should also include single-color stained control samples, which would be necessary for calculating the compensation matrix. Stained cells should be washed and resuspended in 50–100 (l of PBS or other media in optimal concentration of 2–4 $\times$ 10^7/ml (George *et al.*, 2006b; Ortyn *et al.*, 2006). The media containing EDTA and other antiaggregating agents are recom-mended for the acquisition to support smooth flow and acquisition.

IV. Applications of ImageStream System

Multiple fluorescence-based and morphological parameters may be quantita-tively measured by IC analytical software based on the collected images. The features analyzed by ISS software IDEAS include cellular size, shape, texture, shape changes, and location of probes within, on or between cells. Since the measurements may be performed for individual cells as well as selected populations, it became possible to execute complex assays such as evaluating transcription factor (TF) translocation in rare subsets of cells, correlating apoptosis to the phase of the cell cycle or distinguishing the cell type that has particles taken up by phagocytosis.

Thus, multiple applications of imaging cytometry have been already established by several investigators and the list of publications is still growing (Arechiga *et al.*, 2005; McGrath *et al.*, 2008a; Nobile *et al.*, 2010; Parsons *et al.*, 2006; Pawluczkowycz *et al.*, 2009; Riddell *et al.*, 2010; Zuba-Surma *et al.*, 2008a, 2009, 2010). The examples of selected and well-established applications of the ISS are presented below and summarized in Table IV.

A. Cell Death – Apoptosis Quantification

One of the first well-established and vigorously developing application of imaging cytometry was studying of cell death and survival. Since Kerr and colleagues described the biological phenomenon of apoptosis (Kerr *et al.*, 1972), various methods of apoptotic cell detection have been described. Most of the detection assays rely on the cytochemical features characterizing apoptotic cells including morphology and composition of the surface membrane, nuclear events and DNA fragmentation, cellular and nuclear dissolution, activation of cytoplasmic enzymes as well as functionality, and integrity of mitochondria (Barrett *et al.*, 2001; Bedner *et al.*, 1999; Darzynkiewicz *et al.*, 1992, 2006; Kohler *et al.*, 2002; Lovborg *et al.*, 2005; Micoud *et al.*, 2001; Sgonc and Wick, 1994; Willingham, 1999; Wlodkowic *et al.*, 2010; Yasuhara *et al.*, 2003).

Striking changes in nuclear morphology related to DNA fragmentation and condensation are hallmarks of apoptosis. Such changes make the apoptotic cells possible to be identified by different technologies including microscopy, LSC. However, apoptotic cells undergo several other morphological, structural and biochemical changes which may also be detected by electrophoresis, ELISA, Western blotting, and fluorescence cytometry (Barrett *et al.*, 2001; Bedner *et al.*, 1999; Darzynkiewicz *et al.*, 1992, 2006; Kohler *et al.*, 2002; Lovborg *et al.*, 2005; Micoud *et al.*, 2001; Sgonc and Wick, 1994; Willingham, 1999; Wlodkowic *et al.*, 2010; Yasuhara *et al.*, 2003).

The imaging cytometry may be employed for detection and quantification of apoptotic, necrotic and recently autophagic cells by applying assays originally described for fluorescence microscopy and cytometry (Darzynkiewicz *et al.*, 1992, 2004, 2006; Glisic-Milosavljevic *et al.*, 2005; Vermes *et al.*, 2000; Wlodkowic *et al.*, 2010).

One convenient cytometry-based assay for measuring cell death uses fluorescently labeled Annexin V (AnV) and a cell-impermeant nuclear dye such as 7-aminoactinomycid D (7-AAD) or PI (George *et al.*, 2004b). Live and early apoptotic cells exclude the DNA dye, and only late apoptotic or necrotic cells stain with 7-AAD or PI. Annexin V binds to phosphatidylserine, either exposed on the extracellular side of the plasma membrane early, during the apoptotic process, or on the intracellular side of the membrane in permeable necrotic cells. However, there are some limitations of such AnV-binding assay assessed with conventional FC

Table IV

Selected applications of ImageStream X and crucial features employed for each type of analysis by IDEAS software

Application	Features calculated by IDEAS
Apoptosis quantification based on changes in the nuclear morphology and structure (George *et al.*, 2004b; Khuda *et al.*, 2008)	Two features calculated on nuclear image: *"spot small total"* – total intensity of "island" sites within image (higher in fragmented apoptotic nuclei) *"area"* of nucleus computed from special mask (30% threshold mask), that is, smaller for apoptotic cells;
Autophagy based on the presence specific proteins in vacuole (Degtyarev *et al.*, 2008; Lee *et al.*, 2007)	*"Spot count"* – counts the number of masks covering the brightest points in cell compartments – here spots of LC3 protein indicating autophagy;
Cell cycle, mitotic index based on chromatin condensation (Gillard and Farr, 2006; Hall *et al.*, 2010)	*"Spot small total"* and *"spot medium total"* – quantify the fine and coarse nuclear textures, respectively. High values for one or both of these features are indicative of mitosis; *"intensity"* – measured for nuclear image;
DNA repair based on presence proteins of repair complexes (Decaluwe *et al.*, in press)	*"Spot count"* – counts the number of masks covering the brightest points in cell compartments – here the spots in nucleus that are related to nuclear foci containing proteins involved in DNA repair;
Protein translocation analysis (exp. translocation of TFs from cytoplasm to nucleus) (Fanning *et al.*, 2006; George *et al.*, 2006b)	*"Similarity score"* – derived from correlation analysis of intensities of pixels pairs on two different images (exp. cytoplasmic and nuclear) of the same cell (high when TF is localized in nucleus);
Co-localization and trafficking of molecules in various cellular compartments (Amnis, 2006; George *et al.*, 2006a)	*"Similarity bright detail score"* – measures similarity of the small bright details in the image pairs excluding impact of background staining (high when two molecules are co-localized);
Shape change assay. (Tan *et al.*, 2006b) (Amnis materials, www.amnis.com)	*"Aspect ratio"* – computed based on brightfield image as the ratio of the minor axis (width) to major axis (height) of each object ($\sim$1.0 for round objects, $>$1.0 for elongated cells); *"circularity"* – measured as average cellular radius divided by the variance in the radii;
Pseudopod formation and polarization of molecules in one cellular area (Nobile *et al.*, in press; Tan *et al.*, 2006a)	*"Radial delta centroid"* – measured as the distance from the center of the nuclear image to the center of the image of protein involved in pseudopod formation using the pythagorean theorem (larger when protein is capped to one side of the membrane);
Cell classification and morphological identification (Morrissey *et al.*, 2004; Parsons *et al.*, 2006)	Various features including: *"intensity"* – measured for different channels related to the expression of various markers; *"area"* of brightfield and nucleus related to different sizes and types of cells; *"nuclear to the cytopalsmic ratio"* – computed from nuclear and brightfield images, distinguishes types of cells from mixture (blood, bone marrow), related to primitivity of some cell types;
Stem and progenitor cell identification (McGrath *et al.*, 2008a; McGrath *et al.*, 2008b; Zuba-Surma *et al.*, 2008a, 2008d, 2010)	*"Delta centroid XY"* – quantified to establish the content of enucleating mature cells and nucleated erythroid progenitors (McGrath *et al.*, 2008a, 2008b); *"minor axis"* and *"nuclear to cytoplasmic ratio"* – quantified to establish the content of pluripotent stem cells with very small size and high N/C ratio (Zuba-Surma *et al.*, 2010, 2008a, 2008d);

including difficulties in distinguishing AnV$^+$/7-AAD$^+$ necrotic and late apoptotic cells as well as the presence of falsely "AnV-positive" cells which in fact binds microvesicles and represent normal healthy cells (Liu *et al.*, 2009).

Based on imaging capabilities of the ISS, other morphological parameters related predominantly to nuclear structure may distinguish real apoptotic cells from necrotic and falsely "positive" populations.

The chromatin of apoptotic cells condenses and fragments forming the characteristic spots of condensed heterochromatin. Apoptotic nuclei produce small, fragmented, highly textured nuclear images. This causes nuclear dyes to be localized to small punctate regions of the fragmenting nucleus. The brightest 30% of the pixels in the apoptotic cell are in a small area of nuclei that can be identified by these analytical features. Because of the condensed chromatin, apoptotic cells have brighter fluorescence sequestered in small regions and thus higher quantitative scores with this classifier in IDEAS (George *et al.*, 2004b). By quantitatively measuring the area and the intensities of the brightest parts of the nuclear image, the intense and spotted nuclear images of apoptotic cells may be automatically distinguished from the uniformly stained nuclei of normal healthy cells.

On the other hand, late apoptotic and necrotic cells exhibit defective membrane integrity, which permits the entry of nuclear dyes leading often to false impression of the larger nuclear area of these cells. However, necrotic cells appear in imaging cytometry as objects with less complicated and more uniform SSC (darkfield) images when compared to the apoptotic cells which eventually allows for distinguishing of these two fraction of cells. Additional inspection of image galleries associated with all identified populations of cells can confirm the classification of each population (George *et al.*, 2004b).

The IFC may also find its application in assessing apoptosis with techniques such as BrdU staining and nuclear fragmentations in a TUNEL assay. TUNEL has become well validated and typically used by either a fluorescent microscope or a standard flow cytometer (Hewitson *et al.*, 2006; Hunter *et al.*, 2005; Otsuki *et al.*, 2003), but recently the protocols relying on imaging cytometry were also established (Hall *et al.*, 2006; Henery *et al.*, 2008).

Based on extended morphological characterization of cells by IFC, it becomes possible to discriminate true TUNEL positive apoptotic cells from healthy cells with an attached TUNEL positive fragment. Such discrimination may be performed based on visual the inspection of image galleries of cells and thus has not been possible with classical FC. Moreover, the IC allows for further analysis of nuclear morphology and other features of cells identified by TUNEL-positive staining resulting in important correlations of different parameters in detected apoptotic cells (Hall *et al.*, 2006; Henery *et al.*, 2008). Of particular importance is the possibility to discriminate between the genuine apoptotic cells having typical TUNEL-positive nucleus and cells such as macrophages that ingested apoptotic bodies containing TUNEL-positive chromatin fragments and when analyzed by FC are identified also as apoptotic cells (Bedner *et al.*, 1999). The analytical approaches described above have already found a well-established place in analysis and quantification of apoptosis and cell death. The

IFC really allows for the detailed classification of live and apoptotic cells as well as the enumeration of the yield of true as well as false positive objects. Moreover, the capability of ImageStream to compare and combine morphometric and photometric features of cells, which is not possible for classical FC, leads to more detailed classification of cells in terms of different stages of cell death. Importantly, IC allows for further modifications of established protocols by combining for instance nuclear fragmentation analysis with caspase activation, and cellular blebbing, and a more comprehensive picture of the apoptotic process may be created with this technology.

B. Cell Signaling – Protein Translocation Analysis

Imaging cytometry has also been successfully employed in studying of cell signaling especially in quantization of TFs transfer from cytoplasm to nucleus that plays a pivotal role in cellular activation, differentiation, and interaction. Cell activation and translocation of signaling proteins and TFs in different compartments may be assessed by various methods including fluorescence microscopy and classical FC as well as several biochemical and molecular assays (Deptala *et al.*, 1998, 2001; Krutzik and Nolan, 2003; Krutzik *et al.*, 2005).

LSC with its high ability for analyzing morphology of cells has also been effectively employed for measuring translocation of proteins between cellular compartments in adherent cells (Deptala *et al.*, 1998, 2001). IFC becomes a supplemental method for LSC analyzing cells in suspension (George *et al.*, 2006b; Zuba-Surma *et al.*, 2007).

Using the ISS, protein/TF translocation for instance from cytoplasm to nucleus, may be measured by cross-correlation analysis of intensities of fluorescence between the same pixels on two different images stained in two different colors images: (1) related to protein/TF distribution and (2) representing the nuclear image. The "similarity score" may be derived from the Pearson's correlation coefficient, which is computed based on a regression analysis of pairs of intensity values taken from protein/ TF and nuclear fluorescence images from single cellular images. The ISS has been already employed for measuring of cytopalsmic-to-nuclear translocation of several TFs including NF-κB, Interferon Regulatory Factor 7 (IRF-7), Forkhead-box (FOX) p3 (Fox-p3), and p65 in different cell types (Arechiga *et al.*, 2005; Fanning *et al.*, 2006; George *et al.*, 2006b).

The application of IFC occurs to be especially valuable in difficult analysis of protein translocation in (i) cells with small cytoplasmic areas, (ii) in experiments employing a dose- and time-dependent treatment of cells, and (iii) in the analysis of population with a wide range of response for stimulation that results in heterogeneity regarding the stage of the translocation process (George *et al.*, 2006b). Importantly, IFC may also provide valid information regarding potential correlations between the stage of translocation processes and morphological status of cells including shape change, nuclear to cytopalsmic ratio, etc. (Fanning *et al.*, 2006; George *et al.*, 2006b). The analyzed cells me be displayed on the histogram based on their "similarity score" values and further quantitative analyzes may be performed on such selected populations. Moreover, the translocation event can further be confirmed using image galleries and visualizing cells with low- and high-similarity scores (Arechiga *et al.*, 2005; Fanning *et al.*, 2006; George *et al.*, 2006b).

C. Molecule Colocalization and Trafficking

Colocalization of specific molecules, proteins and receptors in different cellular compartments has been previously described as critical for understanding several biological processes. Molecular colocalization or coassociation of proteins may be investigated using methods including, e.g., coprecipitation, yeast two-hybrid analysis and multiple fluorescent microscopy techniques (Kim *et al.*, 2000; Miyashita, 2004; Swick and Kapatos, 2006; Voss *et al.*, 2005; Yu *et al.*, 2002). Fluorescence-based techniques employing fluorescence-labeled antibodies may be especially useful to estimate colocalization of molecules including surface membrane receptors (co-capping) (Cherry *et al.*, 2003; Nohe and Petersen, 2004).

Images collected by IC may be quantitatively analyzed to determine protein colocalization using the "Similarity Bright Detail Score" (SBDS) feature calculated by analytical ISS software. SBDS is computed as similarity between any two channel images on a pixel-by-pixel and cell-by-cell basis. The three-step process of SBDS calculation using a nonmean normalized version of Pearson's correlation coefficient has been provided by Beum and colleagues (George *et al.*, 2006a). The authors employed SBDS to determine potential interactions of Rituxan (RTX), a widely used oncology therapeutic monoclonal antibody against CD20 molecule, with other cellular proteins such as C3b complement fragments to establish potential mechanism of action and trafficking of the drug (George *et al.*, 2006a). Both molecules were identified by staining with specific antibodies conjugated with distinct fluorochromes. Quantitative analysis of similarity between RTX and C3b images was performed postacquisition and was presented as SBDS value (George *et al.*, 2006a). Similar studies showing the binding and co-localization of the ofatumumab (OFA) antibody against CD20 molecule that is clinically more effective than RTX, to C1q complement fragment, have been performed employing the ISS and recently published in (Pawluczkowycz *et al.*, 2009).

The similar approach has also been used to investigate the trafficking of proteins into different cellular compartments including endosomes and lysosomes. By employing compartment-specific staining against EEA-1 (early endosomal antigen 1) and Lamp1, respectively, both compartments were identified and co-localized with other protein of interest such as here RTX (George *et al.*, 2006a). By employing similar methodology, Xu *et al.* studied the endosomal and lysosomal localization of selected therapeutic antibodies targeting human carbonic anhydrase IX (CAIX) – biomarker of various human epithelial tumors associated with tumor resistance to chemotherapy and radiotherapy – to select the ones with the most promising CAIX blocking reactivity (Xu *et al.*, 2010). Recently, other investigators have also used the ISS for examining the localization of other molecules including localization of Nrf-2 in nuclei of stressed human dermal fibroblasts (Gruber *et al.*, 2010), interaction of peroxiredoxin 1 (Prx-1) – tumor-derived molecular chaperone – with TLR-4 receptor on macrophages and dendritic cells which may lead to secretion of proinflammatory cytokines by these immune cells (Riddell *et al.*, 2010), co-localization of programmed death-1 (PD-1) protein with CD95/Fas receptor on the surface of Jurkat cells to understand the mechanisms of apoptotic events undergoing in CD8+ human LyT infected with HIV (Petrovas *et al.*, 2009).

D. Activation – Change of the Cell Shape

The ISS has also been employed to quantify and morphologically analyze cells during the process of their shape changing or pseudopod formation (Tan *et al.*, 2006b). Imaging cytometry may relatively easily distinguish rounded and elongated cells based on calculation of "aspect ratio" value (AR). AR is typically calculated based on the brightfield image of cells and is defined as the ratio of minor cellular axis (width) to the major axis (height). AR close to 1.0 indicates circular objects, while AR below 1.0 represents elongated cells.

More advanced features provided by IFC analytical software allow for distinguishing cells forming pseudopods based on the instance polarization of proteins involved in this process (e.g. Podo) (Tan *et al.*, 2006b). The specific distribution of such proteins during cell migration or pseudopod formation allows for the calculation of "Radial Delta Centroid" feature (RDC) that represents the radial distance between the geometrical centers of the nuclear image and the fluorescent image of Podo (Tan *et al.*, 2006b). If the protein is uniformly distributed throughout the membrane and is visible around the nucleus, the RDC value is lower than in the case when the protein is capped to the one side of the membrane during pseudopod formation. Plotting the RDC between the nuclear and Podo image distribution together with the AR allows for identification of three cellular populations: (i) nonelongated rounded population (with Podo protein uniformly distributed in the membrane); (ii) nonelongated capped cells before migration (with Podo on one pole of the cell); (iii) elongated pseudopod forming cells with high RDC and polarized Podo (Tan *et al.*, 2006b). The classification may be always confirmed by representative image galleries of each identified population (Tan *et al.*, 2006b).

Recently, the application specialists working in the field of imaging cytometry provided a new feature, which may be calculated by IFC software –"Circularity". Circularity parameter is calculated for each cell as the average radius divided by the radial variance. Typical round cells exhibit low radial variance, while ruffled or elongated cells high variance of radii that results in lower circularity parameter. Analyzed cells may be plotted on the histogram based on that value and further subpopulations of round and elongated cells may be identified, gated and further analyzed [Amnis data; www.amnis.com].

The multiple features computed by the ISS provide a wide range of applications for cellular phenotyping and characterization, while combining photometric features with various morphological parameters surely creates further future prospects in use of imaging cytometry.

E. Identification of Stem and Progenitor Cells

IFC may also be applied for identification and characterization of different cellular subpopulations including stem cells. The stem cell field focusing on subfractions of the unique cells with specific functional and morphological features and the cells that are predominantly very rare looks for analytical methods ensuring the presence of such stem/progenitor cell in different materials. The first method that has been commonly used for identification of distinct fractions of bone marrow (BM)-derived primitive cells based on their phenotype and specific antigens expression was classical FC. Recently, the ISS has been also applied for the identification of stem and progenitor cells including hematopoietic BM-derived progenitors from erythroid lineage during their maturation (McGrath *et al.*, 2008a) and adult very small embryonic-like stem cells (VSELs) in different animal and human specimens (Zuba-Surma *et al.*, 2008a, 2008d, 2010).

In the first elegant study, McGrath and colleagues employed IC for identification of erythroid progenitors and to follow their maturation into erythrocytes. Based on the abilities of IFC including correlative analysis of the surface antigen expression, changes in morphology of the nuclei and the cells themselves, it was possible to distinguish six independent stages of erythroid maturation. Moreover, the novel morphological features of these cells identified by ISS analysis including surface unevenness and unusually high contrast in brightfield images allowed for the discrimination of some rare populations of cells appearing during the erythroid development (McGrath *et al.*, 2008a, 2008b).

In the other studies, the ISS occurred to be very helpful for the identification of pluripotent population of VSEL stem cells in adult animal and human tissues (Kucia *et al.*, 2006, 2007; Zuba-Surma *et al.*, 2008a, 2008d, 2010). VSELs have been initially described in adult murine BM as very rare population and small cells with embryonic-like features. These unique adult stem cells were identified and purified by classical flow cytometric analysis and sorting based on the expression of Sca-1 (stem cell antigen 1) expression and lack of panleukocytic marker CD45 and hematopoietic lineages markers (Lin) (Kucia *et al.*, 2006; Ratajczak, 2008; Ratajczak *et al.*, 2004; Zuba-Surma *et al.*, 2008a, 2008d). However, the ISS for the first time allowed for the confirmation of the real existence of such very small Sca-1+/Lin-/CD45- cells and provided their more detailed morphological characterization (Zuba-Surma *et al.*, 2008a, 2008d, 2010). Based on the morphological analysis by IFC, we have confirmed that both murine and human VSELs are the cells that are smaller than erythrocytes, but larger than most of platelets (Ratajczak *et al.*, 2009; Zuba-Surma *et al.*, 2008a, 2010) (Fig. 1). The ISS has been also applied as a method supporting classical FC in identification of VSELs in other murine adult organs (Zuba-Surma *et al.*, 2008d). Based on the ability of the system and direct visualization of cells, we could distinguish real Sca-1+/Lin-/CD45- cells in murine tissues from artifacts, debris, and damaged cells which often imitated "positive" cells (Zuba-Surma *et al.*, 2008d) (Fig. 2). Moreover, for the first time we could confirm the existence of poluripotent Oct-4+ VSELs in

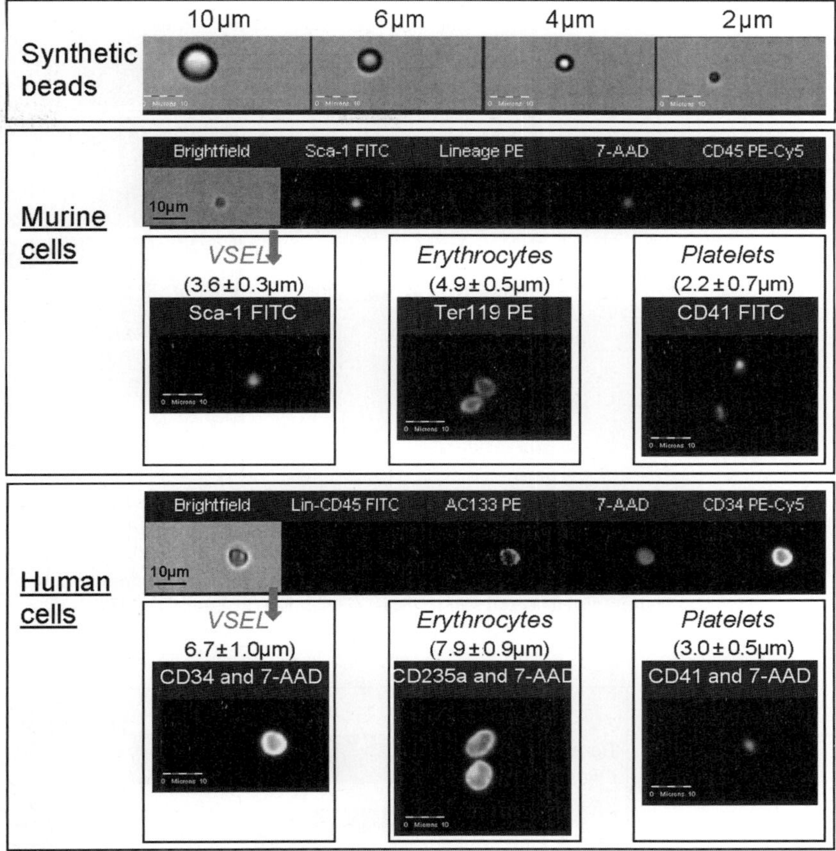

Fig. 1 Representative images presenting the morphology of murine and human VSELs by imaging flow cytometry (ImageStream system). Upper panel: brightfield images of beads with predefined sizes showed as size standards. Middle panel: murine BM-derived Sca-1+/Lin-/CD45- nucleated VSELs (Sca-1 [FITC], Lin [PE], CD45 [PE-Cy5], nucleus [7-aminoactionomycin D]) compared with murine erythrocytes (Ter119+ [PE]) and platelets (CD41+ [FITC]). Lower panel: human cord blood-derived CD133+/CD34 +/Lin-/CD45- nucleated VSELs (Lin and CD45 [FITC], CD133 [PE], CD34 [PE-Cy5], nucleus [7-AAD]). All of the images are shown at the same magnification. Scale bar = 10 (m. VSELs are distinguished from erythrocytes and platelets not only based on distinct surface markers, but also on the presence of nuclei in VSELs. The average sizes of murine and human cells are provided under the respective images. The size of cells has been calculated by employing the "minor axis" feature calculated in IDEAS and presented in micrometers. The same analytical masks have been applied for all analyzed objects.

adult tissues based on the capacity of the ISS to the localization of the fluorescent signal in a particular cellular compartment (Zuba-Surma *et al.*, 2008d). Thus, Oct-4+ VSELs were identified as very small Sca-1+/Lin-/CD45- cells with an intra-nuclear expression of Oct-4 (Fig. 3). We have also used the imaging cytometry for identification and characterization of VSELs in human umbilical cord blood

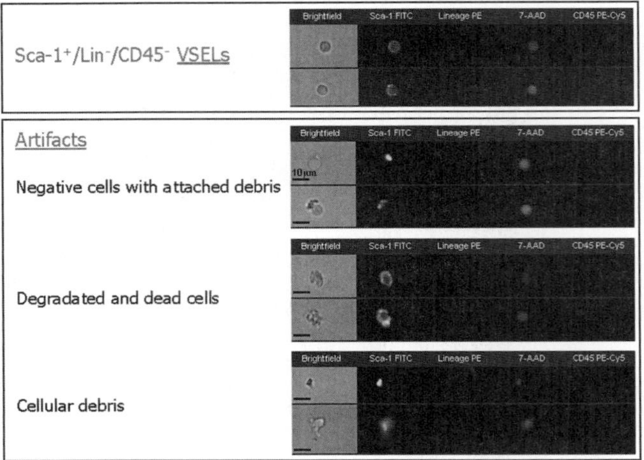

Fig. 2   VSEL stem cells and artifacts distinguished in the ISS. Upper panel: representative image of real Sca-1+/Lin-/CD45- VSEL detected in adult murine tissues. Lower panel: selected example images of falsely "positive" artifacts including (i) negative cells with attached debris fragment, (ii) damaged/ degradating cells, and (iii) cellular debris. Each photograph presents a brightfield image and fluorescent images related to expression of Sca-1 (FITC), Lin (PE), and CD45 (PE-Cy5) as well as nuclear image (7-AAD). The scale bar indicates 10 μm.

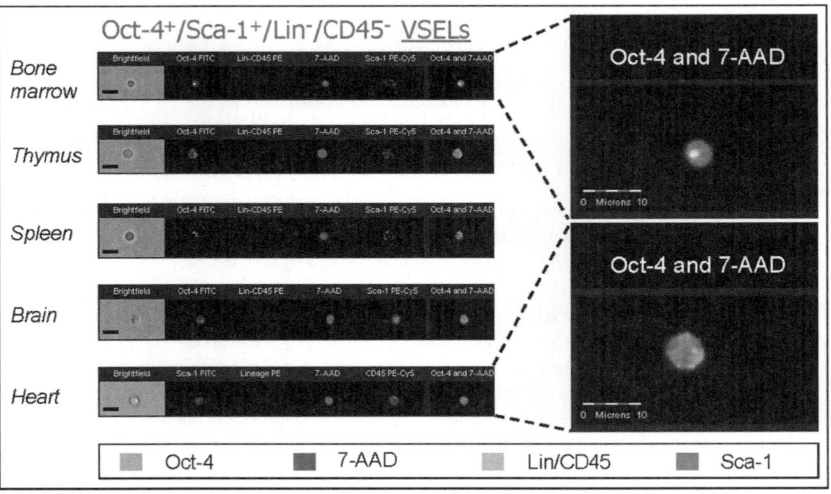

Fig. 3   Images of Oct-4+/Sca-1+/Lin-/CD45- VSEL stem cells detected in selected murine organs by imaging flow cytometry (ImageStream system). Cells were isolated from solid organs that were enzymatically digested to obtain the cell suspension and further stained for pluripotent marker, Oct-4 (FITC), CD45 and hematopoietic lineages markers (Lin, PE), and Sca-1 (PE-Cy5). Nuclei were stained with 7-AAD and cells were analyzed by the ISS to detect intranuclear expression of Oct-4 as shown in magnified, combined images. The scale bars indicate 10 μm.

(Zuba-Surma *et al.*, 2010) and circulating in peripheral blood of patients with tissue injury including acute myocardial infarction (Abdel-Latif *et al.*, 2010; Wojakowski *et al.*, 2009) as well as in embryonic tissues in a mouse model (Zuba-Surma *et al.*, 2009).

F. Other Applications

The ISS has also been employed for: (i) cell classification and identification (Morrissey *et al.*, 2004; Parsons *et al.*, 2006); (ii) identification of fluorescence *in situ* hybridization (FISH) positive cells (Brawley *et al.*, 2004); (iii) identification of mitotically active cells characterized by expression of various specific markers (Gillard and Farr, 2006; Hall *et al.*, 2004); (iv) identification of immune synapses between T cells and antigen presenting cells (APC) (Hall *et al.*, 2004).

V. Future Directions

The IFC has been developed as the next stage of modern flow cytometric technology. Based on combined analysis of photometric as well as morphometric features of cells, the system can be uniquely employed to a multitude of research and clinical applications. The ImageStream technology gives the unique opportunity to combine statistical analysis of large number of cells while discerning multicolor images through combining the features of modern FC and fluorescence microscopy in one system. In the future, when additional lasers and optical parts are widely available, the IFC brings new possibilities for using various combinations of dyes and fluorescent probes.

The potential application of IFC in diagnostics and other clinical assays became next important challenge for this technology and this goal has already high chance to be achieved soon. The important expectation to the system, which would require long-term extended development of the system, includes potential sorting of different cellular populations selected based on morphological and photometric features. Such advancement would equip the system with an option that would make it the most powerful imaging technology and bring it to the next level of scientific applications.

References

Abdel-Latif, A., Zuba-Surma, E. K., Ziada, K. M., Kucia, M., Cohen, D. A., Kaplan, A. M., Van Zant, G., Selim, S., Smyth, S. S., Ratajczak, M. Z. (2010). Evidence of mobilization of pluripotent stem cells into peripheral blood of patients with myocardial ischemia. *Exp. Hematol.* **Aug 25,**.

Amnis, C. (2006). Analysis of Complement 3b deposition on Rituximab opsonized cells using ImageStream® Multispectral Imaging Cytometry. www.amnis.com.

Arechiga, A. F., Bell, B. D., Solomon, J. C., Chu, I. H., Dubois, C. L., Hall, B. E., George, T. C., Coder, D. M., Walsh, C. M. (2005). Cutting edge: FADD is not required for antigen receptor-mediated NF-kappaB activation. *J. Immunol.* **175**, 7800–7804.

Barrett, K. L., Willingham, J. M., Garvin, A. J., and Willingham, M. C. (2001). Advances in cytochemical methods for detection of apoptosis. *J. Histochem. Cytochem.* **49**, 821–832.

Basiji, D., Ortyn, W., Zimmerman, C., Bauer, R., Perry, D., Esposito, R., George, T., Hall, B., and Frost, K. (2006). Image data exploration and analysis software. ISAC XXIII, Quebec, Canada, (conference proceeding).

Basiji, D. A., Ortyn, W. E., Liang, L., Venkatachalam, V., and Morrissey, P. (2007). Cellular image analysis and imaging by flow cytometry. *Clin. Lab. Med.* **27**, 653–670.

Baumgarth, N., and Roederer, M. (2000). A practical approach to multicolor flow cytometry for immunophenotyping. *J. Immunol. Methods* **243**, 77–97.

Bedner, E., Li, X., Gorczyca, W., Melamed, M. R., and Darzynkiewicz, Z. (1999). Analysis of apoptosis by laser scanning cytometry. *Cytometry* **35**, 181–195.

Bonetta, L. (2005). Flow cytometry smaller and better. *Nat. Method* **2**, 785–793.

Brawley, J., George, T., Hall, B., Jorgenson, R., Frost, K., Perry, D., Zimmerman, C., Seo, M., Basiji, D., Ortyn, W., and Morrissey, P. (2004). Fluorescent In Situ Hybridization In Suspension Analysis Using ImageStream® Multispectral Imaging Flow Cytometry. ISAC XII. (conference proceeding).

Cherry, R. J., Morrison, I. E., Karakikes, I., Barber, R. E., Silkstone, G., Fernandez, N. (2003). Measurements of associations of cell-surface receptors by single-particle fluorescence imaging. *Biochem. Soc. Trans.* **31**, 1028–1031.

Conchello, J. A., and Lichtman, J. W. (2005). Optical sectioning microscopy. *Nat. Methods* **2**, 920–931.

Darzynkiewicz, Z., Bedner, E., Li, X., Gorczyca, W., and Melamed, M. R. (1999). Laser-scanning cytometry: a new instrumentation with many applications. *Exp. Cell. Res.* **249**, 1–12.

Darzynkiewicz, Z., Bruno, S., Del Bino, G., Gorczyca, W., Hotz, M. A., Lassota, P., Traganos, F. (1992). Features of apoptotic cells measured by flow cytometry. *Cytometry* **13**, 795–808.

Darzynkiewicz, Z., Huang, X., and Okafuji, M. (2006). Detection of DNA strand breaks by flow and laser scanning cytometry in studies of apoptosis and cell proliferation (DNA replication). *Methods Mol. Biol.* **314**, 81–93.

Darzynkiewicz, Z., Huang, X., Okafuji, M., and King, M. A. (2004). Cytometric methods to detect apoptosis. *Methods Cell. Biol.* **75**, 307–341.

De Rosa, S. C., Brenchley, J. M., and Roederer, M. (2003). Beyond six colors: a new era in flow cytometry. *Nat. Med.* **9**, 112–117.

De Rosa, S. C., and Roederer, M. (2001). Eleven-color flow cytometry. A powerful tool for elucidation of the complex immune system. *Clin. Lab. Med.* **21**, 697–712.

Decaluwe, H., Taillardet, M., Corcuff, E., Munitic, I., Law, H. K., Rocha, B., Riviere, Y., and Di Santo, J. P. (2010) Gamma(c) deficiency precludes CD8+ T cell memory despite formation of potent T cell effectors. *Proc Natl Acad Sci U S A.* **107**, 9311-9316.

Degtyarev, M., De Maziere, A., Orr, C., Lin, J., Lee, B. B., Tien, J. Y., Prior, W. W., van Dijk, S., Wu, H., Gray, D. C., Davis, D. P., Stern, H. M., Murray, L. J., Hoeflich, K. P., Klumperman, J., Friedman, L. S., Lin, K. (2008). Akt inhibition promotes autophagy and sensitizes PTEN-null tumors to lysosomotropic agents. *J. Cell. Biol.* **183**, 101–116.

Deptala, A., Bedner, E., and Darzynkiewicz, Z. (2001). Unique analytical capabilities of laser scanning cytometry (LSC) that complement flow cytometry. *Folia Histochem. Cytobiol.* **39**, 87–89.

Deptala, A., Bedner, E., Gorczyca, W., and Darzynkiewicz, Z. (1998). Activation of nuclear factor kappa B (NF-kappaB) assayed by laser scanning cytometry (LSC). *Cytometry* **33**, 376–382.

Fanning, S. L., George, T. C., Feng, D., Feldman, S. B., Megjugorac, N. J., Izaguirre, A. G., Fitzgerald-Bocarsly, P. (2006). Receptor cross-linking on human plasmacytoid dendritic cells leads to the regulation of IFN- production. *J. Immunol.* **177**, 5829–5839.

George, T., Hall, B., Taylor, R., Beum, P., Cathleen Zimmerman, C., David Basiji, D., Philip Morrissey, P., Keith Frost, K., and William Ortyn, W. (2006). Assessment of Protein Co-localization with an Imaging Flow Cytometer and novel Analytical Method. ISAC XXIII, Quebec, Canada (conference proceeding).

George, T., Hall, B., Zimmerman, C., Frost, K., Seo, M., Perry, D., Ortyn, W., Basiji, D., Morrissey, P. (2004a). Quantitation of nuclear translocation events using ImageStream multispectral imaging cytometry. *Cytometry A* **59**, 121–1121.

George, T. C., Basiji, D. A., Hall, B., Lynch, D. H., Ortyn, W. E., Perry, D. J., Seo, M. J., Zimmerman, C. A., Morrissey, P. J. (2004b). Distinguishing modes of cell death using the ImageStream multispectral imaging flow cytometer. *Cytometry A* **59A**, 237–245.

George, T. C., Fanning, S. L., Fitzgeral-Bocarsly, P., Medeiros, R. B., Highfill, S., Shimizu, Y., Hall, B. E., Frost, K., Basiji, D., Ortyn, W. E., Morrissey, P. J., Lynch, D. H. (2006b). Quantitative measurement of nuclear translocation events using similarity analysis of multispectral cellular images obtained in flow. *J. Immunol. Methods* **311**, 117–129.

Gillard, G. O., and Farr, A. G. (2006). Features of medullary thymic epithelium implicate postnatal development in maintaining epithelial heterogeneity and tissue-restricted antigen expression. *J. Immunol.* **176**, 5815–5824.

Glisic-Milosavljevic, S., Waukau, J., Jana, S., Jailwala, P., Rovensky, J., Ghosh, S. (2005). Comparison of apoptosis and mortality measurements in peripheral blood mononuclear cells (PBMCs) using multiple methods. *Cell. Prolif.* **38**, 301–311.

Gruber, F., Mayer, H., Lengauer, B., Mlitz, V., Sanders, J. M., Kadl, A., Bilban, M., de Martin, R., Wagner, O., Kensler, T. W., Yamamoto, M., Leitinger, N., Tschachler, E. (2010). NF-E2-related factor 2 regulates the stress response to UVA-1-oxidized phospholipids in skin cells. *FASEB J.* **24**, 39–48.

Hall, B., George, T., Basiji, D., Frost, K., Zimmerman, C., and Ortyn, W. (2006). Automated Classification of Apoptosis and Artifact Rejection of TUNEL Positive Cells. ISAC XXIII, Quebec, Canada (conference proceeding).

Hall, B., Kong, R., Mordecai, S., Friend, S., and George, T. C. (2010). Classification of Interphase and Mitotic Phases of the Cell Cycle Using Image Cytometry. *Amnis-materials* (www.amnis.com). Program # 139/P53.

Hall, B., Perry, D., Brawley, J., George, T., Zimmerman, C., Frost, K., Basiji, D., Seo, M., Ortyn, W., Morrissey, P. (2004). Multispectral high content cellular analysis using a flow based imaging cytometer. *Cytometry A* **59**, 152–1152.

Hansma, P. K., Elings, V. B., Marti, O., and Bracker, C. E. (1988). Scanning tunneling microscopy and atomic force microscopy: application to biology and technology. *Science* **242**, 209–216.

Haupt, B. J., Pelling, A. E., and Horton, M. A. (2006). Integrated confocal and scanning probe microscopy for biomedical research. *Sci. World J.* **15**, 1609–1618.

Henderson, R. (2004). Realizing the potential of electron cryo-microscopy. *Q. Rev. Biophys.* **37**, 3–13.

Henery, S., George, T., Hall, B., Basiji, D., Ortyn, W., Morrissey, P. (2008). Quantitative image based apoptotic index measurement using multispectral imaging flow cytometry: a comparison with standard photometric methods. *Apoptosis* **13**, 1054–1063.

Hewitson, T. D., Bisucci, T., and Darby, I. A. (2006). Histochemical localization of apoptosis with in situ labeling of fragmented DNA. *Methods Mol. Biol.* **326**, 227–234.

Hunter, A. L., Choy, J. C., and Granville, D. J. (2005). Detection of apoptosis in cardiovascular diseases. *Methods Mol. Med.* **112**, 277–289.

Jaroszeski, M. J., and Radcliff, G. (1999). Fundamentals of flow cytometry. *Mol. Biotechnol.* **11**, 37–53.

Jones, C. W., Smolinski, D., Keogh, A., Kirk, T. B., and Zheng, M. H. (2005). Confocal laser scanning microscopy in orthopaedic research. *Prog. Histochem. Cytochem.* **40**, 1–71.

Kamentsky, L. A. (2001). Laser scanning cytometry. *Methods Cell. Biol.* **63**, 51–87.

Kamentsky, L. A., Burger, D. E., Gershman, R. J., Kamentsky, L. D., and Luther, E. (1997). Slide-based laser scanning cytometry. *Acta Cytol.* **41**, 123–143.

Kerr, J. F., Wyllie, A. H., and Currie, A. R. (1972). Apoptosis: a basic biological phenomenon with wide-ranging implications in tissue kinetics. *Br. J. Cancer* **26**, 239–257.

Khuda, S. E., Loo, W. M., Janz, S., Van Ness, B., and Erickson, L. D. (2008). Deregulation of c-Myc Confers distinct survival requirements for memory B cells, plasma cells, and their progenitors. *J. Immunol.* **181**, 7537–7549.

Kim, J., Kim, D., and Chung, J. (2000). Replication protein a 32 kDa subunit (RPA p32) binds the SH2 domain of STAT3 and regulates its transcriptional activity. *Cell. Biol. Int.* **24**, 467–473.

Klein, A. B., Witonsky, S. G., Ahmed, S. A., Holladay, S. D., Gogal, R. M. J., Link, L., Reilly, C. M. (2006). Impact of different cell isolation techniques on lymphocyte viability and function. *J. Immunoassay Immunochem.* **27**, 61–76.

Kohler, C., Orrenius, S., and Zhivotovsky, B. (2002). Evaluation of caspase activity in apoptotic cells. *J. Immunol. Methods* **256**, 97–110.

Krutzik, P. O., Clutter, M. R., and Nolan, G. P. (2005). Coordinate analysis of murine immune cell surface markers and intracellular phosphoproteins by flow cytometry. *J. Immunol.* **175**, 2357–2365.

Krutzik, P. O., and Nolan, G. P. (2003). Intracellular phospho-protein staining techniques for flow cytometry: minitoring single cell signaling events. *Cytometr A* **55**, 61–70.

Kucia, M., Halasa, M., Wysoczynski, M., Baskiewicz-Masiuk, M., Moldenhawer, S., Zuba-Surma, E., Czajka, R., Wojakowski, W., Machalinski, B., Ratajczak, M. Z. (2007). Morphological and molecular characterization of novel population of CXCR4(+) SSEA-4(+) Oct-4(+) very small embryonic-like cells purified from human cord blood - preliminary report. *Leukemia* **21**, 297–303.

Kucia, M., Reca, R., Campbell, F. R., Zuba-Surma, E., Majka, M., Ratajczak, J., Ratajczak, M. Z. (2006). A population of very small embryonic-like (VSEL) CXCR4(+)SSEA-1(+)Oct-4+ stem cells identified in adult bone marrow. *Leukemia* **20**, 857–869.

Lee, H. K., Lund, J. M., Ramanathan, B., Mizushima, N., and Iwasaki, A. (2007). Autophagy-dependent viral recognition by plasmacytoid dendritic cells. *Science* **315**, 1398–1401.

Liu, R., Klich, I., Ratajczak, J., Ratajczak, M. Z., and Zuba-Surma, E. K. (2009). Erythrocyte-derived microvesicles may transfer phosphatidylserine to the surface of nucleated cells and falsely "mark" them as apoptotic. *Eur. J. Haematol.*

Lovborg, H., Gullbo, J., and Larsson, R. (2005). Screening for apoptosis – classical and emerging techniques. *Anticancer Drugs* **16**, 593–599.

Lucitti, J. L., and Dickinson, M. E. (2006). Moving toward the light: using new technology to answer old questions. *Pediatr. Res.* **60**, 1–5.

Martin, R. M., Leonhardt, H., and Cardoso, M. C. (2005). DNA labeling in living cells. *Cytometry A* **67**, 45–52.

McGrath, K. E., Bushnell, T. P., and Palis, J. (2008a). Multispectral imaging of hematopoietic cells: where flow meets morphology. *J. Immunol. Methods* **336**, 91–97.

McGrath, K. E., Kingsley, P. D., Koniski, A. D., Porter, R. L., Bushnell, T. P., Palis, J. (2008b). Enucleation of primitive erythroid cells generates a transient population of "pyrenocytes" in the mammalian fetus. *Blood* **111**, 2409–2417.

Micoud, F., Mandrand, B., and Malcus-Vocanson, C. (2001). Comparison of several techniques for the detection of apoptotic astrocytes in vitro. *Cell. Prolif.* **34**, 99–113.

Miyashita, T. (2004). Confocal microscopy for intracellular co-localization of proteins. *Methods Mol. Biol.* **261**, 399–410.

Morrissey, P., George, T., Basiji, D., Frost, K., Ortyn, W., Williams, D. (2004). Cell classification in human peripheral blood using the amnis ImageStream system. *Blood* **104**, 43B–143B.

Njoh, K. L., Patterson, L. H., Zloh, M., Wiltshire, M., Fisher, J., Chappell, S., Ameer-Beg, S., Bai, Y., Matthews, D., Errington, R. J., Smith, P. J. (2006). Spectral analysis of the DNA targeting bisalkyla-minoanthraquinone DRAQ5 in intact living cells. *Cytometry A* **69**, 805–814.

Nobile, C., Rudnicka, D., Hasan, M., Aulner, N., Porrot, F., Machu, C., Renaud, O., Prevost, M. C., Hivroz, C., Schwartz, O., Sol-Foulon, N. (2010). HIV-1 Nef inhibits ruffles, induces filopodia, and modulates migration of infected lymphocytes. *J Virol.* **84**, 2282–2293.

Nobile, C., Rudnicka, D., Hasan, M., Aulner, N., Porrot, F., Machu, C., Renaud, O., Prevost, M. C., Hivroz, C., Schwartz, O., Sol-Foulon, N. (2010). HIV-1 Nef inhibits ruffles, induces filopodia, and modulates migration of infected lymphocytes. *J. Virol.* **84**, 2282–2293.

Nohe, A., and Petersen, N. O. (2004). Analyzing for co-localization of proteins at a cell membrane. *Curr. Pharm. Biotechnol.* **5**, 213–220.

Ortyn, W. E., Hall, B. E., George, T. C., Frost, K., Basiji, D. A., Perry, D. J., Zimmerman, C. A., Coder, D. C., Morrissey, P. J. (2006). Sensitivity measurement and compensation in spectral imaging. *Cytometry A* **69A**, 852–862.

Ortyn, W. E., Perry, D. J., Venkatachalam, V., Liang, L., Hall, B. E., Frost, K., Basiji, D. A. (2007). Extended depth of field imaging for high speed cell analysis. *Cytometry A* **71**, 215–231.

Otsuki, Y., Li, Z., and Shibata, M. A. (2003). Apoptotic detection methods–from morphology to gene. *Prog. Histochem. Cytochem.* **38**, 275–339.

Parsons, C. H., Adang, L. A., Overdevest, J., O'Connor, C. M., Taylor Jr., J. R., Camerini, D., Kedes, D. H. (2006). KSHV targets multiple leukocyte lineages during long-term productive infection in NOD/SCID mice. *J. Clin. Invest.* **116**, 1963–1973.

Pawluczkowycz, A. W., Beurskens, F. J., Beum, P. V., Lindorfer, M. A., van de Winkel, J. G., Parren, P. W., Taylor, R. P. (2009). Binding of submaximal C1q promotes complement-dependent cytotoxicity (CDC) of B cells opsonized with anti-CD20 mAbs ofatumumab (OFA) or rituximab (RTX): considerably higher levels of CDC are induced by OFA than by RTX. *J. Immunol.* **183**, 749–758.

Perfetto, S. P., Chattopadhyay, P. K., and Roederer, M. (2004). Seventeen-colour flow cytometry: unravelling the immune system. *Nat. Rev. Immunol.* **4**, 648–655.

Petrovas, C., Chaon, B., Ambrozak, D. R., Price, D. A., Melenhorst, J. J., Hill, B. J., Geldmacher, C., Casazza, J. P., Chattopadhyay, P. K., Roederer, M., Douek, D. C., Mueller, Y. M., Jacobson, J. M., Kulkarni, V., Felber, B. K., Pavlakis, G. N., Katsikis, P. D., Koup, R. A. (2009). Differential association of programmed death-1 and CD57 with ex vivo survival of CD8+ T cells in HIV infection. *J. Immunol.* **183**, 1120–1132.

Pozarowski, P., Holden, E., and Darzynkiewicz, Z. (2006). Laser scanning cytometry: principles and applications. *Methods Mol. Biol.* **319**, 165–192.

Radcliff, G., and Jaroszeski, M. J. (1998). Basics of flow cytometry. *Methods Mol. Biol.* **91**, 1–24.

Ratajczak, M. Z. (2008). Phenotypic and functional characterization of hematopoietic stem cells. *Curr. Opin. Hematol.* **15**, 293–300.

Ratajczak, M. Z., Kucia, M., Ratajczak, J., and Zuba-Surma, E. K. (2009). A multi-instrumental approach to identify and purify very small embryonic like stem cells (VSELs) from adult tissues. *Micron* **40**, 386–393.

Ratajczak, M. Z., Kucia, M., Reca, R., Majka, M., Janowska-Wieczorek, A., Ratajczak, J. (2004). Stem cell plasticity revisited: CXCR4-positive cells expressing mRNA for early muscle, liver and neural cells 'hide out' in the bone marrow. *Leukemia* **18**, 29–40.

Riddell, J. R., Wang, X. Y., Minderman, H., and Gollnick, S. O. (2010). Peroxiredoxin 1 stimulates secretion of proinflammatory cytokines by binding to TLR4. *J. Immunol.* **184**, 1022–1030.

Roederer, M. (2001). Spectral compensation for flow cytometry: visualization artifacts, limitations, and caveats. *Cytometry A* **45**, 194–205.

Sgonc, R., and Wick, G. (1994). Methods for the detection of apoptosis. *Int. Arch. Allergy Immunol.* **105**, 327–332.

Shapiro, H. M. (1983). Multistation multiparameter flow cytometry: a critical review and rationale. *Cytometry* **3**, 227–243.

Shapiro, H. M. (2005). *Practical Flow Cytometry.* John Wiley & Sons, Inc, New York.

Spibey, C. A., Jackson, P., and Herick, K. (2001). A unique charge-coupled device/xenon arc lamp based imaging system for the accurate detection and quantitation of multicolour fluorescence. *Electrophoresis* **22**, 829–836.

Subramaniam, S., and Milne, J. L. (2004). Three-dimensional electron microscopy at molecular resolution. *Annu. Rev. Biophys. Biomol. Struct.* **33**, 141–155.

Swick, L., and Kapatos, G. (2006). A yeast 2-hybrid analysis of human GTP cyclohydrolase I protein interactions. *J. Neurochem.* **97**, 1447–1455.

Tan, P. C., Kelly M. McNagny, K. M., and Hall, B. (2006a). Quantitative analysis of pseudopod formation with the ImageStream cell imaging system. *http://www.amnis.com/docs/notes/Quantitative-PseudopodFormation.pdf.*

Tan, P. C., McNagny, K. M., and Hall, B. (2006b). Quantitative analysis of pseudopod formation with the ImageStream cell imaging system.

Tarnok, A., and Gerstner, A. O. (2002). Clinical applications of laser scanning cytometry. *Cytometry* **50**, 133–143.

Totino, P. R., Riccio, E. K., Corte-Real, S., Daniel-Ribeiro, C. T., and de Fatima Ferreira-da-Cruz, M. (2006). Dexamethasone has pro-apoptotic effects on non-activated fresh peripheral blood mononuclear cells. *Cell. Biol. Int.* **30**, 133–137.

Vermes, I., Haanen, C., and Reutelingsperger, C. (2000). Flow cytometry of apoptotic cell death. *J. Immunol. Methods* **243**, 167–190.

Voss, T. C., Demarco, I. A., and Day, R. N. (2005). Quantitative imaging of protein interactions in the cell nucleus. *Biotechniques* **38**, 413–424.

Wheeless, L. L., Reeder, J. E., O'Connell, M. J., Robinson, R. D., Cosgriff, J. M., Fradet, Y., Frank, I. N., Cockett, A. T. (1991). DNA slit-scan flow cytometry of bladder irrigation specimens and the importance of recognizing urothelial cells. *Cytometry* **12**, 140–146.

Willingham, M. C. (1999). Cytochemical methods for the detection of apoptosis. *J. Histochem. Cytochem.* **47**, 1101–1110.

Wlodkowic, D., Skommer, J., and Darzynkiewicz, Z. (2010). Cytometry in cell necrobiology revisited. Recent advances and new vistas. *Cytometry A* **77**, 591–606.

Wojakowski, W., Tendera, M., Kucia, M., Zuba-Surma, E., Paczkowska, E., Ciosek, J., Halasa, M., Krol, M., Kazmierski, M., Buszman, P., Ochala, A., Ratajczak, J., Machalinski, B., Ratajczak, M. Z. (2009). Mobilization of bone marrow-derived Oct-4+ SSEA-4+ very small embryonic-like stem cells in patients with acute myocardial infarction. *J. Am. Coll. Cardiol.* **53**, 1–9.

Xu, C., Lo, A., Yammanuru, A., Tallarico, A. S., Brady, K., Murakami, A., Barteneva, N., Zhu, Q., Marasco, W. A. (2010). Unique biological properties of catalytic domain directed human anti-CAIX antibodies discovered through phage-display technology. *PLoS One* **5**, e9625.

Yasuhara, S., Zhu, Y., Matsui, T., Tipirneni, N., Yasuhara, Y., Kaneki, M., Rosenzweig, A., Martyn, J. A. (2003). Comparison of comet assay, electron microscopy, and flow cytometry for detection of apoptosis. *J. Histochem. Cytochem.* **51**, 873–885.

Yu, Z., Zhang, W., and Kone, B. C. (2002). Signal transducers and activators of transcription 3 (STAT3) inhibits transcription of the inducible nitric oxide synthase gene by interacting with nuclear factor kappaB. *Biochem. J.* **367**, 97–105.

Yuan, C. M., Douglas-Nikitin, V. K., Ahrens, K. P., Luchetta, G. R., Braylan, R. C., Yang, L. (2004). DRAQ5-based DNA content analysis of hematolymphoid cell subpopulations discriminated by surface antigens and light scatter properties. *Cytometry B Clin. Cytom.* **58**, 47–52.

Zuba-Surma, E. K., Klich, I., Greco, N., Laughlin, M. J., Ratajczak, J., Ratajczak, M. Z. (2010). Optimization of isolation and further characterization of umbilical-cord-blood-derived very small embryonic/epiblast-like stem cells (VSELs). *Eur. J. Haematol.* **84**, 34–46.

Zuba-Surma, E. K., Kucia, M., Abdel-Latif, A., Dawnn, B., Hall, B., Singh, R., Lillard, J. W., Ratajczak, M. Z. (2008a). Morphological characterization of very small embryonic-like stem cells (VSELs) by ImageStream system analysis. *J. Cell. Mol. Med.* **12**, 292–303.

Zuba-Surma, E. K., Kucia, M., Abdel-Latif, A., Lillard, J. J., and Ratajczak, M. Z. (2007). The ImageStream system: a key step to a new era in imaging. *Folia Histochem. Cytobiol.* **45**, 279–290.

Zuba-Surma, E. K., Kucia, M., and Ratajczak, M. Z. (2008b). "Decoding of Dot": The ImageStream system (ISS) as a supportive tool for flow cytometric analysis. *Cent. Eur. J. Biol.* **3**, 1–10.

Zuba-Surma, E. K., Kucia, M., Rui, L., Shin, D. M., Wojakowski, W., Ratajczak, J., Ratajczak, M. Z. (2009). Fetal liver very small embryonic/epiblast like stem cells follow developmental migratory pathway of hematopoietic stem cells. *Ann. NY Acad. Sci.* **1176**, 205–218.

Zuba-Surma, E. K., Kucia, M., Wu, W., Klich, I., Lillard Jr., J. W., Ratajczak, J., Ratajczak, M. Z. (2008c). Very small embryonic-like stem cells are present in adult murine organs: ImageStream-based morphological analysis and distribution studies. *Cytometry A* **73A**, 1116–1127.

Zuba-Surma, E. K., Kucia, M., Wu, W., Klich, I., Lillard Jr., J. W., Ratajczak, J., Ratajczak, M. Z. (2008d). Very small embryonic-like stem cells are present in adult murine organs: ImageStream-based morphological analysis and distribution studies. *Cytometry A.* **Oct 24**, 2008.

CHAPTER 9

Laser Scanning Cytometry: Capturing the Immune System *In situ*

Mairi A. McGrath, Angela M. Morton and Margaret M. Harnett

Institute of Infection, Immunity and Inflammation, College of Medical Veterinary & Life Sciences, Glasgow Biomedical Research Centre, University of Glasgow, Scotland, UK

Abstract
- I. Introduction
- II. Background: Laser Scanning Cytometry Technology for Quantitatively Imaging and Analyzing Immune Responses *In situ*
- III. Rationale for LSC Analysis of Antigen–Specific T Cell Responses *In vitro* and *In vivo*
- IV. Detailed Protocols for Tracking Antigen–Specific T Cell Responses
    - A. Analysis of Antigen–Specific T Cells *In vitro*
    - B. Analysis of Antigen–Specific T Cells in Tissue *In situ*
- V. Acquisition and Analysis of Data Using WinCyte Software
    - A. Acquisition and Analysis of Antigen Specific T Cells *In vitro*
    - B. Acquisition and Analysis of Antigen Specific T Cells *In situ*
- VI. Results: Analysis of the Role of pERK Signaling in Antigen–Specific Priming of T Cells
- VII. Application of LSC Technology to Analysis of the Immune System in Health and Disease
- VIII. Concluding Remarks and Future Directions
    Acknowledgments
    References

Abstract

Until recently, it has not been possible to image and functionally correlate the key molecular and cellular events underpinning immunity and tolerance in the intact immune system. Certainly, the field has been revolutionized by the advent of tetramers to identify physiologically relevant specificities of T cells, and the introduction of models in which transgenic T-cell receptor and/or B-cell receptor-bearing lymphocytes are adoptively transferred into normal mice and can then be identified by clonotype-specific antibodies using flow cytometry *in vitro*, or immunohistochemistry *ex vivo*. However, these approaches do not allow for quantitative analysis of the

231

0091-679X/10 $35.00
DOI 10.1016/B978-0-12-374912-3.00009-2

precise anatomical, phenotypic, signaling, and functional parameters required for dissecting the development of immune responses in health and disease *in vivo*. Traditionally, assessment of signal transduction pathways has required biochemical or molecular biological analysis of isolated and highly purified subsets of immune system cells. Inevitably, this creates potential artifacts and does not allow identification of the key signaling events for individual cells present in their microenvironment *in situ*. These difficulties have now been overcome by new methodologies in cell signaling analysis that are sufficiently sensitive to detect signaling events occurring in individual cells *in situ* and the development of technologies such as laser scanning cytometry that provide the tools to analyze physiologically relevant interactions between molecules and cells of the innate and the adaptive immune system within their natural environmental niche *in vivo*.

I. Introduction

Understanding the molecular and cellular interactions that regulate the development and phenotype of immune responses is central to the development of safe novel therapies to combat autoimmune and allergic inflammatory disorders as well as the production of efficacious vaccines to fight infection. However, until recently it has not been possible to analyze physiologically relevant interactions *in situ* as the technology to directly image, quantitatively analyze, and functionally correlate the key molecular and cellular events underpinning immunity and tolerance in the intact immune system did not exist. Thus, for example, delineation of the molecular mechanisms underpinning lymphocyte responses traditionally involved cell-free, biochemical assays such as Western blot analysis following polyclonal or mitogenic stimulation of immortalized cell lines or large populations of purified cells, or alternatively, restimulation of antigen-specific clones. Such signals, however, do not necessarily reflect the responses of naive antigen-specific cells found at physiological frequencies within their specialized environmental niche in primary or secondary lymphoid tissue. In addition, such biochemical analysis only represents the "average" of the summed responses of the population at any one time and therefore does not provide any information on the differential kinetics, amplitude or subcellular localization of signals generated by individual cells or functionally distinct subgroups within the population. Moreover, the tissue disruption involved in cell purification inevitably will create potential artifacts and prevent identification of signaling events between functionally or lineage-distinct cells occurring as a consequence of their microenvironment. Although the advent of genetically modified mice expressing transgenic (Tg) antigen receptors or immunoregulatory molecules has resolved some of these problems, functional or signaling analysis of such genetically modified animals has generally still been carried out on distinct cell populations purified *ex vivo* and using classical biochemical methodology, which does not allow preservation of functional cells or cell–cell interactions within physiological microenvironments.

Importantly, however, the recent development of antibodies that can detect post-translational modifications, such as phosphorylation of specific regulatory sites, has now allowed quantitative flow cytometric analysis (FCM) of the activation status of particular signaling elements in individual cells (Krutzik and Nolan, 2006). These types of reagents provided a key breakthrough in signaling technology as the ability of FCM to correlate such signals with cell lineage and functional responses, such as proliferation (carboxyfluorescein succinimidyl ester (CFSE) analysis of cell division), mitochondrial potential integrity (3,3′-dihexyloxacarbocyanine (DiOC6) uptake) and intracellular cytokine production, abrogates the need to purify individual populations of immune system cells (Fang *et al.*, 1998; Krutzik and Nolan, 2006; Marshall *et al.*, 2005b, 2008; Wilson *et al.*, 2003). Furthermore, the new generation of FCM platforms like the ImageStream cytometer (www.Amnis.com) combines the population statistics capabilities of FCM with quantitative image analysis to provide rapid, high-throughput analysis of the morphological parameters or subcellular localization of fluorescence staining of isolated cells, in addition to the immuno-phenotypic information provided by classical FCM (Arechiga *et al.*, 2005; Fanning *et al.*, 2006; Parsons *et al.*, 2006). Even such advanced FCM platforms, however, cannot answer the key questions relating to cell–cell interactions, either in terms of relevant cells involved or indeed, their site of action in tissue microenvironments. In contrast, the recent advances in solid-phase quantitative imaging technology such as laser scanning cytometry (LSC; www.compucyte.com) now allow the detection of signaling and functional events that occur during cell–cell interactions within the intact immune system *in situ* (Harnett, 2007).

This chapter therefore focuses on LSC technology and its potential to provide the tools for quantitative analysis of the precise anatomical, phenotypic, signaling, and functional parameters that are required for dissecting the development of immune responses in health and disease *in vivo*. For example, we have used LSC to investigate the role of antagonistic ERK MAP kinase (ERK) and Rap1 signals in governing antigen (Ag)-specific CD4 T cell responses by tracking expression of activated (dually phosphorylated Thr202Tyr204) ERK and Rap1 in Ag-specific Tg T-cell receptor (TCR)-bearing T cells both *in vitro* and, following adoptive transfer of such Ag-specific T cells, within their physiological environment of the lymph nodes consequent to induction of priming and tolerance *in vivo* (Adams *et al.*, 2004b; Harnett, 2007; Morton *et al.*, 2007). The increasing range of antibodies available for detecting site-specific phosphorylation-associated activatory and inhibitory signals means that this approach can easily be extended to the analysis of the activities of candidate downstream signals such as c-Myc, Rb, cdc2, and p27^{kip} that are also modified by phosphorylation. Similarly, antibodies specific for other regulatory post-translational modifications such as acetylation, ubiquitination, and sumoylation can also be exploited for quantitative LSC analysis of such signaling at the single cell level. Moreover, the ability of LSC to track differential subcellular localization of signals now allows visualization and quantitation of the recruitment and activation of signaling pathways, which, due to problems in radiolabeling primary lymphocytes to high specific activities, have traditionally been difficult to analyze in the immune system by classical biochemical methodologies. For example, lipid-directed signaling enzymes such as phospholipase C (PLC)γ, phospholipase D (PLD), cytoplasmic phospholipase A_2 (cPLA$_2$), protein kinase C, and sphingosine kinase all translocate to membranes on activation, a property that has been widely used as an indicator of their activation (Katz *et al.*, 2001, 2004; Melendez *et al.*, 2007). Such translocation/ activation can be easily quantified by LSC either by setting subcellular localization gates and/or using the max pixel facility. Similarly, for GTPases such as Rap and Ras, LSC-trackable assays based on the principle underlying the well-established pull-down biochemical assays where, for example, exploiting the binding of Rap1 to a RAL-binding domain GST fusion protein, allows an *in situ* Rap activity assay to be developed. Alternatively, antibodies recognising the active form of Ras-related GTPases and similarly, antibodies that can detect the active, DNA-bound forms of transcription factors are now commercially available and can be adapted for quantitative analysis by LSC. In this way, LSC can directly define the functional consequences of particular signals within individual cells or between cells *in situ* and in concert with the use of the reiterative staining/relocalization facility, now allows quantitative imaging of the recruitment and identity of multiple components of signaling pathways (signalsome) within individual cells and analysis of their cellular functional responses.

II. Background: Laser Scanning Cytometry Technology for Quantitatively Imaging and Analyzing Immune Responses *In situ*

The LSC combines an optics unit that generates the laser scanning beam with an upright epifluorescence microscope containing a motorized stage to allow scanning and imaging of samples: scan data are acquired and analyzed using WinCyte software (Harnett, 2007). Computer-controlled shuttering of the main dichroic mirror directs the beams from a range of lasers (argon: 488 nm; helium–neon: 633 nm; and a violet diode laser: 405 nm) to the scan mirror to generate a beam across the scan lens, which is then directed down the microscope objective onto the focal plane of the sample to excite fluorescence. Scattered light is collected by the condenser lens and directed to the forward-scatter photodiode, while emitted fluorescent light is collected by the objective and returned back to the main dichroic mirror and then down through a series of optical filter cubes that reflect selected light frequencies to the appropriate photomultiplier tubes (PMTs). As standard, 463/DF50 (blue), 530/DF30 (green), 580/DF30 (orange), and a 650 nm long-pass filter (long red) band pass filters are used but these can be replaced and/or augmented by additional filters to optimize analysis of tandem dyes or quantum dots (Qdots). Emitted light signals are digitized to create pixel values for each PMT. In addition, the x and y coordinates of the individual cells within a slide are recorded allowing relocalization for restaining, reanalysis, and/or generation of tissue maps (Grierson *et al.*, 2005c; Harnett, 2007). The new generation of LSC, the 4-laser iCys (with additional

532 nm laser), provides a versatile inverted microscope format that allows the use of various sample holders such as multiwell plates and multislide carriers that permit live-cell assays and walk-away analysis. Here, detectors above the samples can collect either transmitted light (to allow quantitation and visualization of light absorption) or scattered light (to obtain brightfield-like laser scatter images) permitting detection of a combination of scatter, absorption, and fluorescent data in real time and hence allowing quantitative analysis of chromatic and fluorescent staining in the same samples (Harnett, 2007).

LSC can analyze the fluorescence and morphological and subcellular features of large numbers of cells, either isolated or within tissue in a slide-based format, with good spatial resolution as it can scan relatively large areas of slides/sections of tissue without the need for refocusing (Gerstner *et al.*, 2004; Grierson *et al.*, 2005a; Harnett, 2007; Kamentsky and Kamentsky, 1991; Luther *et al.*, 2004; Mittag *et al.*, 2005, 2006b; Pozarowski *et al.*, 2006; Tarnok and Gerstner, 2002). This is because the collimated laser beam permits quantification of all the emitted light from cells (depth of field typically 20–30 μm) and hence, this technology is distinct but complementary to that of laser confocal microscopy (LCM) which, although providing highly detailed structural information, can only analyze fluorescence that is emitted close to the focal plane and therefore, can only address small numbers of cells. Moreover, although fluorescence image analysis (FIA) systems, in which cells are excited by highly stabilized mercury or xenon burners and emitted fluorescence is imaged at high resolution by a sensitive, color CCD video camera (e.g., ScanR; www.olympus-europa.com), can replicate some of the features of LSC (Table I), the intense monochromatic excitation that is generated by laser light in LSC not only offers higher detection sensitivity but additionally provides light scatter and absorption and brightfield visualization capabilities (Harnett, 2007; Mittag *et al.*, 2006b).

Thus although LSC, LCM, FIA, and FCM should be viewed and used as complementary quantitative platforms (Table I), only LSC provides all the tools to analyze the intact immune system *in situ* (Dey, 2006; Gerstner *et al.*, 2006a; Grierson *et al.*, 2005a; Mittag *et al.*, 2006a; Mittag *et al.*, 2005, 2006b; Rew *et al.*, 2006; Tarnok and Gerstner, 2002). This is because its capacity to quantitate events in individual cells within tissue sections not only prevents generation of potential signaling artefacts caused by isolation of cells from disrupted tissues but also allows for the identification and dissection of responses occurring due to interactions between subsets of cells of the immune system in their physiological microenvironment (Adams *et al.*, 2004b, 2004c; Gerstner *et al.*, 2004; Grierson *et al.*, 2005b; Harnett, 2007; Marshall *et al.*, 2005b; Morton *et al.*, 2007; Smith *et al.*, 2004a; Taatjes *et al.*, 2001). For example, because it records the precise x and y coordinates of each detected cell on the slide, merging of data files resulting from repeated scans by the LSC operating and analysis software (WinCyte) generates virtual data files where all cells with identical x–y coordinates in the different analyses are identified as being the same cell. Thus, when applied to tissue sections, the x, y-relocation facility permits construction of "tissue maps" which allow visualization and quantitative analysis

Table I

Complementary properties of laser scanning cytometry (LSC), flow cytometry (FCM), and fluorescence image analysis (FIA) systems

Properties	LSC	FCM	FIA
Imaging and quantitative analysis of • tissue sections • adherent cells • cytospins	Yes Brain sections up to 120 μm have been analyzed (Mosch et al., 2006, 2007) but for tissues such as lymph nodes, sections of ∼6 μm are optimal.	No	Yes lower detection sensitivity than lasers and image resolution is limited by the resolution capabilities of the camera. Image acquisition time is dependent on signal level but high magnification (100×) analysis can be performed.
Relationship between quantitative assessment and tissue architecture	Yes Tissue maps (*x*, *y*-plots) can be constructed and the iCys provides instantaneous relocation and image capture generating linked image mosaics alongside tissue maps.	No	Yes Relocation feature provides images of tissue and quantitative analysis of signal, but at present no tissue maps
Imaging and quantitative analysis of • morphology • subcellular localization/ translocation	Yes Contouring (Segmentation) of irregular cell shapes and densely packed cells greatly improved by Watershed feature.	No. ImageStream cytometer provides images and analysis (6-channel CCD) equivalent to 40–60× microscopy. In addition, can analyze *in vitro*-generated cell conjugates	Yes Can contour irregular cell morphology
Relocation Facility • restain • rescan • reanalyze • archive • live cell assays • kinetic assays.	Yes Allows archiving and restain/ reanalysis of samples, live cell and kinetic assays. Generation of *virtual* colors greatly increases number of parameters phenotyped. Automated plate loading and high throughput modules	No Samples are discarded and hence cannot be archived for reanalysis with additional probes Time resolution events cannot be analyzed on individual cells	Yes Allows archiving and restain/ reanalysis of samples, live cell, and kinetic assays. Automated plate loading and high throughput modules

Multicolor analysis	Yes Rescan/reanalysis facilities can generate many *virtual* colors iCys can measure chromatic and fluorescent staining as well as laser light scatter	Yes Up to 17-color plus forward and side scatter	Yes [8] colors per scan. ScanR has eight filter wheel positions and filters can be changed for additional pass if required.
Acquisition and storage acquisition time depends on • area of image • signal brightness • resolution of image • number of channels of data acquired	Lasers provide best resolution of emission and excitation fluorescence and hence higher detection sensitivity than FIA. Slower than FCM and FIA but as iCys analyzes the images as they are acquired this reduces processing time. However, many parameters can be set post acquisition as only scan area, laser, and PMT settings are now required for acquisition. Typically, for a $20\times$ scan field of 500×384 μm, 200 cells can be imaged in 4 s whilst a 5-fold reduction in resolution would typically image 800 cells in a 2.5 mm $\times$ 384 μm scan field in 4 s. iCys data files (.fcs ~2 MB; images ~1–20 MB) Hard drive 1 TB plus 8 GB RAM.	Fast $<$10,000 cells/min High cell capacity allows quantitation of rare events and weak antigens but analysis of small, rare cell populations in clinical samples, such as fine needle aspirates, is difficult FCM data files (fcs: $<$20 MB) but not generally possible to generate imaging files. The exception is the high speed, high cell capacity ImageStream cytometer that uses 6-channel-based CCD for generation of images	Fast Image acquisition time is variable depending on signal (msec-s). FIA data files (fcs. ~GB; requires two hard drives (80 GB + 120 GB) plus 2 GB RAM). acquire and analyze simultaneously.

of the precise molecular and cellular interactions of individual immune system cells within their physiological niche *in situ* (Gerstner *et al.*, 2004; Grierson *et al.*, 2005b; Harnett, 2007).

Importantly, because the sample is not discarded, this capacity of LSC to record the precise *x* and *y* coordinates of individual cells also makes it is possible to relocate to, and perform iterative staining/tracking of cells and should allow development of protocols for the temporal analysis of signalsome recruitment, performing kinetic analyses on individual cells in real time. Indeed, a wide range of live cell assays which can be measured in real time such as analysis of apoptosis, cell cycle status, and proliferation; fluorescence resonance energy transfer (FRET) assays of protein interactions; kinetic assays including translocation of signaling elements from the cytoplasm to the nucleus; intracellular calcium, pH, and membrane potential assays; and functional assays including chemotaxis, phagocytosis, endocytosis, and cell spreading can be analyzed by LSC technology exploiting the relocation facility (Butt *et al.*, 2005; Doyle *et al.*, 2004; Koo *et al.*, 2007; Mital *et al.*, 2006).

In addition, the relocation facility enables archiving of precious samples for future staining and analysis (Gerstner *et al.*, 2004, 2006a; Holme *et al.*, 2007; Laffers *et al.*, 2006; Luther *et al.*, 2004; Mittag *et al.*, 2006a, 2006b; Tarnok and Gerstner, 2002). This feature has proved particularly valuable in the analysis of rare clinical samples, such as fine needle aspirates (Bocsi *et al.*, 2006; Dey, 2006; Gerstner *et al.*, 2005, 2006a, 2006b; Kornblau *et al.*, 2006; Laffers *et al.*, 2006; Rew *et al.*, 2006; Taatjes *et al.*, 2006; Tarnok and Gerstner, 2002), where the iterative staining approach has been used to effectively increase the range of antigens detected in the immunophenotyping of rare cell populations in solid tumors, fine-needle-aspirate biopsies, patient swabs, and peripheral blood leuko-cytes (Gerstner *et al.*, 2005, 2006b; Laffers *et al.*, 2006; Mittag *et al.*, 2005, 2006a, 2006b; Tarnok and Gerstner, 2002).

Pertinent to this, the recent development of 17-color FCM (Perfetto *et al.*, 2004) has highlighted the value of precise phenotyping of distinct effector cell subsets for the consequent unequivocal delineation of their role in the immune system. Although it is only possible at present to detect some eight to nine fluorochromes simultaneously by LSC, this has led to the development of several creative strategies (Mittag *et al.*, 2006b), which exploit the restain, relocate, and reanalysis capabilities of LSC, to generate a large repertoire of "Virtual Colors" by a combining existing polychromatic cytometry, iterative restaining, and differential photobleaching, photoactivation, and photodestruction methodologies (Laffers *et al.*, 2006; Mittag *et al.*, 2006a, 2006b) using a range of conventional and Alexa-based dyes, Qdots, and FRET dyes, respectively (Table II). Although careful optimization and validation of such strategies is required, results to date have indicated that potentially, analysis of up to 100 antigens is feasible (Schubert, 2007), making such approaches extremely valuable, particularly in patient sam-ples in which rare cell populations are present in vanishingly small numbers and hence are not amenable to FCM.

Table II
Hyperchromatic Cytometry: generation of ``virtual colors"

Iterative restaining	Sequential staining using different antibodies but with the same fluorochrome allows identical colors to appear as distinct parameters, that is, "virtual colors" when newly labeled cells appear during sequential scans. During iterative staining, each cell serves as its unstained control and a bleaching step before each restaining can be included to improve sensitivity. Moreover, use of the same fluorochrome reduces the need for compensation.
Differential photobleaching of dyes of similar emission spectra	Differential photostability can be exploited to generate virtual colors. Thus, in the first scan, photostable (e.g., Alexa 532), and conventional (e.g., PE) stains on different antigens cannot be distinguished but following photobleaching two populations appear that can be identified by reference to pre- and postbleaching analysis.
Photoactivation	Qdot fluorescence intensity can be increased by laser exposure allowing generation of novel virtual colors by analogous analysis to that used for differential photobleaching.
Photodestruction of tandem dyes	Virtual colors can be generated by laser-induced photodestruction of the FRET between the donor and acceptor fluorochromes of tandem dyes.

III. Rationale for LSC Analysis of Antigen–Specific T Cell Responses *In vitro* and *In vivo*

The ability of LSC to correlate the precise signature (kinetics, amplitude, and subcellular localization) of intracellular signals with the functional phenotype of individual cells (Fig. 1) has the capacity to make a substantial impact on our understanding of the molecular and cellular mechanisms underpinning development of immune responses in health and disease. For example, it provides a powerful tool for visualizing and quantitating antigen-specific responses following adoptive transfer of Tg TCR- and/or B-cell receptor-bearing lymphocytes in numbers large enough to trace (with antigen receptor-specific antibodies or peptide-tetramers) *in vivo* but small enough to reflect, and indeed not interfere with, the normal physiological response to antigen (Adams *et al.*, 2004b; Garside *et al.*, 1998; Grierson *et al.*, 2005a; Harnett, 2007; Marshall *et al.*, 2005a; Morton *et al.*, 2007; Pape *et al.*, 1997; Smith *et al.*, 2004b). This is because such cells can be distinguished readily from bystander lymphocytes using standard integration contours (which generate data that are directly representative of individual cells in tissue) to detect staining of the antigen receptors, as these low-frequency cells will be distributed sporadically throughout the lymph node (Fig. 2). Here it is possible, by analyzing sections taken throughout the course of the immune response, to identify the key cellular interactions of antigen-specific lymphocytes, potentially dissecting the functional relationship between intracellular signals and effector responses in terms of cytokines, costimulatory molecules or effector-lineage signatures (such as Treg: Foxp3, Th1: T-bet, Th2: GATA-3, or Th17: RORγt). It is also possible to address issues concerning

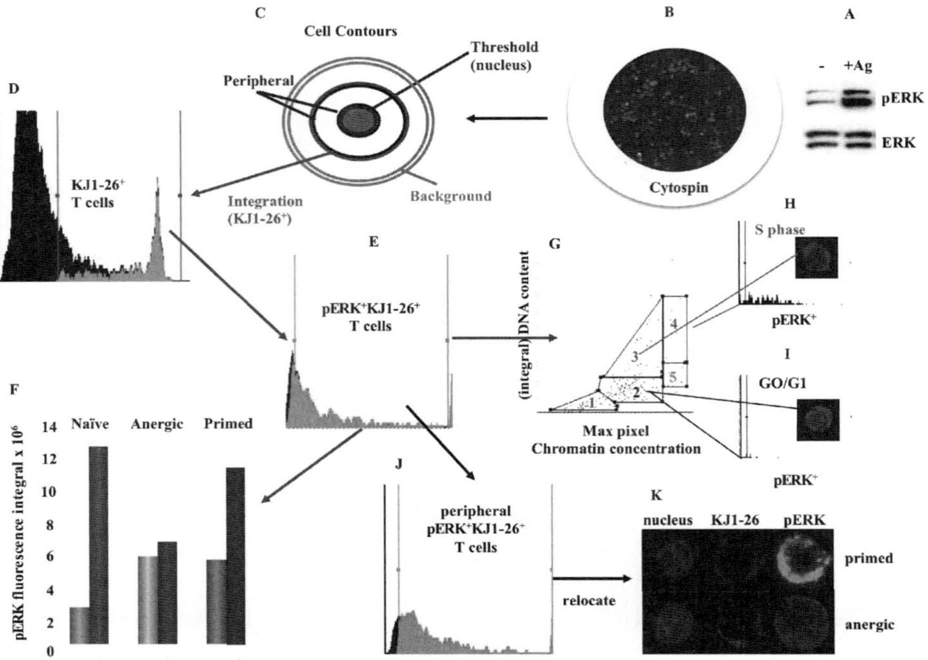

Fig. 1 Analysis of ERK signaling in OVA-specific Tg TCR T cells. The activation of ERK(pERK) MAP kinase in lymph node Tg TCR T cells in response to antigen ($\pm$ Ag) has traditionally been determined by Western blotting (A). This approach represents the average response of the population and does not take into account contributions to the signal from bystander cells. By contrast, LSC analysis can provide information on ERK activation in individual Ag-specific T cells. For example, cells in cytospin preparations (B) of lymph-node cells can be defined and detected by laser scanning cytometry (LSC) on the basis of their nuclear staining (4′,6-diamidino-2-phenylindole (DAPI), blue) which sets the threshold contour (**C**). Ag-specific Tg TCR T cells within this population can then be defined by standard integration contours (**C**) based on staining of the cell surface expression of the clonotypic Tg TCR (KJ1-26; red) and/or an optimal number of pixels (pixel size 0.5 μm × 0.5 μm) that reflect the size of lymphocytes. In this case, to examine pERK expression at the periphery of the T cells, peripheral contours are also set 1 pixel from the threshold and integration contours (**C**). Ag-specific cells are identified by staining of their Tg TCR by the clonotypic antibody KJ1-26 (red) and the levels of receptor expression on individual Tg TCR T cells quantitated (histogram; **D**). Panel *E*, total cellular levels of activated, pERK expression (green staining) in KJ1-26$^+$ Tg TCR T cells are quantitated (integral values) using standard integration contours. Using such analysis, the differential levels of ERK activation in naive, anergic, and primed populations in response to antigen ($\pm$ Ag) can be determined (F). In panel **G**, the cell cycle status of KJ1-26$^+$ Tg TCR T cells is assessed on the basis of DAPI nuclear staining (blue) that allows analysis of both DNA content (*y*-axis; integral value is proportional to DAPI-DNA binding) and also chromatin concentration (*x*-axis; max pixel). Thus, as DNA is more condensed in mitotic versus interphase cells, mitotic cells have a higher max pixel value than other cells during the cell cycle (gate 4). Similarly, new daughter cells exhibit more condensed DNA (gate 5; higher max pixel values) than other GO/G1 cells (gate 2) and hence these two populations can be discriminated by this parameter. Sub-diploid apoptotic cells are found in gate 1. Analysis of the individual phases of the cell cycle showed that while transgenic TCR T cells in S phase (gate 3) expressed varying levels of pERK staining (panel **H**), little or no pERK

lymphocyte migration (in terms of chemokine receptor expression) and T cell help for B cells within lymph nodes (follicular or paracortical localization of T cells) or in clinically relevant tissues such as inflamed joints in arthritis or lungs in asthma (Grierson *et al.*, 2005a; Harnett, 2007; Marshall *et al.*, 2005a; Morton *et al.*, 2007; Smith *et al.*, 2004b).

For example, peripheral tolerance is a state of antigen-specific hyporesponsiveness, which once established, can suppress many aspects of the Ag-specific immune response to subsequent challenge, including lymphocyte proliferation, cytokine production, *in vivo* delayed-type hypersensitivity, and Ab production (Fathman and Lineberry, 2007). The molecular mechanisms underlying induction and maintenance of such tolerance remain unclear although it is well established that TCR ligation, in the absence of costimulatory signals like those provided by CD80/CD86 on the Ag-presenting cell (APC) interacting with CD28 on the T cell, induces such long-lasting unresponsiveness (anergy) in T cells (Fathman and Lineberry, 2007). This is typically evidenced by a lack of IL-2 production and consequent T cell proliferation in response to subsequent challenge with Ag. Consistent with this, in anergic T cells there is reduced recruitment and activation of the MAP kinase signaling cascades resulting in defective activation of transcription factors, such as c-Jun/c-Fos, that are involved in formation of the AP-1 complex required for IL-2 gene induction (Fathman and Lineberry, 2007). However, such signaling defects in anergic T cell populations have generally been identified using immortalized T cell lines or T cell clones, which are unlikely to truly represent the responses of primary Ag-specific T cells, *in vivo*. To address these issues, we therefore attempted to quantitate differential ERK signaling events occurring during priming and tolerance in murine primary Ag-specific T cells on an individual cell basis, using LSC (Adams *et al.*, 2004b; Harnett, 2007; Morton *et al.*, 2007).

IV. Detailed Protocols for Tracking Antigen-Specific T Cell Responses

A. Analysis of Antigen-Specific T Cells *In vitro*

Lymphocytes from a mouse (DO11.10 strain) Tg for an ovalbumin (OVA)-specific TCR are analyzed as cytospins (Fig. 1) for surface expression of the Tg TCR and activated pERK (Adams *et al.*, 2004b; Morton *et al.*, 2007) by LSC as follows:

(panel **I**) was detected in KJ1-26$^+$ cells in the G0/G1 phase (gate 2) of the cell cycle. Representative cells from these two phases of the cell cycle were relocated to and visualized for nuclear (blue) and transgenic TCR (KJ1-26; red) staining (inserts in panels **H** and **I**). In panel **J**, analysis of the pERK expression within the peripheral contour gates of KJ1-26$^+$ T cells quantitated the pERK fluorescence staining associated with the cell periphery (histogram of peripheral integral values; panel **J**). Such peripheral localization of pERK staining was validated by the relocation and visualization of gated cells (panel **K**). Here, representative transgenic TCR T cells demonstrating differential pERK signals in terms of intensity and localization in the primed and anergic populations, with nuclear (blue), surface transgenic TCR (KJ1-26; red) and peripheral pERK (green) staining are shown (panel **K**). (See plate no. 14 in the color plate section.)

i. All samples should be kept in a darkened, humidified chamber at room temperature (RT) throughout.

ii. Place the slide, filter card, and cytofunnel onto the cytoclip in that order and secure with clasp. Add 75 µl of lymphocytes at 2×10^6 cells/ml to cytofunnel. Cytocentrifuge the cells for 4 min at 600 rpm in a Cytospin3 (ThermoShandon, Runcorn, Cheshire, UK). Repeat this step with the slide and filter card turned 180° inside the cytoclip. This will generate duplicate samples on each slide, one of which will be used as a negative staining control.

iii. Fix cells in 4% formaldehyde in PBS in a Coplin jar on ice for 15 min. Wash cells with PBS for 5 min. Draw around samples with wax pen to prevent overspill of antibodies/reagents between samples and with black marker pen for ease of location under microscope.

iv. Incubate in 1% Blocking reagent (BR; Tyramide signal amplification (TSA™) Kit # 12 with Alexa Fluor 488 tyramide, Molecular Probes) for 10 min.

v. Add 1° Ab, biotinylated anti-clonotypic anti-TcR Ab, KJ1-26, diluted 1:250 (stock 1.6 mg/ml) in 1% BR for 30 min. Use 100 µl/cytospin. Wash in TNT wash buffer (0.1 M Tris–HCl pH 7.5, 0.15 M NaCl, and 0.05% Tween 20) for 3 min. Repeat this step twice.

vi. Add streptavidin-horseradish peroxidase (SA-HRP), diluted 1:100 in 1% BR for 25 min. Use 50 µl/cytospin. Wash in TNT wash buffer for 3 min. Repeat this wash step twice.

vii. Add biotinylated-tyramide, diluted 1:50 for 10 min. Use 50 µl/cytospin. Wash in TNT wash buffer for 3 min. Repeat this wash step twice.

viii. Add Streptavidin-Alexa Fluor 647, diluted 1:500 (stock 1 mg/ml) in 1% BR for 30 min. Use 50 µl/cytospin. Wash in TNT wash buffer for 3 min. Repeat this wash step twice.

ix. Quench excess HRP with 0.1% azide/3% H_2O_2 in PBS for 5 min. Repeat this step twice. Wash in TNT wash buffer for 3 min, three times.

x. Permeabilize cells in 50 µl permeabilization buffer A (2% foetal calf serum (FCS), 2 mM EDTA pH 8.0, 0.1% w/v saponin) for 5 min. Wash cells in PBS for 10 s. Repeat this wash step twice.

xi. Incubate cells in 50 µl 1% BR/0.1% w/v saponin for 15 min.

xii. Incubate cells in anti-pERK, diluted 1/250 (anti-Phospho-p44/42 MAP kinase (Thr202Tyr204) New England Biolabs (UK) Ltd.) in 1% BR/0.1% w/v saponin for 30 min, using 50 µl/cytospin. At the same time add rabbit IgG diluted to match concentration of anti-pERK antibody in 1% BR/0.1% w/v saponin for 30 min. Use 50 µl/cytospin. Wash in TNT wash buffer for 3 min. Repeat this wash step twice.

xiii. Add anti-rabbit IgG-HRP conjugate diluted 1:100 in 1% BR/0.1% saponin for 25 min, using 50 µl/cytospin. Wash in TNT wash buffer for 3 min and repeat wash step twice.

xiv. Add Alexa Fluor 488-labelled tyramide, diluted 1:100 in 0.0015% H_2O_2/amplification buffer for 10 min. Use 50 µl working solution/sample. Wash in TNT wash buffer for 3 min. Repeat this wash step twice.

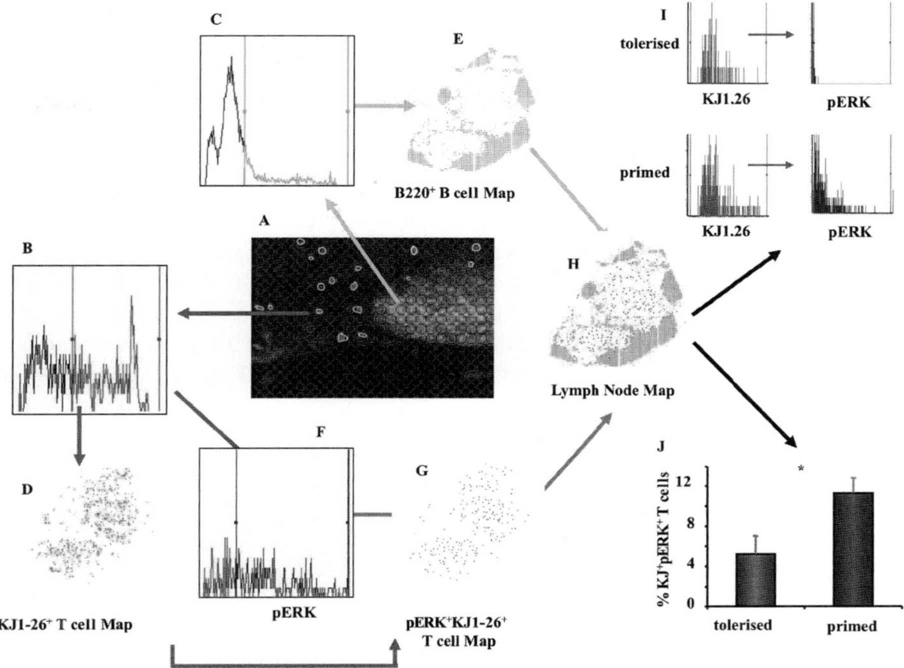

Fig. 2   Tissue map analysis of OVA-specific Tg TCR T cells *in vivo* Tg TCR T cell (KJ1-26[+]; red), B cell (B220[+], green) and activated, dually pERK (blue) staining of a lymph-node section from a mouse that received adoptively transferred Tg TCR T cells and was immunized with OVA in complete Freund's adjuvant *in vivo* (A). Individual OVA-specific T cells were identified by setting standard integration contours (yellow) based on the staining (red) of the Tg TCR by the clonotypic antibody KJ1-26 and the levels of Tg TCR determined on individual cells determined (B). By contrast, B-cell follicles were located by random sampling (phantom contours) set to detect green fluorescence representing B220[+] B cells throughout the tissue (C). Such phantom contours (radius of 6 μm and with a minimal distance between phantom centers of 20 μm) generate fluorescence values that represent the B cell follicles as a whole rather than the individual densely packed B cells. Plotting of the x and y coordinates of fluorescence allows generation of tissue maps identifying the localization of transgenic TCR T cells (red, panel D) and B cells (green, panel E) within the lymph node whilst the expression levels of KJ1.26 and B220 are quantified by histograms (panels B and C respectively). Further analysis of the Tg TCR (KJ1-26[+]) gate allows analysis of the levels of activated pERK expression (integral values) in individual antigen-specific cells (F) and the localization of pERK[+] transgenic T cells within the tissue (blue, panel G). Merging of the individual tissue maps (H) and selection of regional gates allows quantitation of both the number of Tg TCR T cells and also the levels of pERK expression in such T cells in the follicular (green) and paracortical (white) microenvironments within the lymph node. Analysis of tissue from mice primed or tolerized and then rechallenged with antigen *in vivo* show higher levels of pERK in Tg TCR T cells from primed tissue both in terms of levels of expression (I) and also the proportion of cells exhibiting activated ERK (J). (See plate no. 15 in the color plate section.)

xv. Allow to air dry for 5–10 min. Mount in Vectashield with 4′,6-diamidino-2-phenylindole (DAPI) to stain nuclei. Seal cover glass onto slide with nail varnish and store in aluminum foil at 4 °C.

B. Analysis of Antigen–Specific T Cells in Tissue *In situ*

Adoptive transfer allows individual Ag-specific Tg TCR T cells to be tracked (Fig. 2) in different anatomical regions of the draining lymph node, as well as identifying their activation status and expression of signaling molecules (Harnett, 2007; Marshall *et al.*, 2005b, 2008; Morton *et al.*, 2007). Mice homozygous for the cOVA peptide$_{323-339}$/I-A^d-specific DO11.10 Tg TCR (detected using the clonotypic mAb KJ1-26) on the BALB/c background are used as donors and BALB/c mice are used as recipients (Harnett, 2007; Marshall *et al.*, 2005b, 2008; Morton *et al.*, 2007).

1. Preparation of Cell Suspensions for Adoptive Transfer and Immunization

i. Lymph nodes (axillary, brachial, inguinal, cervical, and mesenteric) are pooled and forced through nylon mesh (40 μm) to obtain cell suspensions in sterile RPMI 1640 medium.

ii. Incubate cells with PE-conjugated anti-CD4 and biotinylated clonotypic anti-TCR antibody, KJ1.26 for 20 min at 4 °C. Wash cells in FACS buffer (PBS, 2% FCS, 0.2% NaN$_3$) and then incubate with FITC-conjugated streptavidin for 10 min at 4 °C. Perform two-color FACS analysis on 20,000 events to determine the percentage of KJ1-26$^+$ CD4$^+$ DO11.10 T cells.

iii. Inject a cell suspension containing 3×10^6 Tg TCR T cells/mouse intravenously into BALB/c recipients. Immunize mice 24 h later with OVA$_{323-339}$ (100 μg) with or without LPS (1 μg) in 200 μl PBS intravenously, for priming and tolerance respectively, and seven days later challenge with OVA$_{323-339}$ (100 μg)/LPS (1 μg) in 200 μl PBS intravenously.

2. Harvesting and Preparation of Tissue

Inguinal lymph nodes draining the site of immunization are removed, fixed in 1% paraformaldehyde at 4 °C for 24 h, and then incubated in 30% sucrose in PBS for 16 h. LNs are then embedded in OCT medium and snap-frozen in liquid nitrogen and stored at −70 °C. Sections (6 μm) are cut using a Cryotome$^®$.

3. Staining of Tissue Sections

i. All samples should be kept in a darkened, humidified chamber at RT unless otherwise specified, throughout.

ii. Fix slides in acetone for 10 min and allow to air dry. Mark areas to be stained with wax pen and draw around samples with black marker before rehydrating samples with PBS for 5 min.

iii. Incubate in 100 μl 0.1% azide/3% H_2O_2 in PBS for 15 min. Repeat this step twice. Wash tissue in TNT for 3 min. Repeat this step twice.

iv. Add 100 μl Avidin (four drops in 1 ml of 3% BSA/PBS) for 12 min. Wash tissue in PBS for 5 min. Add 100 μl Biotin (4 drops in 1 ml of 3% BSA/PBS) for 12 min. Wash tissue in PBS for 5 min.

v. Add 1° Ab, biotinylated anti-TcR Ab, KJ1-26, diluted 1:250 (stock 1.6 mg/ml), or isotype antibody to the same concentration in 3% BSA/PBS for 30 min. Use 100 μl/sample. Wash tissue in TNT for 3 min. Repeat this wash step twice.

vi. Add SA-HRP, diluted 1:100 in 3% BSA/PBS for 25 min. Use 100 μl/sample. Wash tissue in TNT for 3 min. Repeat this step twice.

vii. Add biotinylated-tyramide, diluted 1:50 for 10 min. Use 100 μl/sample. Wash tissue in TNT for 3 min. Repeat this wash step twice.

viii. Add Streptavidin-Alexa Fluor 647 (1 mg/ml) diluted 1:500 in 3% BSA for 30 min. Use 50 μl/sample. Wash tissue in TNT for 3 min. Repeat this wash step twice.

ix. Quench excess peroxidase activity with 100 μl 0.1% azide/3% H_2O_2 for 10 min. Repeat this step twice. Wash tissue in TNT for 10 s

x. Permeabilize tissue in permeabilization buffer (3% BSA, 0.1% triton-x-100) for 1 hour. Wash tissue in PBS for 10 s. Repeat this wash step twice.

xi. Incubate tissue in 100 μl 0.1% azide/3% H_2O_2 for 10 min. Repeat this step twice. Wash tissue in PBS for 3 min. Repeat this step twice.

xii. Incubate tissue in anti-pERK diluted 1/250 in 3% BSA/0.1% Triton X-100 at 37 °C for 16 h. Use 100 μl/sample. To control sections, add rabbit IgG diluted to the same concentration in 3% BSA/0.1% Triton X-100 at 37 °C for 16 h. Use 100 μl/sample.

xiii. Wash tissue in TNT for 3 min. Repeat this step twice. Add anti-rabbit IgG-HRP conjugate diluted 1:100 together with anti-B220-FITC diluted 1:250 in 3% BSA/0.1% Triton X-100 for 30 min. Use 50 μl/sample. Wash tissue in TNT for 3 min. Repeat this wash step twice.

xiv. Add Pacific Blue-labeled tyramide diluted 1:100 in 0.0015% H_2O_2/amplification buffer for 10 min. Use 50 μl/sample. Wash tissue in PBS for 3 min. Repeat this step twice.

xv. Allow to air dry for up to 10 min. Mount in Vectashield. Seal cover glass onto slide with nail varnish and store in aluminum foil at 4 °C.

V. Acquisition and Analysis of Data Using WinCyte Software

A. Acquisition and Analysis of Antigen Specific T Cells *In vitro*

Prior to acquisition of data, a number of parameters need to be addressed using the WinCyte software. The first of these is the threshold setting, upon which all events are contoured, or segmented on, as this threshold value allows discrimination of cells or events from background fluorescence levels. The integration contour, is next set a number of pixels out from the threshold contour, and is designed to identify the edge or surface of the cell, and allows for the detection of total fluorescence within the cell. Furthermore, to delineate the subcellular localization of signals with the cell, peripheral contours can be set in order to detect, for example, staining in the region between the nucleus and the cell surface, that is, between the threshold and integration contours. Finally, background contours are set out with cellular areas to allow background fluorescence signal subtraction (Adams *et al.*, 2004a; Gerstner *et al.*, 2004; Grierson *et al.*, 2005a). In this analysis, individual cells are typically detected by identifying their nuclei with DNA binding dyes, such as DAPI, to provide a threshold contour of detection (Fig. 1). This cell recognition or "triggering" approach is simple and straightforward and can be used to analyze cells in suspension that have been cytocentrifuged, or grown in chamber slides or in microtitre plates.

In order to analyze cells that have been labeled with several fluorochromes such as DAPI, FITC, and AF647, it is important to set-up the instrument settings (.PRO) file in the WinCyte software. To do this, in the *Parameters* submenu of the *Instrument settings* menu, the correct lasers and sensors are selected to enable detection of the fluorochromes, for example, labeling pERK (FITC: green), Ag-specific Tg TcR (AF647: red), and nuclear DNA (DAPI: blue) expression, respectively. To contour on cells by identifying their nuclei, in the *Computation* submenu, set the contouring on blue, as this is the color of the DAPI stain following its binding to DNA in the nucleus. Next, the threshold value is set to the appropriate level that is determined by placing the cursor over the DAPI stained nuclei in the scan data display window and noting the pixel value of positive cells. Removal of the background pixel value (an area with no cells present) from the DAPI positive value then allows an appropriate threshold value for the sample to be set. A further minimum area restriction can be placed on this contouring process. This area is the sum of the pixels within the threshold contour. For our analysis, the minimum area was set to 5 μm^2, enabling detection of DAPI-stained nuclei that are sized 5 μm^2 and above as this is the optimal minimum area for the Tg T cells described here. Enable the *peripheral contouring feature* to depict the peripheral area of the cell. Peripheral contours are set between the threshold contour (defined by the nucleus when contouring on DAPI) and the integration contour, for example, one pixel out from the threshold and one pixel in the from integration contours (defined by the edge of the cell; Fig. 1). Then using the *Scan Area* option, highlight an area of the slide to be scanned, corresponding to the location of the cells on the slide. Next set the PMT voltage, offset, and gain settings to the appropriate values that give the optimum signal intensity for each of the fluorochromes being analyzed. This is done during a scan run using the set sensors menu. Optimal settings are indicated by the presence of dark blue lines in the upper third of the PMT scale with very little or no saturation. If this is not the case, the PMTs should be adjusted to increase the signal, or decrease saturation. The contours should be set precisely to the cell, and the threshold value must be set accurately using the *Scan Data Display* window to ensure that all individual cells are being contoured on. Scan the area and save the data file.

In order to analyze the data generated, create a .DPR (template) file which includes the following histograms: (a) KJ1-26 Max Pixel (Max Pixel being the value of the most highly fluorescent pixel in the cell) versus Count, to allow identification of the Ag-specific Tg T cells; (b) pERK Integral (Integral being the total value of fluorescence within the cell) versus Count, to provide a value for the sum of all the emitted fluorescence relating to pERK; and (c) pERK Peripheral Integral (Peripheral Integral being the total value of fluorescence at the periphery of the cell) versus Count, to obtain the levels of peripheral pERK (Fig. 1). Next, relative to control samples, set a region on the KJ1-26 Max Pixel plot denoting the area containing the antigen-specific Tg TcR (KJ1-26$^+$) T cells and then set regions on both of the pERK plots to gate on pERK$^+$ cells. Now connect the KJ1-26$^+$ region to both pERK plots depicting the total and peripheral pERK expression by the Ag-specific Tg T cells on the pERK Integral versus Count and the pERK Peripheral Integral versus Count plots, respectively. Calculate the percentage and number of cells together with mean fluorescence value for every region using the region statistics generated by WinCyte.

B. Acquisition and Analysis of Antigen Specific T Cells *In situ*

The quantitative analysis of individual cells within large phenotypically identical subsets in tissue sections is more problematic than analysis of cells in suspension due to their high cellular density and overlapping nature. To address these problems, Tarnok *et al.* developed and validated the technique of multiple thresholding (Gerstner *et al.*, 2004), an algorithm in which the merging of data files, of varying threshold levels of DNA staining, compensates not only for the presence of nuclei with different diameters in cross sections but also for differential cell densities across sections due to varying tissue microarchitecture (Gerstner *et al.*, 2004).

Alternatively, tissue sections can be analyzed by random sampling, termed phantom contouring, a WinCyte/iGeneration software tool that creates a lattice of cell contours in a random predefined pattern across the area to be scanned (Gerstner *et al.*, 2004; Grierson *et al.*, 2005a; Luther *et al.*, 2004), treating these contours in the same manner as cells (Fig. 2). Although this approach is not directly representative of individual cells but rather an estimate of the mean fluorescence intensity of all such cells within a section, it is ideal, for example, for imaging and analyzing the microarchitecture of B-cell follicles within a lymph node (Fig. 2). By contrast, it is not suitable for the quantitative analysis of individual cell subsets in tissue sections as cells overlap, integral measurement resulting from the use of multiple fluorescence channel-based phantoms is likely to result in the overestimation of the frequency of the particular cell population. Thus at present, despite the benefits of "multiple thresholding" and "phantom contouring", the best way to discriminate individual members of a densely packed cell subset in tissue sections by LSC is by measuring max pixel values (Fig. 2) rather than by the integral fluorescence intensity value (Gerstner *et al.*, 2004; Grierson *et al.*, 2005a). However, the ongoing development of improved contouring software such as the "Watershed" feature now allows the new generation of LSC (iCys) to perform segmentation of even densely packed cells, such as lymphocyte subpopulations in lymph nodes, using either standard integration or phantom contours.

An advantage of using the adoptive transfer of Tg TCR T cells for this type of analysis is therefore that it generates a relatively low and physiologically relevant frequency of Ag-specific Tg T cells, thus overcoming the above problems normally encountered with identifying and contouring on individual cells in densely packed tissue. In this example, contouring of individual Ag-specific cells was set using the long red sensor that detects the surface expression of Tg TcR by AF647-linked KJ1-26 staining. By contrast, to collect data on B cells in the follicles, phantom contours were generated (Fig. 2) and the follicles defined by staining the densely packed B cells as a population. As described above, such phantom contours comprise a lattice of contours which is placed over the tissue section, thus generating fluorescence values which represent the tissue section as a whole, rather than individual cells (Dong *et al.*, 2002). On the *Phantoms* tab, enable phantom contouring, select *lattice* pattern and *allow overlap of events*. Set radius to 6 μm and minimal distance between phantom centers to 20 μm, the settings that are optimal for this type of lymphocyte analysis. Generate a scan area corresponding to the location of the tissue section on the slide. Next set the PMT voltage, offset, and gain to the optimal settings for analysis of such samples.

To quantitate the number of Ag-specific T cells and their level of pERK expression *in situ*, as well as identifying the anatomical location of these cells within the lymph node, generate a DPR file consisting of the following: (a) a histogram of KJ1-26 Max Pixel versus Count to identify the Ag-specific T cells; (b) a histogram of pERK Integral versus Count in order to display the levels of pERK expressed by the Ag-specific T cells; (c) a histogram of B220 Integral versus Count with the *Phantoms only* option selected to identify the B220$^+$ stained B cell follicles; and (d) four *x*-position versus *y*-position plots (tissue maps). Two of these tissue maps should have the *cells only* option selected to depict the location of KJ1-26$^+$ and KJ1-26$^+$ pERK$^+$ Tg T cells. One should have the *phantoms only* option selected to show location of the B220$^+$ stained B cell follicles and one should have the *phantoms and cells option* selected to identify the location of the KJ1-26$^+$ pERK$^+$ Tg T cells in relation to the B220$^+$ stained B cell follicles. Set a region on the KJ1-26 Max Pixel plot denoting the area where KJ1-26$^+$ cells are. Link a tissue map, which has the *cells only* option selected, to this region, thus creating a tissue map depicting the location of KJ1-26$^+$ cells in the tissue section. Next set regions on the pERK plot to gate on pERK$^+$ cell and link the KJ1-26$^+$ region to the pERK plot. This will show the pERK expression of the KJ1-26$^+$ cells on these plots. Link the pERK$^+$ region to the second tissue map plot that has the *cells only* option selected, which will show the location of KJ1-26$^+$ cells that are pERK$^+$. Calculate the percentage and number of cells that are KJ1-26$^+$ pERK$^+$, together with mean pERK fluorescence value for each sample. Set a region on the B220 Integral plot and link this region to the *phantoms only* tissue map to generate a tissue map showing the B220$^+$ B cell follicles. Link both the B220$^+$ region and the pERK$^+$ region to the tissue map with the *phantoms and cells option* selected to locate the KJ1-26$^+$ pERK$^+$ Tg T cells in relation to the B220$^+$ B cell follicles.

VI. Results: Analysis of the Role of pERK Signaling in Antigen–Specific Priming of T Cells

Figure 1 illustrates LSC analysis of a mixed population of lymph-node cells containing a low frequency of antigen (OVA)-specific Tg TCR T cells from DO11.10 mice that have been primed (anti-CD3/CD28) or tolerized (anti-CD3) *in vitro* and then rechallenged with Ag (LPS-matured DC loaded with OVA peptide). This highlights the differential intensities of ERK activation (pERK) in such individual OVA-specific T cells and demonstrates that the primed T cells display higher levels of pERK than tolerized T cells upon rechallenge with Ag. In addition, it also shows that the pERK signal is predominantly localized (approx 80%) to the environment of the TCR at the periphery of primed but not tolerized cells (Fig. 1). Moreover, the functional relevance of such signaling can be assessed in terms of cell cycle progression and proliferation as such analysis showed that while the Tg TCR T cells transiting S-phase expressed pERK, those in the G0/G1 phase of the cell cycle did not (Fig. 1).

Therefore the above approach was used to show that there are marked differences in the amplitude and cellular localization of phosphorylated ERK MAP kinase signals when naive, primed, and anergic T cells are challenged with immunogenic antigen (Adams *et al.*, 2004b; Morton *et al.*, 2007). Thus, primed T cells display more rapid kinetics of phosphorylation and activation of ERK than naive T cells, whereas anergic T cells display a reduced ability to activate ERK upon challenge. In addition, the low levels of pERK found in anergic T cells are distributed diffusely throughout the cell, whereas in primed T cells, pERK appears to be targeted to the same regions of the cell as the TCR (Adams *et al.*, 2004b; Morton *et al.*, 2007). Moreover, the GTPase Rap1, which can antagonize the generation of such pERK signals and has been reported to accumulate in tolerant cells, exhibits an inverse pattern of expression to pERK in individual Ag-specific primed and tolerized T cells (Morton *et al.*, 2007). Although pERK is expressed by more primed than tolerized T cells when rechallenged with Ag *in vitro*, Rap1 is expressed by higher percentages of tolerant compared with primed Ag-specific T cells. Moreover, whereas pERK localizes to the TCR and lipid rafts in primed cells, but exhibits a diffuse cellular distribution in tolerized cells, Rap1 colocalizes with the TCR and lipid raft structures under conditions of tolerance, but not priming, *in vitro* (Morton *et al.*, 2007).

Such analysis was not restricted to *in vitro* studies but was also be extended to *in situ* analysis of OVA-specific cells following induction of priming and tolerance *in vivo* (Fig. 2). Here, tissue maps provide statistical information relating to, for example, the percentage of Tg TCR T cells expressing activated pERK, the differential levels of ERK activity within such cells and the relevant abilities of OVA-specific cells expressing activated ERK or not, to migrate into B-cell follicles to provide T-cell help (Fig. 2). This *in situ* analysis indicated that inverse relationship between Rap1 and pERK expression suggested by the *in vitro* studies was likely to be physiologically relevant, given that we observed the same patterns in Ag-specific T cells *in situ*, following induction of priming and tolerance *in vivo* (Morton *et al.*, 2007). Together, these data suggest that the maintenance of tolerance of individual Ag-specific T cells may reflect the recruitment of upregulated Rap1 to the immune synapse, potentially resulting in sequestration of Raf-1 and uncoupling of the TCR from the Ras-ERK MAP kinase cascade.

VII. Application of LSC Technology to Analysis of the Immune System in Health and Disease

In addition to advancing our understanding of fundamental aspects of the immune response such as priming or tolerance, LSC allows the dissection of the molecular and cellular events underpinning dysfunction of the immune response in autoimmune and allergic inflammatory diseases and well as how key immunoregulatory events are subverted by pathogens to evade the immune response.

For example, filarial nematodes such as *Wuchereria bancrofti*, *Brugia malayi*, and *Onchocerca volvulus* represent major causes of morbidity in the tropics. Infection of humans with these parasitic worms is long-term and the longevity of mature worms (>5 y) appears to be promoted by their secretion of immunomodulatory molecules that act to suppress inflammation, at least in part by modulating effector Th cell responses (Harnett *et al.*, 2010; Harnett and Harnett, 2010). In addition to promoting parasite survival, this immunomodulatory action is beneficial to the host as it also prevents/limits the extreme pathology like elephantiasis that can potentially result from aggressive immune responses to such parasites and hence, the therapeutic potential of these immunomodulatory capabilities in inflammatory disease are now being explored (Harnett *et al.*, 2010; Harnett and Harnett, 2010). One such immunomodulatory molecule, the phosphorylcholine-containing glycoprotein ES-62, acts directly to induce hyporesponsiveness in a number of cells of the immune response including macrophages, dendritic cells, mast cells, and B cells. However, it can also inhibit the development of Th1 phenotype by modulating the maturation of dendritic cells such that they prime T cells that induce Th2/anti-inflammatory responses (Harnett *et al.*, 2010; Harnett and Harnett, 2010). To address identifying the mechanisms underpinning such immunomodulation *in vivo* using LSC, we have exploited the above approaches (Sections III–VI) to show that the decreased clonal expansion of OVA-specific Tg TCR T cells and consequent suppression of the Th1 response to this Ag by ES-62 *in vivo* reflected modulation of the kinetics and extent of antigen-specific T-cell migration into follicles to provide B-cell help (Marshall *et al.*, 2005a, 2008).

There is increasing awareness, however, that whilst T cells play important roles in orchestrating the phenotype of immune responses in autoimmune and allergic inflammatory diseases, many innate cells, in addition to their innate functional responses such as phagocytosis and degranulation, can produce and secrete cytokines important to the pathogenic process at the site of inflammation (Hueber *et al.*, 2010; McInnes and Schett, 2007; Melendez *et al.*, 2007). To date, investigation of pro-inflammatory responses of innate effector cells such as macrophages, neutrophils, and mast cells from animal models of inflammatory disease, as well as patient samples, has generally focused on the responses of such cells purified from peripheral blood samples and these may not reflect either the cell phenotype or functionality at the site of inflammation. Alternatively, in the cases where tissue could be isolated from the site of inflammation, cell purification resulted in the loss of relevant cell–cell interactions and tissue architecture. However, the use of LSC is not restricted to the analysis of antigen-specific B and T cell responses in the lymphoid organs but can also be applied to tracking innate inflammatory responses at the site of inflammation as the new generation 4-laser iCys LSC technology now allows simultaneous analysis of cell morphology, fluorescence and chromatic staining of cells and tissue in a solid-phase format. Hence, uniquely, the iCys can image and quantitatively analyze cell phenotype and signaling in terms of subcellular localization and functionality, in the context of severity of tissue pathology in intact clinically relevant tissue, such as synovial joints. Importantly, the relocation feature permits real-time analysis of individual cells from rare samples from patients without cell purification, as well as their extensive phenotyping by reiterative staining/use of virtual colors and also reanalysis of archived tissue/cells. Collectively, these unique features therefore now allow translation and *in situ* validation of therapeutic targets in rare patient samples, such as fine needle aspirates, to human disease.

Therefore recently our investigation on the role of innate cells, such as mast cells, macrophages, and neutrophils, in the pathology of mouse models of autoimmune diseases such as systemic lupus erythematosus (SLE) and rheumatoid arthritis has involved LSC analysis of their recruitment to, and functional responses in, clinically relevant tissue such as the kidney and paw joints, respectively. A number of lupus-prone mouse strains have been used extensively as models of human SLE, with each displaying the characteristic features of increased circulating autoantibodies, immune complex deposition, and kidney disease (Fairhurst *et al.*, 2006; Liu and Mohan, 2006). However, the joint, skin, and CNS diseases displayed by MRL/lpr mice represent features seen in human SLE but rarely noted in the other models. Moreover, although, the lymphoproliferation of "double negative" T cells exhibited by MRL/lpr mice is not a typical feature of human patients, there is increasing recent evidence for a pathogenic role of such cells in human disease (Crispin *et al.*, 2008). As the major cause of death in MRL/lpr mice, as with human patients, is glomerulonephritis, our major focus has been to study the progressive changes in kidney architecture due to the inflammatory course of disease. Imaging of kidney tissue using traditional techniques has proved difficult as, due to its size, it is generally impossible to examine an entire tissue section in one field. Moreover, the kidney tends to be highly autofluorescent and so it can be problematic to quantify cellular populations using fluorescent dyes. Most importantly, scoring of cellular infiltration and discrimination of the nuclei of the glomeruli or kidney epithelial cells from those nuclei of infiltrating cells by hematoxylin and eosin (H&E) staining usually requires involvement of skilled pathologists. However, the low and high resolution capabilities of the iCys coupled with its ability to discriminate on the basis of size and morphology and analyze both fluorescent and chromatic dyes, overcomes many of these problems and allows objective and quantitative scoring without risk of prejudice. For example, analysis of kidney tissue from MRL/lpr mice throughout the course of the disease has allowed quantitation of the infiltrating cells into the tissue and their discrimination from resident kidney cells (Fig. 3) and this can be combined with the immunophenotypic discrimination of individual cell populations and their functional roles.

Collagen-induced arthritis (CIA) in the DBA/1 mouse is a well-established model of human rheumatoid arthritis (Harnett *et al.*, 2008; McInnes *et al.*, 2003) in which iCys analysis of synovial sections from arthritic paws can provide unique insight into the pathological processes ongoing in inflamed joints. For example, as with the

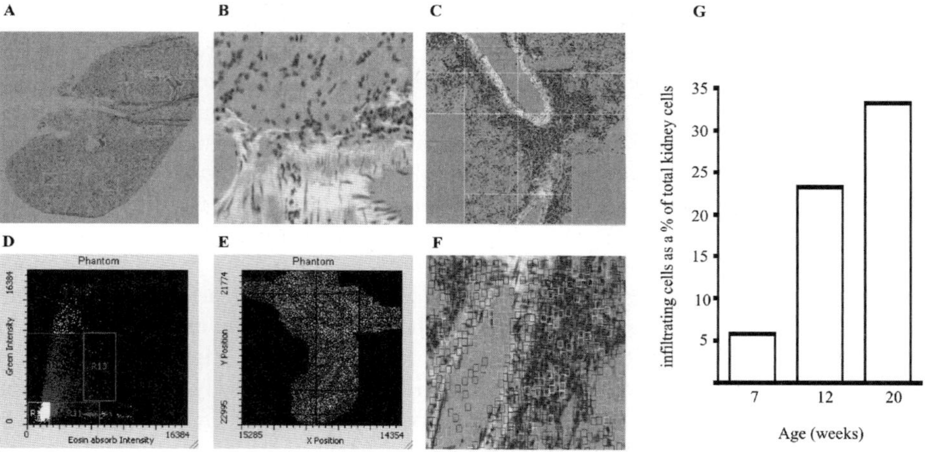

Fig. 3   LSC analysis of cellular infiltration of the kidneys in the MRL/lpr mouse. MRL/lpr mice develop extensive inflammatory cellular infiltration of the kidney and this is most evident around blood vessels. H&E-stained kidney sections from MRL/lpr mice were analyzed by iCys, in the first instance using a low-resolution scan ($\times$10 objective) to produce an image of the entire kidney section (A). Nine areas of interest per kidney section were selected and contoured for high resolution scanning using the 40$\times$ objective (A). The resultant high-resolution scans not only provide qualitative data in terms of high-resolution images (B and C) of the areas of interest, but also allow quantitative analysis of the proportion of infiltrating pro-inflammatory cells relative to the resident kidney cells. In order to discriminate the infiltrating cells from the resident kidney cells, a number of parameter gates based on the H&E staining were applied: thus, infiltrating cells were identified by their high-hemotoxylin intensity and low-eosin staining. A typical scattergram with differential population gates is shown in panel D. In addition, an *xy* tissue map of the data was generated (E) that not only reflected the original image (C) but allowed validation of the analysis as superimposition of the scattergram gates (F) verified that only infiltrating cells had been identified. Analysis of kidney sections from mice of increasing age allowed quantitative scoring of kidney infiltration that correlated with disease progression (G).

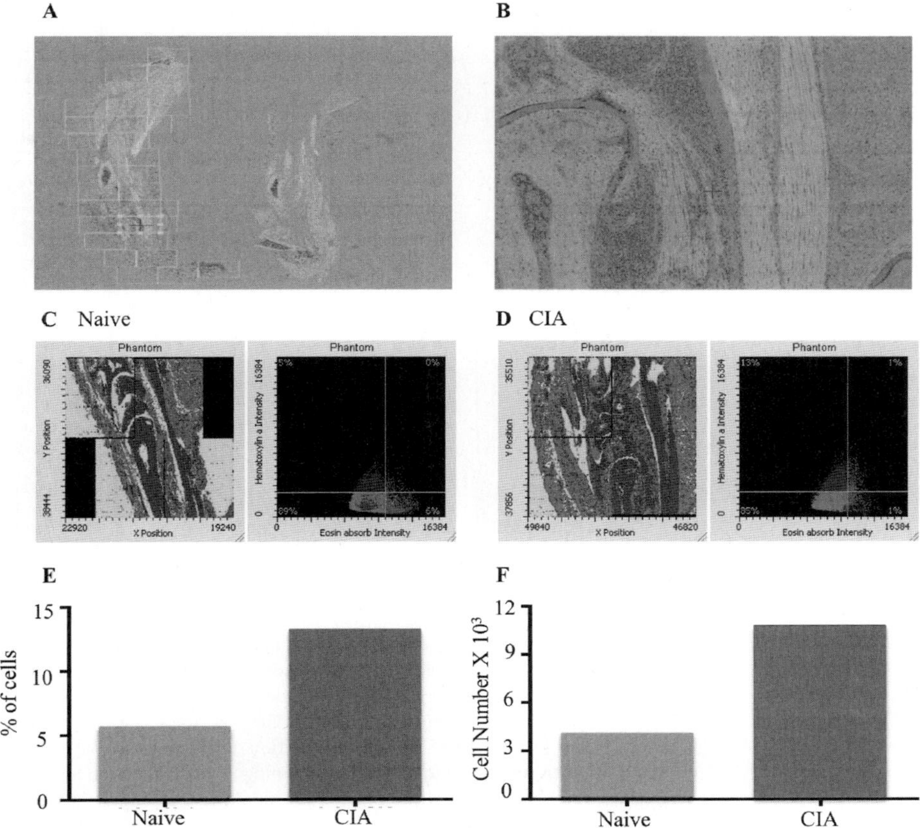

Fig. 4   LSC analysis of the cell infiltration of CIA joints. H&E stained paw-joint sections from naive and CIA mice were scanned by iCys at low (A) and high (B) resolution, the latter providing high quality images of the tissue architecture and how this changes after disease induction (B). To quantitate the infiltrating inflammatory cells surrounding the joint area, phantom contours were applied as this approach not only provided information on infiltrating cells, but also on the surrounding bone and cartilage structure, which could be discriminated on the basis of its differential chromatic absorbance properties. For example, whilst infiltrating cells had low eosin intensity and high hemotoxylin intensity (annotated in blue), cartilage exhibited an inverse pattern of staining (annotated in red). Bone (green) and muscle (pink) cells are also shown. The data from these phantom contours are represented as *xy* tissue maps and scattergram (eosin vs. hemotoxylin intensity) plots to discriminate the different populations in both naive (C) and CIA (D) mice. Quantitative analysis of the data allowed objective scoring of infiltration of pro-inflammatory cells both as a proportion (E) of the joint tissue and also in absolute numbers (F) in naive and CIA mice. (For interpretation of the references to color in this figure legend, the reader is referred to the Web version of this chapter.)

kidney sections above, H&E staining allows imaging and quantitative analysis of the cellular infiltration of the joints occurring in naive and disease-associated mice (Fig. 4) that clearly shows significantly higher cellular infiltration and pathology in the disease state. Such analysis will allow future assessment of the efficacy of immunomodulatory agents to block such pro-inflammatory infiltrates and, when

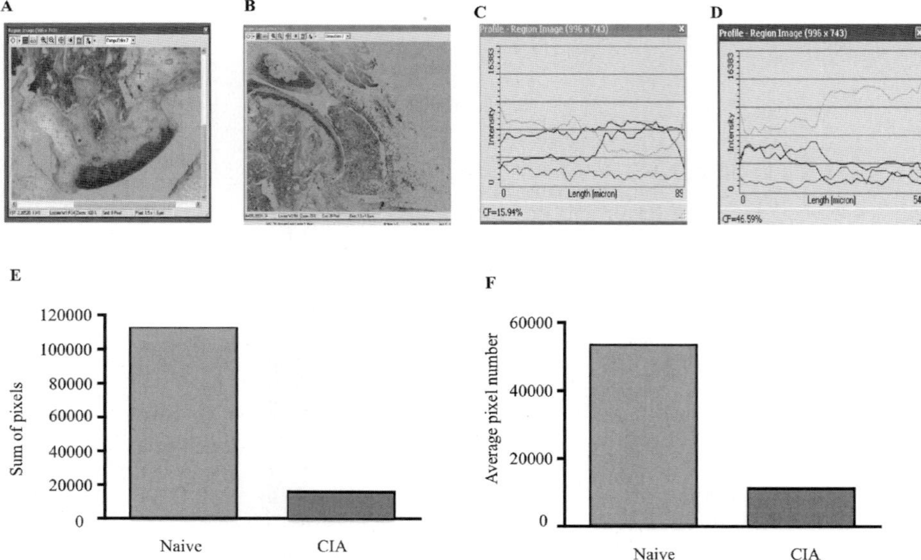

Fig. 5 Analysis of synovial cartilage damage in CIA mice. Cartilage damage in CIA has traditionally been assessed by qualitatively scoring toluidine blue staining of the proteoglycans in cartilage in joint sections. Loss of synovial cartilage can be assessed quantitatively by iGeneration software both in terms of toluidine blue intensity and, by using the profile tool, thickness (μM) of the cartilage layer. The high-resolution images of joints from naive (A) and CIA (B) mice show both of these parameters that are quantitated by the profile tool (C and D). In addition to defining the thickness of the cartilage layer, this feature allows analysis of total toluidine blue staining (E) as well as the intensity across defined sections (F) of the cartilage layer (as indicated by red profile lines in images A and B). (For interpretation of the references to color in this figure legend, the reader is referred to the Web version of this chapter.)

coupled with immunophenotyping, to identify the cellular and molecular targets of such agents. Accompanying the pro-inflammatory cellular infiltration of the synovial joints observed during CIA, a major pathological event is the loss of surface cartilage from the synovial membrane as evidenced by the loss of intensity of toluidine blue staining of the proteoglycans in cartilage (Fig. 5). Moreover, by use of the profiling tool of the iGeneration software, it is possible to generate a quantitative estimate of the thickness of this cartilage layer in addition to simply measuring the intensity of the chromatic dye. Again, this type of parameter will be invaluable in the assessment of the efficacy of novel anti-arthritic therapies and their translation and validation in human disease.

Finally, it has recently emerged that mast cells may play a key pathogenic role in the progression of joint disease (Hueber *et al.*, 2010) and fortunately, in this light, mast cells can be easily identified in tissue using toluidine blue, which stains the heparin in mast cell granules. This has not only allowed us to image, track and quantify the numbers of mast cells in arthritic joints but also, because on degranulation mast cells exhibit a longer, thinner shape with lower dye intensity than resting mast cells which

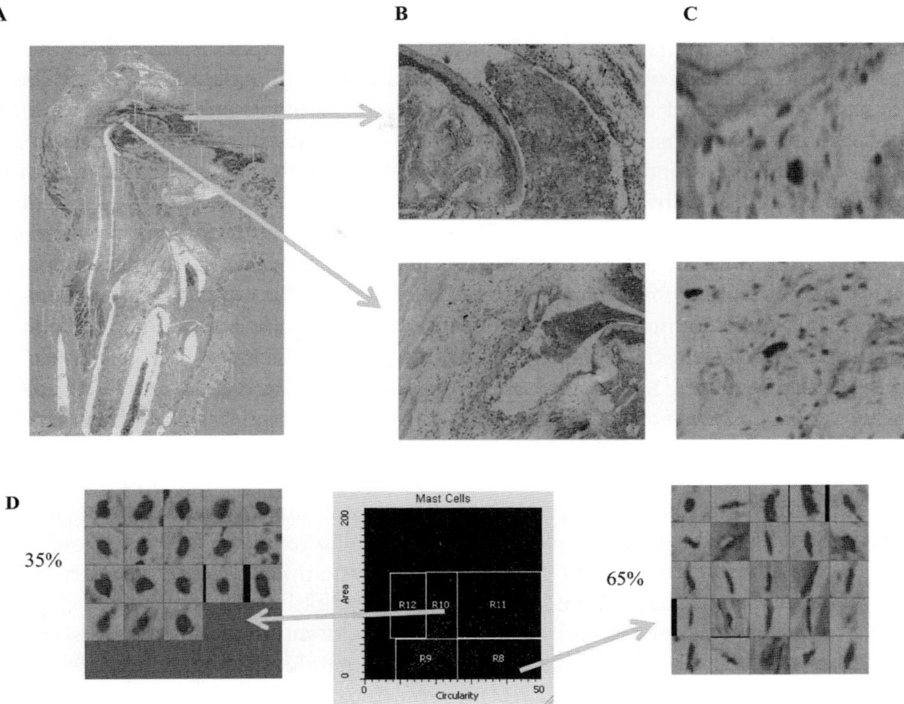

Fig. 6   Analysis of mast cells in the joints of CIA mice. Mast cells were identified in tissue sections from mouse paws from CIA mice by staining with the chromatic dye toluidine blue, which stains the heparin in their granules. Following low resolution scans (A), high resolution scanning of identified regions of interest (B and C) produced images of mast cells (B and C). Analysis of toluidine blue intensity, cell size, and circularity allowed quantitation of the number and activation status of mast cells in this joint tissue (D). This is because degranulated mast cells exhibit a longer, thinner shape of lower toluidine blue intensity than resting mast cells that are smaller, rounder, and more intensely stained. Creation of the gallery images from each region (D) validates the discrimination of resting and degranulated mast cells using these parameters. This application of the iGeneration software therefore allows correlation between the status of mast cell activation and disease progression in individual joint sections and, in this case, indicated that 65% of the total mast cell population was degranulated in this CIA joint tissue. (For interpretation of the references to color in this figure legend, the reader is referred to the Web version of this chapter.)

are smaller, round, and highly stained (Melendez *et al.*, 2007), the segmentation parameters of the iCys now allow correlation of the functional status of joint-based mast cells with the inflammatory course of the disease (Fig. 6).

VIII. Concluding Remarks and Future Directions

Following on from the advances outlined above in analyzing the pathological responses of various types of immune system cells in clinically relevant sites of inflammation, LSC should directly define the functional consequences of particular signals within these individual cells, or between such cells, *in situ*. Indeed, use of the reiterative staining/relocalization facility now provides the capacity to progressively and quantitatively image the recruitment and identity of multiple components of signaling pathways (signalsome) within individual cells and analyze their cellular functional responses. By delineating differential recruitment of downstream effector pathways in response to distinct signal signatures, LSC analysis therefore allows *in situ* dissection of the molecular mechanisms by which a single signaling element can act as a checkpoint to direct distinct cell fate decisions such as apoptosis or survival. Moreover, the ability to perform this analysis on tissue sections provides, at present, the only way of dissecting functionally relevant signals within their physiological environments *in vivo*.

LSC analysis therefore allows direct demonstration of cause (signal) and effect (functional response) within individual cells and when coupled with viral instant transgenesis gene delivery systems (Hurez *et al.*, 2002a, 2002b; Nakagawa *et al.*, 2006) that allow direct comparison between wild type and modified adoptively transferred cells within a single animal, will provide definitive corroborative evidence that such signal signatures are necessary and sufficient for normal development of the immune system *in vivo* and that subversion of these leads to disease. Thus, by allowing quantitative analysis of the effects of signal strength/localization of wild type and mutant signals on signalsome formation and their functional consequence within single cells, LSC provides answers not achievable by traditional investigation. In this way, it allows us to exploit the advantages of instant transgenesis over traditional transgenic or knockout technology in that, by quantitatively imaging the effect of various levels of overexpression of constitutively active/dominant negative constructs on signal signature, downstream effectors and functional responses in individual cells within a mixed population of wild type and gene-modified cells under identical experimental conditions, many of the artefactual all or none phenotypes are eliminated. Moreover, it allows the unmasking of multiple roles for a single signaling element within a single cell that are normally refractory to identification by either traditional or conditional transgenic/knockout technology.

Finally, the advent of an ever-increasing repertoire of commercially available antibodies/reagents highly specific for precise regulatory post-translational modifications of individual signaling elements proven to be key to their biological function ultimately provides tools for the complete delineation of signaling pathways regulating development of the immune response in health and disease *in situ*. Similarly, the recent development of adoptive transfer systems in which the responses of near-physiological levels of transgenic antigen-receptor-bearing B and T lymphocytes can be tracked in animal models of inflammatory autoimmune diseases, such as arthritis, asthma, and multiple sclerosis (Bettelli *et al.*, 2006; Croxford *et al.*, 2006; Jarman *et al.*, 2005; Maffia *et al.*, 2004; Osman *et al.*, 1998) or alternatively, animal models in which physiological autoantigen specificities can be detected by tetramer analysis (Amend *et al.*, 2006; Huang *et al.*, 2004;

Korn *et al.*, 2007)) has now provided the tools to enable LSC-driven dissection of the key events *in situ* in the pathogenesis of such diseases and the translation and validation of candidate therapeutic targets in clinically relevant human disease tissues.

Acknowledgments

The authors would like to thank the Medical Research Council, the Wellcome Trust, the Biotechnology and Biological Sciences Research Council and the Nuffield Foundation for funding this research. They would also like to thank Compucyte for their assistance with the development and analysis of the iCys-based assays of tissue pathology in MRL/Lpr and CIA mice. The authors declare no competing financial interests.

References

Adams, C. L., Grierson, A. M., Mowat, A. M., Harnett, M. M., and Garside, P. (2004a). Differences in the kinetics, amplitude, and localization of ERK activation in anergy and priming revealed at the level of individual primary T cells by laser scanning cytometry. *J. Immunol. Methods* **173**, 1579–1586.

Adams, C. L., Grierson, A. M., Mowat, A. M., Harnett, M. M., and Garside, P. (2004b). Differences in the kinetics, amplitude, and localization of ERK activation in anergy and priming revealed at the level of individual primary T cells by laser scanning cytometry. *J. Immunol.* **173**, 1579–1586.

Adams, C. L., Kobets, N., Meiklejohn, G. R., Millington, O. R., Grierson, A. M., Rush, C. M., Smith, K. M., Garside, P. (2004c). Tracking lymphocytes *in vivo. Archivum immunologiae et therapiae experimentalis* **52**, 173–187.

Amend, B., Doster, H., Lange, C., Dubois, E., Kalbacher, H., Melms, A., Bischof, F. (2006). Induction of autoimmunity by expansion of autoreactive CD4+CD62L low cells *in vivo. J. Immunol.* **177**, 4384–4390.

Arechiga, A. F., Bell, B. D., Solomon, J. C., Chu, I. H., Dubois, C. L., Hall, B. E., George, T. C., Coder, D. M., Walsh, C. M. (2005). Cutting edge: FADD is not required for antigen receptor-mediated NF-kappaB activation. *J. Immunol.* **175**, 7800–7804.

Bettelli, E., Baeten, D., Jager, A., Sobel, R. A., and Kuchroo, V. K. (2006). Myelin oligodendrocyte glycoprotein-specific T and B cells cooperate to induce a Devic-like disease in mice. *J. Clin. Invest.* **116**, 2393–2402.

Bocsi, J., Mittag, A., Sack, U., Gerstner, A. O., Barten, M. J., Tarnok, A. (2006). Novel aspects of systems biology and clinical cytomics. *Cytometry A* **69**, 105–108.

Butt, O. I., Krishnan, P., Kulkarni, S. S., Moldovan, L., and Moldovan, N. I. (2005). Quantification and functional analysis of chemotaxis by laser scanning cytometry. *Cytometry A* **64**, 10–15.

Crispin, J. C., Oukka, M., Bayliss, G., Cohen, R. A., Van Beek, C. A., Stillman, I. E., Kyttaris, V. C., Juang, Y. T., Tsokos, G. C. (2008). Expanded double negative T cells in patients with systemic lupus erythematosus produce IL-17 and infiltrate the kidneys. *J. Immunol.* **181**, 8761–8766.

Croxford, J. L., Miyake, S., Huang, Y. Y., Shimamura, M., and Yamamura, T. (2006). Invariant V(alpha) 19i T cells regulate autoimmune inflammation. *Nature Immunol.* **7**, 987–994.

Dey, P. (2006). Role of ancillary techniques in diagnosing and subclassifying non-Hodgkin's lymphomas on fine needle aspiration cytology. *Cytopathology* **17**, 275–287.

Doyle, S. E., O'Connell, R. M., Miranda, G. A., Vaidya, S. A., Chow, E. K., Liu, P. T., Suzuki, S., Suzuki, N., Modlin, R. L., Yeh, W. C., Lane, T. F., Cheng, G. (2004). Toll-like receptors induce a phagocytic gene program through p38. *J. Exp. Med.* **199**, 81–90.

Fairhurst, A. M., Wandstrat, A. E., and Wakeland, E. K. (2006). Systemic lupus erythematosus: multiple immunological phenotypes in a complex genetic disease. *Adv. Immunol.* **92**, 1–69.

Fang, W., Weintraub, B. C., Dunlap, B., Garside, P., Pape, K. A., Jenkins, M. K., Goodnow, C. C., Mueller, D. L., Behrens, T. W. (1998). Self-reactive B lymphocytes overexpressing Bcl-xL escape negative selection and are tolerized by clonal anergy and receptor editing. *Immunity* **9**, 35–45.

Fanning, S. L., George, T. C., Feng, D., Feldman, S. B., Megjugorac, N. J., Izaguirre, A. G., Fitzgerald-Bocarsly, P. (2006). Receptor cross-linking on human plasmacytoid dendritic cells leads to the regulation of IFN-alpha production. *J. Immunol.* **177**, 5829–5839.

Fathman, C. G., and Lineberry, N. B. (2007). Molecular mechanisms of CD4+ T-cell anergy. *Nature Rev.* **7**, 599–609.

Garside, P. E., Ingulli, E., Merica, R. R., Johnson, J. G., Noelle, R. J., Jenkins, M. K. (1998). Visualization of specific B and T lymphocyte interactions in the lymph node. *Science* **281**, 96–99.

Gerstner, A. O., Mittag, A., Laffers, W., Dahnert, I., Lenz, D., Bootz, F., Bocsi, J., Tarnok, A. (2006a). Comparison of immunophenotyping by slide-based cytometry and by flow cytometry. *J. Immunol. Methods* **311**, 130–138.

Gerstner, A. O., Thiele, A., Tarnok, A., Machlitt, J., Oeken, J., Tannapfel, A., Weber, A., Bootz, F. (2005). Preoperative detection of laryngeal cancer in mucosal swabs by slide-based cytometry. *Eur. J. Cancer* **41**, 445–452.

Gerstner, A. O., Thiele, A., Tarnok, A., Tannapfel, A., Weber, A., Bootz, F. (2006b). Prediction of upper aerodigestive tract cancer by slide-based cytometry. *Cytometry A* **69**, 582–587.

Gerstner, A. O., Trumpfheller, C., Racz, P., Osmancik, P., Tenner-Racz, K., Tarnok, A. (2004). Quantitative histology by multicolor slide-based cytometry. *Cytometry A* **59**, 210–219.

Grierson, A. M., Mitchell, P., Adams, C. L., Mowat, A. M., Brewer, J. M., Harnett, M. M., Garside, P. (2005a). Direct quantitation of T cell signaling by laser scanning cytometry. *J. Immunol. Methods* **301**, 140–153.

Grierson, A. M., Mitchell, P., Adams, C. L., Mowat, A. M., Brewer, J. M., Harnett, M. M., Garside, P. (2005b). Direct quantitation of T cell signaling by laser scanning cytometry. *J. Immunol. Methods* **301**, 140–153.

Grierson, A. M., Mitchell, P., Adams, C. L., Mowat, A. M., Brewer, J. M., Harnett, M. M., Garside, P. (2005c). Direct quantitation of T cell signaling by laser scanning cytometry. *J. Immunol. Methods* **301**, 140–153.

Harnett, M. M. (2007). Laser scanning cytometry: understanding the immune system *in situ. Nature Rev. Immunol.* **7**, 897–904.

Harnett, M. M., Kean, D. E., Boitelle, A., McGuiness, S., Thalhamer, T., Steiger, C. N., Egan, C., Al-Riyami, L., Alcocer, M. J., Houston, K. M., Gracie, J. A., McInnes, I. B., Harnett, W. (2008). The phosphorycholine moiety of the filarial nematode immunomodulator ES-62 is responsible for its anti-inflammatory action in arthritis. *Ann. Rheum. Dis.* **67**, 518–523.

Harnett, M. M., Melendez, A. J., and Harnett, W. (2010). The therapeutic potential of the filarial nematode-derived immunodulator, ES-62 in inflammatory disease. *Clinical Exp. Immunol.* **159**, 256–267.

Harnett, W., and Harnett, M. M. (2010). Helminth-derived immunomodulators: can understanding the worm produce the pill? *Nature Rev.* **10**, 278–284.

Holme, A. L., Yadav, S. K., and Pervaiz, S. (2007). Automated laser scanning cytometry: a powerful tool for multi-parameter analysis of drug-induced apoptosis. *Cytometry A* **71**, 80–86.

Huang, J. C., Vestberg, M., Minguela, A., Holmdahl, R., and Ward, E. S. (2004). Analysis of autoreactive T cells associated with murine collagen-induced arthritis using peptide-MHC multimers. *Int. Immunol.* **16**, 283–293.

Hueber, A. J., Asquith, D. L., Miller, A. M., Reilly, J., Kerr, S., Leipe, J., Melendez, A. J., McInnes, I. B. (2010). Mast cells express IL-17A in rheumatoid arthritis synovium. *J. Immunol.* **184**, 3336–3340.

Hurez, V., Dzialo-Hatton, R., Oliver, J., Matthews, R. J., and Weaver, C. T. (2002a). Efficient adenovirus-mediated gene transfer into primary T cells and thymocytes in a new coxsackie/adenovirus receptor transgenic model. *BMC Immunol.* **3**, 4–17.

Hurez, V., Hautton, R. D., Oliver, J., Matthews, R. J., and Weaver, C. K. (2002b). Gene delivery into primary T cells: overview and characterization of a transgenic model for efficient adenoviral transduction. *Immunol. Res.* **26**, 131–141.

Jarman, E. R., Tan, K. A., and Lamb, J. R. (2005). Transgenic mice expressing the T cell antigen receptor specific for an immunodominant epitope of a major allergen of house dust mite develop an asthmatic phenotype on exposure of the airways to allergen. *Clin. Exp. Allergy* **35**, 960–969.

Kamentsky, L. A., and Kamentsky, L. D. (1991). Microscope-based multiparameter laser scanning cytometer yielding data comparable to flow cytometry data. *Cytometry* **12**, 381–387.

Katz, E., Deehan, M. R., Seatter, S., Lord, C., Sturrock, R. D., Harnett, M. M. (2001). B cell receptor-stimulated mitochondrial phospholipase A2 activation and resultant disruption of mitochondrial membrane potential correlate with the induction of apoptosis in WEHI-231 B cells. *J. Immunol.* **166**, 137–147.

Katz, E., Lord, C., Ford, C. A., Gauld, S. B., Carter, N. A., Harnett, M. M. (2004). Bcl-(xL) antagonism of BCR-coupled mitochondrial phospholipase A(2) signaling correlates with protection from apoptosis in WEHI-231 B cells. *Blood* **103**, 168–176.

Koo, M. K., Oh, C. H., Holme, A. L., and Pervaiz, S. (2007). Simultaneous analysis of steady-state intracellular pH and cell morphology by automated laser scanning cytometry. *Cytometry A* **71**, 87–93.

Korn, T., Reddy, J., Gao, W., Bettelli, E., Awasthi, A., Petersen, T. R., Backstrom, B. T., Sobel, R. A., Wucherpfennig, K. W., Strom, T. B., Oukka, M., Kuchroo, V. K. (2007). Myelin-specific regulatory T cells accumulate in the CNS but fail to control autoimmune inflammation. *Nature Med.* **13**, 423–431.

Kornblau, S. M., Qiu, Y. H., Bekele, B. N., Cade, J. S., Zhou, X., Harris, D., Jackson, C. E., Estrov, Z., Andreeff, M. (2006). Studying the right cell in acute myelogenous leukemia: dynamic changes of apoptosis and signal transduction pathway protein expression in chemotherapy resistant ex-vivo selected "survivor cells". *Cell Cycle* **5**, 2769–2777.

Krutzik, P. O., and Nolan, G. P. (2006). Fluorescent cell barcoding in flow cytometry allows high-throughput drug screening and signaling profiling. *Nature Methods* **3**, 361–368.

Laffers, W., Mittag, A., Lenz, D., Tarnok, A., and Gerstner, A. O. (2006). Iterative restaining as a pivotal tool for n-color immunophenotyping by slide-based cytometry. *Cytometry A* **69**, 127–130.

Liu, K., and Mohan, C. (2006). What do mouse models teach us about human SLE? *Clinical Immunol. (Orlando Fla)* **119**, 123–130.

Luther, E., Kamentsky, L., Henriksen, M., and Holden, E. (2004). Next-generation laser scanning cytometry. *Methods Cell Biol.* **75**, 185–218.

Maffia, P., Brewer, J. M., Gracie, J. A., Ianaro, A., Leung, B. P., Mitchell, P. J., Smith, K. M., McInnes, I. B., Garside, P. (2004). Inducing experimental arthritis and breaking self-tolerance to joint-specific antigens with trackable, ovalbumin-specific T cells. *J. Immunol.* **173**, 151–156.

Marshall, F. A., Grierson, A. M., Garside, P., Harnett, W., and Harnett, M. M. (2005a). ES-62, an immunomodulator secreted by filarial nematodes, suppresses clonal expansion and modifies effector function of heterologous antigen-specific T cells *in vivo*. *J. Immunol.* **175**, 5817–5826.

Marshall, F. A., Grierson, A. M., Garside, P., Harnett, W., and Harnett, M. M. (2005b). ES-62, an immunomodulator secreted by filarial nematodes, suppresses clonal expansion and modifies effector function of heterologous antigen-specific T cells *in vivo*. *J. Immunol.* **175**, 5817–5826.

Marshall, F. A., Watson, K. A., Garside, P., Harnett, M. M., and Harnett, W. (2008). Effect of activated antigen-specific B cells on ES-62-mediated modulation of effector function of heterologous antigen-specific T cells *in vivo*. *Immunology* **123**, 411–425.

McInnes, I. B., Leung, B. P., Harnett, M., Gracie, J. A., Liew, F. Y., Harnett, W. (2003). A novel therapeutic approach targeting articular inflammation using the filarial nematode-derived phosphorylcholine-containing glycoprotein ES-62. *J. Immunol.* **171**, 2127–2133.

McInnes, I. B., and Schett, G. (2007). Cytokines in the pathogenesis of rheumatoid arthritis. *Nature Rev.* **7**, 429–442.

Melendez, A. J., Harnett, M. M., Pushparaj, P. N., Wong, W. S., Tay, H. K., McSharry, C. P., Harnett, W. (2007). Inhibition of FcepsilonRI-mediated mast cell responses by ES-62, a product of parasitic filarial nematodes. *Nature Medicine* **13**, 1375–1381.

Mital, J., Schwarz, J., Taatjes, D. J., and Ward, G. E. (2006). Laser scanning cytometer-based assays for measuring host cell attachment and invasion by the human pathogen *Toxoplasma gondii*. *Cytometry A* **69**, 13–19.

Mittag, A., Lenz, D., Bocsi, J., Sack, U., Gerstner, A. O., Tarnok, A. (2006a). Sequential photobleaching of fluorochromes for polychromatic slide-based cytometry. *Cytometry A* **69**, 139–141.

Mittag, A., Lenz, D., Gerstner, A. O., Sack, U., Steinbrecher, M., Koksch, M., Raffael, A., Bocsi, J., Tarnok, A. (2005). Polychromatic (eight-color) slide-based cytometry for the phenotyping of leukocyte, NK, and NKT subsets. *Cytometry A* **65**, 103–115.

Mittag, A., Lenz, D., Gerstner, A. O., and Tarnok, A. (2006b). Hyperchromatic cytometry principles for cytomics using slide based cytometry. *Cytometry A* **69**, 691–703.

Morton, A. M., McManus, B., Garside, P., Mowat, A. M., and Harnett, M. M. (2007). Inverse Rap1 and phospho-ERK expression discriminate the maintenance phase of tolerance and priming of antigen-specific CD4+ T cells *in vitro and in vivo. J. Immunol.* **179**, 8026–8034.

Mosch, B., Mittag, A., Lenz, D., Arendt, T., and Tarnok, A. (2006). Laser scanning cytometry in human brain slices. *Cytometry A* **69**, 135–138.

Mosch, B., Morawski, M., Mittag, A., Lenz, D., Tarnok, A., Arendt, T. (2007). Aneuploidy and DNA replication in the normal human brain and Alzheimer's disease. *J. Neurosci.* **27**, 6859–6867.

Nakagawa, R., Mason, S. M., and Michie, A. M. (2006). Determining the role of specific signaling molecules during lymphocyte development *in vivo*: instant transgenesis. *Nature Protocols* **1**, 1185–1193.

Osman, G. E., Cheunsuk, S., Allen, S. E., Chi, E., Liggitt, H. D., Hood, L. E., Ladiges, W. C. (1998). Expression of a type II collagen-specific TCR transgene accelerates the onset of arthritis in mice. *Int. Immunol.* **10**, 1613–1622.

Pape, K. A., Kearney, E. R., Khoruts, A., Mondino, A., Merica, R., Chen, Z. M., Ingulli, E., White, J., Johnson, J. G., Jenkins, M. K. (1997). Use of adoptive transfer of T-cell antigen-receptor-transgenic T cell for the study of T-cell activation *in vivo. Immunol. Rev.* **156**, 67–78.

Parsons, C. H., Adang, L. A., Overdevest, J., O'Connor, C. M., Taylor Jr., J. R., Camerini, D., Kedes, D. H. (2006). KSHV targets multiple leukocyte lineages during long-term productive infection in NOD/SCID mice. *J. Clin. Inv.* **116**, 1963–1973.

Perfetto, S. P., Chattopadhyay, P. K., and Roederer, M. (2004). Seventeen-colour flow cytometry: unravelling the immune system. *Nat. Rev. Immunol.* **4**, 648–655.

Pozarowski, P., Holden, E., and Darzynkiewicz, Z. (2006). Laser scanning cytometry: principles and applications. *Methods Mol. Biol.* **319**, 165–192.

Rew, D. A., Woltmann, G., and Kaur, D. (2006). Near-clinical applications of laser scanning cytometry. *Methods Mol. Biol.* **319**, 193–212.

Schubert, W. (2007). A three-symbol code for organized proteomes based on cyclical imaging of protein locations. *Cytometry A* **71**, 352–360.

Smith, K. M., Brewer, J. M., Rush, C. M., Riley, J., and Garside, P. (2004a). *In vivo* generated Th1 cells can migrate to B cell follicles to support B cell responses. *J. Immunol.* **173**, 1640–1646.

Smith, K. M., Brewer, J. M., Rush, C. M., Riley, J., and Garside, P. (2004b). *In vivo* generated Th1 cells can migrate to B cell follicles to support B cell responses. *J. Immunol.* **173**, 1640–1646.

Taatjes, D. J., Palmer, C. J., Pantano, C., Hoffmann, S. B., Cummins, A., Mossman, B. T. (2001). Laser-based microscopic approaches: application to cell signaling in environmental lung disease. *Biotechniques* **31**, 880–882 884, 886–888, 890, 892–884..

Taatjes, D. J., Zuber, C., and Roth, J. (2006). The histochemistry and cell biology vade mecum: a review of 2005–2006. *Histochem. Cell Biol.* **126**, 743–788.

Tarnok, A., and Gerstner, A. O. (2002). Clinical applications of laser scanning cytometry. *Cytometry* **50**, 133–143.

Wilson, E. H., Deehan, M. R., Katz, E., Brown, K. S., Houston, K. M., O'Grady, J., Harnett, M. M., Harnett, W. (2003). Hyporesponsiveness of murine B lymphocytes exposed to the filarial nematode secreted product ES-62 *in vivo. Immunology* **109**, 238–245.

CHAPTER 10

Image Cytometry Analysis of Circulating Tumor Cells

Lori E. Lowes,[*,§,¶] **David Goodale,**[*] **Michael Keeney**[†,‡] **and Alison L. Allan**[*,‡,§,¶]

[*]London Regional Cancer Program, London Health Sciences Centre, London, Ontario, Canada

[†]Special Hematology/Flow Cytometry, London Health Sciences Centre, London, Ontario, Canada

[‡]London Health Sciences Centre, Lawson Health Research Institute, London, Ontario, Canada

[§]Department of Anatomy & Cell Biology, University of Western Ontario, London, Ontario, Canada

[¶]Department of Oncology, University of Western Ontario, London, Ontario, Canada

Abstract

I. Introduction
    A. Cancer Metastasis
    B. Metastasis and Circulating Tumor Cells (CTCs)
    C. Preclinical Models of Metastasis and CTCs

II. Background and Technical Considerations
    A. General Considerations
    B. Enrichment: Negative versus Positive Selection Approaches
    C. Choice of Tumor Markers for CTC Analysis
    D. Techniques for CTC Enrichment and Isolation
    E. Techniques for CTC Detection and Analysis

III. Image Cytometry: Methods and Results
    A. Laser Scanning Cytometry
    B. The CellSearch® System
    C. Other Image Cytometry Approaches for CTC Analysis

IV. Conclusions and Future Directions
    Acknowledgments
    References

Abstract

The majority of cancer-related deaths are as a result of metastatic disease, which has been correlated with the presence of circulating tumor cells (CTCs) in the bloodstream. Therefore the ability to reliably enumerate and characterize these cells could provide useful information about the biology of the metastatic cascade; facilitate patient prognosis; act as a marker of therapeutic response; and/or aid in novel anticancer drug development. Several different techniques have been utilized for the enrichment and detection of these rare CTCs, each having their own unique advantages and disadvantages. In this chapter we will briefly discuss each of these techniques as well as the pros and cons of each approach. In particular, we will provide a comprehensive examination of two image cytometry approaches for CTC analysis that are in routine use in our laboratory; the iCys Laser Scanning Cytometer (Compucyte, Cambridge, MA), and the CellSearch® system (Veridex, North Raritan, NJ). The ability to detect, enumerate, and characterize CTCs is an important tool for the study of the metastatic cascade and the improved clinical management of cancer patients. These rare cells could shed light on the basic biology behind this highly lethal process and ultimately change current patient treatment guidelines.

I. Introduction

A. Cancer Metastasis

It has been estimated that 789,620 men and 739,940 women in the United States will develop cancer and that 299,200 men and 270,290 women will die from cancer in 2010 (Jemal *et al.*, 2010). The majority of these deaths are as a result of metastatic disease (Chambers *et al.*, 2002). Metastasis begins when cells are shed from the primary tumor and enter the bloodstream and lymphatic system through a process called intravasation. The cells that travel through the lymphatic system can metastasize to the lymph nodes or re-enter the bloodstream. Cells in the bloodstream can re-enter (extravasate) into surrounding tissue and form secondary tumors in distant organs. Most current therapies are ineffective at treating metastatic disease and are toxic to patients, in part due to a lack of understanding of the metastatic cascade (Pantel *et al.*, 2008). Future research is therefore needed to address (1) the basic biology of metastasis; (2) drug development and/or treatments targeting metastasis; and (3) development and optimization of techniques to track disease progression and ultimately improve strategies to prevent this deadly process.

B. Metastasis and Circulating Tumor Cells (CTCs)

Metastatic disease is often correlated with the presence of CTCs in the blood (Pantel *et al.*, 2008). In metastatic breast cancer, Cristofanilli *et al.* (2004) have shown that the presence of five or more CTCs in 7.5 ml of blood correlates with significantly reduced progression-free survival and overall survival compared to those individuals who have less than five CTCs in the same blood volume (Cristofanilli *et al.*, 2004). The same has been shown to be true by de Bono *et al.* (2008) with respect to overall survival in metastatic prostate cancer (de Bono *et al.*, 2008). Cohen *et al.* (2009) have also shown that metastatic colorectal cancer patients with three or more CTCs in 7.5 ml of blood have a significantly reduced overall survival compared to patients with less than three CTCs in the same blood volume (Cohen *et al.*, 2009). Therefore, the detection and quantification of CTCs can be used as a method to measure and predict disease progression in the metastatic setting of these three cancer types.

In addition to detection and enumeration of CTCs, the ability to characterize these cells on a cellular and molecular level holds great potential for cancer research and treatment. The importance of CTC characterization is well illustrated with the example of Her2, a member of the epidermal growth factor receptor (EGFR) family. Her2 is a cell surface protein that has been shown to be over-expressed in a subset of breast cancer patients. Exploitation of this marker has allowed for the development of targeted therapies that are "tailor-made" for Her2 + cancer patients, such as the Her2 receptor-interfering monoclonal antibody Herceptin (Stojadinovic *et al.*, 2007). Therefore, other markers of interest expressed by tumor cells have the potential to provide additional new targets for novel anticancer therapies. In addition, examination of the molecular characteristics of CTCs would enable a better understanding of the mechanisms that allow these cells to escape into the circulation, extravasate into distant tissue, and form clinically relevant secondary metastases.

C. Preclinical Models of Metastasis and CTCs

Although *in vitro* assays significantly aid in the study of metastasis, they do not accurately reflect the complexity of the human body. Therefore, the current gold standard for validation of experimental data regarding the metastatic cascade requires the use of *in vivo* preclinical animal models (Welch, 1997). In general, there are two types of models used for the study of the metastatic process, including spontaneous metastasis models and experimental metastasis models, each having their own unique advantages and disadvantages (Welch, 1997). Spontaneous metastasis models allow for the study of all the steps in the metastatic cascade, as tumor cell injections are made into the appropriate orthotopic location (i.e., injection of breast cancer cells into the mammary fat pad for studying breast cancer) and cells are given time to form primary tumors and spontaneously metastasize to secondary sites such as the lymph node and lung. In contrast, experimental metastasis models involve injection directly into the bloodstream (i.e., the tail vein to target the lung) and therefore skip the initial steps of intravasation and dissemination to secondary organs but still allow for the study of later steps in the metastatic process such as extravasation and secondary tumor formation (Welch, 1997). Work from our group and others (Ameri *et al.*, 2010;

Goodale *et al.*, 2009) has demonstrated that cytometric analysis of CTCs in these mouse models can provide important information regarding the kinetics and patterns of metastatic tumor cell dissemination and the relationship of CTCs to the subsequent development of metastases (Fig. 1).

Several methods for the detection and enumeration of CTCs in human patient samples and mouse models have been developed. In this chapter, we will briefly review these techniques, with a focus on image cytometry. In addition, we will examine two image cytometry approaches in detail that are frequently utilized in our laboratory for CTC detection and enumeration, including the iCys Laser Scanning Cytometer (Compucyte, Cambridge, MA), and the CellSearch® system (Veridex, North Raritan, NJ).

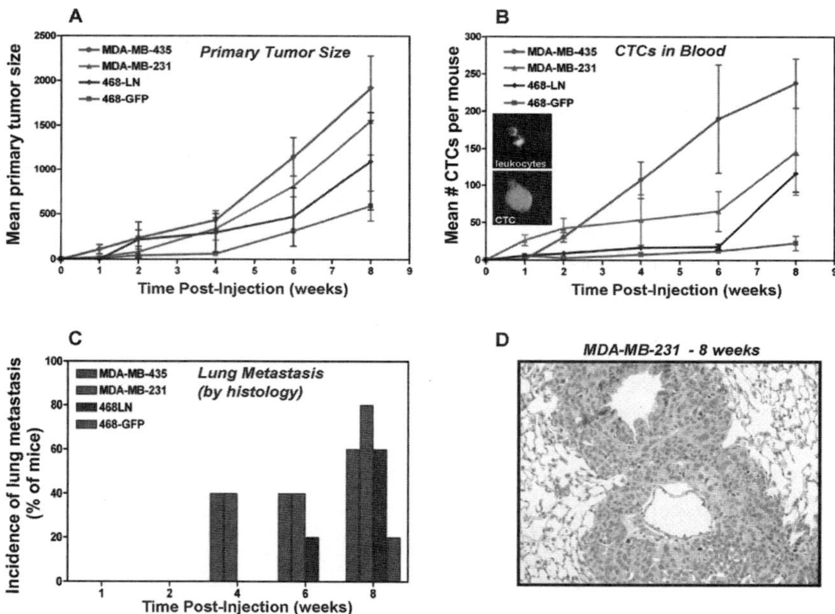

Fig. 1 Kinetics of tumor cell dissemination following mammary fat pad injection. Human breast cancer cell lines of differing metastatic abilities were injected into female nude (*nu/nu*) mice or NOD/SCID mice via the mammary fat pad ($n = 5$ per time point per cell line). MDA-MB-435 cells [red] are highly metastatic; MDA-MB-231 cells [brown]) and 468-LN cells [blue] are moderately metastatic, and 468-GFP cells [green]) are weakly metastatic. At several time points postinjection (1–8 weeks), mice were sacrificed and blood was collected. Fresh blood was processed for analysis by flow cytometry and laser scanning cytometry (LSC). Distant tissues were processed for analysis by histology. (A) Primary tumor growth kinetics as measured weekly by calipers, (B) Circulating tumor cell (CTC) kinetics in blood as measured by flow cytometry and confirmed by LSC (*inset images*; leukocytes stained with CD45⁻PE, CTCs stained with HLA-FITC), (C) Incidence of lung metastasis (% of mice in group) as measured by H&E staining, and (D) histopathology. (See plate no. 16 in the color plate section.)

II. Background and Technical Considerations

A. General Considerations

It has been estimated that in metastatic cancer patients, CTCs are present at a frequency of approximately 1 CTC per 10^{5-7} white blood cells, and in localized disease this frequency may be even lower ($\sim$1 CTC per 10^8 white blood cells) (Allan and Keeney, 2010; Ross *et al.*, 1993; Tibbe *et al.*, 2007). The rare nature of these cells can make it difficult to accurately and reliably detect and analyze CTCs. It is therefore necessary to utilize highly standardized and validated techniques for CTC analysis in order to ensure that results can be compared across laboratories, particularly in the clinical setting (Allan and Keeney, 2010). Because of this necessity for standardization, the majority of techniques currently available for CTC detection cannot be applied in the clinic at the present time, although many are useful for basic science/translational research applications.

The majority of CTC analysis methods require two steps; an enrichment step and a detection step. The enrichment step can be subdivided into two major classes of techniques; cell size/density-based approaches and immunomagnetic separation-based approaches. Similarly, the detection/analysis step can also be subdivided into two classes, nucleic acid-based approaches and cytometric-based approaches. There are also techniques that attempt to improve assay sensitivity, specificity, and reproducibility by using the best combination of enrichment and detection steps necessary for optimal results. We will refer these techniques as combination approaches. The advantages and disadvantages of each technique presented in this chapter are summarized in Tables I–III.

B. Enrichment: Negative versus Positive Selection Approaches

Enrichment approaches for CTCs can either be positive or negative in nature. Positive selection utilizes markers expressed by the target cells (i.e., epithelial markers for tumor cell selection), whereas negative selection utilizes markers expressed by nontarget cells (i.e., CD45 for leukocyte depletion). Positive selection approaches usually result in fairly high purity of the sample recovered, since only CTCs will be selected. However, the recovery of CTCs may be lower than in negative selection approaches, since CTCs that have low or absent expression of epithelial markers (or other positive selection markers) may be lost (Allan and Keeney, 2010; Sieuwerts *et al.*, 2009).

C. Choice of Tumor Markers for CTC Analysis

Currently there is no perfect marker for CTC analysis. Instead, CTCs are enriched and detected using markers that are based on our current understanding of the characteristics of these cells. In general, there are two types of markers that can be utilized for CTC enrichment and/or detection, including epithelial-specific

Table I

Techniques for CTC enrichment

		Advantages	Disadvantages	References
Morphology	ISET/Nucleopore Assay	• Fast, simple, and inexpensive • Can be applied to all tumor types • Can be applied to very small blood volumes (as little as 1 ml) • Additional molecular analysis possible following enrichment • Cells are viable following enrichment	• Small CTCs may be missed • Not an automated system	De Giorgi *et al.*, 2010; Ntouroupi *et al.*, 2008; Pinzani *et al.*, 2006; Vona *et al.*, 2000, 2004
	Density gradient centrifugation (Ficoll/ Lymphoprep/ OncoQuick)	• Fast, simple, and inexpensive • Can be applied to all tumor types • Additional molecular analysis possible following enrichment • Cells are viable following enrichment	• Low sensitivity • Optimal isolation depends on length of time and temperature during centrifugation • Low sample purity • Not an automated system	Rosenberg *et al.*, 2002; Baker *et al.*, 2003; Müller *et al.*, 2005; Balic *et al.*, 2005
	RARE	• Fast, simple, and relatively inexpensive • Can be applied to all tumor types • Additional molecular analysis possible following enrichment	• Low percent recovery of cells following enrichment • Not an automated system	Naume *et al.*, 2004; Alunni-Fabbroni and Sandri, 2010
Immunomagnetic Separation	MACS/EasySep	• Fast and simple • Additional molecular analysis possible following enrichment • Semiautomated systems available for sample processing (autoMACS/ RoboSep)	• A proportion of CTCs may be lost during enrichment due to a lack of target protein expression (false negatives)	Engel *et al.*, 1999; Iinuma *et al.*, 2000; Taubert, 2004; Alunni-Fabbroni and Sandri, 2010; Mostert *et al.*, 2009

markers and tumor-associated markers. Examples of epithelial-specific markers include epithelial cell adhesion molecule (EpCAM) and various cytokeratin (CK) isoforms. The tumor-associated class is much broader and often depends on the tissue of origin of the primary tumor (i.e., breast, prostate, colon, etc.). Therefore, for this class we will discuss some of the more commonly utilized markers as examples.

Table II

Techniques for CTC detection

		Advantages	Disadvantages	References
Cytometric	FAST	• No enrichment required • Quantitative • Rapid sample analysis of large blood volumes (300,000 cells/sec)	• Sample labeling is not an automated process. • CTC determination is subjective • Lack of clinical validation studies	Alunni-Fabbroni and Sandri, 2010; Hsieh *et al.*, 2006; Krivacic *et al.*, 2004; Marrinucci *et al.*, 2010; Mostert *et al.*, 2009
	Flow Cytometry	• High specificity • Quantitative • Highly flexible for research purposes • Multiparameter analysis	• Visual confirmation not possible • Unless fluorescence-activated cell sorting (FACS) is used, CTCs are not collected for downstream analysis	Alunni-Fabbroni and Sandri, 2010; Beitsch *et al.*, 2000; Cruz *et al.*, 2005; Hu *et al.*, 2010; Mostert *et al.*, 2009; Racila *et al.*, 1998
	Laser scanning cytometry	• High specificity • Quantitative • Highly flexible for research purposes • Multiparameter analysis • Visual confirmation of CTCs	• CTCs are not collected for downstream analysis	Balic *et al.*, 2005; Goodale *et al.*, 2009; Pachmann *et al.*, 2008; Ring *et al.*, 2005; Zabaglo *et al.*, 2003; Mostert *et al.*, 2009; Alunni-Fabbroni and Sandri, 2010
	EPISPOT	• Allows detection of only viable cells. • High sensitivity • High specificity	• Not an automated system • CTCs are not collected and therefore subsequent cellular analysis is not possible. • CTC determination is subjective • Lack of clinical validation studies	Alunni-Fabbroni and Sandri, 2010; Mostert *et al.*, 2009; Alix-Panabières *et al.*, 2005, 2007, 2009
Nucleic Acid-Based	RT-PCR	• High sensitivity • Positive results likely indicate viable CTCs due to instability of RNA in the circulation	• Low specificity • Not quantitative • Cells destroyed during processing • Not an automated system • No standardization of marker/primer choice and number of PCR cycles utilized	Alunni-Fabbroni and Sandri, 2010; Hu et al., 2010; Mostert *et al.*, 2009; Paterlini-Brechot and Benali, 2007
	qRT-PCR	• High sensitivity (higher than RT-PCR)	• Cells destroyed during processing	Slade *et al.*, 1999; Van der Auwera *et al.*, 2010;

(Continued)

Table II   *(Continued)*

	Advantages	Disadvantages	References
	• Positive results likely indicate a viable CTC due to instability of RNA in the circulation • Quantitative	• Not an automated system • No standardization of marker/primer choice and number of PCR cycles utilized	Iakovlev *et al.*, 2008; Hu *et al.*, 2000; Paterlini-Brechot and Benali, 2007; Mostert *et al.*, 2009; Alunni-Fabbroni and Sandri, 2010

Table III

Combined techniques for CTC enrichment and detection

	Advantages	Disadvantages	References
AdnaTest	• High sensitivity • Fast and simple • The use of multiple tumor-associated markers for isolation ensures that CTCs that do not express EpCAM can be isolated	• A proportion of CTCs may be lost during enrichment due to a lack of target protein expression (false negatives) • Not an automated system • Lower specificity due to illegitimate expression of target genes by contaminating cells	Alunni-Fabbroni and Sandri, 2010; Cristofanilli, 2009; Lankiewicz *et al.*, 2006; Mostert *et al.*, 2009; Pinzani *et al.*, 2006; Van der Auwera *et al.*, 2010; Zieglschmid *et al.*, 2005
CTC chip	• Very high sensitivity and specificity • Semiautomated system • Visual confirmation of CTCs • High sample purity • High recovery of viable cells following enrichment • Additional molecular analysis possible • No preprocessing of the sample is required thereby reducing the risk of CTC loss due to sample handling	• A proportion of CTCs may be lost during enrichment due to a lack of EpCAM expression (false negatives) • Processing is very slow • CTC determination is subjective • More clinical validation studies needed	Alunni-Fabbroni and Sandri, 2010; Helzer *et al.*, 2009; Mostert *et al.*, 2009; Nagrath *et al.*, 2007; Sequist *et al.*, 2009
CellSearch® System	• High sensitivity • Semiautomated system • Visual confirmation of CTCs • FDA approved for clinical use in metastatic breast, prostate, and colorectal cancer • Has one extra fluorescent channel for additional molecular characterization	• A proportion of CTCs may be lost during enrichment due to a lack of EpCAM expression (false negatives) • Very inflexible for research purposes • CTC determination is subjective • Cells fixed during processing and are therefore unable to be further analyzed or grown in culture.	Cristofanilli, 2009; Cristofanilli *et al.*, 2004

1. Epithelial-Specific Markers

EpCAM and various CK isoforms have been used extensively for CTC enrichment and/or detection. EpCAM is a transmembrane glycoprotein that is a member of the cell adhesion molecule (CAM) family of proteins. This marker is expressed by normal epithelial cells and has been found to be expressed in up to 70% of all primary carcinoma tumor types (Went *et al.*, 2004). EpCAM has been utilized as a tool for CTC enrichment, as epithelial cells should not normally circulate in the bloodstream of healthy individuals. However, this marker has been criticized for its lack of ability to detect the subset of tumor cells that may have a down-regulation in EpCAM expression due to the epithelial-to-mesenchymal transition (EMT), a process that is believed to be necessary for tumor cell invasion and migration from the site of the primary tumor (Sieuwerts *et al.*, 2009; Thiery, 2002). Various CK isoforms have also been utilized as markers for CTC enrichment and detection. These proteins are intermediate filaments found in the cytoskeleton of epithelial cells. Several criticisms of using CKs for CTC detection exist. Firstly, similar to EpCAM, CKs may be downregulated in cells undergoing EMT. Secondly, the appropriate CK isoform(s) for optimal CTC detection have not yet been identified. Finally, depending on the detection technique utilized, CKs can be detected in the blood of healthy individuals (Ko *et al.*, 2000; Vlems *et al.*, 2002).

2. Tumor-Associated Markers

An example of a tumor-associated marker for breast cancer is mammaglobin, is a member of the secretoglobin family of proteins. This protein was discovered in 1996 by Watson and Fleming, and is therefore a relatively new marker (Watson and Fleming, 1996). Originally mammaglobin expression was thought to be confined to the mammary tissue; however, subsequent studies have found mammaglobin expression in normal and malignant ovarian, uterine, and cervical tissue, as well as in sweat glands (Grünewald *et al.*, 2002; Sjödin *et al.*, 2003). In breast cancer, this protein has been found in primary and metastatic lesions, lymph node metastases, and CTCs (Watson *et al.*, 1999). The protein has been suggested as a good marker of micrometastatic disease in the lymph node that is not detectable by routine histological examination (Gillanders *et al.*, 2004). Currently the study of mammaglobin's functional role in breast cancer has not been an area of active research and the literature that is available is highly conflicting, showing mammaglobin as being associated with both high grade breast cancer and less aggressive tumors (Mikhitarian *et al.*, 2008; Núñez-Villar *et al.*, 2003; Sjödin *et al.*, 2008).

Another example of a tumor-associated marker is prostate-specific antigen (PSA), a serine protease. Transcription of this protein is androgen-regulated and therefore production of high levels of PSA is confined to the prostate (Young *et al.*, 1991). It is important to note that PSA is not a specific marker of prostate malignancy, as this protein is also produced in the healthy prostate, as well as in benign prostatic disease (Herrala *et al.*, 2001; Lintula *et al.*, 2005). Therefore it is not the presence of PSA per se that is indicative of prostate cancer, but rather the level of PSA that acts as an indicator for disease presence and progression. PSA is currently utilized as a marker for prostate cancer screening and early diagnosis, however in this setting the use of PSA is highly controversial as many other factors beyond prostate cancer can affect PSA levels (Lintula *et al.*, 2005; Marks *et al.*, 2006; Verhamme *et al.*, 2002). PSA is less controversial when utilized for the detection of prostate cancer recurrence and monitoring of response to therapy. In addition to PSA's use in the clinic, is has also been implicated as a functional contributor to prostate cancer progression (Rubin, 2003; Williams *et al.*, 2007) and as a marker for prostate CTCs (Stott *et al.*, 2010).

Finally, carcinoembryonic antigen (CEA), a tumor-associated marker that has been shown to be associated with colorectal cancer and expressed by CTCs detected in the bloodstream of metastatic colorectal cancer patients (Jonas *et al.*, 1996; Thomson *et al.*, 1969). This glycoprotein is a member of the immunoglobulin super family of proteins and has been implicated in cancer invasion and metastasis due to its structural similarity to other adhesion molecules such as Intercellular Adhesion Molecule-1 (ICAM-1) (Duffy, 2001; Hostetter *et al.*, 1990; Jessup and Thomas, 1989). This marker appears to be inappropriate for screening or diagnostic purposes but has been shown to have utility as a prognostic marker, a marker of tumor recurrence at distant sites, and as a marker for monitoring therapy response (Duffy *et al.*, 2007; Fletcher, 1986).

One criticism of all of these markers is that they are merely tumor-associated but not tumor-specific proteins, and therefore some CTCs may be missed by these techniques and/or benign cells may be erroneously identified as CTCs.

D. Techniques for CTC Enrichment and Isolation

1. Cell Size and Density–Based Enrichment

This type of enrichment relies on size and density differences between blood mononuclear cells and CTCs. In general, there are three techniques that make up this category of enrichment, (1) Isolation by Size of Epithelial Tumor cells (ISET), (2) density gradient centrifugation, and (3) RosetteSep-Applied Imaging Rare Event (RARE).

ISET (Metagenex, Paris, France) is an enrichment technique that separates CTCs based on the assumption that they are larger in size then leukocytes (Vona *et al.*, 2000). Whole blood is diluted and passed through a filter with a pore diameter of 8 µm via vacuum filtration. The larger CTCs are held back by the filter, while smaller leukocytes pass through. This technique is very gentle and does not cause cell damage; therefore cells can be further analyzed following enrichment (Vona *et al.*, 2000). The Nucleopore assay (Whatman International Ltd., UK) is another vacuum filtration technique that utilizes the larger size of CTCs to enrich blood samples (Ntouroupi *et al.*, 2008). However, unlike the ISET technique, subsequent follow-up studies with large patient groups have not been performed to validate the usefulness of the nucleopore assay in patient prognostication (De Giorgi *et al.*, 2010; Ntouroupi *et al.*, 2008; Pinzani *et al.*, 2006; Vona *et al.*, 2004).

Density gradient centrifugation techniques such as Ficoll, Lymphoprep (Axis-Shield PoC, Oslo, Norway), and OncoQuick (Greiner BioOne, Frickenhausen, Germany) separate mononuclear cells from other cells found in whole blood based on differences in blood cell densities (Balic *et al.*, 2005; Naume *et al.*, 2004; Rosenberg *et al.*, 2002). Whole blood is layered on top of a density gradient and centrifuged to separate out distinct cell populations. Erythrocytes and granulocytes will pellet at the bottom of the tube while plasma and mononuclear cells (including CTCs), due to their lower buoyant densities, will remain above the gradient and are readily accessible for subsequent collection and analysis of CTCs. The commercially available OncoQuick has been shown to significantly improve enrichment over the standard Ficoll separation (Rosenberg *et al.*, 2002), likely due to a porous barrier that prevents mixing of cells following separation.

RARE (StemCell Technologies, Vancouver, BC) is a negative selection technique. Tetrameric antibody complexes crosslink CD45-expressing leukocytes to red blood cells present in the sample and form complexes known as ImmunoRosettes. The cross-linked blood sample is then layered on a density gradient (such as Ficoll) and centrifuged. The CD45-expressing cells will then pellet to the bottom of the tube due to their increased density, thereby increasing the enrichment of CD45-negative cells (CTCs) (Alunni-Fabbroni and Sandri, 2010; Naume *et al.*, 2004).

2. Enrichment by Immunomagnetic Separation

Immunomagnetic separation techniques for CTC enrichment depend on the expression of epithelial-specific and/or tumor-associated protein markers discussed in Section II.C. For example, MACS (Miltenyi Biotec, Bergisch Gladbach, Germany) utilizes a magnetic column apparatus that captures cells which have been bound by magnet bead-conjugated antibodies targeted against various cellular antigens (i.e., EpCAM, pan-CK, HER2/neu, or CD45). Labeled cells are held back by the column and unlabeled cells freely pass through. The majority of these kits utilize biodegradable reagents and do not require a fixation or permeabilization step to isolate CTCs (Alunni-Fabbroni and Sandri, 2010; Mostert *et al.*, 2009). Therefore, cells are available for subsequent analysis following enrichment (Engel *et al.*, 1999). EasySep^{TM} (StemCell Technologies) is a very similar technique except that it is not performed in a column, but instead in a standard test tube. Antibody-bound target cells (EpCAM or CD45) are complexed with magnetic nanoparticles and subsequently incubated in a magnet, and unlabeled cells are poured off (Alunni-Fabbroni and Sandri, 2010; Mostert *et al.*, 2009). Both of these techniques can be performed manually or by a semiautomated system (autoMACS and RoboSep, respectively).

3. Combination Approaches

The AdnaTest (Adnagen, Langenhagen, Germany) is a two-component enrichment and detection technique integrating the AdnaSelect and AdnaDetect procedures, respectively. The AdnaSelect uses multiple epithelial-specific and/or tumor-associated targets chosen based upon the cancer type being selected (i.e., breast, prostate, or colon). Magnetic selection beads are incubated with the blood sample and the target cells are labeled. The sample is incubated in a magnet and unlabeled cells are removed. The labeled cells are then lysed, and RNA is isolated using the AdnaDetect kit and subjected to reverse transcription polymerase chain reaction (RT-PCR) for CTC detection. Target RNA depends on cancer type (Lankiewicz *et al.*, 2006; Van der Auwera *et al.*, 2010; Zieglschmid *et al.*, 2005).

The CTC Chip is another combined enrichment and detection technique that utilizes a slide-based chip covered in 78,000 microposts, each coated with EpCAM targeted antibodies (Helzer *et al.*, 2009; Nagrath *et al.*, 2007; Sequist *et al.*, 2009). Unprocessed whole blood is pumped across the surface of the coated chip and EpCAM+ cells bind to the microposts. The captured cells are then subsequently stained using anti-CK/anti-CD45 antibodies and DAPI, and an image-based system is used to detect cells that are CK+/DAPI+/CD45⁻ and present these events to users for qualitative validation as CTCs.

The CellSearch® system (Veridex) is a two-component system that isolates CTCs based on positive EpCAM expression (Cristofanilli *et al.*, 2004; Cristofanilli, 2009). This technique is discussed in detail in Section III.B.

E. Techniques for CTC Detection and Analysis

1. Nucleic Acid–Based Detection

Nucleic acid-based CTC detection techniques depend on the transcription of epithelial-specific and/or tumor-associated RNA in cancerous cells that is not present or is differentially expressed in noncancerous cells. This can be done using RT-PCR or quantitative reverse transcription polymerase chain reaction (qRT-PCR).

Prior to RT-PCR analysis of CTCs, peripheral blood is collected and usually enriched for mononuclear cells using one of the above mentioned enrichment techniques, after which subsequent RT- or qRT-PCR analysis is performed. Both of these techniques utilize RNA transcripts for CTC detection. However, unlike RT-PCR, qRT-PCR allows for semiquantitative information to be obtained (Baker *et al.*, 2003; De Giorgi *et al.*, 2010; Hu *et al.*, 2000; Iakovlev *et al.*, 2008; Pinzani *et al.*, 2006; Ring *et al.*, 2005; Slade *et al.*, 1999; Van der Auwera *et al.*, 2010; Zieglschmid *et al.*, 2005). Using this technique, the relative amount of expression can be determined when compared to a known standardized control curve. This technique is thought to improve specificity over traditional RT-PCR by reducing false positives. However, standardized control curves with appropriate cutoff values to distinguish CTC positive from CTC negative samples are difficult to generate, as the number of

CTCs present in a particular patient population is unknown. Nucleic acid approaches of CTC detection have several advantages, including high sensitivity (Traystman *et al.*, 1997; Wulf *et al.*, 1997) and the ability to analyze very small sample volumes because of the amplification-based nature of PCR (Datta *et al.*, 1994; Hu *et al.*, 2000). However, these approaches also have several disadvantages, including low specificity and a tendency towards false positives. This false positivity can result from illegitimate transcription of target mRNA by nontarget cells, DNA contamination of blood samples, and/or unknown pseudogenes that are not controlled for during the primer design process (Hu *et al.*, 2000). False negatives can also result if mRNA is of poor quality when isolated. Another disadvantage is that mRNA transcription may not be equivalent to protein expression, and therefore may not accurately reflect CTC characteristics. Finally, it is impossible for RT- or qRT-PCR to link transcription of a particular mRNA to a particular cell, which is an important consideration based on the inherent heterogeneity of the blood sample being analyzed, as well as the cancer cells themselves. Many of these issues can be successfully addressed by the use of cytometric detection techniques, which allows for highly specific CTC analysis and characterization on an individual cell-by-cell basis.

2. Cytometric Detection

Cytometric detection of CTCs distinguishes rare CTCs from other cells in the blood based on differential antigen expression or other cellular characteristics, as measured by immunofluorescent labeling of cells. There are three general cytometric detection techniques, (1) epithelial immunospot (EPISPOT), (2) flow cytometry, and (3) image cytometry.

The EPISPOT assay is based on the same principles as the more commonly used enzyme-linked immunosorbant assay (ELISA). In this technique, preenriched cells (using one of the techniques described previously) are added to the EPISPOT plate which has been precoated with antibodies targeted against secreted cellular proteins either of epithelial-specific or tissue-associated origins(Alix-Panabières *et al.*, 2005, 2007, 2009). Viable cells will secrete these target proteins, which will subsequently bind to the EPISPOT plate. Following several days in culture, cells are removed from the plate and antibodies are added which are targeted against a second epitope on the secreted protein. The antibodies can be subsequently labeled or cleaved for detection via immunocytochemical or flow cytometric approaches.

Using flow cytometry, cells are fluorescently labeled with antibodies targeted against epithelial-specific and/or tumor-associated antigens and are subsequently passed single-file through a fine tube. Lasers are then utilized to excite fluorescent particles and, based on scatter and fluorescence analysis; can determine cell shape, size, granularity, and marker expression for appropriate cellular classification. However, flow cytometry does not allow for visual confirmation of identified cells, a problem which is resolved by image cytometry approaches, discussed in detail below.

III. Image Cytometry: Methods and Results

Image cytometry utilizes the same cell-by-cell analysis principle as flow cytometry, except that samples are scanned in a visual plane instead of in suspension (Balic *et al.*, 2005; Goodale *et al.*, 2009; Pachmann *et al.*, 2008; Ring *et al.*, 2005; Zabaglo *et al.*, 2003). For the purposes of this chapter, we will examine two image cytometry approaches in detail that are frequently utilized in our laboratory for CTC detection and enumeration, including the iCys Laser Scanning Cytometer (Compucyte, Cambridge, MA), and the CellSearch® system (Veridex, North Raritan, NJ). Both of these systems provide not only the opportunity to identify and enumerate these rare cells, but also the ability to further characterize CTCs based on markers of particular interest. This additional capacity could provide important molecular and biological information about these cells and facilitate a greater understanding of metastatic cascade and treatments to target metastasis.

A. Laser Scanning Cytometry

Laser scanning cytometry (LSC) uniquely combines the advantages of flow cytometry, image analysis, and automated fluorescence microscopy such that a large amount of multiparameter data can be simultaneously gathered and quantified for individual events in a heterogeneous population of cells, including visualization of fluorescence and morphology (Pozarowski *et al.*, 2006). Specifically, LSC allows for rapid and serial sample analysis; the ability to analyze tissue specimens without disaggregation and therefore without sample loss; the ability to provide unbiased quantitative information; visual confirmation of the sample including cellular localization of target proteins; and the ability for cell positions to be recorded and individual cells relocated for subsequent analysis (Darzynkiewicz *et al.*, 1999; Mach *et al.*, 2010; Tárnok and Gerstner, 2002). In our laboratory, we routinely utilize the iCys LSC (Compucyte) for detection and analysis of CTCs in preclinical models of breast cancer metastasis.

1. Methodology: Detection and Analysis of CTCs in Preclinical Models of Metastasis Using LSC (Allan *et al.*, 2005; Goodale *et al.*, 2009)

Preclinical in vivo breast cancer metastasis models

All animal procedures should be conducted under a protocol approved by institutional Councils on Animal Care. For analysis of human cancer cells, female immunocompromised mice (i.e., athymic nude (*nu/nu*) or NOD/SCID) are required. Depending on the type of metastasis assay used (see Section 1), cells should be injected orthotopically into the thoracic mammary fat pad (spontaneous metastasis model, 2×10^6 cells/mouse); the lateral tail vein (experimental metastasis model to target lung, 1×10^6 cells/mouse); or intracardiac via the left ventricle (experimental metastasis model to target bone, 2×10^5 cells/mouse). At various time points postinjection, mice can be sacrificed and blood collected for CTC analysis. Fresh whole blood should be collected via terminal cardiac puncture of the right ventricle using a 22G needle attached to a 1 ml syringe precoated with heparin (10,000 I.U./ml). All samples should be processed within 2 h of collection.

Labeling and immunomagnetic enrichment procedure

Blood samples should be subjected to red blood cell lysis in $1\times$ NH_4Cl and washed with phosphate-buffered saline (PBS) prior to labeling with mouse antihuman leukocytic antigen (HLA) antibody (clone W6/32) conjugated to fluorescein isothiocyanate (FITC) and rat antimouse pan-leukocytic CD45 antibody (clone 30-F11) conjugated to phycoerythrin (PE). In this protocol, we have chosen HLA as a human-specific marker, since the only human cells present in the mouse blood will be the injected tumor cells. This approach obviously cannot be translated for use in analyzing human CTCs in human blood samples; however, it does serve as a highly specific and reproducible marker for analyzing human CTCs in mouse models. Following labeling, samples can be immunomagnetically enriched using the EasySep^{TM} PE Selection Kit (Stem Cell Technologies; for negative selection/depletion of $CD45^-PE^+$ mouse leukocytes) as per the manufacturer's instructions, using 2×5 min incubations in the EasySep^{TM} magnet. After incubation, the fraction containing the tumor cells (supernatant) should be fixed and permeabilized, i.e. using the IntraPrep^{TM} Fix/Perm Kit (Beckman Coulter, Fullerton, CA).

Laser scanning cytometry analysis

Enriched and labeled blood samples should be placed onto a glass slide and covered with a glass cover slip just prior to LSC analysis. Setup and compensation can be adjusted on a spiked and labeled $\sim$10% mixture of human tumor cells and mouse leukocytes. The LSC acquisition protocol should be configured with primary contour set on light scatter. Green fluorescence (HLA-FITC) and orange fluorescence (CD45$^-$PE) are excited with a 488 nm argon ion laser and measured using standard filter settings. The intensities of maximal pixel (pixel size 0.5×0.5 μm) and integrated fluorescence can be measured and recorded for each event. Cell morphology and fluorescence parameters can be confirmed by visualizing microscopy images through the gallery function in the iCys software. Events which are confirmed to be HLA$^+$CD45$^-$ are considered to be positive CTC events.

2. Typical Results: Detection and Analysis of CTCs in Preclinical Models of Metastasis Using LSC

The LSC acquisition protocol is configured as shown in Fig. 2A, with primary contour set on light scatter. Fig. 2B shows representative analysis of green integral (HLA-FITC) versus orange integral (CD45$^-$PE) of $\sim$0.01% MDA-MB-435 human breast cancer cells spiked in mouse blood. Events which fall within region R1 are counted as meeting the criteria for mouse leukocytes if they are CD45$^+$HLA$^-$, and

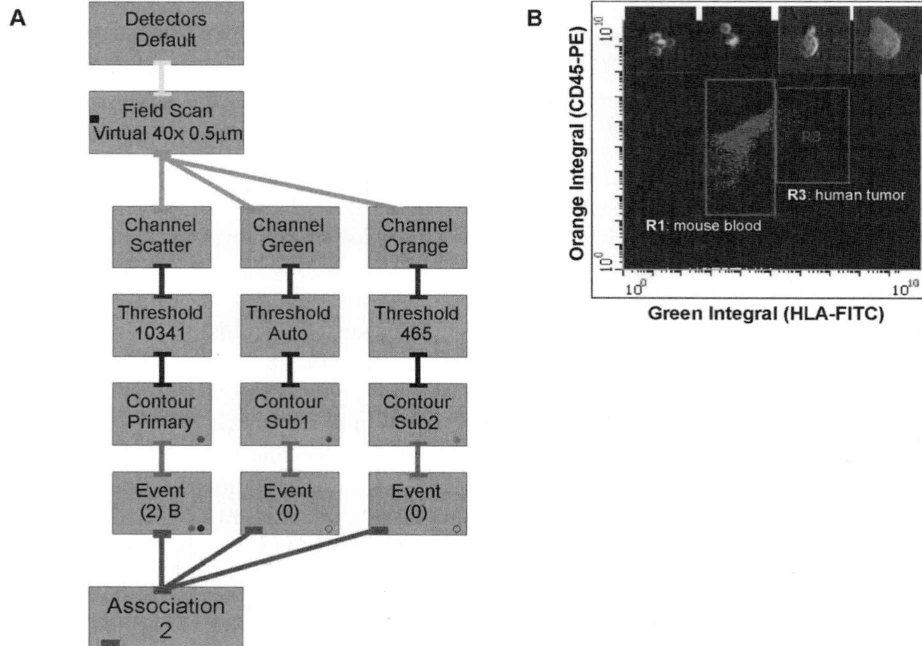

Fig. 2   Laser scanning cytometry (LSC) acquisition setup and representative analysis of MDA-MB-435 human cancer cells in mouse blood. Samples were stained with antimouse CD45⁻PE and antihuman HLA-FITC antibodies, enriched using the EasySep™-PE immunomagnetic selection kit, fixed, and placed on a glass slide with a cover slip for analysis on a Compucyte iCys LSC. For all experiments, one scan field [40 × 0.5 µm] per sample was scanned and analyzed. (A) Acquisition protocol. (B) Representative LSC analysis of ∼0.01% MDA-MB-435 human tumor cells in whole mouse blood. Cell morphology and fluorescence parameters for both populations were confirmed by visualizing microscopy images through the gallery function in the iCys software *(B, top insets)*. Events which fell within region **R1** (orange) and confirmed to be CD45⁺HLA⁻ were counted as meeting the criteria for mouse leukocytes, and events which fell within region **R3** (green) and confirmed to be HLA⁺CD45⁻ were counted as meeting the criteria for human tumor cells. Adapted from Goodale *et al.*, *Cytometry A*. 2009; 75: 344–355. (For interpretation of the references to color in this figure legend, the reader is referred to the Web version of this chapter.)

events which fall within region R3 are counted as meeting the criteria for human tumor cells if they are HLA⁺CD45⁻. Cell morphology and fluorescence parameters of both populations are confirmed by visualization through the gallery function in the iCys software (Figures 2B, top insets).

The use of this LSC assay allows characterization of *in vivo* CTC dissemination patterns and kinetics in the blood of mice injected with human breast cancer cells. The data presented in Fig. 1 suggest that, following mammary fat pad injection, CTC kinetics are related to the metastatic aggressiveness of different human breast cancer cell lines. Interestingly, it can be observed that the timing of micrometastatic incidence to distant organs such as lung is consistent with but slightly delayed relative to CTC kinetics. For example, CTCs can be detected in the blood as early as 1 week post-injection of human breast cancer cells (Fig. 1B), whereas lung micrometastases are only observed starting at 4 weeks post-injection (Fig. 1C, D). This timing corresponds to detection of increasing levels of CTCs, suggesting that, in addition to shedding from the primary tumor, CTCs may also be disseminating from secondary metastatic tumors in the lung. Interestingly, while no correlation was seen between the number of CTCs and primary tumor size for any of the breast cancer models examined, the mice that have the highest number of CTCs in the blood are typically the mice that develop lung metastases (*data not shown*).

3. Practical Application of the Methodology: Detection and Analysis of CTCs in Preclinical Models of Metastasis Using LSC

The ability to quantify CTCs following injection of human breast cancer cells via different routes provides valuable information with regards to the pattern and kinetics of CTC dissemination in spontaneous and experimental metastasis models. Furthermore, since the selection and labeling strategy is designed to specifically identify human cancer cells derived from a variety of sources (Allan *et al.*, 2005), this method has broad application for characterizing the CTC dissemination patterns and metastatic ability of other human cancer cell lines in preclinical mouse models of metastasis. Although a number of quantitative tools have been previously developed to study *in vivo* metastasis (Jenkins *et al.*, 2003; Li *et al.*, 2002; Naumov *et al.*, 1999), the detection and quantification of rare metastatic events such as CTCs has remained challenging. This method addresses this need, and has future potential for helping to elucidate the mechanistic details of early steps in metastasis and how these steps relate to the development of life-threatening macrometastases. In addition, this method could be used for the future identification, development, and testing of new therapeutic strategies to combat cancer.

B. The CellSearch® System

A second image cytometry approach that is utilized in our laboratory for CTC analysis is the CellSearch® system (Veridex). This system is currently the only CTC assay that is approved by the United States Food and Drug Administration (FDA) for clinical use in the metastatic breast, prostate, and colorectal cancer settings using CTC enumeration as a measure of patient prognosis (de Bono *et al.*, 2008; Cohen *et al.*, 2008; Cristofanilli *et al.*, 2004). The CellSearch® system consists of two components; (1) the CellTracks AutoPrep™ system, which automates the blood sample preparation, and (2) the CellTracks Analyzer™ II, which scans the prepared samples. There are currently three unique reagent kits that are available for use with the CellSearch® AutoPrep, including the CellSearch® Epithelial Cell Kit (clinical and research use), the CellSearch® Profile Kit (research use only), and the CellSearch® Circulating Endothelial Cell Kit (research use only). The CellSearch® Epithelial

Cell Kit is used to isolate and enumerate CTCs from human blood samples. The details of this reagent kit are discussed in detail in the Methodology section below. The CellSearch® Profile Kit utilizes the same principle as the CellSearch® Epithelial Kit, except that CTCs are only isolated from blood samples using the AutoPrep, but are not enumerated using the Analyzer. Cells isolated using the Profile kit can be used for downstream applications such as microarray or fluorescence *in situ* hybridization (FISH) analysis of CTCs (Flores *et al.*, 2010). Finally, the CellSearch® Endothelial Cell Kit is utilized to isolate and enumerate $CD146^+$ circulating endothelial cells (CECs) from human blood samples that are believed to be shed into the bloodstream during times of inflammation and/or angiogenesis (Goon *et al.*, 2006).

1. Methodology: Detection and Analysis of CTCs in Human Blood Samples Using the CellSearch® System

Blood sample collection and preparation

Following informed consent, human blood samples should be collected into 10 ml CellSave tubes (Veridex) containing EDTA and a proprietary cellular preservative. To avoid sample degradation, blood must be stored at room temperature and processed within 96 h of collection. After mixture of the blood sample by inversion, 7.5 ml of blood is added to a CellSearch® processing tube along with 6.5 ml of Dilution Buffer (from the CellSearch® reagent kit). The sample is mixed by inversion and centrifuged at $800 \times g$ for 10 min with the brake in the off position, to ensure separation of the blood into three distinct layers.

Detection and enumeration of CTCs using the CellSearch® system

The CellSearch® Epithelial Cell Kit uses an antibody-mediated, ferrofluid-based magnetic separation technique and differential staining with fluorescent particles to distinguish CTCs from contaminating leukocytes in blood samples. Prepared blood samples should be loaded into the AutoPrep system as per the manufacturer's guidelines. Initially, the AutoPrep system performs a positive selection step using anti-EpCAM antibodies conjugated to iron nanoparticles incubated in a magnet. The remainder of the fluid is aspirated from the sample, the selected tumor cells are resuspended, and differential staining antibodies are added to the sample. Samples are then incubated in a magnetic cartridge, called a Magnest™, and scanned using the CellSearch® Analyzer II.

The Analyzer scans the sample using three different filters, each with the exposure time optimized to the appropriate fluorescent particle. CTCs are identified by staining with anti-pan- CK- PE (CK 8/18/19 are characteristic of epithelial cells), and the DNA stain 4′, 6-diamidino-2-phenylindole (DAPI). Leukocytes are identified by staining with anti-CD45⁻ allophycocyanin (APC) and DAPI. After the scan is complete, a gallery of computer-defined, potential tumor cells is presented, from which the user must select, via qualitative review, which cells are CTCs based on the differential staining discussed above.

Molecular characterization of CTCs using the CellSearch® system

Clinically, the CellSearch® system has only been used for the detection and enumeration of CTCs (Cristofanilli *et al.*, 2004; Dawood *et al.*, 2008; Riethdorf *et al.*, 2007). However, the system also allows for additional molecular characterization of CTCs based on molecular markers of interest via differential staining analysis with a fourth filter. This filter allows for the imaging of fluorescein isothiocynate (FITC) or Alexafluor488 particles conjugated to a molecular marker of interest. For example, EGFR is a cell surface protein that has been shown to be over-expressed in a subset of cancer patients. Exploitation of this marker has allowed for targeted therapies that are "tailor-made" for EGFR+ cancer patients, such as the small molecule inhibitor Iressa (Lurje and Lenz, 2009). Analysis of molecular markers such as EGFR is accomplished by adding the antibody of interest to the AutoPrep reagents discussed above. The antibody will then be added during processing. In addition, CTCs can be isolated for downstream molecular characterization using other techniques such as microarray gene expression analysis or FISH (Leversha *et al.*, 2009; Shaffer *et al.*, 2007). This is done via the CellSearch® Profile Kit, which utilizes the same techniques as the CellSearch® Epithelial Kit, except that CTCs are only isolated from blood samples using the AutoPrep, but are not differentially labeled with CK/CD45/DAPI and are not enumerated using the Analyzer.

2. Typical Results: Detection and analysis of CTCs in Human Blood Samples Using the CellSearch® System

CTC detection and enumeration results obtained by the CellSearch® system can vary significantly dependent on many factors such as disease site, disease stage, whether blood samples were spiked with cells from culture, and/or the cell line used for spiking experiments. For example, CTCs from spiked samples (Fig. 3A, B) mainly differ from CTCs detected in patient population in that they appear to be larger, of more consistent morphology, and have more intense staining for CK8/18/19. Variations in CK staining are also observed between different cell lines, presumably based on variable basal expression of CK (Fig. 3A, B). It should be noted that highly aggressive cell lines that are more mesenchymal in nature tend to have reduced or absent expression of EpCAM and/or CK, and therefore cannot be analyzed using the CellSearch® system (Sieuwerts *et al.*, 2009).

From the perspective of molecular characterization of CTCs using the CellSearch® system, a few validated and optimized antibodies are commercially available from Veridex, including EGFR (Fig. 4A). This marker can therefore act as the "gold standard" and/or positive control for the development and optimization of other user-defined markers. Examples of such markers include CD44, a cancer stem cell marker (Fig. 4B) and M30 (Fig. 4C), a novel apoptosis marker, currently being optimized in our laboratory for use in clinical research studies. User-defined

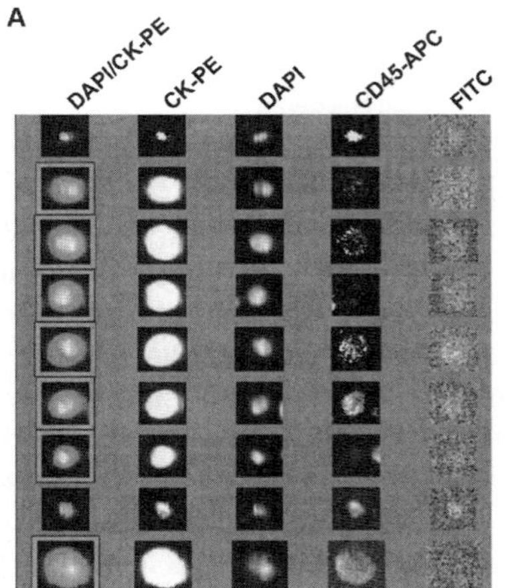
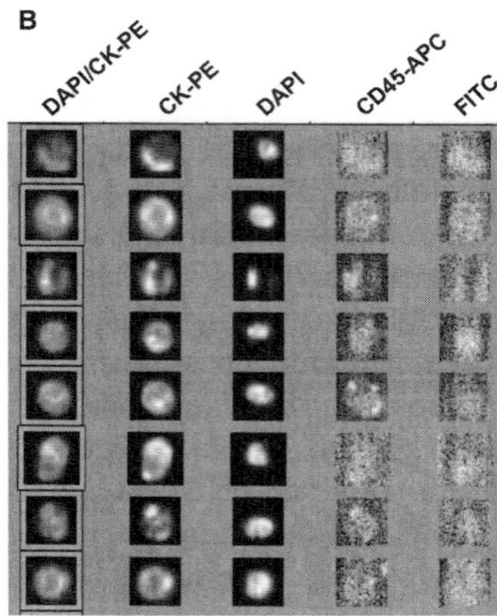
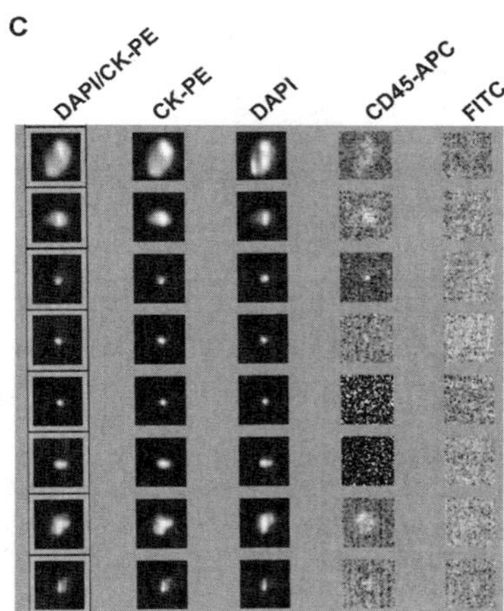

Fig. 3 Detection and enumeration of human CTCs using the CellSearch® system. (A) CellSearch® gallery image of spiked positive control sample for breast cancer. Seven and a half (7.5) ml of blood was collected from a healthy volunteer donor following informed consent, spiked with 1000 MDA-MB-468 human breast cancer cells, and processed on the CellSearch® system as per the manufacturer's instructions. (B) CellSearch® gallery image of spiked positive control sample for prostate cancer. Seven and a half (7.5) ml of blood was collected from a healthy volunteer donor following informed consent, spiked with LNCaP human prostate cancer cells, and processed on the CellSearch® system as per the manufacturer's instructions. (C) CellSearch® gallery images of a prostate cancer patient sample. Orange squares indicate positive CTCs, identified by two independent operators as CK⁺, DAPI⁺, CD45⁻, and FITC⁻ with a round to oval morphology that are at least 4 μm in size. (For interpretation of the references to color in this figure legend, the reader is referred to the Web version of this chapter.)

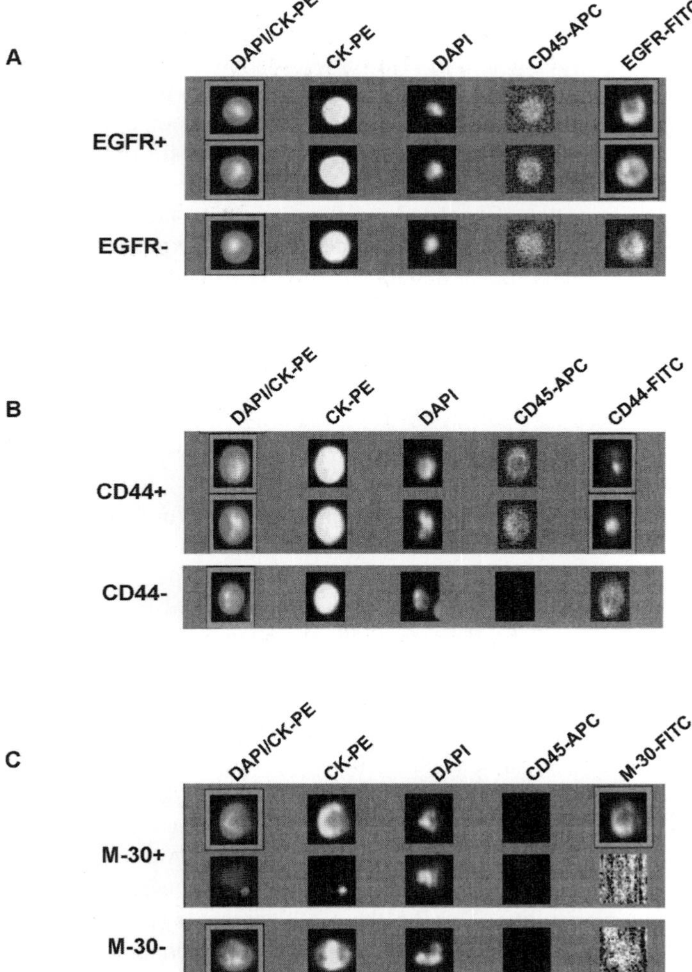

Fig. 4 Molecular characterization of human CTCs using the CellSearch® system. (A) CellSearch® gallery image of 7.5 ml of blood collected from a healthy volunteer donor following informed consent, spiked with 1000 MDA-MB-468 human breast cancer cells, and subsequently incubated with anti-EGFR-FITC (Veridex). Orange squares on the left indicate positive CTCs, identified as CK^+, $DAPI^+$, and $CD45^-$. Orange squares on the right indicate $EGFR^+$ CTCs. (B) CellSearch® gallery image of 7.5 ml of blood collected from a healthy volunteer donor following informed consent, spiked with 1000 MDA-MB-468 human breast cancer cells, and subsequently incubated with 4.0 µg/ml of anti-CD44-FITC (BD Biosciences, Canada). Orange squares on the left indicate positive CTCs, identified as CK^+, $DAPI^+$, and $CD45^-$. Orange squares on the right indicate $CD44^+$ CTCs. (C) CellSearch® gallery image of 7.5 ml of blood collected from a healthy volunteer donor, spiked with 4000 MDA-MB-468 human breast cancer cells treated with 0.1 µg/ml of paclitaxel chemotherapy, and subsequently incubated with 3.7 µg/ml of anti-M30-FITC (Enzo Life Sciences, Lausen, Switzerland). Orange squares on the left indicate positive CTCs, identified as CK^+, $DAPI^+$, and $CD45^-$. Orange squares on the right indicate $M30^+$ CTCs. (See plate no. 17 in the color plate section.)

markers require considerable work-up in order to determine the appropriate antibody concentration and exposure time to avoid false positive and false negative identification of CTCs. Appropriate negative and positive controls for optimization and validation of new user-defined markers include cell lines that are known nonexpressers (negative) and cell lines that are known low- and high- antigen density expressers (positive) of the protein of interest. These cell lines must first be processed without the antibody against the user-defined marker in order to determine the amount of background that can be visualized in the FITC channel. Based on the background observed, an appropriate exposure time can be chosen to ensure that all signals appear negative in the FITC channel. These cell lines are then processed with the antibody for the user-defined marker, and the antibody is titrated such that the low antigen density cell line shows a clear positive signal in the FITC channel that is easily distinguishable from the negative cell line (Veridex White Pages, 2008).

3. Practical Application of the Methodology: Detection and Analysis of CTCs in Human Blood Samples Using the CellSearch[®] System

The CellSearch[®] system is the first (and currently only) standardized and regulatory approved system for detecting and quantifying CTCs in the clinical setting. As such, the assay is straightforward to perform and highly reproducible. However, among its many advantages and clinical utilities, the CellSearch[®] system also has several disadvantages. Firstly, because the system utilizes a positive selection technique for EpCAM, some CTCs will inevitably be lost in the enrichment process due to a lack of/loss of EpCAM expression. As previously mentioned, the transition of highly invasive cells from an epithelial to mesenchymal phenotype (EMT) is a potential biological mechanism by which this could occur (Polyak and Weinberg, 2009; Thiery and Sleeman, 2006; Yang and Weinberg, 2008). Secondly, due to the qualitative nature of the gallery review step for CTC evaluation, the determination of a cell as a positive CTC is highly subjective. However, this problem can be partially resolved by averaging the review results from multiple users to reduce the impact of inter-operator variability. Thirdly, the CellSearch[®] system occasionally presents multiple CTC images within a single frame. The manufacturer's instructions dictate that these images with multiple CTCs be quantified as single events, thereby often underestimating the number of CTCs. Finally, although very useful as a clinical instrument, the CellSearch[®] system is highly inflexible as a research tool, with very few variables that the user can manipulate. For example, an increased number of fluorescent channels for enhanced multiparameter characterization of these cells would be a welcome addition.

C. Other Image Cytometry Approaches for CTC Analysis

Although LSC and the CellSearch[®] system are two of the more common image cytometry approaches for analyzing CTCs, other techniques have been reported. For example, Fibre-optic Array Scanning Technology (FAST) is a scanning technology that allows for rapid analysis of large blood volumes without prior CTC enrichment (Hsieh *et al.*, 2006; Krivacic *et al.*, 2004; Marrinucci *et al.*, 2010). Using this technology, immunofluorescently labeled cells are analyzed by FAST at a rate of up to 300,000 cells/sec. This rapid scanning ability of the FAST system is due to its large field of view (50 mm) that allows a sample to be scanned in one continuous motion, as opposed to traditional automated digital microscopy that requires time-consuming slide-stepping. FAST acts as an image based "enrichment" technique, in which all cells are analyzed but only those that meet minimum CTC criteria are presented to the user for subsequent qualitative analysis. The Automated Cellular Imaging System (ACIS) and ARIOL are other similar systems that utilize automated scanning microscopy for CTC analysis and enumeration (Borgen *et al.*, 2006; Deng *et al.*, 2008).

IV. Conclusions and Future Directions

Although CTCs were described more than a century ago, it has only been in the past few decades that techniques have become available for the detection and enumeration of these rare cells. In particular, in the past 10 years there has been renewed interest in the clinical significance of these cells and an explosion of new techniques for CTC detection and enumeration. Currently the only system that is FDA approved for clinical use is the CellSearch® system, and although additional optimized molecular characterization markers exist for use with this system, they are not yet approved for use in a clinical setting. Therefore the only information regarding CTCs that is currently in clinical use is that of CTC enumeration. We propose that in the future, in addition to simple detection and enumeration of these cells, molecular characterization of CTCs will provide important information with regards to the basic biology and mechanisms of the metastatic cascade; a patient's response to therapy; tracking of clinical disease progression or regression; and the identification of novel therapeutic targets.

The image cytometry techniques discussed in this chapter and others under development will be integral to moving CTC research forward and implementing CTC characterization into the clinic, as these techniques not only allow for the identification of CTCs but also visual confirmation of these cells and their molecular characteristics. As with all techniques, there is always room for improvement and CTC analysis by image cytometry is no exception. Ideally, the optimal solution for all CTC techniques would be the identification of a tumor-specific marker that is expressed by all tumor cells, of all origins, and not by other nontumor cells. However, because of the inherent genetic instability and heterogeneity of cancer, such a perfect marker is highly unlikely to exist and therefore other improvement options must be explored. In particular, two things would significantly enhance CTC detection, enumeration, and clinical utility. First, the epithelial markers EpCAM and CK are currently the most highly utilized markers for CTCs. However, the additional inclusion of mesenchymal markers to aid in CTC detection must be considered due to the

EMT transition observed in highly invasive cancers. Secondly, for any technique to be utilized in the clinic it must meet high quality control standards. In the area of rare event detection this implies that all systems must be standardized and at a minimum semiautomated, although complete automation (including CTC identification) would be optimal.

In conclusion, the ability to detect, enumerate, and characterize CTCs is an important tool in the study of the metastatic cascade and the improved clinical management of cancer patients. These rare cells could shed light on the basic biology behind this highly lethal process and ultimately change current patient treatment guidelines.

Acknowledgments

We thank members of our laboratory and our collaborators for their research work, in particular Tracy Sexton and Benjamin Hedley. The authors' work on CTCs has been supported by grants from the Canada Foundation for Innovation (#13199, to A.L.A); and the Canadian Breast Cancer Research Alliance "Special Competition in New Approaches to Metastatic Disease" with special funding support from the Canadian Breast Cancer Foundation and The Cancer Research Society (#016506, to A.L.A. and M. K.). L.E.L. is supported by a graduate scholarship from the Canadian Institutes for Health Research (CIHR). A.L.A. is supported by a CIHR New Investigator Award and an Early Researcher Award from the Ontario Ministry of Research and Innovation.

References

Alix-Panabières, C., Brouillet, J. P., Fabbro, M., Yssel, H., Rousset, T., Maudelonde, T., Choquet-Kastylevsky, G., Vendrell, J. P. (2005). Characterization and enumeration of cells secreting tumor markers in the peripheral blood of breast cancer patients. *J. Immunol. Methods* **299**, 177–188.

Alix-Panabières, C., Vendrell, J. P., Pellé, O., Rebillard, X., Riethdorf, S., Müller, V., Fabbro, M., Pantel, K. (2007). Detection and characterization of putative metastatic precursor cells in cancer patients. *Clin. Chem.* **53**, 537–539.

Alix-Panabières, C., Vendrell, J. P., Slijper, M., Pellé, O., Barbotte, E., Mercier, G., Jacot, W., Fabbro, M., Pantel, K. (2009). Full-length cytokeratin-19 is released by human tumor cells: a potential role in metastatic progression of breast cancer. *Breast Cancer Res. BCR* **11**, R39.

Allan, A. L., and Keeney, M. (2010). Circulating tumor cell analysis: technical and statistical considerations for application to the clinic. *J. Oncol.* **2010**, 426218.

Allan, A. L., Vantyghem, S. A., Tuck, A. B., Chambers, A. F., Chin-Yee, I. H., Keeney, M. (2005). Detection and quantification of circulating tumor cells in mouse models of human breast cancer using immunomagnetic enrichment and multiparameter flow cytometry. *Cytom. Part. A* **65**, 4–14.

Alunni-Fabbroni, M., and Sandri, M. T. (2010). Circulating tumour cells in clinical practice: methods of detection and possible characterization. *Methods* **50**, 289–297.

Ameri, K., Luong, R., Zhang, H., Powell, A. A., Montgomery, K. D., Espinosa, I., Bouley, D. M., Harris, A. L., Jeffrey, S. S. (2010). Circulating tumour cells demonstrate an altered response to hypoxia and an aggressive phenotype. *Brit. J. Cancer* **102**, 561–569.

Baker, M. K., Mikhitarian, K., Osta, W., Callahan, K., Hoda, R., Brescia, F., Kneuper-Hall, R., Mitas, M., Cole, D. J., Gillanders, W. E. (2003). Molecular detection of breast cancer cells in the peripheral blood of advanced-stage breast cancer patients using multimarker real-time reverse transcription-polymerase chain reaction and a novel porous barrier density gradient centrifugation technology. *Clin. Cancer Res.* **9**, 4865–4871.

Balic, M., Dandachi, N., Hofmann, G., Samonigg, H., Loibner, H., Obwaller, A., van Der Kooi, A., Tibbe, A. G. J., Doyle, G. V., Terstappen, L. W. M. M., Bauernhofer, T. (2005). Comparison of two methods for enumerating circulating tumor cells in carcinoma patients. *Cytom. Part B – Clin. Cy.* **68**, 25–30.

Beitsch, P. D., and Clifford, E. (2000). Detection of carcinoma cells in the blood of breast cancer patients. *Am. J. Surg.* **180**, 446–448.

de Bono, J. S., Scher, H. I., Montgomery, R. B., Parker, C., Miller, M. C., Tissing, H., Doyle, G. V., Terstappen, L. W. W. M., Pienta, K. J., Raghavan, D. (2008). Circulating tumor cells predict survival benefit from treatment in metastatic castration-resistant prostate cancer. *Clin. Cancer Res.* **14**, 6302–6309.

Borgen, E., Pantel, K., Schlimok, G., Mu, P., Otte, M., Renolen, A., Ehnle, S., Coith, C., Nesland, J. M., Naume, B. (2006). A European interlaboratory testing of three well-known procedures for immuno-cytochemical detection of epithelial cells in bone marrow. Results from analysis of normal bone marrow. *Cytometry* **409**, 400–409.

Chambers, A. F., Groom, A. C., and MacDonald, I. C. (2002). Dissemination and growth of cancer cells in metastatic sites. *Nat. Rev. Cancer* **2**, 563–572.

Cohen, S. J., Punt, C. J. A., Iannotti, N., Saidman, B. H., Sabbath, K. D., Gabrail, N. Y., Picus, J., Morse, M. A., Mitchell, E., Miller, M. C., Doyle, G. V., Tissing, H., Terstappen, L. W. M. M., Meropol, N. J. (2009). Prognostic significance of circulating tumor cells in patients with metastatic colorectal cancer. *Ann. Oncol.* **20**, 1223–1229.

Cohen, S. J., Punt, C. J. A., Iannotti, N., Saidman, B. H., Sabbath, K. D., Gabrail, N. Y., Picus, J., Morse, M., Mitchell, E., Miller, M. C., Doyle, G. V., Tissing, H., Terstappen, L. W. M. M., Meropol, N. J. (2008). Relationship of circulating tumor cells to tumor response, progression-free survival, and overall survival in patients with metastatic colorectal cancer. *J. Clin. Oncol.* **26**, 3213–3221.

Cristofanilli, M. (2009). The biological information obtainable from circulating tumor cells. *The Breast* **18**, S38–S40.

Cristofanilli, M., Budd, G. T., Ellis, M. J., Stopeck, A., Matera, J., Miller, M. C., Reuben, J. M., Doyle, G. V., Allard, W. J., Terstappen, L. W. M. M., Hayes, D. F. (2004). Circulating tumor cells, disease progression, and survival in metastatic breast cancer. *New Engl. J. Med.* **351**, 781–791.

Cruz, I., Ciudad, J., Cruz, J. J., Ramos, M., Gómez-Alonso, A., Adansa, J. C., Rodríguez, C., Orfao, A. (2005). Evaluation of multiparameter flow cytometry for the detection of breast cancer tumor cells in blood samples. *Am. J. Clin. Pathol.* **123**, 66–74.

Darzynkiewicz, Z., Bedner, E., Li, X., Gorczyca, W., and Melamed, M. R. (1999). Laser-scanning cytometry: a new instrumentation with many applications. *Exp. Cell Res.* **249**, 1–12.

Datta, Y. H., Adams, P. T., Drobyski, W. R., Ethier, S. P., Terry, V. H., Roth, M. S. (1994). Sensitive detection of occult breast cancer by the reverse-transcriptase polymerase chain reaction. *J. Clin. Oncol.* **12**, 475–482.

Dawood, S., Broglio, K., Valero, V., Reuben, J., Handy, B., Islam, R., Jackson, S., Hortobagyi, G. N., Fritsche, H., Cristofanilli, M. (2008). Circulating tumor cells in metastatic breast cancer: from prognostic stratification to modification of the staging system. *Cancer* **113**, 2422–2430.

De Giorgi, V., Pinzani, P., Salvianti, F., Panelos, J., Paglierani, M., Janowska, A., Grazzini, M., Wechsler, J., Orlando, C., Santucci, M., Lotti, T., Pazzagli, M., Massi, D. (2010). Application of a filtration- and isolation-by-size technique for the detection of circulating tumor cells in cutaneous melanoma. *J. Invest. Dermatol.* 1–8.

Deng, G., Herrler, M., Burgess, D., Manna, E., Krag, D., Burke, J. F. (2008). Enrichment with anti-cytokeratin alone or combined with anti-EpCAM antibodies significantly increases the sensitivity for circulating tumor cell detection in metastatic breast cancer patients. *Breast Cancer Res. BCR* **10**, R69.

Duffy, M. J. (2001). Carcinoembryonic antigen as a marker for colorectal cancer: is it clinically useful? *Clin. Chem.* **47**, 624–630.

Duffy, M. J., van Dalen, A., Haglund, C., Hansson, L., Holinski-Feder, E., Klapdor, R., Lamerz, R., Peltomaki, P., Sturgeon, C., Topolcan, O. (2007). Tumour markers in colorectal cancer: European Group on Tumour Markers (EGTM) guidelines for clinical use. *Eur. J. Cancer* **43**, 1348–1360.

Engel, H., Kleespies, C., Friedrich, J., Breidenbach, M., Kallenborn, A., Schöndorf, T., Kolhagen, H., Mallmann, P. (1999). Detection of circulating tumour cells in patients with breast or ovarian cancer by molecular cytogenetics. *Brit. J. Cancer* **81**, 1165–1173.

Fletcher, R. H. (1986). Carcinoembryonic antigen. *Ann. Intern. Med.* **104**, 66–73.

Flores, L. M., Kindelberger, D. W., Ligon, A. H., Capelletti, M., Fiorentino, M., Loda, M., Cibas, E. S., Jänne, P. A., Krop, I. E. (2010). Improving the yield of circulating tumour cells facilitates molecular characterisation and recognition of discordant HER2 amplification in breast cancer. *Brit. J. Cancer* **102**, 1495–1502.

Gillanders, W. E., Mikhitarian, K., Hebert, R., Mauldin, P. D., Palesch, Y., Walters, C., Urist, M. M., Mann, G. B., Doherty, G., Herrmann, V. M., Hill, A. D., Eremin, O., El-Sheemy, M., Orr, R. K., Valle, A. A., Henderson, M. A., Dewitty, R. L., Sugg, S. L., Frykberg, E., Yeh, K., Bell, R. M., Metcalf, J. S., Elliott, B. M., Brothers, T., Robison, J., Mitas, M., Cole, D. J. (2004). Molecular detection of micrometastatic breast cancer in histopathology-negative axillary lymph nodes correlates with traditional predictors of prognosis. *Ann. Surg.* **239**, 828–840.

Goodale, D., Phay, C., Postenka, C. O., Keeney, M., and Allan, A. L. (2009). Characterization of tumor cell dissemination patterns in preclinical models of cancer metastasis using flow cytometry and laser scanning cytometry. *Cytom. Part A.* **75**, 344–355.

Goon, P. K. Y., Lip, G. Y. H., Boos, C. J., Stonelake, P. S., and Blann, A. D. (2006). Circulating endothelial cells, endothelial progenitor cells, and endothelial microparticles in cancer. *Neoplasia* **8**, 79–88.

Grünewald, K., Haun, M., Fiegl, M., Urbanek, M., Müller-Holzner, E., Massoner, A., Riha, K., Propst, A., Marth, C., Gastl, G. (2002). Mammaglobin expression in gynecologic malignancies and malignant effusions detected by nested reverse transcriptase-polymerase chain reaction. *Lab. Invest.* **82**, 1147–1153.

Helzer, K. T., Barnes, H. E., Day, L., Harvey, J., Billings, P. R., Forsyth, A. (2009). Circulating tumor cells are transcriptionally similar to the primary tumor in a murine prostate model. *Cancer Res.* **69**, 7860–7866.

Herrala, A. M., Porvari, K. S., Kyllönen, A. P., and Vihko, P. T. (2001). Comparison of human prostate specific glandular kallikrein 2 and prostate specific antigen gene expression in prostate with gene amplification and overexpression of prostate specific glandular kallikrein 2 in tumor tissue. *Cancer* **92**, 2975–2984.

Hostetter, R. B., Campbell, D. E., Chi, K. F., Kerckhoff, S., Cleary, K. R., Ullrich, S., Thomas, P., Jessup, J. M. (1990). Carcinoembryonic antigen enhances metastatic potential of human colorectal carcinoma. *Arch. Surg – Chicago* **125**, 300–304.

Hsieh, H. B., Marrinucci, D., Bethel, K., Curry, D. N., Humphrey, M., Krivacic, R. T., Kroener, J., Kroener, L., Ladanyi, A., Lazarus, N., Kuhn, P., Bruce, R. H., Nieva, J. (2006). High speed detection of circulating tumor cells. *Biosens. Bioelectron.* **21**, 1893–1899.

Hu, X. C., and Chow, L. W. (2000). Detection of circulating breast cancer cells by reverse transcriptase polymerase chain reaction (RT-PCR). *Eur. J. Surg. Oncol.* **26**, 530–535.

Hu, Y., Fan, L., Zheng, J., Cui, R., Liu, W., He, Y., Li, X., Huang, S. (2010). Detection of circulating tumor cells in breast cancer patients utilizing multiparameter flow cytometry and assessment of the prognosis of patients in different CTCs levels. *Cytom. Part A.* **77**, 213–219.

Iakovlev, V. V., Goswami, R. S., Vecchiarelli, J., Arneson, N. C. R., and Done, S. J. (2008). Quantitative detection of circulating epithelial cells by Q-RT-PCR. *Breast Cancer Res. Tr.* **107**, 145–154.

Iinuma, H., Okinaga, K., Adachi, M., Suda, K., Sekine, T., Sakagawa, K., Baba, Y., Tamura, J., Kumagai, H., Ida, A. (2000). Detection of tumor cells in blood using CD45 magnetic cell separation followed by nested mutant allele-specific amplification of p53 and K-ras genes in patients with colorectal cancer. *Int. J. Cancer* **89**, 337–344.

Jemal, A., Siegel, R., Xu, J., and Ward, E. (2010). Cancer statistics, 2010. *CA – Cancer J. Clin.* **60**, 277–300.

Jenkins, D. E., Oei, Y., Hornig, Y. S., Yu, S. F., Dusich, J., Purchio, T., Contag, P. R. (2003). Bioluminescent imaging (BLI) to improve and refine traditional murine models of tumor growth and metastasis. *Clin. Exp. Metastas.* **20**, 733–744.

Jessup, J. M., and Thomas, P. (1989). Carcinoembryonic antigen: function in metastasis by human colorectal carcinoma. *Cancer Metastasis Rev.* **8**, 263–280.

Jonas, S., Windeatt, S., O-Boateng, A., Fordy, C., and Allen-Mersh, T. G. (1996). Identification of carcinoembryonic antigen-producing cells circulating in the blood of patients with colorectal carcinoma by reverse transcriptase polymerase chain reaction. *Gut* **39**, 717–721.

Ko, Y., Grünewald, E., Totzke, G., Klinz, M., Fronhoffs, S., Gouni-Berthold, I., Sachinidis, A., Vetter, H. (2000). High percentage of false-positive results of cytokeratin 19 RT-PCR in blood: a model for the analysis of illegitimate gene expression. *Oncology* **59**, 81–88.

Krivacic, R. T., Ladanyi, A., Curry, D. N., Hsieh, H. B., Kuhn, P., Bergsrud, D. E., Kepros, J. F., Barbera, T., Ho, M. Y., Chen., Lan, B., Lerner, R. A., Bruce, R. H. (2004). A rare-cell detector for cancer. *Proc. Natl. Acad. Sci. USA* **101**, 10501–10504.

Lankiewicz, S., Rivero, B. G., and Bócher, O. (2006). Quantitative real-time RT-PCR of disseminated tumor cells in combination with immunomagnetic cell enrichment. *Mol. Biotechnol.* **34**, 15–27.

Leversha, M. A., Han, J., Asgari, Z., Danila, D. C., Lin, O., Anand, A., Lilja, H., Heller, G., Fleisher, M., Scher, H. I. (2009). Fluorescence *in situ* hybridization analysis of circulating tumor cells in metastatic prostate cancer. *Clin. Cancer Res.* **15**, 2091–2097.

Li, X., Wang, J., An, Z., Yang, M., Baranov, E., Jiang, P., Sun, F., Moossa, A. R., Hoffman, R. M. (2002). Optically imageable metastatic model of human breast cancer. *Clin. Exp. Metastas.* **19**, 347–350.

Lintula, S., Stenman, J., Bjartell, A., Nordling, S., and Stenman, U. H. (2005). Relative concentrations of hK2/PSA mRNA in benign and malignant prostatic tissue. *The Prostate* **63**, 324–329.

Lurje, G., and Lenz, H. J. (2009). EGFR signaling and drug discovery. *Oncology* **77**, 400–410.

Mach, W. J., Thimmesch, A. R., Orr, J. A., Slusser, J. G., and Pierce, J. D. (2010). Flow cytometry and laser scanning cytometry, a comparison of techniques. *J. Clin. Monitor. Comp.* **24**, 251–259.

Marks, L. S., Andriole, G. L., Fitzpatrick, J. M., Schulman, C. C., and Roehrborn, C. G. (2006). The interpretation of serum prostate specific antigen in men receiving 5alpha-reductase inhibitors: a review and clinical recommendations. *J. Urology* **176**, 868–874.

Marrinucci, D., Bethel, K., Lazar, D., Fisher, J., Huynh, E., Clark, P., Bruce, R., Nieva, J., Kuhn, P. (2010). Cytomorphology of circulating colorectal tumor cells:a small case series. *J. Oncol.* **2010**, 861341.

Mikhitarian, K., Martin, R. H., Ruppel, M. B., Gillanders, W. E., Hoda, R., Schutte, D. H., Callahan, K., Mitas, M., Cole, D. J. (2008). Detection of mammaglobin mRNA in peripheral blood is associated with high grade breast cancer: interim results of a prospective cohort study. *BMC Cancer* **8**, 55.

Mostert, B., Sleijfer, S., Foekens, J. A., and Gratama, J. W. (2009). Circulating tumor cells (CTCs): detection methods and their clinical relevance in breast cancer. *Cancer Treat Rev.* **35**, 463–474.

Müller, V., Stahmann, N., Riethdorf, S., Rau, T., Zabel, T., Goetz, A., Jänicke, F., Pantel, K. (2005). Circulating tumor cells in breast cancer: correlation to bone marrow micrometastases, heterogeneous response to systemic therapy and low proliferative activity. *Clin. Cancer Res.* **11**, 3678–3685.

Nagrath, S., Sequist, L. V., Maheswaran, S., Bell, D. W., Irimia, D., Ulkus, L., Smith, M. R., Kwak, E. L., Digumarthy, S., Muzikansky, A., Ryan, P., Balis, U. J., Tompkins, R. G., Haber, D. A., Toner, M. (2007). Isolation of rare circulating tumour cells in cancer patients by microchip technology. *Nature* **450**, 1235–1239.

Naume, B., Borgen, E., Tøssvik, S., Pavlak, N., Oates, D., Nesland, J. M. (2004). Detection of isolated tumor cells in peripheral blood and in BM: evaluation of a new enrichment method. *Cytotherapy* **6**, 244–252.

Naumov, G. N., Wilson, S. M., MacDonald, I. C., Schmidt, E. E., Morris, V. L., Groom, A. C., Hoffman, R. M., Chambers, A. F. (1999). Cellular expression of green fluorescent protein, coupled with high-resolution *in vivo* videomicroscopy, to monitor steps in tumor metastasis. *J. Cell Sci.* **112**(Pt 1), 1835–1842.

Ntouroupi, T. G., Ashraf, S. Q., McGregor, S. B., Turney, B. W., Seppo, A., Kim, Y., Wang, X., Kilpatrick, M. W., Tsipouras, P., Tafas, T., Bodmer, W. F. (2008). Detection of circulating tumour cells in peripheral blood with an automated scanning fluorescence microscope. *Brit. J. Cancer.* **99**, 789–795.

Núñez-Villar, M. J., Martínez-Arribas, F., Pollán, M., Lucas, A. R., Sánchez, J., Tejerina, A., Schneider, J. (2003). Elevated mammaglobin (h-MAM) expression in breast cancer is associated with clinical and biological features defining a less aggressive tumour phenotype. *Breast Cancer Res.* **5**, R65–70.

Pachmann, K., Camara, O., Kavallaris, A., Krauspe, S., Malarski, N., Gajda, M., Kroll, T., Jörke, C., Hammer, U., Altendorf-Hofmann, A., Rabenstein, C., Pachmann, U., Runnebaum, I., Höffken, K. (2008). Monitoring the response of circulating epithelial tumor cells to adjuvant chemotherapy in breast cancer allows detection of patients at risk of early relapse. *J. Clin. Oncol.* **26**, 1208–1215.

Pantel, K., Brakenhoff, R. H., and Brandt, B. (2008). Detection, clinical relevance and specific biological properties of disseminating tumour cells. *Nat. Rev. Cancer* **8**, 329–340.

Paterlini-Brechot, P., and Benali, N. L. (2007). Circulating tumor cells (CTC) detection: clinical impact and future directions. *Cancer Lett.* **253**, 180–204.

Pinzani, P., Salvadori, B., Simi, L., Bianchi, S., Distante, V., Cataliotti, L., Pazzagli, M., Orlando, C. (2006). Isolation by size of epithelial tumor cells in peripheral blood of patients with breast cancer: correlation with real-time reverse transcriptase-polymerase chain reaction results and feasibility of molecular analysis by laser microdissection. *Hum. Pathol.* **37**, 711–718.

Polyak, K., and Weinberg, R. A. (2009). Transitions between epithelial and mesenchymal states: acquisition of malignant and stem cell traits. *Nat. Rev. Cancer* **9**, 265–273.

Pozarowski, P., Holden, E., and Darzynkiewicz, Z. (2006). Laser scanning cytometry: principles and applications. *Methods Mol. Biol.* **319**, 165–192.

Racila, E., Euhus, D., Weiss, A. J., Rao, C., McConnell, J., Terstappen, L. W., Uhr, J. W. (1998). Detection and characterization of carcinoma cells in the blood. *Proc. Natl. Acad. Sci. USA* **95**, 4589–4594.

Riethdorf, S., Fritsche, H., Müller, V., Rau, T., Schindlbeck, C., Rack, B., Janni, W., Coith, C., Beck, K., Jänicke, F., Jackson, S., Gornet, T., Cristofanilli, M., Pantel, K. (2007). Detection of circulating tumor cells in peripheral blood of patients with metastatic breast cancer: a validation study of the CellSearch system. *Clin. Cancer Res.* **13**, 920–928.

Ring, A. E., Zabaglo, L., Ormerod, M. G., Smith, I. E., and Dowsett, M. (2005). Detection of circulating epithelial cells in the blood of patients with breast cancer: comparison of three techniques. *Brit. J. Cancer.* **92**, 906–912.

Rosenberg, R., Gertler, R., Friederichs, J., Fuehrer, K., Dahm, M., Phelps, R., Thorban, S., Nekarda, H., Siewert, J. R. (2002). Comparison of two density gradient centrifugation systems for the enrichment of disseminated tumor cells in blood. *Cytometry* **49**, 150–158.

Ross, A. A., Cooper, B. W., Lazarus, H. M., Mackay, W., Moss, T. J., Ciobanu, N., Tallman, M. S., Kennedy, M. J., Davidson, N. E., Sweet, D. (1993). Detection and viability of tumor cells in peripheral blood stem cell collections from breast cancer patients using immunocytochemical and clonogenic assay techniques. *Blood* **82**, 2605–2610.

Rubin, H. (2003). Complementary approaches to understanding the role of proteases and their natural inhibitors in neoplastic development: retrospect and prospect. *Carcinogenesis* **24**, 803–816.

Sequist, L. V., Nagrath, S., Toner, M., Haber, D. A., and Lynch, T. J. (2009). The CTC-chip: an exciting new tool to detect circulating tumor cells in lung cancer patients. *J. Thorac. Oncol.* **4**, 281–283.

Shaffer, D. R., Leversha, M. A., Danila, D. C., Lin, O., Gonzalez-Espinoza, R., Gu, B., Anand, A., Smith, K., Maslak, P., Doyle, G. V., Terstappen, L. W. M. M., Lilja, H., Heller, G., Fleisher, M., Scher, H. I. (2007). Circulating tumor cell analysis in patients with progressive castration-resistant prostate cancer. *Clin. Cancer Res.* **13**, 2023–2029.

Sieuwerts, A. M., Kraan, J., Bolt, J., van Der Spoel, P., Elstrodt, F., Schutte, M., Martens, J. W. M., Gratama, J. W., Sleijfer, S., Foekens, J. A. (2009). Anti-epithelial cell adhesion molecule antibodies and the detection of circulating normal-like breast tumor cells. *J. Natl. Cancer I.* **101**, 61–66.

Sjödin, A., Guo, D., Hofer, P. A., Henriksson, R., and Hedman, H. (2003). Mammaglobin in normal human sweat glands and human sweat gland tumors. *J. Invest. Dermatol.* **121**, 428–429.

Sjödin, A., Ljuslinder, I., Henriksson, R., and Hedman, H. (2008). Mammaglobin and lipophilin B expression in breast tumors and their lack of effect on breast cancer cell proliferation. *Anticancer Res.* **28**, 1493–1498.

Slade, M. J., Smith, B. M., Sinnett, H. D., Cross, N. C., and Coombes, R. C. (1999). Quantitative polymerase chain reaction for the detection of micrometastases in patients with breast cancer. *J. Clin. Oncol.* **17**, 870–879.

Stojadinovic, A., Mittendorf, E. A., Holmes, J. P., Amin, A., Hueman, M. T., Ponniah, S., Peoples, G. E. (2007). Quantification and phenotypic characterization of circulating tumor cells for monitoring response to a preventive HER2/neu vaccine-based immunotherapy for breast cancer: a pilot study. *Ann. Surg. Oncol.* **14**, 3359–3368.

Stott, S. L., Lee, R. J., Nagrath, S., Yu, M., Miyamoto, D. T., Ulkus, L., Inserra, E. J., Ulman, M., Springer, S., Nakamura, Z., Moore, A. L., Tsukrov, D. I., Kempner, M. E., Dahl, D. M., Wu, C. L., Iafrate, A. J., Smith, M. R., Tompkins, R. G., Sequist, L. V., Toner, M., Haber, D. A., Maheswaran, S. (2010). Isolation and characterization of circulating tumor cells from patients with localized and metastatic prostate cancer. *Sci. Transl. Med.* **2**, 25ra23.

Taubert, H. (2004). Detection of disseminated tumor cells in peripheral blood of patients with breast cancer: correlation to nodal status and occurrence of metastases. *Gynecol. Oncol.* **92**, 256–261.

Thiery, J. P. (2002). Epithelial-mesenchymal transitions in tumour progression. *Nat. Rev. Cancer* **2**, 442–454.

Thiery, J. P., and Sleeman, J. P. (2006). Complex networks orchestrate epithelial-mesenchymal transitions. *Nat. Rev. Mol. Cell Bio.* **7**, 131–142.

Thomson, D. M., Krupey, J., Freedman, S. O., and Gold, P. (1969). The radioimmunoassay of circulating carcinoembryonic antigen of the human digestive system. *Proc. Natl. Acad. Sci. USA* **64**, 161–167.

Tibbe, A. G. J., Miller, M. C., and Terstappen, L. W. M. M. (2007). Statistical considerations for enumeration of circulating tumor cells. *Cytometry* **162**, 154–162.

Traystman, M. D., Cochran, G. T., Hake, S. J., Kuszynski, C. A., Mann, S. L., Murphy, B. J., Pirruccello, S. J., Zuvanich, E., Sharp, J. G. (1997). Comparison of molecular cytokeratin 19 reverse transcriptase polymerase chain reaction (CK19 RT-PCR) and immunocytochemical detection of micrometastatic breast cancer cells in hematopoietic harvests. *J. Hematother.* **6**, 551–561.

Tárnok, A., and Gerstner, A. O. H. (2002). Clinical applications of laser scanning cytometry. *Cytometry* **50**, 133–143.

Van der Auwera, I., Peeters, D., Benoy, I. H., Elst, H. J., Van Laere, S. J., Prové, A., Maes, H., Huget, P., van Dam, P., Vermeulen, P. B., Dirix, L. Y. (2010). Circulating tumour cell detection: a direct comparison between the CellSearch System, the AdnaTest and CK-19/mammaglobin RT-PCR in patients with metastatic breast cancer. *Brit. J. Cancer.* **102**, 276–284.

Verhamme, K. M. C., Dieleman, J. P., Bleumink, G. S., van Der Lei, J., Sturkenboom, M. C. J. M., Artibani, W., Begaud, B., Berges, R., Borkowski, A., Chappel, C. R., Costello, A., Dobronski, P., Farmer, R. D. T., Jiménez Cruz, F., Jonas, U., MacRae, K., Pientka, L., Rutten, F. F. H., van Schayck, C. P., Speakman, M. J., Sturkenboom, M. C., Tiellac, P., Tubaro, A., Vallencien, G., Vela Navarrete, R. (2002). Incidence and prevalence of lower urinary tract symptoms suggestive of benign prostatic hyperplasia in primary care – the Triumph project. *Eur. Urol.* **42**, 323–328.

Veridex White Pages. 2008 Available at: http://www.veridex.com/pdf/CXC_Application_Guideline.PDF.

Vlems, F. A., Diepstra, J. H. S., Cornelissen, I. M. H. A., Ruers, T. J. M., Ligtenberg, M. J. L., Punt, C. J. A., van Krieken, J. H. J. M., Wobbes, T., van Muijen, G. N. P. (2002). Limitations of cytokeratin 20 RT-PCR to detect disseminated tumour cells in blood and bone marrow of patients with colorectal cancer: expression in controls and downregulation in tumour tissue. *Mol. Pathol.* **55**, 156–163.

Vona, G., Estepa, L., Béroud, C., Damotte, D., Capron, F., Nalpas, B., Mineur, A., Franco, D., Lacour, B., Pol, S., Bréchot, C., Paterlini-Bréchot, P. (2004). Impact of cytomorphological detection of circulating tumor cells in patients with liver cancer. *Hepatology* **39**, 792–797.

Vona, G., Sabile, A., Louha, M., Sitruk, V., Romana, S., Schütze, K., Capron, F., Franco, D., Pazzagli, M., Vekemans, M., Lacour, B., Bréchot, C., Paterlini-Bréchot, P. (2000). Isolation by size of epithelial tumor cells: a new method for the immunomorphological and molecular characterization of circulating tumor cells. *Am. J. Pathol.* **156**, 57–63.

Watson, M. A., Dintzis, S., Darrow, C. M., Voss, L. E., DiPersio, J., Jensen, R., Fleming, T. P. (1999). Mammaglobin expression in primary, metastatic, and occult breast cancer. *Cancer Res.* **59**, 3028–3031.

Watson, M. A., and Fleming, T. P. (1996). Mammaglobin, a mammary-specific member of the uteroglobin gene family, is overexpressed in human breast cancer. *Cancer Res.* **56**, 860–865.

Welch, D. R. (1997). Technical considerations for studying cancer metastasis *in vivo*. *Clin. Exp. Metastas.* **15**, 272–306.

Went, P. T., Lugli, A., Meier, S., Bundi, M., Mirlacher, M., Sauter, G., Dirnhofer, S. (2004). Frequent EpCam protein expression in human carcinomas. *Hum. Pathol.* **35**, 122–128.

Williams, S. A., Singh, P., Isaacs, J. T., and Denmeade, S. R. (2007). Does PSA play a role as a promoting agent during the initiation and/or progression of prostate cancer? *The Prostate* **67**, 312–329.

Wulf, G. G., Jürgens, B., Liersch, T., Gatzemeier, W., Rauschecker, H., Buske, C., Hüfner, M., Hiddemann, W., Wörmann, B. (1997). Reverse transcriptase/polymerase chain reaction analysis of parathyroid hormone-related protein for the detection of tumor cell dissemination in the peripheral blood and bone marrow of patients with breast cancer. *J. Cancer Res. Clin.* **123**, 514–521.

Yang, J., and Weinberg, R. A. (2008). Epithelial-mesenchymal transition: at the crossroads of development and tumor metastasis. *Dev. Cell* **14**, 818–829.

Young, C. Y., Montgomery, B. T., Andrews, P. E., Qui, S. D., Bilhartz, D. L., Tindall, D. J. (1991). Hormonal regulation of prostate-specific antigen messenger RNA in human prostatic adenocarcinoma cell line LNCaP. *Cancer Res.* **51**, 3748–3752.

Zabaglo, L., Ormerod, M. G., Parton, M., Ring, A., Smith, I. E., Dowsett, M. (2003). Cell filtration-laser scanning cytometry for the characterisation of circulating breast cancer cells. *Cytom. Part A.* **55**, 102–108.

Zieglschmid, V., Hollmann, C., Gutierrez, B., Albert, W., Strothoff, D., Gross, E., Böcher, O. (2005). Combination of immunomagnetic enrichment with multiplex RT-PCR analysis for the detection of disseminated tumor cells. *Anticancer Res.* **25**, 1803–1810.

CHAPTER 11

Preclinical Applications of Quantitative Imaging Cytometry to Support Drug Discovery

David L. Krull and Richard A. Peterson

GlaxoSmithKline, Safety Assessment, Investigative Pathology Laboratory, Research Triangle Park, North Carolina, USA

Abstract

Preclinical drug development is actively involved in testing compounds to find cures or to manage the effects of disease, such as diabetes. Animal models, such as the Zucker diabetic fatty (ZDF) rat, are used to measure efficacy of candidate drugs. This animal model was selected because of its clinical and pathological similarities to diabetic human patients. A method using immunofluorescence and laser scanning cytometry (LSC) technology has been used to measure the development of diabetic phenotype in the ZDF rat during a 17-week time course. The expression levels of insulin, glucagon, voltage-dependent anion channel (VDAC), and Ki67 were quantified. Insulin and VDAC expression were reduced in the ZDF animals in

0091-679X/10 $35.00
DOI 10.1016/B978-0-12-374912-3.00011-0

comparison to the lean control rats, while no significant change was seen in glucagon and Ki67 expression at week 17. This information is useful in the design of studies to test experimental compounds in this model.

Screening drug targets or biomarkers in tissue sections is another important activity in drug development. Tissue microarrays (TMAs) are composed of 60 or more tissue cores from humans or animal models and may contain healthy and/or diseased tissues. Antibodies against target proteins are applied to TMAs using routine immuno-histochemical reagents and protocols. The protein expression across the cores, as labeled by immunohistochemistry, is measured using LSC technology. The process provides an efficient and cost-effective method for evaluating multiple targets in a large number of tissue samples. More recently, IHC and LSC have been taken to the next level to quantify biopharmaceutical drug and target co-localization in tissue sections.

I. Introduction

Within GSK Safety Assessment, toxicology studies are performed on novel compounds using animal models such as mice and rats. Tissues are collected at the end of study, fixed and processed, embedded in paraffin blocks, then sectioned and stained with hematoxylin and eosin. The prepared samples are evaluated by a pathologist who identifies lesions or abnormal changes. Additional follow-up studies may be performed based on the drug target or in order to understand mechanisms leading to the abnormal change. Several laboratory methods, such as Western blots, PCR, ELISA, and flow cytometry, are often employed. While these methods are invaluable to the assessment of novel compounds, there remains an increasing demand for high-content data analysis of protein expression in tissue sections while retaining their complex, heterogeneous amalgamation of structures and cell types. Sample preparation for several of these techniques requires destruction of the tissue architecture and homogenization of the complex cellular environment. While the resulting information is useful for revealing the presence of a particular protein in the tissue sample, localization to a particular cell type is rarely possible.

For this reason, a nondestructive method, such as immunohistochemistry using chromagenic or fluorescent dyes, is preferred. Moreover, with this approach, the samples can be evaluated using multiple imaging platforms to examine protein distribution and expression level. Traditional bright field or fluorescent microscopes and charge-coupled device (CCD) cameras are used to view and capture images. The pathologist typically assigns a numerical value to the level of protein expression − + 1 for mild, + 2 for moderate, and + 3 for severe. This manual analytic method is labor-intensive, inefficient for large studies, and subjective; it can be difficult to differentiate subtle changes in protein expression.

An improved method involves the use of image analysis software such as ImagePro® (Media Cybernetics, Bethesda, MD) to collect quantitative data. It is still inefficient, however, because images must be collected manually and then exported into the analysis software. Another popular method is confocal microscopy, which is useful for collecting high-resolution images and for producing z-stacks, which can be reconstructed into 3D images to allow visualization of protein distribution and co-localization. However, this approach is impractical for quantitation because it is limited to fluorescently stained samples, and the data collection still remains a largely manual process.

There is an increasing need in preclinical drug development for quantitative data collection and analysis, beginning at the earliest stages of the process. Currently, pharmaceutical research operations have a need for increased flexibility with respect to sample type while at the same time shortening development timelines. These requirements limit the number of platforms that combine full automation with the ability to measure chromagenically and fluorescently labeled proteins in cells and tissue sections, and also provide flexible software for data collection and analysis. For the past 6 years, the iCyte®Automated Imaging Cytometer (CompuCyte Corporation, Westwood, MA) has proven to be an ideal platform for measurement of protein expression in tissue sections because of its nondestructive method that preserves tissue architecture (Darzynkiewicz *et al.*, 1999; Harnett, 2007; Luther *et al.*, 2004). In addition, multiple proteins can be labeled in the same section and the relationships among them can be studied. Furthermore, tissue sections and cell cultures can be labeled using standard immunohistochemistry methods and the chromagenic and fluorescent products can be simultaneously imaged and measured (Peterson *et al.*, 2008; Pozarowski *et al.*, 2006).

The iCyte can be equipped with up to four solid-state lasers for excitation at 405, 488, 532, and 633 nm. In addition, two scatter detectors can be set to measure light loss (absorbance) or to show a shaded-relief image. The latter feature is useful for visualizing unstained cells and tissue or to provide contrast to fluorescently or chromagenically stained proteins. An optional robotic arm automatically loads carriers onto the sample stage for scanning and quantitation. This allows for loading 45 carriers with a standard 96-well plate footprint or 180 slides when using the four-position slide carriers.

Here, we demonstrate the use of this technology for evaluation of an animal model of human diabetes and in the quantification of biomarkers in tissue microarrays (TMAs).

II. Specific Examples: Example 1 – High–content automated tissue analysis of ZDF rat pancreas

A. Rationale

Type 2 diabetic patients suffer from insulin resistance, which affects metabolism and several other functions in the body. Drugs are in development to treat diabetes or prevent its onset or progression (Wright *et al.*, 2010), many of which aim to increase insulin production to normal levels by improving the function of insulin-producing cells and/or decreasing insulin resistance.

Animal models of human disease are frequently used to evaluate novel compounds (Chatzigeorgiou *et al.*, 2009). A common model of type 2 diabetes mellitus is the Zucker diabetic fatty (ZDF) rat. Many studies of investigative drugs targeting the pancreas for treatment of diabetes use the ZDF model to assess efficacy. These animals lose their ability to produce adequate quantities of the insulin necessary to keep glucose levels under control due to defective leptin receptors (Nugent *et al.*, 2008). Blood glucose levels can be measured by serum chemistry, but these levels do not reveal information on beta cell morphology or insulin expression.

The islets of Langerhans are a cluster of cells within the pancreas that are responsible for the production and release of hormones that regulate glucose levels. In this study, we focus on the insulin-producing beta cells and the glucagon-producing alpha cells because of their hormones' direct role in glucose regulation. (Insulin decreases blood glucose levels by stimulating the liver to convert glucose into stored glycogen and by stimulating most cells to insert the insulin-responsive glucose transporter into cells so they can take up glucose. Glucagon increases blood glucose levels by stimulating the liver to convert stored glycogen into glucose.) Normal healthy beta cells exhibit homogeneous, cytoplasmic insulin staining in contrast to more heterogeneous staining seen in diseased or malfunctioning beta cells.

Immunohistochemistry was used to label insulin with a guinea pig anti-insulin antibody. Glucagon expression in the alpha cells is labeled using a mouse anti-glucagon antibody. These data may answer questions concerning changes in the ratio of glucagon to insulin. Cell health and activity can be assessed by measuring cell proliferation and the capacity for energy production. The nuclear protein Ki67 is associated with cellular proliferation; it is present during all phases of the cell cycle, except resting (G_O) cells. A rabbit anti-Ki67 antibody is used to label actively dividing cells. Proliferation rates can be determined by both cellular morphology and Ki67 expression.

To investigate the role of mitochondria in diabetes, an antibody against the voltage-dependent anion channel (VDAC) is used to label and identify mitochondria. As a type of mitochondrial porin, VDAC forms an aqueous pathway for metabolites to enter or exit the mitochondria, which is open at low voltages and closed at high voltages. VDAC expression may be associated with beta cell energetics that regulate release of insulin (McCabe, 1994; Muhammed *et al.*, 2010). DAPI is a DNA stain and is used to identify cell nuclei.

B. Materials and methods

1. Animal model: Zucker diabetic fatty (ZDF) rat

ZDF lean/Crl-Lepr$^{+/+}$ and ZDF obese diabetic/Crl-Lepr^{fa} rats (Charles River Laboratories, Inc., Raleigh, NC) were housed one per cage in polycarbonate solid-bottom cages with Bed-O'Cobs^{TM} (The Andersons, Maumee, OH) in an environment with a temperature of 64–79 °F, 30–70% relative humidity, and a 12-h light/dark cycle, and fed Purina LabDiet^{TM} Formulab Diet 5008 (pellets) (PMI^{TM} Nutrition International, Richmand, IN) *ad libitum*. Municipal water supply with additional treatment by reverse osmosis was available to all rat strains *ad libitum* from an automated watering system. All animal handling and treatment in these animal studies was performed in accordance with the GlaxoSmithKline Animal Care and Use Committee (ACUC) guidelines.

2. Tissue processing

ZDF and lean control rats were euthanized at 6, 8, 11, and 17 weeks for tissue collection. The pancreas was cut in half and placed in 10% neutral buffered formalin to fix for 48 h at room temperature. Fixed tissue samples were cut into five to seven pieces to facilitate a more uniform, random islet sampling. Samples were processed in the Leica ASP300S automated tissue processor for ethanol dehydration, xylene clearing, and paraffin infiltration. Pancreas was placed in tissue-embedding molds with the cut surface down. Paraffin sections were cut at 4 μm for each sample, mounted on positively charged slides, and air-dried overnight.

3. Immunohistochemistry

The following dual immunohistochemistry protocol was optimized and run using the Bond^{TM} Max Immunohistochemistry and *in situ* Hybridization Automated Stainer (Leica Microsystems, Bannockburn, IL). Table I shows the list of antibodies used and associated detection reagents.

i. Sections were deparaffinized in Bond Dewax followed by heated epitope retrieval for 20 min in Bond ER2 (EDTA-based solution).
ii. Endogenous peroxidase was blocked using a dual endogenous enzyme block (BioFX) for 10 min.
iii. Endogenous rat IgG was inhibited by Rodent Block R (Biocare Medical) for 15 min.
iv. A rabbit anti-Ki67 primary diluted 1:1000 in Bond Primary Antibody Diluent was applied for 15 min, and then detected with a rabbit-on-rodent HRP polymer (Biocare) for 15 min.
v. The chromagenic product was developed using Refine DAB for 10 min and Bond DAB enhancer for 5 min.

Table I

Antibodies and dectection chemistry applied to pancreas samples

Antibody	Vendor	Dilution	Detection/label
Rabbit anti-Ki67	Neomarkers	1:1000	Rabbit-on-rodent HRP polymer/DAB
Guinea pig anti-insulin	Dako	1:1000	Goat anti-guinea pig AlexaFluor 647
Mouse anti-glucagon	Abcam	1:500	Goat anti-mouse AlexaFluor 532
Rabbit anti-VDAC	Abcam	1:50	RTU biotinylated goat anti-rabbit/SA AlexaFluor 488

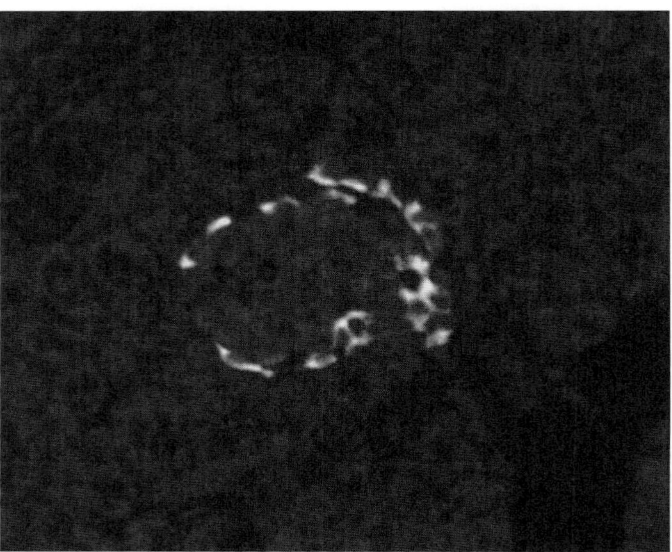

Fig. 1   Representative islet from a 6-week lean rat showing immunofluorescent labeling of insulin (red), glucagon (yellow), and VDAC (green). Ki67 is labeled with DAB chromagen and is displayed as magenta. Nuclei are stained with DAPI (blue). (See plate no. 18 in the color plate section.)

   vi. Chromagen-stained sections were incubated for 30 min in a primary antibody cocktail containing guinea pig anti-insulin at 1:1000, mouse anti-glucagon at 1:500, and rabbit anti-VDAC in Bond Primary Antibody Diluent. Primary antibodies were detected with a secondary antibody cocktail containing RTU biotinylated goat anti-rabbit, goat anti-mouse AlexaFluor 532 at 1:100, and goat anti-guinea pig AlexaFluor 647 at 1:100. The sections were incubated in the secondary antibody cocktail for 30 min.

  vii. Streptavidin-AlexaFluor 488 at 1:200 was applied and incubated for 30 min to label the biotinylated secondary.

 viii. Lastly, the nuclei were stained with DAPI 1:500 for 10 min.

   ix. Excess water was removed from sections by tapping the edge of the slide on absorbent material. Four to five drops of VectaMount were then applied at evenly spaced intervals across the slide surface. A 24 × 50 × 1 mm coverglass was applied to each slide, taking care to avoid air bubbles. Prior to scanning, slides were dried overnight at room temperature, protected from light. Fig. 1 shows the four-color image of the stained islet.

4. Laser scanning cytometry method

The laser scanning cytometer (LSC) scans the samples, and acquired data for each of the fluorescent or chromagenic labels are assembled into a set of images for each

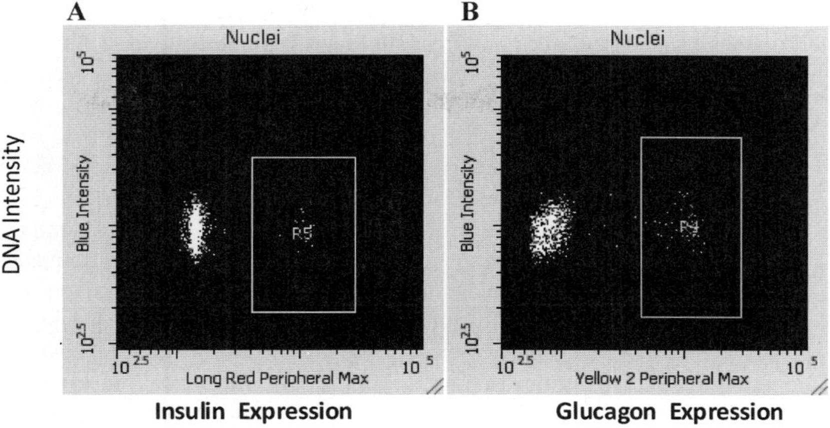

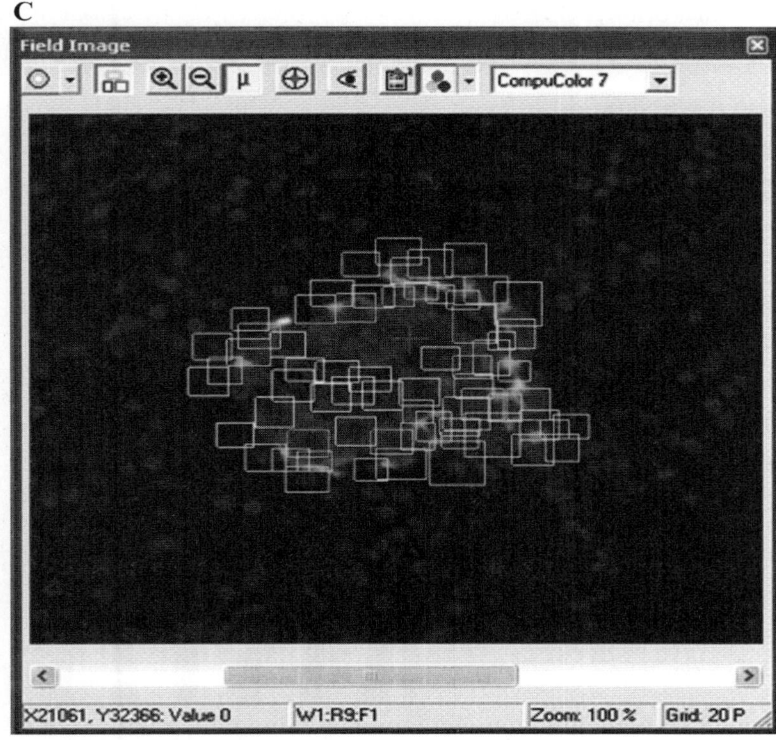

Fig. 2   (A) Beta cell nuclei are separated from other nuclei according to the amount of glucagon staining within the peripheral contour. (B) Alpha cell nuclei are separated from other nuclei according to the amount of insulin staining within the peripheral contour. A region is placed on each scattergram to capture statistical information for each group of cells. (C) Cells that fall within each gated region are displayed on the field image.

label or "channel." Images are "segmented" to automatically identify cells and measure cell characteristics (e.g., cell number, size, and label intensity value). Both nuclear and cytoplasmic expression levels may be measured. Segmentation can be applied using a manually or automatically set threshold value, based on the fluorescent or absorbance intensity signal. In this way, DAPI and Ki67 are used to identify nuclei and measure DNA content and Ki67 expression. Cell cytoplasm can also be segmented in order to measure insulin and glucagon expression. An alternative method of measuring cytoplasmic expression was employed in which a set of two "peripheral contours" were drawn external to the nuclear boundary and used to measure cytoplasmic protein expression associated with the nucleus. DNA intensity was plotted against cytoplasmic protein expression to display populations of cells based on their VDAC, glucagon, and insulin expression. In this way, islet nuclei were separated from acinar nuclei. The separate cell populations were identified by applying gating regions to these scatter plots. The correct placement of these gating regions was verified visually by overlaying colored rectangles on cells in the field images. The color of the rectangles is indicative of the gating region in which data for that cell lies (Fig. 2).

C. Results and discussion

1. Insulin expression in beta cells

There is a clear reduction in the amount of insulin in ZDF islets, even at the first time point (6 weeks). By 8 weeks, insulin expression is reduced by more than half. By the third time point (11 weeks), insulin levels appear to reach a steady state (Fig. 3).

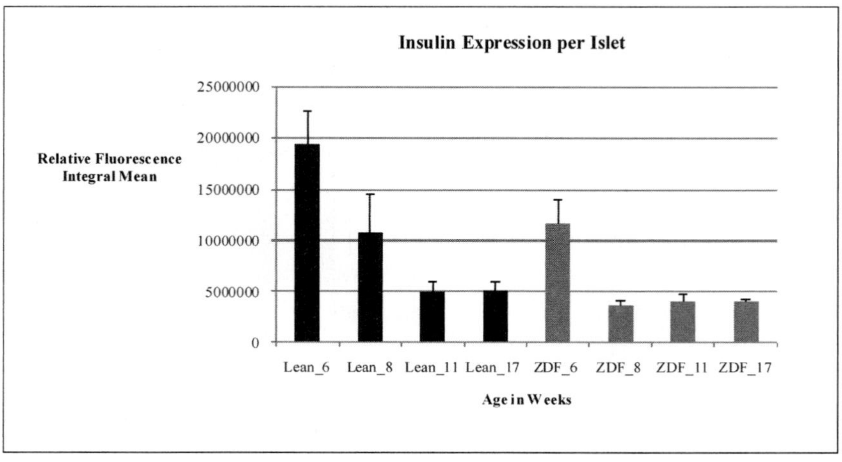

Fig. 3   Mean insulin expression.

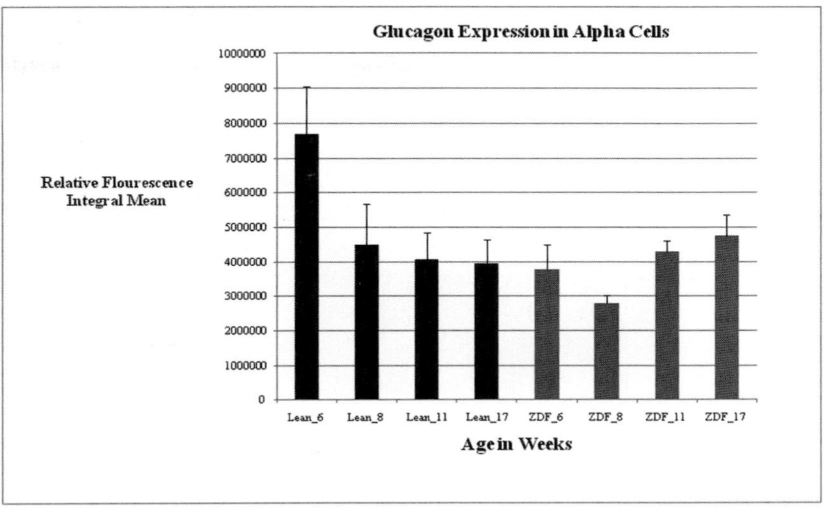

Fig. 4   Mean glucagon expression.

2. Glucagon expression in alpha cells

The glucagon-producing alpha cells are a sub-population of cells in the islets closely associated with beta cells, responsible for raising blood glucose levels by increasing the conversion of glycogen stores, primarily in the liver, to glucose. Glucagon may act to balance the effects of insulin and, if so, should track closely with insulin levels. This appears to be the trend in the lean animals, where both insulin and glucagon are elevated at 6 weeks. After that, the glucagon level decreases as insulin decreases to a steady state at weeks 11 and 17.

The ZDF animals exhibited a trend in which glucagon levels continued to rise in spite of decreased levels of insulin. These data suggest that higher levels of glucagon, in addition to decreased insulin, may be contributing to the diabetic condition in the ZDF rat (Fig. 4).

3. VDAC – mitochondrial expression in beta cells

As the cell's energy producers, mitochondria are vital for all activities requiring ATP. Because VDAC is located within the mitochondrial membrane, they were used in this experiment as a marker to measure mitochondrial biogenesis. Any change in the expression level of the VDAC protein may reveal important

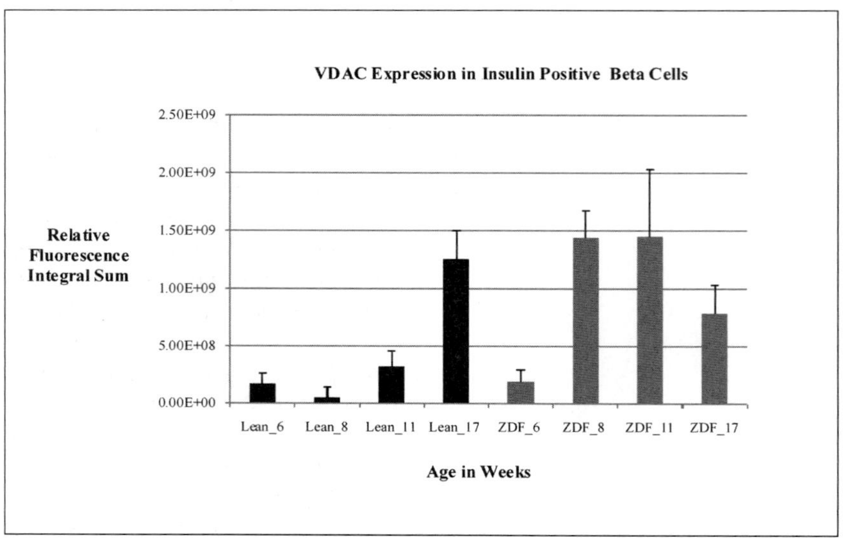

Fig. 5   Total amount of VDAC expression.

information regarding the potential energy of the beta cells. The amount of fluorescence was measured in areas of the islet where segmentation of insulin expression was used. The sum of the VDAC expression in the islets for each group, and the standard error, were calculated. The quantity of VDAC expression in lean and ZDF animals at 6 weeks is closely aligned. There were considerably lower levels of VDAC in the lean animals at 8 and 11 weeks, however, compared to the same time points in the ZDF animals. At 17 weeks, the lean animals show an increase in VDAC expression, in contrast to the ZDF animals which show a decrease (Fig. 5).

4. Beta cell proliferation

The proliferation index is a measure of islet growth and development. Ki67 is a marker of cells that have entered the cell cycle and have committed themselves to division. The proliferation index of the beta cells was calculated by determining the ratio of the number of nuclei with cytoplasmic insulin expression that are Ki67-positive to the total number of cytoplasmic insulin-expressing cells. This ratio was multiplied by 100 to obtain the percentage of Ki67-positive beta cells. Based on these data there does not appear to be a significant difference in the proliferation index between the two groups (Fig. 6).

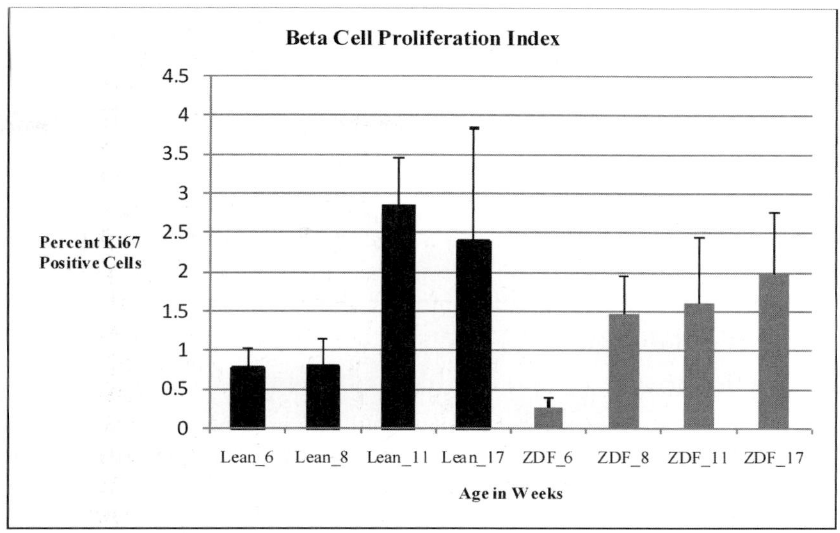

Fig. 6   Mean beta cell proliferation index.

III. Example 2 – Analysis of Biomarkers in Tissue Microarrays

A. Rationale

Preclinical drug development studies are conducted to evaluate various endpoints, such as efficacy, specificity, and safety, by examining protein expression and changes in tissue morphology. There is often an interest in evaluating targeted proteins across multiple tissue types. TMAs containing normal or tumor samples from animal and human donors are an efficient tool for looking at selected markers across multiple tissues using one slide. Here, we demonstrate the use of dual IHC protocols on human multi-tumor arrays. Selected markers include CD31/CD34, cytokeratin 7/CD31, S100/GFAP, kappa/lambda (*in situ* hybridization), and Ki67/CD68. Protein expression across these arrays is measured using the iCyte Automated Imaging Cytometer, which automatically identifies core elements, performs spectral deconvolution to isolate markers of interest, and quantifies the amount of marker expression. High-resolution images of the cores are obtained simultaneously with quantitative analytical data. The development of protocols using TMAs and LSC quantification provides a useful standard for confirming antibody specificity in human TMAs with cross-reactivity and specificity in rodent models.

Table II
Antibodies and chromagens used in dual TMA staining

Multitumor MaxArray™	DAB chromagen	Permanent red chromagen
TMA 1	Kappa *ISH*	Lambda *ISH*
TMA 2	CD31	CD34
TMA 3	Cytokeratin 7	CD31
TMA 4	S100	GFP
TMA 5	Ki67	CD68
TMA 6	Negative control	Negative control

B. Materials and methods

1. Tissue microarrays

MaxArray™ human multitumor TMAs were purchased from Invitrogen (Camarillo, CA). Each array contains 60 formalin-fixed, paraffin-embedded tissue cores from malignant tumors.

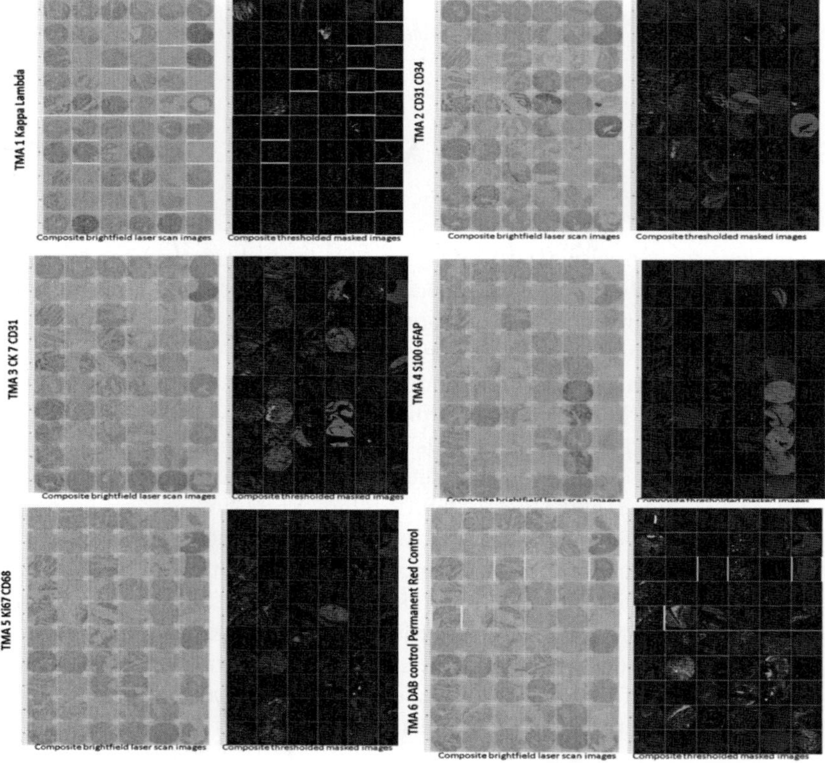

Fig. 7 Representative LSC images of each TMA displayed as a bright field (A) or fluorescent (B) CompuColor image. (For interpretation of the references to color in this figure legend, the reader is referred to the Web version of this chapter.)

2. Immunohistochemistry

Bond Ready-To-Use Primary Antibodies and Refine Polymer Detection kits were purchased from Leica Microsystems (Bannockburn, IL). A dual-IHC protocol using the Refine Polymer DAB and Refine Polymer Red was applied to each TMA. The list of TMA and associated antibodies and chromagens is given in Table II. An *in situ* hybridization protocol for kappa and lambda was run on TMA 1 using the same

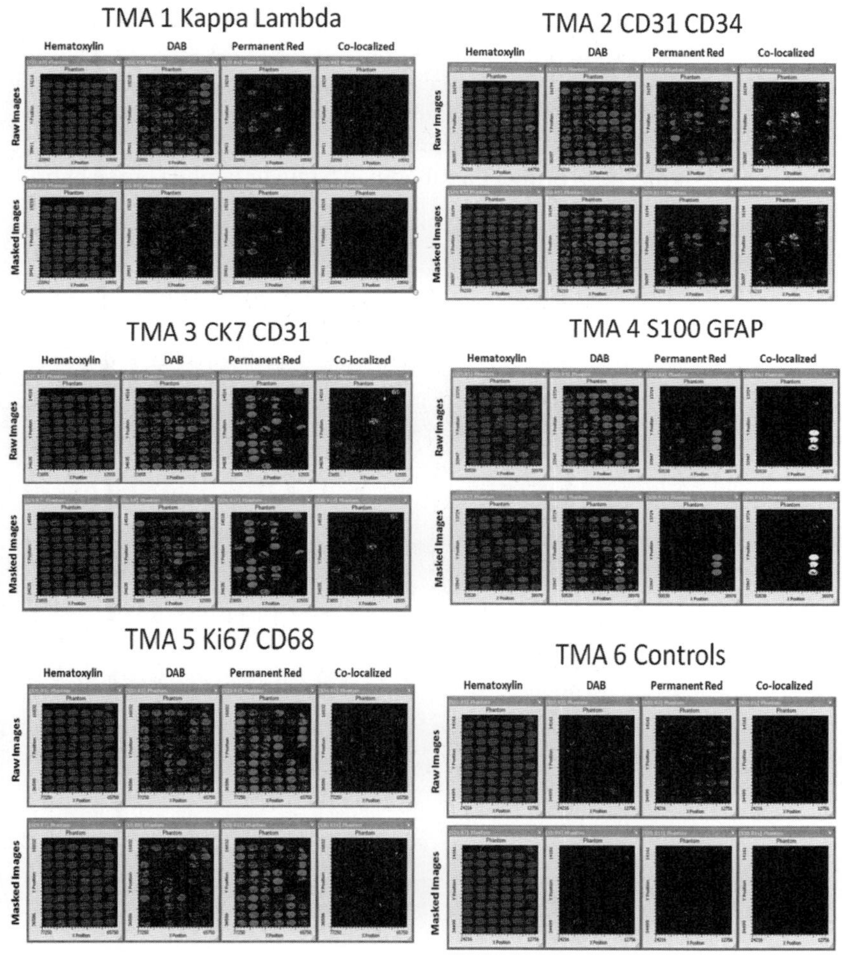

Fig. 8 X–Y mapping of expression data – The X–Y locations of hematoxylin, DAB+, permanent red+, and co-localization of DAB and permanent red are each plotted separately. Data from unmasked images are compared to data from masked images. The masked images show lower levels of background staining. (For interpretation of the references to color in this figure legend, the reader is referred to the Web version of this chapter.)

detection kits. The following dual IHC protocol was optimized and run using the Bond Max IHC/ISH Automated Stainer.

   i. Sections were deparaffinized in Bond Dewax followed by heated epitope retrieval for 20 min in Bond ER2 (EDTA-based solution) or in Bond Enzyme 1 for 10 min @ 37 °C for ISH.
   ii. Endogenous peroxidase block – 5 min.
   iii. [1st] primary antibody – 15 min or kappa probe hybridization for 2 h @ 37 °C for ISH.
   iv. For ISH protocol, a mouse anti-FITC antibody was applied and incubated for 15 min.
   v. Postprimary reagent – 8 min.
   vi. HRP-labeled polymer – 8 min.
   vii. Refine DAB chromagen – 10 min.
   viii. [2nd] primary antibody – 15 min or lambda probe hybridization for 2 h @ 37 °C for ISH.
   ix. Postprimary AP – 20 min.

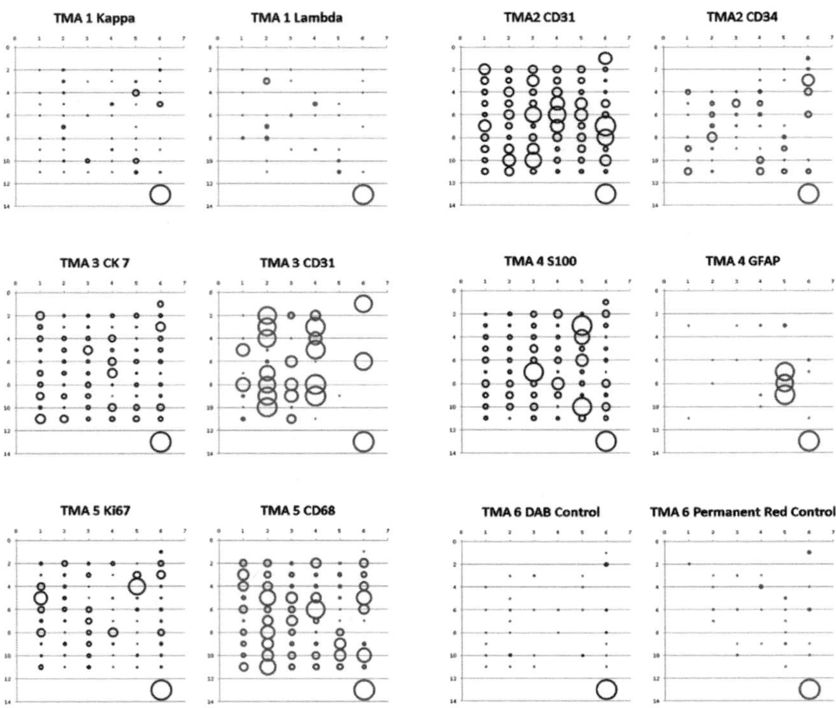

Fig. 9 Bubble plots are shown for percent area of positive expression of DAB and permanent red. The size of the bubble is proportional to the percentage of the core that is covered by the constituents. The lower right bubble is set at 100% as a visual reference for the data sets.

  x. AP-labeled polymer – 30 min.

  xi. Refine Red chromagen – 15 min.

  xii. Hematoxylin – 5 min.

  xiii. Slides were dehydrated in 95% ETOH for 2 min followed by three changes of 100% ETOH for 2 min each, then three changes of xylene for 3 min each. Three drops of Cytoseal 60 were applied to each section, which was then mounted with $24 \times 50 \times 1$ mm coverglass. Slides were dried overnight at room temperature before scanning.

3. Laser scanning cytometry method

The TMA sections were quantified using random sampling, in which circular sampling elements of user-defined number and size, arranged in a lattice pattern or randomly distributed, are overlaid on the field image. Each sampling element collects information with respect to intensity or absorbance of the tissue site it overlays. This method is useful when nuclear segmentation is not required or when measuring protein expression across homogeneous tissues such as liver, brain, and

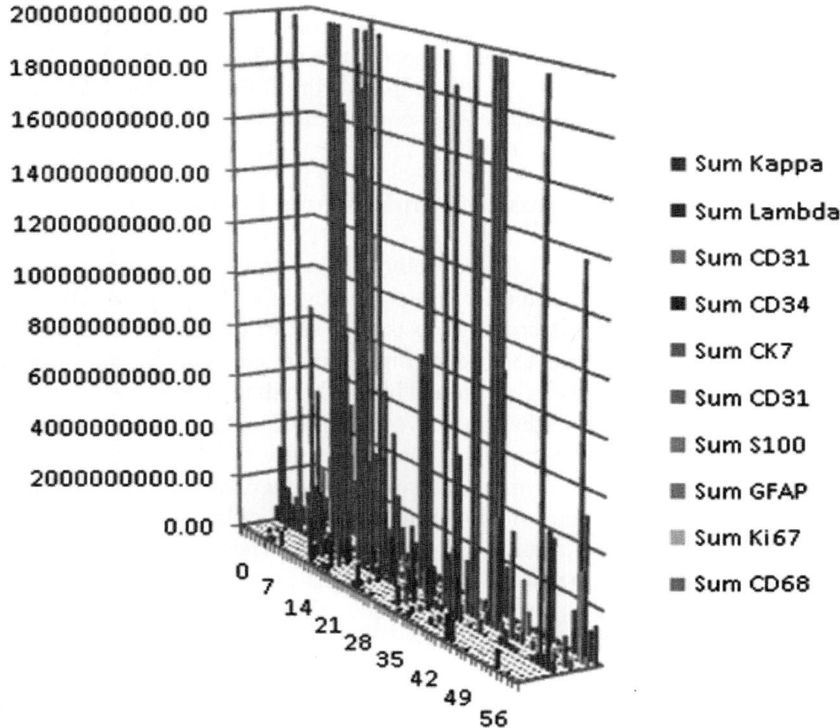

Fig. 10 Total (summed) signal expression per core for all antigens.

muscle. The following parameters for each TMA on a per-core basis were quantified: expression level of each marker, percentage of area of positive expression per core, and percentage of area of co-localization of the two markers per core. Some of the TMA sections had nonspecific background staining that would contaminate measurement of specific markers. A "mask" module, containing only the pixels above the threshold value, was used to measure the TMAs in order to eliminate background signal.

C. Results and discussion

The TMAs were labeled with each of the selected markers and their expression level is displayed in Figure 7. Some of the TMAs had higher than expected background levels, which were corrected using the mask module, which eliminates signal below a specified value Scattergrams were produced for each TMA and antigen and masked-versus-unmasked data is displayed in Fig. 8. The X–Y scattergrams show the expression level of each marker on a per-core basis and provide a visual representation of the distribution of each marker. Quantitative data is plotted as a percentage of area positive for each marker (Fig. 9). These data can also be displayed to show the total expression of the antigens across the entire TMA (Fig. 10). This information reveals which proteins are up-regulated in these cancers; it may be useful for identifying possible drug targets. Since the TMAs represent serial sections, the same core in each TMA can be compared for all the antigens. The use of TMAs allows us to efficiently produce high-content data for 60 tumor types.

IV. Conclusions and Future Directions

Preclinical drug development is under increasing pressures to provide patients with novel and efficacious medicines with exceptional safety profiles, while at the same time decreasing costs. The use of LSC enables screening of multiple tissue or cell elements to produce high-content data that can add value to efficacy testing and early safety prediction. This will enable early decision-making and promote compounds with the most favorable profile. Associated cost savings can be returned to the research and development budget.

The use of immunohistochemical techniques and LSC has shown promise in the development of biopharmaceuticals, including humanized monoclonal antibodies generated against targeted proteins. These molecules are tested in animal models such as the SCID mouse (Bosma *et al.*, 1983, 1988), which has a compromised immune system due to the lack of B and T lymphocytes, and which does not react to the human IgG. Tissues are collected from these animals to assess the expression or activity of the therapeutic antibody. Methods such as TaqMan®, Western blots, clinical pathology, and pharmacokinetic analysis are often performed on the tissues and blood products from these studies. As previously mentioned, however, these methods require destruction of the sample or the tissue morphology, making it impossible to determine the drug distribution in the tissue, or its relationship to the drug's target and other structures. We have developed a method using immuno-histochemistry and LSC to measure therapeutic antibodies and their interaction with their target. The method entails the use of an anti-human IgG antibody or anti-idiotype to detect the drug, followed by a fluorescent or chromagenic detection method. Next, a second antibody directed against the target protein, and the associated detection chemistry, is applied. The amount of target and drug is measured using LSC technology. Using this method we have measured the intensity and integral values for the target protein, drug co-localized with the target and drug in excess. These data are used to examine drug distribution and target saturation.

Additional markers can be applied to study the relationship of the drug to other tissue elements. For example, a drug that targets a protein expressed in skeletal muscle is co-localized with another antibody against myosin heavy chain slow or fast, to determine whether the drug has a preference for type I or type II fibers and the degree of drug affinity. By multiplexing antibodies in this way, we can obtain high-content, multiparameter data in a nondestructive manner. Moreover, the workspace on the iCyte can be reconfigured as necessary by adjusting the thresholds and gating applied to scattergrams, and by adding or subtracting run event statistics to optimize data collection and utility. The quality of the segmentation method used can be evaluated during the data analysis by creating galleries of cells, viewing the specificity of the segmentation contours, and examining gated events in the field images. Adjustments can be made to increase precision, as necessary; then the data is "virtually" rescanned (the saved data is reprocessed with the new settings). Once a workspace has been optimized, it can be routinely applied to future studies with minimal adjustments.

Using LSC technology, we have generated data to support our discovery projects, in addition to regulatory and investigative studies within the Safety Assessment department. These data have added significant value to decision-making and drug development strategies. We will continue to focus our efforts around the quantification of drug and target expression, with the dual goals of reducing attrition and providing patients with high-quality pharmaceuticals that satisfy unmet medical needs.

Reference List

Bosma, G. C., Custer, R. P., and Bosma, M. J. (1983). A severe combined immunodeficiency mutation in the mouse. *Nature* **301**, 527–530.

Bosma, M., Schuler, W., and Bosma, G. (1988). The scid mouse mutant. *Curr. Top. Microbiol. Immunol.* **137**, 197–202.

Chatzigeorgiou, A., Halapas, A., Kalafatakis, K., and Kamper, E. F. (2009). The use of animal models in the study of diabetes mellitus. *In Vivo* **23**, 245–258.

Darzynkiewicz, Z., Bedner, E., Li, X., Gorczyca, W., and Melamed, M. R. (1999). Laser-scanning cytometry: a new instrumentation with many applications. *Exp. Cell Res.* **249**, 1–12.

Harnett, M. M. (2007). Laser scanning cytometry: understanding the immune system in situ. *Nat. Rev. Immunol.* **7**, 897–904.

Luther, E., Kamentsky, L., Henriksen, M., and Holden, E. (2004). Next-generation laser scanning cytometry. *Methods Cell Biol.* **75**, 185–218.

McCabe, E. R. B. (1994). Microcompartmentation of energy metabolism at the outer mitochondrial membrane: role in diabetes mellitus and other diseases. *J. Bioenerg. Biomembr.* **26**, 317–325.

Muhammed, S., Ahmed, M., Kessler, B., and Salehi, A. (2010). Mitochondrial proteome analysis reveals changes in expression of multiple proteins in pancreatic beta cells exposed to high glucose. *Diabetologia* **53**, S200.

Nugent, D. A., Smith, D. M., and Jones, H. B. (2008). 3A review of islet of Langerhans degeneration in rodent models of type 2 diabetes. *Toxicol. Pathol.* **36**, 529–551.

Peterson, R. A., Krull, D. L., and Butler, L. (2008). Applications of laser scanning cytometry in immunohistochemistry and routine histopathology. *Toxicol. Pathol.* **36**, 117–132.

Pozarowski, P., Holden, E., and Darzynkiewicz, Z. (2006). Laser scanning cytometry: principles and applications. *Methods Mol. Biol.* **319**, 165–192.

Wright, E. E., Stonehouse, A. H., and Cuddihy, R. M. (2010). In support of an early polypharmacy approach to the treatment of type 2 diabetes. *Diabetes Obes. Metab.* **12**, 929–940.

CHAPTER 12

Leveraging Image Cytometry for the Development of Clinically Feasible Biomarkers: Evaluation of Activated Caspase-3 in Fine Needle Aspirate Biopsies

Gloria Juan, Stephen J. Zoog and John Ferbas

Department of Clinical Immunology, Amgen, Inc., One Amgen Center Drive, Thousand Oaks, California

Abstract

Quantitation of activated caspases in xenograft models by laser scanning cytometry has demonstrated mechanism-specific biological activity of Anti-Trail Receptor immunoglobulin therapies *in situ.* These preclinical data confirmed that caspase activation is an early event that precedes tumor regression. To apply this platform for clinical monitoring of caspase activation using fine needle aspirate (FNA) biopsies, additional assay feasibility and validation experiments need be addressed. Furthermore, important instrument parameters should be considered including the maintenance and operation of the cytometer in a controlled state to

0091-679X/10 $35.00
DOI 10.1016/B978-0-12-374912-3.00012-2

ensure aspects like data traceability, reliability, and integrity. In the present chapter we describe a method to evaluate caspase activation in Colo205 cells and fine needle aspirate tumors by slide-based, laser scanning cytometry. This approach can be applied to cell cultures, preclinical and clinical fine needle aspirate material.

I. Introduction

Apoptosis could be defined as a physiological cascade that leads to cell death (Darzynkiewicz *et al.*, 1997, 2004). This process can be executed through both the extrinsic and intrinsic apoptotic pathways in addition to the p53 and lysosomal pathways. Initiation of apoptosis leads to activation of effector molecules known as "caspases." These proteases cleave distinct protein substrates, resulting in the morphological changes of apoptosis. Measuring caspase activity may, therefore, constitute a useful diagnostic and prognostic tool in the clinical oncology setting as well as for drug development (Gastman, 2001; Moffitt *et al.*, 2010). However, molecular evidence of drug activity in tumors is often difficult to detect during clinical investigation. Consequently, it is critical to study the on-target pharmacodynamics of drugs using a relevant biopsy method, like fine needle aspirates (FNAs).

The extrinsic apoptotic pathway is initiated at the cell surface by the activation of the death receptors, death receptor 4 (DR4) and death receptor 5 (DR5). DR5, also known as TRAIL receptor 2, is a member of the tumor necrosis factor family of receptors (Pan *et al.*, 1997), and activation by Apo2L/TRAIL stimulation can lead to apoptosis in cancer cells *in vitro* and *in vivo* (LeBlank and Ashkenazi, 2003). Alternatively, agonistic antibodies to DR5 such as conatumumab (Kaplan-Lefko *et al.*, 2010) or others (Adams *et al.*, 2008; Chuntharapai *et al.*, 2001; Georgakis *et al.*, 2005; Guo *et al.*, 2005; Ichikawa *et al.*, 2001; Motoki *et al.*, 2005; Pukac *et al.*, 2005) can be used to bind and activate DR signaling. Since activation of caspases appears to be one of the most specific markers of apoptosis, measuring activated caspase-3 after DR5 signaling constitutes a means of evaluating the pharmacodynamic effect of agonistic drugs targeting this pathway (Adams *et al.*, 2008; Ashkenazi *et al.*, 2008; Herbst *et al.*, 2010; Kaplan-Lefko *et al.*, 2010). Caspase activation can be measured directly by immunocytochemical detection of the neoepitope presented after proteolytic activation of the caspase. Briefly, samples are incubated with antibodies specific to activated caspase-3 and with a fluorescently tagged secondary antibody. Cellular DNA is labeled with DAPI or Hoechst to counterstain the nuclei, and samples are analyzed using cytometric techniques. Morphological visualization and cytometric data of the percentage of caspase positive events are generated. In addition, multiparameter analysis of the cells differentially stained for activated caspase-3 and DNA makes it possible not only to identify and score apoptotic cell populations but also to correlate apoptosis with the cell cycle phase or DNA ploidy (Zoog *et al.*, 2010).

Here we describe a method to evaluate caspase activation in Colo205 cells and FNA tumors by laser scanning cytometry (LSC). This approach can be applied to cell cultures and both preclinical and clinical FNA material.

II. Materials

- Cells to be analyzed
- Phosphate-buffered saline (PBS)
- Bovine serum albumin (BSA)
- Triton X-100 (TX)
- Normal goat serum (NGS)
- Fixation buffer: 2% formaldehyde in PBS (store at 4 °C)
- Polylysine coated slides
- Cytocentrifuge and cytofunnels
- Histology Pap-pen
- Humidifying chamber for overnight slide incubation
- Blocking/permeabilization buffer: PBS/1%BSA/0.2% TX/5%NGS, pH 7.4 (store at 4 °C)
- Rinsing buffer: PBS/1% BSA, pH 7.4 (store at 4 °C)
- Antibody dilution buffer: PBS/1%BSA/0.2%TX
- Anti-EpCAM monoclonal antibody (clone VU1D9; Cell Signaling Technology, Danvers, MA, Cat#2929)
- Anti-caspase-3 antibody (Asp175, rabbit polyclonal antibody, Cell Signaling Technologies, Danvers, MA, Cat#9661L)
- Caspase-3 blocking peptide (Cell Signaling Technologies, Cat#1050)
- Secondary antibodies: goat anti-rabbit IgG Alexa Fluor 647 (Invitrogen, Carlsbad, CA), or goat anti-mouse IgG Alexa Fluor 488 (Invitrogen)
- Hoechst solution; 2 µg/mL Hoechst 33342 (Molecular Probes, Eugene, Oregon) in PBS/1%BSA (Ho342) or DAPI solution: 2 µg/mL in PBS (EMD Chemicals, Gibbstown, NJ)
- Fluorescence mounting media or 50% glycerol–PBS
- iCys Laser Scanning Cytometer (Compucyte Corp., Westwood, MA)

III. Staining and Cytometric Analyses of FNAs or Culture Cell Lines

FNA samples fixed in 2% formaldehyde are gently resuspended by pipetting and washed twice with PBS after centrifugation. The pellets are resuspended in 100 µL of PBS containing 1% BSA and 0.1% sodium azide (PBS–BSA), cytocentrifuged onto polylysine coated slides, and a standard pap-pen is used to encircle the sample with a thin wax ring. The sample is blocked for 30 min with PBS–BSA containing 5% NGS, followed by overnight incubation in the dark at 4 °C with 100 µL of PBS–BSA containing 0.25 µg/mL mouse anti-EpCAM monoclonal antibody.

Slides are then washed in PBS–BSA and incubated overnight in the dark at 4 °C with 100 µL of PBS–BSA containing 0.2% TX (PBS-BSA-TX) and anti-caspase-3 antibody. As a staining control for specificity, a molar excess of blocking peptide may be included in this solution and added to a duplicate slide specimen.

After, the slides are incubated with 100 μL of PBS–BSA–TX containing secondary antibodies conjugated to Alexa488 and Alexa647 for 1 h in the dark at room temperature.

Finally, the cells are stained with 100 μL Hoechst or DAPI solution and incubated for 10 min in the dark.

The slides are then mounted and coverslipped for analysis by LSC. Morphological visualization and cytometric data of the percentage of caspase positive events are generated.

Cytometric measurements can be performed using an iCys Laser Scanning Cytometer with excitation wavelengths of 405, 488, and 633 nm, generated by the violet, argon, and HeNe lasers, respectively. Blue (Hoechst 33342), green (Alexa488), and long red (Alexa647) fluorescence are then measured by separate photomultipliers. Adjacent field images are scanned using 40× magnification to determine the percentage of individual nucleated caspase-3 positive cells. Cellular image quantification is based on nuclear thresholding, and populations are gated to exclude large clumps of cells. The activated caspase-3-positive population is identified based on the intensity of the Alexa647 signal after exclusion of spectral overlap and comparison with control samples (secondary antibody alone and/or peptide blocked primary antibody). The integrity of the population is verified by relocating positive events into an image gallery and visually confirming the morphology of positive staining cells. Additionally, epithelium-derived tumor cells are identified based on the intensity of the EpCAM positive Alexa488 signal (Zoog *et al.*, 2010).

IV. Critical Aspects of the Procedure

A. Fixation and Stability Assessments

Cell fixation and permeabilization are critical steps for immunocytochemical detection of intracellular proteins and often must be customized for particular antigens. The fixative is expected to stabilize the antigen *in situ* and preserve its epitope in a state where it continues to remain reactive with the available antibody. The cell must be permeable to allow access of the antibody to the epitope (Clevenger *et al.*, 1987). It should be noted that extensive DNA–DNA or DNA–protein crosslinking occurs as a result of high concentrations or long fixation times in formaldehyde. Such overfixation can impair staining with intercalating DNA dyes such as propidium iodide. Consequently, if accurate DNA distributions are required, then mild formaldehyde fixation, as described above, is necessary. Similarly, stability assessments to evaluate how reliably the material can be stored in formaldehyde without adversely affecting analyte detection (caspase-3) should be performed. This is particularly a concern when samples need to be shipped from clinical sites to the analytical laboratory. For this assay, colo205 cells stored in 10% neutral buffer formalin at 4 °C are stable for at least 4 months with no observed change in the percentage of cells containing activated caspase-3.

B. Assay Validation

Fine-needle aspiration has been used reliably for clinical diagnosis (Boeddinghaus and Johnson, 2006; Saleh and Masood, 1995), as well as for immunocytochemical analysis of signal-transduction intermediates (Brotherick *et al.*, 1995; Nizzoli *et al.*, 2000). As with any pharmacodynamic assay intended for clinical implementation, preanalytical variability of FNA sampling should be addressed for each analyte.

Perhaps the most important assay performance parameters to be assessed in the FNA assay are the variability and relative specificity of the staining. Precision can be determined by triplicate measurements of nontreated and treated colo205 cells. Staining specificity can be confirmed by including an epitope peptide that blocks caspase-3 immunodetection to levels observed with the secondary antibody alone. Human mock FNAs prepared from resected human colorectal tumor samples can be utilized to determine baseline levels of activated caspase-3 in human tumors. Total variability can then be determined by assessing the contribution to the variability from interpatient, staining replicates within patient (intrapatient), repeat image acquisition on the same instrument.

C. Instrument Validation

Because relative precision is critically important to assessing pharmacodynamic impact in the clinic, it is imperative that the instrument be maintained and operated in a controlled state. An assay is only as good as the instrument used for detection. Therefore, using Good Laboratory Practice guidelines, critical aspects of the LSC that influence the ability to reconstruct experimental data can be investigated. These aspects should include data traceability, reliability, and integrity. The investigation should consist of a validation strategy that defines the system components (including software), the data flow, and the validation deliverables. Requirements can be specified based on the intended use of the system and the Code of Federal Regulations (CFR 21 Part 11). Requirements Specifications include LSC functionality (hardware and software), data acquisition/storage/security, analyst training and environmental/ room conditions. Written test cases should be designed in a Qualification Protocol to challenge each functional requirement defined in the Requirements Specification. A traceability matrix can be generated to map each Requirements Specification to a specific written test step in the Qualification Protocol. This matrix is essential for auditing the validation package. Prior to executing the Qualification Protocol, the system should be placed under change control, which limits access and administrative privileges. The Qualification Protocol is then executed and test results are documented including computer screen shots when appropriate. Finally, a validation report should be generated that describes any test failures and/or deviations from the validation plan or test protocols that may have occurred during execution, how these issues were addressed, and their impact on the validation. A standard operating procedure should govern the access to the LSC as well as its operation, performance testing, and routine maintenance schedule. Because of this demonstrated control over the system, and the assay validation parameters described earlier, an association between data integrity and clinical sample analysis can be made.

V. Results and Discussion

A. Bivariate Analyses

Figure 1 illustrates cytometric bivariate distributions of colo205 cells in which cellular DNA content is plotted versus activated caspase-3 immunoreactivity in asynchronously growing cells (left panel) and cells exposed to DR5 agonist (conatumumab, 1 μg/mL) for 2 h (right panel). In untreated cell cultures, typically 2–5% of cells have detectable levels of activated caspase-3, which corresponds to the percentage of apoptotic cells in these cultures determined microscopically. Cells incubated with a DR5 agonist have a dramatically increased percentage of cells with activated caspase-3 as determined by immunofluorescence.

Likewise, increased caspase-3 activation can be detected in tumor FNAs prepared from freshly resected human colorectal tumors (Fig. 2). The specificity of detection is confirmed by including a blocking peptide that abrogates caspase-3 signal. Cytometric quantitation is then performed in FNAs by gating DNA content to restrict analysis to single cells. A caspase-3 positive threshold is set based on the blocking peptide control for each sample, such that fewer than 2% of cells are positive in the blocking peptide sample. The integrity of the population is then verified by relocating positive events into an image gallery and visually confirming the morphology of positively stained cells. Apoptotic debris is excluded from sample analysis because the sub-G1 peak biases the cell count and skews the apparent percentage of caspase-positive cells, and therefore should be excluded from the calculation of an apoptotic index (Darzynkiewicz *et al.*, 1997). In summary, using quantitative imaging cytometry, caspase-3 activation can be adequately measured in intact cells, rather than relying on lysates or proteins released into the circulation by dying cells.

B. Multiparametric Analysis

The multiparametric analysis of activated caspase-3 in colo205 cells is shown in Fig. 3. By simultaneously staining cells with three different fluorochromes, one reporting the status of caspase-3 activation, another, DNA content (Ho342, cell cycle position), and still another, expression of EpCAM to identify epithelial cells, all three parameters could be evaluated and correlated with each other, on a cell-by-cell basis. It is possible, therefore, to obtain information regarding the proportion of cells

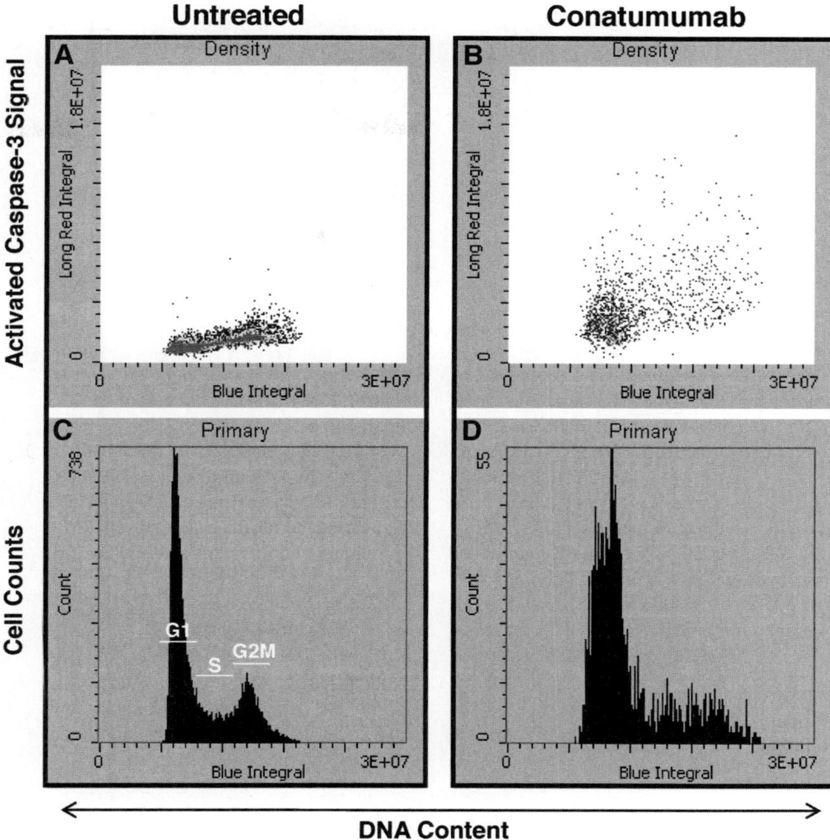

Fig. 1   Untreated (A, C) or conatumumab-treated (B, D) colo205 cells were harvested and stored in 10% neutral buffered formalin. Cells were cytospun onto slides and dual stained with an antibody specific for activated caspase-3 and DAPI. Samples were quantified by LSC for intensity of activated caspase-3 detection (long red integral) and DNA content (blue integral). Data were analyzed in a scatter plot to illustrate activated caspase-3 as a function of DNA content (A, B) or as a histogram to depict cell cycle profiles (C, D). Conatumumab induced caspase-3 activation in all phases of the cell cycle. (For interpretation of the references to color in this figure legend, the reader is referred to the Web version of this chapter.)

expressing EpCAM that are actively undergoing apoptosis. This becomes of particular importance when assaying heterogeneous samples (i.e., tissue sections or tumor biopsies like FNAs), where only sensitive cells will respond to treatment. Selective separation and identification of different cell populations for analysis becomes critical to deconvolute the sample complexity. Figure 3A shows an image field example of human peripheral blood mononuclear cells (PBMCs) with spiked colo205 cells as evaluated using LSC. Finally, imaging cytometry allows confirmation of the typical morphological changes characteristic of apoptosis, which define this mode of cell death and are still considered "the gold standard" (Darzynkiewicz *et al.*, 2004),

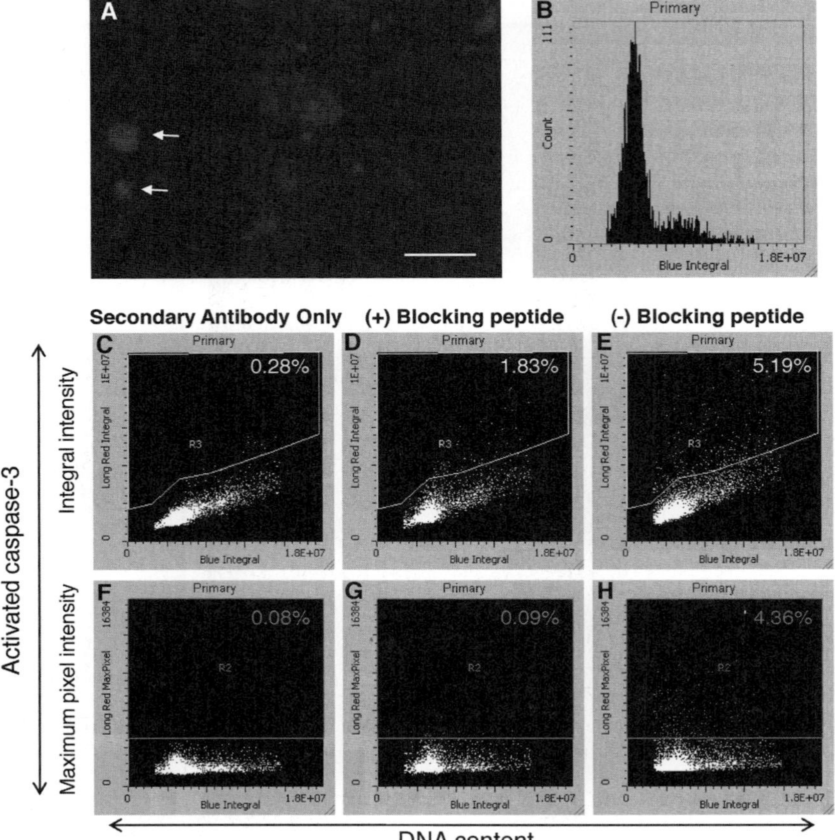

Fig. 2 (A) A representative 40× magnification field image of a mock FNA obtained from a freshly resected human tumor was stained for activated caspase-3 (red) and DNA (blue). FNAs contained a detectable baseline level of cells that contain activated caspase-3 even in the absence of DR5 activation (white arrows); scale bar = 50 μm. (B) DAPI staining enabled DNA content analysis (blue integral) of the single contoured cells in the cytospot. (C–H) Bivariate analysis was then used to analyze staining for activated caspase-3 as a function of DNA content. This analysis uses gating to enumerate the percentage of cells containing activated caspase-3 based on the intensity of staining (integral intensity, C–E) or the maximum pixel intensity in a contoured event (G–H). The gating strategy is based on sample controls that included staining the sample with secondary antibody alone (C, F) and the inclusion of a specific peptide that blocks immunodetection (D, G). (For interpretation of the references to color in this figure legend, the reader is referred to the Web version of this chapter.)

together with the cytometric quantitation (Fig 3B). Interestingly, as a cell advances through cell death, proteases become more abundant, the cell body is fragmented into apoptotic debris, and the intensity of EpCAM becomes diminished (Fig 3C). These data suggest that reliance on EpCAM staining, while necessary to deconvolute

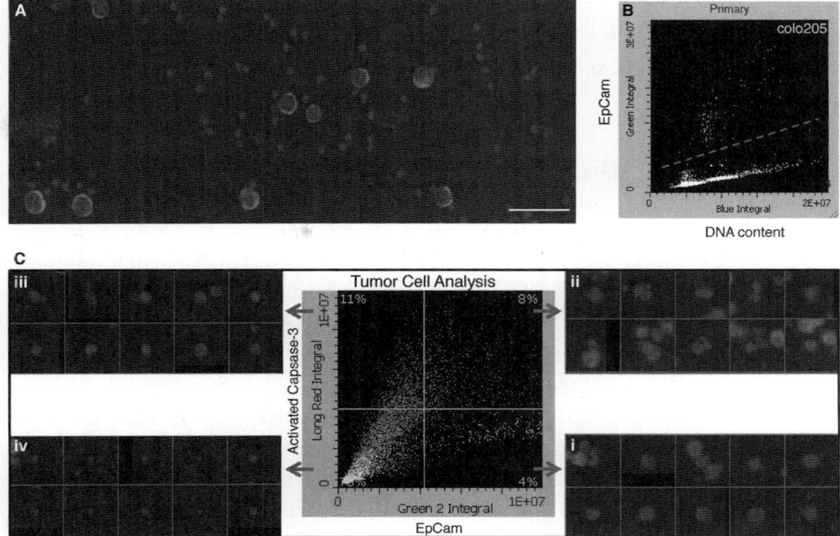

Fig. 3 (A) Colo205 cells were spiked into PBMCs, spun onto slides and stained with anti-EpCAM (green) and DAPI (blue). A representative 40× magnification field image indicates discretely EpCAM positive cells; scale bar = 50 μm. (B) Colo205 cells were readily distinguished from PBMCs based on their EpCAM intensity and higher DNA content. Cells above the dashed red line were verified as colo205 cells by image relocation. (C) Unspiked colo205 cells undergoing apoptosis after conatumumab treatment were stained with anti-EpCAM (green), anti-activated caspase-3 (red), and DAPI (blue) and analyzed by LSC. Image relocation analysis of EpCAM intensity (green integral) as a function of activated caspase-3 intensity (long red integral) indicated that not only viable cells have a distinct phenotype (EpCAM[+] Caspase3[−] shown in panel i) but also dying cells can be observed at various postmortem stages including early death (EpCAM[+] Caspase3[+] shown in panel ii), late death (EpCAM[−] Caspase3[+] shown in panel iii), and apoptotic debris ((EpCAM[−] Caspase3[−] shown in panel iv). (See plate no. 19 in the color plate section.)

heterogeneous biopsy samples, may underreport overall cell death since excluded cells (including late apoptotic cells) could have lost EpCAM expression.

C. Comparison of Caspase-3 Activation in FNAs with Other Methods

The feasibility of using FNAs to evaluate overall pharmacodynamic impact could be assessed in mice by comparing caspase-3 activation in FNAs to xenograft sections and serum caspase-3/7 activity as published (Zoog *et al.*, 2010). Early apoptotic activity (24 h after dosing with DR5 agonistic antibody), as measured by the percentage of activated caspase-3 positive cells in FNAs, correlates significantly both with the amount of activated caspase-3 measured within a tumor section (tumor load) and with serum caspase-3/7 activity. In addition, activated caspase-3 levels can also be compared to the appearance of a downstream product of caspase-3, such as the cleavage product of cytokeratin, M30 (Bivén *et al.*, 2003; Kaplan-Lefko *et al.*, 2010).

VI. Biological Information and Future Directions

Slide-based cytometry of immunostained FNAs is emerging as an application to help clinical scientists better understand the pharmacodynamic activity of antitumor drugs (Assersohn *et al.*, 2003; Gerstner *et al.*, 2009; Nizzoli *et al.*, 2003; Schwock *et al.*, 2007; Verstovsek *et al.*, 2002; Yoder *et al.*, 2004; Zabaglo *et al.*, 2000; Zimmerman *et al.*, 2005). Novel cytometric imaging assays utilizing technologies like LSC, similar to the one described here, allow the detection and quantitation of apoptosis *in situ* by immunocytochemically measuring the activated caspase-3 in the tumor tissue. Moreover, these methods can be modeled in animals in order to closely mimic clinical samples for assay development. A future improvement of this pre-clinical assay will include longitudinal FNA sampling in animals instead of terminal studies. Two key highlights of such approaches include the detection of target activity in small amounts of tumor tissue, rather than in peripheral surrogate compartments (e.g. blood or serum), and early evidence of drug activity prior to signs of tumor regression – both desirable characteristics of biomarkers.

Obviously, the number of serial or longitudinal biopsies that can be procured from a single patient are very limited, in contrast to serum-based assays where repeated samples are easy to acquire. Consequently, it is important to consider that an FNA will provide a static view of a dynamic process, where ongoing clearance of apoptotic cells from the tumor *in vivo* will likely confound absolute quantitation. Additionally, tumor cells can begin to lose their EpCAM signal as they become fragmented into apoptotic debris, restricting the utility of the assay to the evaluation of early cell death. Furthermore, to assess molecular activity of apoptosis-inducing drugs in humans, intrapatient FNAs should be compared. To help compensate for these variables, sophisticated imaging techniques such as endoscopic ultrasound-guided or mammography-guided FNAs may be required to reduce sampling bias (Kulesza and Eltoum, 2007; Stomper *et al.*, 2000;). Further improvements on FNA methodology are also currently being explored to help ensure high-quality sample procurement.

The method presented in this chapter takes full advantage of the cell suspension nature of an FNA, by fixing the aspirates in suspension immediately upon collection (2% formaldehyde) and without embedding them into a cell block, hence avoiding future processing and sectioning of the sample together with preserving cell integrity for DNA content analysis, which ultimately allows measurement of caspase-3 activation relative to the DNA profiles on a single-cell basis. Moreover, multiplexing with tumor cell markers, such as EpCAM or cytokeratin (Brotherick *et al.*, 1995), allows enrichment for tumor cells during analysis and therefore partially compensates for discrepancies in sample heterogeneity.

References

Assersohn, L., Salter, J., Powles, T. J., A'Hern, R., Makris, A., Gregory, R. K., Chang, J., Dowsett, M. (2003). Studies of the potential utility of Ki67 as a predictive molecular marker of clinical response in primary breast cancer. *Breast Cancer Res. Treat.* **82**, 113–123.

Ashkenazi, A., Holland, P., and Eckhardt, S. G. (2008). Ligand-based targeting of apoptosis in cancer: the potential of recombinant human apoptosis ligand 2/tumor necrosis factor-related apoptosis-inducing ligand (rhApo2L/TRAIL). *J. Clin. Oncol.* **26**, 3621–3630.

Adams, C., Totpal, K., Lawrence, D., Marsters, S., Pitti, R., Yee, S., Ross, S., Deforge, L., Koeppen, H., Sagolla, M., Compaan, D., Lowman, H., Hymowitz, S., Ashkenazi, A. (2008). Structural and functional analysis of the interaction between the agonistic monoclonal antibody Apomab and the proapoptotic receptor DR5. *Cell Death Differ.* **15**, 751–761.

Bivén, K., Erdal, H., Hägg, M., Ueno, T., Zhou, R., Lynch, M., Rowley, B., Wood, J., Zhang, C., Toi, M., Shoshan, M. C., Linder, S. (2003). A novel assay for discovery and characterization of pro-apoptotic drugs and for monitoring apoptosis in patient sera. *Apoptosis* **8**, 263–268.

Boeddinghaus, I., and Johnson, S. R. (2006). Serial biopsies/fine-needle aspirates and their assessment. *Methods Mol. Med.* **120**, 29–41.

Brotherick, I., Shenton, B. K., and Lennard, T. W. (1995). Are fine-needle breast aspirates representative of the underlying solid tumour? A comparison of receptor levels, ploidy and the influence of cytokeratin gates. *Br. J. Cancer* **72**, 732–737.

Chuntharapai, A., Dodge, K., Grimmer, K., Schroeder, K., Marsters, S. A., Koeppen, H., Ashkenazi, A., Kim, K. J. (2001). Isotype-dependent inhibition of tumor growth in vivo by monoclonal antibodies to death receptor 4. *J. Immunol.* **166**, 4891–4898.

Clevenger, C. V., Epstein, A. L., and Bauer, K. D. (1987). Quantitative analysis of a nuclear antigen in interphase and mitotic cells. *Cytometry* **8**, 280–286.

Darzynkiewicz, Z., Juan, G., Li, X., Gorczyca, W., Murakami, T., Traganos, F. (1997). Cytometry in cell necrobiology: analysis of apoptosis and accidental cell death (necrosis). *Cytometry* **27**, 1–20.

Darzynkiewicz, Z, Huang, X., Okafuji, M., and King, M.A. (2004). Cytometric methods to detect apoptosis. *In* Methods in Cell Biology, Vol. 75, pp. 307–341.

Gastman, B. R. (2001). Apoptosis and its clinical impact. *Head Neck* **23**, 409–425.

Georgakis, G. V., Li, Y., Humphreys, R., Andreeff, M., O'Brien, S., Younes, M., Carbone, A., Albert, V., Younes, A. (2005). Activity of selective fully human agonistic antibodies to the TRAIL death receptors TRAIL-R1 and TRAIL-R2 in primary and cultured lymphoma cells: induction of apoptosis and enhancement of doxorubicin- and bortezomib-induced cell death. *Br. J. Haem.* **130**, 501–510.

Gerstner, A. O., Laffers, W., and Tarnok, A. (2009). Clinical applications of slide-based cytometry – an update. *J. Biophotonics* **2**, 463–469.

Guo, Y., Chen, C., Zheng, Y., Zhang, J., Tao, X., Liu, S., Zheng, D., Liu, Y. (2005). A novel anti-human DR5 monoclonal antibody with tumoricidal activity induces caspase-dependent and caspase-independent cell death. *J. Biol. Chem.* **280**, 41940–41952.

Herbst, R., Kurzrock, R., Hong, D., Valdivieso, M., Hsu, C. -P., Goyal, L., Juan, G., Hwang, Y., Wong, S., Hill, J., Friberg, G., LoRusso, P. (2010). A first-in-human study of conatumumab in adult patients with advanced solid tumors. *Clin. Cancer Res.* **16**, 5883–5891.

Ichikawa, K., Liu, W., Zhao, L., Wang, Z., Liu, D., Ohtsuka, T., Zhang, H., Mountz, J. D., Koopman, W. J., Kimberly, R. P., Zhou, T. (2001). Tumoricidal activity of a novel anti-human DR5 monoclonal antibody without hepatocyte cytotoxicity. *Nat. Med.* **7**, 954–960.

Kaplan-Lefko, P. J., Graves, J. D., Zoog, S. J., Pan, Y., Wall, J., Branstetter, D. G., Moriguchi, J., Coxon, A., Huard, J. N., Xu, R., Peach, M. L., Juan, G., Kaufman, S., Chen, Q., Bianchi, A., Kordich, J. J., Ma, M., Foltz, I. N., Gliniak, B. C. (2010). Conatumumab, a fully human agonist antibody to death receptor 5, induces apoptosis via caspase activation in multiple tumor types. *Cancer Biol. Ther.* **9**, 616–629.

Kulesza, P., and Eltoum, I. A. (2007). Endoscopic ultrasound-guided fine-needle aspiration: sampling, pitfalls, and quality management. *Clin. Gastroenterol. Hepatol.* **5**, 1248–1254.

LeBlanc, H. N., and Ashkenazi, A. (2003). Apo2L/TRAIL and its death and decoy receptors. *Cell Death Differ.* **10**, 66–75.

Motoki, K., Mori, E., Matsumoto, A., Thomas, M., Tomura, T., Humphreys, R., Albert, V., Muto, M., Yoshida, H., Aoki, M., Tamada, T., Kuroki, R., Yoshida, H., Ishida, I., Ware, C. F., Kataoka, S. (2005). Enhanced apoptosis and tumor regression induced by a direct agonist antibody to tumor necrosis factor-related apoptosis-inducing ligand receptor 2. *Clin. Cancer Res.* **11**, 3126–3135.

Nizzoli, R., Bozzetti, C., Naldi, N., Guazzi, A., Gabrielli, M., Michiara, M., Camisa, R., Barilli, A., Cocconi, G. (2000). Comparison of the results of immunocytochemical assays for biologic variables on preoperative fine-needle aspirates and on surgical specimens of primary breast carcinomas. *Cancer* **90**, 61–66.

Moffitt, K. L., Martin, S. L., and Walker, B. (2010). From sentencing to execution – the processes of apoptosis. *J. Pharm. Pharmacol.* **62**, 547–562.

Nizzoli, R., Bozzetti, C., Crafa, P., Naldi, N., Guazzi, A., Di Blasio, B., Camisa, R., Cascinu, S. (2003). Immunocytochemical evaluation of HER-2/neu on fine-needle aspirates from primary breast carcinomas. *Diagn. Cytopathol.* **28**, 142–146.

Pan, G., O'Rourke, K., Chinnaiyan, A. M., Gentz, R., Ebner, R., Ni, J., Dixit, V. M. (1997). The receptor for the cytotoxic ligand TRAIL. *Science* **276**, 111–113.

Pukac, L., Kanakaraj, P., Humphreys, R., Alderson, R., Bloom, M., Sung, C., Riccobene, T., Johnson, R., Fiscella, M., Mahoney, A., Carrell, J., Boyd, E., Yao, X. T., Zhang, L., Zhong, L., Von, K. A., Shepard, L., Vaughan, T., Edwards, B., Dobson, C., Salcedo, T., Albert, V. (2005). HGS-ETR1, a fully human TRAIL-receptor 1 monoclonal antibody, induces cell death in multiple tumour types *in vitro* and *in vivo*. *Br. J. Cancer* **92**, 1430–1441.

Saleh, H., and Masood, S. (1995). Value of ancillary studies in fine-needle aspiration biopsy. *Diagn. Cytopathol.* **13**, 310–315.

Schwock, J., Ho, J. C., Luther, E., Hedley, D. W., and Geddie, W. R. (2007). Measurement of signaling pathway activities in solid tumor fine-needle biopsies by slide-based cytometry. *Diagn. Mol. Pathol.* **16**, 130–140.

Stomper, P. C., Budnick, R. M., and Stewart, C. C. (2000). Use of specimen mammography-guided FNA (fine-needle aspirates) for flow cytometric multiple marker analysis and immunophenotyping in breast cancer. *Cytometry* **42**, 165–173.

Verstovsek, G., Chakraborty, S., Ramzy, I., and Jorgensen, J. L. (2002). Large B-cell lymphomas: fine-needle aspiration plays an important role in initial diagnosis of cases which are falsely negative by flow cytometry. *Diagn. Cytopathol.* **27**, 282–285.

Yoder, M., Zimmerman, R. L., and Bibbo, M. (2004). Two-color immunostaining of liver fine needle aspiration biopsies with CD34 and carcinoembryonic antigen. Potential utilization in the diagnosis of primary hepatocellular carcinoma vs. metastatic tumor. *Anal. Quant. Cytol. Histol.* **26**, 61–64.

Zabaglo, L., Ormerod, M. G., and Dowsett, M. (2000). Measurement of markers for breast cancer in a model system using laser scanning cytometry. *Cytometry* **41**, 166–171.

Zimmerman, R. L., Logani, S., and Baloch, Z. (2005). Evaluation of the CD34 and CD10 immunostains using a two-color staining protocol in liver fine-needle aspiration biopsies. *Diagn. Cytopathol.* **32**, 88–91.

Zoog, S. J., Ma, C., Kaplan-Lefko, P. J., Hawkins, J., Moriguchi, J., Zhou, L., Pan, Y., Hsu, C. -P., Friberg, G., Herbst, R., Hill, J., Juan, G. (2010). Measurement of Conatumumab-induced apoptotic activity in tumors by fine needle aspirate sampling. *Cytometry* **77A**, 849–860.

CHAPTER 13

Automation of the Buccal Micronucleus Cytome Assay Using Laser Scanning Cytometry

Wayne R. Leifert,[*] **Maxime François,**[*,†] **Philip Thomas,**[*] **Ed Luther,**[‡] **Elena Holden**[§] **and Michael Fenech**[*]

[*]CSIRO Food and Nutritional Sciences, Genome Health Nutrigenomics, Adelaide, SA, Australia

[†]Edith Cowan University, Centre of Excellence for Alzheimer's Disease Research and Care, Joondalup, WA, Australia

[‡]Independent LSC Consultant, Wilmington, Massachusetts, USA

[§]CompuCyte Corporation, Westwood, Massachusetts, USA

Abstract

I. Introduction

II. Rationale

III. Methods

    A. Buccal Cell Sampling and Preparation

    B. Buccal Cell Fixation and Staining

    C. Laser Scanning Cytometry

    D. Low-Resolution Scan

    E. High-Resolution Scan

    F. Virtual Channels and Compensation

    G. Segmentation of Events

    H. Identification of Buccal Cell Types

    I. Nucleus and Micronucleus

    J. DNA Content

IV. Summary

References

METHODS IN CELL BIOLOGY, VOL 102

0091-679X/10 $35.00
DOI 10.1016/B978-0-12-374912-3.00013-4

Abstract

Laser scanning cytometry (LSC) can be used to quantify the fluorescence intensity or laser light loss (absorbance) of localized molecular targets within nuclear and cytoplasmic structures of cells while maintaining the morphological features of the examined tissue. It was aimed to develop an automated LSC protocol to study cellular and nuclear anomalies and DNA damage events in human buccal mucosal cells. Since the buccal micronucleus cytome assay has been used to measure biomarkers of DNA damage (micronuclei and/or nuclear buds), cytokinesis defects (binucleated cells), proliferative potential (basal cell frequency), and/or cell death (condensed chromatin, karyorrhexis, and pyknotic and karyolytic cells), the following automated LSC protocol describes scoring criteria for these same parameters using an automated imaging LSC. In this automated LSC assay, cells derived from the buccal mucosa were harvested from the inside of patient's mouths using a small-headed toothbrush. The cells were washed to remove any debris and/or bacteria, and a single-cell suspension prepared and applied to a microscope slide using a cytocentrifuge. Cells were fixed and stained with Feulgen and Light Green stain allowing both chromatic and fluorescent analysis to be undertaken simultaneously with the use of an LSC.

I. Introduction

The buccal mucosa is an easily accessible tissue for sampling cells in a minimally invasive manner and does not cause undue stress to study subjects. Buccal cells can be used to study the regenerative capacity of the buccal mucosa that is dependent on the number and division rate of the proliferating basal cells, their genomic stability, and their propensity for cell death. This approach is increasingly being used in molecular epidemiological studies to investigate the impact of nutrition, lifestyle factors, genotoxin exposure, and genotype on DNA damage and cell death (Thomas *et al.*, 2009). Since the buccal mucosa is of ectodermal origin, defects in buccal mucosa cells may allow it to act as a surrogate tissue to reflect potential physiological changes that occur in other ectoderm-derived tissues such as fibroblasts and nervous tissue. A method utilizing light and fluorescence microscopy has previously been developed to study DNA damage events such as micronucleus frequency in buccal cells adopting a buccal micronucleus cytome approach (Darzynkiewicz *et al.*, 2011; Thomas *et al.*, 2007, 2008). Furthermore, the presence of micronuclei in epithelial cells is of particular interest because micronuclei are one of the best established biomarkers of DNA damage, representing chromosome breakage and mal-segregation events (Fenech and Crott, 2002).

The buccal micronucleus cytome assay has also been used to measure distinct differences between the cytome profiles associated with normal ageing relative to that for premature ageing clinical outcomes such as Down syndrome and

Alzheimer's disease. These studies highlight the potential diagnostic value of the cytome approach for determining genome instability events (Thomas *et al.*, 2007, 2008). Biomarkers that may identify individuals who are at increased risk or are at an early stage of age-related diseases such as Alzheimer's disease would be valuable to the community since it would be possible not only to monitor the progress of the disease but also to determine the effectiveness of potential therapeutic strategies.

II. Rationale

Laser scanning cytometry (LSC) is a new technology that combines the principles of flow cytometry, quantitative imaging, and immunohistochemistry with high-content, multicolor fluorescence analysis, and can be used to identify specific cells in a heterogeneous population as well as scoring unique molecular events within them (Luther *et al.*, 2004; Pozarowski *et al.*, 2006). Additionally, LSC provides intracellular and within-tissue localization of specific protein targets, generating data that offers the advantage of high-throughput analysis without sample loss. The buccal micronucleus cytome assay is well validated by our group using visual scoring by light microscopy; however, applicability on a large scale for appropriate biomonitoring is hampered by lack of automated high-throughput technology. Visual scoring of the buccal cytome and micronuclei can be time consuming and large numbers of cells and/or donors need to be analyzed to obtain statistically relevant data. This is particularly important when scoring micronuclei due to the low baseline frequencies observed (Ceppi *et al.*, 2010).

The LSC protocol developed here could also be adapted to make use of molecular probes for DNA adducts, aneuploidy, chromosome break measures (Ramirez *et al.*, 1999; Schwartz *et al.*, 2003; Van Schooten *et al.*, 2002), DNA double-strand break (e.g., γH2AX (Tanaka *et al.*, 2007, 2009; Zhao *et al.*, 2009), and measures of oxidative damage to DNA (e.g., 8-oxo-dG) within the nuclei of buccal cells (unpublished observations).

III. Methods

The buccal mucosa is a stratified squamous epithelium. The bottom layer of this epithelium contains actively dividing basal cells and basal stem cells, which produce progeny that differentiate and maintain the structural profile and integrity of the buccal mucosa. The time frame for cellular migration from the basal layer to the keratinized surface layer is thought to range from 7 to 21 days; however, there are only limited data investigating migration rates in buccal mucosa (Bjarnason *et al.*, 1999; Squier and Kremer, 2001). The various cell types and nuclear anomalies among the various cell types in the buccal mucosa, which are observed and scored in a buccal micronucleus cytome assay, are shown schematically in Fig. 1.

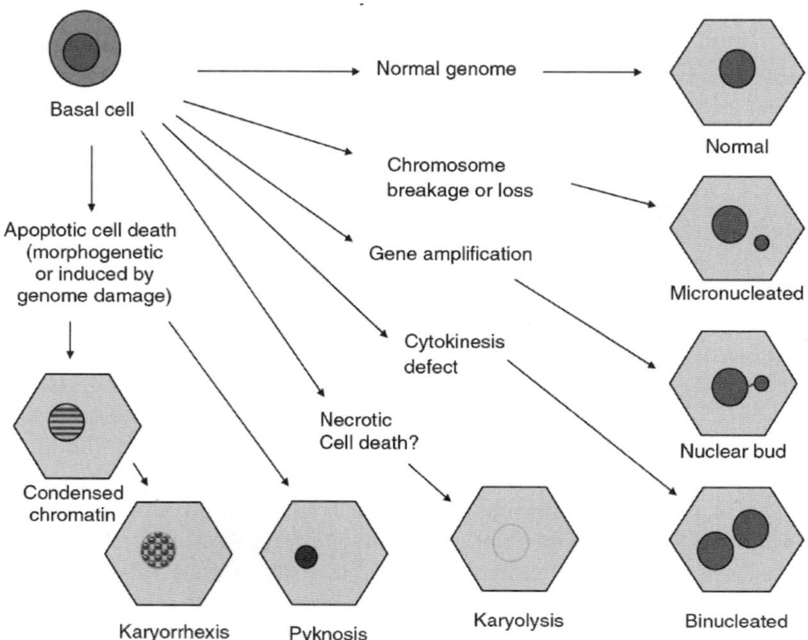

Fig. 1   The various cell types scored in the buccal micronucleus cytome assay (adapted from Thomas *et al.* (2009)).

A. Buccal Cell Sampling and Preparation

Buccal cell isolation and preparation are as previously described by Thomas *et al.* (2009). Human research ethics approval was obtained from CSIRO Food and Nutritional Sciences, Adelaide, South Australia, Adelaide University and Southern Cross University human experimentation ethics Committees. Before sampling, the inside of the mouth was rinsed gently with 30 mL distilled water to remove debris. Buccal cells were sampled using a soft bristle, flat headed toothbrush rotated 20 times against one cheek in a circular motion, and then the toothbrush containing cells was transferred to 30 mL tubes containing "buccal cell buffer" (0.01 M tris (hydroxymethyl)aminomethane, 0.1 M ethylenediaminetetraacetic acid, 0.02 M NaCl, pH 7.0) and cells were dislodged from the toothbrush by agitation of the toothbrush in the buffer. A new toothbrush was used to take the sample from the contralateral cheek, as above and placed in the same buccal cell buffer. The suspension was then centrifuged 10 min at 581*g* at room temperature. Supernatant was discarded and 10 mL of fresh buccal cell buffer was added. Cells were centrifuged twice more and finally resuspended into 5 mL of fresh buccal cell buffer, separated using a syringe with an 18-gauge needle, and then filtered with a 100 μm nylon filter. The cell concentration was determined using a Coulter counter and adjusted to

80,000 cells/mL. Buccal cells were cytocentrifuged (using a Shandon cytocentrifuge) for 5 min at 600 rpm onto microscope slides and air-dried for 10 min.

B. Buccal Cell Fixation and Staining

Cells were fixed in a slide-staining rack containing 50 mL of ethanol:acetic acid mix (3:1) for 10 min at room temperature followed by further air-drying for 10 min at room temperature. Microscope slides containing the fixed cells were immersed for 1 min each in Coplin jars containing 50% (v/v) and then 20% (v/v) ethanol. Cells were washed for 2 min in a Coplin jar containing purified (Milli-Q) water. Slides were placed in a Coplin jar containing 5 M HCl for 30 min and then rinsed in running tap water for 3 min. Slides were drained and placed in a Coplin Jar containing Schiff's reagent for 60 min in the dark at room temperature, and then rinsed for 5 min in tap water and then in Milli-Q water. The cells were counterstained by immersing in Coplin jars containing 0.2% (w/v) Light Green for 30 s and rinsed in Milli-Q water. Slides were then air-dried for at least 45 min before coverslips were applied with DePex mounting medium.

C. Laser Scanning Cytometry

Microscope slides containing fixed/stained buccal cells were inserted into a standard four-slide carrier and analyzed by iCyte® Automated Imaging Cytometer (CompuCyte Corporation, Westwood, MA) with full autofocus function, inverted microsope, three laser excitation (Argon 488 nm, Helium-Neon 633 nm, and Violet 405 nm), four photomultiplier tubes (PMTs) for the quantitation of blue, green, orange, and red fluorescence and dual channel absorption/scatter detector. It is important to select the appropriate excitation lasers and PMT detectors for the analysis of different chromatic or fluorescent probes. In this study, excitation was at 488 and 633 nm, a Long Red emission filter was used for fluorescence, and 488 light loss and 633 light loss photodetectors for absorption were used (Table I). Typically 1000–3000 cells were analyzed using iCyte cytometric analysis software version 3.4.10. The "CompuColor" feature in iCyte was used to provide a green pseudocolor in the cytoplasm (as it is observed when visualized under light microscopy); additionally, nuclei were colored orange.

Table I
Laser and detector selection for buccal cells

Target	Dyes	Excitation lasers	Detectors
Nuclei	Feulgen	488	488 LL (Absorbance) + Long Red (Fluorescence)
Micronuclei	Feulgen	488	488 LL (Absorbance) + Long Red (Fluorescence)
Cytoplasm	Light Green	633	633 LL (Absorbance) + Long Red (Fluorescence)

Abbreviations; LL, light loss (absorption)

D. Low-Resolution Scan

Routinely, a rapid overview scan is initially performed at low resolution using a 20× objective to locate and capture the entire area of the sample (cytospot) that was subsequently analyzed in greater detail. The resolution of an overview scan is low due to large (10 µm) step increments used to acquire the image of the entire sample.

E. High-Resolution Scan

To obtain high-resolution images for analysis, smaller individual (rectangular) scan areas are defined for the high-resolution scan using a 20× objective, outlined in Fig. 2B. In high-resolution scans, small (0.5 µm) laser increment steps are used thus yielding higher resolution detailed "images." It was found that a 20× objective was sufficiently adequate for both the low- and high-resolution scans to analyze the buccal cell cytome.

The user typically defines the size and shape of these regions (shown as user-defined rectangular regions in Fig. 2A,B) and where they will be placed. We routinely place these rectangular regions randomly over the cytospot within the defined low-resolution region being careful not to overlap these scan regions. If there was an obvious artifact present, for example, an air bubble, then this area was excluded from analysis. The size of the high-resolution scan regions was always set to 1500×1110 pixels (or multiples thereof). We have found empirically that this size accommodates the most optimal scanned image size for buccal cells contained within the "field images." Furthermore, by doing this, the LSC will automatically refocus at the start of each scanned 1500×1110 pixel region. This conveniently allows for any refocusing corrections that might be required if a larger single scan region was used. Additionally, if there is a particular reason that a scan region should be excluded from analysis in the main data set, it can easily be excluded later when analyzing data or defining the scattergrams. The other advantage of having multiple analysis scan regions is that each 1500×1110 pixel region represents replicates within the sample being scanned.

In the protocol currently described here, a multipass scan was performed to increase the range of signals that can be detected and optimize image quality. The blue (488 nm) and red (633 nm) excitation lasers are used separately to allow separation of fluorescence from dyes that have similar emission spectra, but different excitation spectra. The two component dyes in this analysis that fall into that category are as follows: Feulgen targets the DNA of cells and fluoresces in the long red region when excited with a 488 nm laser light source, whereas Light Green targets the cytoplasm of cells that also fluoresces in the long red region when excited with the red laser (633 nm). Additionally, the absorbance (light loss) can also be detected using the 488 and 633 light loss photodetectors.

When quantification of the fluorescence signal was required, the photomultiplier voltages should be set so that the brightest pixel value (equivalent to 16,000 units) is

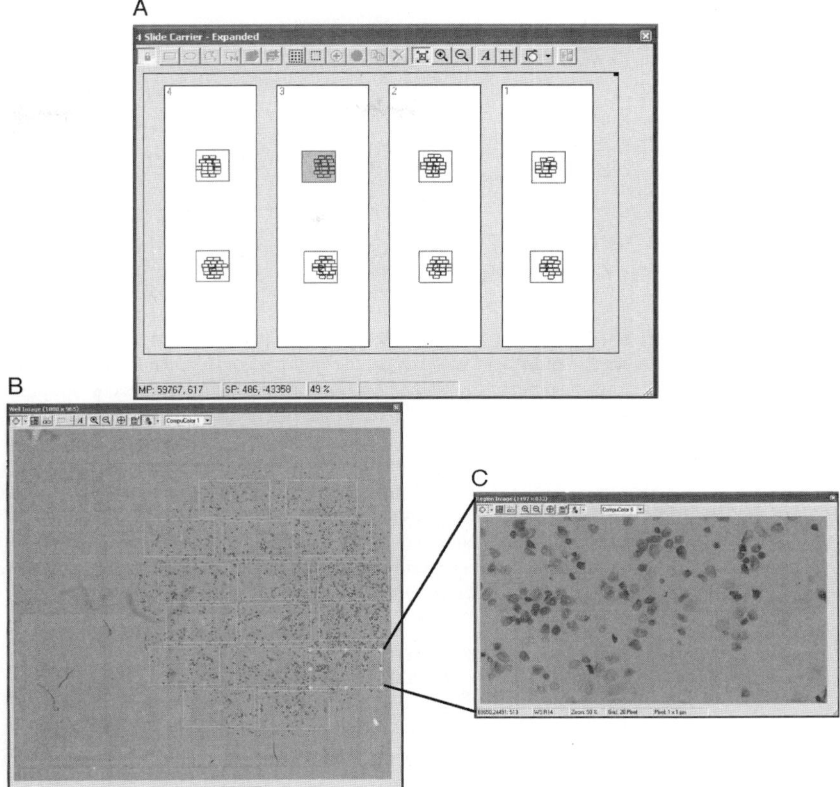

Fig. 2 (A) Shows a diagram of four slides to be analyzed by LSC from right to left. Typically, two cytospots containing buccal cells are prepared on a microscope slide. The large square boxes are the regions in which a "low-resolution" scan of the cytospots are initially performed. This allows the user to define the cytospot region containing the cells. The smaller rectangles within the larger boxes are regions that are scanned at higher resolution for data analysis (see text for full explanation). (B) A typical low-resolution scan "well image" of buccal cells on an entire cytospot also showing the 1500×1110 pixel size rectangular regions to be analyzed at high resolution. (C) An example of a "region image" that consists of a mosaic image showing individual buccal cells stained with Light Green (cytoplasm) and Feulgen (nuclei). (See plate no. 20 in the color plate section.)

just below saturation (e.g., 15,000 units). In the example shown (Fig. 3), this was set to 38 for the blue laser with excitation in the long red channel. The signal intensity can be viewed using the profile feature in the profile window. The "Offset" values (which are used to set the background fluorescence) were set to decrease the background to a pixel value between 200 and 400 units; in this case, for long red (with blue laser excitation), the offset setting was –0.03. By carrying out the above procedure, this will ensure the maximum dynamic range of fluorescence data that can be obtained, hence this will be ideal for the quantification and comparison of data between samples.

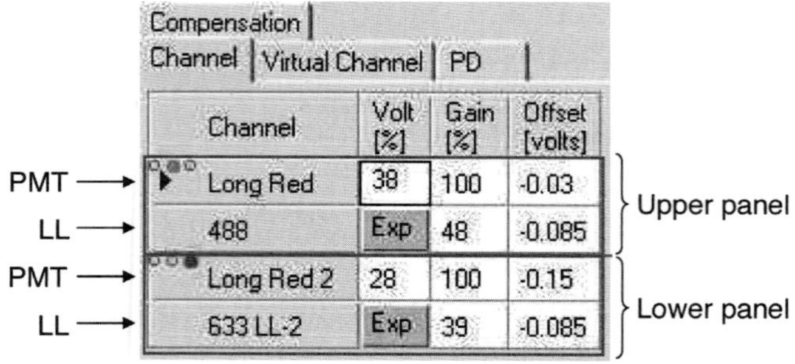

Fig. 3 Setting the channels for excitation and emission: the settings show that in this example the 488 nm excitation laser was used in the first pass of the scan (as indicated in the upper panel), while the text in the upper panel shows the PMT settings and light loss detector settings used for detecting emission, that is, "Long Red" and "488" (which are filters for fluorescence at 647 nm and absorbance at 488 nm, respectively). These settings were used to quantify Feulgen fluorescence and absorbance (light loss) for nuclei and micronuclei. The lower panel shows that the red laser was used to excite the sample on the second pass of the scan. In this instance, the fluorescence emission was at long red ("Long Red 2"), and light loss (absorbance) at 633 (red) was also being recorded ("633 light loss-2"). These settings were used to quantify red fluorescence and absorbance (light loss) of Light Green stain (for cytoplasm). The "volt" and "offset" features are described in the text. (For interpretation of the references to color in this figure legend, the reader is referred to the Web version of this chapter.)

F. Virtual Channels and Compensation

Virtual Channels are used to perform mathematical operations on originally acquired channels to create new "virtual" channels. They are used to increase a weak signal, to add signals together or to isolate the individual fluorescence signals when two or more may overlap in one or more channels (typically termed "compensation"). In our experiment, several virtual channels were created to allow compensation of both absorbance and fluorescence as shown in Table II.

Table II

Creation of virtual channels to compensate for fluorescence and absorbance

Virtual channel	Input channel	Operator	Purpose
		Fluorescence compensation	
LR Fluor M	Long Red	Multiply 0.25	Adjustment factor for Light Green compensation
LR2 Fluor M	Long Red 2	Multiply 0.3	Adjustment factor for Feulgen compensation
Feulgen	Long Red	Subtract LR2 Fluor M	Compensated for Feulgen stain
Light Green	Long Red 2	Subtract LR Fluor M	Compensated for Light Green stain
		Absorbance compensation	
Blue I	488 LL	Invert	Convert from bright field to dark field
Red I	633 LL2	Invert	Convert from bright field to dark field
Blue M	Blue I	Multiply 0.05	Adjustment factor for Light Green compensation
Red M	Red I	Multiply 0.05	Adjustment factor for Feulgen compensation
Blue C	Blue I	Subtract Red M	Compensated for Feulgen stain
Red C	Red I	Subtract Blue M	Compensated for Light Green stain

To properly define and evaluate the compensation settings, it was necessary to monitor the distribution of events in scattergrams using the random segmentation "phantom" feature (Fig. 4). In the scattergram shown in Fig. 4, a region (R35) was drawn around events that fall on areas of the slide where there are no cells and a complementary region (R25) was defined around the events that fall on cells. Events from R35 are excluded and region 25 was used as a gate for further compensation. Both the fluorescence (Fig. 4C) and the absorbance (Fig. 4E) scattergrams of the uncompensated events show a slant in the Y-direction toward the X-direction. In the compensated scattergrams (Fig. 4D,F), that line moves toward a more vertical positioning, indicating that proper compensation has been achieved.

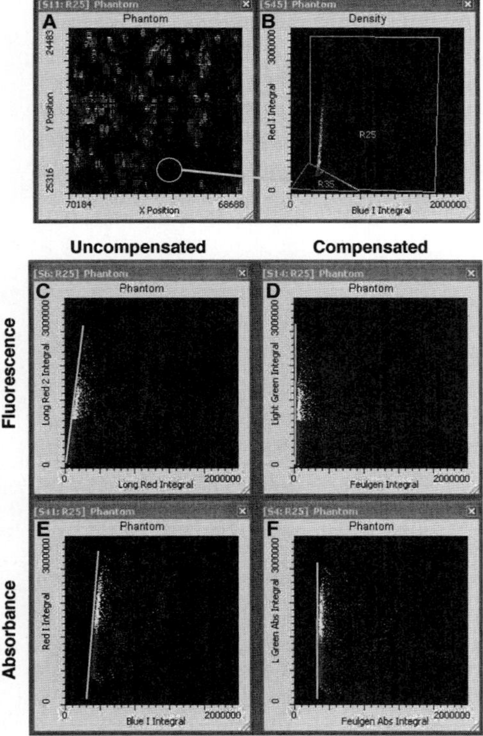

Fig. 4 (A) Phantom contours were generated using the "phantom" feature in iCyte which shows the location of cells (highlighted spots) and where there are no cells (black). This allows the user to define the compensation parameters as described in detail in the text and shown in (B). Uncompensated (C) and compensated (D) fluorescence while uncompensated (E) and compensated (F) absorbance data are shown. The "integral" data was defined as fluorescence per event for the selected channel.

G. Segmentation of Events

In this study, one of the aims was to capture three main buccal cell events that could be analyzed further for scoring and quantification, namely, the cell boundary for identifying and scoring whole cells as well as the nucleus boundary and micronuclei boundary (Fig. 5). Following a high-resolution scan using a $20\times$ objective and using the "protocol" settings as shown in Fig. 6, the clear segmentation lines for cell periphery, nuclei, and micronuclei were generated in iCyte as shown in Fig. 5. The contour lines are automatically drawn around an event such as the cytoplasmic boundary, nucleus, or micronucleus using a user-defined threshold for the pixel values for a particular fluorescent, absorption or virtual channel. Buccal cells are large in diameter and occasionally the cells overlap over two scan fields. In our version of the iCyte software, cells falling on the scan boundaries were excluded from the analysis.

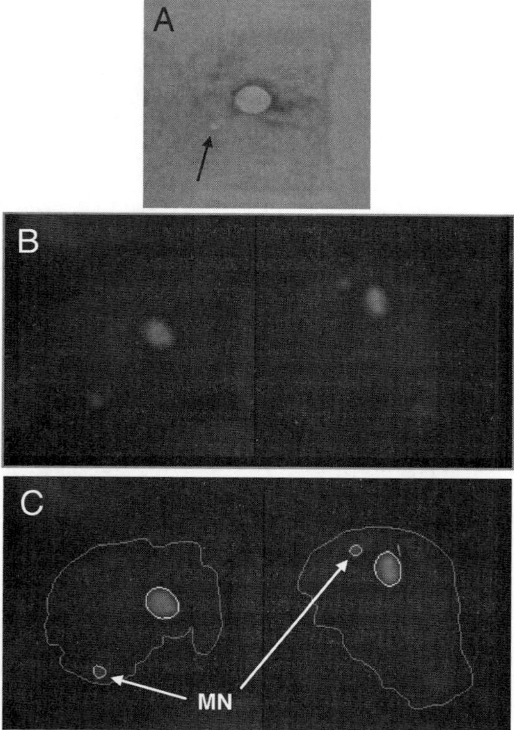

Fig. 5 LSC generated images of buccal cells showing micronuclei. (A) High-resolution image of buccal cells showing a single micronucleus within the cytoplasm. (B) "CompuColor"-generated gallery images of two buccal cells showing distinct micronuclei, and (C) the same cells shown in (B) demonstrating the "segmentation" feature of the iCyte-generated contour lines around the cytoplasmic periphery, nucleus, and micronucleus. (See plate no. 21 in the color plate section.)

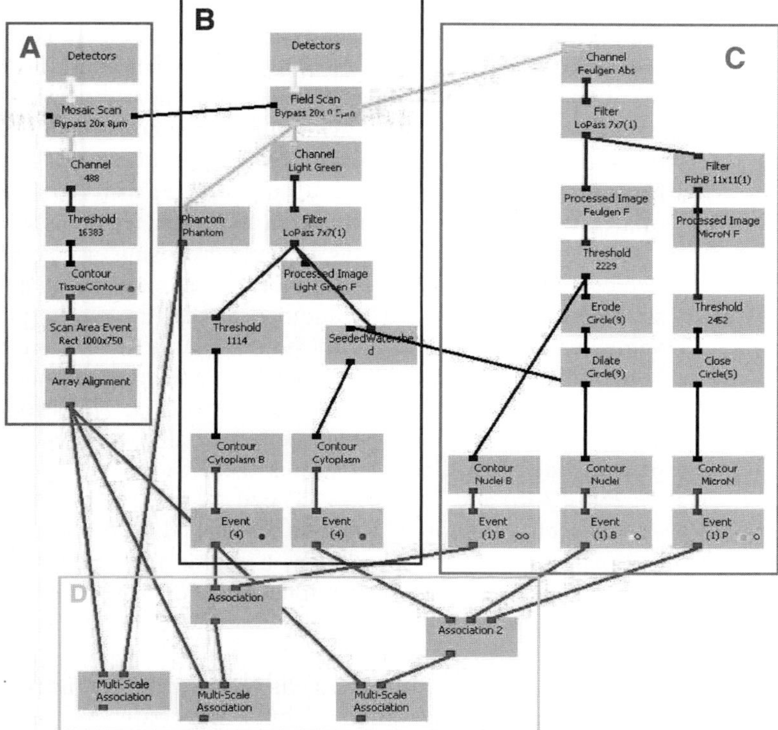

Fig. 6 The iCyte "protocol" was separated into four parts (A, B, C, and D) with all parts being associated together using the "association" module (part D). The first scale (A) provides the settings used for the high-speed overview scan using a 20× objective. The resulting mosaic scan (shown in Fig. 2B) was associated with the second scale settings including parts B and C that contain all the settings for the "high-resolution" scan (as described in text). Part D was the individual component association. All events are associated with each other, which provides a very powerful analysis tool that link parent events to subevents.

H. Identification of Buccal Cell Types

To classify all buccal cells, we generated a scoring system similar to that used previously (Thomas *et al.*, 2009), which consisted of the following cell types: basal, transitional, and differentiated normal viable cells, karyolytic cells (i.e., lacking a nucleus), dead/dying cells ($<2N$), and hyperdiploid cells ($>4N$) using the protocol pathways, as shown in Fig. 6.

Ideally, the iCyte-identified events are defined by a single segmentation of the cytoplasmic periphery. Since buccal cells are occasionally grouped together, it was necessary to use the iCyte algorithm "seeded watershed." This feature divides the groups of cells into individual cells using nuclei as the basis for segmentation. The assumption was that each cell segmented from a group of cells will contain a single nucleus. As a result, however, karyolytic cells, which do not contain nuclei, are eliminated from the segmentation. However, it is rare to see karyolytic cells in groups. To fully score all cells on the slides, the scores obtained from the two segmentation scales, that is, "Contour Cytoplasm B" (which is used only to identify the number of cells without nuclei, i.e., karyolytic) and "Contour Cytoplasm" (which identifies cell types with nuclei) are combined, one without and one with the seeded watershed feature, respectively (also see Fig. 6B). The score of the karyolytic cells obtained from "Contour Cytoplasm B" are then added to the scores of all other cell types obtained from "Contour Cytoplasm."

Cell type segregation was defined by using a scattergram to separate cell aggregates (Fig. 7A) (that could not be separated adequately by the seeded watershed algorithm) followed by another scattergram plotting the Light Green integral value against the circularity of the cytoplasm (a measure of the roundness of the object), where a lower circularity value indicates a higher roundness for the event measured allowing identification of debris (Fig. 7B). From the gate R1, a scattergram was designed to separate cells based on differences in nuclear staining by plotting their DNA content versus the area of the cytoplasm. Figure 7C shows "<2N" (R45) and ">4N" (R13) cells, while regions 46 and 47 were defined as euploid cells. Region 47 was defined as "differentiated cells" due to their large cytoplasmic area. Region 46 was a source for a new scattergram of the area of cytoplasm versus Light Green intensity (Fig. 7D). The following regions are then defined; intensely stained green "basal" cells (R6) and lighter stained "transitional" cells (R4), and are also shown in Fig. 8.

Identification of karyolytic cells (cells without a nucleus) is determined based on the original segmentation (cytoplasm B) pathway (i.e., "no seeded watershed" algorithm applied) shown in Figs. 6B and 9A and B. The percentage of karyolytic cells was obtained with gating region "R23" in Fig. 9B. An example gallery of the cell types scored is shown in Fig. 8. The results of buccal cells obtained from normal healthy "young" (mean age = 22.5 years, $n = 10$) or "old" (mean age = 68.7 years, $n = 10$) volunteers that were scored using the LSC protocol are shown in Fig. 10.

I. Nucleus and Micronucleus

In order to identify and score nuclei and micronuclei (Fig. 6C), various input parameters are used in iCyte such as Feulgen absorption and area of the Feulgen-stained event. The identification of nuclei can be used in conjunction with the cellular segmentation. Both features become associated with the "Cell event," and data obtained from identified nuclei can be correlated to the data obtained from identified cells. The total amount of signal detected (usually the "Integral") in nuclei will define the "DNA content" and hence the ploidy status of that cell. The two modules in Fig. 6C labeled "Contour Nuclei B" and "Contour Nuclei" have identical settings. However, "Contour Nuclei B" was associated with "Contour Cytoplasm B" (in part B); and "Contour Nuclei" events were associated with the

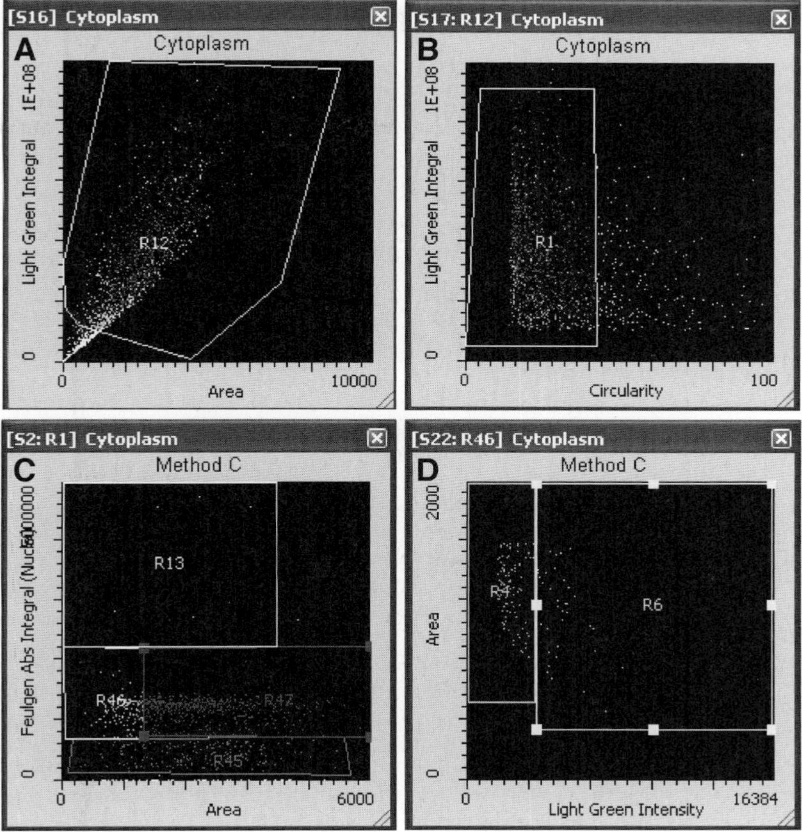

Fig. 7 The identification and scoring of basal, transitional, and differentiated buccal cell types. The scoring of the buccal cell types was achieved by the following criteria: (A) excluding "events" that are either too small or too large to be a buccal cell, (B) have a high "cicrcularity" feature (i.e., are not round in shape), (C) cells that have abnormally high or low nuclear content (i.e., >4N or <2N as shown in R13 and R45, respectively), and euploid cells shown in R46 and R47, with R47 containing the differentiated cells, and (D) was the final stage of cellular classification of basal (R6) and transitional (R4) cells (obtained from scattergram region R46 in (C)). Karyolytic cells are not scored in this set of gating procedures; however, Fig. 9A,B demonstrates the scoring procedure for karyolytic cells.

"Contour Cytoplasm" segmentation in part B. It was necessary to create these two linked events to allow association of nuclei detection with each cell segmentation pathway. The "micronuclei segmentation" was based on the nuclei segmentation; however, a smaller size (area) restriction was defined, since buccal cell micronuclei are typically 1/16 to 1/3 of the main nucleus size (Thomas *et al.*, 2009). A "FISH B" filter was added to the micronucleus segmentation to enhance the spatial resolution of the images, highlight small spots, and therefore increases detection of micronuclei. A peripheral contour around the micronuclei was

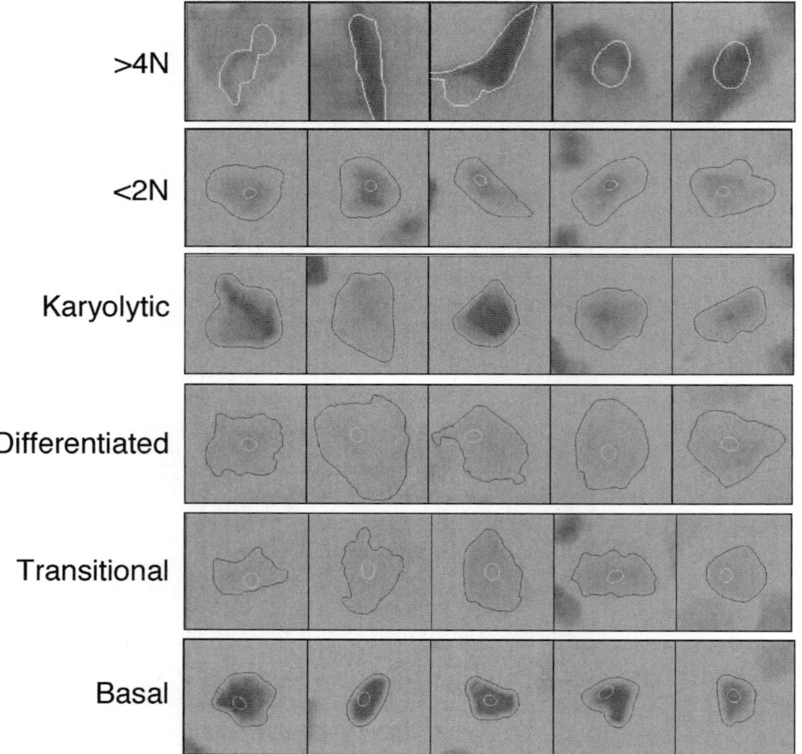

Fig. 8   Gallery images of buccal cells showing the various cell types scored using the automated human buccal cell micronucleus cytome assay by LSC.

routinely applied to segregate the micronuclei that are located within a cell from those that are not (by quantifying the Light Green intensity of the peripheral contour), as shown in Fig. 9. Using these approaches, it was possible to accurately score micronuclei in human buccal cells using LSC, and indeed there was a significantly ($P < 0.001$) higher score of micronuclei in a Down syndrome cohort compared with age-matched controls (Fig. 11). This result compares favorably with our visual scoring observations on an elevated micronucleus frequency in Down syndrome (Thomas *et al.*, 2008).

J. DNA Content

The total DNA content of the cells was based on the Feulgen Absorbance Integral (Fig. 12). The Feulgen Absorbance Max Pixel, a feature that is closely related to the condensation state of the chromatin, is plotted as a scattergram in Fig. 12A. The total

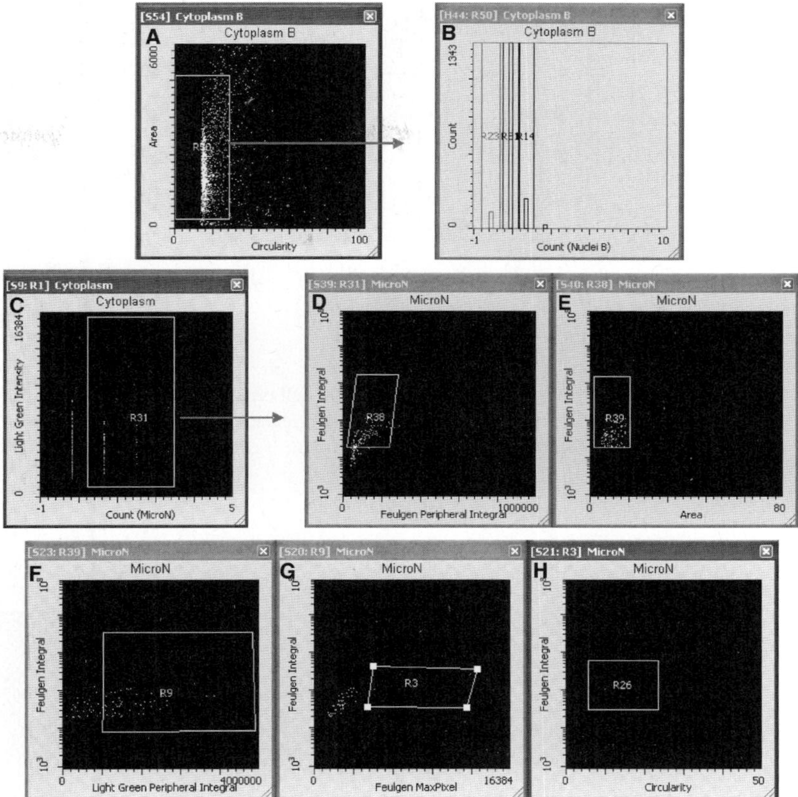

Fig. 9 (A,B) Identification of karyolytic cells (cells without a nucleus) is determined based on the segmentation (cytoplasm B) pathway (i.e., "no seeded watershed" algorithm applied). The percentage of karyolytic cells was obtained with gating region "R23" in panel B. The micronuclei segmentation pathway yields many events that are not micronuclei, and the process of filtering through the events to define true micronuclei entails several steps. The micronuclei identification starts with the same two scattergrams as for cell differentiation status (Fig. 7A,B) and a further scattergram is gated on region R1 (of Fig. 7B). Following this step, all cells that contain a nucleus and with potential micronuclei are identified from the region 31 (C) and several criteria are applied to the micronuclei; an initial gate (R38 in D) was defined to eliminate candidate events that have Feulgen staining surrounding them; this precludes counting bright spots in nuclei. From region 38, another gate (R39 in E) restricts candidates to those with a predetermined area. Then a scattergram showing the Light Green stain (fluorescence) peripheral integral value of the micronuclei versus the Feulgen integral (DNA ploidy) of the micronuclei (F) was used to differentiate candidate micronuclei with no Light Green staining around them and those with (green) cytoplasm surrounding them (R9). Candidates not having the proper intensity of the Feulgen staining are excluded by plotting the Feulgen MaxPixel value of micronuclei and defining a gate (R3 in G). The final step in the micronuclei process was to use a morphology based "circularity feature" to eliminate very irregular candidates from the scoring (H). The circularity feature was plotted against the Feulgen integral (DNA content) of the cells. Lower circularity values translate to round objects. The region was defined around low-circularity objects (R26). Micronuclei detected in region 26 can further be associated to their cell type. (For interpretation of the references to color in this figure legend, the reader is referred to the Web version of this chapter.)

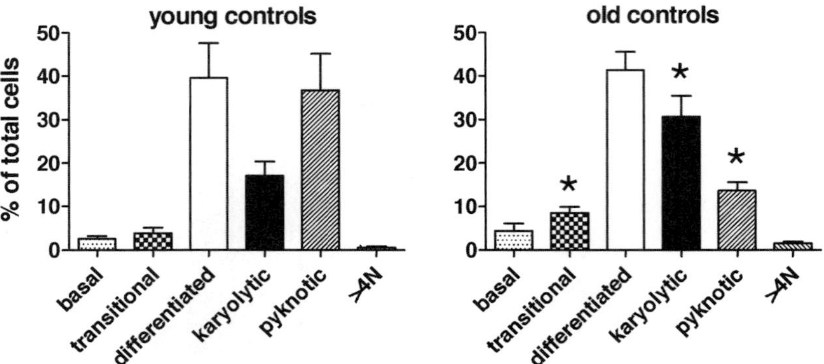

Fig. 10 Classification of the buccal cell types as measured by LSC from "young" (mean age = 22.5 years, $n = 10$) or "old" (mean age = 68.7 years, $n = 10$) healthy volunteers. Data shown are mean ± SEM. [*]$P < 0.05$ compared with young controls.

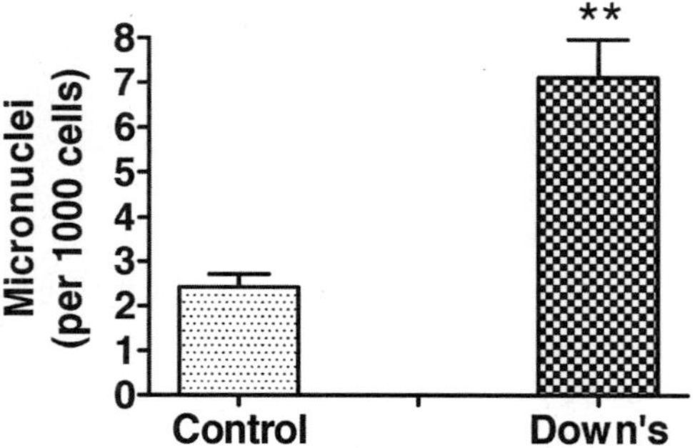

Fig. 11 Frequency of micronuclei in human buccal cells. Human buccal cells on microscope slides were scanned by LSC using the features described in the text. Micronuclei were identified and scored in a Down syndrome cohort ($n = 10$) and an age-matched control group ($n = 10$). Data shown are mean ± SEM, [**]$P < 0.001$.

DNA content (Feulgen Absorbance Integral) is plotted as a histogram in Fig. 12B. Several regions were gated defining different nuclei states; <2N, 2N, 2N–4N, 4N, and >4N. Additionally, we split the Feulgen Absorbance MaxPixel into two groups, that is, Low MaxPixel and High MaxPixel, where "MaxPixel" is the brightest pixel value per event; in this case, it is the brightest pixel value scored within the nuclei. This allowed us to differentiate subtle changes in DNA content of buccal cells (particularly for other studies where we compared buccal cell DNA content of individuals with Down syndrome, which is characterized by trisomy 21, as well as other age-related diseases such as Alzheimer's disease).

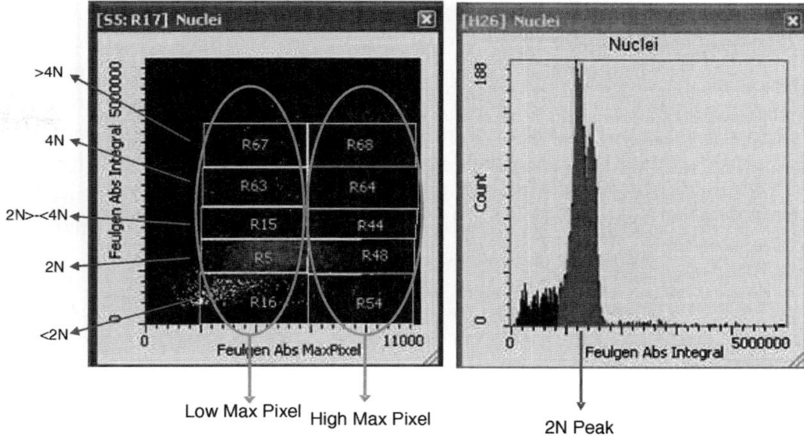

Fig. 12 DNA content of human buccal cells. Scoring criteria were based on Feulgen Absorbance Integral (integrated fluorescence per event) and Feulgen Absorbance MaxPixel (brightest pixel value per event). (Left panel) Nuclear content can be subdivided into two categories "low" and "high" Feulgen Absorbance MaxPixel to allow the detection of subtle changes in DNA staining intensity. (Right panel) A histogram plot of the same data from the left panel showing the delineation of <2N and 2N and the frequency of DNA content "events" scored, with most cells being scored as 2N.

IV. Summary

Our previous studies (Thomas *et al.*, 2009) have shown that the buccal cell micronucleus cytome assay could be used for identifying changes in buccal cell morphology and nuclear parameters in age-related diseases such as Alzheimer's disease and Down syndrome (Thomas *et al.*, 2007, 2008). In this study, a proof-of-principle for an automated LSC approach was used for determining differences in the buccal micronucleus cytome assay. Furthermore, simultaneous scoring of the frequency of buccal cell micronuclei (which are biomarkers for whole chromosome loss and chromosome breakage) was included in the protocol. Indeed, we have also presented preliminary data ($n = 10$ per group) showing an increase in the frequency of micronuclei in a Down syndrome cohort compared with age-matched controls, confirming our previous data using visual scoring techniques.

The automated LSC buccal cell micronucleus cytome assay developed here will be useful for future studies investigating the buccal cell maturation status, cell death, and micronuclei frequency in population-based studies. The nonbiased and automated nature of the protocol will be useful for screening populations of individuals at risk of age-related diseases. Furthermore, this protocol may be extended with standard immunohistochemistry techniques to investigate more specific markers of DNA damage (e.g., histone H2AX/ATM phosphorylation), cell proliferation (Ki67 or cytokeratin(s) expression), and to compare the nutriome, transcriptome, proteome, and cytome status in human buccal epithelial cells that may reveal predictive markers of mild cognitive impairment and Alzheimer's disease risk in asymptomatic individuals. Our aim is to use LSC for "high-content" analysis in single cells as a tool for diagnostic biomarker discovery using a nutrigenomic approach. These biomarkers in combination with the nutriome profile and life-style data may yield valuable information for designing diet and life-style interventions aimed at preventing DNA damage, accelerated ageing, the initiation of mild cognitive impairment, and its progression to Alzheimer's disease.

References

Bjarnason, G. A., Jordan, R. C., and Sothern, R. B. (1999). Circadian variation in the expression of cell-cycle proteins in human oral epithelium. *Am. J. Pathol.* **154**(2), 613–622.

Ceppi, M., Biasotti, B., Fenech, M., and Bonassi, S. (2010). Human population studies with the exfoliated buccal micronucleus assay: statistical and epidemiological issues. *Mutat. Res.* **705**(1), 11–19.

Darzynkiewicz, Z., Smolewski, P., Holden, E., Luther, E., Henriksen, M., François, M., Leifert, W., Fenech, M. (2011). Laser scanning cytometry for automation of the micronucleus assay. *Mutagenesis* **26**(1), 153–161.

Fenech, M., and Crott, J. W. (2002). Micronuclei, nucleoplasmic bridges and nuclear buds induced in folic acid deficient human lymphocytes-evidence for breakage-fusion-bridge cycles in the cytokinesis-block micronucleus assay. *Mutat. Res.* **504**(1–2), 131–136.

Luther, E., Kamentsky, L., Henriksen, M., and Holden, E. (2004). Next-generation laser scanning cytometry. *Methods Cell Biol.* **75**, 185–218.

Pozarowski, P., Holden, E., and Darzynkiewicz, Z. (2006). Laser scanning cytometry: principles and applications. *Methods Mol. Biol.* **319**, 165–192.

Ramirez, M. J., Surralles, J., Galofre, P., Creus, A., and Marcos, R. (1999). FISH Analysis of 1cen-1q12 breakage, chromosome 1 numerical abnormalities and centromeric content of micronuclei in buccal cells from thyroid cancer and hyperthyroidism patients treated with radioactive iodine. *Mutagenesis* **14**(1), 121–127.

Schwartz, J. L., Muscat, J. E., Baker, V., Larios, E., Stephenson, G. D., Guo, W., Xie, T., Gu, X., Chung, F. L. (2003). Oral cytology assessment by flow cytometry of DNA adducts, aneuploidy, proliferation and apoptosis shows differences between smokers and non-smokers. *Oral Oncol.* **39**(8), 842–854.

Squier, C. A., and Kremer, M. J. (2001). Biology of oral mucosa and esophagus. *J. Natl. Cancer. Inst. Monogr.* **29**(29), 7–15.

Tanaka, T., Halicka, D., Traganos, F., and Darzynkiewicz, Z. (2009). Cytometric analysis of DNA damage: phosphorylation of histone H2AX as a marker of DNA double-strand breaks (DSBs). *Methods Mol. Biol.* **523**, 161–168.

Tanaka, T., Huang, X., Halicka, H. D., Zhao, H., Traganos, F., Albino, A. P., Dai, W., Darzynkiewicz, Z. (2007). Cytometry of ATM activation and histone H2AX phosphorylation to estimate extent of DNA damage induced by exogenous agents. *Cytometry A.* **71**(9), 648–661.

Thomas, P., Holland, N., Bolognesi, C., Kirsch-Volders, M., Bonassi, S., Zeiger, E., Knasmueller, S., Fenech, M. (2009). Buccal micronucleus cytome assay. *Nat. Protoc.* **4**(6), 825–837.

Thomas, P., Harvey, S., Gruner, T., and Fenech, M. (2008). The buccal cytome and micronucleus frequency is substantially altered in Down's syndrome and normal ageing compared to young healthy controls. *Mutat. Res.* **638**(1–2), 37–47.

Thomas, P., Hecker, J., Faunt, J., and Fenech, M. (2007). Buccal micronucleus cytome biomarkers may be associated with Alzheimer's disease. *Mutagenesis* **22**(6), 371–379.

Van Schooten, F. J., Besaratinia, A., De Flora, S., D'Agostini, F., Izzotti, A., Camoirano, A., Balm, A. J., Dallinga, J. W., Bast, A., Haenen, G. R., Van't Veer, L., Baas, P., Sakai, H., Van Zandwijk, N. (2002). Effects of oral administration of N-Acetyl-L-cysteine: a multi-biomarker study in smokers. *Cancer Epidemiol. Biomarkers Prev.* **11**(2), 167–175.

Zhao, H., Albino, A. P., Jorgensen, E., Traganos, F., and Darzynkiewicz, Z. (2009). DNA damage response induced by tobacco smoke in normal human bronchial epithelial and A549 pulmonary adenocarcinoma cells assessed by laser scanning cytometry. *Cytometry A.* **75**(10), 840–847.

CHAPTER 14

Laser Scanning Cytometry of Mitosis: State and Stage Analysis

Tammy Stefan and James W. Jacobberger

Case Comprehensive Cancer Center, Case Western Reserve University, Cleveland, Ohio, USA

METHODS IN CELL BIOLOGY, VOL 102

0091-679X/10 $35.00
DOI 10.1016/B978-0-12-374912-3.00014-6

Abstract

Here we use a concept of cell state, which can be defined as the conjunction of expression levels of an arbitrary number of biomolecules or modifications thereof that oscillate, to classify mitotic cells. We describe detection of cell states with quantitative immunofluorescence measurements performed by laser scanning cytometry. This platform allows both measurement of the cell states, capture of cell images within those states, and subsequent analysis of each image to classify by traditional mitotic stages based on nuclear morphology.

I. Introduction

It is not clear that distinctions are made when investigators or scholars refer to phases or stages of a process. Cell division is logically divided into three visually defined periods – interphase, mitosis, and cytokinesis. Mitosis is divided into stages, although each "stage" is termed a phase (prophase, prometaphase, metaphase, anaphase, and telophase). Recently, Pines and Rieder (2001) suggested "restaging" mitosis as a series of transition phases. In this scheme, mitosis would be divided into a series of periods characterized by enzymatic activities that sequentially promote movement through mitosis. As outlined in this chapter, the transition phases parallel and correlate with stages but do not map one-to-one. The advantage to this parsing scheme is that it is independent of structural elements like condensed chromatin, the nuclear membrane, and the metaphase plate, all of which are variably present as mitotic stage landmarks in different cell types.

Starting more than 10 years ago, we began putting together cytometric protocols that included two or more cell-cycle regulated epitopes. Two of these epitopes were on cyclins A2 and B1. Both of these proteins oscillate within the cell cycle, and both are substrates of the anaphase-promoting complex/cyclosome (APC/C), an E3 ubiquitin ligation complex that targets these proteins for proteosomal degradation (Pines, 2006). Cyclin A2 is degraded rapidly after nuclear membrane breakdown, while cyclin B1 is degraded rapidly at the onset of anaphase (Peters, 2006; Pines, 2006). Because these proteins are degraded in mitosis, a bivariate display of a mitotic marker, for example, phospho-S10-histone H3, and either cyclin A2 or B1 for an asynchronous population of mitotic cells, has a characteristic shape (Fig. 1). Further, because each is degraded sequentially, the regions within that data space are enriched sequentially for each succeeding stage of mitosis.

This is where definitions of phases, stages, and states come in. The Oxford English Dictionary defines one meaning of the word "phase" as follows (the *italics* are ours):

2. a. A definite or distinct *state*, *stage*, or *period* in a process of change or development, as the life cycle of an organism; a period marked by a particular characteristic, activity, etc. Also: any particular aspect of a thing of varying aspects; = PHASIS *n.* 2. Now freq. in *phase one* (also *phase two*, etc.): the first (second, etc.) planned stage of a process, project, series of events, etc.

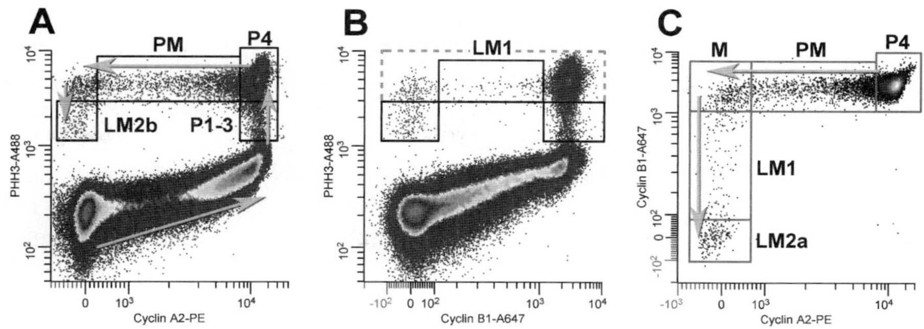

Fig. 1 Exponentially growing Molt4 cells were fixed and stained as described by Jacobberger, Sramkoski, and Stefan, 2011. Dark regions are on unambiguous states. A: P1-3, P4, PM, and LM2b are unamin this plot. B: LM1 is unique to this plot; P1-3 and LM2b are unlabeled. C: displayed data were gated on the high expression cluster of pHH3 (dashed, gray box in B). M is unique to this plot.

Thus, it is not incorrect to equate stages and phases from an English language perspective.

A search of PubMed on "Mitosis [title] AND stages [title]" produces 22 citations, four of which are trivial correlations of the two words, four of which refer to all subdivisions of the cell cycle as stages (Hozak *et al.*, 1986; Leblond and El-Alfy, 1998; Walter *et al.*, 2003; Zatsepina *et al.*, 1988), 13 of which refer to mitotic or mitosis stages (Chakrabarti *et al.*, 2007; D'Amato, 1954; del Campo *et al.*, 1997; Hara *et al.*, 1999; Langendorff *et al.*, 1958; Lopes *et al.*, 2005; Manton *et al.*, 1969, 1970; Rosdahl and Lindstrom, 1980; Santamaria *et al.*, 2007; Soerenby and Lindahl, 1964; Starosvetskaia, 1970; Zosimovskaia and Liapunova, 1966), and one that refers to both mitotic stages and phases within a 36-word title (el-Alfy and Leblond, 1987). A search on "Mitosis [title] AND phases [title]" produces 18 citations, 11 of which refer to mitotic phases (Bozhkova *et al.*, 1978; Delone and Vysotskii, 1964; Dobrokhotov and Valvas, 1981; Gross, 1975; Grosset and Odartchenko, 1975; Hong *et al.*, 2008; Kuz'mina, 1972; Siracky *et al.*, 1984; Weicker, 1954; Zak, 1970; Zak and Cerny, 1966). Of note, eight of those references are from non-English journals. The dates of relevant articles stretch from 1954 to 2008 without a historical bias for both searches. Therefore, we conclude that in native English bioscience language, "stage" is the preferred term, but overall, either of the terms would be technically correct. On the other hand, a search on "Mitosis [title] AND (state [title] OR states [title])" produces 11 citations. One is trivial; two refer to phases or stages (kinetically defined); two refer to the condition of Cdk1; one refers to a cytogenetic state; three refer to either DNA or chromatin states; and two refer to cell states (physiological and mitosis-like states). Clearly, here, states do not fit the definition given above. Rather, "state" refers to the biochemical configuration of a dynamic system.

Given the foregoing discussion, here we refer to cell-cycle phases (G_0, G_1, S, G_2, and M) as periods of variable length, defined by enzymatic synthesis of DNA, any mitotic marker other than morphology, and cell kinetics. We refer to mitotic stages as periods in which structural processes take place, defined visually. Finally, cell states are the biochemical states of the cell, characterized by enough biochemical markers (proteins, epitopes, enzymes, substrates, etc.) such that the state identifies that cell and similar cells as distinctly differentiated from immediately preceding and succeeding cell states. This leads to a somewhat unique view of the cell cycle as a finite-state machine. In this chapter, we will explore analysis of mitosis by measuring phospho-S10-histone H3, cyclin A2, cyclin B1, and DNA content. Quantification of each of these for each cell in an asynchronous cycling population provides cell-state information and the cell cycle can be viewed as a series of unidirectional states. Quantification by laser scanning cytometry (LSC) provides the ability to additionally analyze mitotic stages and determine the correlation between stages and states.

The purpose of this chapter is to present and discuss mitotic analysis by LSC. The single commercially available LSC is an automated microscope developed with imperatives and experience gained by developing flow cytometers (Kamentsky, 2001; Kamentsky and Kamentsky, 1991). Chief among the imperatives are that the output is quantitative and precise. To be specific, this means that two particles, one containing $1x$ of something and one containing $2x$ of the same thing, will be measured as $1 + E_1$ and $2 + E_2$ with E equal to the error that largely comes from the particles themselves (inherent variation) and sample preparation (largely, variation due to probing indirectly). This means that the instrument adds very little noise to the measurement. For example, it is not unusual to measure cycling mammalian nuclei in the G_1 and G_2 phases of the cell cycle by flow cytometry with precision equal to or less than three parts in 100, that is, $G_1 = 1.00$ and $G_2 = 0.97$, with coefficients of variation (CV) of 1–3% for lymphocyte nuclei. Nearly this same level of precision is obtained with current LSCs from Compucyte. The G_2/G_1 ratio of the example data (endothelial cells) in this chapter is 1.99 and the CV is 6.6%.

Most applications that can be done by flow cytometry can be done by LSC. However, there are some applications that require a fixed-substrate approach like LSC. These are applications in which the architecture of the sample needs to remain intact, for example, substrate-dependent cells fixed *in situ* or tissue sections. LSC makes sense for small samples where loss of cells due to the geometry of a flow cytometer is unacceptable, for example (Schwock *et al.*, 2007); when subcellular location is important as a function of something else, for example, movement of NFκB to the nucleus after activation (Deptala *et al.*, 1998; Mercie *et al.*, 2000); and when the image is important for additional analysis by visual inspection. In our case, we needed to account for and evaluate the images of all the measureable mitotic cells within a specific data space. We could not easily do this by conventional flow cytometry and cell sorting.

Widely available LSC is relatively new. Commercialization of the first instrument from CompuCyte occurred in the 1990s and availability of a second-generation instrument came in 2004. A search of PubMed for "'laser scanning cytometry' AND mitosis" returned 23 hits. Four papers labeled mitotic cells and used the

LSC images for additional analysis (Huang *et al.*, 2006; Jacobberger *et al.*, 2008; Kawamura *et al.*, 2006; Mak and Kultz, 2004). Two of those performed some form of mitotic stage analysis (Jacobberger *et al.*, 2008; Mak and Kultz, 2004). Six papers used the CompuCyte software "MaxPixel" feature, which is related to chromatin condensation under normal circumstances (Akimitsu *et al.*, 2003; Gorczyca *et al.*, 2001; Kawamura *et al.*, 2004; Luther and Kamentsky, 1996; Ohshima, 2008; Ohshima and Seyama, 2010), and two of those used the images for analysis (Luther and Kamentsky, 1996; Ohshima, 2008). Two papers described labeled mitotic cells but did not use the images (Kammerer *et al.*, 2009; Maruvada *et al.*, 2004). Eight papers may have used LSC but did not use it for mitotic analysis (Dodson *et al.*, 2004; Fortunato *et al.*, 2002; Guarguaglini *et al.*, 2005; Haider *et al.*, 2003; Kurose *et al.*, 2001; Pachmann *et al.*, 2001; Schwartz *et al.*, 2003; Yamamoto *et al.*, 2006). One paper was part of this series and about detection of mitotic cells (Juan and Darzynkiewicz, 2004). Finally, one paper was in a Japanese journal and we did not analyze its contents. Thus, this is not an overworked area, and while the technical aspect (fixing and staining cells) has been with us for some time, the results presented here are relatively new.

In this chapter, we will demonstrate and discuss measuring with LSC M phase and mitotic cell states based on DNA, phospho-S10-histone H3, and either cyclin A2 or cyclin B1 expression, then analyzing the LSC images of each cell within each state to determine mitotic stage.

II. Background

A. Mitotic Markers

Basic cytometric cell-cycle analysis based on DNA content measurements is common and has been reviewed often. For a historical perspective, see Darzynkiewicz *et al.* (2004). Antibody-based mitotic markers have only recently become popular. Using them depends on the ability to fix and stain cells for intracellular antigens. For a thorough discussion of intracellular antigen staining, see the references in Jacobberger *et al.* (2011); and for a flow cytometry counterpart to this chapter, consult the same reference (Jacobberger *et al.*, 2011). Mitotic markers are abundant. The rates of epitope phosphorylation often increase as cells enter mitosis (Olsen *et al.*, 2010). Much of this is regulatory. For example, histone H3 is highly phosphorylated on T3, S10, T11, and S28. Phosphorylation at T3 affects localization of Aurora B at centromeres (Chen *et al.*, 2010) and phospho-S10 and -S28 are involved in chromatin condensation. For a review of histone H3 phosphorylation, see Perez-Cadahia *et al.* (2009). Because of this, many useful antibodies for detecting mitotic cells react with specific phosphorylated sites. For a quick review and pointers to papers on mitotic markers used in cytometry, see Darzynkiewicz (2008).

B. Analytical Logic and History

The simplest cell-cycle analysis beyond DNA content is DNA coupled with a single marker detected with antibodies. If that marker is a mitotic marker and the cells under study replicate "normally," we can quantify G_1, S, G_2, and M. However, most (if not all) cultured cell lines produce $4C \rightarrow 8C$, $8C \rightarrow 16C,\dots$ $nC \rightarrow 2nC$ subcycles, either through endoreduplication or failure to complete cytokinesis. In that case, a one marker assay is limited to G_1, S, $G_2 + G_{1,4C}$, and M. The remaining subline(s) are analyzable in a similar manner. (Here, we will ignore the desire to analyze the subline(s), since it is repetitious.) If we add cyclin B1 or cyclin A2 to the mix, resolving G2 from $G_{1,4C}$ is solved because $G_{1,4C}$ cells are negative for mitotic markers and negative for cyclin A2 or B1. This was our thinking when we started examining multiparametric protocols with cyclin B1, and mitotic markers (Sramkoski *et al.*, 1999). However, thanks to our friend Vince Shankey, who sent us some PE-labeled anti-cyclin A2, we could put a mitotic marker, cyclin A2, cyclin B1, and DNA content together. We used mouse monoclonal antibodies reactive with cyclin B1 and cyclin A2, conjugated with Alexa Fluor 647 (A647) and R-phycoerythrin (PE), respectively, and with rabbit polyclonal antibodies to phospho-S780-Rb (pRb) or phospho-S10-histone H3 (pHH3) and secondary antibodies labeled with Alexa Fluor 488 (A488) to analyze the complete cell cycle by flow cytometry. One variation was to substitute MPM2-FITC for pHH3 or pRb. For an example of this assay in action, see Fig. 4 in Soni *et al.* (2008). With these four parameters, the cell cycle can be divided into as many as 10 states, with seven of them in mitosis (Fig. 1). See Jacobberger *et al.* (2011) for the flow cytometry protocol and example data. We sorted cells from four of the mitotic regions and convinced ourselves that each state was enriched for mitotic stages. However, sorting was difficult to quantify because cells were lost and those that weren't were widely scattered when sorted onto slides; the sort gates weren't contiguous, and examining the slide by fluorescence after sorting relied on the nuclear DAPI stain. For these reasons, we developed the ability to quantify the states and classify the cell images on the same sample by LSC.

C. Antibody Probes

One of the complications of multiparameter assays is obtaining labeled antibodies or devising an indirect scheme that will be both efficient and of high quality. Quality equals (1) insignificant antibody cross-reactions (specific binding to things that are not targets of the assay); (2) minimized background antibody-binding (nonspecific adsorption), and (3) minimized spectral cross talk. The first two have been reviewed and discussed extensively (Bauer and Jacobberger, 1994; Jacobberger, 2000; Jacobberger, 2001; Jacobberger *et al.*, 2011), but in general, we select antibodies for low cross-reactivity by immunoblot analysis (Jacobberger *et al.*, 1999); we use blocking protein and two or three washes for at least 15 min each to reduce background; we titer our antibodies (see next section), and we choose our probes with the idea of putting bright labels on antibodies to low-abundance antigens and dim labels on high-abundance antigens. (For the meaning of brightness and detail about multicolor experiments in general, see Baumgarth and Roederer, 2000.)

D. Antibody Titers

For optimal signal-to-noise ratio, we titer our antibodies. Titers are first determined conventionally by flow cytometry. We usually choose to operate just at saturation if we can, but for some antibodies that do not appear to saturate, we use the concentration where the signal and the background increase at the same rate. We then re-titer for LSC. The reason for doing this is that we use a higher reaction volume for LSC. For example, in the protocol presented below, we use GNS1-A647 (cyclin B1) at twice the concentration that we use for flow cytometry.

E. Flow Cytometry Protocol Conversion to LSC Protocol

Typical results for Molt4 cells, as shown in Fig. 1, set the "gold standard" for mitotic analysis in our lab. The features we look for are an orthogonal pattern for mitotic cyclin A2 versus cyclin B1; CVs for the cyclin A2/cyclin B1-positive cluster, P4, approaching 20% for both parameters; CVs of approximately 40% for the negative staining cluster, LM2a, and for the cyclins, P4 mean values of 1000–10,000 fluorescence units with the negative, LM2a, set to approximately 0,0. Results by flow cytometry for cells with large cytoplasms (epithelial, fibroblast, etc.) are more variable – broader CVs and less distance between negatives and positives. This is largely due to the increased background antibody-binding in the cytoplasm. The decreased quality of the signals (compared to results like Fig. 1) is compromised further by applying compensation when we use Alexa Fluor 488 (A488) or FITC on pHH3 and R-PE on cyclin A2.

When we implemented our "standard" protocol and reagents for LSC analysis of substrate-dependent cells, we were able to reproduce high-quality data patterns for DNA, pHH3, and cyclin A2, but measurement of cyclin A2-PE was problematic in two areas. First, compensation of pHH3-A488 from cyclin A2-PE broadened CVs that were already broad. Second, PE is very photolabile and scanning the cells with a blue laser produces adequate but weak signals. There is not enough room in this chapter to discuss all the combinations, optimizations, and protocol modifications that we have tried to get around these issues, but as an example, we removed the bright, interfering A488 label from pHH3 and replaced it with Pacific Orange (PacO). pHH3 is a very abundant antigen, and therefore, the pHH3–PacO signal can be detected in combination with DAPI despite spectral overlap, but the resolution of the measurements is compromised by the large offset imposed by spectral overlap, and detection of P1-3 is problematic.

In LSC, to detect cyclin B1 completely, it is necessary to include a cytoplasmic marker. Cyclin B1 expression begins at some point in G_1 and accumulates in the cytoplasm until some point in prophase, at which time it accumulates in the nucleus. Therefore, detecting cyclin B1depends on detecting cyclin B1-related immunofluorescence in the entire cell. The flow cytometry platform works on isolated cells, for which detection of the entire cell is usually coincident with detecting the cell *per se*. The most common approach in LSC is to detect the nucleus first, using the intense signal and well-circumscribed circular contour as the starting point. For refractile cells, light scatter could be used to detect the cytoplasm first, but our cells are flat and spread out, and methanol-fixed cells are essentially transparent and not refractile. The cyclin B1 signal itself can be used if it is scanned multiple times and each scan summed; however, that approach has limited utility, primarily because it will fail for late mitotic cells in which cyclin B1 is very low or undetectable. Therefore, we currently use phosphorylated ribosomal S6 protein (pS6). Of the many cytoplasmic possibilities, we chose this one because we are exploring its use as a marker for cycling cells. Since this phosphoepitope is a response to cytokine signaling, noncycling cells should be positive for this marker briefly, whereas cycling cells should be positive throughout the cell cycle. To detect pS6, we have used a biotin-conjugated rabbit polyclonal antibody and PacO-conjugated streptavidin. This is a very abundant epitope since it is on a ribosomal protein. The choice of this marker and fluorochrome and the aforementioned problems with PE necessitate dividing the assay into two assays: one based on cyclin A2 and one based on cyclin B1. The reasoning is, again, the spillover from blue-excited PacO into the PE channel. There may be a way to recombine the assays (see Section IV.I), but currently we would recommend starting with a two-component assay for any cells other than lymphocytes.

Thus, the current markers in our two-component assay are: (1) cyclin A2-A647, pS6-biotin-streptavidin-PacO, pHH3-A488, and DNA (DAPI); (2) cyclin B1-A647, pS6-biotin-streptavidin-PacO, pHH3-A488, and DNA (DAPI). With this setup, we can detect the states defined in Table I. In this scheme, we do not have the "M" state isolated. Otherwise, the two assays provide the same state information as an assay that would include all markers. (See Section IV for the reasoning behind the state names.) All stages can be analyzed, since that analysis is based on visual inspection of each mitotic image – and stage analysis is the primary purpose for using LSC.

Table I

Mitotic states

	DNA	pHH3	Cyclin A2	Cyclin B1
P1-P3	4C	↑	Max	Max
P4	4C	Max	Max	Max
PM	4C	Max	↓	Max
LM1	4C	Max	Min	↓
LM2a	4C	Max	Min	Min
LM2b	4C	↓	Min	Min

↑ = net accumulation; ↓ = net degradation; Max = expressed high and ∼steady state; Min = absent or lowest levels.

F. Lasers and Data Collection Filters

Laser excitation and emission filters for the CompuCyte iCyte® LSC that we use are standard. We have not tried to optimize at this end. Most likely, we could gain some advantage by optimizing for these fluorochromes; however, like many instruments of this nature, this is a shared instrument and the filters are likely created by the manufacturer to provide the best average signal for many fluorochromes. The normalized spectra, laser lines, and filter bands for our iCyte are shown in Fig. 2.

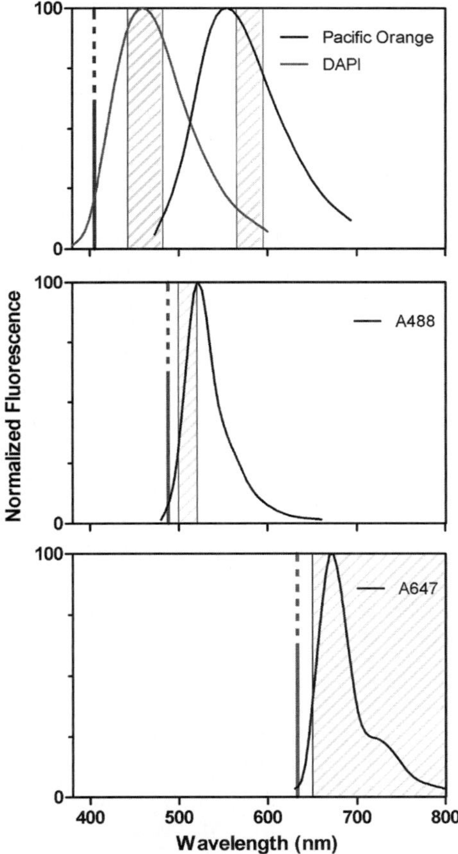

Fig. 2 Laser lines, fluorophore spectra, and filter bands. Normalized spectra were obtained from Invitrogen web site, copies made of the output, and the distances from baseline to key points were measured; the distances were plotted with GraphPad Prism 5.0 after smoothing with a Cubic Spline algorithm (GraphPad Software, San Diego, CA). The purpose of the transformations was to obtain output that could be manipulated for acceptable graphics. Laser line pointers begin as dashed lines.

III. Methods

We present the method by example. We use a specific run and the jargon and specifics of a single instrument. The overall strategy can be retooled for any set of markers or to modify for specific purposes. The concentrations and labels would change, but the protocol is our standard operating procedure for mitotic analysis by LSC. If a home-built instrument or instrument from another manufacturer were used, the jargon and layout might change, but the logic and principles should not.

A. Sample Preparation

iHUVEC cells (hTert-immortalized human umbilical vein endothelial cells) were grown in 35 mm, #1.5/14 mm glass bottom Petri dishes (MatTek Corp., Ashland, MA), with endothelial cell basal medium-2 and SingleQuots® endothelial cell growth medium-2 (Lonza, Walkersville, MD). The dishes (shown in Fig. 3) were either purchased coated with collagen or were coated in the lab with rat tail collagen type I (Invitrogen, Carlsbad, CA) at 10 µg/cm^2 in 0.02 M acetic acid for 1 h. Prior to seeding, the plates were washed twice with sterile PBS, followed by 2 ml of culture medium containing bovine serum albumin (BSA), which helped improved cell distribution prior to attachment. The cells were seeded at 50,000/dish for 3 – 4 days' growth in 5% CO_2. We aim for optimal density and an even distribution after two to three cell cycles. We need the cells to be far enough apart so that they are able to be segmented (contoured), yet dense enough so that we have enough mitotic cells on the plate for a good analysis. Fig. 4 shows a region of a plate, grown as indicated, with what we consider optimal density. The actual parameters of plating and growth will be variable properties of specific cells. The culture medium was replaced with fresh once or twice during the growth period.

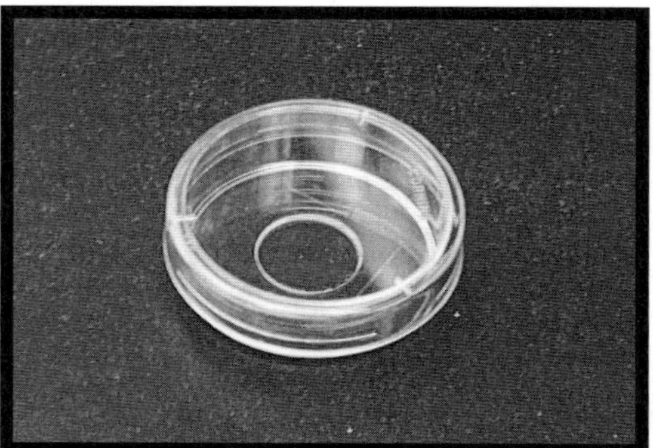

Fig. 3   An image of a 35-mm Petri dish with coverglass insert (microwell), purchased from MatTek Corp. The microwell holds approximately 100 µL.

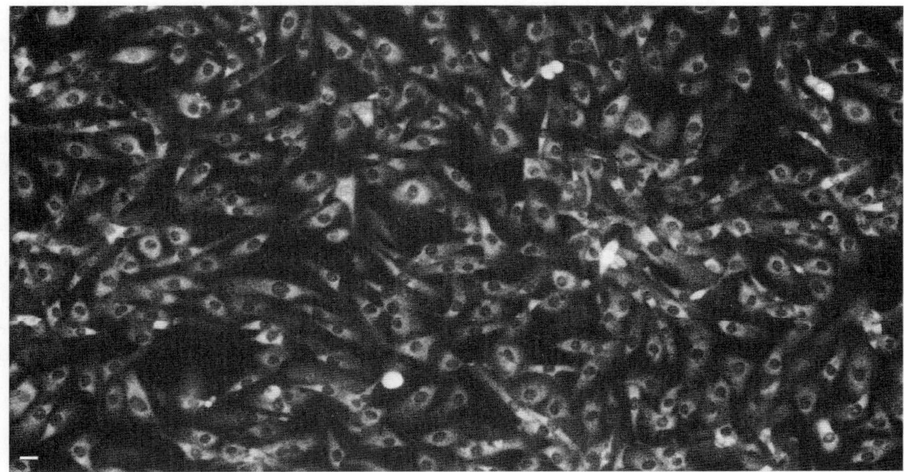

Fig. 4 iHUVEC cells were plated at 50,000/dish, grown for 3 days in 2 mL of growth medium, and fixed and stained for pS6, and cyclin B1-A647 (all colors shown as grayscale). DNA was not stained. The dish was scanned with a 40X objective and 1.0 μm X step. The image is corrected to an aspect ratio of 1:1. Three scan fields are shown, stacked vertically.

B. Fixation, Permeabilization, and Staining

Example cells were fixed in dishes with culture medium by adding formaldehyde (methanol-free, EM grade; Polysciences, Warrington, PA) to a final concentration of 0.13% for 10 min at 37 °C. The dishes were moved to the cold room and media removed. Cells were then rinsed once with PBS and 2 mL 100% methanol at −20 °C were added. All washes and staining were at 4 °C. After methanol fixation, two washes with PBS & one wash with PBS-BSA (10 mM sodium phosphate, 150 mM NaCl, 2% BSA, 0.1% sodium azide) were performed. Nonspecific staining was blocked with 2 ml PBS-BSA for 10–15 min. Antibody cocktails were prepared in 100 μL aliquots for staining within the microwell of the 35 mm dish (∼100 μL). The primary cocktail composition for cyclin A2 analysis was 0.25 μg pS6-biotin (see

Table II

Summary information

Target	Fluor	1° Ab	2°Ab	Excitation (nm)	Emission filter band
DNA	DAPI			405	Blue: 443–483 nm
pS6	PacO	Clone D57.2.2, CST		405	Orange: 565–595 nm
pHH3	A488	Cat #9708, CST	SA	488	Green: 500–521 nm
Cyclin A2	A647	Clone 11B2G3, BCI	GAM, INV	488	Long red: 650 LP
Cyclin B1	A647	Clone GNS1, BD		633	Long red: 650 LP

CST, Cell Signaling Technology, Danvers, MA; BD, BD Biosciences, San Diego, CA; BCI, Beckman Coulter, Incorporated, Miami, FL; INV, Invitrogen; GAM, goat anti-mouse IgG (Fab'2) conjugated to Alexa Fluor 647 (A647), SA: streptavidin conjugated to Pacific Orange

TableII for antibody sources) and 0.31 µg cyclin A2 per 100 µL PBS-BSA. Primary staining time was 1 h followed by two 15-min washes with 2 mL PBS-BSA. The secondary antibody cocktail contained 0.5 µg streptavidin-PacO (SA-PacO), 0.625 µg goat anti-mouse-IgG-A647, and 0.0125 µg anti pHH3-A488 per 100 µL PBS-BSA. Secondary staining was 1 h followed by two 15-min final washes with 2 mL PBS-BSA. To analyze cyclin B1, the primary cocktail contained pS6-biotin. The secondary cocktail contained 0.18 µg anti-cyclin B1-A647, conjugated in our lab with kits from Invitrogen and the same amounts of pHH3-A488 and SA-PacO. DNA was stained with 0.5 µg/mL DAPI in 2 mL PBS containing 1% glycerol and 0.02% *n*-propyl gallate. DAPI solution was added to the plates just prior to scanning. Samples that were scanned for very long periods were scanned with 3 mL DAPI solution to counteract evaporation. During scans, the dish lids are removed to improve scatter images.

Table II lists the summary information for each target in the assay. Filter information is also displayed in Fig. 2.

C. Instrument Setup and Data Acquisition

1. Laser Scanning Cytometer

The samples were scanned on an iCyte® Automated Imaging Cytometer (CompuCyte Corp., Westwood, MA) with iNovator software version 3.4.2.52. The iCyte was equipped with three lasers with excitation at 405, 488, and 633 nm.

2. Filters

Filters are as supplied by the manufacturer. See Fig. 4 and Table II for excitation and emission filter bands used for each dye.

3. Instrument Settings, Data Acquisition, and Segmentation

The settings and logical sequence of processing information comprising data acquisition and segmentation are set using a block-diagram interface, with each block representing an information module, and the connectors representing the relationships between modules (Fig. 5). The setup used here was designed to quantify expression of cyclins A2, B1, and pHH3 in the entire cell or the nucleus of interphase and mitotic cells, using bivariate patterns obtained by flow cytometry as a guide (e.g., Fig. 1). The modifications that are perhaps different from everyday protocols are that (1) we scan each field with each laser independently, in the order blue, violet, then red. This is set up in the "Detector" module of an acquisition protocol; the purpose is to prevent cross talk. (2) We scan for cyclin B1 three to five times, that is, repeated scans of each field. The purpose is to give us the option for averaging the scans (to improve signal-to-noise ratio) and the option to add scans to increase the intensity of low-level signals.

Magnification is set in the Field Scan module (Fig. 5). We use the 60X objective to obtain high-resolution images that increase our chances of classifying mitotic stages

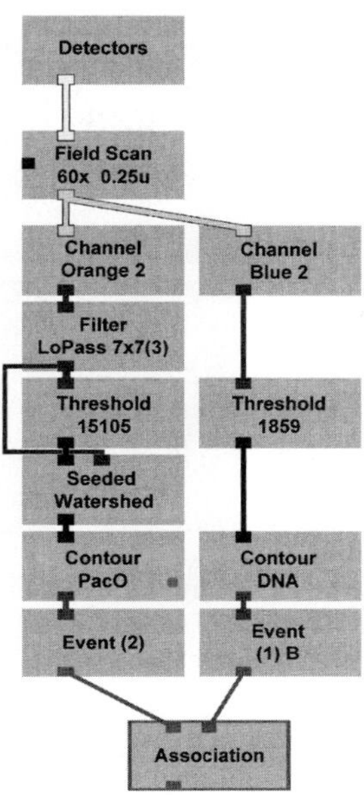

Fig. 5 **Workspace**. The acquisition and analytical workspace as displayed on the iCyte instrument.

correctly. We use a 0.25 μm step size, which produces an image field of 1000 × 768 pixels equal to 250 μm × 119 μm. This is a distorted image that is corrected in software.

Segmentation is the process of identifying events, in this case each cell and the nucleus/nuclei within each cell. This process is determined by several modules in the "workspace" (Fig. 5). To segment the cell, we used a seeded watershed algorithm, executed on the pS6 data. The data source is identified in the "Channel" module (Orange 2, Fig. 5), and the primary data are first processed with a low-pass filter that smoothes the output (LoPass 7 × 7(3), Fig. 5). We set a high threshold on these data (e.g., 15105 in the example output in Fig. 5). This threshold limits detection to the 565–595 nm fluorescence from DAPI and identifies the nucleus. This is used as the seed that is passed to the watershed module. Another threshold is set low within the seeded watershed module so that the algorithm "moves" from the seed to the trough of low-background fluorescence that surrounds each cell. This has the effect of acting on the pS6-specific, or cytoplasmic data. With this approach, we obtain single-cell segmentation on the majority of cells in samples grown at suitable densities. The process then extends to Contour and Event modules in which coding of contour outlines, added distances, and background subtraction features are set. The seeded watershed provides the "primary" segmentation and is supposed to identify cells. A secondary segmentation is performed on the blue fluorescence from DAPI. This is the second column of modules in Fig. 5. This is a much simpler, straightforward single-threshold detection; the module logic is the same. This

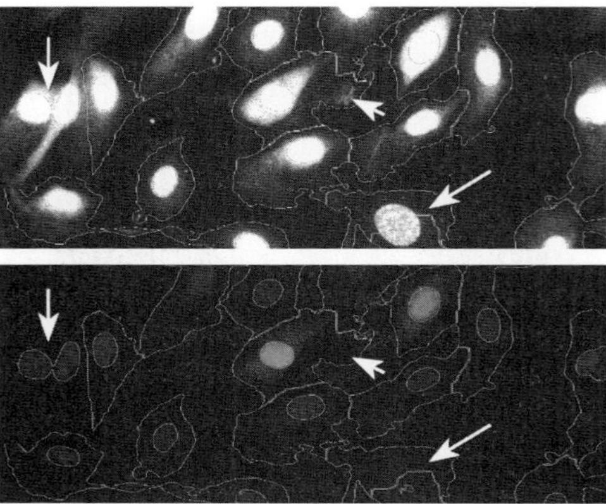

Fig. 6   Illustration of contours. Two scan fields are shown. (Top) pS6 in grayscale. Note the nuclear saturation. (Bottom) cyclin A2 and DAPI fluorescence (two brightest cells are cyclin A2 positive). Nuclear and cytoplasmic contours are thin traces.

feature is subordinate to the primary contour, that is, the software is set up to consider the secondary event as within or contained by the primary event. The signals for pHH3 and cyclin A2 are not processed further except background subtraction (set up in the Contour modules). A segmentation example is shown in Fig. 6.

4. Plate–Specific Settings

We necessarily scan large areas because mitotic cells are infrequent. To minimize out-of-focus issues during the run, we set up the scanning logic as shown and in a software interface as shown in Fig. 7. A number of scan areas are set. The autofocus

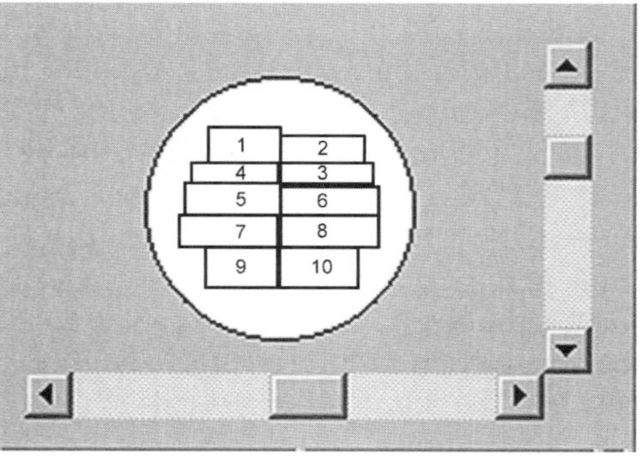

Fig. 7   Scan Areas. Example of the interface for setting multiple scan areas. These are used for data acquisition over large areas to reduce out of focus events. Autofocus readjusts at the start of each new scan field.

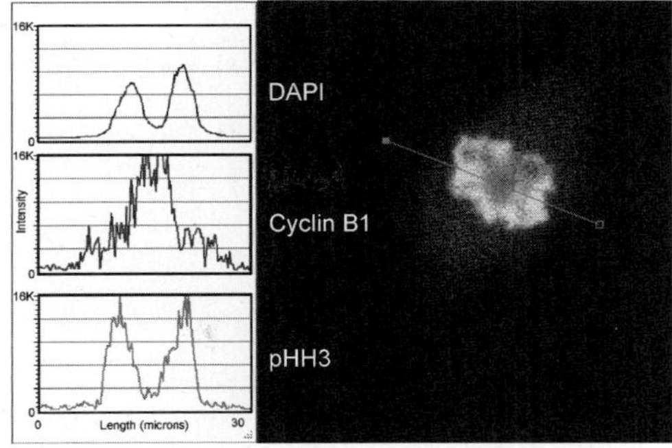

Fig. 8   **Use of the profile tool to set voltages**. The profile tool in the iGeneration software helps set PMT voltages. In the example, the trace lines represent the DNA (top), cyclin B1 (middle), and pHH3 (bottom) signals across the middle of a prometaphase/metaphase cell. A high density of cyclin B1 is evident on the mitotic spindle in the center of the annular chromosomes.

feature is enabled at runtime and serves to refocus the exciting light at the top right of each scan area. We use a "three-point" focus feature that is designed to minimize errors in plate/coverslip distance from the lens.

5. Runtime-Specific Adjustments

Photomultiplier (PMT) voltages are set using the profile tool in the field image window. The highly condensed nuclei of mitotic cells had the brightest signals for all markers, so those cells are singled out for measurement with the profile tool. We aim to set the PMT voltages to just below saturation (Fig. 8).

6. Comment

Because we scan each laser independently, with multiple passes for at least one laser, at high resolution, our scan time per sample is long. To decrease scan time, we use a step size that produces an aspect ratio of 2:1 (0.25 μm rather than 0.10 μm), which requires software adjustment to observe an image with a normal x–y relationship. This adjustment does not negatively impact the resulting images.

D. Data Analysis

1. Preprocessing

There are a series of plots that we examine first. These are used to reduce the event set to those in which we are interested. The first bivariate plot is shown in Fig. 9A. We plot a parameter called "Scan," which is the relative position to the scanline for each field, versus DNA content (Blue 2 Integral). This provides an indication of good laser alignment relative to the bottom of the plate, slide, or coverslip, and whether any

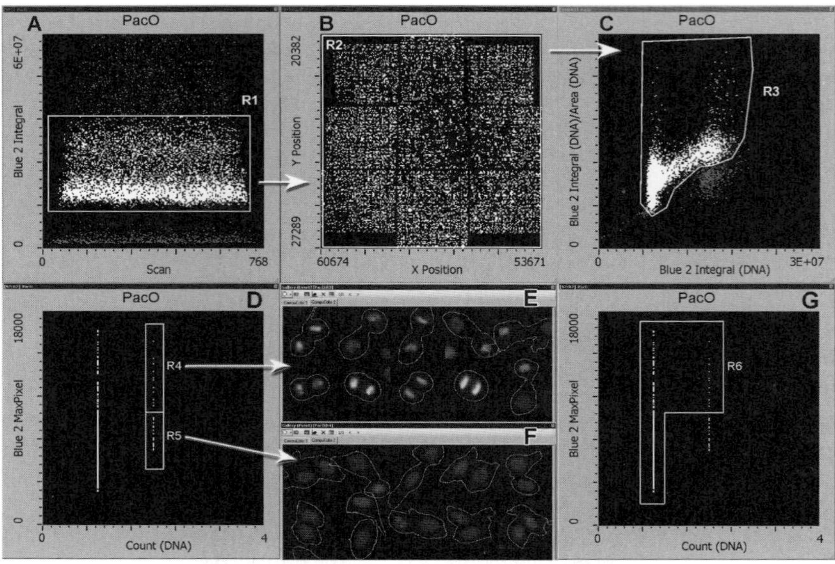

Fig. 9   Data preprocessing. The data are processed from A → G as described in the text. Blue 2 = DAPI florescence channel; count (DNA) = count of the number of subcontours of nuclei within a primary cytoplasmic contour. (B) X and Y position within scan areas. The scan areas depicted by data density are a reflection of the scan areas set prior to running (Fig. 7). (For interpretation of the references to color in this figure legend, the reader is referred to the Web version of this chapter.)

obstruction (like a piece of dust) might have reduced the quality of the measurement. At the analysis stage we cannot correct any problems, but by selective gating, we can remove cells that were not measured well from the analysis. We look at this parameter for each laser (not shown). We also take advantage of the scan plot for the DAPI signal to gate the 2C → 4C population if we are not interested in higher order cells or cell clusters. We next look at an X-position versus Y-position plot (Fig. 9B). This provides a high-level look at cell density for the entire scanned sample. Each scanned area is obvious. From this, we can determine if there are areas of concern vis-à-vis density, or by color-coding (not shown) or gating (shown), we can determine whether gating out cell clusters and higher order cycles appears biased with respect to position on the substrate. Finally, we can scan this plot with a smaller gate and look at a single-parameter histogram for DNA content to determine areas that were not focused well. These areas will generally have a reduced G_1 peak intensity and an increased G_1 CV. If there are troublesome areas, we draw a tortured, amorphous gate around all well-focused areas. In this example, all areas were good enough, and the gate (R2) was set around all areas. Remaining doublets are removed with a plot of DNA "concentration" (DNA content divided by nuclear area) versus DNA content (Fig. 9C). This is based on the idea that the change in the area of the nucleus as cells move from G_1 to G_2 is less than 2X the area of a G_1 cell, and thus doublets fall largely below the G_2 cells in this plot. Empirically, we can set a region on each, look at the images, and confirm that the idea is correct on a sample-to-sample basis. Next, we use the "count" parameter that is unique to the secondary segmented event- the nucleus (DNA). This parameter is the count of the secondary events inside the primary event. We plot it versus the MaxPixel parameter for the DAPI channel (Fig. 9D). In this plot, there are two groups: events with one nucleus and events with two nuclei. We set regions on the "two nuclei" group, and empirically check images (Fig. 9E, 9F) for the break between events with two nuclei and high MaxPixel values due to condensed chromatin, and events with two G_1 nuclei. This information is then used to set an inclusive gate for downstream analysis (Fig. 9G). Note that in the iGeneration software, Boolean gating operates as a function of descendence. Therefore, any data gated from R6 in Fig. 9G are also gated by R3, R2, and R1, since 9G descends from 9C, which descends from 9B, etc.

2. Misclassified Anaphase, Telophase, and Cytokinetic Cells

With our present scheme, some fraction of telophase and cytokinetic cells are segmented as single events per nucleus. These cluster in any display using the MaxPixel parameter for the DAPI signal, e.g., Fig. 9G. At present, we identify these using the same process as doublet discrimination, then color-code and track them (Fig. 9G). Analytically, we divide the count by 2 to obtain the frequency of cells within the LM states.

3. State and Stage Analysis with Cyclin B1

Bivariate plots of pHH3 versus Bvariate plots of pHH3 versus cyclin B1 from LSC data (Fig. 10B) show approximately the same patterns that we observe with flow cytometry (compare to Fig. 1B). The pHH3 versus DNA (Figs. 10A, 13A) or versus cyclin A2 (Fig. 13B) plots are very similar. The LSC CVs for pHH3, cyclin A2, and cyclin B1 are equivalent to flow cytometry and indicate that the quality of this system of measures is high. Plots involving cyclin B1 appear to be more spread than flow cytometry (compare Fig. 1B to Fig. 10B), but calculations of CVs indicate otherwise. Therefore, we can use rules to draw regions around the data that define the same states that we have identified by flow cytometry. The rules are to place boundaries where there are significant changes in frequency (fraction of cells per unit of data space) and/or data direction. What we mean by this is that a virtual or ideal cell can be thought of as moving from state to state in one direction. For example, the arrows in Figs. 1 and 10 point the directions that the ideal cell would move. In reality, it is the time basis of expression programs within cells that create these patterns, but the concept of a period of defined states is equally valid from that point of view. The goal of an analysis as in Fig. 10B is to quantify the cell frequency in P1-3, the aggregate (P4 + PM + M), LM1, LM2a, and LM2b.

LSC offers additional information. In Fig. 10D, we plot mitotic cells excluding those in state P1-3. The plot is the area of the nucleus versus cyclin B1 divided by the area of the cell. This latter transformation should be proportional to cyclin

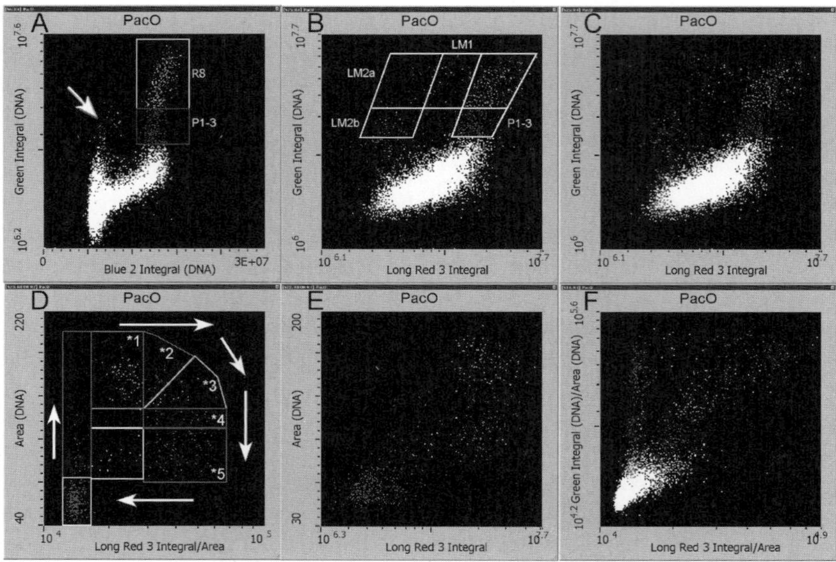

Fig. 10   Cyclin B1 state analysis. Green Integral (DNA) = pHH3 from the nuclear contour; Blue2 Integral (DNA) = DAPI from the nuclear contour; Long Red 3 Integral = cyclin B1 from primary contour; Area = cell footprint calculated from the primary contour. *1 *5 are five new states defined by the footprint of the nucleus and cyclin B1 divided by the footprint of the cells (proportional to concentrations). Regions and events in A and D are color coded to other panels and Fig. 11 panel borders. (For interpretation of the references to color in this figure legend, the reader is referred to the Web version of this chapter.)

B1 concentration. This plot displays the "rounding up" process of the cell, which increases left to right (and thus, the amount of cyclin B1 per footprint increases). It also displays the process of the nuclear material taking up less space as the cell undergoes the prophase through metaphase transition, and then increases again late in mitosis. Because these measures are not biochemical but can be viewed as either a process or morphology (both of which are the result of biochemistry), the image validation is perhaps especially important. This is shown in Figs. 11 and 12. Figure 11 also represents the raw material for stage analysis. This could be done without respect to state, that is, one could simply use a mitotic marker for detection, capture all the images, classify, sort, count, then calculate the frequencies. Here we use states defined by expression of DNA, pHH3, cyclin B1, and DNA and cyclin B1 "concentrations." The states defined by these measures are then further classified based on "image analysis" and assignment of seven categories to prophase (six based on cyclin B1-positive centrosome counts and pHH3 expression pattern, and one based on cyclin B1 localization to the nucleus), an indeterminate breakdown (BD) stage in which we have inferred that cyclin B1 is in flux back to the cytoplasm, and then the remaining stages of mitosis. The criteria for this analysis are listed in Table III. Stage analysis results for one sample are shown in Fig. 12. This analysis suggests that the states as we have defined them are related to mitotic stage but not synonymous.

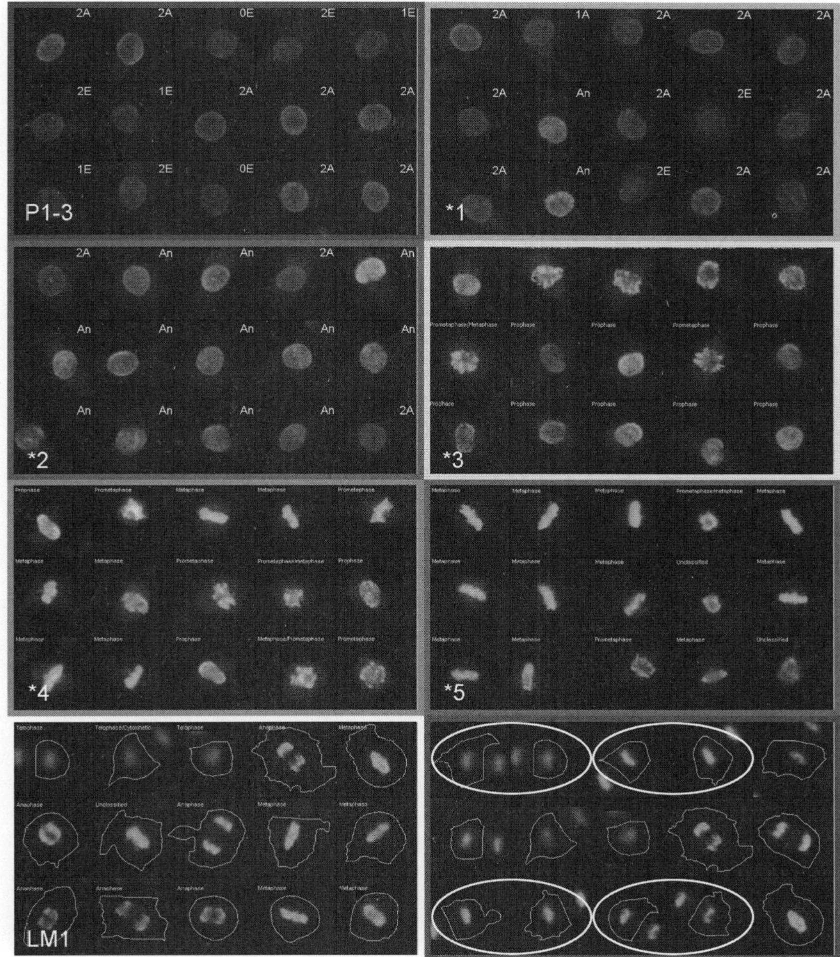

Fig. 11 LSC images of cyclin B1, pHH3, and DNA. Red = cyclin B1; green = pHH3, and blue = DNA. Contours are drawn on the bottom two panels showing successful capture of the binuceate stages in LM1 but often failure to capture these stages correctly in final state gates (green dots and region in Fig. 10). To correct for this problem, the tally should be divided by two when calculating percentages. (See plate no. 22 in the color plate section.)

4. State and Stage Analysis with Cyclin A2

Bivariate plots of pHH3 versus cyclin A2 for data obtained by LSC (Fig. 13B) are very similar to plots of flow cytometry data (Fig. 1A). Presumably, this is due to the more exact ability to specify the location of cyclin A2 by nuclear contouring. After nuclear membrane breakdown, cyclin A2 redistributes throughout the cell. Because of this, capturing the whole cell fluorescence is still important in cyclin A2 analysis;

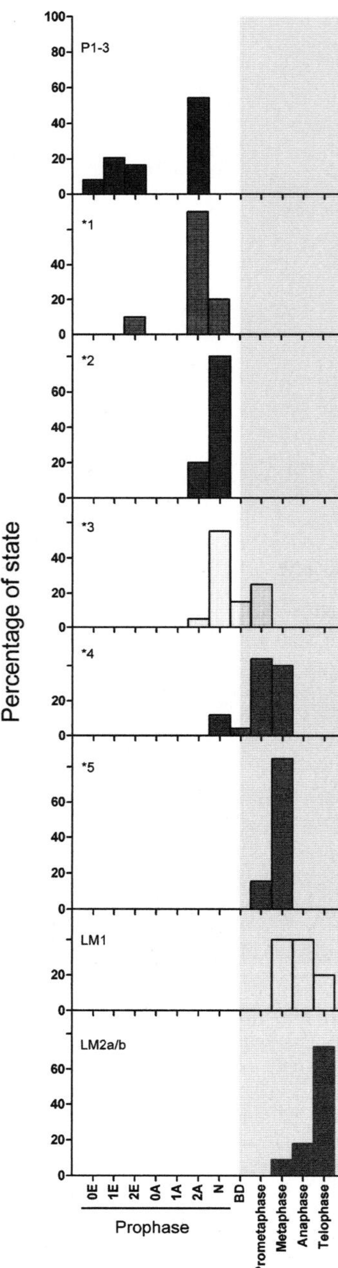

Fig. 12　Stage analysis using LSC images of Cyclin B1, pHH3, and DNA. Bar graphs of the frequencies (%) of cells classified by morphology as discussed in the text (X axis) for each of the states identified in Fig. 11 by criteria in Table III.

Table III

Criteria for stage classification

		Centrosomes	pHH3 pattern	Cyclin B1	DNA	Example
	0E	0	Dim, diffuse	Predominantly cytoplasmic	Circumscribed and spherical	
	1E	1	Dim, diffuse	Predominantly cytoplasmic	Circumscribed and spherical	
	2E	2	Dim, diffuse	Predominantly cytoplasmic	Circumscribed and spherical	
Prophase	**0A	0	Bright, patched, or distinct rim	Predominantly cytoplasmic	Circumscribed and spherical	
	**1A	1	Bright, patched, or distinct rim	Predominantly cytoplasmic	Circumscribed and spherical	
	2A	2	Bright, patched, or distinct rim	Predominantly cytoplasmic	Circumscribed and spherical	
	N		Bright overall, distinct rim	Predominantly nuclear	Circumscribed and spherical	
BD			Bright overall	Cytoplasm, nucleus rebalance	Rough perimeter and spherical	
Prometaphase			Tracks chromatin	Cytoplasmic, mitotic spindle	Chromosomes	
Metaphase			Tracks chromatin	Cytoplasmic, mitotic spindle	Band, or annular	
Anaphase			Tracks chromatin	Mainly absent	Two bands	
Telophase			Tracks chromatin, can be dim	Absent	Separate, smooth	

Note: *It is possible that 0A or 1A exist, either as artifact or perhaps biologically, but in this data set they were not detected. Cyclin B1 image in BD Telophase is the most common image.

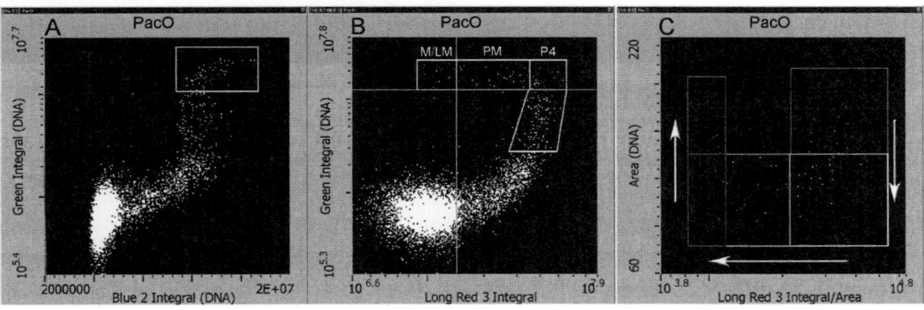

Fig. 13 Cyclin A2 state analysis.

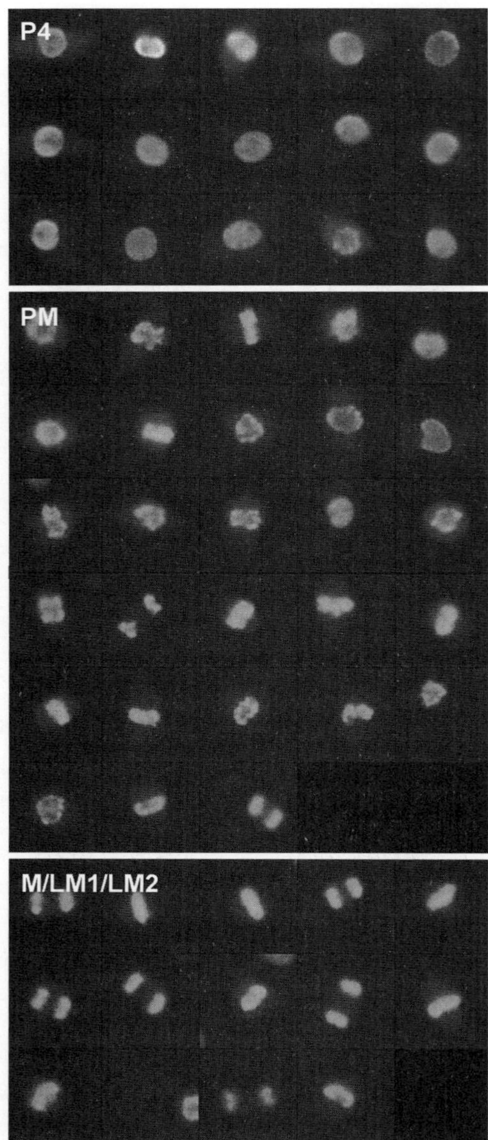

Fig. 14 LSC images of cyclin A2, pHH3, and DNA.

however, at the time of cyclin A2 redistribution, cells are "rounding up" and our ability to detect all the cytoplasm is much better than when cells are flat and irregular. For cyclin A2 analysis, the preprocessing is the same, except that we use one additional scattergram to exclude any endoreduplicated or binucleate G_1 cells (not shown). The basis is that cyclin A2-negative mitotic cells are pHH3-positive and $G_{1,4C}$ cells are cyclin A2-negative, and with Boolean logic, the $G_{1,4C}$ cells can be removed from the analysis.

The analysis itself, with four states, is easier than that for cyclin B1 (Fig. 13B). The two states that are important are P4 and PM, because these describe mitosis from the period when cyclin B1 enters the nucleus (this can be inferred from the analysis for cyclin B1) and when degradation of cyclin A2 by the APC/C gets serious. LSC images from the P1-3 state, defined by DNA, pHH3, and cyclin A2, are unremarkable and cannot be differentiated from interphase by additional image analysis except by difficult parsing of pHH3. However, we know from cyclin B1 analysis that these cells are in the early stage of mitosis, or prophase. The P4 state is dominated by prophase cells (Fig. 14). Within the PM state, we find prometaphase, metaphase, and even anaphase in sharply decreasing frequencies. Finally, the "M/LM" state, which we know from cyclin B1 analysis contains the states LM1, LM2a, and LM2b, is composed of metaphase, anaphase, and telophase cells. Thus, in terms of mitotic stages, cyclin A2 analysis is most informative for early mitosis. Secondly, it should be apparent that, as with cyclin B1 analysis, state and stage are highly correlated but not neatly synonymous. In the data shown in Fig. 14, prometaphase constitutes 63% of the cells in PM and metaphase is 22%, with the rest made up of small numbers of prophase and anaphase cells.

Although not necessary, it is instructive to ask whether we will get any benefit from measuring cyclin A2 "concentration" and nuclear size changes during mitosis. Fig. 13C shows the pattern for the mitotic cells (excluding P1-3) and color-coded regions set at boundaries. Following the colors in Fig. 13B, we can see that the progression through P4 $\rightarrow$ PM $\rightarrow$ M/LM is reflected in the changes in nuclear area, that is, P4 cells with large nuclear area precede cells with smaller area, which precede PM cells. We have not explored this further. It is included for symmetry.

IV. Discussion

A. Brief Summary

This chapter is about a specific LSC approach to multiparametric cell-cycle analysis that was relatively easy to implement using flow cytometry as the platform but required more thought to implement on the LSC platform. However, if we had not had access to flow cytometry patterns as "gold standards," we would not have questioned the LSC data that we were getting early on. Additionally, we knew from cell sorting, drug treatment, and siRNA experiments that there should be a relatively strong correlation between state and stage progression. Our early LSC results did not support this. The imperfect patterns were largely a function of cyclin B1, and our inability to obtain measurements from the entire cytoplasm appears to be the reason that we could not get the patterns that we knew were possible. Secondly, we suspected that spectral overlap, something that is not usually considered a problem in cytometry, was intolerable when the rigor of directly examining the cell stages and states, made possible by LSC, was applied to what had become a standard flow cytometry assay in the lab. While we have not proved that this was the case, breaking the assay up into two separate assays using fluorophores for the cyclins and pHH3 that are without spectral overlap, and obtaining the kind of results that we knew or suspected we should obtain, we conclude that we are right about this aspect as well. Thus, at the present time, if we are interested in cell state and/or mitotic stage analysis, we employ the two components described here.

B. State Names

The naming of cell states is problematic. We have gone through several iterations without finding a really satisfactory convention. The mitotic state names, P1-3, P4, PM, M, LM1, LM2a, LM2b, are the number of states that are possible from an analysis of pHH3 coupled with cyclins A2 and B1 in human somatic cells. They were named P, PM, and M based on dominant stages (prophase, prometaphase, and metaphase) within the states. It was reasonable that the period when pHH3 was ascending was a mitotic state, although microscopic analysis indicates that detection begins in G_2 (Perez-Cadahia *et al.*, 2009). Our idea is now supported by the analysis of centrosome numbers in cells in this state (unpublished). If we coupled other mitotic phospho-markers (e.g., phosho-S780-Rb, MPM2, phospho-T56-Bcl2) and pHH3, we could describe three states within P1-3 in bivariate plots (unpublished). Thus, the name P1-3 was assigned to that state and P4 to the $pHH3_{max}$/cyclin $A2_{max}$/cyclin $B1_{max}$ state. The problem with this is that it is obvious that we can subdivide states again and again. States are created by the timing differences between two or more parameters at basal levels, going up, at max, and returning to baseline. If we add informative markers for a state, we will break that state up into at least three to five new states. Thus, the naming convention has to accommodate constant change/modification and a marker-specific aspect. Some of this can be solved by using subscripts, for example, LM2a and LM2b, which leave room for expansion. *LM* was named after "late mitosis"; *2* is a second state based on cyclin B1, and the *a* and *b* designations describe two more states based on pHH3 expression within the LM2 state. This approach can become tedious, however. At present, we do not have any better ideas.

C. Concordance

We developed this approach to mitotic cell analysis by LSC to correlate mitotic stages and states. We have investigated whether or not we would find prophase, prometaphase, metaphase, anaphase, and telophase cells in specific areas of bivariate patterns by plotting cyclin B1 or cyclin A2 versus pHH3 for populations of asynchronous cells (unpublished). We expected that the "P" states would be populated predominantly with prophase cells, that PM would be populated with predominantly prometaphase cells, and that the LM states would be populated with metaphase, anaphase, and telophase cells. The reasoning was based on the idea that the APC/C marks cyclin A2 for degradation at a high rate after nuclear membrane breakdown, which occurs at the prophase $\rightarrow$ prometaphase transition, and that the APC/C marks cyclin B1 for degradation at a high rate after the spindle checkpoint has been extinguished, which is either "in" metaphase, or perhaps has become synonymous with metaphase. Essentially, the results are concordant with our expectations. However, the concordance is not as neat as perhaps narrative models of mitotic progression might lead us to imagine. Part of this lack of concordance is the result of applying absolute boundaries to probabilistic distributions. For example, the P4 state is a two-dimensional Gaussian distributed conjunction of pHH3 and cyclin A2 expression. The PM state can be modeled as a series of overlapping two-dimensional Gaussian distributions. Partly, this is based on our imperfect ability to measure. However, it is not unreasonable to expect that biologically there will be a mismatch in a process and the levels of the molecules that regulate the process. This is the way biology works, since it is a system that evolved to work under a variety of unexpected conditions, that is, it is robust. Therefore, we presume, for example, that some cells initiate more APC/C activity toward cyclin A2 before nuclear membrane breakdown than the average cell and some cells may be late in this regard. Thus, that states and stages do not match one-for-one is expected both at the experimental/technological and biological levels. In fact, this is the point of the article by Pines and Rieder (2001). We presume that with more parameters, this concordance should become more exact. This is hinted at when the expression levels of the cyclins are divided by the cell area and analyzed with respect to the nuclear area.

D. Critical Aspects of the Methodology

The aspects that we think are critical are as follows:

1. The method requires whole cell contouring, which is challenging, should be carefully thought about, and exactly described. To date, our best results have been obtained with a seeded watershed algorithm.
2. The stage analysis aspect suffers the same problems as any image-based classification system. The classification rests on images that are "ideal," and on either side of this "ideal" we find visual variation that presents difficulties. An example of that is the determination whether a cell belongs to the annular metaphase or prometaphase categories. Therefore, there is always more uncertainty in image classification than in determination of states.

   Related to this, the use of states prior to image analysis and classification is different from, and seems significantly less complicated than, using computer-based image analysis as a primary classification system. It would be interesting to compare the two approaches to test this idea.

3. Because of the emphasis on analysis of images and precise quantification, the requirement for 60X magnification, multiple laser passes, and segmented scan areas leads to long data acquisition times. This is not without problems, for example, evaporation of the sample during the run. We expect that as we become more experienced, we can lower the magnification to 40X, and reduce the x-step to 0.5μ which will reduce the time. Also, once the state and stage correlation is made robustly, we might rely more on state for applications like evaluating drug effects and use visual image inspection as a backup, or to ask specific questions.

4. One significant problem with our approach here is that we would like to be able to say that we measure every mitotic cell that matters over the assay area, that is, those we lose to cell aggregates are randomly lost, so we should not need to worry about those. However, we have not proven this, and it remains to be studied further. Areas of dense cell growth will be more problematic with respect to cell aggregates. At present, if we are satisfied that cell density is under control, then mitotic loss to aggregates should be random. Secondly, the configuration of the platform software is such that cells that are partially outside the boundary of the scan area not included in the analysis. The impact of this loss has not been evaluated. Currently, we rectify the low numbers of mitotic cells with repeated assays. Moving the assay to a lower magnification should reduce this problem.

E. Pitfalls and Misinterpretation of the Data

We have discussed problems throughout the text of this chapter. One comment needs restating here: one should not get into the habit, even in the lab, of calling states "stages." For example, P4 cells are not prophase cells. States and stages are not synonymous. They represent two different systems of measurement.

F. Comparison with Other Methods

Other approaches to this methodology, for example, flow cytometry and sorting cells by FACS, have been mentioned in the text. The approach is tedious and prone to error. A second approach would be to perform imaging flow cytometry with an instrument like the ImageStream (Amnis, Seattle, WA). We do not have experience with this protocol on that instrument, but it should be a viable approach, especially for hematopoietic cells.

G. Applications

One of us (JWJ) embarked on measuring cell-cycle regulating proteins as a function of DNA content with the not-very-well-formed idea that the system of measures would be a tool for relating the quantity of something and the rate at which cells cycle. It turns out that this is true. We did the specific work outlined here to provide a means of correlating state and stage that would be comprehensive, that is, we would measure all the mitotic cells and define the changes in stage as a function of state progression for populations of cells under different conditions. That work is mostly done and will be reported elsewhere. Other obvious applications of this assay are to test drug effects on the cell cycle. For example, 1.5 uM, alsterpaulone causes increased frequencies of PM and M (Soni and Jacobberger, 2004). Sulforaphane causes an increase in M (Pham *et al.*, 2004). This measurement system seems ideal for measuring the combined effects of multiple drug treatments. We have begun preliminary studies in this regard, and the results are promising and intriguingly complex. Other analyses are to monitor the levels of expression of some third marker as a function of these defined states. Surprisingly, even if the state-specific markers are affected, provided they are not reduced to background, the analysis can be performed. For example, see Soni *et al.*, 2008. Finally, evaluating the effects of genetic manipulation of genes that affect the cell cycle, that is, standard cell biology, could be powerfully improved by this approach.

H. Biological/Biomedical Information

This assay provides the frequencies of cells within cell states related to the cell cycle and the frequencies of mitotic stages. It also provides the frequencies of cells in G_1, S, G_2, and M of successive endocycles ($2C \rightarrow 4C \rightarrow 8C \rightarrow 16C$, etc.). Because the analysis provides a closed loop, that is, it includes progressive states from the beginning of the cell cycle to the end, it is possible to extract the cell-cycle expression profile of any epitope, including the ones used to define the states. See Frisa and Jacobberger, 2009, for a crude example of this process.

I. Future Directions

LSC is an attractive approach to the overall desire to express an understanding of the cell cycle in quantitative terms. The cell cycle can be explained well as a series of oscillating biochemical reactions. Measuring any number of these will produce interesting patterns that allow definition of specific states that might be important to monitor, especially in studies of the actions of drugs and multiple combinations of drugs. It is reasonable that any cell process should be similar in biochemical logic. It would be interesting to build a fully automated cell culture, treatment, and staining system, work with 24-well plates, and integrate a series of cell-cycle assays that focus on different parts of the cell cycle with assays that monitor other processes, for example, DNA damage repair or cell death. On a less grand scale, uppermost in our minds at this time is testing the use of an additional (green) laser to the system to address the problem of Alexa Fluor 488 spectral overlap in the PE channel. This may allow us to combine the cyclin A2 and B1 assays into a single assay that would not suffer any degradation in precision.

J. Final Comment

Independently, both of us have noted that when analyzing flow cytometry data without images, we miss the ability to "click" to a gallery of images to support or modify what we may be thinking at the time of analysis. Both of us are experienced flow cytometry experts, very used to dealing with abstract electronic dots and patterns that represent cell or molecular systems behavior, but both of us have become very comfortable with and very attached to having the support of images underlying our analysis. This is more of an emotional than reasoned statement – however, we believe that it is rooted in an improved ability to obtain additional information during an analytical session that would not be available if we were doing non-image-based cytometry.

Acknowledgments

This work was supported by grants from NCI, R01CA73413 to JWJ, and P30CA43703 to Stan Gerson, which supports the Cytometry and Imaging Microscopy Core facility in which the cytometry was performed. Additional thanks go to Vince Shankey (Beckman Coulter), Chuck Goolsby (Northwestern University), David Hedley (Ontario Cancer Institute), and Elena Holden, Mel Henriksen, Ed Luther, and Geoff Westgate (CompuCyte) for advice and comments on this work.

The authors do not have conflicts of interest associated with this work.

References

Akimitsu, N., Adachi, N., Hirai, H., Hossain, M. S., Hamamoto, H., Kobayashi, M., Aratani, Y., Koyama, H., Sekimizu, K. (2003). Enforced cytokinesis without complete nuclear division in embryonic cells depleting the activity of DNA topoisomerase IIalpha. *Genes Cells* **8**, 393–402.

Bauer, K. D., and Jacobberger, J. W. (1994). Analysis of Intracellular Proteins. *Methods Cell Biol.* **41**, 351–376.

Baumgarth, N., and Roederer, M. (2000). A practical approach to multicolor flow cytometry for immunophenotyping. *J. Immunol. Methods* **243**, 77–97.

Bozhkova, V. P., Rott, N. N., and Chailakhian, L. M. (1978). Relationship between the phases of mitosis in the cells of chimeric loach embryos. *Ontogenez.* **9**, 296–300.

Chakrabarti, R., Jones, J. L., Oelschlager, D. K., Tapia, T., Tousson, A., Grizzle, W. E. (2007). Phosphorylated LIM kinases colocalize with gamma-tubulin in centrosomes during early stages of mitosis. *Cell Cycle* **6**, 2944–2952.

Chen, E., Huang, X., Zheng, Y., Li, Y. J., Chesney, A., Ben-David, Y., Yang, E., Hough, M. R. (2010). Phosphorylation of HOX11/TLX1 on Threonine-247 during mitosis modulates expression of cyclin B1. *Mol. Cancer* **9**, 246.

D'Amato, F. (1954). Considerations on selective inhibition of various stages of mitosis. *G Ital Chemioter.* **1**, 129–134.

Darzynkiewicz, Z., Crissman, H., and Jacobberger, J. W. (2004). Cytometry of the cell cycle: cycling through history. *Cytometry A* **58**, 21–32.

Darzynkiewicz, Z. (2008). There's more than one way to skin a cat: yet another way to assess mitotic index by cytometry. *Cytometry A* **73**, 386–387.

del Campo, A., Gimenez-Martin, G., Lopez-Saez, J. F., and de la Torre, C. (1997). Frailty of two cell cycle checkpoints which prevent entry into mitosis and progression through early mitotic stages in higher plant cells. *Eur. J. Cell. Biol.* **74**, 289–293.

Delone, N. L., and Vysotskii, V. G. (1964). Sensitivity to acceleration of various phases of mitosis in the microspores of *Tradescantia paludosa. Izv. Akad. Nauk. SSSR Biol.* **12**, 900–907.

Deptala, A., Bedner, E., Gorczyca, W., and Darzynkiewicz, Z. (1998). Activation of nuclear factor kappa B (NF-kappaB) assayed by laser scanning cytometry (LSC). *Cytometry* **33**, 376–382.

Dobrokhotov, V. N., and Valvas, V. S. (1981). Dynamics of the phases of mitosis and the duration of cell division. *Biull. Eksp. Biol. Med.* **92**, 76–78.

Dodson, H., Bourke, E., Jeffers, L. J., Vagnarelli, P., Sonoda, E., Takeda, S., Earnshaw, W. C., Merdes, A., Morrison, C. (2004). Centrosome amplification induced by DNA damage occurs during a prolonged G2 phase and involves ATM. *EMBO J.* **23**, 3864–3873.

el-Alfy, M., and Leblond, C. P. (1987). Long duration of mitosis and consequences for the cell cycle concept, as seen in the isthmal cells of the mouse pyloric antrum. II. Duration of mitotic phases and cycle stages, and their relation to one another. *Cell Tissue Kinet.* **20**, 215–226.

Fortunato, E. A., Sanchez, V., Yen, J. Y., and Spector, D. H. (2002). Infection of cells with human cytomegalovirus during S phase results in a blockade to immediate – early gene expression that can be overcome by inhibition of the proteasome. *J. Virol.* **76**, 5369–5379.

Frisa, P. S., and Jacobberger, J. W. (2009). Cell cycle-related cyclin b1 quantification. *PLoS One.* **4**, e7064.

Gorczyca, W., Smolewski, P., Grabarek, J., Ardelt, B., Ita, M., Melamed, M. R., Darzynkiewicz, Z. (2001). Morphometry of nucleoli and expression of nucleolin analyzed by laser scanning cytometry in mitogenically stimulated lymphocytes. *Cytometry.* **45**, 206–213.

Gross, W. O. (1975). Mitosis. demonstration of separate phases. *Verh. Anat. Ges.* **69**, 851–854.

Grosset, L., and Odartchenko, N. (1975). Duration of mitosis and separate mitotic phases compared to nuclear DNA content in erythroblasts of four vertebrates. *Cell. Tissue Kinet.* **8**, 91–96.

Guarguaglini, G., Duncan, P. I., Stierhof, Y. D., Holmstrom, T., Duensing, S., Nigg, E. A. (2005). The forkhead-associated domain protein Cep170 interacts with Polo-like kinase 1 and serves as a marker for mature centrioles. *Mol. Biol. Cell.* **16**, 1095–1107.

Haider, A. S., Grabarek, J., Eng, B., Pedraza, P., Ferreri, N. R., Balazs, E. A., Darzynkiewicz, Z. (2003). *In vitro* model of "wound healing" analyzed by laser scanning cytometry: accelerated healing of epithelial cell monolayers in the presence of hyaluronate. *Cytometry A.* **53**, 1–8.

Hara, M., Igarashi, J., Yamashita, K., Iigo, M., Yokosuka, M., Ohtani-Kaneko, R., Hirata, K., Herbert, D. C. (1999). Proteins recognized by antibodies against isolated cytological heterochromatin from rat liver cells change their localization between cell species and between stages of mitosis (interphase vs metaphase). *Tissue Cell.* **31**, 505–513.

Hong, K. U., Choi, Y. B., Lee, J. H., Kim, H. J., Kwon, H. R., Seong, Y. S., Kim, H. T., Park, J., Bae, C. D., Hong, K. M. (2008). Transient phosphorylation of tumor associated microtubule associated protein (TMAP)/cytoskeleton associated protein 2 (CKAP2) at Thr-596 during early phases of mitosis. *Exp] Mol. Med.* **40**, 377–386.

Hozak, P., Zatsepina, O., Vasilyeva, I., and Chentsov, Y. (1986). An electron microscopic study of nucleolus – organizing regions at some stages of the cell cycle (G0 period, G2 period, mitosis). *Biol. Cell.* **57**, 197–205.

Huang, X., Kurose, A., Tanaka, T., Traganos, F., Dai, W., Darzynkiewicz, Z. (2006). Sequential phosphorylation of Ser-10 on histone H3 and ser-139 on histone H2AX and ATM activation during premature chromosome condensation: relationship to cell-cycle phase and apoptosis. *Cytometry A.* **69**, 222–229.

Jacobberger, J. W. (2000). Flow cytometric analysis of intracellular protein epitopes. *In* "Immunophenotyping," (C. A. Stewart, and J. K. A. Nicholson, eds.), pp. 361–405. Wiley-Liss, New York.

Jacobberger, J. W. (2001). Stoichiometry of immunocytochemical staining reactions. *Methods Cell Biol.* **63**, 271–298.

Jacobberger, J. W., Frisa, P. S., Sramkoski, R. M., Stefan, T., Shults, K. E., Soni, D. V. (2008). A new biomarker for mitotic cells. *Cytometry A.* **73**, 5–15.

Jacobberger, J. W., Sramkoski, R. M., and Stefan, T. (2011). Multiparameter cell cycle analysis. *In* "Flow Cytometry Protocols, Methods in Molecular Biology," (T. S. Hawley, and R. G. Hawley, eds.), Vol. 699, Springer Science, New York, p. 21.

Jacobberger, J. W., Sramkoski, R. M., Zhang, D., Zumstein, L. A., Doerksen, L. D., Merritt, J. A., Wright, S. A., Shults, K. E. (1999). Bivariate analysis of the p53 pathway to evaluate Ad-p53 gene therapy efficacy. *Cytometry* **38**, 201–213.

Juan, G., Darzynkiewicz, Z., 2004. Detection of mitotic cells. Curr Protoc Cytom. Chapter 7, Unit 7 24.

Kamentsky, L. A. (2001). Laser scanning cytometry. *Methods Cell Biol.* **63**, 51–87.

Kamentsky, L. A., and Kamentsky, L. D. (1991). Microscope-based multiparameter laser scanning cytometer yielding data comparable to flow cytometry data. *Cytometry.* **12**, 381–387.

Kammerer, B. D., Sardella, B. A., and Kultz, D. (2009). Salinity stress results in rapid cell cycle changes of tilapia (Oreochromis mossambicus) gill epithelial cells. *J. Exp. Zool. A Ecol. Genet. Physiol.* **311**, 80–90.

Kawamura, K., Fujikawa-Yamamoto, K., Ozaki, M., Iwabuchi, K., Nakashima, H., Domiki, C., Morita, N., Inoue, M., Tokunaga, K., Shiba, N., Ikeda, R., Suzuki, K. (2004). Centrosome hyperamplification and chromosomal damage after exposure to radiation. *Oncology.* **67**, 460–470.

Kawamura, K., Morita, N., Domiki, C., Fujikawa-Yamamoto, K., Hashimoto, M., Iwabuchi, K., Suzuki, K. (2006). Induction of centrosome amplification in p53 siRNA-treated human fibroblast cells by radiation exposure. *Cancer Sci.* **97**, 252–258.

Kurose, A., Yoshida, W., Yoshida, M., and Sawai, T. (2001). Effects of paclitaxel on cultured synovial cells from patients with rheumatoid arthritis. *Cytometry.* **44**, 349–354.

Kuz'mina, S. V. (1972). The ratio of the phases of mitosis in decontaminated cell lines and in cell lines contaminated with mycoplasma. *Biull. Eksp. Biol. Med.* **13**, 100–102.

Langendorff, H., Hagen, U., and Ernst, H. (1958). Studies on a biological radiation protection. XXVIII. Effect of cysteamine on various stages of the mitosis cycle. *Strahlentherapie* **107**, 574–579.

Leblond, C. P., and El-Alfy, M. (1998). The eleven stages of the cell cycle, with emphasis on the changes in chromosomes and nucleoli during interphase and mitosis. *Anat. Rec.* **252**, 426–443.

Lopes, C. S., Sampaio, P., Williams, B., Goldberg, M., and Sunkel, C. E. (2005). The Drosophila Bub3 protein is required for the mitotic checkpoint and for normal accumulation of cyclins during G2 and early stages of mitosis. *J. Cell. Sci.* **118**, 187–198.

Luther, E., and Kamentsky, L. A. (1996). Resolution of mitotic cells using laser scanning cytometry. *Cytometry.* **23**, 272–278.

Mak, S. K., and Kultz, D. (2004). Gadd45 proteins induce G2/M arrest and modulate apoptosis in kidney cells exposed to hyperosmotic stress. *J. Biol. Chem.* **279**, 39075–39084.

Manton, I., Kowallik, K., and von Stosch, H. A. (1969). Observations on the fine structure and development of the spindle at mitosis and meiosis in a marine centric diatom (Lithodesmium undulatum). II. The early meiotic stages in male gametogenesis. *J. Cell. Sci.* **5**, 271–298.

Manton, I., Kowallik, K., and von Stosch, H. A. (1970). Observations on the fine structure and development of the spindle at mitosis and meiosis in a marine centric diatom (*Lithodesmium undulatum*). 3. The later stages of meiosis I in male gametogenesis. *J. Cell Sci.* **6**, 131–157.

Maruvada, P., Dmitrieva, N. I., East-Palmer, J., and Yen, P. M. (2004). Cell cycle-dependent expression of thyroid hormone receptor-beta is a mechanism for variable hormone sensitivity. *Mol. Biol. Cell.* **15**, 1895–1903.

Mercie, P., Belloc, F., Bihlou-Nabera, C., Barthe, C., Pruvost, A., Renard, M., Seigneur, M., Bernard, P., Marit, G., Boisseau, M. R. (2000). Comparative methodologic study of NFkappaB activation in cultured endothelial cells. *J. Lab. Clin. Med.* **136**, 402–411.

Ohshima, S. (2008). Abnormal mitosis in hypertetraploid cells causes aberrant nuclear morphology in association with H_2O_2-induced premature senescence. *Cytometry A.* **73**, 808–815.

Ohshima, S., and Seyama, A. (2010). Cellular aging and centrosome aberrations. *Ann. N.Y. Acad. Sci.* **1197**, 108–117.

Olsen, J. V., Vermeulen, M., Santamaria, A., Kumar, C., Miller, M. L., Jensen, L. J., Gnad, F., Cox, J., Jensen, T. S., Nigg, E. A., Brunak, S., Mann, M. (2010). Quantitative phosphoproteomics reveals widespread full phosphorylation site occupancy during mitosis. *Sci. Signal.* **3**, ra3.

Pachmann, K., Zhao, S., Schenk, T., Kantarjian, H., El-Naggar, A. K., Siciliano, M. J., Guo, J. Q., Arlinghaus, R. B., Andreeff, M. (2001). Expression of bcr-abl mRNA in individual chronic myelogenous leukaemia cells as determined by in situ amplification. *Br. J. Haematol.* **112**, 749–759.

Perez-Cadahia, B., Drobic, B., and Davie, J. R. (2009). H3 phosphorylation: dual role in mitosis and interphase. *Biochem. Cell. Biol.* **87**, 695–709.

Peters, J. M. (2006). The anaphase promoting complex/cyclosome: a machine designed to destroy. *Nat. Rev. Mol. Cell Biol.* **7**, 644–656.

Pham, N. A., Jacobberger, J. W., Schimmer, A. D., Cao, P., Gronda, M., Hedley, D. W. (2004). The dietary isothiocyanate sulforaphane targets pathways of apoptosis, cell cycle arrest, and oxidative stress in human pancreatic cancer cells and inhibits tumor growth in severe combined immunodeficient mice. *Mol. Cancer Ther.* **3**, 1239–1248.

Pines, J. (2006). Mitosis: a matter of getting rid of the right protein at the right time. *Trends Cell. Biol.* **16**, 55–63.

Pines, J., and Rieder, C. L. (2001). Re-staging mitosis: a contemporary view of mitotic progression. *Nat. Cell. Biol.* **3**, E3–6.

Rosdahl, I. K., and Lindstrom, S. (1980). Morphology of epidermal melanocytes in different stages of mitosis. *Acta Derm. Venereol.* **60**, 209–215.

Santamaria, A., Neef, R., Eberspacher, U., Eis, K., Husemann, M., Mumberg, D., Prechtl, S., Schulze, V., Siemeister, G., Wortmann, L., Barr, F. A., Nigg, E. A. (2007). Use of the novel Plk1 inhibitor ZK-thiazolidinone to elucidate functions of Plk1 in early and late stages of mitosis. *Mol. Biol. Cell.* **18**, 4024–4036.

Schwartz, J. L., Muscat, J. E., Baker, V., Larios, E., Stephenson, G. D., Guo, W., Xie, T., Gu, X., Chung, F. L. (2003). Oral cytology assessment by flow cytometry of DNA adducts, aneuploidy, proliferation and apoptosis shows differences between smokers and non-smokers. *Oral. Oncol.* **39**, 842–854.

Schwock, J., Ho, J. C., Luther, E., Hedley, D. W., and Geddie, W. R. (2007). Measurement of signaling pathway activities in solid tumor fine-needle biopsies by slide-based cytometry. *Diagn. Mol. Pathol.* **16**, 130–140.

Siracky, J., Sparrow, S., and Jones, M. (1984). Changes in distribution of the phases of mitosis in human tumor xenografts following drug treatment. *Neoplasma.* **31**, 539–544.

Soerenby, L., and Lindahl, P. E. (1964). On the concentrating of ascites tumour cells in stages of pre-mitosis and mitosis by counter-streaming contrifugation. *Exp. Cell Res.* **35**, 214–217.

Soni, D. V., and Jacobberger, J. W. (2004). Inhibition of cdk1 by alsterpaullone and thioflavopiridol correlates with increased transit time from mid G2 through prophase. *Cell Cycle.* **3**, 349–357.

Soni, D. V., Sramkoski, R. M., Lam, M., Stefan, T., and Jacobberger, J. W. (2008). Cyclin B1 is rate limiting but not essential for mitotic entry and progression in mammalian somatic cells. *Cell. Cycle.* **7**, 1285–1300.

Sramkoski, R. M., Wormsley, S. W., Bolton, W. E., Crumpler, D. C., and Jacobberger, J. W. (1999). Simultaneous detection of cyclin B1, p105, and DNA content provides complete cell cycle phase fraction analysis of cells that endoreduplicate. *Cytometry.* **35**, 274–283.

Starosvetskaia, N. A. (1970). Ultrastructural characteristics of the final stages of mitosis in tissue culture cells. *Vestn. Akad. Med. Nauk. SSSR.* **25**, 51–56.

Walter, J., Schermelleh, L., Cremer, M., Tashiro, S., and Cremer, T. (2003). Chromosome order in HeLa cells changes during mitosis and early G1, but is stably maintained during subsequent interphase stages. *J. Cell. Biol.* **160**, 685–697.

Weicker, H. (1954). Erythroblastic mitosis; duration of mitosis and course of phases of the hemi-homo-plastic karyokinesis of the proerythroblasts and of the three erythroblast maturation divisions. *Z. Klin. Med.* **151**, 407–428.

Yamamoto, Y., Matsuyama, H., Kawauchi, S., Matsumoto, H., Nagao, K., Ohmi, C., Sakano, S., Furuya, T., Oga, A., Naito, K., Sasaki, K. (2006). Overexpression of polo-like kinase 1 (PLK1) and chromosomal instability in bladder cancer. *Oncology.* **70**, 231–237.

Zak, M. (1970). Changes i the mitotic activity and the portion of single mitosis-phases of the bone marrow cells after x-ray irradiation and administration of chloroquine. *Strahlentherapie* **139**, 587–592.

Zak, M., and Cerny, M. (1966). A method for the evaluation of the mitotic index and the individual phases of mitosis in bone marrow cells using hypotonized cells. *Cas Lek Cesk.* **105**, 1178–1179.

Zatsepina, O., Hozak, P., Babadjanyan, D., and Chentsov, Y. (1988). Quantitative ultrastructural study of nucleolus-organizing regions at some stages of the cell cycle (G0 period, G2 period, mitosis). *Biol. Cell.* **62**, 211–218.

Zosimovskaia, A. I., and Liapunova, N. A. (1966). The duration of mitosis and stages of the mitotic cycle in primary culture of human embryonic fibroblasts. *Tsitologiia.* **8**, 208–215.

Instrumentation, new probes and methods

CHAPTER 15

Lasers in Flow Cytometry

William G. Telford

Experimental Transplantation and Immunology Branch, National Cancer Institute, National Institutes of Health, Bethesda, Maryland, USA

METHODS IN CELL BIOLOGY, VOL 102

0091-679X/10 $35.00
DOI 10.1016/B978-0-12-374912-3.00015-8

Abstract

Laser technology has advanced tremendously since the first gas lasers were incorporated into early flow cytometers. Gas lasers have been largely replaced by solid-state laser technology, making virtually any desirable visible light wavelength available for flow cytometry. Multiwavelength, white light, and wavelength tunable lasers are poised to enhance our analytical capabilities even further. In this chapter, I summarize the role that lasers play in cytometry, and the practical characteristics that make a laser appropriate for flow cytometry. I then review the latest single wavelength lasers available for flow cytometry, and how they can be used to excite the ever-expanding array of available fluorochromes. Finally, I review the contribution and potential of the latest tunable laser technology to flow cytometry, and show several examples of these novel sources integrated into production instruments. Technical details and critical parameters for successful application of these lasers for biomedical analysis are covered in depth.

Keywords: diode; DPSS; flow cytometer; laser; supercontinuum (SC); tunable laser

I. Introduction

Lasers provide the excitation light source for virtually every flow cytometer system available today. The historical development of the practical laser (which celebrated its 50th birthday in 2010) has in many ways paralleled the development of the flow cytometer. Although the earliest cytometer systems were forced to rely on noncoherent light sources like mercury lamps, the developers of the first commercial systems were quick to incorporate lasers as excitation sources when their size and cost made that integration practical (Shapiro, 2003). Lasers provide monochromatic, coherent, stable light at high power densities, in a form that can be easily shaped, directed to, and focused on a sample stream. While a few flow cytometers still use other light sources for purposes of either economy or specialized applications, lasers have become and remain the dominant light source for flow.

The evolution of flow cytometric analysis has in fact been largely dictated by the development of laser technology. The 488 and 514.5 nm lines available from the argon-ion lasers integrated into the earliest commercial systems ensured the utility of fluorescein, rhodamine, and later the phycobiliproteins as some of the earliest fluorochromes used for extracellular and intracellular labeling (Shapiro, 2003). These same laser lines allowed the use of nucleic acid binding dyes like propidium iodide and acridine orange to unravel the mechanisms of cell cycling. Gas laser sources also provided high levels of ultraviolet, violet, and blue light, allowing for the continued use of UV-excited DNA dyes like DAPI for cell cycle analysis, and the use of multiple UV and violet-excited DNA dyes for chromosome analysis and sorting. Red laser lines were one of the first visible products of the laser diode, making multiparametric immunolabeling possible with the monomeric cyanin dye Cy5 and the phycobiliprotein allophycocyanin (Hoffman *et al.*, 1987; Loken *et al.*, 1987). Violet laser light, previously available only from gas sources, can now be inexpensively produced from diode sources, making these fluorescent probes far more accessible on affordable instruments (Shapiro and Perlmutter, 2001; Telford *et al.*, 2003, 2006). The recent development of diode-pumped solid-state (DPSS) sources and fiber lasers have allowed virtually any wavelength to be produced, opening up probes such as the red fluorescent proteins to flow cytometric analysis (Telford, 2004a,b). Flow cytometry used to be limited by available laser wavelengths; this has nearly ceased to be a restriction. A summary of wavelengths available from both earlier gas lasers and more modern solid-state sources is shown in Fig. 1. The visible spectrum is nearly covered even with single wavelength sources, and this list is constantly growing.

While adding additional lasers to a flow cytometer can increase the total number of parameters we can simultaneously measure; this is not the only motivation. A "simple" three-laser cytometer, for example, can theoretically carry out 17-color flow cytometry: six colors using a 488 nm laser, three using a red diode, and eight with a violet laser diode (including quantum nanoparticles). These are more than enough simultaneous parameters for even the serious user. The primary value of additional excitation wavelengths is getting access to fluorochromes previously unavailable on the usual instruments (Shapiro and Telford, 2009). Green and yellow lasers, for example, can excite the broad array of red fluorescent proteins now available, something the usual blue and red lasers cannot do well (Chudakov *et al.*, 2005; Shaner *et al.*, 2004; Telford *et al.*, 2005). Modifying and adding to our laser wavelengths also allows us to fine-tune our excitation conditions to the precise requirements of our assay systems, maximizing sensitivity. We can also design sophisticated fluorescent experiments such as energy transfer systems, enhancing our capabilities in measuring complex biological events.

In this chapter, we first discuss the technical requirements for lasers used in flow cytometers. These specifications remain largely unchanged since the early days of cytometer development, although certain new systems can capitalize on laser technology not normally used for flow. We will then briefly review the history of lasers in flow cytometers, concentrating mainly on gas laser technology that is still used in some legacy systems. We will further review more modern diode and diode-pumped lasers, including the recent exciting developments in fiber laser technology. Lasers ranging from the ultraviolet to near infrared that can be integrated into flow systems will be reviewed. Finally, we will discuss technology for producing multiple wavelengths from a single laser, including merge modules and supercontinuum (SC) laser sources. The latter of these can produce *any* laser wavelength for unparalleled flexibility in fluorochrome excitation, although with important limitations. The next step in this technological evolution, truly tunable lasers, will point the way for eliminating excitation limitations from flow cytometric analysis.

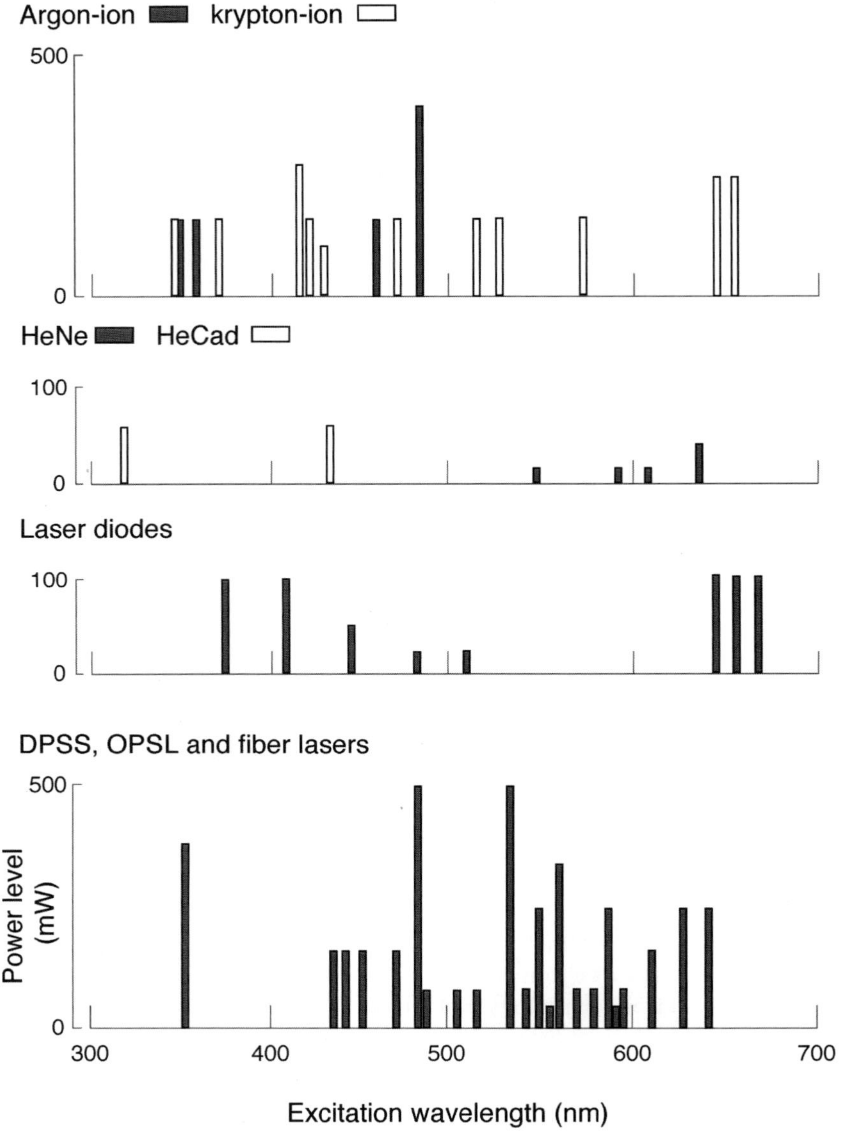

Fig. 1 Emission wavelengths of argon-ion, krypton-ion, HeNe, HeCad, diode, and DPSS/OPSL lasers.

II. Laser Characteristics for Flow Cytometry

Lasers remain to be one of the most expensive components of a modern flow cytometer, despite the dramatic decrease in manufacturing cost per unit in recent years. Lasers have long been incorporated into a variety of consumer goods, a phenomenon that generally assures a decrease in manufacturing cost as production volume increases. This is particularly true in the development of the laser diode, which has been incorporated into everything from laser pointers to CD and DVD players. Red laser diodes suitable for laser pointers are now available for pennies, a few dollars for a diode array in a CD player. Violet laser diodes have made the precision read/write characteristics of DVD and Blu-Ray players possible; while more expensive than red diodes, the cost for these units has dropped dramatically as a result of the music and video reproduction and playback market. The manufacturing industry, particularly semiconductor production has also driven the development of low-cost lasers than could conceivably be used for other purposes. The biomedical analysis field has directly benefited from these economies of production; lowering the cost of lasers we incorporate it into our instruments (Shapiro, 1993 and Perlmutter, 2001).

However, the lasers we incorporate into our Blu-Ray players and laser pointers do not meet the technical specifications required for flow cytometry for a variety of reasons. The laser requirements for flow cytometry remain relatively unchanged since their first introduction into flow cytometers, despite the dramatic advances in the technology. An instrument user who is contemplating the integration of a novel laser source into their system needs to keep these specifications in mind. While this responsibility is usually relegated to the instrument manufacturer, modern flow cytometers can incorporate as many as seven or more lasers; the end user should therefore have a good idea of what is going into their instrument. Some specifications are also affected by having many sources present simultaneously.

Beam shape remains a critical specification for lasers in flow cytometers. Early gas laser manufacturers traditionally classified their bead shapes as circular or noncircular, with a circular Gaussian TEM_{00} (transverse electronic mode) single mode being specified for most instruments. This circular beam can be predictably focused onto a sample stream, and the beam shaped using prisms or cylindrical lenses. Mode mixing can sometimes produce a flat-top or "top hat" beam profile, which has sharper edges than a typical Gaussian beam and takes better advantage of the total laser output. Diode and DPSS lasers, while relying on different mechanisms to produce the beam output, are still typically specified as having TEM_{00} output for flow cytometry.

Gas lasers applicable for flow cytometry usually have excellent Gaussian TEM_{00} single mode characteristics (Shapiro, 2003). Keep in mind that diode laser usually possess multiple modes (and are called multimode) along one axis. Rather than producing a single spot, these lasers produce multiple spots along a single axis. Laser diode beams can therefore be highly irregular and are usually not suitable for flow stream focusing without some type of postlaser modification. Beams can be optically modified with cylindrical lenses or anamorphic optics to produce a single roughly circular or elliptical beam spot, usually with a good bit of power loss. The resulting "single mode" beams are generally applicable for flow, although their often uneven beam profiles can sometimes produce focusing problems. Several enterprising manufacturers have shaped these multimode laser diode beams into elliptical beam spots for flow cytometry with less power loss. These lasers can sometimes be integrated into instruments with less downstream beam shaping. DPSS lasers generally have cleaner Gaussian TEM_{00} beams and require less post-laser beam shaping. Fiber optic coupling (used on many commercial flow cytometers) will generally render any laser beam into a Gaussian beam spot, albeit with a large loss of power. However, successful and efficient fiber optic coupling is very dependent on the beam shape and focus; large power losses can result in a laser that is not optically coupled.

A. Continuous Wave (CW) Versus Pulsed

Lasers can emit continuously (continuous wave (CW)), or they can be pulsed. In pulsed lasers, the laser energy is released at discrete intervals, with a discrete pulse length. Lasers intended for surgery, manufacturing, etc. are frequently pulsed; the individual pulses can be generated at higher power levels than a continuous emission, and the material being worked on can cool between pulses. Generally, most lasers on flow cytometers are CW. However, mode-locked DPSS lasers with extremely fast repetition rates (>10 MHz) and long enough pulse intervals (so-called quasi-CW) lasers can be used for flow cytometry; from the perspective of a cell, these lasers function like CW sources. Quasi-CW lasers will be discussed in more detail in the solid-state UV laser section.

B. Laser Noise

Laser noise refers to the stability of the laser in the interval required for the laser to do its job. For flow cytometry, this interval is the time in which a cell passes through the beam, and allow subsequent cells to pass through the beam as well. In a typical cytometer a cell will pass through the beam in approximately 1 μs. The laser therefore needs to be *quiet* in this interval, and is typically evaluated between 20 Hz and 20 MHz. Lasers applicable for flow cytometry generally exhibit 0.5% peak-to-peak noise in this range, and often much less for high quality units. Achieving low noise is not easy, and distinguishes cheaper lasers from more expensive ones. Red diode laser pointers, for example, can exhibit greater than 25% peak-to-peak noise, fine for pointing but not for most types of flow cytometry. Laser noise can originate from many sources, including the laser medium itself, the self-regulation of the laser, and the power source. Some older gas laser and most diodes have a light control mode which self-regulates the output, reducing laser noise. Temperature control is critical for diode and DPSS operation and can also be a source of laser noise.

Beam shape, laser noise, and stability are usually carefully dictated by the instrument manufacturer prior to integration. However, the end user should still be aware of these factors, as they can result in instrument problems in complex instruments with multiple laser sources.

Nevertheless, extremely low noise levels are not always necessary for flow cytometry. If cells are being analyzed for DNA content, a measurement where considerable precision is required, a "noisy" laser will show a dramatic loss of peak resolution and an increase in peak coefficient of variation generally measured for G_1 cell subpopulation having uniform DNA content. If measuring a surface marker with a broad range of variation, however, a noisy laser might not produce a significantly different profile from a quiet one. Furthermore, phenotyping introduces error from a variety of sources, and laser noise may be a minor contributor relative to other factors. Instrument developers have therefore experimented with reliable but cheap solid-state laser sources for inexpensive instruments designed for one task, such as immunophenotyping T cells from HIV patients. Inexpensive diodes and DPSS lasers used for industrial applications often achieve $\sim$5% peak-to-peak noise, high but acceptable for some applications. Laser noise in this case is a minor component of the overall measurement, and accurate phenotyping can still be carried out. A simple flow cytometer has been constructed using a green laser pointer using this principle. Some advanced multiwavelength laser sources (such as SC lasers) also have somewhat higher noise levels than the traditional specification. Noise levels of 1–2% in these cases are probably acceptable for some applications.

When looking at laser noise on instrument with many lasers, the overall *noise budget* is often taken into account. This is defined as the sum of the RMS noise from all system lasers. Many manufacturers attempt to keep the overall noise budget below a certain level (i.e., 3%); the typical noise levels of modern diodes and DPSS lasers are typically low enough to meet these specifications, in spite of the larger number of lasers available on high-end instruments.

C. Laser Power Stability

Laser stability can be measured in minutes or hours, as opposed to noise, which is typically measured over milliseconds to microseconds. While laser instability on this scale might not affect cell-to-cell measurement as peak-to-peak noise would, it can have a dramatic effect on measurement of an entire sample, or on multiple experiments over hours, days or weeks. Laser manufacturers will usually specify stability over an 8-h period following warm-up; this should be 1% RMS or less. As with laser noise, this specification can be difficult to achieve at the engineering stage, and separates the inexpensive units from the expensive ones. As with laser noise, laser stability can be affected by a variety of factors, including the laser cavity, the power supply and temperature regulation. Unlike laser noise, laser stability can be easily measured by the end user using a recording power meter.

D. Pointing Stability

Laser pointing stability over an extended period of time is an issue with many early laser diodes, but is generally not a major issue with modern modules, where temperature fluctuations (a cause of drift) are minimal. It is expressed in $\mu rad/°C$.

E. Module Size

Although there is an excessively large number of size specifications for laser modules, market dominance by several large manufacturers has resulted in several standard form factors that have been informally adopted by the industry, and are now present in many cytometers. Cylindrical HeNe lasers are typically 45 mm in diameter, varying in length depending on the plasma tube. Small laser diodes are typically mounted in cylindrical housings at 0.5 in./12.5 mm, although this can vary between manufacturers. Larger cylindrical diodes are quite variable in diameter. Coherent, Inc. has produced both direct diode and DPSS/ optically pumped semiconductor laser (OPSL) modules in the "Sapphire" ($125 \times 70 \times 34$ mm, without heat sink) and more recently the "Cube" casing (100 mm $\times 40$ mm $\times 40$ mm, without heat sink), size standards that are now being frequently emulated by other manufacturers. These size specifications may dictate whether a laser can be easily integrated into an existing system.

Keep in mind that many inexpensive solid-state lasers are produced that physically resemble more expensive units, but are intended for applications much less rigorous than flow cytometry. Examine the specifications carefully. If some specifications are omitted (i.e., peak-to-peak noise level), they are not likely suitable for flow cytometry.

III. Laser Safety

A. Gas Lasers for Flow Cytometry

The earliest flow cytometers utilized mercury arc lamps, and were able to take advantage of the ultraviolet, blue, green, and yellow lines these lamps produced (Shapiro, 2003). However, lasers were rapidly integrated into flow cytometer designs during their early development at Los Alamos, Stanford University and other pioneering institutions, and formed the core of the earliest commercial cell sorters and analyzers. These lasers were almost exclusively gas-filled ion lasers, including argon, krypton, and helium–neon (Fig. 2). The argon- and krypton-ion lasers tasked for flow cytometry were large and typically water cooled; later on, lower power air-cooled variants were available, particularly argon. Water-cooled argon- and krypton-ion are still found on some older sorting systems, although they are being rapidly superseded by the solid-state laser source (Fig. 2a and 2b). Air-cooled argon-ion lasers were incorporated into an enormous number of benchtop analyzers, and are still present in many legacy systems (Fig. 2c).

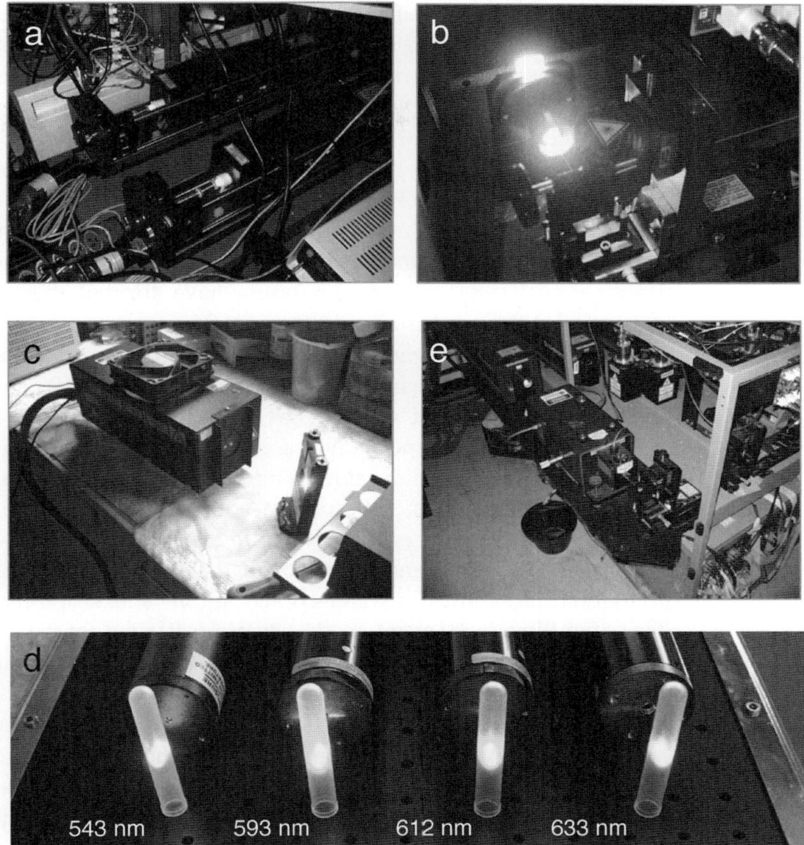

Fig. 2 Examples of gas lasers. (a) Open water-cooled argon-ion and krypton-ion lasers (Coherent, Santa Clara, CA); (b) krypton-ion laser emitting at 568 nm (Coherent); (c) air-cooled argon-ion 488 nm laser (Spectra-Physics, Mountain View, CA); (d) HeNe lasers emitting at 543, 594, 612, and 633 nm (JDS Uniphase, Milpitas, CA); (e) dye head laser with argon-ion pump laser (Coherent).

1. Argon- and Krypton-Ion Lasers

Argon- and krypton-ion lasers make excellent laser sources for flow cytometry. They produce high-quality Gaussian TEM_{00} beams with low noise and good power stability, particularly in the higher-power water-cooled units. They are large lasers and produced copious amounts of waste heat, requiring a complex cooling system. However, they have relatively long lifetimes, often operating for several thousand hours before requiring gas tube replacement. Both argon- and krypton-ion lasers emit several monochromatic wavelengths simultaneously, which can be isolated using a Littrow prism inserted between the laser mirrors flanking the gas tube. Argon-ion lasers emit powerful lines at 488 and 514.5 nm, allowing initial excitation of fluorescein and rhodamine (and later emission of phycobiliproteins and their tandem dyes) (Shapiro, 2003). Argon-ion lasers also produce a relatively strong series of lines in the ultraviolet, particularly at 351 and 364 nm (Fig. 1). This allows the use of UV-excited dyes such as DAPI, which had first been used in mercury arc lamp systems, and also allowed the use of UV physiological probes such as the calcium indicator indo-1. Water-cooled argon lasers also produce a relatively strong line at 457 nm; this wavelength permits such applications as chromosome sorting using the DNA dyes chromomycin A3 and mithramycin, and simultaneous excitation of green, yellow, and cyan fluorescent protein (Gray and Cram, 1990; Van den Engh *et al.*, 1984). Argon-ion lasers have formed the excitation core of most commercial flow cytometers in the past 25 years.

Krypton-ion lasers, more expensive and less common than argon, produce a wider variety of wavelengths, including UV lines at 351 and 356 nm, violet lines at 407 and 413 nm, a blue line at 476 nm, green and yellow lines at 520, 531 and 568 nm, respectively, and red lines at 647 and 676 nm (Fig. 1) (Shapiro, 2003). They do not however produce a 488 nm line. While less powerful than argon-ion lasers and more maintenance intensive, krypton-ion sources allow for the excitation of a broad variety of fluorescent probes, including the earliest violet-excited fluorochromes such as Cascade Blue and yellow-excited probes such as Texas Red. Power levels for individual laser lines in water-cooled argon- and krypton-ion sources range from tens of milliwatts to several watts, usually sufficient for the jet-in-air sample delivery systems and low numerical aperture optics of earlier instruments. Mixed-gas argon–krypton lasers are less common but have been incorporated into high-end cell sorters; the coherent spectrum is a well-known example of this, and can produce a wide range of wavelengths from the UV to the red, albeit sometimes at low power levels. Air-cooled argon-ion lasers are smaller and can be integrated into small benchtop instruments such as the Becton–Dickinson FACScan and FACSCalibur and the Beckman-Coulter XL (Fig. 2c). Their power levels can approach 50 mW, making them very applicable for cuvette-based cytometers with higher numerical apertures. These lower power argon lasers are restricted to the 488 and 514.5 nm lines, unable to produce useful levels of UV light. Many of these lasers remain in service.

2. Helium–Neon (HeNe) Red Lasers

Helium–neon (HeNe) lasers have been a mainstay of flow cytometry almost since their introduction (Hoffman *et al.*, 1987; Shapiro, 1986; Shapiro and Stephens, 1986). Also as gas lasers, HeNe lasers also have excellent TEM_{00} beam characteristics, low noise, and good long-term stability. They were crucial in the introduction of multilaser flow cytometers, allowing additional fluorochromes to be used beyond those accessible at 488 nm (Fig. 2d). The dominant red 633 nm HeNe wavelength was ideal for exciting flow molecular weight fluorochromes such as Cy5, and the brighter phycobiliprotein allophycocyanin. By the early 1990s, investigators with high-end instruments could excite three fluorochromes using their 488 nm source (typically fluorescein, phycoerythrin (PE), and a PE tandem) and APC with a red HeNe or diode, permitting four-color flow cytometry. While high-power HeNe lasers (over 30 mW) were nearly as large as their water-cooled counterparts, they could be air-cooled and thus easier to operate. Lower power red HeNe lasers (5 to 20 mW) were much smaller and could be integrated into benchtop instruments. HeNe laser also emits other laser lines, including green (546 nm), yellow (594 nm), and orange (612 nm) (Fig. 4d). These lines are far less efficient than the red line and emit at considerably lower power levels. As a result, they are not often used in flow cytometry. However, some early benchtop analyzers and sorters, including the Becton-Dickinson FACS Analyzer and the Beckman-Coulter Epics Elite have used green and yellow HeNe lasers, respectively, as low-power excitation sources. While being replaced in many cases with solid-state sources, red HeNe lasers are still reliable, inexpensive laser sources with excellent characteristics for flow cytometry.

3. Helium–Cadmium (HeCad) Laser

HeCad lasers produce wavelengths in the UV at 325 and in the blue at 442 nm and are uncommon on commercial flow cytometers. They rely on the generation of cadmium vapor in a helium lasing tube, constituting an interesting variation on the typical gas laser. They have been used for a few flow cytometric applications, including DAPI excitation at 325 nm and chromosome analysis using the 442 nm line (Frey *et al.*, 1993; Shapiro and Perlmutter, 1993; Snow and Cram, 1993). They have also made brief appearances on benchtop flow cytometers, including the original BD LSR. HeCad lasers tend to become very noisy over their tube life, and can be very maintenance intensive (Shapiro, 2003). As of this writing, these lasers are essentially no longer in production in formats suitable for flow cytometry.

4. Dye Head Lasers

While not technically a gas laser source, these lasers rely on gas lasers or another powerful source to "pump" a circulating laser dye in a quartz cuvette, emitting wavelengths dependent on the dye used (Fig. 2e). Until recently, dye head lasers were the only practical method of producing laser lines not easily available from other lasers, including the yellow and orange lines (580 to 610 nm). A dye head loaded with the laser dye rhodamine 6G could produce powerful laser lines in the 570 to 620 nm range for excitation of Texas Red, Cy5 or allophycocyanin (Loken *et al.*, 1987). Many laser dyes are available, although only a few have been applied to flow cytometry. Recent advances in solid-state laser technology have largely replaced the dye head laser for flow cytometry.

Gas lasers in general are becoming much less common on commercial instrumentation. Most manufacturers of the large-frame argon- and krypton-ion lasers have begun to phase out production of these units, and support for these modules from the original manufacturers will likely cease in the near future. Several third-party companies, including Evergreen Laser (Durham, CT, USA, http://www.evergreenlaser.com) still provide support for these units, including tube fabrication and replacement. Air-cooled argon lasers are still being manufactured for legacy cytometers utilizing these units, but are rapidly being replaced by solid-state substitutes.

IV. Laser Diodes

Laser diodes are essentially light emitting diodes (LEDs) fabricated in the form of a laser cavity (Fig. 3). The first commercial laser diodes emitted in the infrared range and were generally not useful for flow cytometry. A breakthrough came with the development of the first truly visible red laser diode, with monochromatic wavelengths in the 635–645 nm range (Fig. 3a). These small red lasers were introduced into several commercial cytometers in the early 1990s, substituting

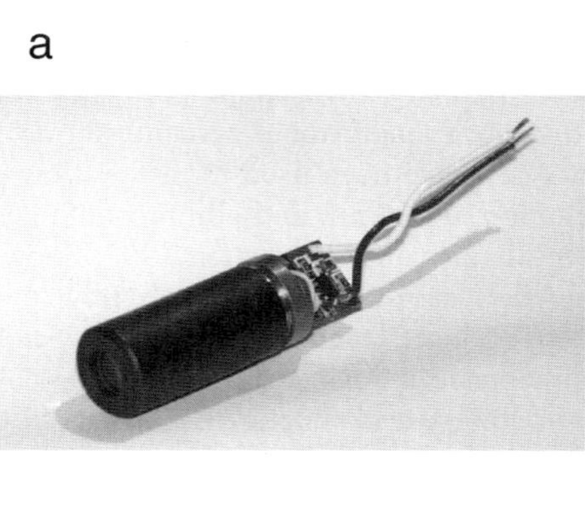
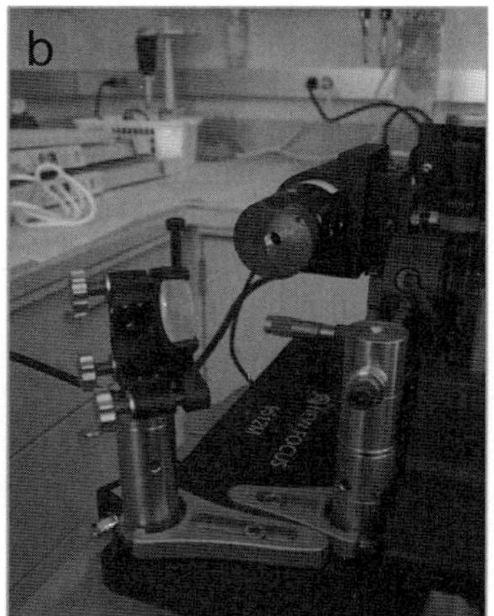

Fig. 3 Diode, DPSS, and fiber laser modules. (a) Small red diode laser module; (b) larger red laser diode (coherent); (c) DPSS laser emitting at 561 nm (Cobolt AB, Solna, Sweden); (d) fiber laser emitting at 542 nm. (For interpretation of the references to color in this figure legend, the reader is referred to the Web version of this chapter.)

for a larger HeNe red laser but exciting the same fluorochromes (Doornbos *et al.*, 1993, 1994). Red diodes therefore made practical benchtop multilaser flow cytometry possible.

Red diodes were the only laser diodes applicable for flow cytometry until the turn of the century, when Shuji Nakamura at Nichia developed the first true blue and violet LEDs and laser diodes, based on gallium arsenide semiconductor chemistry (Nakamura and Fasol, 1997). Howard Shapiro tested a very early generation violet laser for flow cytometry, and found it to be very useful for doing DNA content analysis using the DNA binding dye Hoechst 34580 (Shapiro and Perlmutter, 2001). Like red diodes, violet laser diodes are monochromatic but may emit in a range from 395 to 415 nm, depending on the specifications of the individual diode. Violet laser diodes have since become common fixtures on flow cytometers, useful for exciting a variety of fluorescent probes, including low molecular weight immunolabeling dyes (i.e., Cascade Blue, Pacific Blue, Pacific Orange) and quantum nanoparticles (Hoffman, 2002; Telford *et al.*, 2002, 2003, 2006, 2007) (Fig. 3b). The combination of a blue–green 488 nm, red HeNe or diode laser and a violet laser diode has become a standard multilaser configuration on many commercial instruments, allowing the simultaneous analysis of fourteen or more judiciously selected fluorochromes.

Other laser diodes have also found uses in flow cytometry, although to a lesser degree. Blue laser diodes emitting in the 440 to 450 nm range can excite DNA dyes such as mithramycin and chromomycinone, both used in chromosome sorting. However, their typical power levels are usually insufficient for this application, and they see little use in flow cytometry compared to more powerful blue DPSS lasers. Near UV or UV laser diodes emitting in the 370–390 nm range have been used for DAPI and Hoechst dye cell cycle analysis, as well as for Hoechst side population analysis of stem cells (Cabana *et al.*, 2006; Telford, *et al.*, 2004, 2004a; Telford and Frolova, 2004). Their relatively low power levels (typically less than 20 mW) have similarly limited their usage compared to more powerful solid-state UV sources. Recent development of a 488 nm laser diode (not to be confused with DPSS sources at the same wavelength) should make a useful replacement for argon-ion 488 nm lasers in cytometers. Green laser diodes, also distinct from DPSS modules, are also in development and should be similarly applicable for cytometry.

Laser diodes make good laser sources for flow cytometry. They are relatively long-lived (5000 to 10,000 h is typical), and their relative simplicity of design makes them inexpensive to manufacture. They are small, have low power requirements, and relatively nondemanding from an integration standpoint, as long as they can operate in a controlled temperature environment. High-end modules also have good long-term stability and low noise levels. They require good temperature control to maintain both power and wavelength, and use light control mode to maintain their power level. Their beam shapes can be problematic, since they are usually multimodal in the uncorrected state, with multiple beam spikes along one axis. This multimodal pattern can be corrected using pairs of cylindrical lenses or anamorphic corrective optics to convert their beam pattern into a circular or elliptical spot. This correction is often imperfect, and "hot spots" can form that can make beam focusing difficult. Laser manufacturers have refined these methods of beam correction, making the resulting beam spots applicable for flow in most cases. The range of applications for laser diodes is huge, with highly variable quality of manufacture and specification depending on the intended purpose of the laser. Only high-end modules (usually apparent in the cost) should be used for most flow cytometry applications. The size of a fully assembled laser diode has decreased considerably, with some lasers applicable for flow contained in a "lipstick" sized module As discussed earlier, however, less expensive modules can be used for some applications where a certain amount of laser noise can be tolerated.

V. Diode-Pumped Solid State (DPSS) Lasers

Most of the solid -state lasers now installed on flow cytometers are DPSS or OPSL modules (Perfetto and Roederer, 2007; Telford *et al.*, 2005). DPSS lasers consist of a powerful diode infrared laser that "pumps" a crystalline lasing medium, resulting in laser output (Fig. 3c). For example, an 808 nm infrared pump laser can be used to pump an yttrium aluminum garnet (YAG) or yttrium vanadate (YVO_4) lasing cavity, resulting in a 1064 nm laser emission. This wavelength can be frequency doubled down to 532 nm, the familiar green laser line, or frequency tripled down to 355 nm (Shapiro, 2003). Green Nd:YAG and Nd:YVO_4 lasers are becoming common installations on flow cytometers, and can be used to excite PE and its tandems with lower autofluorescence than the usual 488 nm sources. They also give us access to rhodamine-based probes, as well as many red fluorescent proteins. The frequency-tripled 355 nm modules can be used for UV applications.

While 532 nm modules were the first to see widespread use, virtually any laser wavelength is now available in a DPSS or OPSL format (Shapiro and Telford, 2009). By modifying the pump laser wavelength and/or the chemical or physical structure of the lasing cavity, almost any wavelength can in theory be generated. Modifications to laser cavity design have resulted in a plethora of wavelengths that can be applied to flow cytometry. Both mode-locked quasi-CW and CW UV lasers are now available, as are blue DPSS lasers emitting in the 440 to 480 nm range. DPSS 488 nm lasers have now been available for some time, at both low and high power levels. Green DPSS lasers now range from 505 to 561 nm, with 532 and 561 nm sources becoming a popular option for both PE and red fluorescent protein excitation (Perfetto and Roederer, 2007; Telford *et al.*, 2005; Telford and Huber, 2005). Yellow and orange sources ranging from 570 to 610 nm have been recently developed, and red sources ranging from 625 to 650 nm have recently been produced (Kapoor *et al.*, 2007; 2008). Many of these lasers closely match the legacy wavelengths previously generated from argon and krypton ion and dye head sources, including 488, 515, 530, 568, and 594 nm (Shapiro and Telford, 2009) (Fig. 1).

As with diodes, DPSS lasers are reliable with long potential lifetimes. They are small and relatively easy to integrate into existing instrumentation. They can be built at power levels ranging from milliwatts to watts, for incorporation into small cuvette instruments or larger jet-in-air sorters. They produce clean Gaussian TEM_{00} beams. Since they are available in a rainbow of colors, a DPSS laser can usually be found for virtually any application. Their main disadvantage is their cost, which remains considerably higher than diodes. DPSS lasers are more complex than diodes and are thus more expensive to manufacture. Economies of scale have reduced their cost somewhat, but they remain pricey relative to other laser sources. Newly developed diodes emitting at 488 and 532 nm are therefore seen as economically viable alternatives to DPSS units at these two wavelengths. They have also not replaced diodes in the violet range, a relatively unimportant issue since diodes fill this niche quite well and at a reasonable cost.

A. Fiber Lasers

An exciting offshoot of DPSS laser development is the fiber laser. One of the difficulties in DPSS laser development and manufacture is the need to fabricate highly pure, uniform crystals as the lasing media. These crystals must be grown under highly controlled conditions, a difficult manufacturing process. Doping these crystals with impurities to modify their characteristics and potentially generate new wavelengths is similarly difficult.

Fiber lasers utilize a linear or nonlinear fiber optic as a lasing cavity, instead of a crystal (Fig. 3d). This technology has many advantages. Optical fibers are far easier and cheaper to fabricate than crystals. In addition, it is much easier to modify the composition of a fiber, allowing more rapid development of new fiber constructions and easier subsequent manufacture. The resulting laser consists of the traditional pump laser coupled to a fiber optic; laser light is emitting directly from the fiber end. The resulting lasers are usually smaller than their traditional DPSS counterparts, and are easier to integrate into existing instrumentation. Fiber lasers can now be constructed that emit at virtually any wavelength in the visible spectrum (Kapoor *et al.*, 2007; 2008; Telford *et al.*, 2009a). They also form the basis for some intriguing advances in nonlinear laser technology, including SC white light lasers, and continuously tunable fiber lasers (Telford *et al.*, 2009b). Since the ultimate laser output emits from a fiber optic, beam quality is Gaussian TEM_{00} and remains polarized, with noise and stability characteristics similar to DPSS lasers.

VI. Lasers by Wavelength

The following section will describe a variety of laser diodes, DPSS, and fiber lasers in order of ascending wavelength, ranging from the ultraviolet to the far red. All of these lasers represent the state-of-the-art in flow cytometry laser excitation at the time of writing. Specific manufacturers will be indicated only for unique modules not available from multiple sources.

A. Ultraviolet (355 nm)

As described above, the early availability of ultraviolet light for flow cytometry from arc lamps and ion lasers resulted in the development of several useful applications, including cell cycle analysis using the DNA binding dyes DAPI and the Hoechst dye, as well as intracellular calcium measurement using the chelating probe indo-1. More recently, efflux of the DNA dye Hoechst 33342 by stem cells (the so-called Hoechst side population technique) has proven critical for stem cell identification (Goodell *et al.*, 1996). Quantum nanocrystals or quantum dots are also well excited by UV lasers (Telford *et al.*, 2004). Chromosome sorting also requires a high-power UV laser source (Van den Engh *et al.*, 1984). While not essential on many cytometers, UV lasers are still required for these and other techniques, and are frequently installed on high-end instruments.

Examples of modern solid-state UV sources are shown in Fig. 4. This requirement is now largely filled by DPSS mode-locked frequency-tripled Nd:YVO$_4$ lasers, two examples shown in Fig. 4a and 4b. These lasers are quasi-CW, with repetition rates in the 10 to 100 MHz range. Several commercial units are available that are applicable for flow cytometry, including the Newport Spectra-Physics Vanguard (Irvine, CA) and the JDS Uniphase Xcyte (Milpitas, CA) series. The Fianium UVPower is similarly mode locked but built as a fiber laser (Fig. 4c). All three units are available in power levels ranging from several hundred milliwatts to several watts; the lower end of this range is certainly adequate for most flow applications including chromosome sorting. The Xcyte units are available in power levels ranging from 20 to 160 mW, the lower powers which might be more cost-effective for benchtop instrumentation. All of these lasers are relatively large compared with the typical small DPSS laser.

CW UV lasers are also available for flow cytometry. The Coherent Genesis 355 nm laser emits at 100 mW in CW mode. Very small UV lasers have also become available; an example is the Cobolt Zouk (Solna, Sweden), which occupies a much small footprint than the units described above and emits 10 × 10 mW (Fig. 4d). This lower power would be applicable for cuvette instruments. This unit was able to analyze both sensitivity microsphere arrays and mouse bone marrow labeled for side population very effectively (Fig. 4e and 4f).

Many Q-switched 355 nm lasers are commercially available, generally built for industrial purposes. The repetition rate of these lasers is in the 1 to 100 kHz range, much lower than a quasi-CW unit and too low for flow cytometry. However, a Q-switched laser with a modified lasing cavity was recently demonstrated to produce CW 355 nm laser light at over 100 mW, at noise and stability levels comparable to quasi-CW units (DPSS Laser, Inc., Santa Clara, CA) (Fig. 4g). Since industrial lasers are built in economies of scale, this may result in a less expensive UV laser source suitable for flow cytometry.

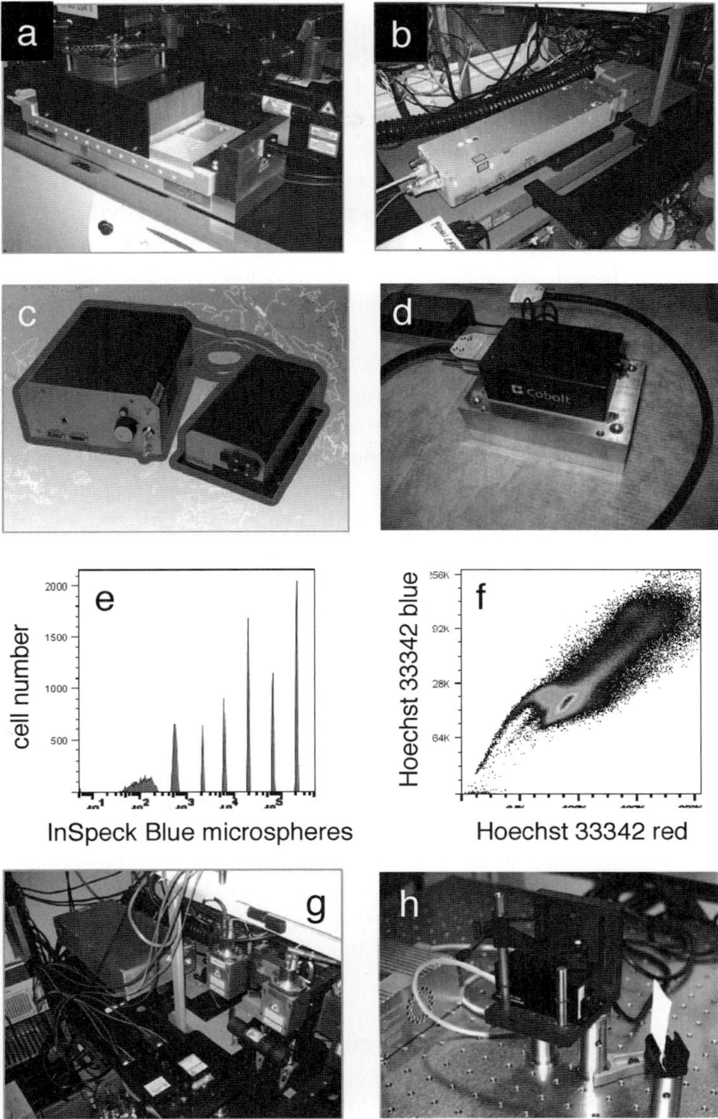

Fig. 4 Ultraviolet and near-UV lasers. (a and b) Nd:YVO$_4$ frequency tripled quasi-CW lasers (Xcyte 355 nm at 20 mW, JDS Uniphase and Vanguard 355 nm at 350 mW, Newport Spectra-Physics); (c) UV fiber quasi-CW laser (UVPower, Fianium, Southampton, UK); (d) small UV CW laser, 355 nm at 10 mW (Cobolt Zouk, Cobolt AB); (e) analysis of InSpeck Blue seven-population microsphere array using a Cobolt Zouk 355 nm laser at emitting at 10 mW mounted on a BD Biosciences LSR II (San Jose, CA); (f) mouse bone marrow labeled with UV-excited DNA dye Hoechst 33342, and analyzed for Hoechst side population using a Cobolt Zouk 355 nm laser emitting at 10 mW on a BD Biosciences LSR II (San Jose, CA); (g) modified Q-switched UV laser emitting CW UV light at 355 nm (DPSS Lasers, Inc., Santa Clara, CA); (h) UV laser diode, 375 nm emitting at 100 mW (Pavilion Integration Corporation, San Jose, CA). (For interpretation of the references to color in this figure legend, the reader is referred to the Web version of this chapter.)

B. Near Ultraviolet (370–390 nm)

Ultraviolet or near-ultraviolet laser diodes are a very different technology than the 355 nm lasers described above, consisting of a direct diode. They can be used for all the UV applications described above with the exception of indo-1 and chromosome analysis. The calcium-bound form of the chelating dye indo-1 emits at ~390 nm, too close to the emission of the laser to allow for signal discrimination. UV diodes also emit at considerably lower power levels than the DPSS laser, making them unsuitable for chromosome discrimination. However, they are far less expensive than Nd:YVO$_4$ lasers, and work very well for whole cell applications like DAPI cell cycle analysis and Hoechst side population (Cabana *et al.*, 2006; Telford, 2004; Telford and Frolova, 2004). While the final units emit in the 370–390 nm range, most modules tend to center around 375 nm.

Several manufacturers produce UV laser diodes, ranging in power from 10–20 mW. UV diodes now use anamorphic optics to circularize their beams, which are generally acceptable for flow cytometry. One exception to this power level is a unit produced by Pavilion Integration Corporation (San Jose, CA), which produces a 100 mW UV laser diode unit (Fig. 4h). This module uses an unusual beam modification technique, which combines multiple modes from the laser diode into a blade-like beam pattern with a top hat profile. This beam profile has proven adaptable to get-in-air instruments like the BD Biosciences FACSVantage.

C. Violet (395–410 nm)

Violet laser diodes have become important tools in flow cytometric analysis (Hoffman, 2002; Shapiro and Perlmutter, 2001; Telford *et al.*, 2002, 2003, 2006, 2007). Several low molecular weight phenotyping dyes have been developed that excite at this wavelength, including Cascade Blue, Alexa Fluor 405, Pacific Blue, Pacific Orange (Invitrogen Life Technologies, Carlsbad, CA) and V450 and V500 (BD Biosciences, San Jose, CA). Quantum nanocrystals or dots also excite very well at this wavelength. Coumarin-based viability and proliferation dyes are now in common use, and the violet-excited expressible fluorescent protein CFP and its derivatives (i.e., Cerulean) are often used as fluorescent resonance energy transfer (FRET) donors for intermolecular measurement studies (He *et al.*, 2004; Lelimousin *et al.*, 2009). Violet laser diodes are now available from a variety of manufacturers in single mode configurations at power levels at 100 mW and above. This power level should be adequate for most flow cytometers. Violet laser diodes with multiple modes and higher power levels have been used in imaging cytometer like the Amnis ImageStream (Seattle, CA). To date, no DPSS laser with violet emission has been used in flow cytometry, and is probably not necessary given the reliability of diode sources.

D. Blue (430–480 nm)

Prior to solid-state laser development, only HeCad and argon lasers could produce emission in this range (442 and 457 nm, respectively) (Shapiro, 2003; Snow and

Cram, 1993). While not a widely used wavelength, the argon 457 nm has been used to excite the DNA binding dyes chromomycin A3 in combination with Hoechst 33258 or DAPI for chromosome analysis and sorting. The 457 nm argon line has also been used to simultaneously excite GFP, YFP, and CFP, although this has been largely discontinued due to the prevalence of multilaser cytometers (Beavis and Kalejta, 1999).

Blue DPSS lasers emitting at 435 nm have been developed but are no longer manufactured at specifications appropriate for cytometry. Blue laser diodes emitting at ~445 nm are available from Coherent (Santa Clara, CA) at power levels up to 50 mW. This may not be powerful enough for chromosome sorting on a jet-in-air sorter. DPSS lasers have been developed at 442, 447, and 457 nm, with high power 457 nm units available from several manufacturers; the 457 nm lasers are of sufficient power for chromosome analysis and sorting. Several manufacturers now produce 473 nm lasers as well although the cytometric applications for this wavelength are likely to be limited.

E. Blue–Green (480–515 nm)

DPSS 488 nm modules at relatively low power levels (10–50 mW) are available from a variety of manufacturers. These lasers are largely replacing the air-cooled argon lasers typically found in benchtop cytometers. They are small, have excellent beam characteristics, are quiet, and can have long lifetimes. They are also essentially maintenance free, not requiring the tube changes necessitated by ion lasers. Higher power units for jet-in-air instruments are less common; these include the Coherent Sapphire series, which range from 20 to several watts in power, and Newport Spectra Physics, which produces a 100 mW module. MPB Communications also manufacturers the first commercially available 488 nm fiber laser, which peaks at several hundred milliwatts. The 488 nm laser as a core excitation source is a legacy requirement that is unlikely to change despite advances in laser wavelength availability.

Following the development of a 488 nm diode by Nichia (Tokyo, Japan), several manufacturers are planning production of low-power 488 nm laser diodes, intended to be inexpensive alternatives to DPSS lasers. Power Technology (Alexander, AR) is an early producer of this technology in a format suitable for flow cytometry.

An interesting 491 nm DPSS laser manufactured by Cobolt emits at up to 50 mW, and could substitute for 488 nm. They also have an unusual dual emission 491/532 nm laser that might find interesting applications in multilaser cytometry. Newport manufactures a 505 nm laser, Coherent a 514 nm laser, and both Newport Spectra Physics and Cobolt a 515 nm source. These are probably intended as legacy replacements for the argon-ion 514.5 nm line. While not particularly useful for flow cytometry, they may have utility for some confocal microscopy applications.

F. Green to Yellow (520–565 nm)

Several examples of green to yellow DPSS lasers are shown in Fig. 5. Green DPSS frequency-doubled 532 nm lasers are now common fixtures on flow cytometers

(Perfetto and Roederer, 2007) (Fig. 5a). They often substitute for the 488 nm line for excitation of PE and its tandems. PE tandem dyes in particular benefit from this laser line, which conforms better to the excitation maxima of PE. In addition, 532 nm laser light results in less cellular autofluorescence, further improving the resolution of signal to background. DPSS 532 nm lasers are also useful for exciting the shorter wavelength red fluorescent proteins, including DsRed and dTomato (Chudakov *et al.*, 2005; Shaner *et al.*, 2004; Telford *et al.*, 2005). A variety of rhodamine-based

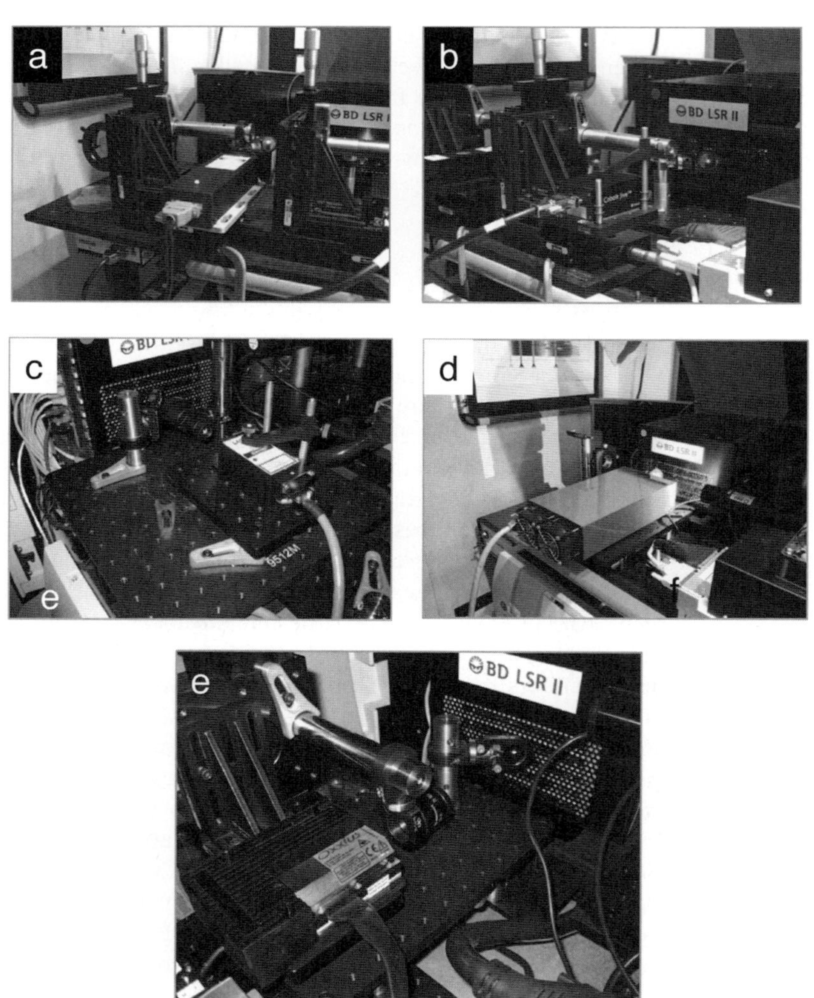

Fig. 5 Green to yellow lasers. (a) DPSS 532 nm, 50 mW (Laser-Compact, Russia); (b) DPSS 561 nm emitting at 50 mW (Cobolt Jive, Cobolt AB); (c) DPSS 542 nm emitting at 50 mW (Lasos, Germany); (d) fiber laser 550 nm emitting at 150 mW (Zecotek Photonics Ltd., Singapore); (e) DPSS 553 nm emitting at 200 mW (Oxxius, France). (For interpretation of the references to color in this figure legend, the reader is referred to the Web version of this chapter.)

dyes and fluorescent probes are also available that could be exploited with a green laser line. These lasers have also found their way into several small cytometers for simultaneous excitation of PE and the viability dye 7-aminoactinomycin D.

Green 532 nm lasers are mainly available as DPSS units, and now range from several milliwatts to several watts in power. A 50 mW unit should be adequate for most applications on cuvette instruments, with higher powers recommended for jet-in-air instruments. A 532 nm fiber laser emitting at several hundred milliwatts is now in production (MPB Communications, Quebec, Canada).

The first direct diode green lasers are now becoming available as well. The firm Kaai (partly founded by Shuji Nakamura, the original developer of the blue and violet gallium nitride diode) has produced a 523 nm green diode, and diodes up to the 532 nm range are in development. Like the 488 nm direct diode laser, this may prove to be reliable, an inexpensive alternative to DPSS lasers.

Modifications to the pump laser and the lasing cavity have produced additional useful laser lines in the green range. DPSS 561 nm lasers have also been integrated into many flow cytometers (Telford *et al.*, 2005; Telford and Huber, 2006) (Fig. 5b). Like the 532 nm unit, they can be used to excite PE and its tandems with even lower autofluorescence background. They are also more optimal for exciting the longer red fluorescent proteins, including mCherry, mStrawberry, and mKate2. Green–yellow 561 nm lasers are now available at power levels up to several hundred milliwatts, and a high power 560 nm fiber laser is also available (MPB Communications). The 561 nm overcomes one disadvantage of the 532 nm laser, namely its overlap into the fluorescein detection bandwidth (Telford *et al.*, 2009). Flow cytometers cannot completely exclude laser light from its detector paths; as a result, green laser light can impinge on the fluorescein filter and make fluorescein or GFP detection difficult or impossible. Green laser installations require the modification of the fluorescein filter to a shorter bandwidth, usually 510/20 nm or similar. Installing a 561 nm in place of the 532 nm line is another approach to the problem, since this laser line does not overlap the fluorescein filter range. However, 561 nm can impinge on the PE filter bandwidth (usually 575/26 or 585/42 nm on most instruments), similarly necessitating a modification of the PE filter to a 590/20 nm or similar.

The close proximity of the 532 and 561 nm lines to these bandwidths suggests that a laser line intermediate to these two would make a good compromise for PE, PE tandem, and red fluorescent protein detection, while avoiding the fluorescein and PE filter ranges. While lagging behind the development of these two frequency-tripled units, several options are now available. DPSS 542 nm units suitable for flow cytometry are now available from Newport and Lasos, probably intended as legacy replacements for the green HeNe 543 nm laser (Fig. 5c). These lasers still slightly overlap into the traditional fluorescein band, but filter modification is easier. A green 550 nm fiber laser is available from Zecotek Photonics Ltd. (Singapore), and a DPSS 553 nm laser from Oxxius (France) (Fig. 5d and 5e). This wavelength range is almost ideal for positioning between fluorescein and PE. The 550 nm has been shown to provide excitation of PE and the red fluorescent proteins as levels nearly equal to 561 nm, while not overlapping into the traditional PE range (Telford *et al.*, 2009). The flexibility of design associated with fiber lasers suggests that any monochromatic wavelength in the 520 to 560 nm lines should be possible, if an application can be found.

G. Yellow to Orange (568–610 nm)

Laser in this range have not been heavily exploited for flow cytometry, mainly due to the unavailability. However, they have several useful applications. The longer expressible red fluorescents such as mCherry, mStrawberry, mPlum, and mKate possess excitation maxima in this region (Chudakov *et al.*, 2005; Shaner *et al.*, 2004; Kapoor *et al.*, 2008; 2009). Texas Red and its more recent modification Alexa Fluor 594 is a very bright fluorescent probe, widely used in epifluorescence and confocal microscopy by use of the yellow 577 nm line of an arc lamp, or a yellow HeNe 594 nm laser (Telford, 2003; Titus *et al.*, 1982). This probe has seen some use in flow cytometers equipped with dye head lasers, but has declined as instruments have become smaller and dye heads retired (Loken *et al.*, 1987). Yellow HeNe lasers emit at very low power levels (<5 mW) and have not been widely integrated into cytometers. This range of laser light has also proven difficult to generate from a solid-state source.

Fortunately there are now several alternatives that are suitable for flow cytometry. Examples are shown in Fig. 6. Coherent has tested a DPSS 570 nm module, and

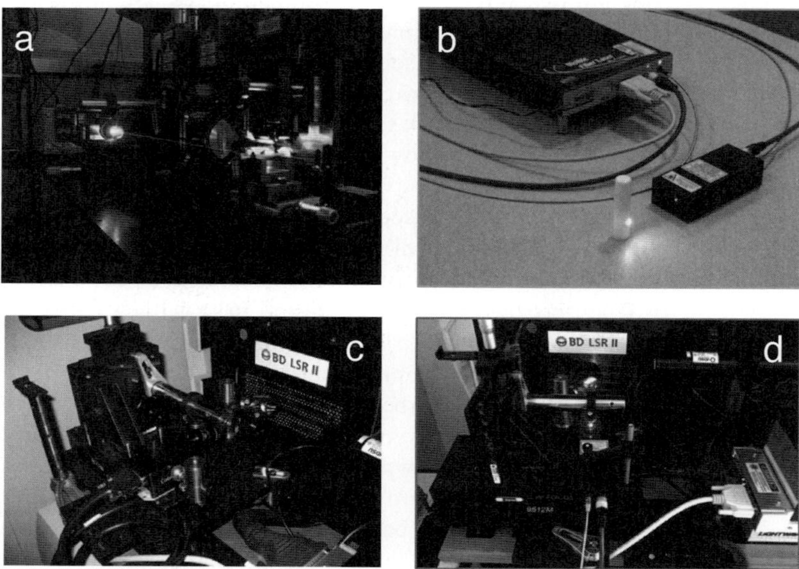

Fig. 6 Yellow to red lasers. (a) Fiber laser 580 nm emitting at 200 mW (MPB Communications); (b) fiber laser 592 nm emitting at 200 mW (MPB Communications); (c) DPSS 594 nm laser emitting at 50 mW (Cobolt AB); (d) fiber laser 628 nm emitting at 200 mW (MPB Communications). (For interpretation of the references to color in this figure legend, the reader is referred to the Web version of this chapter.)

manufactures both 568 and 577 nm modules that would be suitable for flow cytometry. A 580 nm fiber laser is available that provides excellent excitation of long-wavelength red fluorescent proteins, and has been demonstrated on a jet-in-air cell sorter (MPB Communications) (Fig. 6a). Both DPSS and fiber lasers are now available from several manufacturers, particularly at the yellow HeNe legacy wavelength range of 592 to 594 nm (Fig. 6b and 6c). A 594 nm laser can excite Texas Red, APC, and two APC tandem dyes simultaneously, adding one additional parameter over a traditional red laser. Orange 594 nm lasers are also excellent for exciting the longest red fluorescent proteins, including mPlum, HcRed, mKate2, and E2 Crimson (Kapoor *et al.*, 2008; 2009).

Coherent manufactures a high power 607 nm laser, fairly close to the 612 nm orange HeNe line. Orange HeNe lasers were rarely used in flow cytometry, and a potential application for a line in this range remains to be seen.

H. Red (620–650 nm)

These include the familiar red laser sources, namely the HeNe 633 nm and the red laser diode. Red HeNes are still excellent lasers for many applications, possessing good beam characteristics and being stable, long-lived and inexpensive (Hoffman *et al.*, 1987; Shapiro, 1986; Shapiro and Stephens, 1986). They are now a little large and fragile relative to solid-state sources, and are thus being replaced in my systems with much smaller red diodes (Doornbos *et al.*, 1993, 1994). One good advantage of a HeNe was its fixed wavelength at 633 nm (really 632.8 nm), which is the same from laser to laser. Red laser diodes using gallium arsenide phosphide chemistry on a GaP substrate can range anywhere from 635 to 645 nm, usually centering in the middle (Shapiro, 2003). As a result, a red diode is not an identical replacement for a 633 nm HeNe; if the diodes emit at over 640 nm, it might overlap slightly into the typical range for an APC filter, usually at 660/20 nm. When red diodes are used for flow cytometry, special care should be taken to use a detection filter that is blocked for their emission. A narrow bandwidth "clean-up" filter (~640/8 nm) is often used in front of the laser to similarly block any diode "glow" that might occur above 640 nm.

Red HeNe lasers in small packages (the standard 44 mm cylinder) are generally limited in power to less than 25 mW). Higher power NeNe lasers in larger packages such as the Spectra-Physics Model 127 are increasingly unavailable. Red laser diodes in the 640 nm range are now quite powerful, exceeding 100 mW in larger packages and reaching 25 mW in "lipstick" modules. Several red DPSS lasers have also been developed, including a high-power fiber laser emitting at 642 nm (MPB Communications).

While red laser diodes are frequently replacing red HeNe lasers in flow, their often undesirable increased wavelength has driven the development of a true red HeNe replacement. A red 628 nm fiber laser is now available (MPB Communications) that provides comparable excitation to a red HeNe, at much higher power levels (Fig. 6d). This laser is now being integrated into many flow cytometer systems.

I. Long Red to Near Infrared (650 nm and Beyond)

The number of fluorescent dyes available for flow cytometry that excite in the long red region of the spectra is relatively small, and few instruments are equipped with the long red lasers to excite them. However, extending detection into the infrared is one strategy for increasing our multiparametric analysis capability (Lawrence *et al.*, 2008). Some of the earliest red diodes possesses emission wavelengths ranging from 650 to 780 nm, and would be applicable for flow cytometry. At least one instrument manufacturer offers a 780 nm laser as an option.

VII. Multiwavelength Sources for Flow Cytometry

All of the lasers discussed above can potentially be integrated into flow cytometers. There is a growing need to carry out high-order multiparametric analysis (10–20 simultaneous parameters), as well as the desire to analyze many novel fluorochromes with unusual excitation/emission properties. Flow cytometers with several lasers in now the norm, and commercial instruments with 7 to 10 laser sources are now available.

Nevertheless, there are considerable technical challenges in integrating so many lasers into a single instrument. In addition to the usual space requirements, the problem of dealing with so many excitation wavelengths projecting into "detection space" is a big one. The filter and dichroic mirror elements must be properly chosen, blocked, and arrayed to maximize sensitivity in the face of a complex selection of laser light. It is also likely that not all lasers will be needed all the time; while a 580 nm yellow laser might be necessary for mCherry excitation, it is not essential for all analysis, and could be switched on or off as the experiment requires.

Being able to choose the most optimal wavelength for excitation is also an important consideration in flow cytometry. While the lasers discussed above cover a wide swath of the visible spectrum, there are still gaps that might correspond to an existing or future fluorochrome. We might also need to modify our excitation wavelengths to conform to a particular detector arrangement (as the 550 nm laser does in the case of simultaneous fluorescein and PE detection). Fixed laser wavelengths may therefore not provide the flexibility we need for some applications.

Several new technologies are currently under development to extend the flexibility of laser excitation for flow cytometry. *Laser merge modules* allow for the integration of many laser wavelengths into a single fiber output, reducing the complexity of the instrument design while maintaining laser flexibility. *Supercontinuum white light lasers* emit *all* laser wavelengths simultaneously in the visible and infrared spectra, allowing desired bandwidths to be extracted for excitation purposes. Tunable fiber lasers emit a continuous series of monochromatic wavelengths within a particular range, allowing fine-tuning of laser wavelength for particular applications. All of these technologies can potentially improve the flexibility and precision of our flow cytometry. While most of these applications have not yet been widely applied to flow cytometers, they have been used in confocal microscopy systems.

A. Laser Merge Modules

Laser-merged modules allow multiple lasers to be installed on a cytometer in a minimal space. They use fiber optic coupling, a technique now widely used to deliver laser power to a device in situations where a traditional "air-launched" laser is inconveniently large. Merge modules take this technology one-step further. Multiple lasers are assembled on an optical table or box, and their beams merged using dichroic mirrors to a single collinear beam spot. This combined beam spot is then coupled to a single mode fiber optic that is optimized for a broad range of wavelengths. The fiber output can then be integrated into a cytometer, with the appropriate collimation lens to reconstruct the beam geometry at the fiber output. The mounted lasers can therefore be placed at some distance from the instrument, reducing the size of the portion of the cytometer containing the detection elements. Optical shutters can then be employed to turn on or off the instruments needed for a particular application (Fig. 7a).

This approach to integrating multiple lasers in a small space has been employed on a Cytek Development modified FACSCalibur, where a merge module containing 405, 440, 532, 561, and 640 nm lasers was coupled to one of the instrument detection

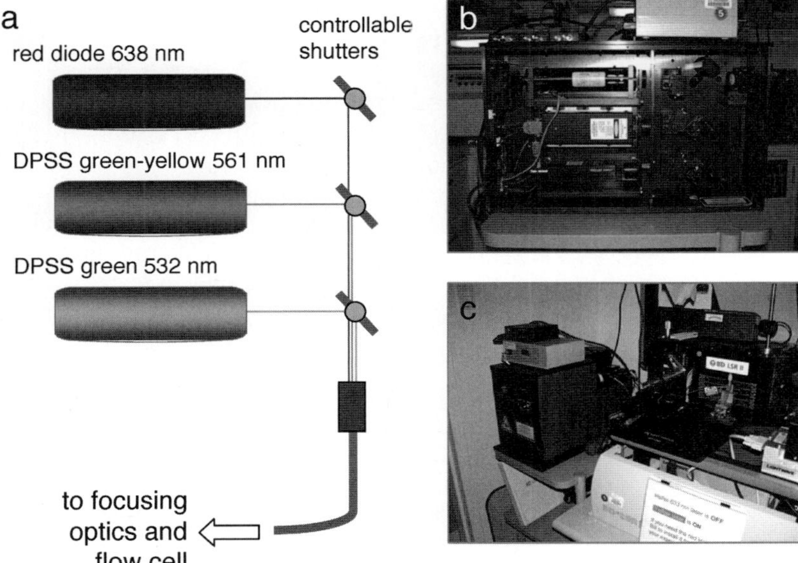

Fig. 7 Laser merge module. (a)Schematic of laser merge module; (b) interior of laser merge module, showing lasers and steering mirrors (Spectral Applied Research, Toronto, CA); (c) laser merge module mounted on a LSR II.

paths (Fig. 7b and 7c). It is also employed commercially in the Beckman-Coulter MoFlo XDP and Astrios cell sorters, where one or two merge modules allow for the integration of up to seven lasers simultaneously.

B. Supercontinuum White Light Lasers

SC white light lasers are solid-state nonlinear fiber lasers that emit at all visible wavelengths simultaneously and continuously within a certain range (Birks *et al.*, 2000; Gmachl *et al.*, 2002; Nowak *et al.*, 1999; Sun *et al.*, 2007). The commercial units available today generally emit in a range from approximately 450 to 2200 nm, providing a more or less continual curve of emission over the entire range (Fig. 8a and 8b) (Sun *et al.*, 2007). When viewed (carefully, with proper protection) by eye, this laser light is white in appearance; it in fact contains almost the entire visible spectrum, with a large component of infrared light (Fig. 8c). In theory, wavelength ranges desired for flow cytometry could be filtered out of this continuous spectrum, and employed the way monochromatic laser light is employed. The implications of this technology are enormous; if we can selectively filter wavelength or wavelengths

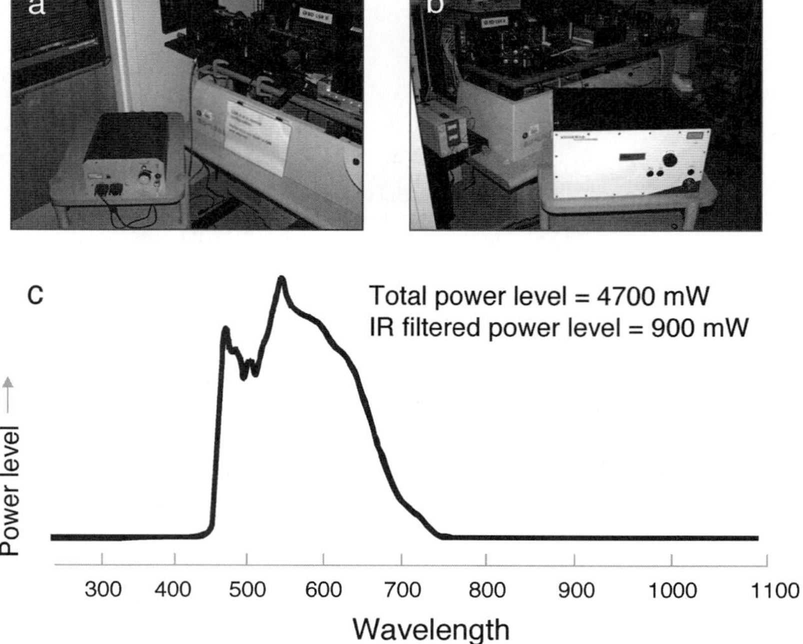

Fig. 8   Supercontinuum lasers. (a) SC450 6W SC laser (Fianium, Southampton, UK); (b) SuperXtreme 6W SC laser (NKT Photonics, Birkerod, Denmark); (c) visible emission spectrum of SC source after removal of IR component using two dielectric cold mirrors.

we need for our probes, we can use a single laser to provide precise wavelength ranges of light, fine-tuned to the needs of the experiment and the detection system.

Supercontinuum lasers are already being used in other biomedical detection systems, including spectroscopy, spectrofluorimetry, and confocal microscopy. Leica has released a laser confocal system based on a single SC laser source, filtering the total emission for up to eight distinct wavelength bandwidths. Pilot experiments have also been carried out using SC laser light flow flow cytometry (Telford *et al.*, 2009b). These experiments demonstrated some of the limitations that SC sources have for this particular application. SC lasers are very powerful, emitting up to 6 W total power output. However, this output is distributed over a wide wavelength range. The output in the visible spectrum is about 30% of the total. The resulting output of visible light is therefore between 2 and 7 mW/nm. Filtering of single nanometer "windows" of laser light is therefore not likely to provide enough power for flow cytometer. We instead need to filter multinanometer bandwidths of SC light and use them as we would traditional monochromatic laser beams.

In a SC configuration for flow cytometry, the total SC emission is reflected off two "cold mirrors" to remove the infrared portion of the light (above 750 nm) (Fig. 9a and 9b). The remaining laser light is then filtered through a heat-resistant narrow bandpass filter, not unlike those used for fluorescence detection. The resulting laser light will include all wavelengths transmitted by the bandpass filter (Fig. 9c). If the filter is sufficiently narrow (~10 nm), the beam properties will be similar to a traditional laser.

SC laser excitation has been demonstrated for a variety of bandwidths corresponding to traditional laser lines, including ~530, 580, 590, and 630 nm (Telford *et al.*, 2009b). The power level was roughly equal to that derived from a traditional laser, albeit spread out over a larger spectral range. Sensitivity for a variety of fluorochromes was found to be comparable with traditional lasers (Fig. 9d) (Telford *et al.*, 2009b). Care was required to ensure that the laser bandwidth did not overlap into any detection filter bandwidths; this is an optical design issue that would need to be addressed in any commercial system employing a SC laser.

Using this system, any laser wavelength range can be produced on a flow cytometer, assuming the user has the appropriate bandpass filter. Use of bandpass filters is an easy and relatively efficient way to select the desired wavelength range, but does not provide continuous tuning of the laser. There are other methods by which SC laser light could be selectively tuned across the visible spectrum. Computer-controlled acoustic-optical tuning filters (AOTFs) can also be employed to separate wavelengths of interest; this is the technology employed in laser confocal microscopy. Unfortunately AOTFs are not particular efficient, and generate considerable noise in their operation. This is not a major problem for imaging but presents difficulties for flow cytometry. In another approach, monochromators or prisms could also be used to split the laser light into a multispectral band; monochromators are used for white-light-laser-based spectrometers and spectrofluorimeters (Fig. 10a and 10b). This approach has been tried using both single and paired equilateral dispersive prisms and an optical slit, allowing very narrow bandwidths to be isolated from the SC laser (Fig. 10c and 10d). The resulting

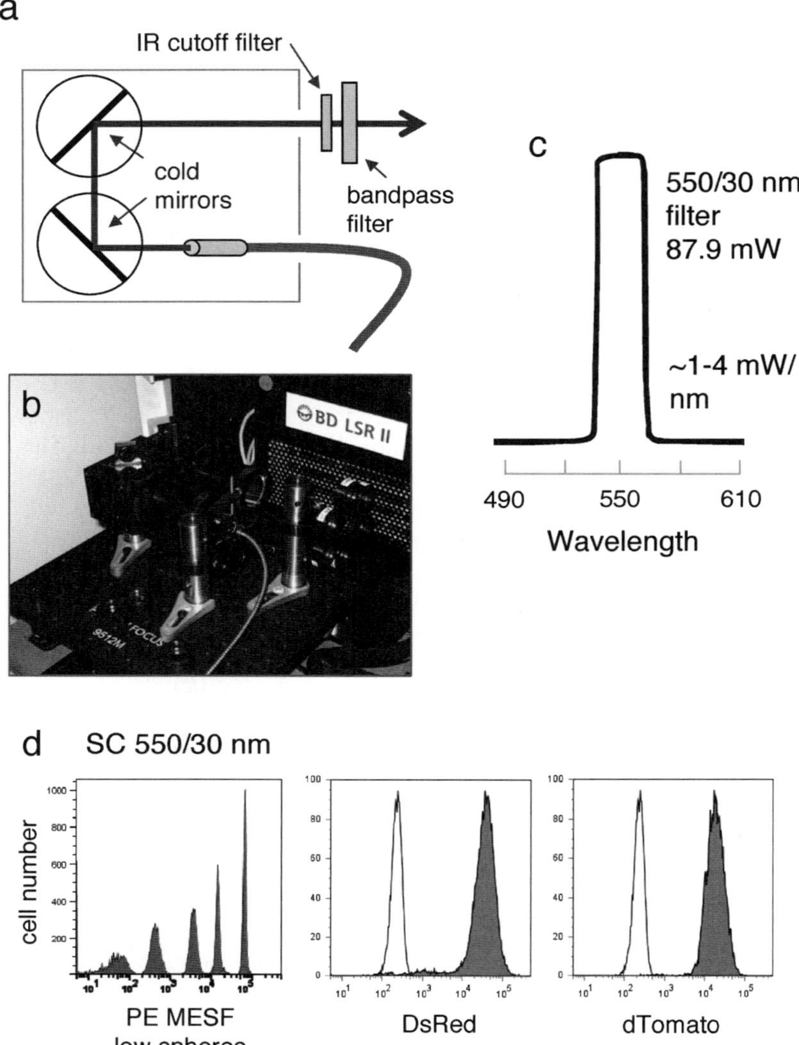

Fig. 9 Integrating SC lasers into flow cytometers. (a) Optics to remove infrared component from SC light, including two reflecting cold mirrors, and IR cutoff filter (700 nm shortpass) and the desired narrow bandpass filter; (b) SC filter components on a BD Biosciences LSR II flow cytometer; (c) emission spectra of SC laser light filtered with a 550/30 nm narrow bandpass filter; (d) flow cytometric analysis of a PE MESF low-fluorescence microsphere array and SP2/0 cells expressing the red fluorescent proteins DsRed and dTomato, using the above SC laser 550/30 nm emission on a LSR II flow cytometer. (For interpretation of the references to color in this figure legend, the reader is referred to the Web version of this chapter.)

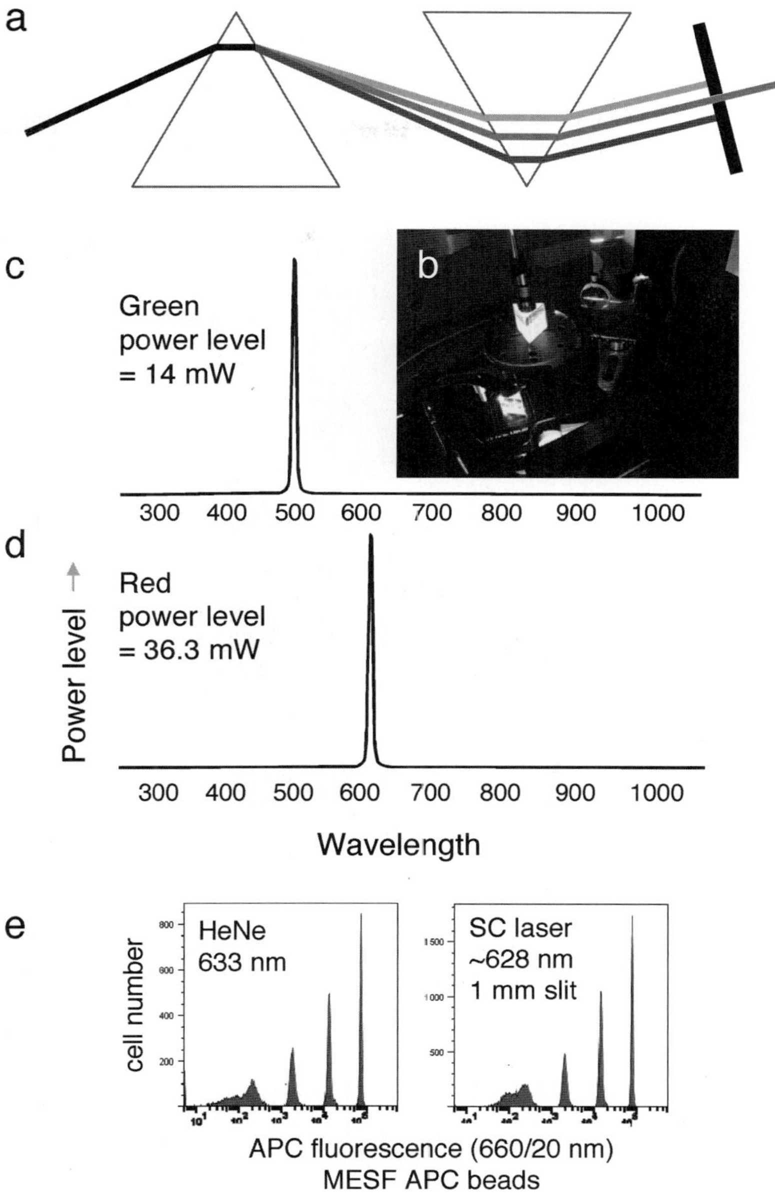

Fig. 10 Spectral separation of SC laser light. (a) Schematic of equilateral dispersive prism pair used for dispersing and realigning the separated light, with an adjustable slit for isolating the desired bandwidth; (b) actual prism assembly on LSR II; (c and d) spectroradiometric scans green and red wavelength bandwidths separated by a prism pair; (e) flow cytometric analysis of an allophycocyanin (APC) MESF low-fluorescence microsphere array using either a conventional HeNe laser or the SC emission shown in Fig. 10d on an LSR II flow cytometer. (For interpretation of the references to color in this figure legend, the reader is referred to the Web version of this chapter.)

beam patterns can be used as a conventional laser, with sensitivity comparable to traditional sources (Fig. 10e). Finally, continuously tunable filters that smoothly increase or decrease their transmission bandwidth depending on their angle of incidence can be used to isolate SC bandwidths. VersaChrome^{TM} filters from Semrock allow gradually increasing 20 nm bandwidths of light to pass through them as they are rotated in the laser path (Fig. 11a, 11b and 11c). These bandwidths can also be used

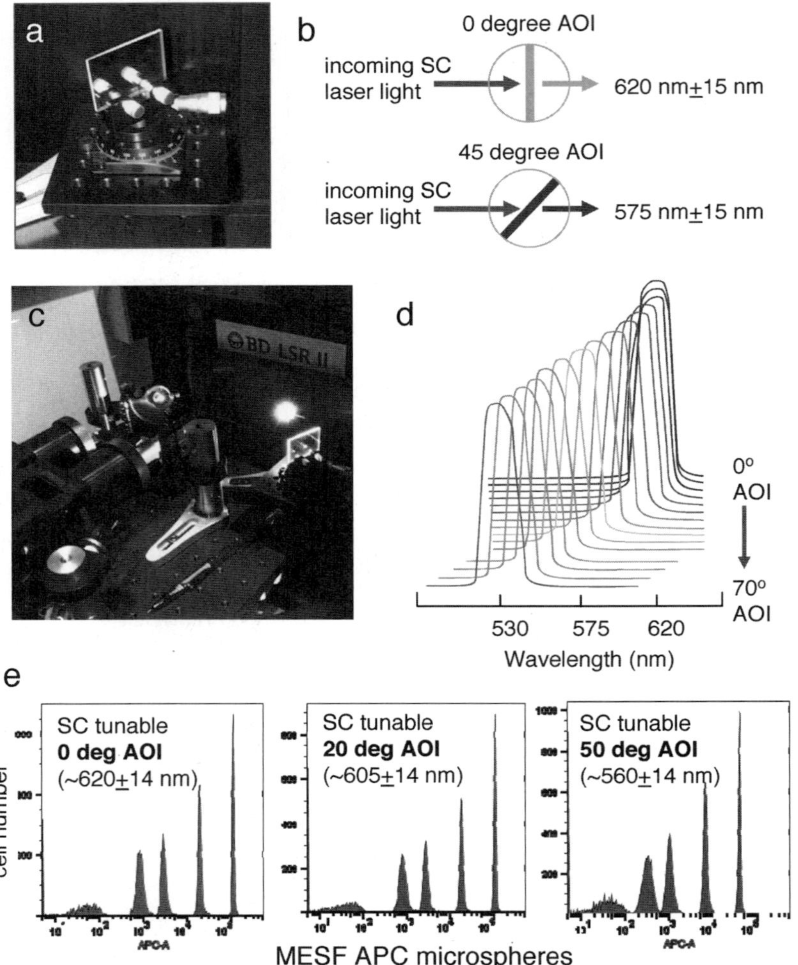

Fig. 11 Tunable filters with SC laser light. (a) Semrock VersaChrome tunable filter; (b) angle of incidence for VersaChrome light transmission ranging from 620 nm at 0° to 575 nm at 45°; (c) VersaChrome^{TM} filter mounted in path of the SC laser with reflected light baffle; (d,) spectroradiography of tunable filter transmitted SC laser light, ranging from 530 to 620 nm in roughly 14 nm bandwidths; (e) flow cytometric analysis of an APC MESF low-fluorescence microsphere array using tunable filter transmitted SC laser light at different filter angles of incidence.

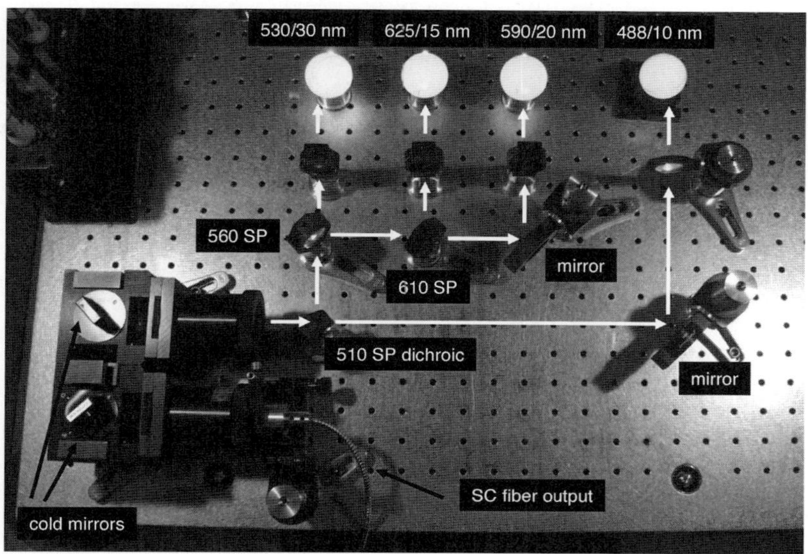

Fig. 12 Multiple wavelengths from SC laser light. Filter configuration for splitting SC laser light into four useful wavelengths. Specifications of dichroics and bandpass filters are indicated. Final power levels for 530/30, 615/25, 590/20, and 488/10 nm were 37.5, 81.8, 29.5, and 17.4 mW, respectively.

as a conventional laser, with similar results (Fig. 11d). This provides a very elegant and efficient way to separate the desired wavelengths of light from a SC laser.

In addition to continuous tuning, SC lasers could provide all the necessary laser lines required for a single instrument, using a single laser. A standard array of dichroics and filters can be set up to isolate several wavelength ranges relevant for flow cytometry, at power levels sufficient for cuvette-based instruments (Fig. 12).

C. Tunable Fiber Lasers

While SC white light lasers may are intriguing due to their ability to "tune" to wavelength ranges of interest, they have some limitations. Considerable power is "lost" to most cytometrists in the infrared range, and must be removed prior to introduction into a standard flow cytometer. Considerable research and development is being exerted toward concentrating SC emission more strongly in the visible range. SC lasers are class IV lasers and operate at very high power levels, posing a degree of user risk due to these high emissions (particularly in the infrared range). Nevertheless, the power level per nanometer is not particularly high, and most of this energy is discarded in attempt to provide reasonably narrow bandwidths for fluorochrome excitation.

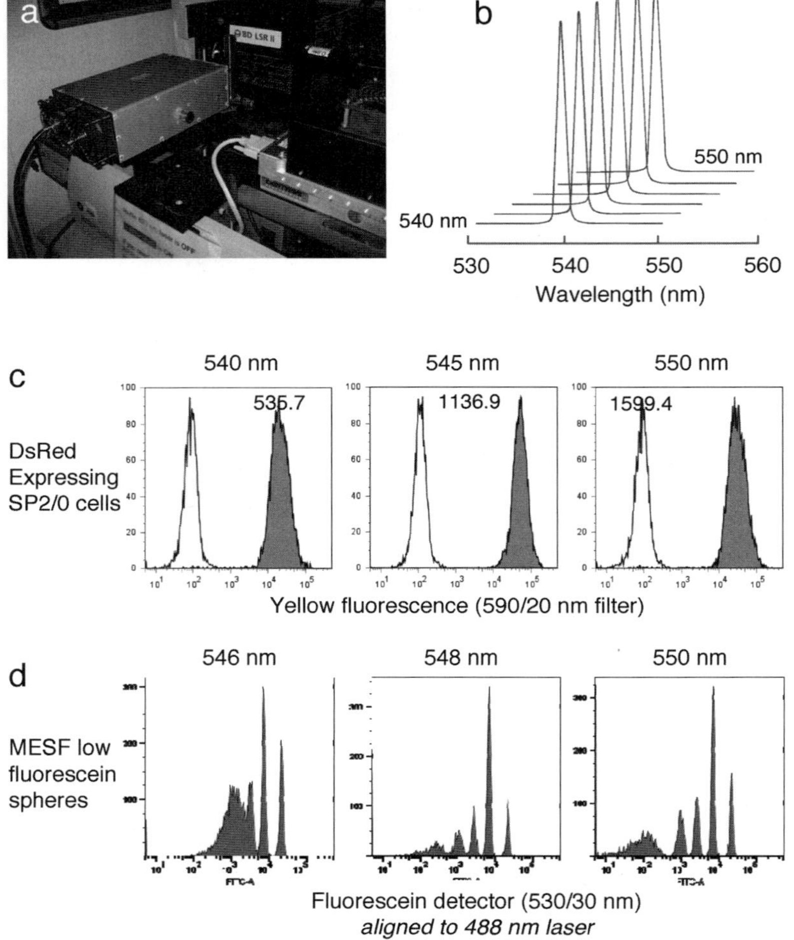

Fig. 13 True tunable lasers. (a) Tunable green fiber laser, with an adjustable range of 540 to 550 nm (Zecotek); (b) spectroradiography of tunable green laser at 540, 542, 544, 546, 548, and 550 nm intervals. Banwidths were approximately 1 nm; (c) flow cytometric detection of DsRed expression in SP2/0 cells at 540, 545, and 550 nm; (d) overlap of green laser light into the fluorescein detector of a LSR II, using a FITC MESF low-fluorescence microsphere arrays to measure low-signal discrimination. At 550 nm, almost no green laser light reaches the detector, and all spheres can be resolved. At 546 nm, some laser light reaches the detector, and some resolution is lost. (For interpretation of the references to color in this figure legend, the reader is referred to the Web version of this chapter.)

The ideal tunable laser source would emit at monochromatic wavelengths ($\sim$1 nm bandwidths), one or more than one at a time, with essentially no emission in between. The power level of the tunable wavelengths should be sufficient for most flow cytometers, between 10 and 100 mW and more if possible. Such a system would address the concerns found with SC lasers, and would provide flow cytometers with complete tuning flexibility.

True tunable lasers fulfilling this description have existed for some time in the infrared range (>1200 nm). Several recent innovations in nonlinear fiber lasers have recently produced the first true tunable solid-state lasers in the visible range with specifications appropriate for flow cytometry. A SC nonlinear quasi-CW fiber laser from Toptica Photonics (Munich, Germany), the iChrome can produce any 1 nm wavelength from 488 to 640 nm. However, the power output is roughly 1 to 3 mW per nanometer, adequate for imaging but not yet optimal for flow cytometry. A fiber laser module produced by Zecotek Photonics Ltd. (Singapore) can be continuously tuned in 1 nm intervals from 540 to 550 nm, producing between 50 and 150 mW/nm interval (Fig. 13a and 13b) (Akulov *et al.*, 2007). This laser has been mounted on a flow cytometer and produces data comparable with a single wavelength laser source, requiring minimal effort to change the wavelength (Fig. 13c and 13d) (Telford *et al.*, 2009a). While this range is still very limited, a module that can tune from 525 to 540 nm is also available, and future units should be able to cover the entire green range of 515 to 560 nm. Such a unit should have applications in fine-tuning excitation to the broad range of red fluorescent proteins, for example. The ultimate goal will have a handful of laser, or perhaps a single unit, that can provide the entire range of wavelengths necessary for flow cytometry.

VIII. Summary

The single laser 488 nm flow cytometers of 30 years have become dual laser cytometers (blue and red), and subsequently blue, red, and violet. The standard benchtop cytometer now has at least three lasers, and often far more; high-end instruments with 7 to 10 laser modules are not uncommon. Increase in the number of lasers can increase the number of simultaneous parameters we can measure, but only up to a point; a three-laser cytometer can excite and analyze up to 16 markers simultaneously, for example. The main advantage of increased laser number lies in their ability to excite fluorochromes that were previously inaccessible to flow cytometers (e.g., the red fluorescent proteins), and to fine-tune our excitation capabilities to maximize detection sensitivity. It also allows us to design complex experiments that is FRET, to investigate cell function beyond the simple presence or absence of a protein. Increasing the number and flexibility of our excitation options greatly enhances the analytical "vision" of flow cytometry.

References

Akulov, V. A., Afanasiev, D. M., Babin, S. A., Churkin, D. V., Kablukov, S. I., Rybakov, M. A., Vlasov, A. A. (2007). Frequency tuning and doubling in Yb-doped fiber lasers. *Laser Phys.* **17**, 124–129.
Beavis, A. J., and Kalejta, R. F. (1999). Simultaneous analysis of the cyan, yellow and green fluorescent proteins by flow cytometry using single-laser excitation at 458 nm. *Cytometry* **37**, 68–73.

Birks, T. A., Wadsworth, W. J., and Russell, P. S. (2000). Supercontinuum generation in tapered fibers. *Opt. Lett.* **25**, 1415–1417.

Cabana, R., Frolova, E. G., Thomas, R. A., Krishan, A., and Telford, W. G. (2006). The minimal instrumentation requirements for Hoechst side population analysis: Stem cell analysis on low-cost flow cytometry platforms. *Stem Cells* **24**, 2573–2581.

Chudakov, D. M., Lukyanov, S., and Lukyanov, K. A. (2005). Fluorescent proteins as a toolkit for *in vivo* imaging. *Trends Biotechnol.* **23**, 605–613.

Doornbos, R. M. P., De Grooth, B. G., Kraan, Y. M., Van Der Poel, C. J., and Greve, J. (1994). Visible diode lasers can be used for flow cytometric immunofluorescence and DNA analysis. *Cytometry* **15**, 267–271.

Doornbos, R. M. P., Hennink, E. J., and Putman, C. A. J. (1993). White blood cell differentiation using a solid state flow cytometer. *Cytometry* **14**, 589–594.

Frey, T., Stokdijk, W., and Hoffman, R. A. (1993). Bivariate flow karyotyping with air-cooled lasers. *Cytometry Suppl.* **6**, 71.

Gmachl, C., Sivco, D. L., Colombelli, R., Capasso, F., and Cho, A. Y. (2002). Ultra-broadband semiconductor laser. *Nature* **415**, 883–887.

Goodell, M. A., Brose, K., Paradis, G., Conner, A. S., and Mulligan, R. C. (1996). Isolation and functional properties of murine hematopoietic stem cells that are replicating in vivo. *J. Exp. Med.* **183**, 1797–1806.

Gray, J. W., and Cram, L. S. (1990). Flow karyotyping and chromosome sorting. In *Flow cytometry and sorting*, 2nd Edition, Wiley-Liss, New York, pp. 503–529.

He, L., Wu, X., Meylan, F., Olson, D. P., Simone, J., Hewgill, D., Siegel, R., Lipsky, P. E. (2004). Monitoring caspase activity in living cells using fluorescent proteins and flow cytometry. *Am. J. Pathol.* **164**, 1901–1913.

Hoffman, R. A. (2002). Multicolor immunofluorescence flow cytometry using 400 nm laser diode excitation. *Cytometry Suppl.* **11**, 124.

Hoffman, R. A., Reinhardt, B. N., and Stevens, F. E. (1987). Two-color immunofluorescence using a red helium neon laser. *Cytometry Supp* **1**, 103.

Kapoor, V., Karpov, V., Linton, C., Subach, F. V., Verkhusha, V. V., Telford, W. G. (2008). Solid state yellow and orange lasers for flow cytometry. *Cytometry* **73**, 570–577.

Kapoor, V., Subach, F. V., Kozlov, V. G., Grudinin, A., Verkhusha, V. V., Telford, W. G. (2007). New lasers for flow cytometry: filling the gaps. *Nature Methods* **4**, 678–679.

Lawrence, W. G., Varadi, G., Entine, G., Podniesinski, E., and Wallace, P. K. (2008). Enhanced red and near infrared detection in flow cytometry using avalanche photodiodes. *Cytometry A.* **73**, 767–776.

Lelimousin, M., Noirclerc-Savoye, M., Lazareno-Saez, C., Paetzold, B., Le Vot, S., Chazal, R., Macheboeuf, P., Field, M. J., Bourgeois, D., Royant, A. (2009). Intrinsic dynamics in ECFP and Cerulean control fluorescence quantum yield. *Biochemistry* **48**, 10038–10046.

Loken, M. R., Keij, J. F., and Kelley, K. A. (1987). Comparison of helium–neon and dye lasers for the excitation of allophycocyanin. *Cytometry* **8**, 96.

Nakamura, S., and Fasol, G. (1997). *The blue laser diode. GaN based light emitters and lasers.* Springer, Berlin.

Nowak, G. A., Kim, J., and Islam, M. N. (1999). Stable supercontinuum generation in short lengths of conventional dispersion-shifted fiber. *Appl. Opt.* **38**, 7364–7369.

Perfetto, S. P., and Roederer, M. (2007). Increased immunofluorescence sensitivity using 532 nm excitation. *Cytometry* **71A**, 73–79.

Shaner, N. C., Campbell, R. E., Steinbach, P. A., Giepmans, B. N. G., Palmer, A. E., Tsien, R. Y. (2004). Improved monomeric red, orange and yellow fluorescent proteins derived from Discosoma sp. red fluorescent proteins. *Nature Biotechnol.* **22**, 1567–1572.

Shapiro, H. M. (1986). The little laser that could: applications of low power lasers in clinical flow cytometry. *Ann. N. Y. Acad. Sci.* **468**, 18.

Shapiro, H. M. (1993). Trends and developments in flow cytometry instrumentation in clinical flow cytometry. *Ann. N. Y. Acad. Sci.* **677**, 155–166.

Shapiro, H. (2003). How flow cytometers work. *In* "Practical flow cytometry," (and H. Shapiro, ed.), pp. 124–149. John Wiley and Sons, New York, NY.

Shapiro, H. M., and Perlmutter, N. G. (1993). Bivariate chromosome flow cytometry using single-laser instruments. *Cytometry Suppl.* **6**, 71.

Shapiro, H. M., and Perlmutter, N. G. (2001). Violet laser diodes as light sources for cytometry. *Cytometry* **44**, 133–136.

Shapiro, H. M., and Stephens, S. (1986). Flow cytometry of DNA content using oxazine 750 or related laser dyes with 633 nm excitation. *Cytometry* **7**, 107.

Shapiro, H. M., and Telford, W. G. (2009). Lasers for flow cytometry. *In* "Current Protocols in Cytometry," (J. P. Robinson, Z. Darzynkiewicz, J. Dobrucki, R. A. Hoffman, J. P. Nolan, A. Orfao, and P. S. Rabinovitch, eds.), pp. 1.9. John Wiley and Sons, New York, NY.

Snow, C., and Cram, L. S. (1993). The suitability of air-cooled helium cadmium (HeCad) lasers for two color analysis and sorting of human chromosomes. *Cytometry Suppl.* **6**, 20.

Sun, J. H., Gale, B. J., and Reid, D. T. (2007). Composite frequency comb spanning 0.4–2.4 micron from a phase-controlled femtosecond Ti:sapphire laser and synchronously pumped optical parametric oscillator. *Opt. Lett.* **32**, 1414–1416.

Telford, W. G. (2004a). Analysis of UV-excited fluorochromes by flow cytometry using a near-UV laser diode. *Cytometry* **61A**, 9–17.

Telford, W. G. (2004b). Small lasers in flow cytometry (invited book chapter). *In* "Methods in Molecular Biology Volume 263, Flow Cytometry Protocols," (T. S. Hawley, and R. G. Hawley, eds.), pp. 399–418. Humana Press, London, UK.

Telford, W. G., Babin, S. A., Khorev, S. V., and Rowe, S. H. (2009a). Green fiber lasers: An alternative to traditional DPSS green lasers for flow cytometry. *Cytometry A* **75**, 1031–1039.

Telford, W. G., Bradford, J., Godfrey, W., Robey, R. W., and Bates, S. E. (2007). Side population analysis using a violet-excited cell permeable DNA binding dye. *Stem Cells* **25**, 1029–1036.

Telford, W. G., and Frolova, E. G. (2004). Discrimination of Hoechst side population in mouse bone marrow with violet and near-UV laser diodes. *Cytometry* **57A**, 45–52.

Telford, W. G., and Huber, C. (2006). Novel solid-state lasers in flow cytometry. *Biophotonics Int.* **13**, 50–53.

Telford, W. G., Hawley, T. S., and Hawley, R. G. (2002). Analysis of violet-excited fluorochromes by flow cytometry using a violet laser diode. *Cytometry Suppl.* **11**, 123.

Telford, W. G., Hawley, T. S., and Hawley, R. G. (2003). Analysis of violet-excited fluorochromes by flow cytometry using a violet laser diode. *Cytometry* **54A**, 48–55.

Telford, W. G., Kapoor, V., Jackson, J., Burgess, W., Buller, G., Hawley, T., Hawley, R. (2006). Violet laser diodes in flow cytometry: an update. *Cytometry* **69**, 1153–1160.

Telford, W. G., Murga, M., Hawley, T., Hawley, R. G., Packard, B. Z., Komoriya, A., Haas, R., Hubert, C. (2005). DPSS yellow-green 561 nm lasers for improved fluorochrome detection by flow cytometry. *Cytometry* **68A**, 36–44.

Telford, W. G., Subach, F. V., and Verkhusha, V. V. (2009b). Supercontinuum white light lasers for flow cytometry. *Cytometry* **75A**, 450–459.

Titus, J. A., Haugland, R., Sharrow, S. O., and Segal, D. M. (1982). Texas Red, a hydrophilic, red-emitting fluorophore for use with fluorescein in dual parameter flow microfluorometric and fluorescence microscopic studies. *J. Immunol. Methods* **50**, 193–204.

Van den Engh, G., Trask, B., Cram, S., and Bartholdi, M. (1984). Preparation of chromosome suspensions for flow cytometry. *Cytometry* **5**, 108–117.

CHAPTER 16

The Use of Hollow Fiber Membranes Combined with Cytometry in Analysis of Bacteriological Samples

Jerzy Kawiak,[*,‡] **Radosław Stachowiak,**[†] **Marcin Lyżniak,**[*] **Jacek Bielecki**[†] **and Ludomira Granicka**[‡]

[*]Department of Clinical Cytology, Medical Center Postgraduate Education, Warsaw

[†]Department of Applied Microbiology, Warsaw University, Warsaw

[‡]Institute of Biocybernetics and Biomedical Engineering PAS, Warsow/Poland

411

0091-679X/10 $35.00
DOI 10.1016/B978-0-12-374912-3.00016-X

Abstract

To avoid destruction of the implanted biological material it may be separated from host immunological system by enclosure within a permiselective membrane. Two-directional diffusion through the membrane of nutrients, metabolic products, as well as bioactive products of encapsulated cells is required to ensure their survival and functional activities.

The system of cells encapsulated within the membrane releasing the biologically active substance may be applied either locally to give an opportunity of therapeutic agent activity in the specified place and/or at some convenient site (tissue) for a prolonged period of time. The novel system of bacteria bio-encapsulation using modified membranes, and its assessment by flow cytometry is described and discussed.

The encapsulated in membrane bacteria, functioning and releasing their products were evaluated in the systems in vitro and in vivo. The bacteria cells products impact on Eukariotic cells was evaluated. The cytometric evaluation demonstrates the membrane ability to avoid the release of bacteria enclosed within the membrane wall. In experiments with treatment of the bacteria with antibiotic to release products from damaged bacteria it was possible to distinguish stages of the applied antibiotic impact on encapsulated bacteria cells. In *E. coli* following stages were distinguished: induction of membrane permeability to PI, activation of proteases targeting GFP (protein) and subsequent nucleic acids degradation. In the another experiment the evidence was presented of the cytotoxic activity of live *Bacillus subtilis* encapsulated within the membrane system. The *Bacilus* products mediated by secreted listeriolysin O (LLO) on the chosen eukaryotic cells was evaluated. Similar systems releasing bacterial products locally and continuously may selectively affect different types of cells and may have possible application in the anticancer treatment at localized sites.

I. Introduction

To avoid destruction *in vivo* the biological material has to be separated from host immunological system by enclosure within a permeant-selective membrane. Two-directional diffusion over the membrane of small molecular weight solutes such as O_2, glucose, metabolic products, as well as bioactive products of encapsulated cells is required to ensure their survival and functional activities. The biological material such as live cells encapsulated in membranes producing biologically active substances allow for constant, potentially long-term systemic production of regulatory substances. The use of encapsulated cells is of particular importance when the half-time of the synthetic factor is too short to give therapeutic function after disposable injection. For instance, following intravenous systemic injection of $25–100$ $\mu g/dm^2$ of purified TNF-α its half-time in the human plasma is only $14–18$ min. The system of cells encapsulated in the membrane releasing the biologically active substance may be applied either locally to give an opportunity of therapeutic agent activity in the specified place and/or at some convenient site (tissue) for a prolonged period of time.

Contemporary biotechnology gives opportunity of synthesizing biologically active substances similar to natural ones. Natural products for example recombinant IL-3 produced by bacteria *Escherichia coli (E. coli)* has similar properties as its equivalent produced by lymphocytes and leukemic cell lines. However, to obtain the comparable biological response sometimes 50 times higher amount of such recombinant IL-3 must be applied (Robak, 1991).

A. Encapsulation of Bacteria

Encapsulation of the bacteria within a semipermeable polymer membrane opens several technological possibilities; it improves the bacteria cells stability during passage under adverse conditions of the gastrointestinal tract in food industry applications, where bacteria, *Lactobacillus acidophilus, Bifidobacterium, Lactobacillus casei*, are added to ice cream, cheese, mayonnaise, and yogurt because of its beneficial action in human organism (Charalampopoulos *et al.*, 2002; Hou *et al.*, 2003; Shah, 2000; Sultana *et al.*, 2000). Bacteria isolated from places contaminated with petrol, when encapsulated in microcapsules, are protected, and degrade hydrocarbons in petrol three times faster with application of encapsulated bacteria as compared to nonencapsulated (Moslemy *et al.*, 2002). Bacteria innoculum encapsulated in alginate microcapsules are used for removal of the components of phenol (Hajii *et al.*, 2000). For the same purpose many bacteria, *Pseudomonas, Acidiomonas, Commamonas, Zooglea Azotobacter,* immobilized in polysulfone or polypropylene hollow fibers' (HF) lumen may be applied. Similar system protects from the attack of the implanted microorganisms to the host animal (Granicka *et al.*, 2005). Some encapsulated microorganisms may carry a transfected gene and express the regulatory molecules, thereby becoming a source of valuable regulating factors. Such factors released in strategic locations may direct or modify the biological processes in the eukaryotic organism (Chang and Prakash, 2001; Prakash and Chang, 2000a; 2000b).

The encapsulation can be used to harness bacteria for longer period release of the therapeutic molecule. The most commonly performed procedure of encapsulation is based on calcium alginate gel capsule formation (Kailasapathy, 2002). Kappa-carrageenan, gellan gum, gelatin, and starch have also been used (Kidchob *et al.*, 1998; Morikawa *et al.*, 1997; Yang *et al.*, 1994; Zimmerman *et al.*, 2003) for the microencapsulation of probiotic organisms. Usually bacteria applied for therapeutic purposes are encapsulated in microcapsules. Genetically modified *E. coli* strain DH5 encapsulated in microcapsules may serve for creatinine, urea, and ammonia level reduction in renal or liver dysfunction. The serum creatinine as well as urea level decline was observed on a rodent model after microencapsulated *E. coli* DH5 oral application (Prakash and Chang, 2000a, 2000b). There are several methods of bacteria encapsulation: spry-drying, extrusion, emulsion, and phase separation.

B. Application of the Capillary Membranes

In our experiments on bacteria systems the capillary membranes were applied. The capillary membranes are typical devices among macrocapsules and they seem to be attractive for transplantation applications. HFs ensure a reproducible composition and shape, smooth surface, and high mechanical properties as well as stability. In preparation of HFs the solvent is removed directly from the membrane during its washing, after the production process (e.g., polypropylene membranes), or it diffuses directly into the coagulation bath (e.g., polysulfone, 2.5-cellulose acetate membranes). These procedures are independent of the encapsulation process. In the case with the alginian microcapsules their preparation proceeds parallel to the cells encapsulation, and the choice of the nontoxic solvent for the encapsulating cells is important. In other cases it is necessary to apply techniques reducing the toxic impact of the solvent (Crooks *et al.*, 1990; Dawson *et al.*, 1987; Zhang *et al.*, 2000).

The polypropylene surface modified hollow fiber (Granicka *et al.*, 2010) was applied for the bacteria *E. coli* isolation. Applied membrane modification allows not only to avoid bacteria adhesion, the phenomenon that was the subject of interest of some authors (Emery *et al.*, 2003; Homma *et al.*, 2006; Lewis *et al.*, 2001; Walker *et al.*, 2005; Wang *et al.*, 2004), but also prevents bacteria escaping from the lumen of the hollow fiber as well as improves membrane biocompatibility (Figs. 1 and 2).

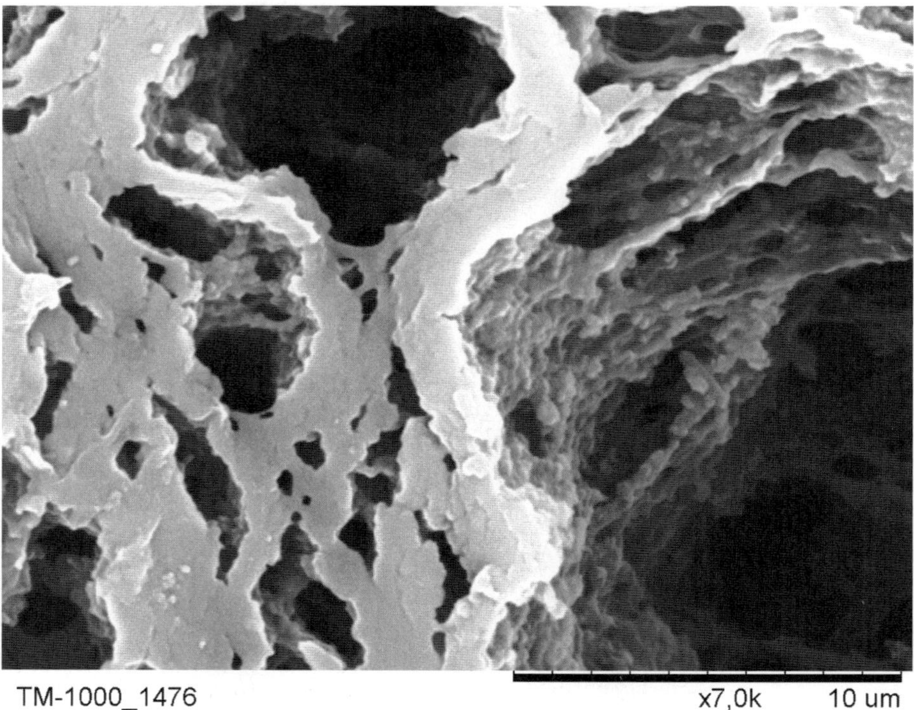

Fig. 1   EM photographic documentation of the HF membrane fragment used for encapsulation of *E. coli*. The membrane was modified and its properties preventing the escape of bacteria depend on the compact structure seen on the right surface of the membrane.

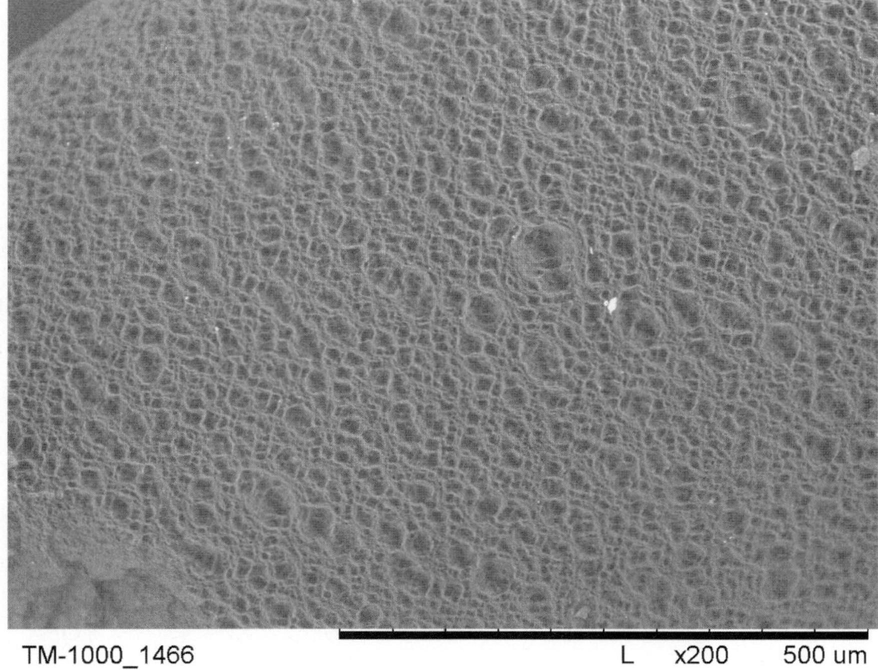

Fig. 2   EM fragment of the modified membrane surface used for *B. subtilis* encapsulation.

The membrane suitability for bacteria encapsulation may be assessed using the microbeads model.

II. Assessment of Membrane Suitability for Encapsulation of Microorganisms

The microbeads of diameter comparable to dimensions of microorganisms may be used for testing the membrane impermeability to the bacteria.

A. Permeability of the Modified Membrane For Microbeads of Diameter 0.2 μm

The HFs original or surface modified were filled up with about 20 μl volume 1% suspension of FluoroSpheres, 0.2 μm microbeads (Invitrogen, Molecular Probes, USA) in physiological saline. The encapsulated FluoroSpheres were incubated for 24 h in 1 ml physiological saline and than the sample of the saline from outer medium was cytometrically tested for the presence of the beads. As a

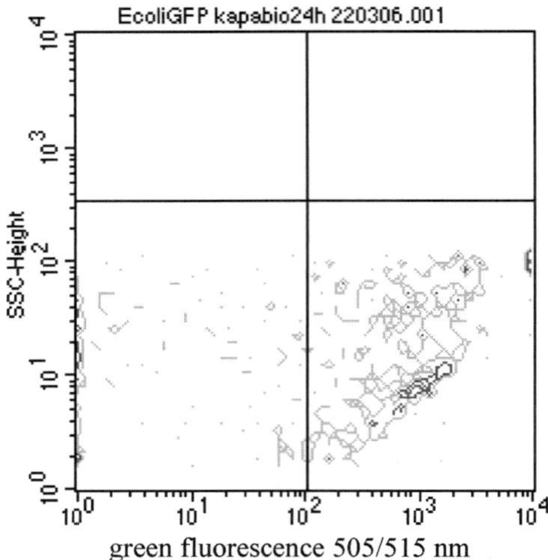

Quad	Events	% Gated	% Total
UL	0	0.00	0.00
UR	0	0.00	0.00
LL	154	27.95	27.95
LR	397	72.05	72.05

Fig. 3 Assessment of beads yellow–green fluorescence (quadrant lower right (LR)) of physiological saline supernatant above the microbeads nonencapsulated in HF after 24 h incubation. (For interpretation of the references to color in this figure legend, the reader is referred to the Web version of this chapter.)

positive control 20 µl nonencapsulated microbeads suspension was added and incubated in 1 ml physiological saline (Fig. 3). The physiological saline alone was incubated as a negative control. No presence of microbead fluorescence events was observed in external physiological saline when beads were enclosed within surface-modified HFs. In the Fig. 4 the bivariate distribution of green fluorescence versus side scatter is presented for the external culture medium in which the encapsulated microbeads were incubated enclosed in surface-modified HFs. However, the presence of microbeads with green fluorescence (FL1) was present in external physiological saline when they were encapsulated in unmodified HFs (Fig. 5). In conclusion, the modified HFs appeared to close safely FluoroSphere microbeads 0.2 µm in diameter for 24 h period as compared to unmodified HFs. The FluoroSphere microbeads diameter 0.2 µm has size comparable to *E. coli* dimensions.

The results are processed by the FACSCalibur flow cytometer (Becton Dickinson Immunocytochemistry Systems, USA) equipped with the argon ion (488 nm) laser and the CellQuest software system (Becton Dickinson, USA). Microbeads and bacteria were separated from other events on light scatter characteristics (the gate of FSC and SSC, log scale) with the proper threshold.

Some model systems of bacteria encapsulated in membranes were verified concerning: (i) the survival of encapsulated bacteria; (ii) the ability of releasing some factors from encapsulated bacteria; (iii) the ability of production and secretion of biologically active factor by HF-encapsulated bacteria.

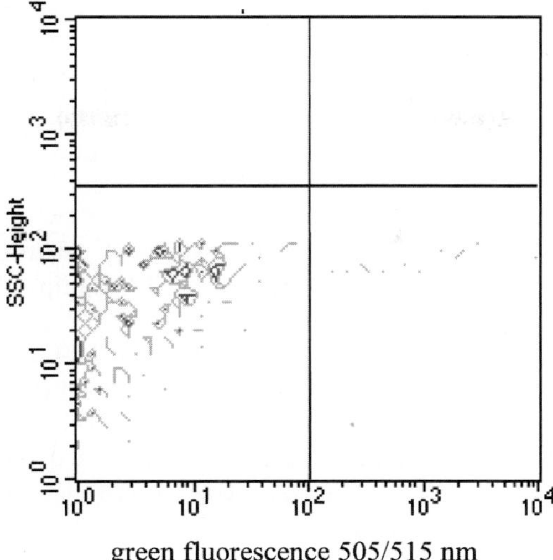

Quad	Events	% Gated	% Total
UL	0	0.00	0.00
UR	0	0.00	0.00
LL	244	95.31	95.31
LR	12	4.69	4.69

Fig. 4   Assessment of FluoroSphere beads green fluorescence (quadrant lower right (LR)) of physiological saline supernatant above the microbeads encapsulated in modified HFs after 24 h incubation (from Granicka *et al.*, 2010). (For interpretation of the references to color in this figure legend, the reader is referred to the Web version of this chapter.)

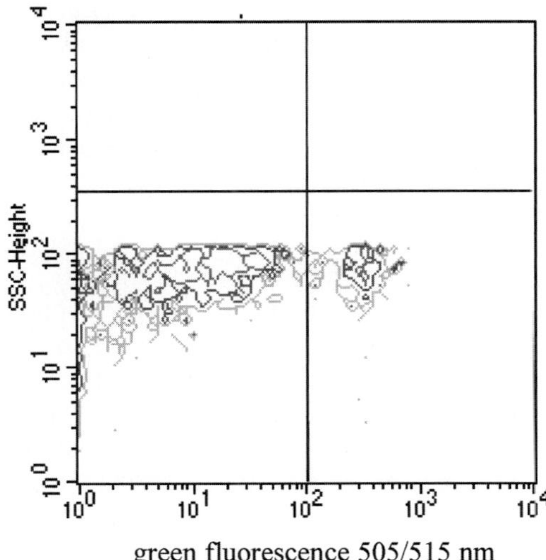

Quad	Events	% Gated	% Total
UL	0	0.00	0.00
UR	0	0.00	0.00
LL	1745	82.23	82.23
LR	377	17.77	17.77

Fig. 5   Assessment of FluoroSphere beads green fluorescence (quadrant lower right (LR)) of physiological saline supernatant above the microbeads encapsulated in unmodified HF after 24 h incubation. (For interpretation of the references to color in this figure legend, the reader is referred to the Web version of this chapter.)

B. The Encapsulated Bacteria Survival

The *E. coli* expressing GFP protein were utilized in this experiment. The suspension of bacteria treated with tested concentration of the antibiotic (tetracycline) does not instantly switch on the death signal in all cells, probably depending on its physiological status and/or microenvironment. At the same time we may observe bacteria with differing cell membrane permeability to propidium iodide (PI) and its binding to nucleic acids. When bacteria are characterized by the expression of the marker such as GFP we may concurrently assess cell membrane permeability, maximal value for PI binding (DNA content) and destruction of the GFP protein in time within the same two-dimensional graph.

E. coli strain SG3103 (Qiagen) was transfected with pQE-GFP (green fluorescent protein) plasmid in Institute of Biophysics and Biochemistry, PAS, Poland as described before (Granicka *et al.*, 2005). Shortly, the pQE-GFP plasmid was created by cloning into the pQE60 (Qiagen) vector. The GFP sequence amplified was: 5'CAT G*CC ATG G*CA ATG AGT AAA GGA GAA GAA CTT-3' and GFP-1 5'CG*G GAT CC*A TGT TTG TAT AGT TCA TCC ATG CC-3'. The expression of GFP in the bacteria for *in vitro* experiments was induced by 2 mM isopropyl β-D-1-thiogalactopyranoside (IPTG) treatment for 3.5 h, before encapsulation (bacteria *E. coli* GFPI). *E. coli* strain DH5 was used as an untransfected GFP control.

The presence of microorganisms is assessed using the FACSCalibur flow cytometer (Becton Dickinson Immunocytochemistry Systems, USA) as described above for FuoroBeads. Using a cytometer scaled for small particles forward light scatter as well as side light scatter may be used to detect *E. coli* population.

Severe combined immune deficiency (SCID) mice, age about 2 months, body weight about 20 g were used in the experiment. The animals were on a special diet supplied with vitamins, received sterile water *ad libitum*, and were bred in a sterile compartment. The protocol for animal experiments was approved by the Local Ethical Committee.

C. Tests *in vitro* and *in vivo*

(1) The initial concentration of bacteria is set spectrophotometrically at the 550 nm wavelength. The absorbance 0.125 was related to the concentration of 1×10^8 bacteria/ml. *E. coli* GFPI at the concentration of 1.5×10^8 bacteria/ml in the Luria–Bertan, Broth, Miller (Difco, USA) (LB) culture medium were encapsulated in HFs of 2 cm length and HFs were placed in the culture medium LB. The culture system in a humidified atmosphere (5% CO_2, 37 °C) was tested for 1, 2, 4, or 5 days. The culture of *E. coli* strain DH5 without GFP was used as described above, as the negative control. After culture the encapsulated bacteria have to be washed out with 0.3 ml of sterile physiological saline from HFs and analyzed in a flow cytometer (Granicka *et al.*, 2005) to assess the presence of GFP fluorescence of organisms present inside. The samples of the culture medium in which the HF-encapsulated *E. coli* were incubated were also analyzed.

(2) *E. coli* GFPIs at the concentration of 1.5×10^8 bacteria/ml in LB/RPMI (1:1) culture medium were encapsulated in HFs of 2 cm length and subcutaneously implanted (3–4 HFs/mouse) into mice at barbiturate anesthesia. Retro-orbital peripheral blood samples of about 0.5 ml volume were taken from mice in barbiturate narcosis after 1, 2, 4, or 5 days (in the 5-day experiment) to assay the presence of bacteria in the peripheral blood. Then the animal was sacrificed and the HFs were explanted. The content of the explanted HFs was washed out with 0.3 ml sterile physiological saline as described above. The bacteria washed out from HFs were analyzed in the flow cytometer to assess the presence of GFP fluorescence of microorganisms. As the negative control served the GFP untransfected *E. coli* strain DH5 at the concentration of 1.5×10^8 bacteria/ml in the LB/RPMI culture medium, encapsulated in 2 cm length HFs and subcutaneously implanted (3 HFs/per animal) into mouse. To control the possible presence of bacteria released from HF, the peripheral blood was added to the LB medium, induced by IPTG for 3.5 h and incubated for 12 h, in 37 °C in 200 rot/min shaker to assess the presence of GFP fluorescence of organisms. Than the GFPI organisms were compared with an uninduced sample by flow cytometry. A parallel microbiological test for the presence of *E. coli* in the peripheral blood of animal was performed as well.

D. Evaluation of GFP Expression *in vitro* and *in vivo*

The GFP expression value of encapsulated *E. coli* GFP during culture *in vitro* is presented in Fig. 6. The obtained values of GFP expression were comparable for 1 to 5 days of culture ($p = 0.32$) estimated by one-factor analysis of variation for obtained values at different days. The mean expression channel value was 869 ± 27 ($n = 12$). Neither the bacteria with GFP expression were noticed in the culture medium outside the encapsulated *E. coli* culture nor GFP expression was observed in the negative control samples.

In summary, the procedure described *in vitro* was used for testing the time of GFP expression after a single (3.5 h) IPTG induction of the GFP-gene expression. Applied bacteria strain sustains the GFP expression over 5 days after a single gene induction with IPTG.

The expression of GFP in *E. coli* GFPI encapsulated HFs and implanted into the mice may be observed. The SCID mice as a host for encapsulated bacteria were used assuming, that the bacteria released from HFs would induce sepsis in the animal. After the implantation of HFs with encapsulated bacteria, the presence of *E. coli* was tested in the blood. In the 5-day experiment neither a septic animal nor the presence of *E. coli* in the blood was observed.

The GFP expression of *E. coli* GFPI encapsulated in HFs after 1, 2, 4, or 5 days' subcutaneous implantation into mice is presented in Fig. 7. Explanted *E. coli* GFPIs exhibited the mean expression value (units) 603 ± 17 ($n = 32$) during 5-day implantation. In the Fig. 8 the representative cytogram is presented, obtained for *E. coli* GFPIs encapsulated in HFs after a subcutaneous implantation for 4 days into a

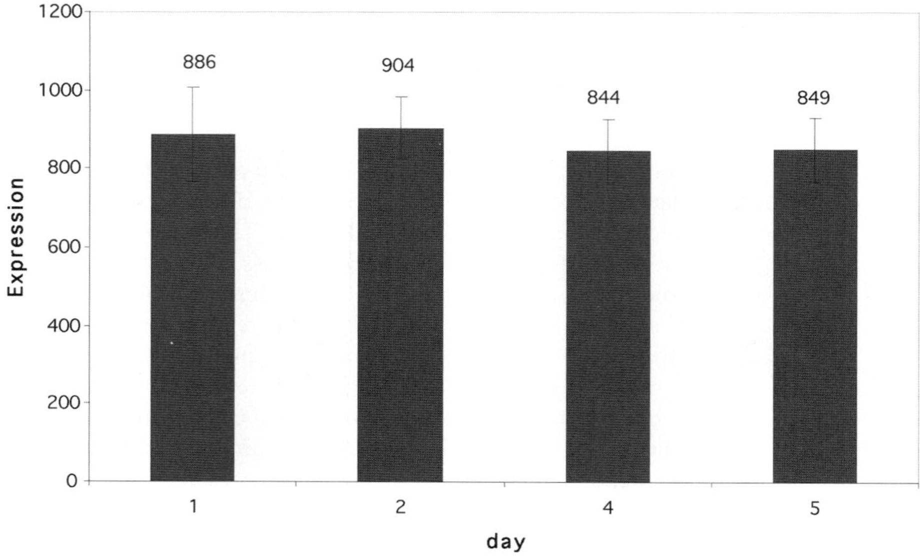

Fig. 6   The GFP expression in 5-day culture of encapsulated in membranes *E. coli* GFPI (the values are presented as mean ± SD, *n* = 12).

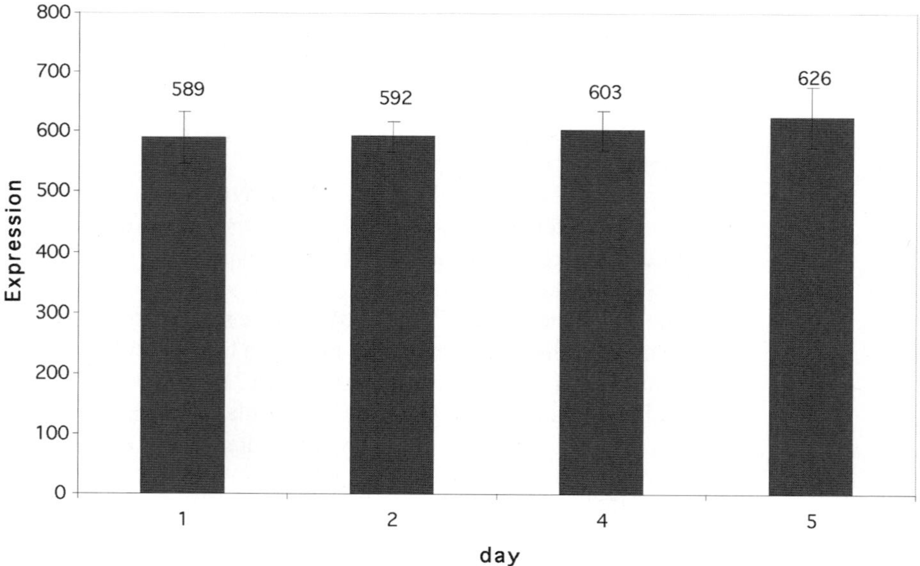

Fig. 7   The GFP expression of *E. coli* GFPI encapsulated in membranes, tested in explants after 1-, 2-, 4-, 5-day subcutaneous implantation into SCID mice (the values are presented as mean ± SD, *n* = 8).

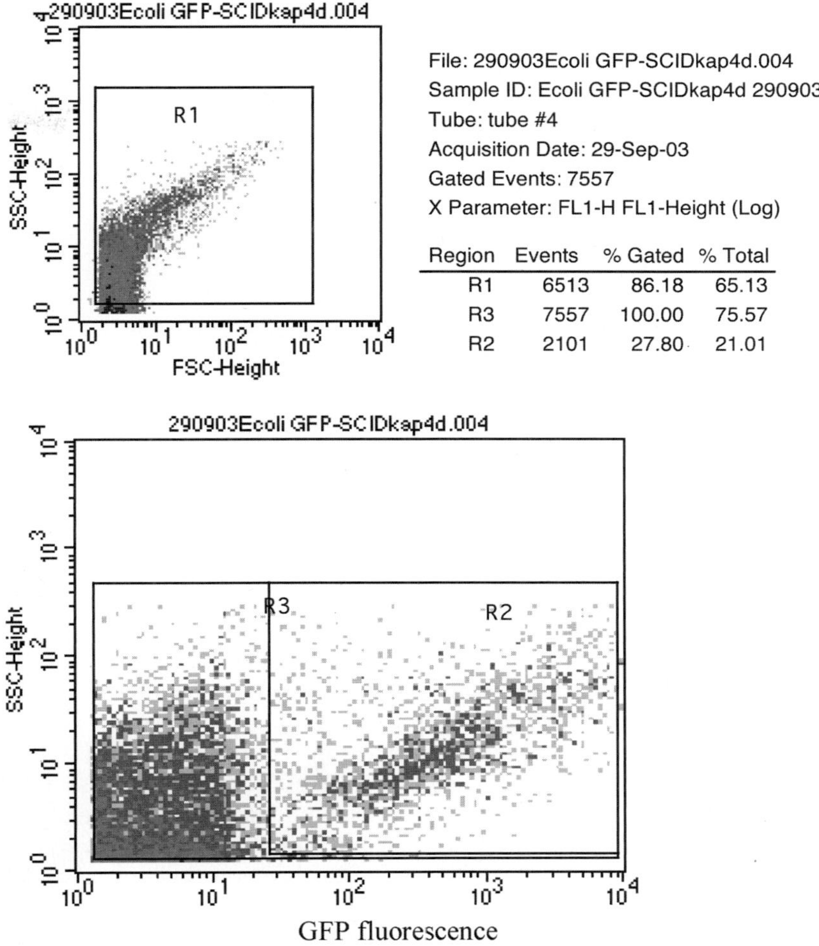

Fig. 8   Assessment of green fluorescence GFP (quadrant lower right LR, gate R2- positive events) of encapsulated *E. coli* GFPI explanted and washed out of the HF after 4-day subcutaneous implantation into mice. It was observed 27.8% organisms GFP positive (gate 2, R2). (For interpretation of the references to color in this figure legend, the reader is referred to the Web version of this chapter.)

mouse. The green fluorescence (FL1) median fluorescence intensity unit readings obtained for *E. coli* DH5 strain without GFP gene were five units, *E. coli* GFP untreated with IPTG (constitutive GFP expression) utilized in the *in vivo* experiments were about 90 units and for *E. coli* GFPI, after treatment *in vitro* with IPTG were 1963 units (Granicka *et al.*, 2010). It was observed, that the expression of GFP in *E. coli*, even after IPTG treatment before HFs implantation is lower *in vivo* than *in vitro*, possibly due to the differences in the surrounding medium.

In summary, the described HFs system ensures the bacteria survival within the HFs space when cultured *in vitro* or implanted into a mouse *in vivo*. A single induction of the GFP gene expression may be satisfactory for the observation of GFP expression for 5 days.

III. The Release of Bacteria Products

The performance of bacteria encapsulated in HFs when treated with chosen antibiotic may be evaluated. The antibiotic application may cause bacteria cytolysis and release of biologically active substance(s) for which production the bacteria was genetically modified. GFP protein was used here as a model biological substance expressed in the bacteria.

A. Evaluation of the Antibiotic–Induced Changes in Bacteria *E. coli*–GFPI *in vitro*

The antibiotic impact on bacteria *E. coli in vitro* may be assayed using established the procedure.

The suspension of *E. coli*-GFPI encapsulated in HFs at the concentration of $1,5 \times 10^8$ bacteria/ml (8.5×10^5/HFs) was cultured for 48 h in the 1 ml LB culture medium with addition the antibiotic, tetracycline at concentration 1 mg/ml (35 °C). As a control the encapsulated in HFs bacteria *E. coli*-GFPI were incubated in LB medium devoid of antibiotic. The HFs content was washed out 1, 2, 24, 48 h from the application of the tetracycline and bacteria were evaluated in the flow cytometer after the cytochemical reaction with PI. The presence of *E. coli*-GFPI fluorescence as well as PI fluorescence of the bacteria (living/dead test) was assessed with this method. The samples of the outside culture medium in which the HFs encapsulated *E. coli*-GFPI were analyzed did not contained bacteria. The effect of flow cytometric assay for encapsulated in modified HFs *E. coli*-GFPI culture treated 2 h with tetracycline is presented in Fig. 9A.

As compared with the negative control (Fig. 9B), the *E. coli*-GFP living cells localized as GFP positive and PI negative (R2) are recognizable from nonfluorizing events. Numerous bacteria containing GFP has a cell membrane with increasing permeability for PI (R3) till some maximal value of PI fluorescence. Some of the bacteria maximally PI positive (PI^+) lost their GFP from the cytosol ($GFP^{+/-}$) (R4) with different fraction of GFP remaining within the cell. The GFP fluorescence decrease was probably an effect of protease activity and diffusion of protein fragments from the PI positive (permeable cell membrane) bacteria. Then the cells start to lose its PI fluorescence ($PI^{+/-}$), DNA content (R5).

It was observed, that the percentage number of PI^+ population with retained GFP (GFP^+) or population with partially lost GFP ($GFP^{+/-}$) increased about 19 times after

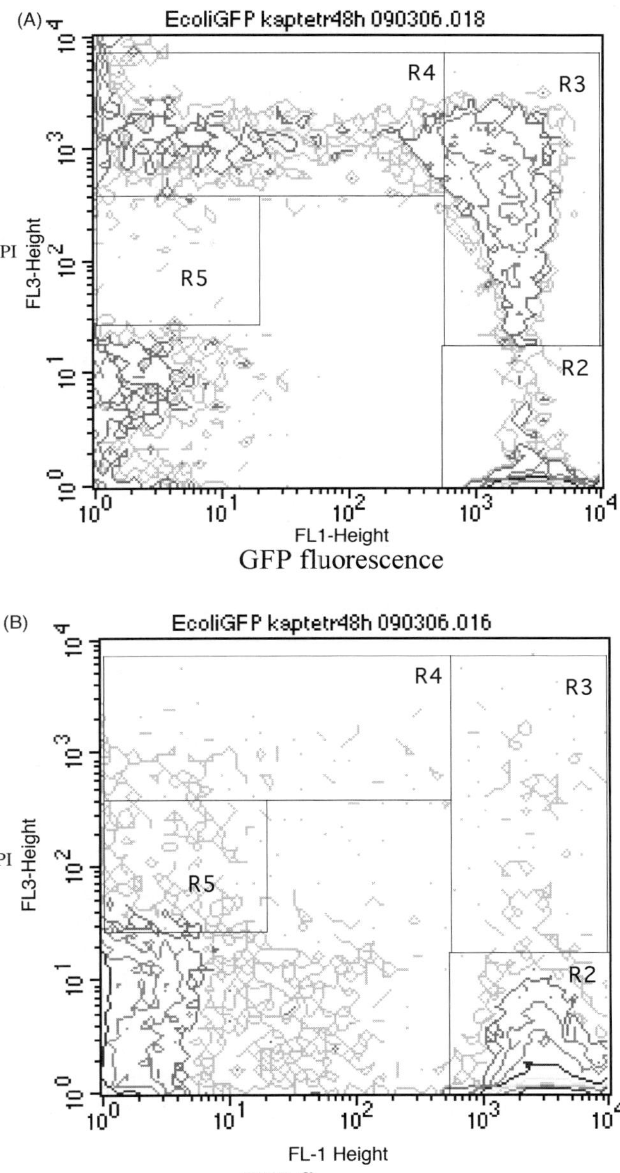

Fig. 9   A. *E. coli*-GFP encapsulated in modified hollow fiber, incubated 2 h, 35 °C, with tetracycline (1 mg/ml). Hollow fiber encapsulated *E. coli*-GFP may be localized as FL1 events. Incubation with PI localize beside living cells (R2, 44.7%), also several dead bacteria (R3, 32.6%), some of them loosing GFP-protein from cytosol (R4, 20.2%), probably due to intracellular protease activation. Some bacteria were loosing DNA as well (R5, 2.4%). Fig. 9B Negative control: E. coli GFP incubated without tetracycline. R2, 96.2%; R3, 0.8%;R4, 0.6%; R5, 2.1%.

2 h from the start of tetracycline treatment and was $18.24 \pm 2.19\%$ ($n = 12$) as compared to a negative control $0.95 \pm 0.78\%$. These proportions changed after 24-h culture in the presence of tetracycline as result of the bacteria repopulation for $61.70 \pm 21.18\%$ and $7.12 \pm 2.05\%$ in negative control, respectively (nine times), and after 48-h culture for $85.78 \pm 12.70\%$ and $11.57 \pm 5.22\%$, respectively (seven times). The percentage of GFP^{+-}/PI^{+} cells changes with time of the tetracycline treatment. There were no microorganisms GFP positive observed within the culture medium outside the HFs loaded with *E. coli*.

B. Evaluation of The Antibiotic Impact on Bacteria *E. coli In vivo*

The suspension of *E. coli*-GFPI in LB/RPMI at the concentration of $1,5 \times 10^8$ bacteria/ml may be encapsulated in HF for implantation to the animal. The HF of 2 cm length containing bacteria was implanted subcutaneously (3–4 HF/mouse) into SCID mice under barbiturate anesthesia as described before. After 2 days the mouse received subcutaneous injection of 1 ml tetracycline solution (1 mg/ml) at the site far from implanted HFs. At a predetermined time the peripheral blood samples were taken, the animal was sacrificed and the HFs were explanted. The content of the explanted HFs was washed out with sterile physiological saline and analyzed in a flow cytometer for presence of GFP and PI fluorescence of microorganisms. As the negative control, SCID mice with encapsulated in HFs *E. coli*-GFPI implanted received 1 ml of physiological saline. The following procedure was as described above.

The mice cannot be effectively treated with IPTG to induce rise of GFP fluorescence within *E. coli*-GFP. However, the *E . coli*-GFP bacteria have constitutive, low expression of GFP fluorescence, which is not observed in GFP untransfected *E. coli* strain DH5. No release of bacteria from HFs, which would induce sepsis in the animal, was observed on second day after the implantation of encapsulated bacteria. The expression of constitutive fluorescence of *E. coli*-GFP encapsulated in HFs after 2 days from implantation and after tetracycline injection: 6, 8, 24 h before HFs explantation was evaluated.

The most substantial necrotic bacteria proportion induced in mouse by single subcutaneous tetracycline treatment was observed after 8 h. The ratio of living to damaged cells decreased after 6 and 8 h as compared to the control ratio values in the untreated animals while they returned to control ratio value at 24 h from start of the tetracycline treatment. The observed rebound may be due to the repopulation of the surviving bacteria within the HFs. The calculated tetracycline concentration applied in the mice was about 20-fold lower compared to our *in vitro* experiment.

This observation of the flow cytometric assessment of *E. coli*-GFP fluorescence during culture or implantation to the animal is interpreted as reflecting the tetracycline effect on *E. coli*-GFP. The following stages of this process can be recognized by flow cytometry. First when the bacteria cell membrane permeability increases, the transport of PI into the cell rises. The PI within the bacteria cell binds to intracellular nucleic acids, DNA and RNA. The binding process progresses and approaches the maximal fluorescence value when the PI-binding sites become occupied; the GFP protein fluorescence does not change at this stage. The next step involves initiation of GFP-fragments loss from the cells, probably due to protease (s) activation within the PI permeable cells (García-Fruitós *et al.*, 2007). This process progresses without significant loss of cellular DNA and RNA, as the PI binding does not change. Finally, when most GFP protein fluorescence disappears from the cells due to loss of proteolytically degraded fragments from the bacteria, the DNA and RNA degradation produces diffusible fragments released from the cells. The DNA fragmentation must be a rapid process compared to GFP degradation, since only few DNA-losing cells (about 2%) are noted. Such distinguishable steps of bacterial death process were not previously described. It should be noted that the model of *E. coli*-GFP process may be similar in the other prokaryotic organisms.

The described observations may be due to the progressive increase of bacteria cell membrane permeability in the first stage. Then in the next step the GFP protein aggregates, the inclusion bodies formed by IPTG-induced overproduction of the protein, are degraded by proteases, possible ATP-dependent proteases like Lon and ClpP. The final fast step is due to the degradation of nucleic acids. The HF-enclosed prokaryotic cells treated with antibiotic may have an application for a biologically active peptide release in a therapeutic program.

IV. Production and Release by Bacteria of Biologically Active Factor(s)

In this experiment *Bacillus subtilis*, a strain producing biologically active substance listeriolysin O (LLO), were encapsulated in HFs to evaluate the effect on chosen eukaryotic cells for future application in the anticancer treatment. It was assumed, that the active substance will be continuously, locally supplied by active bacteria, while the system finely may be removed.

A. *Bacillus Subtilis* Strain BR1–S as a Model Gram–Positive Bacterium

Bacillus subtilis strain BR1-S is a model gram-positive bacterium which, is *generally regarded as safe. B. subtilis* strain BR1-S is a derivative of ZB307 strain (Zuber & Losick, 1987) producing LLO. Gene *hly* encoding LLO from *L. monocytogenes* 10403S strain was cloned in vector pAG58 and introduced into ZB307 similarly as previously described (Bielecki *et al.*, 1990; Wiśniewski and Bielecki, 2004). Bacteria were grown at 37 °C with 120 rpm agitation on LB medium (Sigma) and BHI (Becton–Dickinson) supplemented with erythromycin (1 µg /ml) and chloramfenicol (3 µg/ml). To induce LLO production and release by the live bacteria the HF-encapsulated bacteria were activated for production of LLO in the presence of 1 mM IPTG (Schallmey *et al.*, 2004; Simonen and Palva, 1993).

B. Evaluation of Encapsulated Bacteria Impact on Jurkat Line Cells or on Peripheral Blood Mononuclear Cells Obtained from Leukemia Patients

Two types of target leukemia cells were used: Jurkat – human T-lymphocyte cell line or suspension of peripheral blood mononuclear cells obtained from the peripheral blood of patients with chronic lymphocytic B-cell leukemia. The patients' cells were used after informed consent obtained from the patient in accordance with the Declaration of Helsinki and with the approval by the Local Ethics Committee.

The bacteria *Bacillus subtilis* BR1-S were encapsulated in the HFs membrane system. The HF was then placed in 0.5 ml suspension of Jurkat line cells or in peripheral blood mononuclear cells, with addition of erythromycin (3 µg/ml) and IPTG (1 µg/ml) for 24 h (5% CO_2, 37 °C). As a negative control the cells suspension was cultured with encapsulated bacteria *B. subtilis* BR1-S without IPTG addition or with the empty HF membranes (control). The samples of eukaryotic cells were collected after 24 h, to evaluate the cells viability by assessing their ability to exclude PI by flow cytometry. It was observed that the viability of the Jurkat cells incubated 24 h with the HF-encapsulated bacteria declined (19.5 ± 9.41% and 4.6 ± 2.09% viable Jurkat cells treated with encapsulated BR1-S and BR1-S+IPTG, respectively), as compared to a negative control that had 71% of live cells (Fig. 10). Similar effect of encapsulated bacteria *B. subtilis* BR1-S was observed with

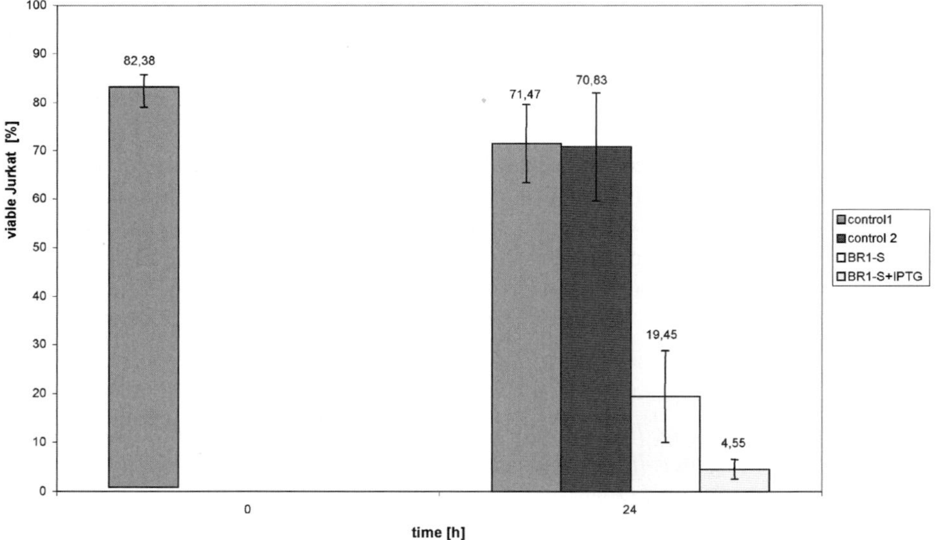

Fig. 10　Evaluation of encapsulated bacteria LLO secreting impact on Jurkat cells viability. The percentage number of living cells after 24 h culture of HF encapsulated bacteria *Bacillus subtilis* strain BR1-S in Jurkat cells suspension. BR1-S+ IPTG – the Jurkat suspension cultured with IPTG induced encapsulated bacteria BR1-S (n = 12); BR1-S – the Jurkat suspension cultured with encapsulated bacteria BR1-S not induced (n = 6); control 1 – the Jurkat cells suspension; control 2 – the Jurkat cells suspension cultured with empty membrane (n = 6).

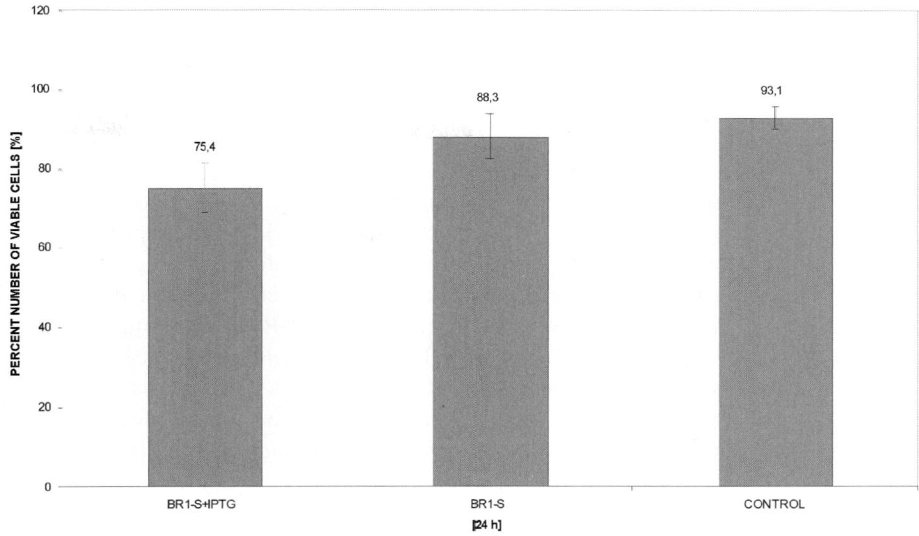

Fig. 11   Evaluation of encapsulated bacteria impact on peripheral blood mononuclear cells obtained from leukemia patients (B-CLL). The percentage number of living cells after 24 h culture with encapsulated bacteria *Bacillus subtilis* BR1-S in MNC suspension. BR1-S+IPTG – the MNC suspension cultured with induced encapsulated bacteria BR1-S ($n = 12$); BR1S – a negative control – the MNC suspension cultured with encapsulated bacteria BR1-S not induced ($n = 6$); control – a second negative control – the MNC suspension cultured with empty HF ($n = 6$).

leukemic mononuclear cells (Fig. 11). However, killing by encapsulated *B. subtilis* BR1-S toxins of B-leukemia cells was less effective than in experiment with Jurkat T-cells. There was the difference in viability of cells tested in the same conditions with IPTG as compared to a negative control. The difference was 18% and 67% for B-cell leukemia and Jurkat T-cells, respectively. This suggests an effective and selective activity of the bacterial toxins produced by HF membrane-encapsulated *Bacillus subtilis* BR1-S cells.

In conclusion, the encapsulated bacteria *Bacillus subtilis*, BRI-S produce and secrete toxins, one of them is LLO, which diffuses to the culture medium and kills the target eukaryotic cells, here cells of Jurkat T-cells line and mononuclear peripheral blood B-cells from the leukemic patient. This effect was selective; more T-cells were killed than B-cells in similar experimental test. The system appeared useful in killing the leukemia cells in the *in vitro* system; however, it needs further improvements to be applied as system *in vivo*.

V. Conclusion

Described and discussed is the novel system of bacteria bio-encapsulation using modified membranes, and its assessment by flow cytometry. The described bio-encapsulation system demonstrates the following advantages: (i) it allows one to avoid the release of objects of bacteria size or bacteria *E. coli* and *Bacillus subtilis* through the membrane wall *in vitro* and *in vivo* as well; (ii) makes it possible to distinguish stages of the tetracycline impact on encapsulated bacteria *E. coli* such as induction of membrane permeability to PI, activation of proteases targeting GFP and subsequent nucleic acids degradation and release; (iii) allows one to observe the tetracycline impact on encapsulated *E. coli* viability *in vivo*, with evidence of increasing the necrotic bacteria share followed by repopulation of live encapsulated bacteria while the they are safely enclosed within HF at site of implantation; (iv) provides evidence of the cytotoxic activity of live *Bacillus subtilis* encapsulated in the membrane system mediated by BR1-S secreting LLO on the chosen eukaryotic cells. Such system releasing bacterial products locally and continuously may selectively affect different types of cells and may have possible application in the anticancer treatment at localized sites. The applied flow cytometric methods made it possible to demonstrate encapsulated bacteria viability, their protein expression as well as cytotoxic activity.

Acknowledgments

This study was partly supported by grant N N-401015936 and grant CMKP 2008.

References

Bielecki, J., Youngman, P., Connelly, P., and Portnoy, D. A. (1990). Bacillus subtilis expressing a haemolysin gene from Listeria monocytogenes can grow in mammlian cells. *Nature (London)* **345**, 175–176.

Chang, T. M., and Prakash, S. (2001). Procedures for microencapsulation of enzymes, cells and genetically engineered microorganisms. *Molec. Biotechnol.* **17**, 249–260.

Charalampopoulos, D., Wang, R., Pandiella, S. S., and Webb, C. (2002). Application of cereals and cereal components in functional foods. *Int. J. Food Microbiol.* **79**, 131–141.

Crooks, C. A., Douglas, A. J., Broughton, R. J., and Sefton, A. V. (1990). Microencapsulation of mammalian cells in a HEMA copolymer: effects on capsule morphology and permeability. *J. Biomed. Mater. Res.* **24**, 1241–1262.

Dawson, R. M., Broughton, R. L., Stevenson, W. T. K., and Sefton, M. V. (1987). Microencapsulation of CHO cells in a hydroxyethyl methacrylate-methyl methacrylate copolymer. *Biomaterials* **8**, 360–366.

Emery, B. E., Dixit, R., Formby, C. C., and Biedlingmaier, J. F. (2003). The resistance of maxillofacial reconstruction plates to biofilm formation *in vitro*. *Laryngoscope* **113**, 1977–1982.

García-Fruitós, E., Martínez-Alonso, M., Gonzàlez-Montalbán, N., Valli, M., Mattanovich, D., Villaverde, A. (2007). Divergent genetic control of protein solubility and conformational quality in *Escherichia coli*. *J. Mol. Biol.* **374**, 195–205.

Granicka, L. H., Wdowiak, M., Kosek, A., Świeżewski, S., Wasilewska, D., Jankowska, E., Weryński, A., Kawiak, J. (2005). Survival analysis of *Escherichia coli* encapsulated in hollow fibre membrane *in vitro* & *in vivo*. preliminary report. *Cell Transplant* **14**, 323–330.

Granicka, L. H., Żołnierowicz, J., Wasilewska, D., Weryński, A., and Kawiak, J. (2010). Induced death of *Escherichia coli* encapsulated in a hollow fiber membrane as observed *in vitro* or after subcutaneous implantation. *J. Microbiol. Biotechnol.* **20**, 224–228.

Hajii, K. T., Lepine, F., Bisaillon, J. G., Beaudet, R., Hawari, J., Guiot, S. R. (2000). Effects of bioaugmentation strategies in UASB reactors with a methanogenic consortium for removal of phenolic compounds. *Biotech. & Bioeng.* **67**, 417–423.

Homma, H., Nagaoka, S., Mezawa, S., Matsuyama, T., Masuko, E., Ban, N., Watanabe, N., Niitsu, Y. (2006). Bacterial adhesion on hydrophilic heparinized catheters compared with adhesion on silicone catheters, in patients with malignant obstructive jaundice. *J. Gastroenterol.* **31**, 836–843.

Hou, R. C., Lin, M. Y., Wang, M. M., and Tzen, J. T. (2003). Increase of viability of entrapped cells of *Lactobacillus delbrueckii* ssp. bulgaricus in artificial sesame oil emulsions. *J. Dairy Sci.* **86**, 424–428.

Kailasapathy, K. (2002). Microencapsulation of probiotic bacteria: technology and potential applications. *Curr. Issues Intestinal. Microbiol.* **3**, 39–48.

Kidchob, T., Kimura, S., and Imashini, Y. (1998). Degradation and release profile of microcapsules made of poly[L-lactic acid-*co*-L-Lysine (Z)]. *J. Contr. Rel.* **54**, 283–292.

Lewis, A. L., Cunning, Z. L., Goreish, H. H., Kirkwood, L. C., Tolhurst, L. A., Stratford, P. W. (2001). Crosslinkable coatings from phosphorylcholine-based polymers. *Biomater.* **22**, 99–111.

Morikawa, N., Iwata, H., Matsuda, S., Miyazaki, J. I., and Ikada, Y. (1997). Encapsulation of mammalian cells into synthetic polymer membranes using least toxic solvents. *J. Biomater. Sci.* **8**, 575–586.

Moslemy, P., Neufeld, R. J., and Guiot, S. R. (2002). Biodegradation of gasoil by gellan gum-encapsulated bacterial cells. *Biotech. & Bioeng.* **80**, 175–184.

Prakash, S., and Chang, T. M. (2000a). Artificial cells microencapsulated genetically engineered E. coli DH5 cells for the lowering of plasma creatinine *in vitro* and *in vivo*. *Artif. Cells, Blood Substitutes & Immobilization Biotechnol.* **28**, 397–408.

Prakash, S., and Chang, T. M. (2000b). *In vitro* and *in vivo* uric acid lowering by artificial cells containing microencapsulated genetically engineered *E. coli* DH5 cells. *Int. J. Artif Organs* **23**, 429–435.

Robak, T. (1991). Hemopoietic growth factors (polish). *Postpy Hig. Med. Dosw.* **45**, 461–469.

Schallmey, M., Singh, A., and Ward, O. P. (2004). Developments in the use of *Bacillus* species for industrial production. *Can. J. Microbiol.* **50**, 1–17.

Shah, N. P. (2000). Probiotic bacteria: selective enumeration and survival in dairy foods. *J. Dairy Sci.* **83**, 894–907.

Simonen, M., and Palva, I. (1993). Protein secretion in *Bacillus* species. *Microbiol. Rev.* **57**, 109–137.

Sultana, K., Godward, G., Reynolds, N., Arumugaswam, R., Peiris, P., Kailasapathy, K. (2000). Encapsulation of probiotic bacteria with alginate-starch and evaluation of survival in simulated gastrointestinal conditions and in yoghurt. *Int. J. Food Microbiol.* **62**, 47–55.

Walker, S. L., Hill, J. E., Redman, J. A., and Elimelech, M. (2005). Influence of growth phase on adhesion kinetics of *Escherichia coli* D21g. *Appl. Environ. Microbiol.* **71**, 3093–3099.

Wang, J., Huang, N., Yang, P., Leng, Y., Sun, H., Liu, Z. Y., Chu, P. K. (2004). The effect of amorphous carbon films deposited on polyethylene terephalate on bacterial adhesion. *Biomater* **25**, 3163–3170.

Wiśniewski, J., and Bielecki, J. (2004). Polymerizer-mediated intracellular movement. *Pol. J. Microbiol.* **53**, 35–38.

Zhang, Z. Y., Ping, Q. N., and Xiao, B. (2000). Microencapsulation and characterization of tramadol-resin complexes. *J. Controll. Rel.* **66**, 107–113.

Zimmerman, H., Hillgartner, M., Manz, B., Feilen, P., Brunnenmeier, F., Leinfelder, U., Weber, M., Cramer, H., Schneider, S., Hendrich, C., Volke, F., Zimmermann, U. (2003). Fabrication of homogeneously cross-linked, functional alginate microcapsules validated by NMR-, CLSM-, and AFM-imaging. *Biomaterials* **24**, 2083–2096.

Zuber, P., and Losick, R. (1987). Role of AbrB in SpoOA- and SpoOB-dependent utilization of a sporulation promoter in *Bacillus subtilis*. *J. Bacteriol.* **169**, 2223–2230.

Yang, H., Iwata, H., Shimizu, H., Takagi, T., Tsuji, T., Ito, F. (1994). Comparative studies of *in vitro* and *in vivo* function of three different shaped bioartificial pancreas made of agarose hydrogel. *Biomaterials* **15**, 113–120.

CHAPTER 17

Guide to Red Fluorescent Proteins and Biosensors for Flow Cytometry

Kiryl D. Piatkevich and Vladislav V. Verkhusha

Department of Anatomy and Structural Biology, and Gruss-Lipper Biophotonics Center, Albert Einstein College of Medicine, Bronx, New York

Abstract

Since the discovery of the first red fluorescent protein (RFP), named DsRed, 12 years ago, a wide pallet of red-shifted fluorescent proteins has been cloned and biotechnologically developed into monomeric fluorescent probes for optical microscopy. Several new types of monomeric RFPs that change the emission wavelength either with time, called fluorescent timers, or after a brief irradiation with violet light, known as photoactivatable proteins, have been also engineered. Moreover, RFPs with a large Stokes shift of fluorescence emission have been recently designed. Because of their distinctive excitation and fluorescence detection conditions developed specifically for microscopy, these fluorescent probes can be suboptimal for flow cytometry. Here, we have selected and summarized the advanced orange, red, and far-red fluorescent proteins with the properties specifically required for the flow cytometry applications. Their effective brightness was calculated for the laser sources available for the commercial flow cytometers and sorters. Compatibility of the fluorescent proteins of different colors in a multiparameter flow cytometry was determined. Novel FRET pairs, utilizing RFPs, RFP-based intracellular biosensors, and their application to a high-throughput screening, are also discussed.

Keywords: DsRed; RFP; mCherry; mOrange; FRET; FACS

I. Introduction

Chalfie *et al.* (1994) showed that a green fluorescent protein (GFP) from the jellyfish *Aequorea victoria* could be used as a marker for protein localization and expression in living bacteria and worm cells. Cloning of GFP and it first application *in vivo* dramaticly altered the nature and scope of the issues that could be addressed by cell biologists. Together with the introduction of new microscopy techniques, fluorescent proteins (FPs) changed the way life science research is performed today.

The next breakthrough in FP technology occurred in 1999 when six new FPs were cloned from nonbioluminescent *Anthozoa* species (Matz *et al.*, 1999). One of the proteins, named drFP583, differed from GFP dramatically in its spectral properties, demonstrating a red fluorescence. The drFP583 protein, the gene for which was optimized for expression in mammalian cells, became the first commercially available red fluorescent protein (RFP), named DsRed for *Discosoma* sp. Recently, the majority of RFPs have been isolated and cloned from *Anthozoa* species living in the Indo-Pacific region (Piatkevich *et al.*, 2010a; Verkhusha *et al.*, 2003a).

Subsequently, the race was on to succeed in monomerization and improvement of wild-type RFPs in order to produce new probes suitable for multicolor imaging of cellular proteins and FRET pairs with emission in the longer wavelength region. A number of desirable changes to the physical and biochemical properties of FPs have been achieved through the intense molecular evolution (Campbell *et al.*, 2002; Shaner *et al.*, 2004). The increasing brightness, maturation efficiency, photostability, pH stability, and minimizing cytotoxicity significantly improved the utilization of RFPs for live-cell microscopy. Beside these enhancements, researches also succeeded in developing new types of monomeric RFPs, such as fluorescent timers (FTs) and photoactivatable FPs (PA-FPs), which are particularly useful for subcellular dynamics studies and superresolution imaging (Piatkevich and Verkhusha, 2010). Recently, engineered RFPs with large Stokes shift (LSS) can serve as additional red colors for multicolor imaging and hold great promises for multiphoton microscopy. At present, near-infrared FPs and FP-based biosensors with red emission are of great interest for FP developers. However, in general, the vast variety of currently available FPs is quite sufficient to study the majority of problems in cell biology.

Genetically encoded markers based on the GFP-like proteins have several advantages over the fluorescent dyes for applications in cell biology. FPs possess a unique ability to produce fluorescence after their expression in any prokaryotic or eukaryotic cell without additional enzymes or cofactors except for molecular oxygen. Moreover, a protein of interest can be easily tagged with an FP on a DNA level without affecting its intrinsic function. Consequently, FPs as reporter markers allow for studying temporal and spatial expression of genes by measuring their fluorescence signal in live cells and tissues, as well as localization and dynamics of cellular proteins, organelles, and virus particles. The recent expansion of spectrally distinct FP variants has enabled multicolor imaging for monitoring several events simultaneously.

Coupled with flow cytometry, the FP technology provides an amazing opportunity to noninvasively differentiate between various cell populations, monitor gene activities, and detect protein–protein interactions and small molecules in individual living cells in a high-throughput manner. Unfortunately, as yet RFPs have been underutilized in the flow cytometry and fluorescence-activated cell sorting (FACS) approaches. Recent advances in FP development and invention of the lasers emitting a variety of wavelengths greatly increase the performance and the capabilities of a multiparameter analysis with the FP-based flow cytometry. This chapter focuses on the novel improved RFPs, which are potentially useful for many FACS applications. We also describe the techniques for the utilization of RFPs as reporters and biosensors for the multiparameter flow cytometry.

II. Major Characteristics of FPs

Since the clonning of GFP, a large amount of practically usefull FPs have been isolated from different organisms or developed on the basis of wild-type FPs. We believe that the number of novel FPs will continue to grow. Not surprisingly, it is not always obvious which FP should be chosen from a nubmer of similar probes for a certain application. Below we describe the key characteristics of FPs that should be considered by a researcher for any applications of those probes in biological systems.

A. Molecular Brightness (Intrinsic Fluorescence Intensity)

Brightness of an FP is one of its most important characteristics. High brightness of an FP is usually an additional advantage for any cell biology application. Brightness

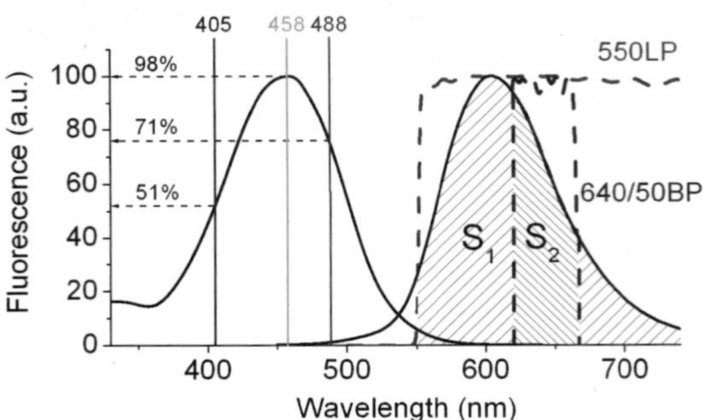

Fig. 1 The fluorescence excitation and emission spectra of LSSmKate2 (solid lines) are shown. The wavelengths of the laser lines, 405, 458, and 488 nm, are shown by vertical lines with indication of excitation efficiency. The emission that passes through the 550 nm long pass (red dash line) and 640/50 nm band pass (blue dash line) filters is shown as cross-hatched region with square S_1 and S_2, respectively. The fluorescence intensity collected through the filters is proportional to the S_1 and S_2 squares. (For interpretation of the references to color in this figure legend, the reader is referred to the Web version of this chapter.)

determines the sensitivity and signal-to-noise ratio for the fluorescence detection. By definition, the molecular brightness of FP is the product of a molar extinction coefficient and quantum yield, measured *in vitro*. Since the extinction coefficient is a function of the wavelength, the effective brightness of an FP depends on the wavelenght of the excitation light. Moreover, the effective brightness is determined by emission filter sets used for detecting fluorescence (Fig. 1). Molecular brightness and effective brightness specified for certain laser excitations for common RFPs are presented in Tables I and II, respectively.

FPs are usually expressed and visualized in cells. Fluorescence brightness of an FP measured *in vitro* does not always correspond to its actual brightness observed in live cells (Kremers *et al.*, 2007). The *in vivo* brightness, beside intrinsic spectral characteristics, is determined by a number of parameters such as expressing cell type, expression efficiency, mRNA and protein stability, efficiency and rate of chromophore maturation, and the fusion protein partner. Therefore, in order to choose the best FP for a certain application in cells, it is recommended to screen several optimal candidates, since it is impossible to predict their behavior theoretically.

For the application of FPs in superresolution microscopy and fluorescent correlation spectoscopy, an important characteristic is the single-molecule brightness, which tends to be higher than that measured for an ensemble of FP molecules (Subach *et al.*, 2009a). The single-molecule brightness is defined as the average fluorescence intensity per light-emitting FP molecule; however, not every FP molecule in the ensemble is fluorescent due to an incomplete formation of the chromophore. Importantly, a two-photon brightness, which corresponds to the two-photon action cross-section, cannot be readily predicted from the common one-photon absorbance spectra (Drobizhev *et al.*, 2009). Measurment of both single-molecule and two-photon brightness is not a trivial task and requires sophisticated equipment.

Table I

Spectroscopic and biochemical properties of red–shifted FPs efficient in flow cytometry

Protein	Oligomeric state	Ex_{max} (nm)	Em_{max} (nm)	ε (M^{-1} cm^{-1})	QY	Molecular brightness	pK_a	$\tau_{1/2}$ maturation 37 °C (h)	References
				Orange fluorescent proteins					
E2-Orange	Tetramer	540	561	36,500	0.54	20	4.5	1.3	Strack *et al.*, 2009a
mOrange	Monomer	548	562	71,000	0.69	49	6.5	2.5	Shaner *et al.*, 2004
mKOκ	Monomer	551	563	105,000	0.61	64	4.2	1.8	Tsutsui *et al.*, 2008
mOrange2	Monomer	549	565	58,000	0.60	35	6.5	4.5	Shaner *et al.*, 2008
tdTomato	Pseudomonomer	554	581	138,000	0.69	95	4.7	1.0	Shaner *et al.*, 2004
TagRFP	Monomer	555	584	98,000	0.41	40	<4.0	1.7	Merzlyak *et al.*, 2007
TagRFP-T	Monomer	555	584	81,000	0.41	33	4.6	1.7	Shaner *et al.*, 2008
				Red fluorescent proteins					
DsRed-Express2	Tetramer	554	591	35,600	0.42	15	–	0.7	Strack *et al.*, 2008
mStrawberry	Monomer	574	596	90,000	0.29	26	<4.5	0.8	Shaner *et al.*, 2004
LSSmKate2	Monomer	460	605	26,000	0.17	4.5	2.7	2.5	Piatkevich *et al.*, 2010b
mRuby	Monomer	558	605	112,000	0.35	39	5	2.8	Kredel *et al.*, 2009
mCherry	Monomer	587	610	72,000	0.22	16	<4.5	0.25	Shaner *et al.*, 2004
mKeima	Monomer	440	620	14,400	0.24	3	6.5	4.7	Kogure *et al.*, 2006
mRaspberry	Monomer	598	625	79,000	0.15	12	5.0	1.0	Wang *et al.*, 2004
				Far-red fluorescent proteins					
mKate2	Monomer	588	633	62,500	0.40	25	5.4	<0.33	Shcherbo *et al.*, 2009a
Katushka2	Dimer	588	633	66,250 × 2	0.37	25 × 2	5.4	<0.33	Shcherbo *et al.*, 2009a
E2-Crimson	Tetramer	605	646	58,500	0.12	7	4.5	0.4	Strack *et al.*, 2009b
mNeptune	Monomer	600	650	67,000	0.20	13	5.4	ND	Lin *et al.*, 2009
TagRFP657	Monomer	611	657	34,000	0.10	3.4	5.0	2.0	Morozova *et al.*, 2010

Ex_{max}, excitation maximum; Em_{max}, emission maximum; ε, molar extinction coefficient; QY, quantum yield; pK_a, pH value, at which protein retains half of its fluorescent intensity; $\tau_{1/2}$, half-time of protein maturation. Molecular brightness is determined as a product of quantum yield and molar extinction coefficient divided by 1000. ND, not determined.

Table II

Effective brightness of red–shifted FPs using common lasers

Protein	Solid-state lasers			Gas lasers		
	Line (nm)	Efficiency of excitation (%)	Effective brightness	Line (nm)	Efficiency of excitation (%)	Effective brightness
mOrange	488	20	10	488	20	10
	532	62	30	514	50	24
	561	47	23	530	58	28
	592	0.4	0.2	568	18	9
mKOκ	488	14	9	488	14	9
	532	55	35	514	43	28
	561	70	45	530	52	33
	592	1	0.6	568	33	21
mOrange2	488	20	7	488	20	7
	532	64	22	514	52	18
	561	53	19	530	60	21
	592	0.6	0.2	568	22	8
tdTomato	488	27	26	488	27	26
	532	64	61	514	55	52
	561	91	86	530	63	60
	592	4	4	568	63	60
TagRFP	488	5	2	488	5	2
	532	36	14	514	22	9
	561	92	37	530	35	14
	592	5	2	568	61	24
TagRFP-T	488	5	1.6	488	5	1.6
	532	36	12	514	22	7
	561	92	30	530	35	12
	592	5	1.6	568	61	20
DsRed-Express2	488	33	5	488	33	5
	532	71	11	514	60	9
	561	87	13	530	70	11
	592	1	0.2	568	60	9
mStrawberry	488	12	3	488	12	3
	532	56	15	514	31	8
	561	81	21	530	54	14
	592	45	12	568	94	24
LSSmKate2	405	51	2	407	43	2
	440	85	4	458	98	4.4
	488	71	3	476	90	4
mCherry	488	8	1	488	8	1
	532	40	6	514	21	3
	561	64	10	530	37	6
	592	93	15	568	72	12

(Continued)

Table II *(Continued)*

Protein	Solid-state lasers			Gas lasers		
	Line (nm)	Efficiency of excitation (%)	Effective brightness	Line (nm)	Efficiency of excitation (%)	Effective brightness
mKeima	405	62	2	407	65	2
	440	100	3	458	94	3
	488	31	0.9	476	60	2
mRaspberry	488	9	1	488	9	1
	532	44	5	514	27	3
	561	74	9	530	42	5
	592	99	12	568	77	9
mKate2	488	8	2	488	8	2
	532	33	8	514	18	4.5
	561	59	15	530	31	8
	592	97	24	568	68	17
tdKatushka2	488	4	2	488	4	2
	532	23	11	514	11	6
	561	51	25	530	21	11
	592	97	48	568	60	30
E2-Crimson	532	50	3.5	514	54	4
	561	59	4	530	51	3.5
	592	90	6	568	68	5
	638	19	1.3	633	30	2
mNeptune	488	11	1.4	488	11	1.4
	532	37	5	514	24	3
	561	64	8	530	35	5
	592	97	13	568	69	9
	638	13	2	633	20	3
TagRFP657	561	51	1.7	530	22	0.7
	592	82	3	568	58	2
	638	36	1.2	633	50	1.7

Effective brightness is determined as a product of quantum yield, molar extinction coefficient, and efficiency of excitation divided by 100,000.

B. Maturation: Protein Folding and Chromophore Formation

FP maturation includes two consecutive steps: protein folding and chromophore formation, the latter usually being a rate-limiting step. The tertiary structure of all known FPs is highly conserved and organized as a capped β-barrel that serves as a shell to prevent chromophore quenching by solvent molecules (Day and Davidson, 2009). The main peculiarity of GFP-like fluorescent proteins is that the formation of the chromophore responsible for protein fluorescent properties occurs without any cofactors or enzymes but requires molecular oxygen. The chromophore is generated as a result of several consecutive autocatalytic reactions involving internal amino acid residues of the FP. Even though the first RFPs were cloned in 1999, it was not until recently that the actual mechanism of red chromophore formation has been revealed. It was shown that the posttranslational modifications of RFPs are more complicated than those for GFPs and do not always lead to the red chromophore formation (Piatkevich and Verkhusha, 2010).

A characteristic parameter for estimating the rate, at which an FP becomes fluorescent, is maturation half-time, which is the time required for the FP to reach a half of its maximal fluorescence. The maturation half-time can range from minutes to hours (Table I). Rapid maturation of an FP enables detection of promoter activation and single translational events, while the slow-maturated FPs have rather limited quantitative applications.

Maturation kinetic is usually measured for a rapidly purified FP after expression under anaerobic conditions (Merzlyak *et al.*, 2007). Alternatively, maturation can be monitored in live bacteria after induction of FP expression (Kremers *et al.*, 2007). Maturation rates measured for common RFPs by these approaches are presented in Table I. It should also be kept in mind that the maturation rate and efficiency for an FP *in vivo* depend on oxygen concentration, temperature, expressing cell types, and fusion protein partner. To reduce the negative folding interference in one or both of the fusion proteins, a flexible linker of appropriate length should be used. The Gly-Gly-Ser and Gly-Gly-Thr segments usually provide sufficeint flexibility to the linker (Snapp, 2005).

C. Oligomerization and Cytotoxicity

Oligomerization and cytotoxicity of FPs are among the key deciding issues for their use in cell biology. Depending on protein sequence and cellular environment, FPs can display varying degrees of quatenary structure. To further complicate matters, wild-type Anthozoan FPs may also form nonspecific, high-order aggregates. Both *in vitro* and *in vivo* studies involving a specific FP can be compromised by a tendency of that protein to either oligomerize or aggregate, causing mislocalization of target proteins, disruption of their function, interference with signaling cascades, and cell toxicity of aggregates.

Aggregation of RFPs, which usually results in brightly fluorescent dots in cells and high cytotoxicity, may be caused by electrostatic interactions between positively and negatively charged surfaces. The interactions between macromolecules of FPs can be significantly reduced by the substitution of the positively charged amino acid cluster near the N-terminus with negatively charged or neutral residues (Yanushevich *et al.*, 2002). The first seven amino acids of enhanced GFP (MVSKGEE) followed by a spacer sequence NNMA have been successfully employed for the majority of improved RFPs (Campbell *et al.*, 2002; Shaner *et al.*, 2004). In many respects, optimization of N- and C-termini helps to reduce the cytotoxicity of RFPs (Strack *et al.*, 2008).

In contrast to GFPs from *Hydrozoa* species, which are dimers, virtually all wild-type Anthozoan FPs form stable terameric complexes even in diluted solutions and dissociate only in harsh conditions, which cause irreversible denaturation of polypeptide chains (Vrzheshch *et al.*, 2000). Oligomerization tendency increases the rate constants of the thermal and guanidinium chloride-induced denaturation and extends intracellular life span of FPs (Stepanenko *et al.*, 2004; Verkhusha *et al.*, 2003b). Monomerization of wild-type RFPs by mutagenesis has proven to be a challenging task in the course of their enhancement, since the loss of ability to form tetramers can lead to substantional decrease of fluorescence brightness and protein maturation rates (Campbell *et al.*, 2002; Kredel *et al.*, 2009). Oligomerization tendency of an FP can be quantitively characterized by the dissociation constant; however, the majority of papers discribing FPs lack this information. An acceptable approach for the determination of an FP oligomeric state in cells would be to express its fusion with α-tubulin or β-actin to check whether cytoskeletal structures form correctly (Rizzo *et al.*, 2009). However, even FPs performing well in actin and tubulin fusions may form oligomers under other physiological conditions due to their high local concentration. High local concentration can be reached by extreme overexpression of certain FP fusions and is manifested by background fluorescence, aggregate formation, distortion of cellular organelles, misslocalization of targeted protein, and false-positive FRET signals (Rizzo *et al.*, 2009; Zacharias *et al.*, 2002).

For fusion proteins, the FP must be truly monomeric to minimize the interference with the normal function and localization of targeted protein. Nevertheless, tetrameric and dimeric FPs are still suitable for labeling of luminal spaces of organelles and whole cells. Oligomeric FPs can be also used to control the expression level of a protein of interest by means of translation from the same mRNA controled by an internal ribosome entry sequence (Kamio *et al.*, 2010). Another approach for creating "pseudo"-monomers involves covalent "head-to-tail" cross-linking of two identical FP molecules that would result in an intramolecular or "tandem" dimer, which usually possesses identical spectral characteristics as the original proteins and performs essentially as a monomer although at twice as bigger molecular weight and size. This technique was successfully applied to dimeric variants of DsRed (Campbell *et al.*, 2002; Shaner *et al.*, 2004) and to Katushka2 (Shcherbo *et al.*, 2009a).

D. Photostability and Photoactivation

Similarly to low-molecular-weight fluorescent dyes, all FPs undergo photobleaching upon extended irradiation with excitation light. In flow cytometry analyzers and FACS instruments, however, photobleaching of FPs at the interrogation point is negligible due to the short duration of laser illumination. Thus, photostability and phototoxicity of FPs are not crucial for their usage in flow cytometry. For example, cells expressing JRed (Shagin *et al.*, 2004) or KillerRed (Bulina *et al.*, 2006), which are genetically encoded photosensitizers, can be applied to flow cytometers without any light-induced cell killing. This feature of flow cytometry instruments, however, limits FACS applications of PA-FPs by their insufficient photoactivation. Although "soft" in terms of photoconversion PA-FPs, such as PAmCherry (Subach *et al.*, 2009a), can be photoactivated using LED arrays in cultured cells prior to FACS. Fortunately, the gap of PA-FPs' limited use in flow cytometry can be filled by recently developed monomeric FTs, which change their spectral properties simply with time (Subach *et al.*, 2009b; Tsuboi *et al.*, 2010).

Photostability of FPs can be important if postsort analysis by fluorescent microscopy is required. A comparison of bleaching half-times revealed that the DsRed, TagRFP-T, and mOrange2 proteins are the most photostable among the RFPs (Shaner *et al.*, 2008). When choosing an FP for a particular experiment, it should be kept in mind that protein photostability strongly depends on the conditions of the protein biosynthesis and photobleaching (Bogdanov *et al.*, 2009).

E. pH Stability of Fluorescence

The FP β-barrel is stable in a wide range of pH, and so the rapid denaturation and chromophore degradation are observed at pH values lower than 2 and above 12. Oligomerization tends to increase the overall pH stability of FPs (Vrzheshch *et al.*, 2000). Nevertheless, the spectroscopic properties of FPs can be affected in the physiological range of pH. The pH value, at which the fluorescence of an FP equals to half of its maximum, is designated as pK_a and is employed for a pH-stability comparison. The majority of RFPs developed to date has an apparent pK_a value ranging from 4.0 to 6.5 (Table I). On one hand, high pH stability is advantageous when using FPs in the Golgi, endosomes, lysosomes, secretory granules, and other acidic organelles, as well as for performing quantitative imaging. On the other hand, ratiometric pH-dependence of the FP spectral properties can be efficiently applied for monitoring pH changes in living cells (Hanson *et al.*, 2002).

F. Optimization of Nucleotide and Amino Acid Sequences

Jellyfish and corals, the original source of majority of FPs, differ from mammals by the DNA codon usage. Codon optimization may significantly increase the fluorescence signal of an FP by improving its expression in mammalian cells (Zolotukhin *et al.*, 1996). Usually, commercially available FPs incorporate mammalian codon usage. However, humanized versions of FPs are not necessarily suitable for expression in other model organisms. For instance, the expression of mRFPmars protein, which was optimized for use in *Dictyostelium* cells, failed in mammalian cells (Fischer *et al.*, 2006).

The secondary structure of mRNA and its stability can have a great impact on the expression level of an FP in cells too. Disruption of a predicted stem-loop involved in the ribosomal binding site in mRNA may account for stronger bacterial expression of some FPs (Strack *et al.*, 2009a). Available software for codon usage analysis (EXPASY) and mRNA structure prediction (Zuker, 2003) can be used for the optimization of FP expression in certain cell types or model organisms.

The analysis of the amino acid sequence of an FP, particularly the amino acid residues external to the protein fold and the amino acid content of the N- and C-termini, can help to prevent FP fusion mislocalization. External cystein resiudes and N-glycosylation sites induce an incorrect folding and mislocalization of FP fusions due to their oligomerization in endoplasmic reticulum (Jain *et al.*, 2001). Sometimes an FP can contain short localization sequences that disturb the original localization of a targeted protein (Kredel *et al.*, 2009). Undesirable FP fusion behavior can be eliminated by a substitution of all external cysteins and glycosilation sites and by truncation of targeting domains.

III. Modern Advanced Red–Shifted FPs

The first-generation monomeric RFPs, such as mFruits (Shaner *et al.*, 2004; Wang *et al.*, 2004), mKO (Karasawa *et al.*, 2004), TagRFP (Merzlyak *et al.*, 2007), and mKate (Shcherbo *et al.*, 2007), can be successfully used as markers for protein localization, dynamics, and interactions. Unfortunately, limited brightness and low photostability of these proteins hampered their applications. That is why, recently, a whole series of FPs with improved brightness and photostability has been developed on the basis of these proteins. The palette of enhanced conventional RFPs has been additionally extended by a number of FTs and RFP variants with a LSS of more than 140 nm. The first far-red FP, which can be efficiently excited by the standard 633 and 638 nm laser lines, has been also created.

All RFPs can be divided into three groups according to the fluorescence emission maximum: orange (with emission maximum from 550 to 590 nm), red (with emission maximum from 590 to 630 nm), and far-red (with emission maximum more than 630 nm) (see Table I and Fig. 2).

A. Orange Fluorescent Proteins

Recently, several potentially useful orange fluorescent proteins (OFPs) have been derived from various Anthozoan FPs. The brightest monomeric OFP available to

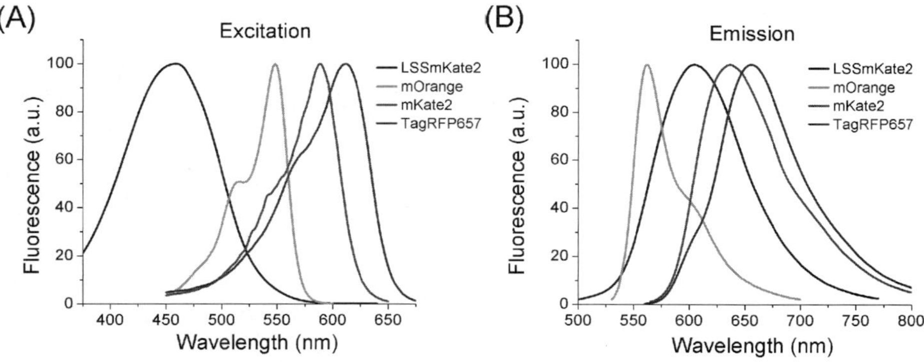

Fig. 2   The excitation (A) and emission (B) spectral profiles of the LSS red (LSSmKate2), orange (mOrange), red (mKate2), and far-red (TagRFP657) FPs are shown. (For interpretation of the references to color in this figure legend, the reader is referred to the Web version of this chapter.)

date is mKOκ (mKusabira Orange kappa), which was derived from mKO (Karasawa *et al.*, 2004), by introducing seven mutations (Tsutsui *et al.*, 2008). The mKOκ protein is characterized by improved pH stability and maturation rate relative to mKO precursor. Photostability of mKOκ under arc lamp illumination is about threefold lower than that for popular mCherry, but both perform similarly under laser illumination. mKOκ has been used for stably expressing cell lines that can indicate its low cytotoxicity (Sakaue-Sawano *et al.*, 2008). Moreover, mKOκ was successfully used as a FRET acceptor (Tsutsui *et al.*, 2008). Finally, poor localization of some mKO fusions could be treated by its replacement with enhanced mKOκ (Rizzo *et al.*, 2009).

The eqFP578 protein turned out to be a rather prospective precursor for a series of monomeric RFPs. Extensive mutagenesis of eqFP578 resulted in a bright monomeric FP, named TagRFP (Merzlyak *et al.*, 2007). In addition to high brightness and fast maturation, TagRFP possesses good pH stability and performs well as a fusion partner for a variety of proteins in different cell types. The main disadvantage of TagRFP is its low photostability under arc lamp and laser illumination.

Tsien and coworkers succeeded in improving the photostability of TagRFP and mOrange significantly (Shaner *et al.*, 2008). Selection of the brightest bacterial colonies, expressing TagRFP and mOrange mutants, under long-term irradiation by intense green-yellow light allowed for the isolation of photostable clones. The resulting FPs, named TagRFP-T and mOrange2, are 9-fold and 25-fold more photostable than the precursors, respectively. mOrange2, however, possesses the reduced brightness and substantially increased maturation time. Photostability of mOrange2 appeared to be almost insensitive to oxygen. In the case of TagRFP-T, photosensitivity to oxygen remained unchanged. Both TagRFP-T and mOrange2 preserve the monomeric state and perform well as tags for cellular proteins. Unfortunately, there is no literature-supported data on the cytotoxicity of these proteins in mammalian cells.

In addition, enhanced tetrameric versions of DsRed, termed DsRed-Express2 (Strack *et al.*, 2008), and E2-Orange (Strack *et al.*, 2009a) have been developed. These FPs yield minimum cytotoxicity and high maturation rates, which makes them more attractive for expression in transgenic animals and stable cell lines as reporter markers for whole cell labeling or for monitoring gene activity. The cytotoxicity of DsRed-Express2 and E2-Orange with respect to mammalian cells is comparable to that of common enhanced GFP. Their photostabilities are comparable to those of many improved RFPs. The tetrameric state of the proteins, however, prevents them from being used in fusion proteins. Another drawback of these proteins is the existence of green species in the absorbance spectra that may limit their applications for multicolor microscopy. Nonetheless, the usage of DsRed-Express2 and E2-Orange simultaneously with EGFP in flow cytometry has been demonstrated (Strack *et al.*, 2008, 2009a).

The tdTomato protein, which is the tandem dimer orange FP, still holds the position of one of the brightest red-shifted FPs yet developed (Shaner *et al.*, 2004). Besides, it is characterized by high photostability and good maturation rate.

While tdTomato can be successfully used for whole-cell and organelle labeling, the large size of the tandem unit may disrupt the function of the target protein.

B. Red Fluorescent Proteins

Continued and intensive optimization of the most red-shifted wild-type RFP, called eqFP611, yielded a bright monomeric mRuby protein (Kredel *et al.*, 2009). Good performance in most fusions and resistance to acidic environments make mRuby potentially useful for many cell biology applications. mRuby turned out to be an effective marker for the visualization of peroxisomes and endoplasmic reticulum in mammalian cells. Another advantage of mRuby is a good separation between excitation and emission spectra (Stokes shift is 47 nm), allowing for better detection of its fluorescence by flow cytometry analyzers and FACS instruments. There are no, however, data regarding mRuby cytotoxicity, and its photostability values in literature are contradictory. The main mRuby runner-ups, mCherry and mStrawberry, have superior maturation rates and exhibit good brightness in mammalian cells. The red-shifted fluorescence spectrum of mRaspberry may allow its good spectral separation with OFPs (Wang *et al.*, 2004).

Recently, a variety of OFPs and RFPs that can be efficiently excited by blue or cyan light have been developed on the basis of conventional FPs (Kogure *et al.*, 2006; Piatkevich *et al.*, 2010b, 2010c). The OFPs and RFPs with an LSS (more than 140 nm) are particularly promising as additional orange and red colors for multiparameter flow cytometry. LSS-RFPs have an absorption maximum at about 440–460 nm, which is ideal for excitation with blue lasers (Table II). The first LSS-RFP, named mKeima, has been already applied to multicolor two-photon microscopy and fluorescence cross-correlation spectroscopy (Kogure *et al.*, 2006). Low brightness and strong pH-dependence of the fluorescence limit its applications though. The presence in Keima samples species with conventional red fluorescence (having the excitation peak at 584 nm) hinders multicolor imaging of common RFPs simultaneously with mKeima (Piatkevich *et al.*, 2010b).

A newly developed LSSmKate2, which was derived from far-red protein mKate, is brighter than mKeima and demonstrates almost pH-independent behavior of the spectral properties in the physiological conditions (Piatkevich *et al.*, 2010b). Its apparent pK_a is 2.7, making LSSmKate2 the most pH-stable among all RFPs reported to date. The several types of mammalian cells stably expressing LSSmKate2 either in cytosol or targeted to the nucleus had normal *in vitro* and *in vivo* proliferation rates and constant expression level of FPs, indicating low cytotoxicity.

The rare examples of RFPs isolated from *Hydrozoa* species are JRed (Shagin *et al.*, 2004) and KillerRed (Bulina *et al.*, 2006). These RFPs are the only known genetically encoded intracellular photosensitizers of reactive oxygen species. In addition, recent attempts to develop an RFP on the basis of a commonly used GFP from *A. victoria* (class Hydrozoa) have finally resulted in the first green-red FP, called R10-3, which has been successfully used in flow cytometry as an additional color (Mishin *et al.*, 2008).

C. Far–Red Fluorescent Proteins

The growing number of researchers is showing interest in far-red FPs due to their superiority over other RFPs for tissue and whole-body imaging, which can make them attractive for application in the biomedical research (Shcherbo *et al.*, 2007, 2009a). Far-red FPs can be simultaneously used for multicolor microscopy and flow cytometry of living cells in conjugation with OFPs and RFPs (Morozova *et al.*, 2010). Monomeric far-red FPs are the proteins of choice for tagging cellular proteins even in strongly autofluorescence conditions. In addition, infrared FPs, obtained on the basis of phytochrome family of proteins, will compete in whole-body and deep-tissue imaging with far-red FPs of the GFP-like family in near future (Shu *et al.*, 2009).

Several far-red FPs have been derived from the wild-type eqFP578 protein. Random mutagenesis of eqFP578 gene yielded a dimeric far-red FP, named Katushka (Shcherbo *et al.*, 2007). A monomeric version of Katushka, called mKate, exhibits high pH sensitivity of fluorescence in the physiological range and a weak photoactivation that complicates quantitative analysis. Subsequent random mutagenesis of mKate has enhanced its brightness, photostability, and pH stability. The improved version, known as mKate2, features rapid maturation with half-time less than 20 min, but it exhibits residual green fluorescence (Shcherbo *et al.*, 2009a). A transgenic expression of mKate2 in *Xenopus* embryos demonstrated reduced cytotoxicity even at high expression levels. Nevertheless, cytotoxicity assay in transiently transfected HeLa cells revealed a noticeable loss of high expression of mKate and Katushka in 24 h after transfection (Strack *et al.*, 2008). Cytotoxicity of mKate and Katushka has been additionally supported by a bacterial expression assay (Strack *et al.*, 2008).

On the basis of mKate, Lin *et al.* (2009) succeeded in engineering the bright far-red protein, named mNeptune, achieving the emission maximum of 649 nm. The red-shifted excitation and emission spectra of mNeptune make it preferable for multicolor imaging with OFPs in live cells. Relative to its parental protein, however, mNeptune was shown to have a tendency to form weak dimers, an increased cytotoxicity in bacterial cells, and residual green fluorescence (Morozova *et al.*, 2010).

Another far-red protein, named E2-Crimson, has been derived from DsRed–Express2 (Strack *et al.*, 2009b). E2-Crimson forms a tetramer and features an excitation peak at 605 nm and an emission peak at 646 nm. E2-Crimson possesses fast maturation and high pH stability. Because of its tetrameric state, E2-Crimson should be used to label whole cells or luminal spaces of organelles. The analysis of HeLa cells transiently transfected with E2-Crimson showed a constant level of protein expression through the 120 h after transfection that indicates its low cytotoxicity (Strack *et al.*, 2009b).

Despite the great effort in the development of far-red FPs, none of them has achieved sufficient excitation using a 633 nm HeNe laser. Only recently, the first monomeric far-red FPs, called TagRFP657, with absorbance and emission maxima at 611 and 657 nm, respectively, have been developed (Morozova *et al.*, 2010). The red-shifted absorbance allows for the excitation of TagRFP657 by the standard 633–640 nm red lasers used in flow cytometry analyzers and FACS instruments. Moreover, TagRFP657 was shown to be an efficient protein tag for the superresolution fluorescence imaging using a stimulated emission depletion microscope, as well as for multicolor wide-field microscopy with OFPs. The good pH stability and photostability coupled with low cytotoxicity make TagRFP657 the far-red probe of choice for multicolor labeling.

IV. Simultaneous Detection of Multiple FPs

One of the main prospects presented by the recent development of the spectrally distinct FPs is multicolor flow cytometry. To a large extent, the instrument configuration and available lasers determine the number of FPs that can be detected simultaneously. However, even with a limited number of available lasers and detectors, there can still be a wide range of possible FP combinations. Despite the possible number of simultaneously used FPs, one should consider two basic rules when selecting fluorescent probes for multicolor flow cytometry. Selected fluorophores should possess high brightness and have minimal emission overlap. We have already defined the brightness for FPs above. In this section, we give general recommendations for drafting FP combinations for multicolor flow cytometry.

At present, there is a vast range of GFP-like FPs with colors spanning the whole visible spectrum of emission wavelengths, peaking from 424 to 670 nm. Based on the value of the emission maximum, all FPs can be divided into the following groups: blue FPs (emission peaking at 420–460 nm), cyan FPs (emission peaking at 470–500 nm), green FPs (emission peaking at 500–520 nm), yellow FPs (emission peaking at 520–540 nm), orange FPs (emission peaking at 550–590 nm), red FPs (emission peaking at 590–630 nm), and far-red FPs (emission peaking at 630–650 nm).

The blue, cyan, green, and yellow FPs have been recently described in detail in a number of reviews (Chudakov *et al.*, 2010; Day and Davidson, 2009). Here we only mention the most efficient for flow cytometry FPs from each spectral region. These include blue FPs: Azurite, EBFP2, and TagBFP; cyan FPs: ECFP, Cerulean, mTurquoise, and mTFP1; green FPs: T-Sapphire, EGFP, spGFP, Emerald, and TagGFP2; and yellow FPs: mAmetrine, EYFP, Venus, Citrine, and YPet (Chudakov *et al.*, 2010; Day and Davidson, 2009).

The appropriate optical filter configuration and real-time electronics or postanalysis software-based compensation provide the ability to distinguish between large numbers of different fluorophores with partially overlapping spectra (Hawley *et al.*, 2004). We list a number of possible combinations of FPs with respect to their spectral properties and laser availability often seen in currently existing flow cytometry analyzers and FACS sorters (Table III). Nonetheless, the emission of FPs from the neighboring cells across Table II generally exhibit noticeable spectral overlap. It should be pointed out that a cross talk between emissions occurs toward the red edge of the spectrum (Fig. 2B). Moreover, by increasing the number of detected colors, the

Table III

Compatibility of various FPs in multiparameter flow cytometry

Laser line (nm)	Blue	Cyan	Green	Yellow	Orange	Red	Far-red
350–356	Azurite EBFP2 TagBFP		T-Sapphire	mAmetrine			
405, 407	Azurite EBFP2 TagBFP	ECFP Cerulean mTurquoise	T-Sapphire	mAmetrine	LSS-mOrange	LSS-mKate1 LSS-mKate2 mKeima	
440, 458		ECFP Cerulean mTurquoise	EGFP spGFP Emerald TagGFP2		LSS-mOrange	LSS-mKate1 LSS-mKate2 mKeima	
488			EGFP spGFP Emerald TagGFP2	EYFP Venus Citrine YPet			
514				EYFP Venus Citrine YPet	mKO mOrange mKOk mOrange2 dTomato		
530, 532					mKO mOrange mKOk mOrange2 dTomato TagRFP TagRFP-T	mStrawberry	
561					mKO mOrange mKOk mOrange2 dTomato TagRFP TagRFP-T	mStrawberry mCherry mRaspberry	mKate mKate2 Katushka mNeptune
568					dTomato TagRFP TagRFP-T	mStrawberry mCherry mRaspberry	mKate mKate2 Katushka E2-Crimson mNeptune
592						mStrawberry mCherry mRaspberry	mKate mKate2 Katushka E2-Crimson mNeptune TagRFP657
633, 638							TagRFP657 E2-Crimson

Table IV
Internet resources providing description and characteristics of FPs

Resource	Web page	Comments
Clontech Laboratories	http://www.clontech.com/clontech/gfp/index.shtml	General description of FPs
Olympus	http://www.olympusconfocal.com/applications/fpcolorpalette.html	Reviews of FPs
Evrogen	http://www.evrogen.com	Detailed information on many FPs
MBL International	http://www.mblintl.com	General description of FPs
Albert Einstein College of Medicine	http://www.einstein.yu.edu/facs/	Recommended for flow cytometry FPs
Florida State University	http://micro.magnet.fsu.edu	List of useful links
MicroscopyU	http://www.microscopyu.com	Interactive applications for selecting filter sets

experiment also gains an additional complexity. Simultaneous detection of several FPs at the single cell level requires extra circumspection. Several flow cytometry approaches have been reported to detect RFPs with up to three GFP variants using single- or dual-laser excitation (Hawley *et al.*, 2001, 2004; Kamio *et al.*, 2010). The strategies utilizing the multicolor flow cytometry with fluorescent dyes can be also applied for FPs. The lack of yellow and orange lasers on some flow cytometers can be compensated by RFPs with increased Stokes shift, which are useful for combinations with conventional FPs. For instance, TagBFP, T-Sapphire, and LSSmKate2 can be excited with 405–407 nm light, but emit in different spectral ranges: blue, green, and red, respectively. The 458 nm laser can also be used for the simultaneous excitation of LSSmKate2 with ECFP and EYFP.

The appropriate choice of FPs with minimal emission overlap and the selection of filter sets that detect the fluorescence from one fluorophore can minimize the need for compensation, making the flow cytometry analysis more accurate. A forward estimate of spectral spillover and optimal filters can be theoretically made by overlaying spectral profiles of FPs. There are a number of web resources presenting valuable information on the spectral properties of FPs, which might be useful for designing multicolor flow cytometry experiments (Table IV).

V. Fluorescent Timers

Multiparameter flow cytometry can be also performed using only one FP that changes its spectral properties with time. The FPs changing their spectral properties during maturation in a time-dependent manner are known as FTs. Usually, at an early stage of maturation, FTs fluoresces in one spectral region, and after complete maturation, they exhibit a red-shifted fluorescence. The predictable time course of changing fluorescent colors allows for a quantitative analysis of temporal and spatial molecular events, based on the ratio between fluorescence intensities of two forms. Therefore, FTs yield temporal and spatial information regarding protein translation and target promoter activity. Moreover, recently developed new subclass of monomeric FTs enables tracking of protein dynamics and determination of half-life of the tagged proteins.

The first reported FT was tetrameric DsRed-Timer (Terskikh *et al.*, 2000). Its color transition from green to red was shown to be due to superefficient FRET between green and red units in the tetramer (Verkhusha *et al.*, 2004). DsRed-Timer has been utilized for monitoring the dynamics of gene expression in various tissues, for separating cells with different gene activity patterns, and for studying intracellular trafficking and organelle aging (Chudakov *et al.*, 2010).

Recently, the first monomeric FTs, which exhibit distinctive fast, medium, and slow blue-to-red chromophore maturation rates, have been developed (Table V) (Subach *et al.*, 2009b). During the maturation of these monomeric FTs, the fluorescence of the blue forms increases to its maximum value, and after that decreases to zero (Fig. 3). The fluorescence of the red forms increases with time with some delay and then reaches a plateau. Maturation of both fluorescent forms is temperature sensitive; however, monomeric FTs exhibit a similar timing behavior in bacteria, insect, and mammalian cells. The maturation times increase at lower temperatures like 16 °C and 25 °C and decrease at higher temperatures like 45 °C (Table V). The higher the temperature, the faster is the growth of the red-to-blue ratio. This observation provided a unique interpretation for the age of the particular FT. The data suggest that calibration curves based on the FT red-to-blue ratio can be used to determine the time from the start of FT production under any specific conditions. In microscopy, the quantitative analysis can be affected by a blue-to-red photoactivation of FTs under prolonged or intense illumination with violet light. However, in flow cytometry, FTs do not show any noticeable photoconversion (Subach *et al.*, 2009b).

High pH stability of blue and red forms makes it possible to use FTs in a wide range of pH values. The blue and red forms of FTs are bright enough to use them either alone, in protein fusions, or together with green FPs for multicolor imaging.

Table V

Spectroscopic and biochemical properties of monomeric fluorescent timers

Protein	Ex_{max} (nm)	Em_{max} (nm)	ε (M^{-1} cm^{-1})	QY	pK_a	Characteristic time (h)	Optimal excitation laser (nm)
Fast-FT	403	466	49,700	0.30	2.8	0.25	405, 407
	583	606	75,300	0.09	4.1	7.1	532, 561, 592
Medium-FT	401	464	44,800	0.41	2.7	1.2	405, 407
	579	600	73,100	0.08	4.7	3.9	532, 561, 592
Slow-FT	402	465	33,400	0.35	2.6	9.8	405, 407
	583	604	84,200	0.05	4.6	28	532, 561, 592
mK-GO	500	509	35,900	ND	6.0	10	458, 488
	548	561	42,000	ND	4.8	ND	514, 532

Ex_{max}, excitation maximum; Em_{max}, emission maximum; ε, molar extinction coefficient; QY, quantum yield; pK_a, pH value, at which protein retain half of its fluorescent intensity. ND, not determined.

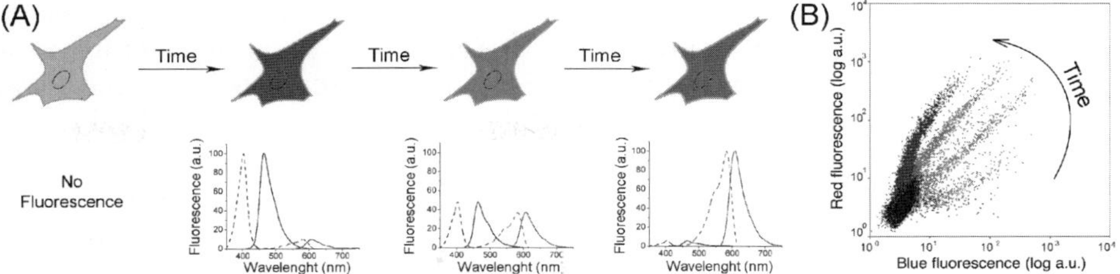

Fig. 3   (A) A cell expressing a fluorescent timer gradually changes fluorescence from blue to red with time. The respective time changes of the excitation and emission spectra for the blue and red forms of the fluorescent timer are shown. (B) A flow cytometry plot shows the same population of cells, which expresses the blue-to-red fluorescent timer, analyzed at different times after its expression. (For interpretation of the references to color in this figure legend, the reader is referred to the Web version of this chapter.)

Availability of three monomeric FTs with distinctive blue-to-red maturation times is useful for studies of intracellular processes with different timescales.

Recently, by introducing six mutations into mKO, Tsuboi *et al.* (2010) generated another monomeric FT, named Kusabira Green-Orange, or mK-GO, that changes fluorescence from green to orange over time (Table V). Absorption of mK-GO is slightly pH sensitive, with the apparent pK_a values at 6.0 and 4.8 for the green and orange forms, respectively. The ratio of orange per green fluorescence, determined by *in vitro* translation, was linearly increased and reached a plateau at approximately 10 h. It is reasonable to expect that the maturation rate is temperature dependent. Monomeric behavior of mK-GO enables its utilization in fusions with the targeting proteins.

The color transition for monomeric FTs is thought to be due to a conversion of one fluorescent species into another. It was discovered that a blue-emitting form of FTs contains the TagBFP-like chromophore, which is converted after autooxidation by molecular oxygen to a red-emitting DsRed-like chromophore (Pletnev *et al.*, 2010; Subach *et al.*, 2010).

Interestingly, color transition of living cell can also be due to regulation of gene expression. For example, spatial and temporal patterns of cell-cycle dynamics can be visualized with a Fucci approach (fluorescent, ubiquitination-based cell cycle indicator) reported by Sakaue-Sawano *et al.* (2008). Fucci is composed of mKO2-hCdt1 (30/120) and mAG-hGem (1/110) fusion proteins that make the expressing cells being yellow at a start of replication, then switch to green during S phase, and to orange during G$_1$ phase. This approach allows for the analysis and sorting of Fucci-expressing cells using single laser excitation at 488 nm.

VI. FRET–Based Genetically Encoded Biosensors

FPs have been successfully employed in engineering genetically encoded biosensors. FP-based biosensors mediate the monitoring and detection of different intracellular events while preserving spatial and temporal resolution. The genetically encoded biosensors possess a unique advantage of being able to be fused to a certain protein or organelle within a cell, facilitating a noninvasive assay of a specific protein activity or cellular signal. The expansion of the FP palette with newly developed RFPs gives unprecedented opportunities to design biosensors with spectrally distinct properties. Moreover, RFPs may serve as templates for engineering biosensors fundamentally different from what already exists with GFP variants.

Detection and quantification of a FRET signal by flow cytometry provides sensitive measurements of protein–protein interactions or protease activity in live cells. The FACS-based FRET can be efficiently applied for cell-based protein library screening and directed evolution of FP biosensors. Upon FRET, a decrease in the donor fluorescence intensity and an increase in the acceptor fluorescence intensity are observed, from either of which the experimental FRET efficiency can be calculated (Domingo *et al.*, 2007). Based on this phenomenon, the FRET positive events can be easily detected with flow cytometry. Cells expressing a donor and an acceptor either in combination (for negative control) or as a fusion protein (for positive control) should be used for establishing an assay to measure FRET signals. The optical configuration of the instruments for FRET measurements should allow for an efficient excitation of both donor and acceptor by two different lasers and for collection of the donor, acceptor, and FRET signals (Chan *et al.*, 2001). Alternatively, the fluorescence of the donor and acceptor upon excitation by a single laser can be measured simultaneously, and the efficiency of the transfer can be judged from the ratio of fluorescence intensities at two wavelengths, corresponding to the emission of the donor and the acceptor or by donor quenching (He *et al.*, 2003a). This strategy directly analyzes the efficiency of FRET pairs for screening of cell-based libraries using flow cytometry.

A. Conventional FRET Pairs

For a long time, EYFP was the most red-shifted monomeric FP available, while mainly the ECFP–EYFP pair has been extensively used in cell biology. It has been shown that flow cytometry can be effectively used to detect the FRET from ECFP to EYFP (Banning *et al.*, 2010; Chan *et al.*, 2001; He *et al.*, 2003a; Siegel *et al.*, 2000). The ECFP–EYFP pair was applied for detection of protein–protein interactions and for monitoring caspase activity in live yeast and mammalian cells by flow cytometry (Banning *et al.*, 2010; Dye *et al.*, 2005; He *et al.*, 2003b; van Wageningen *et al.*, 2006). Enhanced cyan and yellow FPs, such as CyPet and YPet proteins, overcome performance of the ECFP–EYFP pair in FACS analysis by means of a better signal-to-noise ratio and an improved dynamic range (Nguyen and Daugherty 2005; You *et al.*, 2006), possibly caused by their weak dimerizing tendency. However, the considerable spectral overlap between cyan and yellow FPs makes FRET measurements difficult for the purpose of screening molecular libraries with few positive hits. Thus, development of new FRET pairs is highly desirable. One of the options is to utilize the recently created

RFPs as acceptors or even donors. For instance, a combination of GFP and RFP compared to the ECFP–EYFP pair has advantages for flow cytometry analysis due to the minimal emission overlap and superior brightness of GFPs versus ECFP. In addition, orange FPs can be efficient FRET donors for far-red FPs.

Engineering of reliable and efficient FRET pairs requires FPs that combine such properties as true monomeric behavior, high pH stability and photostability, fast maturation, and bright fluorescence. Moreover, in order to achieve a high dynamic range (the donor/acceptor emission ratio change), which is one of most important characteristics of FRET sensors, the donor and the acceptor should ideally have minimal cross talks.

Since FRET is highly distance-sensitive, any oligomerization tendency of the donor and acceptor FP molecules can lead to nonspecific FRET, even if they are expressed in their free, untagged form (Nguyen and Daugherty, 2005). For example, the ECFP–HcRed pair tested for the detection of proteins interacting in the endoplasmic reticulum demonstrated a sufficient FRET signal in FACS screening. However, a high level of nonspecific FRET, which could be attributed to the strong oligomerization of HcRed, noticeably hindered FRET measurements (Johansson *et al.*, 2007). Slowly maturating FPs may demonstrate insufficient or completely abrogated chromophore formation during folding of the biosensor, due to an unpredictable interference with the rest of the biosensor molecule. It can result in a strong variation of measured FRET efficiency (Goedhart *et al.*, 2007). Low pH stability limits the application of FRET pairs in acidic organelles. Sometimes, the postsorting analysis is required to detect FRET in collected cells using fluorescence microscopy. In this case, highly photostable FPs are the proteins of choice for accurate FRET efficiency measurements.

B. Novel Advanced FRET Pairs

A high FRET efficiency has been achieved for the combination of GFPs and recently developed monomeric RFPs. The high extinction coefficient of TagRFP and mCherry makes them attractive FRET acceptors for the GFPs (Shcherbo *et al.*, 2009b; Tramier *et al.*, 2006). Shcherbo *et al.* generated a bright, high contrast apoptosis reporter, named CaspeR3 (*Caspase-3 R*eporter), constructed from TagRFP and TagGFP. Both FPs exhibited profound maturation in CasreR3. A combination of LSS green FPs, such as T-Sapphire, with orange or red FPs may be a promising FRET pair due to almost complete separation of fluorescence profiles that simplifies intensity-based FRET experiments (Bayle *et al.*, 2008; Zapata-Hommer and Griesbeck, 2003).

Yellow FPs, which are red-shifted compared to GFP, can achieve a greater spectral overlap with RFPs and still preserve the separation of donor and acceptor fluorescence. Several bright yellow FPs, namely mCitrine, SYFP2, Venus, and LSS mAmetrine, have quantum yields exceeding 0.58 (Ai *et al.*, 2008; Griesbeck *et al.*, 2001; Kremers *et al.*, 2006; Nagai *et al.*, 2002) and are very promising as fluorescent donors for RFPs. The mCitrine and mKate2 proteins can be excited independently and exhibit a limited cross talk. This pair of yellow and far-red FPs has been employed for the construction of a voltage-sensitive biosensor (Mutoh *et al.*, 2009). The fusion of SYFP2 as a donor and mStrawberry or mCherry as an acceptor can be used for protein–protein interaction studies (Goedhart *et al.*, 2007).

Anthozoan cyan and orange FPs exhibit a close spectral overlap of the emission and absorption spectra, respectively. Their improved versions have been utilized as efficient FRET pairs. For example, the MiCy–mKO pair has a high ratio change (14-fold) and was used to monitor the activity of caspase-3 during apoptosis (Karasawa *et al.*, 2004). Another pair, mUKG–mKOk, was optimized for membrane voltage measurements (Tsutsui *et al.*, 2008). Both pairs exhibited reduced cross-excitation and pH sensitivity. All proteins are quite bright and should be easily detected in flow cytometry analysis. However, the performance of the MiCy–mKO pair can be limited by the MiCy dimerization tendency and the slow maturation of mKO. The problem reported for mUKG–mKOk, in turns, is its weak tendency to form fluorescent aggregates (Mutoh *et al.*, 2009). The enhanced versions of mKO, named mKO2, can be also used as a FRET acceptor in conjugation with cyan FPs (Sakaue-Sawano *et al.*, 2008). Moreover, the mKO protein was used as a donor to the mCherry acceptor for quantitative FRET in live cells (Goedhart *et al.*, 2007). However, a significant overlap of orange fluorescence of the donor with red fluorescence of the acceptor can disable FRET detection by flow cytometry.

C. FRET Biosensors in Multicolor Flow Cytometry

Using a FRET-biosensor in combination with a second biosensor makes it possible to monitor and detect multiple biochemical parameters in a single cell. To make the multiparameter analysis effective, the FRET pairs used simultaneously should fulfill certain criteria. Each FRET pair or additional fluorophore should be excited at a wavelength at which the other FRET pair or fluorophore is not excited or excited insignificantly. Another possibility is to collect emission from each FRET pair at a wavelength where another FRET pair does not emit fluorescence. The engineering of FRET pair combinations that meet these demands became possible only with the introduction of RFPs.

In some cases, a multiparameter analysis is possible by means of adding a monomeric RFP to the ECFP–EYFP pair. Depending on experimental conditions, the measurement of FRET signals within a system of three donor–acceptor pairs, such as ECFP–EYFP, ECFP–mRFP, and EYFP–mRFP, can be performed (Galperin *et al.*, 2004). Additionally, in trimeric complexes, the two-step FRET can be achieved: from ECFP to EYFP and, subsequently, from EYFP to mRFP. The "linked" FRET can be accurately separated from individual steps by using a flow cytometer with the six-color three-laser system, which measures distinct signals from FPs and from four possible FRET signals (He *et al.*, 2005). He *et al.* also described the optical configuration for FRET measurements in details. The linked FRET was used to study the homotrimerization of TRAF2 protein in live cells. This method can be utilized for studying an interaction between three specific proteins. Also, the sequential fusing of ECFP, EYFP, and mRFP using the caspase-3 and caspase-6 recognition motifs yielded a dual parameter FRET sensor (Wu *et al.*, 2006). Flow cytometry facilitated a simultaneous monitoring of the activity of two caspases in apoptotic cells, using the 405 and 488 nm lasers to excite ECFP and EYFP, respectively.

Two spectrally distinct FRET pairs may also enable multiparameter flow cytometry. The cyan-fluorescing mTFP1 and yellow-fluorescing mAmetrine, which possess distinct excitation spectra, were used as donors for yellow FP and bright RFP, named tdTomato, as acceptors (Ai *et al.*, 2008). The mOrange–mCherry and ECFP–EYFP can be excited separately enabling a multiparameter detection. However, the performance of the mOrange–mCherry pair is limited by a strong spectral overlap and a poor-sensitized emission of mCherry (Piljic and Schultz, 2008).

D. Optimizing Biosensors for High-Throughput Screening

Utilization of FP-based sensors in combination with high-throughput techniques has a tremendous potential in basic science research as well as in large-scale approaches for the screening of novel pharmaceuticals or the development of therapeutic strategies. For a sustainable flow cytometry analysis, the FRET-based biosensor should exhibit a reproducible signal and have a wide dynamic range (the difference between maximal and minimal FRET signals). Unfortunately, a poor dynamic range is generally one of the biggest drawbacks for FRET-based biosensors. A biosensor usually consists of a sensing domain, which is fused to FPs, and its dynamic range is highly sensitive to the way in which these two components are connected. The length and the amino acid content of linkers between the sensing domain and FPs have a strong influence on the FRET efficiency and overall dynamic range, because they mainly determine the orientation and the distance between the FPs (van Dongen *et al.*, 2007; Evers *et al.*, 2006; Nagai and Miyawaki, 2004). The process of optimization of a FRET biosensor involves varying the linker length and sequence by generating sensor libraries. Classical linkers in biosensor engineering are $(GGSGGS)_n$ and $(GGSGGT)_n$ that are flexible and reduce interactions between nearby domains (van Dongen *et al.*, 2007; Nagai and Miyawaki, 2004). Constructed libraries, which may contain up to 10^8 independent clones, can be efficiently screened by flow cytometry, in order to choose variants with improved characteristics for further application in high-throughput screening (Nguyen and Daugherty, 2005).

VII. Biosensors Consisting of a Single FP

Spectral properties of some FPs can change in response to a specific stimulus. This feature allows the construction of a single FP-based biosensor, which contains only one engineered FP molecule. In certain cases, it is possible to modulate the optical properties of a single FP. One of the options is a mechanical deformation of the FP's β-barrel. The mechanical deformation can be caused by changing conformation of the inserted into the FP molecule additional peptide or domain or by manipulating the new N- and C-termini in the middle of the barrel after a circular permutation of the FP.

Circularly permutated GFP variants have been widely utilized to design efficient biosensors of hydrogen peroxide, calcium, phosphorylation, and membrane potential (Chudakov *et al.*, 2010). In general, RFPs are less tolerant to the polypeptide or domain insertions and circular permutations in comparison to GFPs, probably due to the more complicated posttranslational chemistry of the chromophore formation (Subach *et al.*, 2010). However, Gautam *et al.* (2009) succeeded in engineering a circularly permuted mKate with new N- and C-termini formed by residues 180 and 182, respectively. The cpmKate180 permutant was shown to serve a promising sensing platform for the development of a series of voltage-sensitive probes, which can be used in live cells. In mCherry, six distinct sites in β-barrel have been identified that were used to create circular permutation variants (Li *et al.*, 2008). Recently, the extensive directed evolution of the most promising cpmCherry variant has yielded a mutant that exhibited 61% of parental mCherry brightness and was found to be highly tolerant of the circular permutation at other locations within the amino acid sequence (Carlson *et al.*, 2010). The main advantage of the circularly permutated FPs as biosensors over the FRET-based biosensors is their dynamic range that can be further improved (Nagai *et al.*, 2004).

Some FP variants exhibit different spectral properties depending on the presence of the particular ions, such as proton, halides, or heavy metal cations. Some examples of biosensors utilizing this feature are YFP-H148Q to detect halides (Jayaraman *et al.*, 2000) and ratiometric deGFPs to measure intracellular pH (Hanson *et al.*, 2002). There are also several examples of a similar behavior among RFPs. A DsRed derivative, known as mNectarine, exhibits a strong pH-dependent fluorescence in physiological pH range with apparent pKa value about 6.9 (Johnson *et al.*, 2009). The mNectarine fusion with human concentrative nucleoside transporter was used to measure pH during the H^+–nucleoside cotransport in live cells. However, the calibration required for intensity-based pH measurements and a nonuniform biosensor distribution complicate mNectarine application as a ratiomentric pH biosensors. It has been shown that mKeima is ratiometric by excitation. Upon alkalization from pH 5.0 to 8.0, the fluorescence of mKeima increased almost 3-fold and decreased more than 12-fold, when excited at 440 and 584 nm, respectively (Piatkevich *et al.*, 2010b). A red fluorescent biosensor based on mKeima may provide a ratiometric readout of intracellular pH for high-throughput assays. A quenching of the red fluorescence of HcRed in the presence of copper ions was employed to develop a soluble copper-sensing system (Rahimi *et al.*, 2007). The same principle can be utilized for the determination of copper concentrations in live cells, based on HcRed as a sensing platform.

Another recently described approach for the detection of protein–protein interactions, as well as for measuring changes in protein complexes upon drug influence and identification of compounds interfering in protein–protein binding, is a bimolecular fluorescent complementation (BiFC) (Kerppola, 2008). The principle of the

BiFC assay is to split an FP into two nonfluorescent fragments that can be fused to two proteins of interest. If the proteins of interest interact, the two nonfluorescent fragments are brought into close proximity, so that an FP barrel is reconstituted, and fluorescence arises. All BiFC systems are irreversible: once a split FP is formed, it stays in conjugation, even if the proteins of interest do not interact any more. In contrast to FRET, an interpretation of the BiFC assay results requires less analysis and fewer corrections. However, temporal resolution is limited to the length of time necessary for the chromophore formation after the nonfluorescent fragments association.

Coupling BiFC to flow cytometry has been proven to be a powerful technique for the validation of weak protein interactions and for a screening and identification of optimal ligands in biologically synthesized libraries (Morell *et al.*, 2008). A novel red fluorescent BiFC system that can support fluorescence complementation in most mammalian cell-based studies have been reported recently (Chu *et al.*, 2009). This advanced BiFC system is based on mLumin, a mKate derivative, and has a good performance and sensitivity at 37 °C. Indeed, previously developed red fluorescent BiFC systems based on mRFP1-Q66T (Jach *et al.*, 2006) or mCherry (Fan *et al.*, 2008) worked at a lower temperature only, possibly due to the misfolding of the RFP fragments at 37 °C. The GFP variants, which support BiFC under physiological conditions, namely, Cerulean, Citrine, and Venus, can be used in combination with the mLumin-based BiFC system to facilitate simultaneous monitoring of multiple intracellular protein–protein interactions (Shyu *et al.*, 2006). Furthermore, a recently described BiFC-based FRET technique enables an identification of the ternary complexes in live cells (Shyu *et al.*, 2008).

VIII. Perspectives

In a near future, we expect to observe significant efforts to design enhanced bright noncytotoxic FPs for the far-red and infrared regions, which can be efficiently excited with conventional red sources such as 633 nm HeNe and 635–640 nm solid-state lasers.

Further development of the commercial imaging flow cytometers and scanning cytometers and equipping them with green and yellow lasers will allow for more extensive use of OFP and RFP to detect subcellular events and intracellular protein localization.

A wider use of orange solid-state lasers and white supercontinuum lasers will enable the closing the gap between 568 and 633 nm in the excitation sources currently available for flow cytometry (Kapoor *et al.*, 2007, 2008; Telford *et al.*, 2009).

Engineering the next generation of the bright and efficiently folding orange, red, and far-red LSS FPs, which can be excited with 405–457 nm laser sources, will provide the additional red-shifted colors for the multiparameter flow cytometry applications. The LSS FPs will also enable the detection of three FRET pairs or three FRET-based biosensors in a live cell.

Acknowledgments

For helpful discussions, the authors thank Drs. Jinghang Zhang and Steven Porcelli of the Einstein Flow Cytometry Core Facility (supported by the Einstein Cancer Center (NIH/NCI grant CA013330) and the Einstein Center for AIDS Research (NIH grant AI-51519)). This work was supported by the NIH/NIGMS grant GM073913 to V.V.V.

References

Ai, H. W., Hazelwood, K. L., Davidson, M. W., and Campbell, R. E. (2008). Fluorescent protein FRET pairs for ratiometric imaging of dual biosensors. *Nat. Methods* **5**, 401–403.

Banning, C., Votteler, J., Hoffmann, D., Koppensteiner, H., Warmer, M., Reimer, R., Kirchhoff, F., Schubert, U., Hauber, J., Schindler, M. (2010). A flow cytometry-based FRET assay to identify and analyse protein–protein interactions in living cells. *PLoS One* **5**, e9344.

Bayle, V., Nussaume, L., and Bhat, R. A. (2008). Combination of novel green fluorescent protein mutant TSapphire and DsRed variant mOrange to set up a versatile in planta FRET-FLIM assay. *Plant Physiol.* **148**, 51–60.

Bogdanov, A. M., Bogdanova, E. A., Chudakov, D. M., Gorodnicheva, T. V., Lukyanov, S., Lukyanov, K. A. (2009). Cell culture medium affects GFP photostability: a solution. *Nat. Methods* **6**, 859–860.

Bulina, M. E., Chudakov, D. M., Britanova, O. V., Yanushevich, Y. G., Staroverov, D. B., Chepurnykh, T. V., Merzlyak, E. M., Shkrob, M. A., Lukyanov, S., Lukyanov, K. A. (2006). A genetically encoded photosensitizer. *Nat. Biotechnol.* **24**, 95–99.

Campbell, R. E., Tour, O., Palmer, A. E., Steinbach, P. A., Baird, G. S., Zacharias, D. A., Tsien, R. Y. (2002). A monomeric red fluorescent protein. *Proc. Natl. Acad. Sci. USA* **99**, 7877–7882.

Carlson, H. J., Cotton, D. W., and Campbell, R. E. (2010). Circularly permuted monomeric red fluorescent proteins with new termini in the beta-sheet. *Protein Sci.* **19**, 1490–1499.

Chalfie, M., Tu, Y., Euskirchen, G., Ward, W. W., and Prasher, D. C. (1994). Green fluorescent protein as a marker for gene expression. *Science* **263**, 802–805.

Chan, F. K., Siegel, R. M., Zacharias, D., Swofford, R., Holmes, K. L., Tsien, R. Y., Lenardo, M. J. (2001). Fluorescence resonance energy transfer analysis of cell surface receptor interactions and signaling using spectral variants of the green fluorescent protein. *Cytometry* **44**, 361–368.

Chu, J., Zhang, Z., Zheng, Y., Yang, J., Qin, L., Lu, J., Huang, Z. L., Zeng, S., Luo, Q. (2009). A novel far-red bimolecular fluorescence complementation system that allows for efficient visualization of protein interactions under physiological conditions. *Biosens. Bioelectron.* **25**, 234–239.

Chudakov, D. M., Matz, M. V., Lukyanov, S., and Lukyanov, K. A. (2010). Fluorescent proteins and their applications in imaging living cells and tissues. *Physiol. Rev.* **90**, 1103–1163.

Day, R. N., and Davidson, M. W. (2009). The fluorescent protein palette: tools for cellular imaging. *Chem. Soc. Rev.* **38**, 2887–2921.

van Dongen, E. M., Evers, T. H., Dekkers, L. M., Meijer, E. W., Klomp, L. W., Merkx, M. (2007). Variation of linker length in ratiometric fluorescent sensor proteins allows rational tuning of Zn(II) affinity in the picomolar to femtomolar range. *J. Am. Chem. Soc.* **129**, 3494–3495.

Domingo, B., Sabariegos, R., Picazo, F., and Llopis, J. (2007). Imaging FRET standards by steady-state fluorescence and lifetime methods. *Microsc. Res. Tech.* **70**, 1010–1021.

Drobizhev, M., Tillo, S., Makarov, N. S., Hughes, T. E., and Rebane, A. (2009). Absolute two-photon absorption spectra and two-photon brightness of orange and red fluorescent proteins. *J. Phys. Chem. B* **113**, 855–859.

Evers, T. H., van Dongen, E. M., Faesen, A. C., Meijer, E. W., and Merkx, M. (2006). Quantitative understanding of the energy transfer between fluorescent proteins connected via flexible peptide linkers. *Biochemistry* **45**, 13183–13192.

Fan, J. Y., Cui, Z. Q., Wei, H. P., Zhang, Z. P., Zhou, Y. F., Wang, Y. P., Zhang, X. E. (2008). Split mCherry as a new red bimolecular fluorescence complementation system for visualizing protein-protein interactions in living cells. *Biochem. Biophys. Res. Commun.* **367**, 47–53.

Fischer, M., Haase, I., Wiesner, S., and Müller-Taubenberger, A. (2006). Visualizing cytoskeleton dynamics in mammalian cells using a humanized variant of monomeric red fluorescent protein. *FEBS Lett.* **580**, 2495–2502.

Galperin, E., Verkhusha, V. V., and Sorkin, A. (2004). Three-chromophore FRET microscopy to analyze multiprotein interactions in living cells. *Nat. Methods* **1**, 209–217.

Gautam, S. G., Perron, A., Mutoh, H., and Knöpfel, T. (2009). Exploration of fluorescent protein voltage probes based on circularly permuted fluorescent proteins. *Front. Neuroengineering.* **2**, 14.

Goedhart, J., Vermeer, J. E., Adjobo-Hermans, M. J., van Weeren, L., and Gadella Jr., T. W. (2007). Sensitive detection of p65 homodimers using red-shifted and fluorescent protein-based FRET couples. *PLoS One* **2**, e1011.

Griesbeck, O., Baird, G. S., Campbell, R. E., Zacharias, D. A., and Tsien, R. Y. (2001). Reducing the environmental sensitivity of yellow fluorescent protein. Mechanism and applications. *J. Biol. Chem.* **276**, 29188–29194.

Hanson, G. T., McAnaney, T. B., Park, E. S., Rendell, M. E., Yarbrough, D. K., Chu, S., Xi, L., Boxer, S. G., Montrose, M. H., Remington, S. J. (2002). Green fluorescent protein variants as ratiometric dual emission pH sensors. 1. Structural characterization and preliminary application. *Biochemistry* **41**, 15477–15488.

Hawley, T. S., Herbert, D. J., Eaker, S. S., and Hawley, R. G. (2004). Multiparameter flow cytometry of fluorescent protein reporters. *Methods Mol. Biol.* **263**, 219–238.

Hawley, T. S., Telford, W. G., Ramezani, A., and Hawley, R. G. (2001). Four-color flow cytometric detection of retrovirally expressed red, yellow, green, and cyan fluorescent proteins. *Biotechniques* **30**, 1028–1034.

He, L., Bradrick, T. D., Karpova, T. S., Wu, X., Fox, M. H., Fischer, R., McNally, J. G., Knutson, J. R., Grammer, A. C., Lipsky, P. E. (2003a). Flow cytometric measurement of fluorescence resonance energy transfer from cyan fluorescent protein to yellow fluorescent protein using single laser excitation at 458 nm. *Cytometry A* **53**, 39–54.

He, L., Olson, D. P., Wu, X., Karpova, T. S., McNally, J. G., Lipsky, P. E. (2003b). A flow cytometric method to detect protein-protein interaction in living cells by directly visualizing donor fluorophore quenching during CFP-YFP fluorescence resonance energy transfer (FRET). *Cytometry A* **55**, 71–85.

He, L., Wu, X., Simone, J., Hewgill, D., and Lipsky, P. E. (2005). Determination of tumor necrosis factor receptor-associated factor trimerization in living cells by CFP-YFP-mRFP FRET detected by flow cytometry. *Nucleic Acids Res.* **33**, e61.

Jach, G., Pesch, M., Richter, K., Frings, S., and Uhrig, J. F. (2006). An improved mRFP1 adds red to bimolecular fluorescence complementation. *Nat. Methods* **3**, 597–600.

Jain, R. K., Joyce, P. B., Molinete, M., Halban, P. A., and Gorr, S. U. (2001). Oligomerization of green fluorescent protein in the secretory pathway of endocrine cells. *Biochem. J.* **360**, 645–649.

Jayaraman, S., Haggie, P., Wachter, R. M., Remington, S. J., and Verkman, A. S. (2000). Mechanism and cellular applications of a green fluorescent protein-based halide sensor. *J. Biol. Chem.* **275**, 6047–6050.

Johansson, D. X., Brismar, H., and Persson, M. A. (2007). Fluorescent protein pair emit intracellular FRET signal suitable for FACS screening. *Biochem. Biophys. Res. Commun.* **352**, 449–455.

Johnson, D. E., Ai, H. W., Wong, P., Young, J. D., Campbell, R. E., Casey, J. R. (2009). Red fluorescent protein pH biosensor to detect concentrative nucleoside transport. *J. Biol. Chem.* **284**, 20499–20511.

Kamio, N., Hirai, H., Ashihara, E., Tenen, D. G., Maekawa, T., Imanishi, J. (2010). Use of bicistronic vectors in combination with flow cytometry to screen for effective small interfering RNA target sequences. *Biochem. Biophys. Res. Commun.* **393**, 498–503.

Kapoor, V., Karpov, V., Linton, C., Subach, F. V., Verkhusha, V. V., Telford, W. G. (2008). Solid state yellow and orange lasers for flow cytometry. *Cytometry A* **73**, 570–577.

Kapoor, V., Subach, F. V., Kozlov, V. G., Grudinin, A., Verkhusha, V. V., Telford, W. G. (2007). New lasers for flow cytometry: filling the gaps. *Nat. Methods* **4**, 678–679.

Karasawa, S., Araki, T., Nagai, T., Mizuno, H., and Miyawaki, A. (2004). Cyan-emitting and orange-emitting fluorescent proteins as a donor/acceptor pair for fluorescence resonance energy transfer. *Biochem. J.* **381**, 307–312.

Kerppola, T. K. (2008). Bimolecular fluorescence complementation (BiFC) analysis as a probe of protein interactions in living cells. *Annu. Rev. Biophys.* **37**, 465–487.

Kogure, T., Karasawa, S., Araki, T., Saito, K., Kinjo, M., Miyawaki, A. (2006). A fluorescent variant of a protein from the stony coral Montipora facilitates dual-color single-laser fluorescence cross-correlation spectroscopy. *Nat. Biotechnol.* **24**, 577–581.

Kredel, S., Oswald, F., Nienhaus, K., Deuschle, K., Röcker, C., Wolff, M., Heilker, R., Nienhaus, G. U., Wiedenmann, J. (2009). mRuby, a bright monomeric red fluorescent protein for labeling of subcellular structures. *PLoS One* **4**, e4391.

Kremers, G. J., Goedhart, J., van den Heuvel, D. J., Gerritsen, H. C., and Gadella Jr., T. W. (2007). Improved green and blue fluorescent proteins for expression in bacteria and mammalian cells. *Biochemistry* **46**, 3775–3783 http://physrev.physiology.org/cgi/external_ref?access_num=10.1021%2Fbi0622874&link_type=DOI.

Kremers, G. J., Goedhart, J., van Munster, E. B., and Gadella Jr., T. W. (2006). Cyan and yellow super fluorescent proteins with improved brightness, protein folding, and FRET Forster radius. *Biochemistry* **45**, 6570–6580.

Li, Y., Sierra, A. M., Ai, H. W., and Campbell, R. E. (2008). Identification of sites within a monomeric red fluorescent protein that tolerate peptide insertion and testing of corresponding circular permutations. *Photochem. Photobiol.* **84**, 111–119.

Lin, M. Z., McKeown, M. R., Ng, H. L., Aguilera, T. A., Shaner, N. C., Campbell, R. E., Adams, S. R., Gross, L. A., Ma, W., Alber, T., Tsien, R. Y. (2009). Autofluorescent proteins with excitation in the optical window for intravital imaging in mammals. *Chem. Biol.* **16**, 1169–1179.

Matz, M. V., Fradkov, A. F., Labas, Y. A., Savitsky, A. P., Zaraisky, A. G., Markelov, M. L., Lukyanov, S. A. (1999). Fluorescent proteins from nonbioluminescent Anthozoa species. *Nat. Biotechnol.* **17**, 969–973.

Merzlyak, E. M., Goedhart, J., Shcherbo, D., Bulina, M. E., Shcheglov, A. S., Fradkov, A. F., Gaintzeva, A., Lukyanov, K. A., Lukyanov, S., Gadella, T. W., Chudakov, D. M. (2007). Bright monomeric red fluorescent protein with an extended fluorescence lifetime. *Nat. Methods* **4**, 555–557.

Mishin, A. S., Subach, F. V., Yampolsky, I. V., King, W., Lukyanov, K. A., Verkhusha, V. V. (2008). The first mutant of the Aequorea victoria green fluorescent protein that forms a red chromophore. *Biochemsitry* **47**, 4666–4673.

Morell, M., Espargaro, A., Aviles, F. X., and Ventura, S. (2008). Study and selection of *in vivo* protein interactions by coupling bimolecular fluorescence complementation and flow cytometry. *Nat. Protoc.* **3**, 22–33.

Morozova, K. S., Piatkevich, K. D., Gould, T. J., Zhang, J., Bewersdorf, J., Verkhusha, V. V. (2010). Far-red fluorescent protein excitable with red lasers for flow cytometry and superresolution STED nanoscopy. *Biophys. J.* **99**, L13–L15.

Mutoh, H., Perron, A., Dimitrov, D., Iwamoto, Y., Akemann, W., Chudakov, D. M., Knöpfel, T. (2009). Spectrally-resolved response properties of the three most advanced FRET based fluorescent protein voltage probes. *PLoS One* **4**, e4555.

Nagai, T., Ibata, K., Park, E. S., Kubota, M., Mikoshiba, K., Miyawaki, A. (2002). A variant of yellow fluorescent protein with fast and efficient maturation for cell-biological applications. *Nat. Biotechnol.* **20**, 87–90.

Nagai, T., and Miyawaki, A. (2004). A high-throughput method for development of FRET-based indicators for proteolysis. *Biochem. Biophys. Res. Commun.* **319**, 72–77.

Nagai, T., Yamada, S., Tominaga, T., Ichikawa, M., and Miyawaki, A. (2004). Expanded dynamic range of fluorescent indicators for Ca(2+) by circularly permuted yellow fluorescent proteins. *Proc. Natl. Acad. Sci. USA* **101**, 10554–10559.

Nguyen, A. W., and Daugherty, P. S. (2005). Evolutionary optimization of fluorescent proteins for intracellular FRET. *Nat. Biotechnol.* **23**, 355–360.

Piatkevich, K. D., Efremenko, E. N., Verkhusha, V. V., and Varfolomeev, S. D. (2010a). Red fluorescent proteins and their properties. *Russ. Chem. Rev.* **79**, 243–258.

Piatkevich, K. D., Hulit, J., Subach, O. M., Wu, B., Abdulla, A., Segall, J. E., Verkhusha, V. V. (2010b). Monomeric red fluorescent proteins with a large Stokes shift. *Proc. Natl. Acad. Sci. USA* **107**, 5369–5374.

Piatkevich, K. D., Malashkevich, V. N., Almo, S. C., and Verkhusha, V. V. (2010c). Engineering ESPT pathways based on structural analysis of LSSmKate red fluorescent proteins with large Stokes shift. *J. Am. Chem. Soc.* **132**, 10762–10770.

Piatkevich, K. D., and Verkhusha, V. V. (2010). Advances in engineering of fluorescent proteins and photoactivatable proteins with red emission. *Curr. Opin. Chem. Biol.* **14**, 23–29.

Piljic, A., and Schultz, C. (2008). Simultaneous recording of multiple cellular events by FRET. *ACS Chem. Biol.* **3**, 156–160.

Pletnev, S., Subach, F. V., Dauter, Z., Wlodawer, A., and Verkhusha, V. V. (2010). Understanding blue-to-red conversion in monomeric fluorescent timers and hydrolytic degradation of their chromophores. *J. Am. Chem. Soc.* **132**, 2243–2253.

Rahimi, Y., Shrestha, S., Banerjee, T., and Deo, S. K. (2007). Copper sensing based on the far-red fluorescent protein, HcRed, from Heteractis crispa. *Anal. Biochem.* **370**, 60–67.

Rizzo, M. A., Davidson, M. W., and Piston, D. W. (2009). Fluorescent protein tracking and detection: applications using fluorescent proteins in living cells. *Cold Spring Harb. Protoc.* 2009: pdb.top64.

Sakaue-Sawano, A., Kurokawa, H., Morimura, T., Hanyu, A., Hama, H., Osawa, H., Kashiwagi, S., Fukami, K., Miyata, T., Miyoshi, H., Imamura, T., Ogawa, M., Masai, H., Miyawaki, A. (2008). Visualizing spatiotemporal dynamics of multicellular cell-cycle progression. *Cell* **132**, 487–498.

Shagin, D. A., Barsova, E. V., Yanushevich, Y. G., Fradkov, A. F., Lukyanov, K. A., Labas, Y. A., Semenova, T. N., Ugalde, J. A., Meyers, A., Nunez, J. M., Widder, E. A., Lukyanov, S. A., Matz, M. V. (2004). GFP-like proteins as ubiquitous metazoan superfamily: evolution of functional features and structural complexity. *Mol. Biol. Evol.* **21**, 841–850.

Shaner, N. C., Campbell, R. E., Steinbach, P. A., Giepmans, B. N., Palmer, A. E., Tsien, R. Y. (2004). Improved monomeric red, orange and yellow fluorescent proteins derived from Discosoma sp. red fluorescent protein. *Nat. Biotechnol.* **22**, 1567–1572.

Shaner, N. C., Lin, M. Z., McKeown, M. R., Steinbach, P. A., Hazelwood, K. L., Davidson, M. W., Tsien, R. Y. (2008). Improving the photostability of bright monomeric orange and red fluorescent proteins. *Nat. Methods* **5**, 545–551.

Shcherbo, D., Merzlyak, E. M., Chepurnykh, T. V., Fradkov, A. F., Ermakova, G. V., Solovieva, E. A., Lukyanov, K. A., Bogdanova, E. A., Zaraisky, A. G., Lukyanov, S., Chudakov, D. M. (2007). Bright far-red fluorescent protein for whole-body imaging. *Nat. Methods* **4**, 741–746.

Shcherbo, D., Murphy, C. S., Ermakova, G. V., Solovieva, E. A., Chepurnykh, T. V., Shcheglov, A. S., Verkhusha, V. V., Pletnev, V. Z., Hazelwood, K. L., Roche, P. M., Lukyanov, S., Zaraisky, A. G., Davidson, M. W., Chudakov, D. M. (2009a). Far-red fluorescent tags for protein imaging in living tissues. *Biochem. J.* **418**, 567–574.

Shcherbo, D., Souslova, E. A., Goedhart, J., Chepurnykh, T. V., Gaintzeva, A., Shemiakina, I. I., Gadella, T. W., Lukyanov, S., Chudakov, D. M. (2009b). Practical and reliable FRET/FLIM pair of fluorescent proteins. *BMC Biotechnol.* **9**, 24.

Shu, X., Royant, A., Lin, M. Z., Aguilera, T. A., Lev-Ram, V., Steinbach, P. A., Tsien, R. Y. (2009). Mammalian expression of infrared fluorescent proteins engineered from a bacterial phytochrome. *Science* **324**, 804–807.

Shyu, Y. J., Liu, H., Deng, X., and Hu, C. D. (2006). Identification of new fluorescent protein fragments for bimolecular fluorescence complementation analysis under physiological conditions. *Biotechniques* **40**, 61–66.

Shyu, Y. J., Suarez, C. D., and Hu, C. D. (2008). Visualization of AP-1 NF-kappaB ternary complexes in living cells by using a BiFC-based FRET. *Proc. Natl. Acad. Sci. USA* **105**, 151–156.

Siegel, R. M., Chan, F. K., Zacharias, D. A., Swofford, R., Holmes, K. L., Tsien, R. Y., Lenardo, M. J. (2000). Measurement of molecular interactions in living cells by fluorescence resonance energy transfer between variants of the green fluorescent protein. *Science STKE* **2000**(38), pl1.

Snapp, E. (2005). Design and use of fluorescent fusion proteins in cell biology. *Curr. Protoc. Cell Biol.* Chapter 21, Unit 21.4, 21.4.1-21.4.13.

Stepanenko, O. V., Verkhusha, V. V., Kazakov, V. I., Shavlovsky, M. M., Kuznetsova, I. M., Uversky, V. N., Turoverov, K. K. (2004). Comparative studies on the structure and stability of fluorescent proteins EGFP, zFP506, mRFP1, dimer2 and DsRed. *Biochemistry* **43**, 14913–14923.

Strack, R. L., Bhattacharyya, D., Glick, B. S., and Keenan, R. J. (2009a). Noncytotoxic orange and red/green derivatives of DsRed-Express2 for whole-cell labeling. *BMC Biotechnol.* **9**, 32.

Strack, R. L., Hein, B., Bhattacharyya, D., Hell, S. W., Keenan, R. J., Glick, B. S. (2009b). A rapidly maturing far-red derivative of DsRed-Express2 for whole-cell labeling. *Biochemistry* **48**, 8279–8281.

Strack, R. L., Strongin, D. E., Bhattacharyya, D., Tao, W., Berman, A., Broxmeyer, H. E., Keenan, R. J., Glick, B. S. (2008). A noncytotoxic DsRed variant for whole-cell labeling. *Nat. Methods* **5**, 955–957.

Subach, F. V., Patterson, G. H., Manley, S., Gillette, J. M., Lippincott-Schwartz, J., Verkhusha, V. V. (2009a). Photoactivatable mCherry for high-resolution two-color fluorescence microscopy. *Nat. Methods* **6**, 153–159.

Subach, F. V., Subach, O. M., Gundorov, I. S., Morozova, K. S., Piatkevich, K. D., Cuervo, A. M., Verkhusha, V. V. (2009b). Monomeric fluorescent timers that change color from blue to red report on cellular trafficking. *Nat. Chem. Biol.* **5**, 118–126.

Subach, O. M., Malashkevich, V. N., Zencheck, W. D., Morozova, K. S., Piatkevich, K. D., Almo, S. C., Verkhusha, V. V. (2010). Structural characterization of acylimine-containing blue and red chromophores in mTagBFP and TagRFP fluorescent proteins. *Chem. Biol.* **17**, 333–341.

Telford, W. G., Subach, F. V., and Verkhusha, V. V. (2009). Super-continuum white light lasers for flow cytometry. *Cytometry A* **75**, 450–459.

Terskikh, A., Fradkov, A., Ermakova, G., Zaraisky, A., Tan, P., Kajava, A. V., Zhao, X., Lukyanov, S., Matz, M., Kim, S., Weissman, I., Siebert, P. (2000). "Fluorescent timer": protein that changes color with time. *Science* **290**, 1585–1588.

Tsutsui, H., Karasawa, S., Okamura, Y., and Miyawaki, A. (2008). Improving membrane voltage measurements using FRET with new fluorescent proteins. *Nat. Methods* **5**, 683–685.

Tsuboi, T., Kitaguchi, T., Karasawa, S., Fukuda, M., and Miyawaki, A. (2010). Age-dependent preferential dense-core vesicle exocytosis in neuroendocrine cells revealed by newly developed monomeric fluorescent timer protein. *Mol. Biol. Cell.* **21**, 87–94.

Tramier, M., Zahid, M., Mevel, J. C., Masse, M. J., and Coppey-Moisan, M. (2006). Sensitivity of CFP/YFP and GFP/mCherry pairs to donor photobleaching on FRET determination by fluorescence lifetime imaging microscopy in living cells. *Microsc. Res. Tech.* **69**, 933–939.

Verkhusha, V. V., Chudakov, D. M., Gurskaya, N. G., Lukyanov, S., and Lukyanov, K. A. (2004). Common pathway for the red chromophore formation in fluorescent proteins and chromoproteins. *Chem. Biol.* **11**, 845–854.

Verkhusha, V. V., Matz, M. V., Sakurai, T., and Lukyanov, K. A. (2003a). GFP-like fluorescent proteins and chromoproteins of the class Anthozoa. *In* "Protein Structures: Kaleidoscope of Structural Properties and Functions," (and V. N. Uversky, ed.), pp. 405–439. Research Signpost, Kerala, India.

Verkhusha, V. V., Pozhitkov, A. E., Smirnov, S. A., Borst, J. W., van Hoek, A., Klyachko, N. L., Levashov, A. V., Visser, A. J. (2003b). Effect of high pressure and reversed micelles on the fluorescent proteins. *Biochim. Biophys. Acta* **1622**, 192–195.

Vrzheshch, P. V., Akovbian, N. A., Varfolomeyev, S. D., and Verkhusha, V. V. (2000). Denaturation and partial renaturation of a tightly tetramerized DsRed protein under mildly acidic conditions. *FEBS Lett.* **487**, 203–208.

van Wageningen, S., Pennings, A. H., van der Reijden, B. A., Boezeman, J. B., de Lange, F., Jansen, J. H. (2006). Isolation of FRET-positive cells using single 408-nm laser flow cytometry. *Cytometry A* **69**, 291–298.

Wang, L., Jackson, W. C., Steinbach, P. A., and Tsien, R. Y. (2004). Evolution of new nonantibody proteins via iterative somatic hypermutation. *Proc. Natl. Acad. Sci. USA* **101**, 16745–16749.

Wu, X., Simone, J., Hewgill, D., Siegel, R., Lipsky, P. E., He, L. (2006). Measurement of two caspase activities simultaneously in living cells by a novel dual FRET fluorescent indicator probe. *Cytometry A* **69**, 477–486.

Yanushevich, Y. G., Staroverov, D. B., Savitsky, A. P., Fradkov, A. F., Gurskaya, N. G., Bulina, M. E., Lukyanov, K. A., Lukyanov, S. A. (2002). A strategy for the generation of non-aggregating mutants of Anthozoa fluorescent proteins. *FEBS Lett.* **511**, 11–14.

You, X., Nguyen, A. W., Jabaiah, A., Sheff, M. A., Thorn, K. S., Daugherty, P. S. (2006). Intracellular protein interaction mapping with FRET hybrids. *Proc. Natl. Acad. Sci. USA* **103**, 18458–18463.

Zacharias, D. A., Violin, J. D., Newton, A. C., and Tsien, R. Y. (2002). Partitioning of lipid-modified monomeric GFPs into membrane microdomains of live cells. *Science* **296**, 913–916.

Zapata-Hommer, O., and Griesbeck, O. (2003). Efficiently folding and circularly permuted variants of the Sapphire mutant of GFP. *BMC Biotechnol.* **3**, 5.

Zolotukhin, S., Potter, M., Hauswirth, W. W., Guy, J., and Muzyczka, N. (1996). A "humanized" green fluorescent protein cDNA adapted for high-level expression in mammalian cells. *J. Virol.* **70**, 4646–4654.

Zuker, M. (2003). Mfold web server for nucleic acid folding and hybridization prediction. *Nucleic Acids Res.* **31**, 3406–3415.

CHAPTER 18

Quantum Dot Technology in Flow Cytometry

Pratip K. Chattopadhyay

ImmunoTechnology Section, Vaccine Research Center, NIAID, NIH, Bethesda, Maryland, USA

Abstract

The development of quantum dot (QD) technology represents one of the most dramatic advances in flow cytometry history, offering the opportunity for highly multiplexed experiments and allowing better resolution of dimly staining markers. Here, we guide users through the technical aspects of using QDs (including instrumentation and antibody conjugation), demonstrate why QDs are useful in multicolor flow cytometry, and describe some of the challenges investigators may face when adopting this technology.

I. Introduction

The development of flow cytometry technology has been propelled by the need to identify and characterize cell populations (Chattopadhyay *et al.*, 2008). As new lasers and fluorochromes have been introduced over the years, researchers have

0091-679X/10 $35.00
DOI 10.1016/B978-0-12-374912-3.00018-3

revealed a wide variety of proteins that are differentially expressed across immune cell subsets (Chattopadhyay and Roederer, 2010a). Often, these proteins identify functionally distinct cell subsets, some of which may be linked to disease pathogenesis or vaccine efficacy (Betts *et al.*, 2006; Bolton and Roederer, 2009; Darrah *et al.*, 2007). With the recent development of quantum dots (QDs) for flow cytometry, the enumeration of new and/or important cell subsets has become considerably easier. Here, we review fundamental properties of QDs, describe how they can be used in flow cytometry, and explore a variety of applications.

II. Fundamental Aspects of QD Flow Cytometry: Fluorescence and Hardware

QDs are derived from semiconductor materials (cadmium, selenium, and tellurium) that assemble into nanometer-scale crystals as a result of a complex production process. By varying a high-temperature (>300 °C) incubation step, or altering the mixture of precursor materials, the size of the QD can be modulated (Peng *et al.*, 1998; Smet *et al.*, 1999). This has important consequences, as the size of the nanocrystal dictates the wavelength of the fluorescence emitted.

Currently, nanocrystals can be produced between 2 to 6 nm in diameter, which emit a wide variety of fluorescent light (Bruchez, 2005). The smallest fluoresce in the green range of the visible light spectrum (emitting 525 nm light), while the largest QDs emit red light (at 800 nm). Commercial vendors offer a number of other QD species, named by the wavelength of their peak emission, including QD545, 565, 585, 605, 625, 650, 655, and 705. Among these, QD605 and QD655 (emitting light at 605 and 655 nm, respectively) are the brightest, while QD525 is the dimmest (Fig. 1A).

Because QDs are available in a wide variety of "colors," multiplexed analysis of protein expression is possible when the flow cytometer is equipped with several photomultiplier tubes (PMTs). For maximum flexibility, an octagon (eight) PMT can be coupled to the laser used for QD excitation, as indicated in Fig. 1B. Each PMT (except the last one) is associated with two filters. The first consists of a dichroic mirror to select light sharply, and is called a "long-pass" (LP) filter. This filter only allows light above a certain wavelength to pass. A second (bandpass) filter then purifies the light further by collecting a range of wavelengths within the fluorochrome's emission spectrum. The wavelengths of light that fall below the first (long-pass) filter's cut-off are reflected to the next detector, where they are queried in a similar fashion (Perfetto *et al.*, 2004).

A number of hardware-related factors are critical to multicolor QD experiments. First, as for any other fluorochromes, appropriate filters must be selected. Long-pass filters must transmit light efficiently, while bandpass filters should collect a sufficiently broad range of wavelengths to maximize QD signal (without collecting light from lasers or other fluorochromes). Filter selection and qualification are discussed in greater detail elsewhere (Perfetto *et al.*, 2006). Second, the intrinsic variation in signal-to-noise ratios between different PMTs may be an important consideration. In theory, PMTs generating more electronic noise can diminish the resolution of dimmer QDs; thus, dim QDs should be associated with PMTs displaying high sensitivity, while brighter fluorochromes can be relegated to less sensitive PMTs. Measures of PMT sensitivity are beyond the scope of this chapter, but are being developed and are discussed elsewhere (Wood and Hoffman, 1998). Finally, voltage settings for QD PMTs should be selected carefully, and monitored over time, as part of an instrument calibration and quality control program (Perfetto *et al.*, 2006). When such systems are not employed, data may not be comparable between experiments, and some of the benefits of using QDs (described later in this chapter) may not be realized.

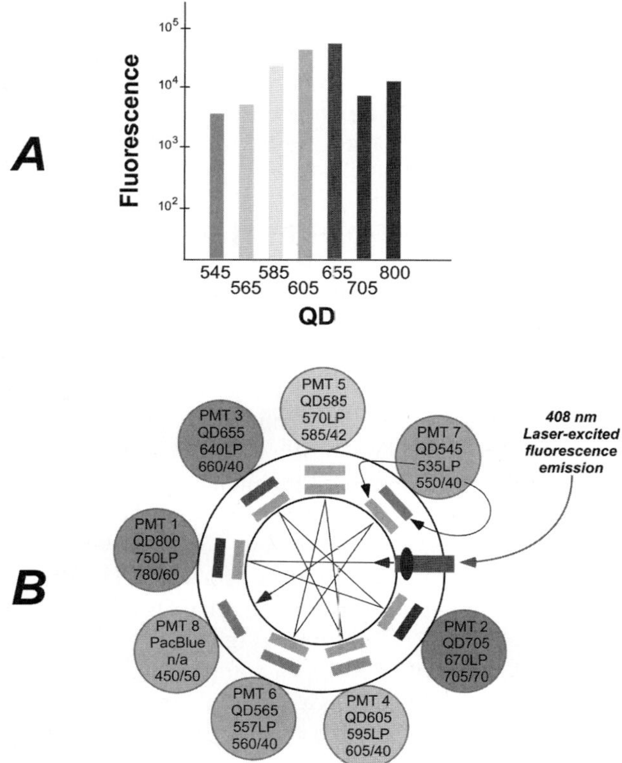

Fig. 1 Detection of QD fluorescence. (A) Fluorescence intensity of anti-IgK beads stained with various QD-labeled CD8 antibodies. QD655 is the brightest, followed by QD605 and QD585. Note that PMT voltages were optimized prior to this analysis, using the technique described in Perfetto *et al.*, 2007. (B) Configuration of the octagonal PMT system in a modified LSRII flow cytometer (described in Chattopadhyay *et al.*, 2006). Dichroic LP filters transmit light at a certain wavelength to band pass filters, which further purify the light before transmission to the detector. Light below a particular wavelength is reflected to the next dichroic filter in the sequence.

QDs have broad excitation spectra, and therefore the PMTs used for QD detection can be coupled to a variety of laser sources. Generally, excitation weakens with increasing laser wavelength (Chattopadhyay *et al.*, 2010b), so ultraviolet lasers provide the strongest excitation. However, these lasers also induce a strong cellular autofluorescence, which may mask dim QD signals and negate the benefits of stronger excitation. For this reason, violet lasers are often the preferred excitation source for QDs in multilaser instruments.

In multicolor settings, it is important to recognize that QDs are excited by any laser below their emission wavelength. Thus, higher wavelength QDs (QDs 655, 705, 800) can be excited by ultraviolet, violet, blue, green, and red lasers. This characteristic suggests a role for QDs as replacements for common organic fluorophores, like Cy5.5PerCP, Cy5PE, APC, Alexa 680, or Cy7APC, although qdots release considerably less fluorescence than organic fluorochromes when excited by higher wavelength lasers. More importantly, though, the broad excitation spectrum of QDs demands significant compensation to remove QD signals from non-QD channels (i.e., removal of QD655 light from the APC detector (Chattopadhyay *et al.*, 2006)).

III. Utility of QDs in Multicolor Flow Cytometry

Multicolor flow cytometry requires careful consideration of the excitation and emission characteristics of fluorochromes in order to identify those that can be measured distinctly. The organic fluorochromes commonly used today have different, and narrow, excitation spectra (Baumgarth and Roederer, 2000); therefore, instruments intended for multicolor analysis typically require three or more lasers for exciting a variety of fluorochromes. This raises the cost of flow cytometers, and complicates calibration, maintenance, and troubleshooting. In contrast, the various species of QDs share relatively broad (and overlapping) excitation spectra so that multicolor fluorescence can be achieved with just one laser.

The fluorescence characteristics of QDs also offer three important advantages over organic fluorochromes. First, QDs are relatively bright, with 90% of the energy they absorb released as light (a characteristic known as quantum yield (Michalet *et al.*, 2005)). In comparison, the organic fluorochrome fluorescein isothiocyanate (FITC) has a quantum yield of only 30% (Haugland, 1994). Second, the emission spectra of the organic fluorochromes used in multicolor flow cytometry can overlap significantly – more so than QDs. For example, four tandem dyes (TRPE, Cy5PE, Cy5.5PE, and Cy7PE) emit significant fluorescent signal between 550 and 575 nm, the same region in which PE is usually detected (Fig. 2A). In contrast, for any given QD channel, light from the two adjacent QDs are the primary sources of spillover fluorescence (Fig. 2B); moreover, the signal contributed by this contaminating light is typically minimal. The third advantage afforded by QDs is that their emission spectra are generally symmetrical. Unlike many organic fluorochromes, no long tails of fluorescence emission are observed in the red region of the spectrum (Fig. 2B).

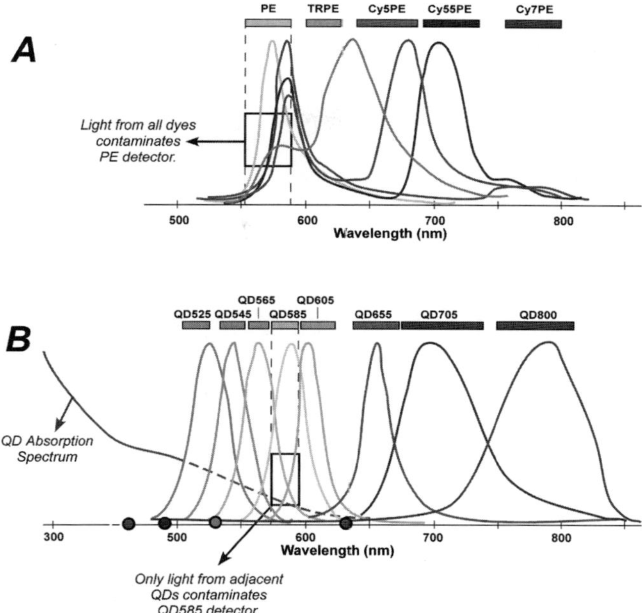

Fig. 2 Emission spectra for organic dyes and QDs. The colored bars indicate the range of wavelengths for filters used to detect each fluorochrome. (A) PE and PE-based tandems overlap significantly, such that all these organic dyes are detected in the PE channel and high compensation values are required. Also, these fluorochromes have long tails of emission in the red region of the spectrum, which can induce significant spreading error. (B) In contrast, QD emission spectra are narrow and symmetrical, resulting in better sensitivity. For example, only light from adjacent QDs contaminates the QD585 channel. QDs also have broad (and roughly equivalent) excitation spectra (solid gray line), so multiple QDs can be excited by the same laser. However, because higher wavelength QDs have broader excitation spectra, these QDs are excited by multiple lasers (depicted as circles in the figure). Note that the dashed gray line roughly describes the absorption spectra; in reality, the absorption spectrum cannot extend past the emission spectrum. (For interpretation of the references to color in this figure legend, the reader is referred to the Web version of this chapter.)

The favorable excitation and emission characteristics of QDs have important implications for multicolor staining panels, since the success of these panels depends heavily on compensation (i.e., subtraction of spillover fluorescence) and spreading error (Roederer, 2001). The latter represents the measurement error associated with low fluorescence events, which appears more prominent as compensation values increase. Thus, when two fluorochromes overlap significantly, high levels of compensation are required; this is associated with increased spreading error, which can hide dimly staining populations. For example, the high compensation between Hilyte 750-APC and APC, coupled with the measurement error associated with low fluorescence red events in the APC channel, gives high levels of spreading error that could mask dimly staining populations (Fig. 3A). In contrast, the minimal compensation between QD800 and QD655 does not reveal any spreading error (Fig. 3A). The difference in spreading error is starker when stained (and compensated) beads

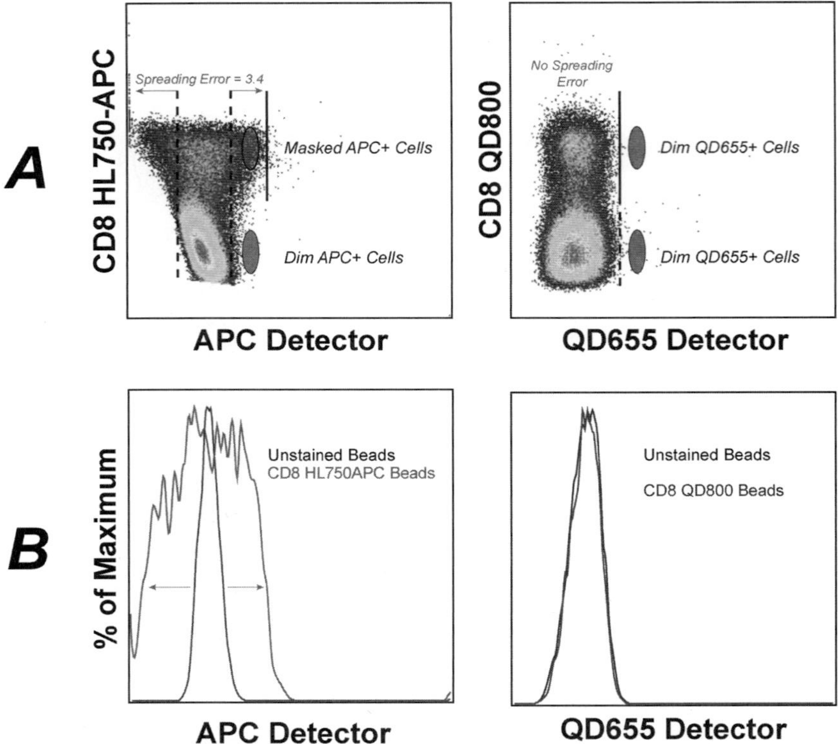

Fig. 3 The effect of spreading error. PBMCs were stained with anti-CD8 HiLyte750-APC (HL750APC) or QD800, and spreading error into other channels was examined after compensation and biexponential transformation of data. HL750APC contributes significant spreading error into APC, which can mask populations exhibiting dim APC fluorescence. In contrast, QD800 contributes no spreading error into the QD655 channel, allowing better discrimination of dim QD655+ cells.

are compared to unstained beads (Fig. 3B). Moreover, QDs are uniquely excited by low-wavelength lasers (e.g., 405 nm violet), which have no effect on organic fluorochromes; therefore, spillover fluorescence *from* organic fluorochromes *into* QD channels is minimal, and spreading error is negligible when QD and organic fluorochromes are combined. These factors allow excellent resolution of dimly staining populations in the channels with the brightest QDs. However, it should be noted that the converse is not true: QDs contribute substantial spillover fluorescence into the channels used to detect organic fluorochromes (Chattopadhyay *et al.*, 2010b). Still, this usually has little practical consequence, as QD spillover is mostly found within the channels used for the brightest organic fluorochromes; so most markers still resolve well when QDs and organic fluorochromes are used together in the same panel. Thus, with the availability of QDs, it becomes (relatively) easy to incorporate new markers into existing multicolor panels.

Interestingly, with the wide variety of QD "colors" available, multicolor panels can be designed even without the use of organic fluorochromes. In fact, it is possible to design a five-color panel with minimal ($<10\%$) compensation requirements (Chattopadhyay *et al.*, 2006). This feature of QDs could propel the future development of simple, single-laser, five-color flow cytometers capable of highly automated data analysis.

IV. QD Conjugation to Antibodies

The major obstacle for incorporating QDs into multicolor experiments is the limited number of reagents commercially available. Currently, very few companies offer antibodies conjugated to QDs. Fortunately, the conjugation of antibodies to QDs is fairly straightforward; it mostly relies on simple equipment and low-cost materials. Thus, the method is accessible to most laboratories.

On the other hand, it is important to note that the production of QD nanocrystals is complex. The assembly of semiconductor materials into nanocrystals requires specialized equipment, and the process must be tightly controlled so that the size of the crystals is uniform and the material is stable (Dabbousi *et al.*, 1997; Wu *et al.*, 2007). Moreover, the quantum yield of raw QD nanocrystals is very low, unless several layers of zinc sulfide (an inorganic shell) are applied (Grabolle *et al.*, 2008; Reiss *et al.*, 2003). To increase solubility, the QD nanocrystals and inorganic shells must also be encrusted with an organic material (polyethylene glycol, or PEG). Finally, functional groups must be linked to the QD for antibody conjugation. Given the complexity of the QD synthesis process, it is best to purchase QDs, functionalized with chemical groups for antibody conjugation, from commercial sources.

To perform conjugations in-house, commercial kits are available; however, a more cost-effective approach requires nothing more than home-made buffers, low-cost columns, and common chemicals (Chattopadhyay *et al.*, 2007). Moreover, this "home-made" conjugation system may have a slightly higher success rate than commercial kits. The process we employ for in-house conjugations begins with highly purified antibody (containing no exogenous protein) at high concentrations (>4 mg/mL). Manufacturers typically do not sell such concentrated material off-the-shelf; custom orders for unconjugated antibodies are usually required. Still, if only dilute antibody is available, it can be concentrated; however, researchers should keep in mind that some antibody is lost in this process. We recommend that users begin by conjugating at least 0.5 mg of antibody, for the sake of reaction efficiency. Notably, commercial kits require much less unconjugated antibody (0.1 mg), and appear to be less affected by diluted antibody preparations.

Generally, the procedure we employ for QD conjugations is very similar to the methods described for making PE-conjugated antibodies (using reductive cross-linking (Chattopadhyay *et al.*, 2007)). The process involves activating the QDs using sulfosuccinimidyl 4-*N*-maleimidomethyl cyclohexane-1-carboxylate (sulfo-SMCC); this generates malemide groups on the surface of the QDs. At the same time, the antibody is reduced by treatment with dithiothreitol, in order to expose free thiol groups. The activated QD and reduced antibody are then mixed (at a 1:1 molar ratio), incubated for 1 h, and the reaction is quenched with *N*-ethylmaleimide (NEM). Finally, the QD-conjugated antibody is exchanged into a storage buffer containing Tris, NaCl, and sodium azide.

Other methods for QD conjugations are available, including those that rely on other functional groups. When QDs with carboxylic acid groups are used, these can be coupled to primary amines to form stable amide bonds (Biju *et al.*, 2008). With mercapto-coated QD, thiolated proteins or oligonucleotides can be attached after an overnight adsorption period (Mitchell *et al.*, 1999; Wu *et al.*, 2003). Similarly, through electrostatic interactions, negatively charged QDs can be adsorbed to poly-lysine chains that have been previously attached to the protein (Michalet *et al.*, 2005). Notably, methods that rely on adsorption are typically not stable in long term, and contain large amounts of unreacted material.

V. Developing Staining Panels with QDs

The creation of new antibody panels has become easier with the availability of QDs; however, a number of overall and QD-specific issues must be considered in development. Generally, the process begins with the identification of proteins of interest, which are then ranked by their importance. Thus, we list antibody targets that are *required* for the experiment, those that are *important but not required*, and those that may be interesting, but are of low priority (i.e., luxury markers; Mahnke and Roederer, 2007).

Next, we characterize expression levels for these markers, and assign each marker to one of three categories (primary, secondary, tertiary) on the basis of this infor-mation. Primary markers are typically expressed by parent populations (e.g., CD3, CD4, CD8) and exhibit on/off patterns (low or intermediate levels of expression are rarely observed). In contrast, secondary markers (e.g., CD45RA) provide meaning-ful information across populations expressing high, intermediate, and low levels of the protein. Finally, tertiary markers are those with poorly characterized or dim expression (Mahnke and Roederer, 2007).

Fluorochrome choice is governed by three factors. First, the expression level of each marker is considered. Tertiary markers are assigned to the fluorochromes with the brightest expression, while secondary markers are assigned to the next tier of fluorochromes. Primary markers can be assigned to any available fluorochrome. The second factor governing fluorochrome–antibody pairing is reagent availabil-ity. In laboratories capable of in-house conjugation, a wide variety of antibody–fluorochrome conjugates can be made for testing. Otherwise, fluorochrome choice may be limited severely by what is commercially available. The third and final factor is the quality of the reagents available. There can be substantial differences among monoclonal antibody clones, and even among fluorochrome–antibody conjugates. These differences can be observed in the proportion of cells identified by the reagent, the brightness of the staining, or the separation between positive and negative populations.

Once markers have been categorized, and fluorochrome–antibody conjugates have been obtained and tested, a number of candidate panels can be proposed and compared. These panels should be judged first for the resolution of required tertiary markers (i.e., those markers that must be included in the panel, and have the dimmest expression), and then for resolution of important (but not necessary) markers and/or secondary markers. An important determinant for success or failure of the candidate panels will be spreading error (described in more detail above). Importantly, where a bright reagent contributes spreading error in another channel, lowering the concentration of the reagent (and therefore the brightness, compensation, and spreading error in the other channel) may improve staining resolution (Mahnke and Roederer, 2007).

So, how can QDs be incorporated into staining panels? The brightest QDs (QD655 and QD605) are particularly suitable for tertiary markers, because they are not susceptible to spreading error and because very little fluorescence from organic fluorochromes contaminates these channels (Chattopadhyay *et al.*, 2010b). This allows better resolution of populations with low expression. Moreover, the dimmest QDs (e.g., QD545) are suitable for primary markers, which do not exhibit dim staining (their expression is on/off). Secondary or luxury markers can be slotted into the other QD channels, based on the quality of the reagent.

In theory, the most powerful implementation of QD technology is to assign tertiary markers to bright QDs; however, in practice, this approach is difficult. Unfortunately, antibodies against tertiary markers are not commercially available, and in-house conjugations for such antibodies have a high failure rate. In contrast, antibodies against primary markers are easily conjugated to QDs, and commercial reagents are available. Therefore, for most researchers, assigning primary markers to QD channels is a simple and useful way to adopt QD technology. For researchers who perform in-house conjugations, it is also relatively easy to assign secondary markers to QDs, since in-house conjugations of such markers (e.g., CD45RA, CD45RO, and CD57) are frequently successful.

VI. Troubleshooting QD Use

The problems associated with using QDs in multicolor flow cytometry are unique; however, these challenges are not necessarily any more complex than those presented by other fluorochromes. Here, we describe the problems we have experienced with QDs, and offer some troubleshooting guidance.

The success rate of our QD–antibody conjugations has been fairly high (Chattopadhyay *et al.*, 2010b), and successful conjugations are extremely reproducible (i.e., subsequent production of the same conjugate is always successful, even if different antibody or fluorochrome lots are used). However, there are some antibodies that do not conjugate well to QDs in our hands. Unfortunately, we have not yet been able to identify a common reason for these failures; they cannot be attributed to the antibody isotype (anti-CD57 IgM works well, but anti-CD27 IgG has been poor) or brightness of the marker (the brightest QDs are as bright as PE or APC, which usually yield successful conjugates). Given this uncertainty, we recommend that laboratories developing protocols for QD–antibody conjugation make a series of "control" conjugates against a brightly expressed marker, present at high frequency within peripheral blood mononuclear (PBMC) samples, which display on/off expression (such as CD8). When attempts to conjugate a new antibody to QDs fail, the success of the control conjugation can serve as confirmation that the method is working. If the control fails, then the reagents used to reduce the antibody and SMCC activate the QD should be replaced (Chattopadhyay *et al.*, 2007).

When using QDs in multicolor flow cytometry, it is also important to consider the stability of antibody conjugates. Over our 7-year experience, the majority of QD reagents have exhibited a long "shelf-life," with some reagents displaying little evidence of degradation even 5 years after production. However, we have noted that some reagents are not stable; as with conjugation failures – the exact cause cannot be ascertained. Nevertheless, we have observed some early warning signs of conjugate degradation. First, over time, higher and higher concentrations of antibody must to be used to achieve the same staining brightness across experiments. Second, inspection of the reagent vial reveals visible particulates (and staining is lost after filtering the antibody conjugate). Third, conjugates often fail when stored in plastic vials (rather than low retention microcentrifuge tubes or glass vials). Finally, as conjugates are used and the volume in the vial decreases, the risk of degradation increases significantly.

The use of QDs in intracellular staining protocols presents a unique set of challenges. First, QDs prepared by in-house conjugation methods may not be suitable for intracellular staining, because they can contain a large amount of unreacted fluorochrome. In surface-staining experiments, this free fluorochrome is removed with washes; however, washes after the intracellular step seem less effective, as free fluorochrome introduces significant background staining. Second, QDs may not disperse evenly throughout the intracellular environment, either because of their propensity to form multivalent complexes that are too big to enter the cell, or because they can become trapped nonspecifically within a cell. Third, the fluorescence of QDs may be strongly diminished by the harsh fixatives used in many intracellular staining products. In some cases, fluorescence can be maintained (Fig. 4) by adjusting the temperature of intracellular incubation steps to 4 °C (Chattopadhyay *et al.*, 2010b).

Finally, we recently found that low levels of heavy metal contaminants in staining or fixation buffers can completely abrogate QD staining (Zarkowsky *et al.*, 2010). The problem was uncovered in animal studies, where some stained PBMC samples contained higher levels of contaminating erythrocytes than others (i.e., the cell pellet was red). The samples with more red blood cells had severely diminished or completely absent QD staining. Given previous reports that heavy metal anions could impair QD fluorescence, we reasoned that the source of the problem was iron from lysed erythrocytes. We noted similar problems with some lots of formaldehyde

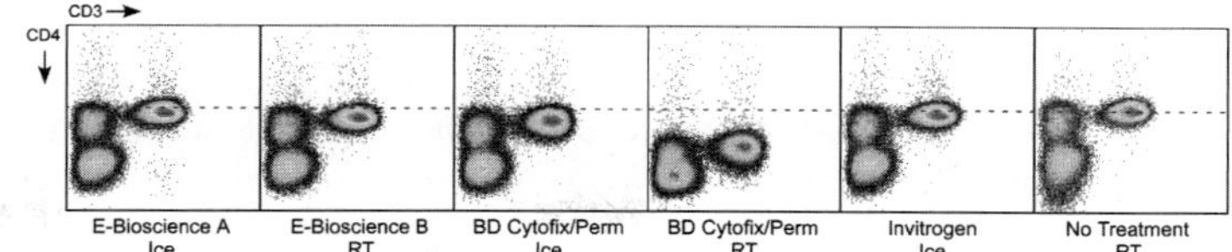

Fig. 4   The effect of various fixation and permeabilization buffers and conditions on QD fluorescence. PBMC were stained with CD3 Cy7APC and CD4 QD565, and then treated with the kits manufactured by e-biosciences, BD, or Invitrogen. To test the effect of temperature, all steps were performed on ice or room temperature. Results were compared to stained, but otherwise untreated, cells.

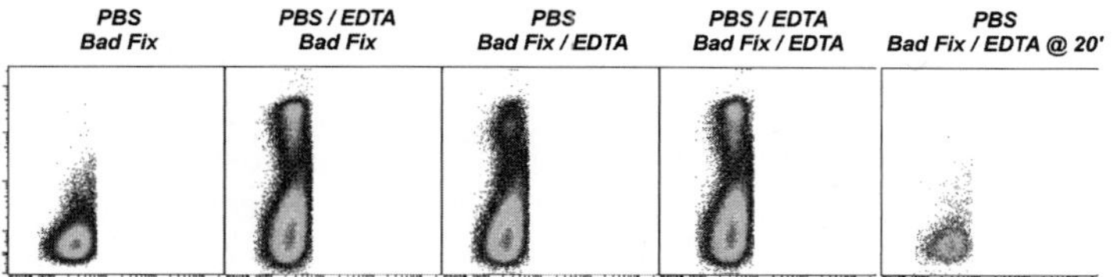

Fig. 5   The effect of fixative with high levels of heavy metals on QD fluorescence. Samples were washed and stained in the specified media, containing various combinations of PBS, fixative, and EDTA. The effect of heavy metals in fixative is shown in the first panel; staining is rescued in the second panel by pretreatment with 1 mM EDTA. When EDTA is added after fixation, staining is not completely rescued (third panel), nor is it rescued when added after a 20-min rest. Cells were stained either with QD655 or QD605 conjugated to anti-CD8 for these tests.

fixative (Fig. 5), which were subsequently tested and found to contain high levels of various metal anions. When these anions were added to samples stained with QD at varying concentrations, we found that ferrous and zinc ions could inhibit staining at concentrations of 0.1 and 0.5 mM, respectively, while cupric ions were much more potent (with a half-maximal inhibitory concentration of 20 nM). Importantly, washing samples with 1 mM EDTA before QD staining prevented this problem (Fig. 5), presumably by chelating the free metal ions. However, EDTA treatment after staining did not rescue QD fluorescence (Fig. 5), demonstrating that the inhibition of fluorescence was irreversible. Interestingly, this property could be beneficial in some experiments, where it may be useful to completely eliminate the cell-associated fluorescence after analysis. For example, in adoptive transfer experiments that use stained, sorted cells, the treatment of samples with cupric sulfate prior to transfer could eliminate the QD fluorescence and allow subsequent *in vivo* tracking or staining on the same QD channel (Zarkowsky *et al.*, 2010).

VII. Applications for QDs

A wide variety of flow cytometry applications have emerged from QD technology. In 2006, we published an early example of the utility of QDs (Chattopadhyay *et al.*,

2006). The study analyzed the maturity of various antigen-specific T-cell populations using a 17-color staining panel. This panel consisted of seven QDs and 10 organic fluorochromes, which were measured simultaneously in a single tube. The QDs were not only conjugated to conventional antibodies, but also incorporated into peptide MHC Class I (pMHCI) multimers designed to detect T-cells specific for HIV, EBV, and CMV antigens. The ability to use QD-multimers was a significant advantage, since multimers were only available in FITC, PE, and APC at the time. By using QD-multimers, we could take advantage of the high resolution of the FITC, PE, and APC channels to measure dimly staining (e.g., tertiary) markers.

The work also demonstrated the utility of multiplexed technologies: by identifying multiple, phenotypically distinct subsets within each antigen-specific T-cell population, the remarkable diversity of T-cell subsets can be appreciated. Moreover, by comparing the frequency of such finely defined subsets between vaccinated, diseased, and healthy individuals, we might reveal the phenotypic traits of the T-cells most capable of conferring immunity (Chattopadhyay *et al.*, 2008). Finally, by using QDs, we are able to measure many more cell types in the same tube, an important consideration when sample availability is limited.

QDs are being used to develop other multiplexed platforms, as well. For example, by carefully varying the brightness of pMHCI multimers, multiple antigen specificities can be queried in the same QD channel (Newell *et al.*, 2009). When multiple QD channels are employed, this combinatorial approach can assay 15 or more antigen specificities simultaneously. Extreme multiplexing is also possible with QD-encoded beads (Gao *et al.*, 2002). In these systems, beads can be produced in two sizes (as large as a cell and as small as a virus) and loaded with six different QDs at 10 intensity levels. When these beads are bound to probes, they can interrogate as many as one million distinct protein or nucleic acid targets (such as secreted cytokine from a culture). The technologies, although nascent, have shown good reproducibility and accuracy (Wang *et al.*, 2006). Thus, the narrow and symmetrical emission spectra of QDs allow for dramatic multiplexing that is not possible with other fluorochromes. Moreover, this multiplexing can be achieved with just one laser, thereby simplifying the high throughput systems that will ultimately be desired for these applications.

QDs have also been used to monitor pathogenic bacteria. In a study of *Escherichia coli* strains (Hahn *et al.*, 2008), QD conjugated to antibodies against pathogenic strains could detect one pathogenic bacterium among 99 harmless ones. The detection limit of this assay was comparable to previous assays using FITC and PE, but QDs were 10-fold brighter and gave more accurate results. Interestingly, when QD-conjugated antibodies bound to bacteria, they exhibited strong shifts in emission, such that their peak emission was blue-shifted by as much as 140 nm (Dwarakanath *et al.*, 2004). These shifts may be due to local pH, charge, or hydrophobicity around the bacterium, or may have been the result of slight changes in the shape of the QD as it bounds. In any event, the blue-shifting of QDs during binding could have application in pathogen detection by flow cytometry.

Finally, even though there are challenges associated with using QDs intracellularly, important advances have been made. One approach is to target the QD (with or without conjugated antibody) into the cell using enzymes or signal peptides (Chattopadhyay *et al.*, 2010b), in order to avoid steric issues or intracellular degradation. This has been demonstrated with matrix metalloproteinases, enzymes that can be used to translocate QDs into cells, and nuclear or mitochondrial signal peptides (Hoshino *et al.*, 2004). The latter can deliver the QDs directly to the intracellular organelle of interest, through simple uptake from cell culture media. When coupled to antibodies, QDs bound to deliver molecules might allow organelle directed, specific intracellular staining without fixation or permeabilization. Additionally, QDs can be used for general (i.e., nonspecific) labeling of cells, yielding results comparable to the organic dyes carboxyfluorescein succinimidyl ester (CFSE) and PKH in terms of stability (Jaiswal *et al.*, 2003). Notably, when QDs replace CFSE and PKH, the FITC and PE channels become available for measuring cell-surface phenotypes, thereby providing information about the nature of dividing or quiescent cells.

VIII. Conclusion

Although specific applications for QDs are still emerging, the basic technology has matured to the point that it can be relatively easily employed for multicolor flow cytometry. Unfortunately, the commercial catalog for QDs remains relatively small, which may be hindering the widespread adoption of this technology. Therefore, researchers wishing to gain the substantial benefits of QDs will need to turn to in-house conjugations. Fortunately, this process is relatively straightforward and cost-effective. However, troubleshooting and monitoring the quality of QD conjugations and staining is necessary, as the use of QD-conjugated antibodies can present unique problems. Nevertheless, the remarkable spectral properties of QDs allow excellent resolution of dimly staining populations and easy multiplexing. The latter is perhaps the most important benefit associated with QD technology, as it allows more information to be acquired from less sample, thereby conserving precious patient material.

Conflict of Interest

The author has no conflicts of interest.

Acknowledgments

The author greatly appreciates the contributions of the following individuals: Dr. Mario Roederer for his leadership and guidance in developing QD technology for flow cytometry, Stephen P. Perfetto for optimization of hardware and assistance troubleshooting QD staining problems, Dr. Joanne Yu for expertise in antibody conjugation, and the QD users at the Vaccine Research Center (and elsewhere) who have provided feedback about their experience.

References

Baumgarth, N., and Roederer, M. (2000). A practical approach to multicolor flow cytometry for immunophenotyping. *J. Immunol. Methods* **243**, 77–97.

Betts, M. R., Nason, M. C., West, S. M., De Rosa, S. C., Migueles, S. A., Abraham, J., Lederman, M. M., Benito, J. M., Goepfert, P. A., Connors, M., Roederer, M., Koup, R. A. (2006). HIV nonprogressors preferentially maintain highly functional HIV-specific CD8+ T cells. *Blood* **107**, 4781–4789.

Biju, V., Itoh, T., Anas, A., Sujith, A., and Ishikawa, M. (2008). Semiconductor quantum dots and metal nanoparticles: syntheses, optical properties, and biological applications. *Ana. Bioanal. Chem.* **391**, 2469–2495.

Bolton, D. L., and Roederer, M. (2009). Flow cytometry and the future of vaccine development. *Expert Rev. Vaccines* **8**, 779–789.

Bruchez, M. (2005). Turning all the lights on: quantum dots in cellular assays. *Curr. Opin. Chem. Biol.* **9**, 533–537.

Chattopadhyay, P. K., Hogerkorp, C., and Roederer, M. (2008). A chromatic explosion: the development and future of multiparameter flow cytometry. *Immunology* **125**, 441–449.

Chattopadhyay, P. K., Perfetto, S. P., Yu, J., and Roederer, M. (2010b). The use of quantum dot nanocrystals in multicolor flow cytometry. *Wiley Interdiscip Rev. Nanomed. Nanobiotechnol.* **2**, 334–348.

Chattopadhyay, P. K., Price, D. A., Harper, T. F., Betts, M. R., Yu, J., Gostick, E., Perfetto, S. P., Goepfert, P., Koup, R. A., De Rosa, S. C., Bruchez, M. P., Roederer, M. (2006). Quantum dot semiconductor nanocrystals for immunophenotyping by polychromatic flow cytometry. *Nat. Med.* **12**, 972–977.

Chattopadhyay, P. K., and Roederer, M. (2010a). Good cell, bad cell: flow cytometry reveals T-cell subsets important in HIV disease. *Cytometry A* **77**, 614–622.

Chattopadhyay, P. K., Yu, J., and Roederer, M. (2007). Application of quantum dots to multicolor flow cytometry. *Methods Mol. Biol.* **374**, 175–184.

Dabbousi, B. O., Rodriguez-Viejo, J., Mikulec, F. V., Heine, J. R., Mattoussi, H., Ober, R., Jensen, K. F., Bawendi, M. G. (1997). (CdSe)ZnS core–shell quantum dots: synthesis and characterization of a size series of highly luminescent nanocrystallites. *J. Phys. Chem. B* **101**, 9463–9575.

Darrah, P. A., Patel, D. T., De Luca, P. M., Lindsay, R. W., Davey, D. F., Flynn, B. J., Hoff, S. T., Andersen, P., Reed, S. G., Morris, S. L., Roederer, M., Seder, R. A. (2007). Multifunctional TH1 cells define a correlate of vaccine-mediated protection against Leishmania major. *Nat. Med.* **13**, 843–850.

Dwarakanath, S., Bruno, J. G., Shastry, A., Phillips, T., John, A. A., John, A., Kumar, A., Stephenson, L. D. (2004). Quantum dot-antibody and aptamer conjugates shift fluorescence upon binding bacteria. *Biochem. Biophys. Res. Commun.* **325**, 739–743.

Gao, X., Chan, W. C., and Nie, S. (2002). Quantum-dot nanocrystals for ultrasensitive biological labeling and multicolor optical encoding. *J. Biomed. Opt.* **7**, 532–537.

Grabolle, M., Ziegler, J., Merkulov, A., Nann, T., and Resch-Genger, U. (2008). Stability and fluorescence quantum yield of CdSe–ZnS quantum dots – influence of the thickness of the ZnS shell. *Ann. N.Y. Acad. Sci.* **1130**, 235–241.

Hahn, M. A., Keng, P. C., and Krauss, T. D. (2008). Flow cytometric analysis to detect pathogens in bacterial cell mixtures using semiconductor quantum dots. *Anal. Chem.* **80**, 864–872.

Haugland, R. P. (1994). Spectra of fluorescent dyes used in flow cytometry. *Methods Cell. Biol.* **42**(Pt B), 641–663.

Hoshino, A., Fujioka, K., Oku, T., Nakamura, S., Suga, M., Yamaguchi, Y., Suzuki, K., Yasuhara, M., Yamamoto, K. (2004). Quantum dots targeted to the assigned organelle in living cells. *Microbiol. Immunol.* **48**, 985–994.

Jaiswal, J. K., Mattoussi, H., Mauro, J. M., and Simon, S. M. (2003). Long-term multiple color imaging of live cells using quantum dot bioconjugates. *Nat. Biotechnol.* **21**, 47–51.

Mahnke, Y. D., and Roederer, M. (2007). Optimizing a multicolor immunophenotyping assay. *Clin. Lab. Med.* **27**, 469–485v.

Michalet, X., Pinaud, F. F., Bentolila, L. A., Tsay, J. M., Doose, S., Li, J. J., Sundaresan, G., Wu, A. M., Gambhir, S. S., Weiss, S. (2005). Quantum dots for live cells, *in vivo* imaging, and diagnostics. *Science* **307**, 538–544.

Mitchell, G. P., Mirkin, C. A., and Letsinger, R. L. (1999). Programmed assembly of DNA functionalized quantum dots. *J. Am. Chem. Soc* **121**, 8122–8123.

Newell, E. W., Klein, L. O., Yu, W., and Davis, M. M. (2009). Simultaneous detection of many T-cell specificities using combinatorial tetramer staining. *Nat. Methods* **6**, 497–499.

Peng, X., Wickham, J., and Alivisatos, A. P. (1998). Kinetics of II–VI and III–V colloidal semiconductor nanocrystal growth: "Focusing" of size distributions. *J. Am. Chem. Soc.* **120**, 5343–5344.

Perfetto, S. P., Ambrozak, D., Nguyen, R., Chattopadhyay, P., and Roederer, M. (2006). Quality assurance for polychromatic flow cytometry. *Nat. Protoc.* **1**, 1522–1530.

Perfetto, S. P., Chattopadhyay, P. K., and Roederer, M. (2004). Seventeen-colour flow cytometry: unravelling the immune system. *Nat. Rev. Immunol.* **4**, 648–655.

Reiss, P., Bleuse, J. L., and Pron, A. (2002). Highly luminescent CdSe ZnSe core/shell nanocrystals of low size dispersion. *Nano. Lett.* **2**, 781–784.

Roederer, M. (2001). Spectral compensation for flow cytometry: visualization artifacts, limitations, and caveats. *Cytometry* **45**, 194–205.

Smet, Y. D., Deriemaeker, L., Parloo, E., and Finsy, R. (1999). On the determination of ostwald ripening rates from dynamic light scattering measurements. *Langmuir* **15**, 2327–2332.

Wang, H. Q., Liu, T. C., Cao, Y. C., Huang, Z. L., Wang, J. H., Li, X. Q., Zhao, Y. D. (2006). A flow cytometric assay technology based on quantum dots-encoded beads. *Anal. Chim. Acta* **580**, 18–23.

Wood, J. C., and Hoffman, R. A. (1998). Evaluating fluorescence sensitivity on flow cytometers: an overview. *Cytometry* **33**, 256–259.

Wu, X., Liu, H., Liu, J., Haley, K. N., Treadway, J. A., Larson, J. P., Ge, N., Peale, F., Bruchez, M. P. (2003). Immunofluorescent labeling of cancer marker Her2 and other cellular targets with semiconductor quantum dots. *Nat. Biotechnol.* **21**, 41–46.

Wu, Y., Lopez, G. P., Sklar, L. A., and Buranda, T. (2007). Spectroscopic characterization of streptavidin functionalized quantum dots. *Anal. Biochem.* **364**, 193–203.

Zarkowsky, D., Lamoreaux, L., Chattopadhyay, P., Koup, R. A., Perfetto, S. P., Roederer, M. (2010). Heavy metal contaminants can eliminate quantum dot fluorescence. *Cytometry A.* **79a**, 84–89.

CHAPTER 19

Background-free Cytometry Using Rare Earth Complex Bioprobes

Dayong Jin

Advanced Cytometry Labs, MQ Photonics Centre, Faculty of Science, Macquarie University, Sydney, Australia

Abstract

In the analytical fields of microbiology, disease diagnosis, and antibioterrorism, there are increasing demands for rapid yet inexpensive quantification of rare cells. This has proven to be challenging by the conventional spectral discrimination of using traditional fluorescent probes, since the strong autofluorescence from background cells or particles overlaps spectrally with the probe fluorescence. This is particularly true when the target cell occurs at very low frequency (one in more than 100,000 background cells) representing a needle-in-a-haystack problem. This chapter describes a low-cost solution to overcome this problem by employing a novel

0091-679X/10 $35.00
DOI 10.1016/B978-0-12-374912-3.00019-5

detection technology, namely the use of rare-earth (lanthanide) complex bioprobes with luminescence lifetimes in the hundreds of microseconds. Due to this long persistence in lifetime, microsecond duration luminescence can be detected under conditions where fluorescent backgrounds would overwhelm the emission of conventional fluorochromes. The nanosecond duration autofluorescence associated with cells can be suppressed by time-gated detection, allowing detection of long lifetime lanthanide-based bioprobes with minimal background interference. This technology is applicable to a broad range of detection technologies in both cytometry and imaging. In this chapter, we highlight a typical application in the monitoring of the rare microbial pathogens *Cryptosporidium parvum* and *Giardia lamblia* against the complex background of concentrated drinking water. We also describe recent nanotechnological developments in the production of rare-earth nanoparticle bioprobes required for this technology. Other applications of rare-earth bioprobes and time-gated flow cytometry will also be discussed.

I. Introduction

A. Fluorescence Detection

Sensitivity is a critical issue in advanced biotechnology, clinical diagnosis, food safety, and antibioterrorism security. In such fields, there are increasing demands in ultrasensitive detection of trace amounts of specific cells or microorganisms within complex biosamples (Gross *et al.*, 1995; Hemmila and Mukkala, 2001; Jin *et al.*, 2009; Nelson *et al.*, 2004; Thibon and Pierre, 2009b; Yao *et al.*, 2006; Zhao *et al.*, 2004). Nowadays, fluorescence-based detection has become a dominant signal transduction strategy to probe cellular or subcellular targets. In this procedure, the target molecules are tagged by a reporter dye featuring distinctive optical properties typically in color or wavelength (Campbell *et al.*, 2002; Chan and Nie, 1998). In conventional immunofluorescence labeling, for example, a fluorescent tag will be conjugated to an antibody, and then react with the specific antigen on the cell, so that only the target antigen-specific cells emit at the desired fluorescence using a fluorescence microscope. To achieve more detailed spatial resolution, the laser scanning confocal microscope (Pawley and Masters, 2008) was developed. In order to analyze a large population of cells, automated cell analysis required laser scanning cytometry (Kamentsky and Kamentsky, 1991), flow cytometry (FCM) (Shapiro, 2003), or automated imaging FCM (George *et al.*, 2004).

Conventional fluorescence bioprobes, including organic dyes and fluorescent proteins are usually sensitive enough to label and detect even subcellular information. However, many naturally occurring substances are autofluorescent under UV or visible wavelength excitation, and traditional fluorescence bioprobes lose much of their relative sensitivity in the presence of strong autofluorescence. This is particularly true in the detection of rare events. Many biological and clinical procedures require accurate detection of rare target cells at frequencies in the range 1 in

1,000,000 or below (Bajaj *et al.*, 2000). For example, residual cancer cells in bone marrow transplantations are a major concern (Deisseroth *et al.*, 1994), and it is vital to be able to remove such cells prior to the procedure (Gross *et al.*, 1995). This requires a detection level of one residual cancer cell per 10,000,000 of bone marrow or peripheral blood stem cells. Fetal cells present in maternal blood during pregnancy are an ideal source of genetic material for noninvasive prenatal diagnosis; however, the target fetal nucleated red blood cells need to be detected against the maternal cells at extremely low frequencies ranging from 1 in 10^7 to 10^9 (Bajaj *et al.*, 2000; Bianchi *et al.*, 1996; Johnson *et al.*, 2007). In water safety inspection, very small number of organisms are capable of causing infection. The methods of analysis must therefore be sufficiently sensitive to detect a single microorganism (e.g., *Cryptosporidium parvum* and *Giardia lamblia*) in as large as 10 L volume of water additionally containing millions of nontarget microorganisms and particles (Veal *et al.*, 2000). The fundamental problem in this last example is autofluorescent noise from nontarget particle/cells in a complex medium containing many objects that can be mistaken for the target of interest. Here we describe a typical problem in waterborne and foodborne pathogen detection as an example to demonstrate the potential power of the lanthanide-based cytometry methods.

B. Microbial Cell Assessment

Monitoring of waterborne and foodborne pathogens is a good application for the emerging field of microbial FCM (Veal *et al.*, 2000). Table I describes several typical waterborne and foodborne pathogenic bacteria which have been responsible for many outbreaks of disease in recent years. Major sources of microbial pathogens come from human and animals waste that find their way into domestic water supplies. Bacteria, viruses, protozoa, and helminths can be carried by water and transmitted to people by direct contact (e.g., water contact disease) or ingestion (e.g., waterborne and foodborne disease) (Lemarchand *et al.*, 2004). Microbial contamination of water arguably represents the most significant health hazard and public concern in both developed and developing nations (Lemarchand *et al.*, 2004; Sharma *et al.*, 2003). There is a tremendous need for real-time and rapid monitoring of water quality for early detection and initiation of treatment to control the spread of infection.

There are basically two categories of techniques used to monitor water quality: population analysis, for example, total target microorganism counts or total subpopulation counts, and rare event detection of specific microbial cells to confirm the presence of pathogens. Traditionally, culture-based methods have been employed to amplify signals by growth of a single organism into colony on a plate, followed by microscopic examination. The culture process is time-consuming and labor intensive, and often requires a day or more to yield a result (Lemarchand *et al.*, 2001). It should also be noted that a large majority of microorganisms cannot be easily cultured and some species of ecological interest, for example, pathogens, cannot be grown outside their natural environments.

Table I

Typical waterborne and foodborne pathogens

Baterium	Size	Disease	Human infection	Infectious dose	Reference
Salmonella	~1 μm width; 2~6 μm long	Typhoid or typhoid-like fever; gastroenteritis; salmonellosis	Ingestion of organisms in contaminated food or water	$>10^5$ cells	(Kothary and Babu, 2001; Ohl and Miller, 2001)
			Escherichia coli		
(OH:157)	0.5~1.5 μm	Bloody diarrhea (hemorrhagic colitis) and Hemolytic Uremic syndrome	Consumption of milk, yoghurt, meats, dry cured salami, raw vegetables, unpasteurized juice, and potable water	10–100 cells, or even lower	(Lekkas *et al.*, 2006; Lemarchand *et al.*, 2001; Sharma *et al.*, 2003; Wilks *et al.*, 2005)
Entamoeba Histolytica/Coli	10~20 μm	Diarrhoea accompanied by nausea, fever, colic and symptomatic disease	Consumption of contaminated water or food or by person to person contact	1 cell	(Kothary and Babu, 2001; Sharp *et al.*, 2001)
Cryptosporidium parvum	4~5 μm	Acute gastrointestinal disease, persistent diarrhoea life-threatening to immunocompromised individuals	Ingestion of water contaminated with the oocysts of cryptosporidium and also by direct contact with the infected persons or animals.	1–10 cells	(Dawson, 2005; Dillingham *et al.*, 2002; Kothary and Babu, 2001; Quintero-Betancourt *et al.*, 2002; Sharma *et al.*, 2003)
Legionella pneumophila	1~3 μm	Legionnaires, a common form of severe pneumonia	Inhalation of aerosols, showers, air conditioning and other aerosol-generating devices	Unofficial threshold 10^2–10^3 cells L^{-1}	(Aurell *et al.*, 2004; Levi, 2001; Veal *et al.*, 2000)
Cyclospora cayetanensis	8~10 μm	Severe diarrhea, weight loss, stomach cramps, nausea, vomiting, fatigue, and fever	Sewage or water contaminated by human sewage effluent can affect humans or contaminate crops	Low and unknown due to less studies	(Dawson, 2005; Quintero-Betancourt *et al.*, 2002; Rose and Slifko, 1999)
Campylobacter jejuni	0.8~5 μm	Acute gastroenteritis	Foodborne pathogen, poultry product, water is also regarded as an important route for the transmission, untreated water, and unpasteurized milk	500 cells	(Kothary and Babu, 2001; Sharma *et al.*, 2003)
Giardia lamblia	9 μm ~12 μm	Watery, foul-smelling diarrhea, often accompanied by nausea, smelling abdominal cramps or gurgling, bloating	Consumption of contaminated water or food or by person to person contact	10–100 cells	(Dawson, 2005; Rose and Slifko, 1999)

Another difficulty in microbial cell detection is that microorganism of interest usually exists in an immense pool of other microorganisms, which constitute an enormous source of background. Microbial cells, typically bacteria, are less than 10 μm in diameter, approximately one thousandth the volume of mammalian cells. They also have considerably lower DNA content, a characteristic often exploited in FCM for detection. For example, the DNA contents of Escherichia coli, being some 1000 times less than the normal diploid human cell. Target microorganisms often occur in complex samples, for example, water, mud, food, and beverages that have high autofluorescence backgrounds, both from organic and inorganic debris and other microorganisms. Conventional microscopy of these samples is frequently difficult, slow, and tedious to identify and enumerate cells of interest from background particles, and may not be possible at all if the mixture is too complex (Connally *et al.*, 2002). Diagnosis via microscopic examination of a single stool specimen has a low sensitivity and has been observed to miss up to 50% of infections caused by *Giardia* or *Entamoeba* (Schunk *et al.*, 2001).

The accurate detection of target microorganisms (<10 μm diameter) within an intrinsically fluorescent matrix of nontarget particles, for example, minerals, plant debris, and algae (Fig. 1), often requires multicolor fluorescence staining followed by multispectral detection. The complexity and cost of such techniques are significant and detection accuracy is not guaranteed, particularly when the target organism is in very low concentration, for example, <10 cells in 1 L of water.

C. Time-Domain Techniques

One promising approach for overcoming these difficulties is the use of time-gated luminescence (TGL) lifetime analysis. In such a scenario, a pulsed excitation source and luminescent probes with long luminescence lifetimes are used to ensure discrimination from each other and background autofluorescence. Fortunately, most naturally occurring autofluorescent sources typically cease their emission within 10 ns following excitation. For example, fluorescein has a fluorescence lifetime as

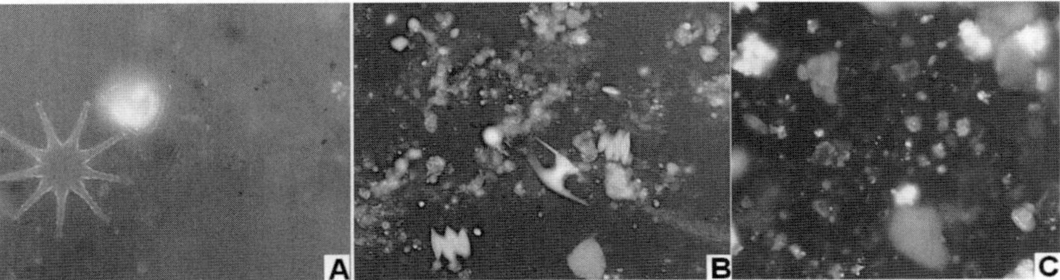

Fig. 1 Images of concentrated environmental backwash water (10,000:1) from various sources in Sydney, Australia, under the UV fluorescence microscope (original figure from (Jin *et al.*, 2007b)). Many organisms, organic and nonorganic debris are visible.

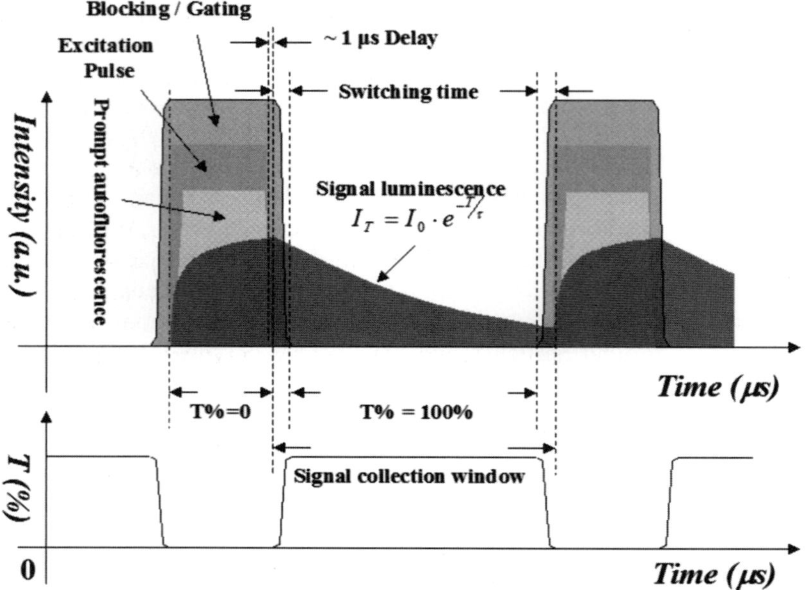

Fig. 2   TGL biosensing using a microsecond-duration luminescence lanthanide bioprobe.

short as 4.5 ns. This provides excellent opportunities in time-resolved detection strategies based on long-lifetime luminescent bioprobes (Fig. 2).

To take advantage of these phenomena, we need to use fluorescent probes with much longer fluorescence lifetimes than those typically used for FCM. Three groups of long-lifetime luminescent bioprobes can be used for this technique (Jin *et al.*, 2007a). The first and most important group is the lanthanide metal complexes, including europium (Eu^{3+}) and terbium (Tb^{3+}), with typical lifetimes ranging from 300 μs to 1 ms(Yuan and Wang, 2006). Recent reports on intensely luminescent lanthanide chelates have prompted their use in clinical immunoassays (Jiang and Luo, 2004; Soukka *et al.*, 2003; Tan *et al.*, 2004a; Wu and Zhang, 2002; Wu *et al.*, 2002; Yuan and Matsumoto, 1997), environmental pathogen detection (Connally *et al.*, 2002), and DNA-protein interaction assays using time-resolved fluorescence energy resonance transfer (FRET) (Heyduk and Heyduk, 2001; Selvin, 2002; Sueda *et al.*, 2000; Xiao and Selvin, 2001). The second group comprises of palladium (Pd) and platinum (Pt) metalloporphyrins with lifetime about 40 μs (Hennink *et al.*, 1996). These probes have relatively high absorptive coefficients at ~400 nm (150,000–200,000 (litre/mole/cm)) and emit in the far-red region of the visible spectrum (640–660 nm). However, they are strongly quenched by the oxygen and other quenching molecules (Hennink *et al.*, 1996). Botchway and coworkers (Botchway *et al.*, 2008) recently reported a stable Pt complex bioprobe (quantum yield: 70%; absorptive coefficient at 350–380 nm: ~10,000 (litre/mole/cm); lifetime: $\sim$ 3 µs) for time-resolved and two-photon emission imaging microscopy. The third potential type of long-lifetime fluorescent dyes are the charge-transfer-excited transition metal chelates, typically ruthenium, osmium, and rhenium chelates, with lifetimes of 0.1 to 10 µs (Hemmila and Mukkala, 2001; Kurner *et al.*, 2001). These chelates are similar to the Pd and Pt metalloporphyrins in their sensitivity to oxygen, requiring suitable buffers and deoxygenation are required during the assays.

Lanthanide (rare-earth element) chelates therefore have the most desirable characteristics for time-resolved bioimaging and FCM, including exceptional long lifetime, large Stokes shifts, sharp emission profiles, and lower environmental sensitivity (Bunzli and Piguet, 2005). Europium, terbium, samarium, and dysprosium are the most popularly used ions for generating these complexes. The excitation of these lanthanide chelates is usually restricted to UV wavelength, typically from 320 to 370 nm. This has been recently extended to the 400 nm violet range for some complexes (Jiang *et al.*, 2010b) with emission in the visible to near IR range at different lifetimes (Fig. 3). Europium chelates have a quantum yield of up to 70%, and terbium chelates up to 100% (Yuan *et al.*, 2001), but the other two have low quantum yields as low as 2% (Hemmila and Laitala, 2005; Hemmila and Mukkala, 2001).

A successful chelate complex (Yuan and Wang, 2006) contains several parts, and the excitation/emission process of the complex involves three steps: photon absorption, energy transfer, and lanthanide fluorescence emission. As illustrated in Fig. 4, the organic chromophore, BHHCT $(4,4'$-bis$(1'',1'',1'',2'',2'',3'',3''$-heptafluoro-$4'',6''$-hexanedion-$6''$-yl)-chlorosulfo-*o*-terphenyl) chelate acts as an antenna or sensitizer, absorbing excitation light and transferring energy to the lanthanide, this overcoming the inherently weak absorption cross-section of the rare-earth ion; the

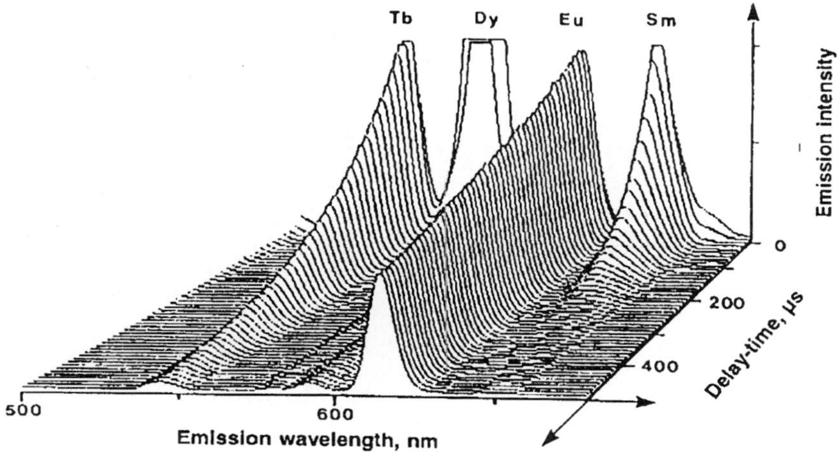

Fig. 3 Time-resolved spectrum decay profiles of a mixture lanthanide chelates including europium, terbium, samarium, and dysprosium (original figure from (Hemmila and Mukkala, 2001)). Europium maintains its emission to well over 400 µs.

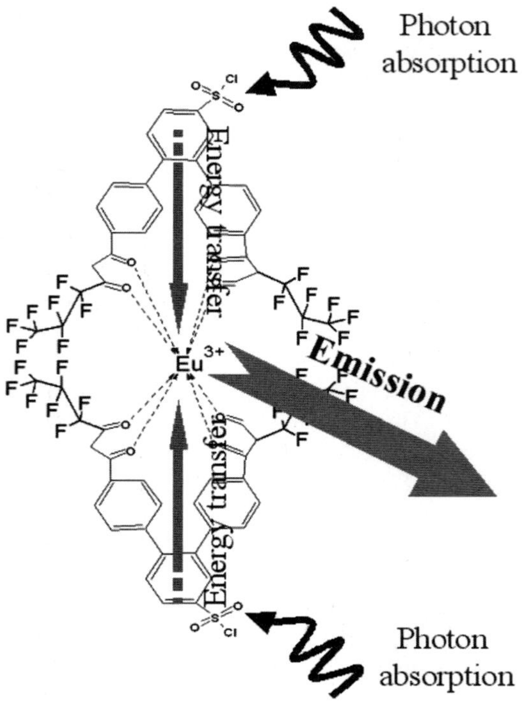

Fig. 4 Structure of europium chelates BHHCT (the –C3F7 groups attaching in close proximity to Eu^{3+}; the –ClSO₂ groups reacting with and attaching to biomolecules). The molecular location of the excitation/emission process is also shown.

–C_3F_7 group displaces water from the primary coordination sphere of the lanthanide, reducing quenching caused by the the O–H vibration. The chlorosulfonate activating group also provides a scaffold for attaching a reactive group for coupling to biomolecules. The strongly fluorescent europium chelate BHHCT uses tetradentate β-diketone chelates that have been demonstrated to be excellent antenna molecules for the excitation of trivalent europium (Connally *et al.*, 2004b, 2005; Yuan and Matsumoto, 1996; Yuan and Wang, 2005). The molecules have two electron-withdrawing heptafluoro groups that flank the β-diketone moiety and strongly bind the lanthanide ion.

II. Instrumentation Development

A. TGL Microscopy

Millisecond TGL imaging was developed in the early 1990s. Phosphorescence/luminescence imaging was pioneered by a number of groups (Beverloo *et al.*, 1990,

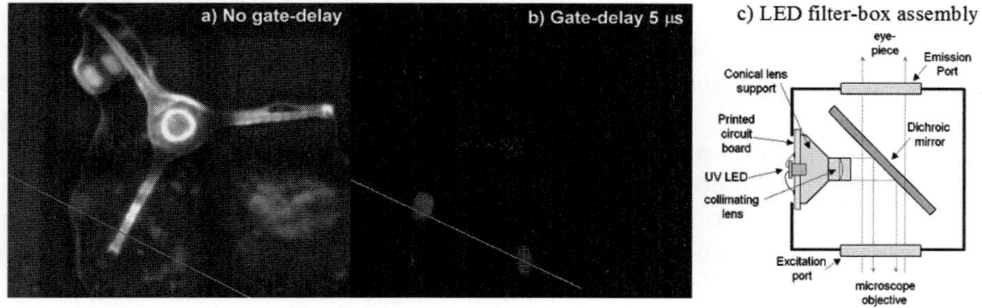

Fig. 5   (a) Two immunfluorescently labeled Giardia cysts embedded in an autofluorescent matrix of algae (line transects cysts) viewed with conventional epifluorescence; (b) time-resolved fluorescence microscopy with UV LED excitation; (c) the small size of the LED permitted the device to be mounted within the filter housing to simplify the optical arrangement. (Original figure was from (Connally *et al.*, 2006).)

1992; Hennink *et al.*, 1996; Marriott *et al.*, 1991, 1994; Phimphivong and Saavedra, 1998; Phimphivong *et al.*, 1995; Seveus *et al.*, 1992; Templeton *et al.*, 1991; Verwoerd *et al.*, 1994). The TGL imaging technique offers effective background suppression for a wide range of applications in both analytical and clinical detection (Bjartell *et al.*, 1999a, 1999b; deHaas *et al.*, 1996; Faulkner *et al.*, 2005; Schaferling *et al.*, 2003; Vaisanen *et al.*, 2000). A multiparameter TGL microscope was also reported (Connally *et al.*, 2006; Soini *et al.*, 2003). Recently, we used TGL microscopy techniques for imaging of individual *in situ* microorganisms including *Giardia* (Fig. 5 a and b) and *Cryptosporidium* cysts (Connally *et al.*, 2002, 2004a, 2004b, 2005). A TGL system was built on a conventional epifluorescence microscope platform equipped with a pulsed excitation source (a flashlamp or a UV LED) and a time-gated image-intensified CCD camera. A microprocessor was used to synchronize the excitation and capture sequences. The microscope was fitted with a UV excitation filter and a long-pass emission filter set typically employed for the examination of the DNA binding dye DAPI (4′,6-diamindino-2-phenylindole) (Connally *et al.*, 2006).

Connally has reviewed the recent progress in TGL microscopy based on lanthanide probes (Connally and Piper, 2008c). In the past decade, there has been a rapid evolution of the technology of TGL microscopy to make it more effective and cheaper; in the case of the pulsed excitation source, the technology has moved from chopper-interrupted continuous Hg lamps or UV lasers to xenon flashlamps, and most recently current-switched UV LEDs. In the case of time-gated detection the technology has also evolved from mechanical chopper or ferroelectric shutters to gated CCDs, gated-intensifier CCDs and most recently to gain-switched electron-multiplying CCDs (Connally and Piper, 2008c). Despite these advances TGL microscopy remains a relatively complex and an expensive tool, restricting application to a relatively small range of specialized bioassays.

Another practical problem arises from the relatively weak luminescence intensity of lanthanide probes and slow TGL cycling rates (maximum at $\sim 10^2$ Hz)

(Jin *et al.*, 2007a). This requires quite long (often 30 s) signal accumulation times to produce quality TGL images. Researchers therefore typically report either grayscale (Connally and Piper, 2008a; Harma *et al.*, 2001) or pseudocolor (Soini *et al.*, 2003; Vereb *et al.*, 1998; Wu *et al.*, 2009) TGL cell images obtained using high-quality monochrome (high-gain CCD) cameras. Ideally, TGL microscopes should permit direct visual (naked eye) observation of true-color TGL images in real time, which would enable practical utilization of TGL microscopy to identify multiplexed target organisms free of background in routine laboratory diagnostics (Vereb *et al.*, 1998).

More recently, we have successfully demonstrated the feasibility of real-time direct visual observation of high-contrast lanthanide stained cells under the background-free condition using a standard epiflourescence microscope incorporating a comparatively simple and low-cost modification (Fig. 6). This development should

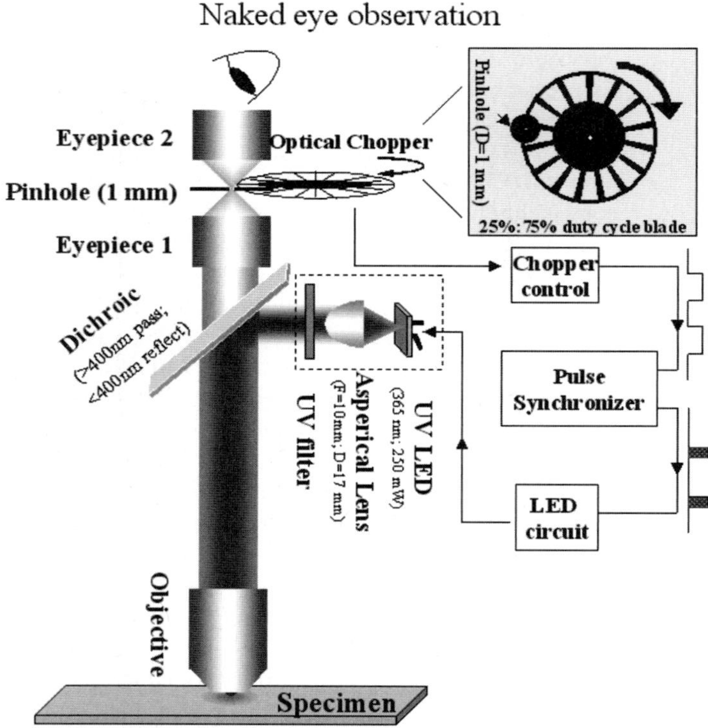

Fig. 6   Schematic layout of the TGL microscope, using chopper gating the imaging plain to achieve much faster chopper switching times, and an additional eyepiece to allow direct observation of time-gated luminescence. The pulse synchronizer accepting monitored TTL signals from the chopper blade was used to trigger the UV LED circuit, so that an appropriate time delay between the excitation pulse to gated-detection phase was achieved (Jin *et al.*, 2006). (Original figure was from (Jin *et al.*, 2010).)

be easily adaptable to a larger number of laboratories since presently there are only expensive time-resolved luminescence microscopes. Briefly, (refer to Fig. 6), we choose a current switched UV LED, since it can be switched up to several MHz rapidly with complete switch-off time <3 μs (Jin *et al.*, 2007b). Though mechanical chopper gating has been reported with slow switching times of 100–200 μs (Connally and Piper, 2008b; Hanaoka *et al.*, 2007; Hashino *et al.*, 2007; Kimura *et al.*, 2008; Soini *et al.*, 2003; Vereb *et al.*, 1998), we employed a new compact fast chopper (167 rotating cycles per second) and a new optics configuration to achieve up to 11 μs switching time and 2.5 kHz repetition rate at a transparent/blocking duty ratio of 75:25%. One simple idea exploited in this technique was to accelerate the switching time, bringing the image to a focus by the eyepiece of the conventional fluorescence microscope, and to project the time-gated image by the second eyepiece onto the naked eye or the digital camera, so that the new system functions without any complicated change of emission pathway. With the improved optics, we achieved a relatively high TGL signal photon collection efficiency of 72.1% (time-gated mode vs. conventional mode for europium emissions), and demonstrated strong europium labeled cryptosporidium oocysts in the TGL collection mode that could be visualized with the naked eye. The image shown in Fig. 7 was collected at

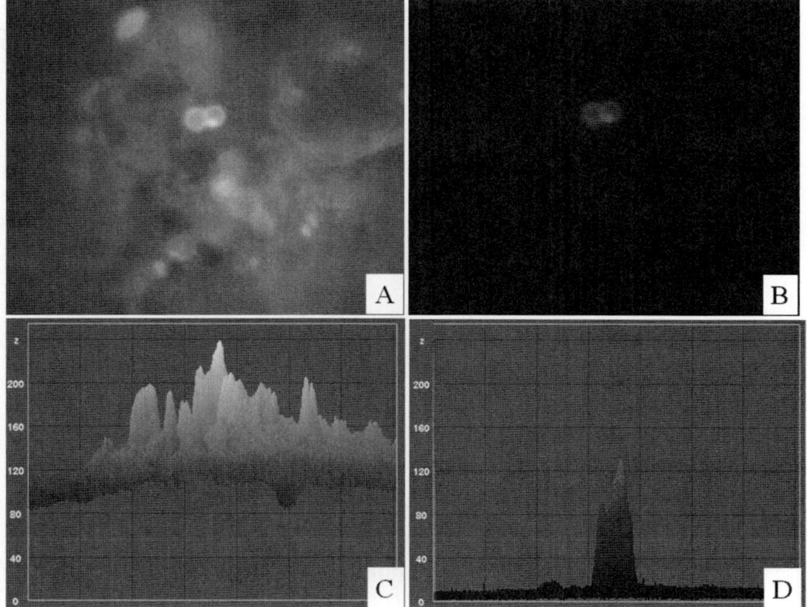

Fig. 7   True color time-gated imaging of two *Cryptosporidium* oocysts labeled by a BHHCT–europium complex within a complex concentrated water sample with significant debris. (A) conventional mode, (B) time-gated mode, (C) and (D) are the 3D intensity-spectrum-spatial image analysis corresponding to (A) and (B), respectively.

1 s exposure times by a color 2 megapixel Nikon digital CCD camera, with an image intensity equivalent to the real-time naked eye observation. The signal to background ratio (luminescence intensity of *Cryptosporidium* vs. autofluorescence background intensity of algae) was improved by 7-fold (132/16 (cryptosporidium/red algae in time-gated mode):238/202 (cryptosporidium/red algae in conventional mode)) with autofluorescence background reduction by 12.6-fold [202/16].

This demonstration of naked eye observation is significant to the biosensing field, since it directly addresses the concern about the signal intensity issue from lanthanide-based probes, due to their much slower excitation-emission cycle rate compared to conventional fluorophores. This improved time-gated technique allows for ultrahigh contrast bioimaging through enhanced temporal, spectral, morphological, and intensity resolution, rendering TGL imaging microscopy superior to the existing advanced microscopy techniques in many ways. Surprisingly, this technique can be easily applied to the normal conventional fluorescence (or light) microscopes at low cost (<$2,000) with very limited modifications to the microscope itself. We envision this technology having a large impact on lanthanide (as well as phosphorescence dye) based advanced biosensing applications, including continuous monitoring of biological cellular processes (e.g., celluar chemical sensing of pH, cations using lanthanide complexes as excellent sensors/switches) in real time (Botchway *et al.*, 2008; Thibon and Pierre, 2009a, 2009b), tissue imaging (Bornhop *et al.*, 1999), lifetime mapping (Beeby *et al.*, 2000), and lanthanide sensitized FRET emission mapping. In this last application, a long-lived chelate would be used as energy transfer donor and a fluorophore with a lifetime in the nanosecond range as the acceptor (Vereb *et al.*, 1998) Since this microscopy system uses true-color imaging, multicolor lanthanided-based TGL imaging also becomes feasible (Petoud *et al.*, 2003). This method should also accelerate the ongoing discoveries of lanthanide-based biochemistry and bio-medical diagnostics toward more advanced background-free imaging, for example, practical simultaneous temporal-spectral-spatial 3D bioimaging.

B. Time–Gated Flow Cytometry

FCM is an automated technique capable of rapidly identifying rare cells contained within a large sample. It relies on scattering or fluorescence measurements that are made while the cells or particles pass, usually in a single file, through a capillary flow cell. Commercial FCM instruments equipped with multiple spectral channels are capable of real-time rapid analysis with rates reaching 10,000 cells s^{-1} (Shapiro, 2003). However, detecting very rare cells poses significant challenges in terms of accuracy. The fundamental challenge is autofluorescence noise from nontarget particle/cells in the complex biological samples (Ferrari *et al.*, 2006; Mcclelland and Pinder, 1994; Shapiro, 2003; Veal *et al.*, 2000). Rare cells are often difficult to resolve from other cell types, and nontarget objects can often mimic the cell type of interest.

The use of lanthanide chelates as fluorescent probes for FCM was first proposed by Leif and Vallarino (1991) and Leif *et al.* (1976). A detailed theoretical analysis and discussion of TGL flow cytometer design concepts was presented by Condrau *et al.* (1994a, 1994b). In Condrau's system, the TGL detector was positioned orthogonally to the excitation beam (a 26 mW CW 325 nm He–Cd laser) and opposite to the standard multicolor detection channels. The TGL detection channel was normally off, but once a forward scattering detection signal indicated the arrival of target, the laser beam was deflected away from the flow cell by an acousto-optic modulator and the photon counting photomultiplier tube was then gated on briefly for TGL analysis of the incoming cell. However, since the long decay over a $\sim$1 ms inhibits the laser excitation, the whole system suffers from the limited cell flow rate by no more than 200 cells s^{-1} (flow velocity of 1.7 m s^{-1}) at counting accuracy of 90% (Condrau *et al.*, 1994a). This flow rate is far slower than most commercial systems, and would not allow sufficient throughput to detect rare events in a reasonable period of time. The laser system coupled with acousto-optic modulation resulted in a bulky, expensive, and complex instrument.

We recently returned to lanthanide-based FCM by studying the temporal-to-spatial transformation effects, which provides many new opportunities to achieve high-speed cell analysis rates at 100% detection efficiency (Jin *et al.*, 2007a, 2007b, 2009). Briefly, when the labeled cells are traveling in a rapid flow stream, following excitation the rapid flow can distribute the long-lived emission spatially. At flow rates of 2 to 10 m s^{-1}, the emission from europium or terbium chelates is detectable for over hundreds of microns, as modeled in Fig. 8.

This is shown in Fig. 9, where a conventional fluorochrome Sulforhodamine 101 is only bright within the excitation spot in the rapid flow cytometer (Fig. 9A). In contrast, the europium complex (Fig. 9B) in the rapid flow system displays a spatially distributed emission "spike". Since other traditional fluorochromes and sources of autofluorescence would not show this signal pattern, it offers the unique opportunity to distinguish target organisms against the autofluorescence backgrounds.

In principle, time gating to discriminate lanthanide emission against short lifetime autofluorescence could be achieved simply by detecting the long-lived luminescence from target organisms at some distance downstream from the (continuously illuminated) excitation spot. In practice however, excitation light can be scattered with autofluorescence background along the flow stream, which may be much brighter than the faint luminescent signal from the target organisms; this is particularly true for a flow-stream in air (which acts as a light guide), rendering such a simple approach impractical. This situation becomes even worse when incoherent excitation sources, for example, flash lamps, or LEDs are used to excite lanthanide bioprobes. It follows that implementation of TGL techniques in flow requires application of pulsed excitation and gated detection after a delay interval. The question then arises of how pulse sequencing is related to the excitation and detection spot dimensions and velocity of the flow.

Fortunately, a good understanding of temporal-spatial transformation allows us to precisely control the time-gated detection period by the size of detection aperture. In another words, it is not necessary for the next TGL cycle to wait until the lanthanide luminescence to cease for more than 1 ms. This should in principle allow for an

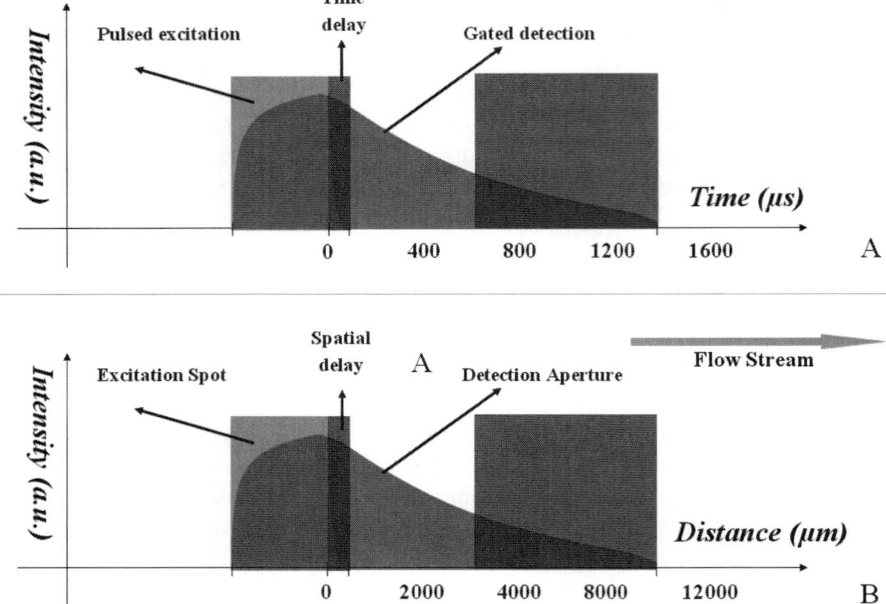

Fig. 8   (A) The general theory of TGL detection operation exploring the large difference in luminescence decay in the temporal domain, where the appropriate time-gated detection window can eliminate
both scattering and the autofluorescence background; (B) long-lifetime biomarkers in rapid flow can
convey the temporal luminescence decay into spatial decay along the flow axis (assuming luminescence
lifetime $\tau = 500$ μs and flow velocity $\upsilon = $ m s^{-1}). Correspondingly, the time-gated detection window in
temporal domain is conveyed into a spatially resolved luminescence aperture.

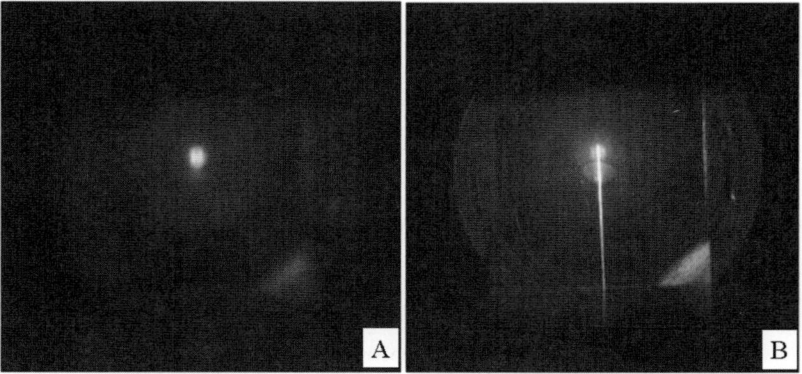

Fig. 9   Experimental demonstration of temporal-spatial-spectral 3D detection opportunities offered by
lanthanide bioprobes in flow cytometry (FCM): (A) a conventional fluorochrome (Sulforhodamine, 101)
probes in a rapid flow system, with fluorescence occurring only at the laser intercept point; (B) the
europium probes in flow cells at flow speed of $\sim$3.2 m s^{-1} downstream. The fluorescence emission
extending well beyond the laser intercept is clearly visible.

increase in the cell flow rate from Condrau's 200 cells s^{-1} to several thousand cells s^{-1} (multikilohertz cell/particle arrival rates). This flow rate is closer to those achieved in typical commercial systems, and would make rare event detection more practical.

Recently we successfully developed two approaches to address these issues (Jin *et al.*, 2007a, 2007b, 2009; Leif *et al.*, 2009). As shown in Fig. 10, labeled target organisms as well as other autofluorescence nontarget cells were firstly illuminated by pulsed excitation in the excitation spot, during which the detector was switched off (Fig. 10 left). Once the excitation pulse has extinguished, the autofluorescence rapidly faded within 1 µs. The fluorescent signal from target organisms/cells however had lost very little of its original intensity, and its emission continued for a given distance downstream of the excitation zone. The opportunity existed for the time-delayed detection of target cells at some point distant beyond the excitation zone, so the signal was both temporally and spatially displaced against a theoretically zero background (Fig. 10 right). To achieve a high target-organism count rate, the observation aperture was situated to permit a much shorter luminescence measurement time (e.g., <50 µs) rather than the typical ∼1 ms for lanthanide chelates luminescence.

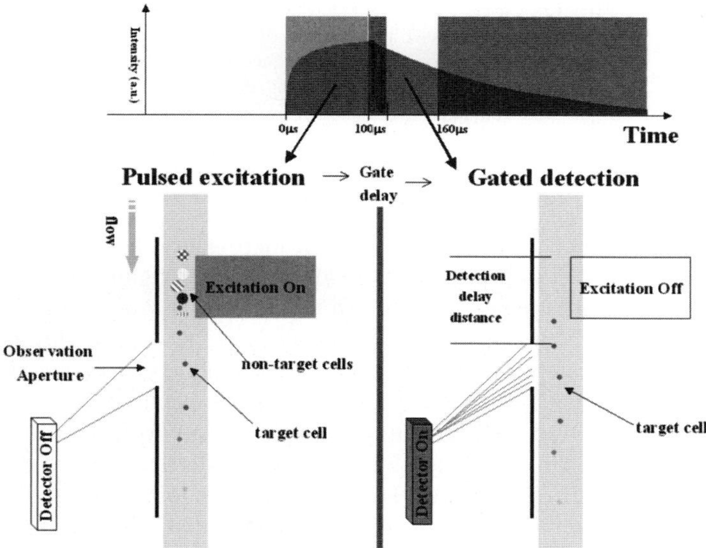

Fig. 10 Schematics of the TGL FCM geometry. The detection of long-lifetime biomarker labeled target cells in flow involved three steps (pulsed excitation, gate delay, and gated detection) with both temporal and spatial considerations: during the period of pulsed excitation (e.g., 0 to 100 µs shown in top), both autofluorescence from nontarget particles and signal luminescence from target cells were detectable at the excitation spot, but the detector was positioned downstream from the excitation spot and switched off (bottom left). After an appropriate gate delay time, both light scattering and autofluorescence ceased and only long-lifetime luminescence from target cells remained to be detected while target cells. (Original figure was from (Jin *et al.*, 2007a).)

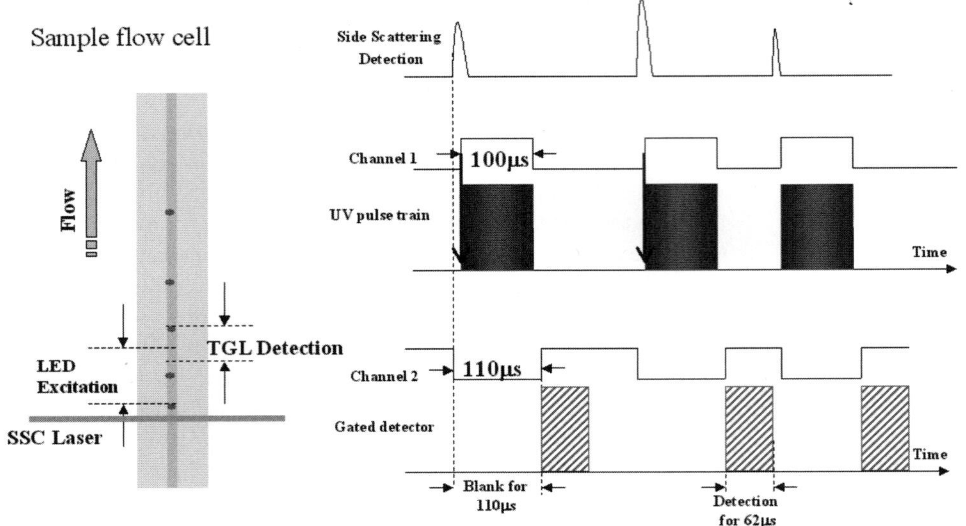

Fig. 11 Schematic of the side-scattering triggered TGL detection model: (A) geometrical positions for the scattering detection channel, the pulsed excitation spot, and the time-gated detection spot; (B) temporal operational triggering sequence showing the appearance of an event (the third pulse) during the TGL analysis of the previous event (the second pulse), which is never excited or counted. (Original figure was from (Leif *et al.*, 2009).)

Our first model was referred to as the "triggered model" (shown in Fig. 11), in which conventional FCM scattering or fluorescence channels were employed to detect an incoming cell, and the TGL excitation and detection channel was positioned downstream and triggered to confirm this event (Jin *et al.*, 2007a; Leif *et al.*, 2009). Although the maximum cell analysis rate (count rate) for which acceptable detection efficiency can be achieved was limited, one of the advantages for the "triggered model" is that every potential positive event can obtain the exact amount of excitation exposure, resulting in good signal sensitivity and resolution reflected in a low sample coefficient of variation. This model was optimized and experimentally demonstrated (Leif *et al.*, 2009) (Fig. 12A). Thanks to its accurate signal resolution; such a flow cytometer allowed an automated statistical analysis of intensity from individual europium-containing microspheres (5 µm in diameter), which facilitated our collaborator Dr. Robert Leif to optimizing his conjugation chemistry and produce extremely high-quality europium calibration microspheres. The measured microsphere intensity variation coefficient was as low as 7%, as shown in Fig. 12B (Leif *et al.*, 2009).

Our second approach was referred to as the "continuous flow-section model" (Fig. 13) (Jin *et al.*, 2007a). This is the only approach to achieve high spatial detection efficiency for multikilohertz particle arrival rates. In such a model, a

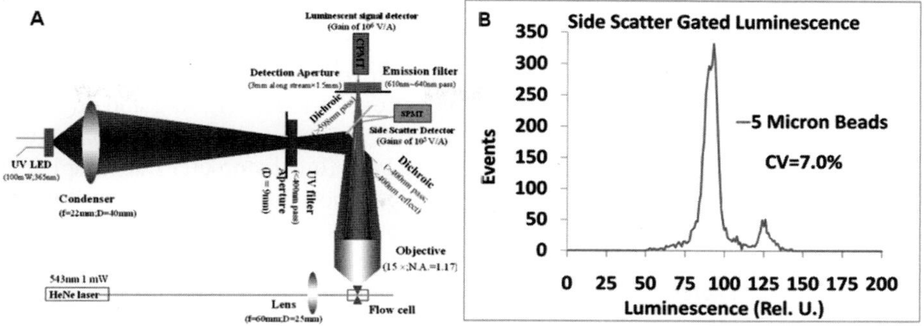

Fig. 12 (A) Optics of the light scatter gated time-delayed luminescence flow cytometer. The terms channel photomultiplier tube emission detector and silicon photomultiplier tube have been abbreviated, respectively, as CPMT and SPMT. (B) Luminescence distribution obtained with a light-scatter-gated flow cytometer. (Original figure was from (Leif *et al.*, 2009).)

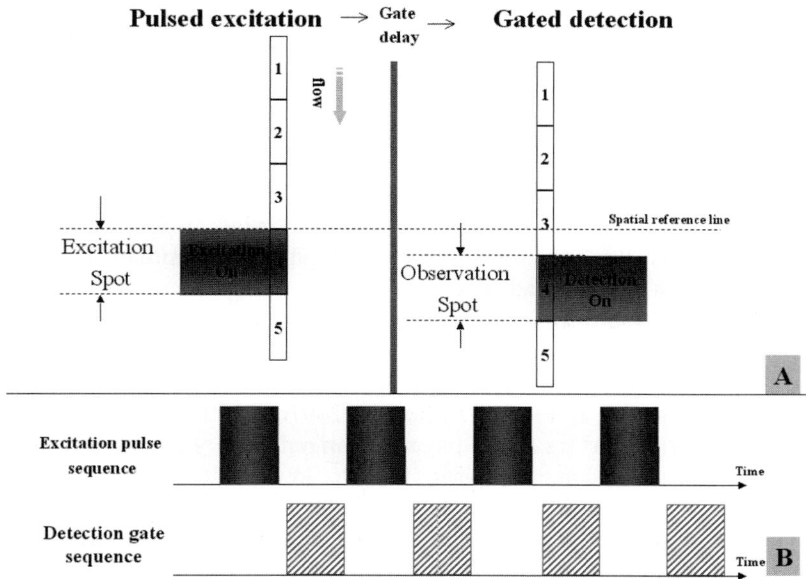

Fig. 13 Flow-section TGL FCM model. The TGL sequence (pulsed excitation, gate delay, and time-gated detection) at a fixed repetition rate (shown in (B)) was applied continuously to the flow sample. The continuous flowing stream can be conceived as adjacent continuous flow sections, and in proper design of sizes and positions of the excitation and detection spots in consideration of the sample flow velocity; one pulsed excitation and detection cycle will be responsible for screening of a TGL event in one corresponding section, so that theoretically all parts of flow will be analyzed sequentially. (Original figure was from (Jin *et al.*, 2007b).)

continuous TGL-cycle rate must be applied where the excitation and detection zone sizes are matched to the sample flow velocity and the TGL temporal pulse sequence. As illustrated in Fig. 13, the flow stream was conceived as continuously adjacent sections, and each TGL pulsed excitation and time-gated detection cycle was responsible for examining the sample within a particular section on the flow.

A continuous flow-section TGL flow cytometer may employ either a brief pulsed excitation source, for example, nanosecond pulsed lasers (Ogilvy and Piper, 2005), or a long-pulsed excitation sources, for example, 100 μs pulsed UV LEDs. In the case of short-pulsed excitation, the distance traveled by a target cell during excitation is less than 1 μm (Laser pulse $\sim$10 ns; $\upsilon\sim$10 m s^{-1}). To properly excite each section, the excitation zone dimensions must be equal to or greater than the distance travelled by the particle during each TGL cycle. Since the traveled distance during the excitation duration is very small in the case of short-pulsed excitation, the size of a flow section is effectively equal to the size of the excitation zone. Experimentally this laser excitation system is currently under our investigation

When the flow section is excited with relatively long pulses (typically >5 μs), the distance traveled by a particle can be significantly large compared to the excitation spot size, and particles/cells would experience different exposure depending upon their location within the excitation zone. For example, the cells entering earlier into the excitation spot will obtain longer exposure than the cells just entering the excitation spot at the end of the excitation pulse (Jin *et al.*, 2007a). To compensate the events with less exposure, we employed a higher excitation pulse repetition rate, so that the events partially excited by the first pulse will have another chance being excited by the second pulse. At a flow rate of 3.2 m s^{-1}, LED excitation pulse for 100 μs, excitation spot diameter size of 530 μm, a repetition rate of 6.45 kHz (Jin *et al.*, 2007b, 2009), the compensation level reached 93.5% of the maximum in theory (Jin *et al.*, 2007a), so that the whole flow stream is covered by UV illumination without "dead time."

Our prototype flow-section flow cytometer used a high-intensity (average power of 250 mW at 365 nm, generating $\sim$15 mW peak intensity at the flow cell) UV LED as an excitation source (schematic layout shown in Fig. 14 and systematic optics shown in Fig. 15 (Jin *et al.*, 2007b, 2009)). At a TGL repetition rate of 6.45 kHz (consisting of 100 μs excitation pulse, a $\sim$10 μs time-resolving period and a $\sim$45 μs gated-detection period), each flow section covers 496 μm, so that the detection region distance is set for 528 μm by detection aperture of 8 mm on the image plan (Fig. 14 B). Thus, each TGL section is monitored through a 528 μm detection window, within which any time-gated signal pulses with the duration of more than 10 μs (filtered by software) can be counted as valid events. A filtering function was also programmed in the software to only count for events appearing >300 μs away from each other; hence, no long-lifetime luminescent target cells/particles were counted twice.

With the prototype flow-section TGL FCM, we have so far done four demonstrations: (1) signal train capture and analysis (Fig. 16 A) (Jin *et al.*, 2007b); (2) real-time TGL event counting (Fig. 16 B) (Jin *et al.*, 2009); (3) ultrarare event real-time

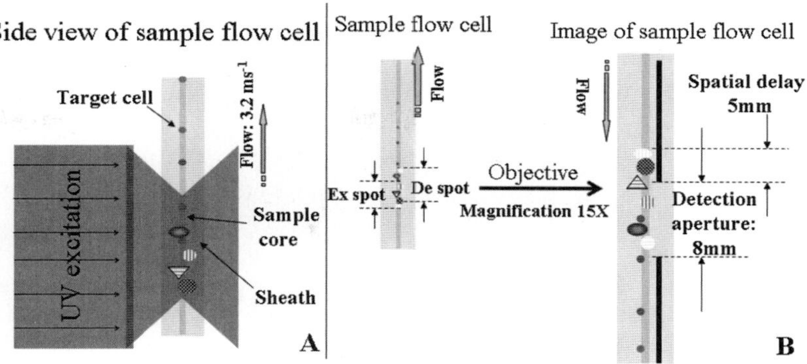

Fig. 14   Schematic geometric layout of TGL FCM. (A) UV LED excitation at the flow cell in the sample flow stream, the round solid (red) particles represent target europium particles/cells and the other larger round (pattern), oval, and triangle particles represent nontarget autofluorescent particles/cells. The smallest illumination spot size from focused UV LED light was 0.53 mm along the flow stream, more than 20 times larger than in conventional flow cytometers that use laser illumination, thus covering many cells instead of the single cell per illumination region. (B) Instead of continuous illumination and detection as in conventional flow cytometers, pulsed excitation and time-gated detection sequence in antiphase were applied to monitor flow sections. This allowed both the position and the size of the detection spot to be related to the flow rate and the TGL repetition rate, in order to discriminate long-lifetime luminescence labeled cells. The image at the flow cell was magnified ($15\times$) via the signal collection objective optics after projection onto the image plane, which is inverted by the objective. (Original figure was from (Jin *et al.*, 2009).)

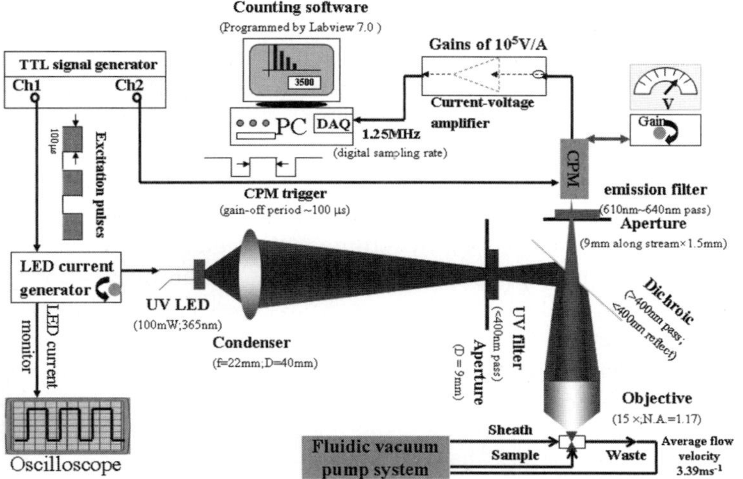

Fig. 15   Layout of the UV LED excited TGL flow cytometer consisting of five key subsystems: UV LED excitation optics, fluidics system including flow cell, signal collection optics, time-gated detection with control electronics, and signal processing unit. The TGL sequence (pulsed excitation and gated detection) was controlled by a dual channel TTL signal generator: channel 1 sent a TTL pulse train to control the periodically pulsed UV excitation and channel 2 was responsible for synchronizing the time-gated detection; the purpose of gating the detector was to prevent the intense LED emission and autofluorescence from reaching the sensitive photodetector during the excitation phase. (Original figure was from (Jin *et al.*, 2007b).)

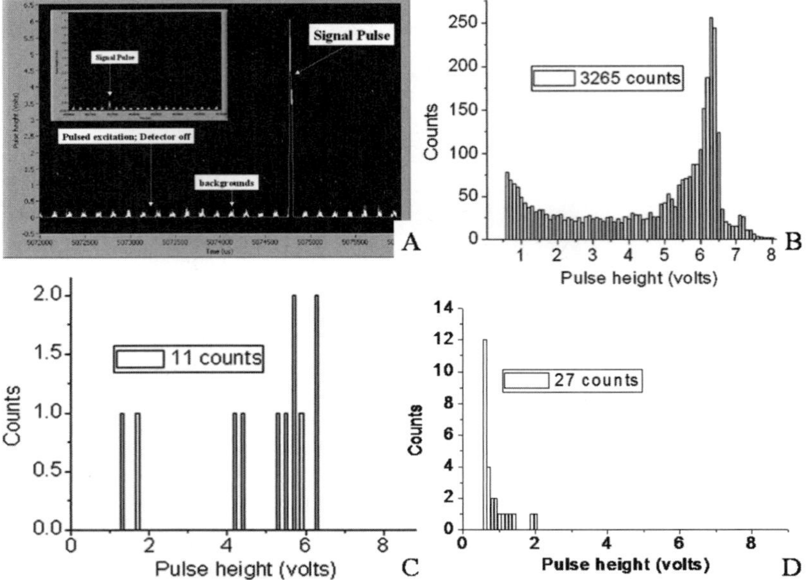

Fig. 16 TGL flow cytometer detection signal sequence observation when europium targets were analyzed in the presence of a strongly fluorescent S101 solution. The signal sequence consisted of three components: the period when the detector was switched off during pulsed excitation, the background pulses (uniform and low-height pulse profile) when no target TGL signal was presented, and the signal pulses when a target event with long-lived luminescence passed the detection spot. The inset shows that one europium target event with low pulse intensity (0.2–0.4 V) was clearly distinguishable from other background pulses (this observation train was resulted from TGL counting of a europium beads sample containing a large number of nontarget particles from complex water samples). (Original figure was from (Jin *et al.*, 2007b, 2009).)

counting (Fig. 16 C) (Jin *et al.*, 2009); (4) biological demonstration (Fig. 16 D) (Jin *et al.*, 2009). According to the signal train sequence (Fig. 16 A), the TGL FCM rendered invisible all the background autofluorescence from the particle/cells, and the captured signal sequence possessed only three components, dramatically reducing the analysis complexity and increasing the detection accuracy. Since the time-gated detection mode also rejected the majority of the scattering noise from excitation, the application of an incoherent low-cost LED source in the flow cytometer became practical. This is a major consideration in conventional FCM where scattering and autofluorescence are directly related to the size of the laser intercept area (Shapiro, 2003). Although the TGL intensity histogram (Fig. 16 B) displayed a large coefficient of variation comparing to the triggered model, the particle/cell arrival rate for the flow-section TGL FCM is extremely high, since this performance parameter is not limited by the overall particle/cell frequency, but by the targeted long-lifetime-labeled event number, thanks to the independence of number of nontarget particles/cells appearing invisible. Clearly, the flow-section detection model is well matched to rare event target detection in a matrix consisting of large amount of nontarget background particles. In Fig. 16 C, we successfully recovered as few as ten 5 μm diameter Europium containing calibration beads from over one million nontarget autofluorescent background particles. The specific detection rates (accuracy) have reached to $100 \pm 30\%$ and $91 \pm 3\%$ for 10 and 100 target beads recovery, respectively (Jin *et al.*, 2009). Shown in Fig. 16 D, limited by the current available power from UV LED source, the low excitation efficiency ($\sim$8%) resulted in a relatively low detection rate of 13% for recovering the targeted *Giardia* cysts in the flow-section TGL FCM. However, the potential for future improvement scope is very large (Jin *et al.*, 2009). For example, recently we have been optimizing our system with a 355 nm Q-switched nanosecond-pulsed (25 kHz) laser (Nd: YAG), and preliminary tests have found the detected TGL intensity of europium labeled *Giardia* was enhanced by approximately threefold (data not shown).

III. Bioprobes Development

A. Nanoparticle Bioprobes

The rapidly evolving field of nanoscience and nanotechnology has also opened up a promising era in developing rare-earth-element-containing luminescence nanoparticles as novel biomarkers. This has been contributing the rapid advancement of the lanthanide-based cytometry. Nanotechnology has so far demonstrated several advantages over the chelate-based lanthanide bioprobes, which include significantly amplified signal intensity, enhanced photostability, and extended color diversity to increase multiplexing detection capacity. We present here one representative core technique, namely encapsulation nanoparticles. They are typically smaller than 50 nm in diameter, highly reproducible, bright with unique optical properties, and can be functionalized to produce nanoparticle-based bioprobes.

B. Encapsulation Nanoparticles

The most popular nanoparticle bioprobes can be categorized into three families: the quantum dots (Gao *et al.*, 2005; Sharrna *et al.*, 2006; Wang *et al.*, 2006; Zhong, 2009), gold/silver nanoparticles, and the silica/polymer nanoparticles. Silica/polymer nanoparticles are a relatively easy concept to understand, since there are mature chemistry technologies capable of encapsulating up to thousands of molecular dyes inside the polymer or the silica network layer. Thus, the immediate benefits include remarkable signal amplification and enhanced photostability. Typically around 40 to 60 nm (Burns *et al.*, 2006; Mukai *et al.*, 2006; Ow *et al.*, 2005) these nanoparticles may show enhanced signal by up to several hundred fold when comparing to traditional organic dyes (Burns *et al.*, 2006; Ow *et al.*, 2005). The polymer or silica network forms a protective shell, limiting access to environmental quenching agents such as oxygen, certain solvents, and other species. This effect is beneficial to protect many lanthanide complexes (Yuan and Wang, 2006) with enhanced photo-stability, which is important for monitoring real-time biological processes that need a long-time continuous excitation, such as microscopy observation.

This encapsulation technology was first applied to lanthanide complexes by a method known as solvent swelling. Polystyrene microspheres become swollen when suspended in an organic solvent, allowing the dye molecules to infiltrate the microsphere. When these microspheres are later transferred to an aqueous solution they shrink and the dye molecules become entrapped (Leif et al., 2009; Wilson et al., 2006). Europium and Pt dyes can be loaded into the polystyrene nanoparticles to make long lifetime fluorescent probes (600 µs for europium and 40 µs for Pt). As shown earlier, we employed this technique in preparation of europium containing microspheres (5 µm diameter) as calibration beads to evaluate TGM instruments (Leif et al., 2009). In 2001 Lovgren and coworkers reported an immunoassay of trace prostate-specific antigen (PSA) using a carboxyl-modified polystyrene nanoparticle (107 nm containing more than 30,000 β-diketones chelate europium molecules). As a result, an extremely low detection limit of 0.38 ng L^{-1}, corresponding to 10 fmol L^{-1} or 60 zeptomoles (60 × 10^{-21} moles) of biotinylated PSA molecules was achieved (Harma et al., 2001). This was an over 70-fold improvement in detection limit compared to a conventional immunoassay. In 2001, Klimant, Wolfbeis and coworkers reported a simple nanoencapsulation technique based on coprecipitation of phosphorescent ruthenium(II) – tris(polypyridyl) complex and polyacrylonitrile (PAN) derivatives from a solution in N,N-dimethylformamide (DMF) (Kurner et al., 2001). Due to the distinctive feature of PAN being impermeable to oxygen and ionic chemical species, the doping dye ruthenium complex inside had a tremendously enhanced stability against oxygen (the oxygen cross-sensitivity $\Delta\tau$ decreased from 75 to 3–5%) increased the lifetime (from 1.2 µs in air, 4.7 µs in nitrogen and as high as 6 µs in nanoparticles) and improved the quantum yield from below 30% in nitrogen to 40% inside nanoparticle (Kurner et al., 2001). The optimized nanoparticles were of spherical shape, 30 ± 7 nm in size, and separated by a high zeta potential of $\xi = -54.0$ mV using carboxylic surface groups. This may provide opportunity to produce smaller lanthanide-containing nanoparticles than polystyrene as the host material. One of the drawbacks, however, is that polymer particles are hydrophobic, tend to agglomerate in an aqueous medium, and swell in organic solvents, resulting in dye leakage (Wang et al., 2006). And when the size is smaller than 100 nm, the polymer nanoparticles are not uniform and less reproducible, which may limit their use for cell labeling and cell imaging applications.

Silica nanoparticles possess several advantages. Thanks to its higher density (e.g., 1.96 g cm^{-3} for silica vs. 1.05 g cm^{-3} for polystyrene), silica nanoparticles are easy to separate via centrifugation during particle preparation, surface modification, and other solution treatment processes. Silica nanoparticles are more hydrophilic and biocompatible with amino groups during synthesis. Silica nanoparticles are not subject to microbial attack, and no swelling or porosity change occurs with changes in pH (7). Surface modification of microemulsion nanoparticles can be achieved via direct hydrolysis and cocondensation of tetraethylorthosilicate (TEOS) and other organosilanes in the microemulsion solution (Bagwe *et al.*, 2006). This results in the relative ease of functionalization of the particles' surface with amines, thiols, carboxyls, and methacrylate groups (Rossi *et al.*, 2005). With an optimum balance of the use of inert and active surface functional groups, the aggregation and nonspecific binding can be minimized (Bagwe *et al.*, 2006).

Two general synthetic routes have been reported for preparation of dye-doped silica nanoparticles: the Stöber synthesis method and water-in-oil microemulsion processes. The Stöber method is a one-step synthesis that involves the condensation of TEOS in ethanol:water mixtures under alkaline conditions at room temperature. This method avoids the use of potentially toxic organic solvents and surfactants (Rossi *et al.*, 2005). However, it is difficult to obtain particle sizes below 100 nm (Bagwe *et al.*, 2004). Alternatively, nanoparticles prepared using the water-in-oil microemulsion (W/O) are monodispersed, and it is relatively easy to control the size.

There have been several innovative techniques proposed to produce microsecond lifetime silica nanoparticles (Jiang *et al.*, 2010; Song *et al.*, 2009; Tan *et al.*, 2004a, 2004b; Wu *et al.*, 2008, 2009; Ye *et al.*, 2004a, 2004b; Yuan and Wang, 2005, 2006). Professor Yuan's group at Dalian University of Technology, P. R. China, has been leading this area of research since their earlier reports in 2004 (Tan *et al.*, 2004a, 2004b; Ye *et al.*, 2004a, 2004b; Yuan and Wang, 2006). Fig. 17 shows ultrabright long-lifetime luminescence nanobioprobes containing europium complex 4,4′-bis (1″,1″,1″,2″,2″,3″,3″-heptafluoro-4″,6″-hexanedion-6″-yl)chlorosulfo-o-terphenyl (BHHCT) and terbium complex $N,N,N1,N1$-[2,6-bis(3′-aminomethyl-1′-pyrazolyl)-phenylpyridine]tetrakis(acetate)(BPTA), respectively.

One such synthesis technique is the microemulsion synthesis method. A reverse microemulsion is an isotropic and thermodynamically stable single-phase system consisting of water, oil, and surfactant (Yao *et al.*, 2006). As shown in Fig. 18, the oil phase is cyclohexane and n-octanol, and the surfactant is Triton X-100. By

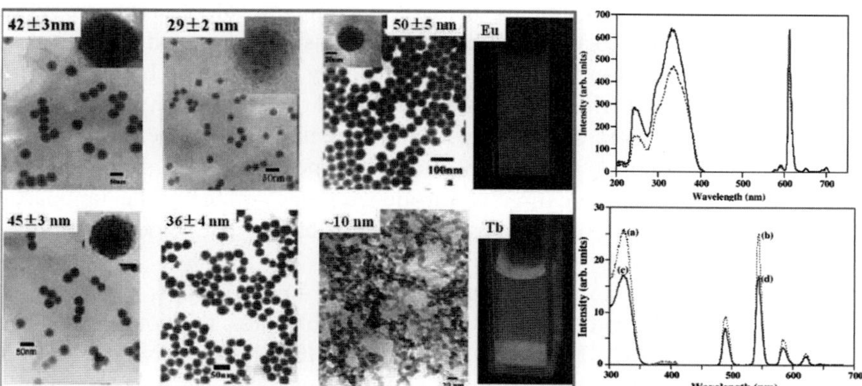

Fig. 17 Silica nanoparticles containing europium complex BHHCT and terbium complex BPTA. Absorption and emission spectra are shown on the right. (Figure courtesy of Professor J. Yuan at Dalian University of Technology, P. R. China.)

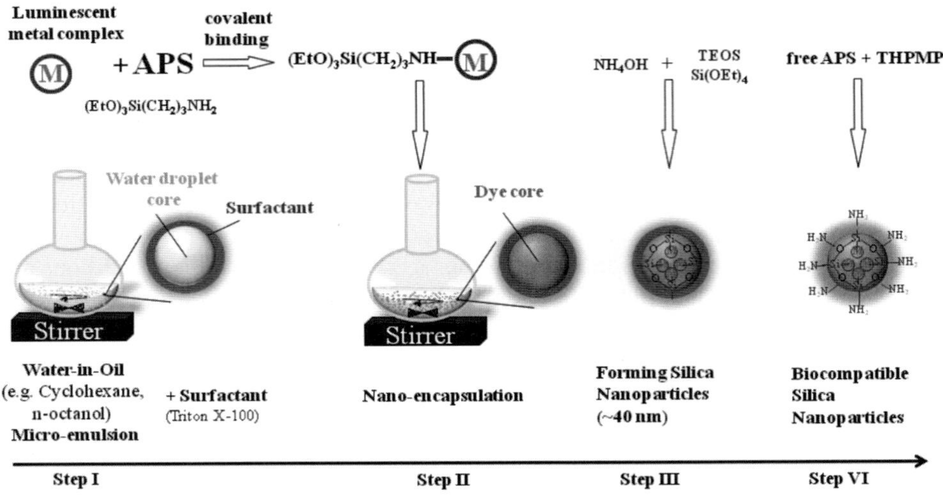

Fig. 18 Schematic of the covalent binding nanoencapsulation method to synthesize silica-based highly luminescent nanoparticles containing a large number of lanthanide metal complexes.

appropriate stirring, nanosized water droplets surrounded by surfactant are generated and dispersed in the continuous bulk oil phase. The nanosized water core serves as a confined media for the formation of discrete nanoparticles. The size of the final silica nanoparticles is typically determined by the size of the water droplet core, which is controlled by the water-to-surfactant molar ratio. The particle size was found to decrease with an increase in either ammonium hydroxide concentration or the water-to-surfactant molar ratio or the cosurfactant-to-surfactant molar ratio. The uniform microemulsion solution forms after 1 h of vigorous stirring. Instead of the direct dye-doping method, a conjugate of (3-aminopropyl)triethoxysilane (APS) bound to a lanthanide complex was first prepared as a precursor, so that the lanthanide complex dye could be firmly (covalently) bound to silicon atoms. This happened when forming silica nanoparticles through ammonia-catalyzed hydrolysis of TEOS in those water-in-oil microemulsions. Since the nanoparticles were prepared by copolymerization of an APS-metal complex, free APS, and TEOS, free amino groups were directly introduced to the surface of the nanoparticles, and these amino groups made the surface modification and bioconjugation of the nanoparticles easier.

Due to the presence of surface amino groups, the nanoparticles can be directly conjugated to streptavidin (SA) molecules. Previous work (Tan *et al.*, 2004a, 2004b; Ye *et al.*, 2004a, 2004b) has shown that the bioactivity of the nanoparticle-labeled SA prepared by the direct conjugation method was rather low. To improve the low SA bioactivity on the nanoparticles, Yuan and coworkers established a BSA (bovine serum albumin) coating the method to prepare the nanoparticle-labeled SA. The BSA-coated nanoparticles were then used for binding to SA through the amino groups with glutaraldehyde. For details, refer to the earlier publications. (Wu *et al.*, 2008, 2009). One of the advantages of the BSA-bridge method is that protein recognition sites are oriented away from the nanoparticle surface to ensure that they do not lose their ability to bind to a target.

Since 2007, our group (the Advanced Cytometry Laboratories at Macquarie University) has been collaborating with Professor Yuan's group and successfully developed a range of novel bioprobes based on lanthanide-containing silica nanoparticles. Our results include radical extension of excitation wavelength from UV to visible (excitation peak from 330 to 406 nm) (Wu *et al.*, 2008, 2009) and demonstration for multicolor lanthanide nanobioprobes (Jiang *et al.*, 2010a). Toward the advancement for lanthanide cytometry, we also succeeded in demonstrating such nanobioprobes for background-free TGL bioimaging of environmental pathogens.

C. Visible–Light–Sensitized Highly Luminescent Europium Nanoparticles

We have also reported two approaches to synthesize visible-light-excited europium containing silica nanoparticles, and have demonstrated their use in bioassay and bioimaging applications (Wu *et al.*, 2008, 2009). The fundamental principle in producing these nanoparticles is to provide shielding for vulnerable europium complexes APS-CDHH-Eu^{3+}-DPBT (Wu *et al.*, 2008) and APS-BHHCT-Eu^{3+}-DPBT (Wu *et al.*, 2009), which are only visible-light-excitable in organic solutions and unstable in aqueous media as shown in Fig. 19.

Fig. 20 shows the bright-field, luminescence, and TGL images (excited with 330–380 and 380–420 nm, respectively) of APS-BHHCT-Eu^{3+}-DPBT nanoparticles labeled *Giardia lamblia* cysts in a water concentrate. TGL imaging technique using as-prepared nanoparticles as a bioprobe can eliminate the interference of short-lived background fluorescence, providing a specific and high-contrast detection of

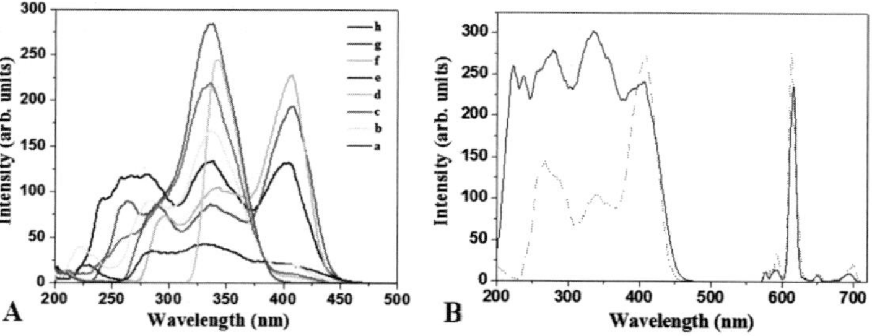

Fig. 19 (A) TGL excitation spectra of the APS-BHHCT-Eu3+-DPBT precursor in different solvents ($\lambda_{em} = 614$ nm). (a) DMF, (b) THF, (c) ethanol, (d) acetone, (e) heptane, (f) toluene, (g) chloroform, and (h) dichloromethane. (B) UV VIS absorption spectra of the APS-BHHCT-Eu3 +-DPBT (dotted line) in chloroform and the APS-BHHCT-Eu3+-DPBT nanoparticles (solid line) in 0.05 M Tris–HCl buffer of pH 7.8. (Reproduced by permission of The Royal Society of Chemistry (RSC) (Wu *et al.*, 2009).)

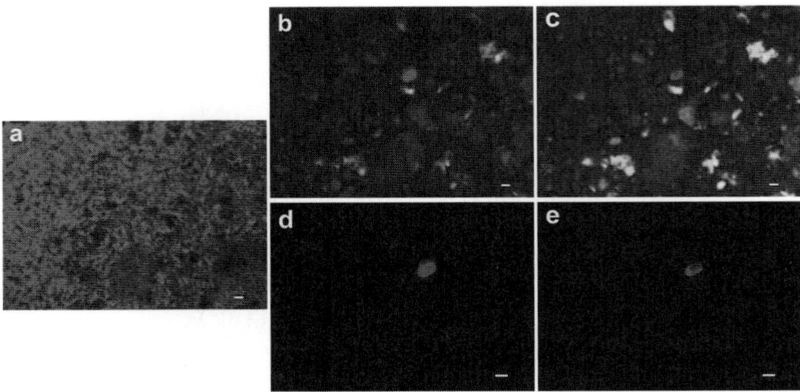

Fig. 20   Bright-field (a), luminescence (b, c, excited with 330–380 and 380–420 nm, respectively) and TGL (d, e, excited with 330–380 and 380–420 nm, respectively) images of *Giardia lamblia* stained by the nanoparticle-labeled SA in the concentrated environmental water sample. Scale bars, 10 μm. The TGL images are shown in pseudo color (wavelength of 615 nm) treated by a SimplePCI software. (Reproduced by permission of The Royal Society of Chemistry RSC (Wu *et al.*, 2009).) (For interpretation of the references to color in this figure legend, the reader is referred to the Web version of this chapter.)

environmental microorganisms. The strong luminescence signals shown in the images with 380–420 nm excitation, which were comparable to 330–380 nm excitation, further demonstrated the efficacy of the nanoparticles for TGL biosensing with a visible light excitation.

This work is important for both developing lanthanide cytometry instruments and their many practical applications, since it overcame many restrictions of ultraviolet excitation (typically <360 nm) requirements. To excite a lanthanide cytometer, the semiconductor diodes are superior to the traditional gas lasers or flash lamps in terms of size, cost, and repetition rates, but are only available in the longer wavelength range. Laser diodes only available at wavelengths no lower than 375 nm, and are not yet ideal to excite the most europium and terbium complex bioprobes at typical excitation peak of 330 nm. Violet laser diodes are now relatively inexpensive, and are common fixtures on flow cytometers. Substitution of visible light for UV sources is also advantageous since a number of clinical and biological systems are sensitive to UV radiations, which is particularly true in fluorescence bioimaging of living cells.

D. Multicolor Luminescent Lanthanide Nanoparticles

Multicolor fluorescence biolabeling requires introduction of two or more fluorescent labels to simultaneously detect multiple biomolecules. Multiplexing detection reduces the time and cost per assay, decreases the sample volumes required, and most importantly, makes the measurement more informative through co-association of multiple labels (Liu *et al.*, 2007; Wang *et al.*, 2005). Although Sm^{3+}, Eu^{3+}, Tb^{3+}, and

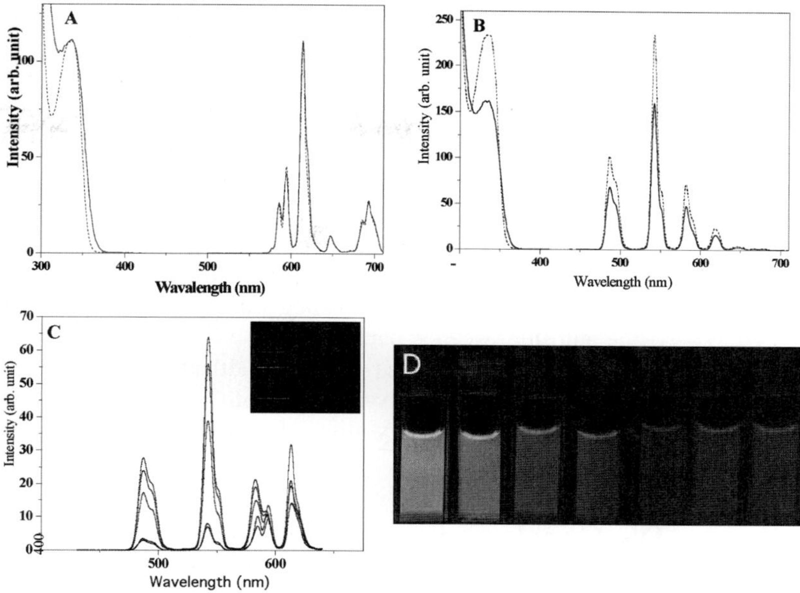

Fig. 21 Time-resolved excitation and emission spectra of (A) the PTTA-Eu^{3+}-nanoparticles (solid lines) and free PTTA-Eu^{3+} complex (dashed lines); (B) the PTTA-Tb^{3+}-nanoparticles (solid lines) and free PTTA-Tb^{3+} complex (dashed lines) in 0.05 M Tris-HCl buffer of pH 7.8; (C) mission spectra of the nanoparticles prepared with different molar ratios of PTTA-Eu^{3+}/PTTA-Tb^{3+} complexes; (D) emission color images of the nanoparticles prepared with different molar ratios of PTTA-Eu^{3+}/PTTA-Tb^{3+} complexes in aqueous solution under irradiation of a 365 nm UV lamp (from left to right, Eu^{3+}:Tb^{3+} = 0:1, 1:5, 1:3, 1:1, 3:1, 5:1, 1:0). (Original figure was from (Jiang *et al.*, 2010a).)

Dy^{3+} complexes have been used as labels for multilabel time-resolved fluoroimmunoassay (Xu *et al.*, 1992), Sm^{3+} and Dy^{3+} complexes are weakly fluorescent with shorter fluorescence lifetime. Only Eu^{3+} and Tb^{3+} complexes with emission maximum wavelengths at ∼615 nm (Eu^{3+} complexes, red emission) and ∼545 nm (Tb^{3+} complexes, green emission) are suitable as the labels. This has limited the use of anthanide cytometry for multiplexing applications.

We succeeded in synthesizing a series of silica nanoparticle biolabels emitting green, yellow, orange, and red colors with an excitation peak at 335 nm (Jiang *et al.*, 2010). The new bioprobes have been prepared by cobinding different molar ratios of fluorescent Eu^{3+} and Tb^{3+} complexes with a ligand N,N,N^1,N^1-(4′-phenyl-2,2′:6′,2″-terpyridine-6,6″-diyl) bis(methylenenitrilo) tetrakis (acetic acid) (PTTA) inside the silica nanoparticles (Jiang *et al.*, 2010), since the PTTA complex can form highly stable and strongly fluorescent complexes with both Eu^{3+} (Fig. 21 A) and Tb^{3+} (Fig. 21 B) ions. As described in the original publication (Jiang *et al.*, 2010), seven kinds of the nanoparticles were prepared with PTTA-Eu^{3+}/PTTA-Tb^{3+} molar ratios of 0:1, 1:5, 1:3, 1:1, 3:1, 5:1, and 1:0, respectively (Fig. 21 C and D). All nanoparticles are monodispersed, spherical, and uniform in size, 56 ± 4 nm in diameter, without the effect of PTTA-Eu^{3+}/PTTA-Tb^{3+} molar ratio (Jiang *et al.*, 2010).

Our ongoing investigation of nanoparticle synthesis techniques has been aimed at developing "ideal" bioprobes for advanced cytometric detection. The uniqueness of these probes could benefit from the following features.

- Amplified strong fluorescence intensity can lower the detection limit. The nanoparticle dye enrichment of lanthanide complex is absent of concentration quenching (Selvin, 2002), while essentially all organic fluorescent dyes at high local concentrations are self-quenching. (Makarova *et al.*, 1999; Wu *et al.*, 2006)
- Microsecond exceptionally-long-lifetime allows for TGL detection to suppress the autofluorescence background.
- Sharp-emission profile ($\sim$ 10 nm width at half-maximum) by lanthanide complex provides multiplexing variety by both spectrum and lifetime combinations.
- Excellent photostability allows for prolonged imaging and observation.

IV. Conclusion

In recent decades, we have witnessed many important advances in fluorescence techniques for modern biochemical and pharmaceutical research and disease diagnosis. The most representative techniques include both instruments (such as FCM, fluorescence activated cytometry sorting, confocal microscopy, and multiphoton microscopy) and fluorescent probes (such as quantum dots, noble metal nanoparticles, and natural fluorescence proteins). They are sensitive enough to explore even subcellular phenomena. However, there are many constraints associated with these typical detection schemes, which have limited the development of advanced cytometry to address particular detection challenges, such as rare event pathogen detection in complex samples. Spectral overlap and high levels of autofluorescence background can reduce target visibility and affecting detection accuracy when using traditional instruments and fluorochromes.

The recent development of rare-earth elements (lanthanides) as fluorescent probes for both imaging and FCM has been proposed as a solution to several problems not adequately addressed by traditional technologies. Time-gated detection of long lifetime decaying lanthanide probes can effectively suppress the autofluorescence background and greatly increase the detection throughput since nontarget events are rendered invisible to the detector. The lifetime tuning techniques, for example, through fluorescence resonance energy transfer and local plasmonics, can provide another dimension to top-up spectral labeling identities (codes) and increase multiplexing capacities. The time gating in microsecond to millisecond region is mechanically easy to detect, which makes the lanthanide cytometry compatible with many existing automated bioinstrumentation. Since the excitation and emission are in antiphase, low-beam-quality but low-cost and compact diode sources are adequate to achieve desired sensitivity, which dramatically benefits the overall cost and size of the lanthanide cytometers. Furthermore, in the era of nanoscience and nanotechnology, fluorescence nanoparticles are starting to replace traditional fluorescent labels. Lanthanide cytometry should therefore be able to take full advantage of this technology. Lanthanide-containing nanoparticles are also improving the characteristics of chelate-based lanthanide probes in terms of brightness, photostability, and spectral and lifetime tuning opportunities. We anticipate that progress in both time-gated instrumentation and rare-earth fluorescent probes will have widespread impact on the advancement of analytical cytology.

Acknowledgments

Dr. D. JIN wishes to acknowledge Mrs. DENG, Wei, Dr. WU, Jing, Dr. YE, Zhiqiang, Prof. WANG, Guilan, Prof. YUAN, Jingli, Mrs. SONG, Cuihong, Mrs. JIANG, Hongfei, Prof. GUAN, Yafeng, Mr. LU, Yiqing, Mr. ZHAO, Jiangbo, Dr. Russell Connally, Mr. Sean Yang, Prof. J. Paul Robinson, Dr. Robert R. Leif, Prof. Ewa M. Goldys, and Prof. James A. Piper for the long-term collaboration and contributions to Advanced Cytometry Laboratories @ Macquarie, and funding schemes: Australian Research Council FABLS ("Fluorescence Applications in Biotechnology and Life Sciences") network, the Australian Research Council Nanoscience and Nanotechnology network, Australian Research Council Discovery Project DP 1095465, Macquarie University Research Fellowship Scheme, the ISAC (International Society for Analytical Cytology) scholar program, and the National Natural Science Foundation of China (no. 20575069).

References

Aurell, H., Catala, P., Farge, P., Wallet, F., Le Brun, M., Helbig, J. H., Jarraud, S., Lebaron, P. (2004). Rapid detection and enumeration of *Legionella pneumophila* in hot water systems by solid-phase cytometry. *Appl. Environ. Microbiol.* **70**, 1651–1657.

Bagwe, R. P., Hilliard, L. R., and Tan, W. H. (2006). Surface modification of silica nanoparticles to reduce aggregation and nonspecific binding. *Langmuir* **22**, 4357–4362.

Bagwe, R. P., Yang, C. Y., Hilliard, L. R., and Tan, W. H. (2004). Optimization of dye-doped silica nanoparticles prepared using a reverse microemulsion method. *Langmuir* **20**, 8336–8342.

Bajaj, S., Welsh, J. B., Leif, R. C., and Price, J. H. (2000). Ultra-rare-event detection performance of a custom scanning cytometer on a model preparation of fetal nRBCs. *Cytometry* **39**, 285–294.

Beeby, A., Botchway, S. W., Clarkson, I. M., Faulkner, S., Parker, A. W., Parker, D., Williams, J. A. G. (2000). Luminescence imaging microscopy and lifetime mapping using kinetically stable lanthanide (III) complexes. *J. Photochem. Photobiol. B-Biol.* **57**, 83–89.

Bianchi, D. W., Zickwolf, G. K., Weil, G. J., Sylvester, S., and DeMaria, M. A. (1996). Male fetal progenitor cells persist in maternal blood for as long as 27 years postpartum. *Proc. Natl. Acad. Sci. U.S.A.* **93**, 705–708.

Bjartell, A., Laine, S., Pettersson, K., Nilsson, E., Lovgren, T., Lilja, H. (1999a). Time-resolved fluorescence in immunocytochemical detection of prostate-specific antigen in prostatic tissue sections. *Histochemical J.* **31**, 45–52.

Bjartell, A., Siivola, P., Hulkko, S., Pettersson, K., Rundt, K., Lilja, H., Lovgren, T. (1999b). Time-resolved fluorescence imaging (TRFI) far direct immunofluorescence of PSA and alpha-1-antichymotrypsin in prostatic tissue sections. *Prostate Cancer Prostatic Dis.* **2**, 140–147.

Bornhop, D. J., Hubbard, D. S., Houlne, M. P., Adair, C., Kiefer, G. E., Pence, B. C., Morgan, D. L. (1999). Fluorescent tissue site-selective lanthanide chelate, Tb-PCTMB for enhanced imaging of cancer. *Analytical Chem.* **71**, 2607–2615.

Botchway, S. W., Charnley, M., Haycock, J. W., Parker, A. W., Rochester, D. L., Weinstein, J. A., Williams, J. A. G. (2008). Time-resolved and two-photon emission imaging microscopy of live cells with inert platinum complexes. *Proc. Natl. Acad. Sci. U.S.A.* **105**, 16071–16076.

Bunzli, J. C. G., and Piguet, C. (2005). Taking advantage of luminescent lanthanide ions. *Chem. Soc. Rev.* **34**, 1048–1077.

Burns, A., Ow, H., and Wiesner, U. (2006). Fluorescent core-shell silica nanoparticles: towards "Lab on a Particle" architectures for nanobiotechnology. *Chem. Soc. Rev.* **35**, 1028–1042.

Campbell, R., Tour, O., Palmer, A., Steinbach, P., Baird, G., Zacharias, D., Tsien, R. (2002). A monomeric red fluorescent protein. *Proc. Natl. Acad. Sci. U.S.A.* **99**, 7877.

Chan, W. C. W., and Nie, S. M. (1998). Quantum dot bioconjugates for ultrasensitive nonisotopic detection. *Science* **281**, 2016–2018.

Condrau, M. A., Schwendener, R. A., Niederer, P., and Anliker, M. (1994a). Time-resolved flow-cytometry for the measurement of lanthanide chelate fluorescence. 1. Concept and theoretical evaluation. *Cytometry* **16**, 187–194.

Condrau, M. A., Schwendener, R. A., Zimmermann, M., Muser, M. H., Graf, U., Niederer, P., Anliker, M. (1994b). Time-resolved flow-cytometry for the measurement of lanthanide chelate fluorescence. 2. Instrument design and experimental results. *Cytometry* **16**, 195–205.

Connally, R., Jin, D., and Piper, J. (2006). High intensity solid-state UV source for time-gated luminescence microscopy. *Cytometry Part A* **69A**, 1020–1027.

Connally, R., D. Jin, and J. A. Piper (2005). BHHST: An improved lanthanide chelate for time-resolved fluorescence applications. *In* "Plasmonics in Biology and Medicine II." (T. L. Vo-Dinh, R. Joseph, Gryczynski, and Zygmunt K., eds.), Vol. 5704, pp. 93–104.

Connally, R., and Piper, J. (2008a). Solid-state time-gated luminescence microscope with ultraviolet light-emitting diode excitation and electron-multiplying charge-coupled device detection. *J. Biomed. Opt.* **13** (3), 034022.

Connally, R., Veal, D., and Piper, J. (2002). High resolution detection of fluorescently labeled micro-organisms in environmental samples using time-resolved fluorescence microscopy. *FEMS Microbiol. Ecol.* **41**, 239–245.

Connally, R., Veal, D., and Piper, J. (2004a). Flashlamp excited time-resolved fluorescence microscope suppresses autofluorescence in water concentrates to deliver 11-fold increase in signal-to-noise ratio. *J. Biomed. Opt.* **9**, 725–734.

Connally, R., Veal, D., and Piper, J. (2004b). Time-resolved fluorescence microscopy using an improved europium chelate BHHST for the *in situ* detection of *Cryptosporidium* and *Giardia*. *Microsc. Res. Tech.* **64**, 312–322.

Connally, R. E., and Piper, J. A. (2008b). Time-gated luminescence microscopy. *In* Fluorescence Methods and Applications: Spectroscopy, Imaging, and Probes. *Ann. N. Y. Acad. Sci.* **1130**, 106–116.

Connally, R. E., and Piper, J. A. (2008c). Time-gated luminescence microscopy. *In* Fluorescence Methods and Applications: Spectroscopy, Imaging, and Probes. *Ann. N. Y. Acad. Sci.* **1130**, 106–116.

Dawson, D. (2005). Foodborne protozoan parasites. *I. J. Food Microbiol.* **103**, 207–227.

deHaas, R. R., Verwoerd, N. P., vanderCorput, M. P., vanGijlswijk, R. P., Siitari, H., Tanke, H. J. (1996). The use of peroxidase-mediated deposition of biotin-tyramide in combination with time-resolved fluorescence imaging of europium chelate label in immunohistochemistry and in situ hybridization. *J. Histochem. Cytochem.* **44**, 1091–1099.

Deisseroth, A. B., Zu, Z. F., Claxton, D., Hanania, E. G., Fu, S. Q., Ellerson, D., Goldberg, L., Thomas, M., Janicek, K., Anderson, W. F., Hester, J., Korbling, M., Durett, A., Moen, R., Berenson, R., Heimfeld, S., Hamer, J., Calvert, L., Tibbits, P., Talpaz, M., Kantarjian, H., Champlin, R., Reading, C. (1994). Genetic marking shows that Ph(+) cells present in autologous transplants of chronic myelogenous leukemia (Cml) contribute to relapse after autologous bone-marrow in Cml. *Blood* **83**, 3068–3076.

Dillingham, R. A., Lima, A. A., and Guerrant, R. L. (2002). Cryptosporidiosis: epidemiology and impact. *Microbes Infection* **4**, 1059–1066.

Faulkner, S., Pope, S. J. A., and Burton-Pye, B. P. (2005). Lanthanide complexes for luminescence imaging applications. *Appl. Spectrosc. Rev.* **40**, 1–31.

Ferrari, B. C., Stoner, K., and Bergquist, P. L. (2006). Applying fluorescence based technology to the recovery and isolation of *Cryptosporidium* and *Giardia* from industrial wastewater streams. *Water Res.* **40**, 541–548.

Gao, X. H., Yang, L. L., Petros, J. A., Marshal, F. F., Simons, J. W., Nie, S. M. (2005). *In vivo* molecular and cellular imaging with quantum dots. *Curr. Opin. Biotechnol.* **16**, 63–72.

George, T., Basiji, D., Hall, B., Lynch, D., Ortyn, W., Perry, D., Seo, M., Zimmerman, C., Morrissey, P. (2004). Distinguishing modes of cell death using the ImageStream multispectral imaging flow cytometer. *Cytometry Part A* **59**, 237–245.

Gross, H. J., Verwer, B., Houck, D., Hoffman, R. A., and Recktenwald, D. (1995). Model study detecting breast-cancer cells in peripheral-blood mononuclear-cells at frequencies as low as 10(-7). *Proc. Natl. Acad. Sci. U.S.A.* **92**, 537–541.

Beverloo, H. B., van Schadewijk, A., Bonnet, J., van der Geest, R., Runia, R., Verwoerd, N. P., Vrolijk, J., Ploem, J. S., Tanke, H. J. (1992). Preparation and microscopic visualization of multicolor luminescent immunophosphors. *Cytometry* **13**, 561–570.

Beverloo, H. B., van Schadewijk, A., van Gelderen-Boele, S., and Tanke, H. J. (1990). Inorganic phosphors as new luminescent labels for immunocytochemistry and time-resolved microscopy. *Cytometry* **11**, 784–792.

Hanaoka, K., Kikuchi, K., Kobayashi, S., and Nagano, T. (2007). Time-resolved long-lived luminescence imaging method employing luminescent lanthanide probes with a new microscopy system. *J. Am. Chem. Soc.* **129**, 13502–13509.

Harma, H., Soukka, T., and Lovgren, T. (2001). Europium nanoparticles and time-resolved fluorescence for ultrasensitive detection of prostate-specific antigen. *Clin. Chem.* **47**, 561–568.

Hashino, K., Ikawa, K., Ito, M., Hosoya, C., Nishioka, T., Makiuchi, M., Matsumoto, K. (2007). Application of a fluorescent lanthanide chelate label on a solid support device for detecting DNA variation with ligation-based assay. *Anal. Biochem.* **364**, 89–91.

Hemmila, I., and Laitala, V. (2005). Progress in lanthanides as luminescent probes. *J. Fluoresc.* **15**, 529–542.

Hemmila, I., and Mukkala, V. M. (2001). Time-resolution in fluorometry technologies, labels, and applications in bioanalytical assays. *Crit. Rev. Clin. Lab. Sci.* **38**, 441–519.

Hennink, E. J., deHaas, R., Verwoerd, N. P., and Tanke, H. J. (1996). Evaluation of a time-resolved fluorescence microscope using a phosphorescent Pt-porphine model system. *Cytometry* **24**, 312–320.

Heyduk, T., and Heyduk, E. (2001). Luminescence energy transfer with lanthanide chelates: interpretation of sensitized acceptor decay amplitudes. *Anal. Biochem.* **289**, 60–67.

Jiang, C. Q., and Luo, L. (2004). Spectrofluorimetric determination of human serum albumin using a doxycycline-europium probe. *Analytica Chimica Acta* **506**, 171–175.

Jiang, H. F., Wang, G. L., Zhang, W. Z., Liu, X. Y., Ye, Z. Q., Jin, D. Y., Yuan, J. L., Liu, Z. G. (2010a). Preparation and time-resolved luminescence bioassay application of multicolor luminescent lanthanide aanoparticles. *J. Fluorescence* **20**, 321–328.

Jiang, L. N., Wu, J., Wang, G. L., Ye, Z. Q., Zhang, W. Z., Jin, D. Y., Yuan, J. L., Piper, J. (2010b). Development of a visible-light-sensitized europium complex for time-resolved fluorometric application. *Anal. Chem.* **82**, 2529–2535.

Jin, D., Connally, R., and Piper, J. (2006). Long-lived visible luminescence of UV LEDs and impact on LED excited time-resolved fluorescence applications. *J. Phys. D – Appl. Phys.* **39**, 461–465.

Jin, D., Connally, R., and Piper, J. (2007a). Practical time-gated luminescence flow cytometry: I. Concepts. *Cytometry Part A* **71A**, 783–796.

Jin, D., Connally, R., and Piper, J. (2007b). Practical time-gated luminescence flow cytometry: II. Experimental evaluation using UV LED excitation. *Cytometry Part A* **71A**, 797–808.

Jin, D., and Piper, J. (2011). Time-gated luminescence microscopy allowing direct visual inspection of lanthanide-stained microorganisms in background-free condition. *Analytical Chemistry*, DOI: 10.1021/ac103207r.

Jin, D., Piper, J., Yuan, J., and Leif, R. (2010). Time-gated real-time bioimaging system using multicolor microsecond-lifetime silica nanoparticles. *In* "Imaging, Manipulation, and Analysis of Biomolecules, Cells, and Tissues VIII," (D. L. Farkas, D. V. Nicolau, and R. C. Leif, eds.), 7568, Proc. Int. Soc. Opt. Eng. (SPIE), San Francisco, CA, USA 756819.

Jin, D. Y., Piper, J. A., Leif, R. C., Yang, S., Ferrari, B. C., Yuan, J. L., Wang, G. L., Vallarino, L. M., Williams, J. W. (2009). Time-gated flow cytometry: an ultra-high selectivity method to recover ultra-rare-event mu-targets in high-background biosamples. *J. Biomed. Opt.* **14**(02), 024023.

Johnson, K. L., Stroh, H., Khosrotehrani, K., and Bianchi, D. W. (2007). Spot counting to locate fetal cells in maternal blood and tissue: A comparison of manual and automated microscopy. *Microsc. Res. Tech.* **70**, 585–588.

Kamentsky, L., and Kamentsky, L. (1991). Microscope-based multiparameter laser scanning cytometer yielding data comparable to flow cytometry data. *Cytometry Part A* **12**, 381–387.

Kimura, H., Mukaida, M., Watanabe, M., Hashino, K., Nishioka, T., Tomino, Y., Yoshida, K. I., Matsumoto, K. (2008). Quantitative evaluation of time-resolved fluorescence microscopy using a new europium label: Application to immunofluorescence imaging of nitrotyrosine in kidneys. *Anal. Biochem.* **372**, 119–121.

Kothary, M. H., and Babu, U. S. (2001). Infective dose of foodborne pathogens in volunteers: a review. *J. Food Safety* **21**, 49–73.

Kurner, J. M., Klimant, I., Krause, C., Preu, H., Kunz, W., Wolfbeis, O. S. (2001). Inert phosphorescent nanospheres as markers for optical assays. *Bioconjugate Chem.* **12**, 883–889.

Leif, R., Clay, S. P., Gratzner, H. G., Haines, H. G., Rao, K. V., Vallarino, L. M. (1976). Markers for instrumental evaluation of cells of the female reproductive tract: Existing and new markers. *In* "The Automation of Uterine Cancer Cytology, Chicago," (G. L. Wied, G. F. Bahr, and P. H. Bartels, eds.), pp. 313–344. Tutorials of Cytology, Chicago.

Leif, R., and Vallarino, L. (1991). Rare-earth chelates as fluorescent markers in cell separation and analysis. *In* "Cell Separation Science and Technology, ACS Symposium Series 464," (D. S. Kompala, and P. W. Todd, eds.), pp. 41–58. American Chemical Society, New York.

Leif, R. C., Yang, S., Jin, D. Y., Piper, J., Vallarino, L. M., Williams, J. W., Zucker, R. M. (2009). Calibration beads containing luminescent lanthanide ion complexes. *J. Biomed. Opt.* **14**(02), 024022.

Lekkas, C., Kakouri, A., Paleologos, E., Voutsinas, L. P., Kontominas, M. G., Samelis, J. (2006). Survival of *Escherichia coli* O157: H7 in Galotyri cheese stored at 4 and 12 degrees C. *Food Microbiol.* **23**, 268–276.

Lemarchand, K., Masson, L., and Brousseau, R. (2004). Molecular biology and DNA microarray technology for microbial quality monitoring of water. *Critical Rev. Microbiol.* **30**, 145–172.

Lemarchand, K., Parthuisot, N., Catala, P., and Lebaron, P. (2001). Comparative assessment of epifluorescence microscopy, flow cytometry and solid-phase cytometry used in the enumeration of specific bacteria in water. *Aquatic Microbial. Ecol.* **25**, 301–309.

Levi, Y. (2001). Ecologie microbienne des reseaux d'eau potable et risque microbiologique: l'exemple de Legionella pneumophila. *Revue Francaise des Laboratoires* **2001**, 33–37.

Liu, J. W., Lee, J. H., and Lu, Y. (2007). Quantum dot encoding of aptamer-linked nanostructures for one-pot simultaneous detection of multiple analytes. *Anal. Chem.* **79**, 4120–4125.

Makarova, O. V., Ostafin, A. E., Miyoshi, H., Norris, J. R., and Meisel, D. (1999). Adsorption and encapsulation of fluorescent probes in nanoparticles. *J. Phys. Chem. B.* **103**, 9080–9084.

Marriott, G., Clegg, R., Arndt-Jovin, D., and Jovin, T. (1991). Time resolved imaging microscopy. Phosphorescence and delayed fluorescence imaging. *Biophys. J.* **60**, 1374–1387.

Marriott, G., Heidecker, M., Diamandis, E., and Yan-Marriott, Y. (1994). Time-resolved delayed luminescence image microscopy using an europium ion chelate complex. *Biophys. J.* **67**, 957–965.

Mcclelland, R. G., and Pinder, A. C. (1994). Detection of low-levels of specific salmonella species by fluorescent-antibodies and flow-cytometry. *J. Appl. Bacteriol.* **77**, 440–447.

Mukai, T., Morita, D., Yamamoto, M., Akaishi, K., Matoba, K., Yasutomo, K., Kasai, Y., Sano, M., and Nagahama, S. (2006). Investigation of optical-output-power degradation in 365-nm UV-LEDs. *Physica Status Solidi C.* 2211–2214.

Nelson, P. T., Baldwin, D. A., Scearce, L. M., Oberholtzer, J. C., Tobias, J. W., Mourelatos, Z. (2004). Microarray-based, high-throughput gene expression profiling of microRNAs. *Nat. Methods* **1**, 155–161.

Ogilvy, H., and Piper, J. A. (2005). Compact, all solid-state, high-repetition-rate 336 nm source based on a frequency quadrupled, Q-switched, diode-pumped Nd: YVO4 laser. *Opt. Express* **13**, 9465–9471.

Ohl, M. E., and Miller, S. I. (2001). Salmonella: A model for bacterial pathogenesis. *Annual Rev. Medicine* **52**, 259–274.

Ow, H., Larson, D. R., Srivastava, M., Baird, B. A., Webb, W. W., Wiesner, U. (2005). Bright and stable core-shell fluorescent silica nanoparticles. *Nano Lett.* **5**, 113–117.

Pawley, J., and Masters, B. (2008). Handbook of biological confocal microscopy. *J. Biomed. Opt.* **13**, 029902.

Petoud, S., Cohen, S. M., Bunzli, J. C. G., and Raymond, K. N. (2003). Stable lanthanide luminescence agents highly emissive in aqueous solution: Multidentate 2-hydroxyisophthalamide complexes of SM3+, Eu3+, Tb3+, Dy3+. *J. Am. Chem. Soc.* **125**, 13324–13325.

Phimphivong, S., Kolchens, S., Edmiston, P. L., and Saavedra, S. S. (1995). Time-resolved, total internal-reflection fluorescence microscopy of cultured-cells using a Tb chelate label. *Analytica Chimica Acta* **307**, 403–417.

Phimphivong, S., and Saavedra, S. S. (1998). Terbium chelate membrane label for time-resolved, total internal reflection fluorescence microscopy of substrate-adherent cells. *Bioconjugate Chem.* **9**, 350–357.

Quintero-Betancourt, W., Peele, E. R., and Rose, J. B. (2002). Cryptosporidium parvum and *Cyclospora cayetanensis*: a review of laboratory methods for detection of these waterborne parasites. *J. Microbiol. Methods* **49**, 209–224.

Rose, J. B., and Slifko, T. R. (1999). *Giardia, cryptosporidium*, and *Cyclospora* and their impact on foods: a review. *In J. Food Protection* **62**, 1059–1070.

Rossi, L. M., Shi, L. F., Quina, F. H., and Rosenzweig, Z. (2005). Stober synthesis of monodispersed luminescent silica nanoparticles for bioanalytical assays. *Langmuir* **21**, 4277–4280.

Schaferling, M., Wu, M., Enderlein, J., Bauer, H., and Wolfbeis, O. S. (2003). Time-resolved luminescence imaging of hydrogen peroxide using sensor membranes in a microwell format. *Appl. Spectrosc.* **57**, 1386–1392.

Schunk, M., Jelinek, T., Wetzel, K., and Nothdurft, H. D. (2001). Detection of *Giardia lamblia* and *Entamoeba histolytica* in stool samples by two enzyme immunoassays. *Eur. J. Clin. Microbiol. & Infect. Dis.* **20**, 389–391.

Selvin, P. R. (2002). Principles and biophysical applications of lanthanide-based probes. *Ann. Rev. Biophys. Biomol. Struct.* **31**, 275–302.

Seveus, L., Vaisala, M., Syrjanen, S., Sandberg, M., Kuusisto, A., Harju, R., Salo, J., Hemmila, I., Kojola, H., Soini, E. (1992). Time-resolved fluorescence imaging of europium chelate label in immunohistochemistry and *in situ* hybridization. *Cytometry* **13**, 329–338.

Sharrna, P., Brown, S., Walter, G., Santra, S., and Moudgil, B. (2006). Nanoparticles for bioimaging. *Adv. Colloid Interf. Sci.* **123**, 471–485.

Sharma, S., Sachdeva, P., and Virdi, J. S. (2003). Emerging water-borne pathogens. *Appl. Microbiol. Biotechnol.* **61**, 424–428.

Shapiro, H. (2003). *Practical flow cytometry.* Wiley-Liss, New York.

Sharp, S. E., Suarez, C. A., Duran, Y., and Poppiti, R. J. (2001). Evaluation of the triage micro parasite panel for detection of *Giardia lamblia*, Entamoeba histolytica/Entamoeba dispar, and *Cryptosporidium parvum* in patient stool specimens. *J. Clin. Microbiol.* **39**, 332–334.

Soini, A. E., Kuusisto, A., Meltola, N. J., Soini, E., and Seveus, L. (2003). A new technique for multiparameter imaging microscopy: Use of long decay time photoluminescent labels enables multiple color immunocytochemistry with low channel-to-channel crosstalk. *Microsc. Res. Tech.* **62**, 396–407.

Song, C. H., Ye, Z. Q., Wang, G. L., Jin, D. Y., Yuan, J. L., Guan, Y. F., Piper, J. (2009). Preparation and time-gated luminescence bioimaging application of ruthenium complex covalently bound silica nanoparticles. *Talanta* **79**, 103–108.

Soukka, T., Antonen, K., Harma, H., Pelkkikangas, A. M., Huhtinen, P., Lovgren, T. (2003). Highly sensitive immunoassay of free prostate-specific antigen in serum using europium(III) nanoparticle label technology. *Clinica Chimica Acta* **328**, 45–58.

Sueda, S., Yuan, J. L., and Matsumoto, K. (2000). Homogeneous DNA hybridization assay by using europium luminescence energy transfer. *Bioconjugate Chem.* **11**, 827–831.

Tan, M. Q., Wang, G. L., Hai, X. D., Ye, Z. Q., and Yuan, J. L. (2004a). Development of functionalized fluorescent europium nanoparticles for biolabeling and time-resolved fluorometric applications. *J. Mater. Chem.* **14**, 2896–2901.

Tan, M. Q., Ye, Z. Q., Wang, G. L., and Yuan, J. L. (2004b). Preparation and time-resolved fluorometric application of luminescent europium nanoparticles. *Chem. Mater.* **16**, 2494–2498.

Templeton, E., Wong, H., Evangelista, R., Granger, T., and Pollak, A. (1991). Time-resolved fluorescence detection of enzyme-amplified lanthanide luminescence for nucleic acid hybridization assays. *Clin. Chem.* **37**, 1506–1512.

Thibon, A., and Pierre, V. C. (2009a). A highly selective luminescent sensor for the time-gated detection of potassium. *J. Am. Chem. Soc.* **131**, 434+.

Thibon, A., and Pierre, V. C. (2009b). Principles of responsive lanthanide-based luminescent probes for cellular imaging. *Analy. Bioanal. Chem.* **394**, 107–120.

Vaisanen, V., Harma, H., Lilja, H., and Bjartell, A. (2000). Time-resolved fluorescence imaging for quantitative histochemistry using lanthanide chelates in nanoparticles and conjugated to monoclonal antibodies. *Luminescence* **15**, 389–397.

Veal, D. A., Deere, D., Ferrari, B., Piper, J., and Attfield, P. V. (2000). Fluorescence staining and flow cytometry for monitoring microbial cells. *J. Immunol. Methods* **243**, 191–210.

Vereb, G., Jares-Erijman, E., Selvin, P. R., and Jovin, T. M. (1998). Temporally and spectrally resolved imaging microscopy of lanthanide chelates. *Biophys. J.* **74**, 2210–2222.

Verwoerd, N. P., Hennink, E. J., Bonnet, J., Vandergeest, C. R. G., and Tanke, H. J. (1994). Use of ferroelectric liquid-crystal shutters for time-resolved fluorescence microscopy. *Cytometry* **16**, 113–117.

Wang, L., Wang, K. M., Santra, S., Zhao, X. J., Hilliard, L. R., Smith, J. E., Wu, J. R., Tan, W. H. (2006). Watching silica nanoparticles glow in the biological world. *Anal. Chem.* **78**, 646–654.

Wang, L., Yang, C. Y., and Tan, W. H. (2005). Dual-luminophore-doped silica nanoparticles for multiplexed signaling. *Nano Lette.* **5**, 37–43.

Wilks, S. A., Michels, H., and Keevil, C. W. (2005). The survival of Escherichia coli O157 on a range of metal surfaces. *Int. J. Food Microbiol.* **105**, 445–454.

Wilson, R., Cossins, A. R., and Spiller, D. G. (2006). Encoded microcarriers for high-throughput multiplexed detection. *Angew. Chem. Int. Ed.* **45**, 6104–6117.

Wu, C. F., Szymanski, C., and McNeill, J. (2006). Preparation and encapsulation of highly fluorescent conjugated polymer nanoparticles. *Langmuir* **22**, 2956–2960.

Wu, F. -B., and Zhang, C. (2002). A new europium b-diketone chelate for ultrasensitive time-resolved fluorescence immunoassays. *Anal. Biochem.* **311**, 57–67.

Wu, F. B., Han, S. Q., Zhang, C., and He, Y. F. (2002). Synthesis of a highly fluorescent beta-diketone-europium chelate and its utility in time-resolved fluoroimmunoassay of serum total thyroxine. *Anal. Chem.* **74**, 5882–5889.

Wu, J., Wang, G. L., Jin, D. Y., Yuan, J. L., Guan, Y. F., Piper, J. (2008). Luminescent europium nanoparticles with a wide excitation range from UV to visible light for biolabeling and time-gated luminescence bioimaging. *Chem. Comm.* **21**(3), 365–367.

Wu, J., Ye, Z. Q., Wang, G. L., Jin, D. Y., Yuan, J. L., Guan, Y. F., Piper, J. (2009). Visible-light-sensitized highly luminescent europium nanoparticles: preparation and application for time-gated luminescence bioimaging. *J. Mater. Chem.* **19**, 1258–1264.

Xiao, M., and Selvin, P. R. (2001). Quantum yields of luminescent lanthanide chelates and far-red dyes measured by resonance energy transfer. *J. Am. Chem. Soc.* **123**, 7067–7073.

Xu, Y. Y., Pettersson, K., Blomberg, K., Hemmila, I., Mikola, H., Lovgren, T. (1992). Simultaneous quadruple-label fluorometric immunoassay of thyroid-stimulating hormone, 17-alpha-hydroxyprogesterone, immunoreactive trypsin, and creatine-kinase Mm isoenzyme in dried blood spots. *Clin. Chem.* **38**, 2038–2043.

Yao, G., Wang, L., Wu, Y. R., Smith, J., Xu, J. S., Zhao, W. J., Lee, E. J., Tan, W. H. (2006). FloDots: luminescent nanoparticles. *Anal. Bioanal. Chem.* **385**, 518–524.

Ye, Z. Q., Tan, M. Q., Wang, G. L., and Yuan, J. L. (2004a). Novel fluorescent europium chelate-doped silica nanoparticles: preparation, characterization and time-resolved fluorometric application. *J. Mater. Chem.* **14**, 851–856.

Ye, Z. Q., Tan, M. Q., Wang, G. L., and Yuan, J. L. (2004b). Preparation, characterization, and time-resolved fluorometric application of silica-coated terbium(ill) fluorescent nanoparticles. *Anal. Chem.* **76**, 513–518.

Yuan, J. L., and Matsumoto, K. (1996). Synthesis of a new tetradentate beta-diketonate-europium chelate that can be covalently bound to proteins in time-resolved fluorometry. *Anal. Sci.* **12**, 695–699.

Yuan, J. L., and Matsumoto, K. (1997). Synthesis of a new tetradentate beta-diketonate-europium chelate and its application for time-resolved fluorimetry of albumin. *J. Pharm. Biomed. Anal. s* **15**, 1397–1403.

Yuan, J. L., and Wang, G. L. (2005). Lanthanide complex-based fluorescence label for time-resolved fluorescence bioassay. *J. Fluoresc.* **15**, 559–568.

Yuan, J. L., and Wang, G. L. (2006). Lanthanide-based luminescence probes and time-resolved luminescence bioassays. *Trac-Trends in Analytical Chemistry* **25**, 490–500.

Yuan, J. L., Wang, G. L., Majima, K., and Matsumoto, K. (2001). Synthesis of a terbium fluorescent chelate and its application to time-resolved fluoroimmunoassay. *Anal. Chem.* **73**, 1869–1876.

Zhao, X. J., Hilliard, L. R., Mechery, S. J., Wang, Y. P., Bagwe, R. P., Jin, S. G., Tan, W. H. (2004). A rapid bioassay for single bacterial cell quantitation using bioconjugated nanoparticles. *Proc. Natl. Acad. Sci. U.S.Am.* **101**, 15027–15032.

Zhong, W. W. (2009). Nanomaterials in fluorescence-based biosensing. *Anal. Bioanal. Chem.* **394**, 47–59.

CHAPTER 20

Surface-Enhanced Raman Scattering (SERS) Cytometry

John P. Nolan[*] **and David S. Sebba**[†]

[*]La Jolla Bioengineering Institute, La Jolla, California, USA

[†]NanoComposix, Inc., San Diego, California, USA

Abstract

Significant advances have been made in the preparation and applications of surface-enhanced Raman scattering (SERS)-active materials for biomolecular analysis. Bright signals, photostability, and narrow spectral features of SERS-active materials offer attractive advantages for cytometric analyses. However, SERS cytometry is still in an early stage of development, and advances in both instrumentation and reagents will be necessary to realize its full potential. In this chapter, we discuss the challenges of expanding the numbers of fluorescent labels that can be measured in cytometry, and introduce SERS tags with extremely narrow spectral peaks as an approach to make more efficient use of the optical spectrum and increase the number of parameters in cytometry.

0091-679X/10 $35.00
DOI 10.1016/B978-0-12-374912-3.00020-1

Keywords: Biomolecular analysis; Data acquisition; Flow cytometry; Fluorescent labels; Optical spectrum; Raman Scattering; Surface-enhanced Raman scattering (SERS)

I. Introduction

A key element of modern cytometry is the ability to measure many parameters simultaneously, with the demand for an increasing number of parameters driving cytometry technology development over the last decade. As new fluorescence probes have been developed and adapted to biological reagents such as antibodies, cytometry instrumentation has expanded the numbers of light sources and detectors to the point where the entire visible spectrum, from the UV to the far red, is now fully employed to measure different fluorescence labels. Further increases in the number of parameters that can be measured simultaneously necessitate one or more of the following paths: (A) the adaptation of nonoptical methods of detection, and thus the development and validation of entirely new instrumentation and reagent sets; (B) expansion of optical methods beyond the visible, requiring new probes, detectors, light sources, and optical components; or (C) more efficient use of the currently accessible spectrum by using alternative optical techniques. Each of these options is the subject of active research and development, and each has specific advantages and challenges. In this chapter, we discuss the challenges of expanding the numbers of fluorescent labels that can be measured in cytometry, and introduce surface-enhanced Raman scattering (SERS) tags with extremely narrow spectral peaks as an approach to make more efficient use of the optical spectrum and increase the number of parameters in cytometry.

II. Multiparameter Fluorescence Measurements

The development of multiparameter flow cytometry has been enabled by increasing the number of lasers, the number of detectors interrogating the sample stream at each laser intersection point, and the availability of fluorophores with appropriate excitation and emission spectra. The number of discrete parameters (or colors) that can be measured with a flow cytometer is determined by the useful spectral range of the instrument, and the absorption (excitation) and emission spectra of available fluorophores. The useful spectral range for a flow cytometer is about 350–900 nm, and is limited by the transmission of optical components at shorter wavelengths and by the poor sensitivity of commonly available detectors at longer wavelengths. Many fluorophores are available with excitation and emission within this range of wavelengths, but the width of a typical emission spectrum is 50–100 nm wide. Some degree of spectral overlap between fluorophores can be accommodated through compensation, but this adds noise to the data and complicates experimental design and analysis.

A common configuration for multiparameter flow cytometry is illustrated in Fig. 1. A blue laser can excite the small green-emitting organic fluorophores such

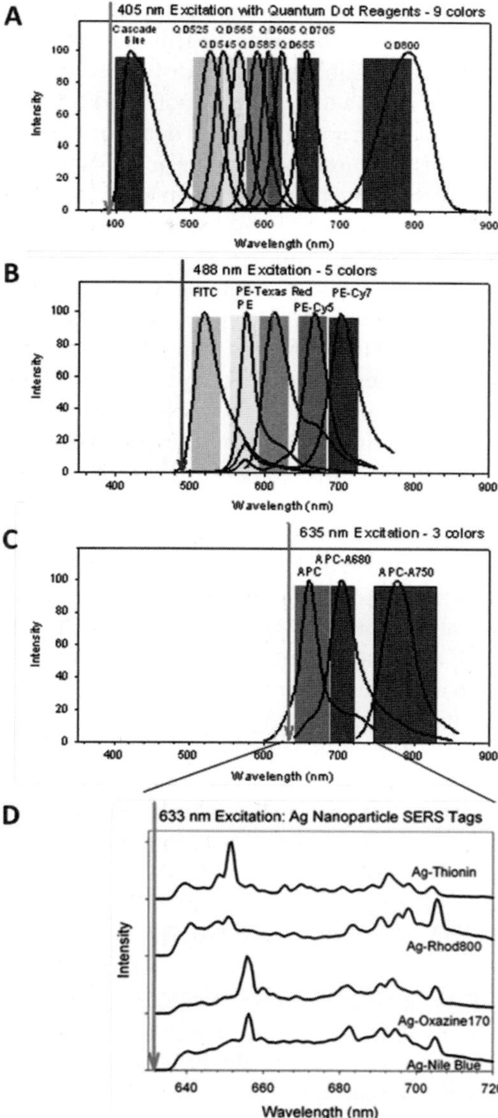

Fig. 1 Use of the optical spectrum in cytometry. (A–C) Conventional fluorescence-based cytometry uses multiple excitation sources to excite multiple dyes. As an example taken from conventional flow cytometry, a violet laser (A) can excite small organic dyes and a range of QDots with emission that span the visible spectrum, a blue laser (B) can excite small organic dyes or PE and its tandem conjugates, and a red laser (C) can excite small organic dyes and allophycocyanin and its tandem conjugates. The number of dyes that can be detected for a given excitation wavelength is ultimately limited by the spectral width of the emission and the available spectral space. (D) Laser-excited Raman scattering features much narrower spectral features, opening the possibility to use more labels in narrower regions of the spectrum. (See plate no. 23 in the color plate section.)

as fluorescein, and the large fluorescent protein phycoerythrin (PE) and its tandem conjugates, and by using appropriate dichroic mirrors, band- pass, and long-pass filters it is possible to distinguish these fluorophores. The addition of UV, violet, or red laser allows the excitation of additional fluorophores that can be measured with additional mirrors, filters, and detectors. However it is easy to see that because of the width of the fluorescence emission spectra, it is difficult to detect more than four or five fluorophores per excitation laser.

One approach to increase the number of parameters is to use fluorophores with more suitable excitation and emission characteristics. A notable recent example is the adaptation of semiconductor quantum dots (QDots), which are efficiently excited in the violet range of the spectrum, and which can be tuned to emit from the blue to the far red. This allows as many as seven different labels to be excited with a single laser. Another approach is to increase the efficient excitation of fluorophores by using additional lasers more closely matched to the absorption spectra of fluorophores of interest. For example, green (532 nm) excitation more efficiently excites PE and its tandem conjugates compared to a 488-nm laser. Thus, the configuration of commercial flow cytometers optimized for highly multiparameter (or polychromatic) applications feature four or more lasers and multiple banks of detectors to allow detection of as many as 17 different fluorochromes. However, these increases in instrument complexity have reached a point of diminishing returns in terms of increased number of parameters, as the fundamental constraints of useful spectral range and fluorescence emission spectral width have not changed.

An alternative approach to increase the number of labels is to make more efficient use of the available spectrum. Raman light scatter, in comparison to fluorescence, has very narrow spectral features (Fig. 1D) that originate from the interaction of the exciting light with molecular vibrations of chemical bonds in a material. Raman scattering has been used for label-free analysis of cells and tissues as well as for the development of bright labels for antibodies or other targeting molecules.

III. Raman Scattering in Cytometry

Raman scattering originates from the interaction of light with molecular vibrations in chemical bonds. The inelastic scattering results in scatter at longer wavelengths proportional to the energy that is lost as a result of the interaction with the scattering material. The correspondence of the Raman spectral fingerprint of a sample with its molecular composition has made Raman spectroscopy a popular tool in chemical analysis, and there has been interest in applying Raman analysis to biological systems as well, although the spectral signatures are typically not specific enough to distinguish individual proteins, for example. From the standpoint of labels for cytometry, Raman is interesting because it offers the possibility of encoding many distinct labels in a relatively narrow region of spectral space.

Like fluorescence, Raman scattering occurs at longer wavelengths from excitation (anti-Stokes Raman scattering at shorter wavelengths also occurs, but we will not focus on this here due to the very low intensity of the anti-Stokes scattering signal). Unlike fluorescence, Raman scattering can be observed with any excitation wavelength and is always shifted the same degree (in terms of frequency, expressed as wave numbers in units of cm^{-1}) relative to the excitation wavelength. Raman scattering is not subject to saturation or photobleaching, making it possible to increase signal by increasing laser power and measurement times.

A. Intrinsic Raman Scattering

There is significant interest in using Raman scattering to detect and quantify different biochemical species from biological samples. For example, protein, DNA, carbohydrate, and lipid have distinct Raman spectra (Mourant *et al.*, 2006), and this has been used to estimate the relative abundance of each of these classes of biological molecules in tumorigenic and nontumorigenic cultured cells (Mourant *et al.*, 2005; Short *et al.*, 2005). While such measurements have the advantage of being label-free and thus relatively noninvasive, they do not have a great deal of molecular specificity compared, for example, to a measurement using a labeled antibody to detect a specific protein. A second disadvantage is that Raman scattering in endogenous compounds produces relatively weak signals that require long measurement times (seconds). Still, there is significant interest in using these approaches to develop diagnostic tools for cancer and other diseases (Kendall *et al.*, 2009; Nijssen *et al.*, 2009; Owen *et al.*, 2006).

Stimulated Raman scattering techniques, including coherent anti-Stoke Raman scattering (CARS) (Krafft *et al.*, 2009; Rodriguez *et al.*, 2006), are multiphoton methods that can be tuned to specific Raman scattering frequencies and can produce significantly stronger signals, allowing label-free live imaging of particular biochemical constituents in cells. A notable example is the real-time imaging of lipid droplets within living cells (Evans *et al.*, 2005; Nan *et al.*, 2003). However, like intrinsic Raman scatting, CARS lacks the molecular specificity of methods that use antibodies or other targeting molecules.

B. Surface-Enhanced Raman Scattering

More than 30 years ago, it was recognized that the intensity of Raman scattering was greatly increased near certain metal surfaces (Jeanmaire and Van Duyne, 1977). In recent years, improvements in the ability to reproducibly make uniform nanostructures have enabled significant advances in the theoretical understanding and experimental control of SERS (Stiles *et al.*, 2008). SERS combines the molecular information provided by Raman scattering with a very bright signal, enabling, in some cases, single molecule analysis (Kneipp *et al.*, 1997; Nie and Emory, 1997). This combination makes SERS an extremely promising analytical technique, and numerous groups are working on the development of substrates for the direct detection of biological analytes. For example, several groups have reported the detection of SERS from bacterial (Jarvis and Goodacre, 2008; Jarvis *et al.*, 2006; Patel *et al.*, 2008) or viral (Driskell *et al.*, 2010; Hoang *et al.*, 2010; Shanmukh *et al.*,

2006, 2008) pathogens, or their metabolites (Daniels *et al.*, 2006; Evanoff *et al.*, 2006; Zhang *et al.*, 2005). In some cases, multivariate analysis approaches such as principal components analysis (PCA) were used to resolve the slight spectral differences among species. While these direct detection approaches have the advantage of enabling label-free analysis, they also used relatively pure samples of cultured organisms. Extending this work to complex environmental samples, with low levels of targets and high levels of irrelevant species or background, is likely to be very challenging (Efrima and Zeiri, 2009; Golightly *et al.*, 2009).

Similarly, direct detection of biomolecules such as proteins and nucleic acids using SERS has been demonstrated. For example, it has been shown that the phosphorylation of a peptide substrate by a kinase can be detected using SERS (Moger *et al.*, 2007; Sundararajan *et al.*, 2006; Yue *et al.*, 2009). Such an approach is feasible for an *in vitro* assay of enzyme activity, but is unlikely to be applicable to cytometry applications using single cells, whether live or fixed. The challenge of interpreting intrinsic SERS spectra directly from a complex sample such as a cell are illustrated by reports of SERS spectra obtained from single cells incubated with gold or silver nanoparticles. Spectra obtained with such approaches allow only the most general components of cells to be resolved, and offer little promise of useful cytometric measurements.

In order to provide information with relevant molecular specificity, as with fluorescence, SERS benefits from the use of a targeting molecule such as an antibody or oligonucleotide. In contrast to the previously described direct detection methods that measure SERS from endogenous compounds, in indirect methods the SERS signal serves as an exogenous label for the targeting model, allowing exploitation of the photostability and narrow spectral features of SERS for multiplexed analysis. The use of SERS as a label can be implemented in many popular detection formats currently used for fluorescence, including solid-phase assays such as immmunoassays, planar microarrays, and lateral flow assays.

Early examples focused on nucleic acid detection, and employed oligonucleotides bearing Raman-active compounds, whose scattering is enhanced upon hybridizing near an appropriate metal surface (Isola *et al.*, 1998; Vo-Dinh *et al.*, 1994), or when metal is deposited over the hybridized probe (Cao *et al.*, 2002). A variation in these two approaches exploits the formation of SERS "hotspots" upon aggregation of silver or gold colloids during binding of Raman-tagged oligonucleotides (Fabris *et al.*, 2007). Similar approaches have been used to detect the binding or Raman-tagged antibodies in immunoassays (Grubisha *et al.*, 2003) and to detect DNA–protein interactions. These detection schemes all require either the assay to be performed on an SERS-active surface, or the aggregation or deposition of SERS-active metal onto the sample.

A further evolution of the indirect or extrinsic SERS-detection approach involves the development of discrete SERS tags, in which the plasmonic nanoparticle, Raman-active compound, and targeting molecule are incorporated into a single entity that can be used to label a sample. This format of labeling is likely to be the most useful for the analysis of single cells, whether by flow or image cytometry, and is the focus of this chapter.

IV. Reagents and Instrumentation

A. Anatomy of an SERS Tag

At the most general level, an SERS tag (Fig. 2) is composed of (A) a plasmonically active nanoparticle with a resonance at a desired excitation wavelength, (B) a Raman-active compound, which is ideally also resonant at the excitation wavelength and gives the SERS tags its distinctive spectral fingerprint, and (C) a coating to stabilize the SERS tag, provide a surface for conjugation to a targeting entity such as an antibody, and passivate the surface to reduce nonspecific binding. In this section, we will discuss each of these components as well as specific examples of SERS tags.

1. Plasmonic Nanoparticles

The starting material for an SERS tag is a plasmonic nanoparticle. Noble metal nanoparticles support localized surface plasmon resonances (LSPRs), a collective oscillation of electrons that strongly couple to light at specific wavelengths, and produce extremely high electromagnetic fields near the nanoparticle surface.

In bulk measurements, the LSPR results in distinctive changes in the extinction of a nanoparticle suspension. The extinction is the result of increases in both nanoparticle absorbance and scatter, and the wavelengths at which plasmonic nanoparticles interact with light can be tuned by varying the composition, shape, and size of the particle (Fig. 3). Gold and silver nanoparticles are the most commonly used plasmonic nanoparticles as they absorb and scatter light in the 350–900 nm spectral range. For example, solid silver spheres exhibit a resonance in the blue, shifting to longer wavelengths as the diameter increases. Solid gold spheres have resonances at green wavelengths, also red shifting as the diameter increases. Thin shells of silver or gold over a nonconducting material such as silica

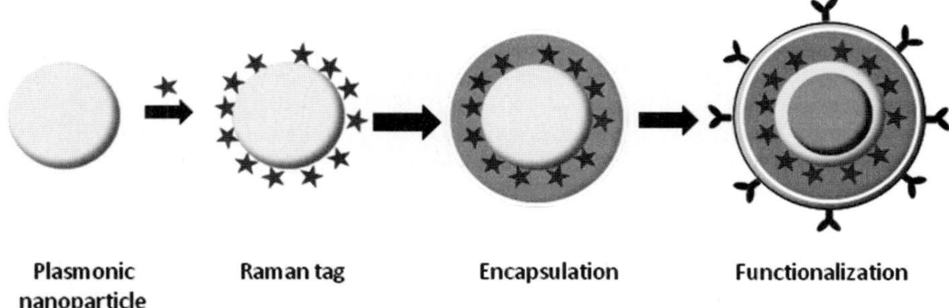

Fig. 2 Anatomy of a SERS tag. A SERS tag generally consists of a plasmonic nanoparticle with the desired resonant wavelength, a Raman active compound that confers a particular spectral signature, a coating to stabilize and protect the SERS tag, and a targeting molecule such as an antibody or nucleic acid that confers molecular specificity to the SERS tag.

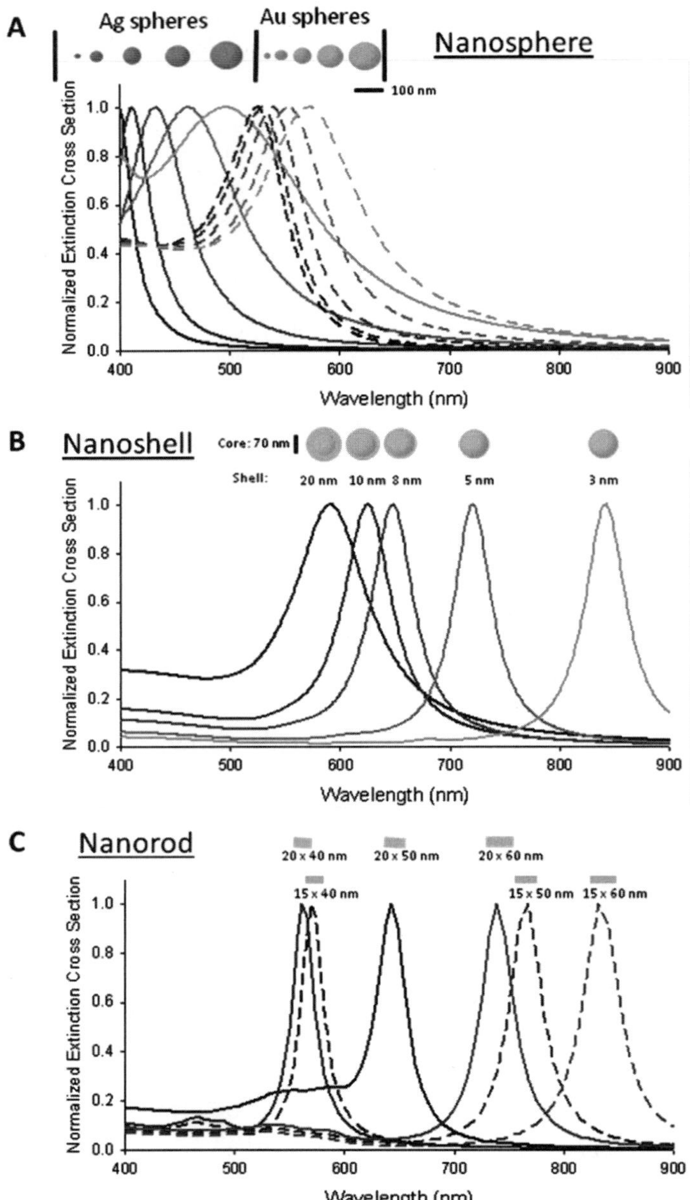

Fig. 3 Wavelength tunability of plasmonic nanoparticles. Gold and silver nanostructures have size- and shape-dependent optical properties that can be predicted from electromagnetic theory or numerical simulations. (A) Solid spheres of silver and gold have resonance wavelengths that shift to longer wavelengths with the sphere radius. (B) Gold nanorod resonance wavelengths shift to longer wavelengths with aspect ratio. (C) Shells of gold over a silica core have resonance wavelengths that shift to longer wavelengths as the ratio of core radius to shell thickness increases.

also support an LSPR that can lead to strong SERS enhancements (Oldenburg *et al.*, 1999). In this case, the resonance wavelength depends on the ratio of the shell thickness to the core diameter. Asymmetric structures such as nanorods and spheroids support strong resonances on their long axes, the wavelength of which varies with the particle's aspect ratio (Hao and Schatz, 2004; Murphy *et al.*, 2008; Orendorff *et al.*, 2006). Beyond spheres, shells, and rods, a variety of other nanoparticle structures have been shown to produce SERS including nanocubes (McLellan *et al.*, 2006), nanostars (Khoury and Vo-Dinh, 2008), nanoflowers (Xie *et al.*, 2008), and nanorice (Wiley *et al.*, 2007).

The electric field intensity resulting from the LSPR varies across the surface of the nanoparticle, and depends on the shape of the nanoparticle, its orientation relative to the exciting light, and the polarization and wavelength of that light. The LSPR is predicted to be highest at particle edges, points, or surface irregularities, which has led to the concept of SERS "hot spots" as being responsible for generating the highest SERS enhancement and signals. Additionally, the LSPRs of adjacent nanoparticles can couple, creating hotspots at the junctions between aggregated particles (Camden *et al.*, 2008). The engineering of metal nanoparticle systems to provide the highest and most uniform LSPRs and SERS enhancement is a key goal of SERS tag design.

2. Raman Tag

The Raman tag is a compound that exhibits a distinctive Raman spectrum, or fingerprint, when adsorbed to the surface of a plasmonic particle that can be distinguished from other Raman tags. The electromagnetic field is strongest nearest the nanoparticle surface, so physical adsorption of the Raman compound directly to the metal surfaces is required for the strongest SERS enhancement. Often, the Raman tag is a thiol-containing compound, which takes advantage of the affinity of that moiety for gold and silver surfaces, although in principle any compound that adsorbs to the metal surface is suitable. Small organic thiols such a mercaptobenzoic acid (MBA) have been the subject of many fundamental SERS studies. As mentioned above, if the Raman tag has a molecular absorbance that is resonant with the exciting light, an additional signal enhancement is obtained. Highly absorbing fluorophores such as rhodamine 6G (R6G) have been used to demonstrate single-molecule SERS detection. Such surface-enhanced resonant Raman scattering (SERRS) is a special case of SERS that produces the strongest Raman scattering enhancement and thus the brightest SERS tags. Isothiocyanate versions of many common fluorophores combine the advantages of a resonant tag with a metal binding sulfur group. While fluorescence is often a significant source of background in Raman measurements, in the case of fluorophores used as resonant Raman tags, the fluorescence is fortuitously quenched by proximity to the metal surface.

3. SERS Tag Coating and Functionalization

SERS tags are generally coated or encapsulated to stabilize and protect the tag from environmental components and to provide functional groups for conjugation to targeting molecules such as antibodies. Silica is a popular coating that provides chemical and mechanical stability, as well as a surface that is readily functionalized for conjugation to targeting molecules. SERS tags have also been coated with polymers such as polyethylene glycol (PEG) to stabilize the tags, provide functional groups for conjugation, and passivate the surface to reduce nonspecific binding.

B. Examples of SERS Tags

1. SERS Tags Based on Metal Spheres

Some of the first SERS tags were based on gold or silver spheres. In early examples based on gold spheres, organic thiols (Mulvaney *et al.*, 2003) or sulfur-containing resonant dyes (Doering and Nie, 2003) served as the Raman tags, and a silica coating encapsulated the tags and served as the surface for functionalization. Subsequent variations employed resonant dyes (Gong *et al.*, 2006) and polymer (Merican *et al.*, 2007) or protein (Lee *et al.*, 2007) coatings. SERS tags based on metal spheres can be made to be fairly uniform, but do not provide the brightest SERS signals compared to SERS tags made with other nanoparticles due to the relatively low electric field intensities found near the nanosphere surface.

2. SERS Tags Based on Aggregated Spheres

Empirical observations led to the recognition that aggregates of nanoparticles produce greater SERS enhancements than individual metal spheres. An early example of this approach (Su *et al.*, 2005) employed organic thiols and heat to trigger the aggregation of silver spheres. The aggregation was slowed by cooling and the addition of protein (bovine serum albumin, BSA), which encapsulated and stabilized the aggregates. The BSA coating also provided a source of amine groups for bioconjugation. Later variants employed resonant dye molecules and a silica (Brown and Doorn, 2008), rather than protein, coating. SERS tags based on aggregated colloids have resonance wavelengths that are red-shifted relative to isolated spheres because of the overall larger size of the aggregates, and because the LSPRs of individual nanoparticles couple, shifting the resonance to a lower energy state. They also can be significantly brighter than SERS tags based on single spheres because of hot spots formed at the interfaces of the aggregated spheres, and there are examples of these types of SERS tags being used in image cytometry (Lutz *et al.*, 2008a; Shachaf *et al.*, 2009; Sun *et al.*, 2007a) and flow cytometry (Watson *et al.*, 2008, 2009). However, the aggregation process is difficult to control, and typically results in highly polydisperse aggregates with low uniformity that have wide range of sizes, resonant wavelengths, and brightness. This makes the scalable and reproducible preparation of aggregate-based SERS tags a significant challenge.

3. Nanoshell-Based SERS Tags

Thin shells of metal over a core of dielectric material such as silica provide a plasmonic particle in which the LSPRs of two edges of the shell can couple providing a strong SERS enhancement (Oldenburg *et al.*, 1999). The resonant wavelength can be tuned by adjusting the thickness of the metal shell and size of the dielectric core, with decreasing shell–core ratios shifting the resonance to longer wavelengths. By building the shell over monodisperse silica cores, very uniform plasmonic particles with high SERS enhancements can be prepared. Silver nanoshells have been labeled with organic thiols and coated in silica for subsequent conjugation and use as an imaging probe (Yu *et al.*, 2007). Gold nanoshells have been labeled with resonant dye, and coated with a thin layer of silver, which provides additional signal enhancement (Sebba *et al.*, 2009) that allows the Raman scattering from individual tags to be detected with submillisecond integration times. These particles can be effectively stabilized and functionalized with a sulfhydral-PEG-based coating. Nanoshell-based SERS tags represent a promising approach to the scalable manufacturing of bright, uniform SERS tags for high-speed imaging and flow cytometry applications.

4. Nanorod–Based SERS Tags

Metal nanorods exhibit LSPR that can be tuned by varying their aspect ratio (length to width) and have strong electric field enhancements at the tips that leads to strong SERS when illuminated by the appropriate wavelength of light (Orendorff *et al.*, 2006). Gold nanorods labeled with an organic thiol and coated in polymer served as antibody label for the detection of surface markers on cultured cells (Park *et al.*, 2009b). Gold rods labeled with resonant compounds were coated with polymer and injected into mice, where their distribution was determined using *in vivo* imaging (von Maltzahn *et al.*, 2009). SERS tags based on nanorods also appear as promising reagents for high-speed imaging and flow cytometry applications.

C. SERS Cytometry Data Acquisition and Analysis

1. General Considerations

A distinguishing feature of Raman scattering compared to fluorescence is its narrow spectral features, which requires higher spectral resolution to resolve than is commonly employed in image or flow cytometry. The approaches to cytometry spectral imaging hardware have been recently reviewed (Lerner *et al.*, 2010). In general, Raman and SERS detection demands the high spectral resolution provided by grating-based dispersive optics and array-type detectors. In addition, because Raman scattering spectral features occur at a fixed frequency shift from the exciting light, the resolution of SERS spectrum also depends on the spectral width of the excitation laser. Many diode laser sources suitable for fluorescence cytometry have spectral widths of 1 nm or more and will produce unacceptably broad Raman peaks. Once continuous spectral data are collected, they are subjected to spectral unmixing to extract the contributions of individual tags and background sources to the spectra. Common approaches include linear least squares unmixing (Lutz *et al.*,

2008b) or multivariate analysis such as PCA (Watson *et al.*, 2008) or multiple curve resolution (MCR) (Haaland *et al.*, 2009) to resolve individual components in mixed spectra.

2. SERS Imaging

Raman imaging systems are commercially available; however most are not designed for high-speed biological imaging. In fluorescence imaging, relatively broad spectral bands are selected with dichroic mirrors and bandpass filters, and the system magnification and pixel size of the CCD camera determine the spatial resolution. Alternatively, laser scanning cytometry couples high-speed scanning optics to render high-resolution information using point detectors such as PMTs. Such approaches have enabled the development of very high speed imaging systems, as employed in the high content analysis field. In Raman imaging, the requirement for high-resolution spectra necessitates the use of a high-performance spectrograph and CCD array detector to collect this spectral information. Making such spectral measurements point-by-point at high spatial resolution from cells on a microscope slide, for example, is very slow by comparison to conventional fluorescence imaging. A more efficient approach is to use line focus excitation to collect one-dimensional spatial information and spectral information simultaneously on a CCD array. This approach is taken by a few commercial systems, and can offer significantly higher imaging throughput.

3. SERS Flow Cytometry

There are no commercial Raman flow cytometers, but recent years have seen a renewed interest in developing high-resolution spectral capabilities in flow cytometry. In contrast to imaging, where recent hardware developments have increased the speed of imaging systems, flow cytometry is intrinsically fast. Conventional flow cytometer analyzers employ measurement integration times on the order of 10 μs to measure hundreds to thousands of cell per second. High-speed cell sorters may use integration times 10-fold faster to reach 10 times higher analysis rates. Specially constructed laboratory instruments have used slower measurement times up to 1 ms, to increase measurement sensitivity. For high-resolution spectral flow cytometry, as for high-resolution spectral imaging, dichroic mirrors and bandpass filters are replaced with a spectrograph. Modern CCD arrays provide high resolution and sensitivity at exposure times as fast as 10 μs; however they have slower readout times compared to PMTs, which can limit the analysis rates to a few hundred cells per second. Lab-built Raman flow cytometers have demonstrated performance suitable for most popular flow cytometry applications (Watson *et al.*, 2008), and the addition of spectral data acquisition capabilities to a commercial flow cytometer has also been demonstrated (Watson *et al.*, 2009). As for SERS imaging, a limiting factor for the future development of SERS flow cytometry is the availability of bright and uniform SERS tags.

V. SERS Cytometry Applications

While the phenomenon of SERS has been known for more than 30 years, and its potential as a bioanalytical tool recognized for some time, it is only in the past decade that techniques for controlled nanoparticle fabrication have become widely disseminated. Thus, from an applications standpoint, SERS is a still young field. As discussed earlier, there have been several implementations of SERS in nucleic acid and protein analysis, and SERS tags have been used as labels in bioassays in a manner analogous to chromogenic or fluorescent labels on a variety of measurement platforms including ELISA plates, planar substrates including microarrays, and lateral flow assays. Our focus in this chapter is on the potential for SERS to extend the multiparameter analysis of cells and cell systems.

A. *In vitro* Measurements

At present, cytometry generally focuses on the study of cells cultured *in vitro*, or *ex vivo* analysis of primary cells or tissues, and most cytometry-related SERS applications have focused on these classes of applications as well. Several groups have used SERS-tagged antibodies or other ligands to detect cell-surface markers (Kim *et al.*, 2006; Nguyen *et al.*, 2010; Noh *et al.*, 2009; Park *et al.*, 2009a; Yu *et al.*, 2007) or intracellular antigens (Lee *et al.*, 2007; Shachaf *et al.*, 2009) on cultured cell lines. SERS tags have also been used as antibody labels to stain tissue sections (Lutz *et al.*, 2008a, 2008b; Sun *et al.*, 2007b). In many cases, these examples were qualitative demonstrations involving single SERS tags, but in a few cases efforts were made to extract quantitative information in multiplexed measurements (Lutz *et al.*, 2008a, 2008b; Shachaf *et al.*, 2009).

B. *In vivo* Measurements

There is significant interest in adapting quantitative cytometry approaches to *in vivo* measurements, and SERS has some potential advantages for enabling this. The tunability of SERS tag resonance allows tags to be developed that are excitable in the far red, allowing greater light penetration into tissue and minimizing many sources of autofluorescence. Several groups have injected untargeted SERS tags into animals and used *in vivo* SERS imaging to determine their distribution and fate (Keren *et al.*, 2008; Matschulat *et al.*, 2010; von Maltzahn *et al.*, 2009; Wang *et al.*, 2010; Zavaleta *et al.*, 2009).

More specific biological information can be obtained when SERS tags are functionalized to allow targeting, for instance with an antibody. Qian *et al.* (2008) used anti-EGFR SERS tags to localize tumors in live mice using *in vivo* SERS imaging. An exciting aspect of this approach to *in vivo* imaging is the use of plasmonic nanoparticles to thermally ablate diseased tissues (Cobley *et al.*, 2010; El-Sayed *et al.*, 2006; O'Neal *et al.*, 2004) presenting the prospect for coordinated cancer diagnosis and therapy using targeted nanoparticle probes (Hirsch *et al.*, 2003; Lal *et al.*, 2008).

VI. Summary and Prospects

The past decade has seen significant advances in the preparation and applications of SERS-active materials for biomolecular analysis. Bright signals, photostability, and narrow spectral features offer attractive advantages for cytometric analyses. Following the path of semiconductor QDots, improvements in nanoparticle synthesis methods and surface chemistry have led to plausible schemes for the scalable and reproducible production of uniform SERS tags, a prerequisite for commercial application. In parallel, developments in high-speed multispectral image and flow cytometry have laid the groundwork for the application of Raman tags application in high speed cell analysis. However, SERS cytometry is still in an early stage of development, and advances in both instrumentation and reagents will be necessary to realize its full potential.

Reagents will be critical to the further development of SERS cytometry. In order to allow researchers to focus on the development of the biological applications of SERS, stable and uniform tags that are readily functionalized with an antibody or other targeting molecule must be available, ideally from commercial sources. Tags must be reproducibly bright and uniform from batch to batch to allow comparison of results over time and between instruments. The status of SERS tags might be compared to that of QDots 5 or 10 years ago, with a handful of lab having extensive experience in making tags and with a limited number of commercial products making their way toward the market, but with further development and optimization being pursued in both academic and commercial labs.

Finally, it is clear that SERS complements, rather than replaces, fluorescence as a biological detection tool. The chief advantages of narrow spectral features and photostability make them very attractive as tags for antibodies or other targeting molecules. However, there are many applications, including cell viability, ion and pH fluxes, and the measurement of enzyme activities, for which fluorescence is currently clearly better suited. Thus, future applications will likely involve the simultaneous measurement of fluorescence and SERS tags. The wavelength tunability of SERS tags makes it possible to engineer SERS tags to be excited in the NIR, leaving the visible wavelengths for the wide variety of visible wavelength-excited currently available. In the future, we may see instruments designed for the simultaneous high speed measurement of both fluorescence and Raman signals.

References

Brown, L., and Doorn, S. (2008). A controlled and reproducible pathway to dye-tagged, encapsulated silver nanoparticles as substrates for SERS multiplexing. *Langmuir* **24**, 2277–2280.

Camden, J. P., Dieringer, J. A., Wang, Y., Masiello, D. J., Marks, L. D., Schatz, G. C., Van Duyne, R. P. (2008). Probing the structure of single-molecule surface-enhanced Raman scattering hot spots. *J. Am. Chem. Soc.* **130**, 12616–12617.

Cao, Y. C., Jin, R., and Mirkin, C. A. (2002). Nanoparticles with Raman spectroscopic fingerprints for DNA and RNA detection. *Science* **297**, 1536–1540.

Cobley, C. M., Au, L., Chen, J., and Xia, Y. (2010). Targeting gold nanocages to cancer cells for photothermal destruction and drug delivery. *Expert Opin. Drug Deliv.* **7**, 577–587.

Daniels, J. K., Caldwell, T. P., Christensen, K. A., and Chumanov, G. (2006). Monitoring the kinetics of *Bacillus subtilis* endospore germination via surface-enhanced Raman scattering spectroscopy. *Anal. Chem.* **78**, 1724–1729.

Doering, W. E., and Nie, S. (2003). Spectroscopic tags using dye-embedded nanoparticles and surface-enhanced Raman scattering. *Anal. Chem.* **75**, 6171–6176.

Driskell, J. D., Zhu, Y., Kirkwood, C. D., Zhao, Y., Dluhy, R. A., Tripp, R. A. (2010). Rapid and sensitive detection of rotavirus molecular signatures using surface enhanced Raman spectroscopy. *PLoS ONE* **5**, e10222.

Efrima, S., and Zeiri, L. (2009). Understanding SERS of bacteria. *J. Raman Spectrosc.* **40**, 277–288.

El-Sayed, I. H., Huang, X., and El-Sayed, M. A. (2006). Selective laser photo-thermal therapy of epithelial carcinoma using anti-EGFR antibody conjugated gold nanoparticles. *Cancer Lett.* **239**, 129–135.

Evanoff, D. D., Heckel, J., Caldwell, T. P., Christensen, K. A., and Chumanov, G. (2006). Monitoring DPA release from a single germinating Bacillus subtilis endospore via surface-enhanced Raman scattering microscopy. *J. Am. Chem. Soc.* **128**, 12618–12619.

Evans, C., Potma, E., Puoris'haag, M., Cote, D., Lin, C., Xie, X. (2005). Chemical imaging of tissue *in vivo* with video-rate coherent anti-Stokes Raman scattering microscopy. *Proc. Natl. Acad. Sci. U.S.A* **102**, 16807–16812.

Fabris, L., Dante, M., Braun, G., Lee, S. J., Reich, N. O., Moskovits, M., Nguyen, T., Bazan, G. C. (2007). A heterogeneous PNA-based SERS method for DNA detection. *J. Am. Chem. Soc.* **129**, 6086–6087.

Golightly, R. S., Doering, W. E., and Natan, M. J. (2009). Surface-enhanced Raman spectroscopy and homeland security: a perfect match? *ACS Nano* **3**, 2859–2869.

Gong, J., Jiang, J., Liang, Y., Shen, G., and Yu, R. (2006). Synthesis and characterization of surface-enhanced Raman scattering tags with Ag/SiO$_2$ core–shell nanostructures using reverse micelle technology. *J. Colloid Interface Sci.* **298**, 752–756.

Grubisha, D., Lipert, R., Park, H., Driskell, J., and Porter, M. (2003). Femtomolar detection of prostate-specific antigen: an immunoassay based on surface-enhanced Raman scattering and immunogold labels. *Anal. Chem.* **75**, 5936–5943.

Haaland, D. M., Jones, H. D. T., Van Benthem, M. H., Sinclair, M. B., Melgaard, D. K., Stork, C. L., Pedroso, M. C., Liu, P., Brasier, A. R., Andrews, N. L., *et al.* (2009). Hyperspectral confocal fluorescence imaging: exploring alternative multivariate curve resolution approaches. *Appl. Spectrosc.* **63**, 271–279.

Hao, E., and Schatz, G. C. (2004). Electromagnetic fields around silver nanoparticles and dimers. *J. Chem. Phys.* **120**, 357–366.

Hirsch, L. R., Stafford, R. J., Bankson, J. A., Sershen, S. R., Rivera, B., Price, R. E., Hazle, J. D., Halas, N. J., West, J. L. (2003). Nanoshell-mediated near-infrared thermal therapy of tumors under magnetic resonance guidance. *Proc. Natl. Acad. Sci. U.S.A* **100**, 13549–13554.

Hoang, V., Tripp, R. A., Rota, P., and Dluhy, R. A. (2010). Identification of individual genotypes of measles virus using surface enhanced Raman spectroscopy. Analyst. Available at: http://www.ncbi.nlm.nih.gov/pubmed/20838669. [Accessed October 6, 2010].

Isola, N. R., Stokes, D. L., and Vo-Dinh, T. (1998). Surface-enhanced Raman gene probe for HIV detection. *Anal. Chem.* **70**, 1352–1356.

Jarvis, R. M., Brooker, A., and Goodacre, R. (2006). Surface-enhanced Raman scattering for the rapid discrimination of bacteria. *Faraday Discuss.* **132**, 281–292 discussion 309–319.

Jarvis, R. M., and Goodacre, R. (2008). Characterisation and identification of bacteria using SERS. *Chem. Soc. Rev.* **37**, 931–936.

Jeanmaire, D. L., and Van Duyne, R. P. (1977). Surface raman spectroelectrochemistry: Part I. Heterocyclic, aromatic, and aliphatic amines adsorbed on the anodized silver electrode. *J. Electroanalytical Chem.* **84**, 1–20.

Kendall, C., Isabelle, M., Bazant-Hegemark, F., Hutchings, J., Orr, L., Babrah, J., Baker, R., Stone, N. (2009). Vibrational spectroscopy: a clinical tool for cancer diagnostics. *Analyst* **134**, 1029–1045.

Keren, S., Zavaleta, C., Cheng, Z., de la Zerda, A., Gheysens, O., Gambhir, S. (2008). Noninvasive molecular imaging of small living subjects using Raman spectroscopy. *Proc. Natl. Acad. Sci.* **105**, 5844–5849.

Khoury, C. G., and Vo-Dinh, T. (2008). Gold nanostars for surface-enhanced Raman scattering: synthesis, characterization and optimization. *J. Phys. Chem. C* **112**, 18849–18859.

Kim, J., Kim, J., Choi, H., Lee, S., Jun, B., Yu, K., Kuk, E., Kim, Y., Jeong, D., Cho, M., *et al.* (2006). Nanoparticle probes with surface enhanced Raman spectroscopic tags for cellular cancer targeting. *Anal. Chem.* **78**, 6967–6973.

Kneipp, K., Wang, Y., Kneipp, H., Perelman, L. T., Itzkan, I., Dasari, R. R., Feld, M. S. (1997). Single molecule detection using surface-enhanced Raman scattering (SERS). *Phys. Rev. Lett.* **78**, 1667.

Krafft, C., Dietzek, B., and Popp, J. (2009). Raman and CARS microspectroscopy of cells and tissues. *Analyst* **134**, 1046–1057.

Lal, S., Clare, S. E., and Halas, N. J. (2008). Nanoshell-enabled photothermal cancer therapy: impending clinical impact. *Acc. Chem. Res.* **41**, 1842–1851.

Lee, S., Kim, S., Choo, J., Shin, S. Y., Lee, Y. H., Choi, H. Y., Ha, S., Kang, K., Oh, C. H. (2007). Biological imaging of HEK293 cells expressing PLCgamma1 using surface-enhanced Raman microscopy. *Anal. Chem.* **79**, 916–922.

Lerner, J. M., Gat, N., and Wachman, E. (2010). Approaches to spectral imaging hardware. *Curr. Protoc. Cytom. Chapter 12*, Unit 12.20.

Lutz, B., Dentinger, C., Nguyen, L., Sun, L., Zhang, J., Allen, A., Chan, S., Knudsen, B. (2008b). Spectral analysis of multiplex Raman probe signatures. *ACS Nano* **2**, 2306–2314.

Lutz, B., Dentinger, C., Sun, L., Nguyen, L., Zhang, J., Chmura, A., Allen, A., Chan, S., Knudsen, B. (2008a). Raman nanoparticle probes for antibody-based protein detection in tissues. *J. Histochem. Cytochem.* **56**, 371–379.

von Maltzahn, G., Centrone, A., Park, J., Ramanathan, R., Sailor, M. J., Hatton, T. A., Bhatia, S. N. (2009). SERS-coded gold nanorods as a multifunctional platform for densely multiplexed near-infrared imaging and photothermal heating. *Adv. Mater.* **21**, 3175–3180.

Matschulat, A., Drescher, D., and Kneipp, J. (2010). Surface-enhanced Raman scattering hybrid nanoprobe multiplexing and imaging in biological systems. *ACS Nano* **4**, 3259–3269.

McLellan, J. M., Siekkinen, A., Chen, J., and Xia, Y. (2006). Comparison of the surface-enhanced Raman scattering on sharp and truncated silver nanocubes. *Chem. Phys. Lett.* **427**, 122–126.

Merican, Z., Schiller, T. L., Hawker, C. J., Fredericks, P. M., and Blakey, I. (2007). Self-assembly and encoding of polymer-stabilized gold nanoparticles with surface-enhanced Raman reporter molecules. *Langmuir* **23**, 10539–10545.

Moger, J., Gribbon, P., Sewing, A., and Winlove, C. P. (2007). Feasibility study using surface-enhanced Raman spectroscopy for the quantitative detection of tyrosine and serine phosphorylation. *Biochim. Biophys. Acta* **1770**, 912–918.

Mourant, J., Dominguez, J., Carpenter, S., Short, K., Powers, T., Michalczyk, R., Kunapareddy, N., Guerra, A., Freyer, J. (2006). Comparison of vibrational spectroscopy to biochemical and flow cytometry methods for analysis of the basic biochemical composition of mammalian cells. *J. Biomed. Opt.* **11**, 064024.

Mourant, J., Short, K., Carpenter, S., Kunapareddy, N., Coburn, L., Powers, T., Freyer, J. (2005). Biochemical differences in tumorigenic and nontumorigenic cells measured by Raman and infrared spectroscopy. *J. Biomed. Opt.* **10**, 031106.

Mulvaney, S., Musick, M., Keating, C., and Natan, M. (2003). Glass-coated, analyte-tagged nanoparticles: a new tagging system based on detection with surface-enhanced Raman scattering. *Langmuir* **19**, 4784–4790.

Murphy, C. J., Gole, A. M., Hunyadi, S. E., Stone, J. W., Sisco, P. N., Alkilany, A., Kinard, B. E., Hankins, P. (2008). Chemical sensing and imaging with metallic nanorods. *Chem. Commun. (Camb.)* 544–557.

Nan, X., Cheng, J., and Xie, X. (2003). Vibrational imaging of lipid droplets in live fibroblast cells with coherent anti-Stokes Raman scattering microscopy. *J. Lipid Res.* **44**, 2202–2208.

Nguyen, C., Nguyen, J., Rutledge, S., Zhang, J., Wang, C., Walker, G. (2010). Detection of chronic lymphocytic leukemia cell surface markers using surface enhanced Raman scattering gold nanoparticles. *Cancer Lett.* **292**, 91–97.

Nie, S., and Emory, S. (1997). Probing single molecules and single nanoparticles by surface-enhanced Raman scattering. *Science* **275**, 1102–1106.

Nijssen, A., Koljenović, S., Bakker Schut, T. C., Caspers, P. J., and Puppels, G. J. (2009). Towards oncological application of Raman spectroscopy. *J. Biophotonics* **2**, 29–36.

Noh, M., Jun, B., Kim, S., Kang, H., Woo, M., Minai-Tehrani, A., Kim, J., Kim, J., Park, J., Lim, H., *et al.* (2009). Magnetic surface-enhanced Raman spectroscopic (M-SERS) dots for the identification of bronchioalveolar stem cells in normal and lung cancer mice. *Biomaterials* **30**, 3915–3925.

Oldenburg, S. J., Westcott, S. L., Averitt, R. D., and Halas, N. J. (1999). Surface enhanced Raman scattering in the near infrared using metal nanoshell substrates. *J. Chem. Phys.* **111**, 4729.

O'Neal, D. P., Hirsch, L. R., Halas, N. J., Payne, J. D., and West, J. L. (2004). Photo-thermal tumor ablation in mice using near infrared-absorbing nanoparticles. *Cancer Lett.* **209**, 171–176.

Orendorff, C. J., Gearheart, L., Jana, N. R., and Murphy, C. J. (2006). Aspect ratio dependence on surface enhanced Raman scattering using silver and gold nanorod substrates. *Phys. Chem. Chem. Phys.* **8**, 165.

Owen, C. A., Notingher, I., Hill, R., Stevens, M., and Hench, L. L. (2006). Progress in Raman spectroscopy in the fields of tissue engineering, diagnostics and toxicological testing. *J. Mater. Sci. Mater. Med.* **17**, 1019–1023.

Park, H., Lee, S., Chen, L., Lee, E., Shin, S., Lee, Y., Son, S., Oh, C., Song, J., Kang, S., *et al.* (2009a). SERS imaging of HER2-overexpressed MCF7 cells using antibody-conjugated gold nanorods. *Phys. Chem. Chem. Phys.* **11**, 7444–7449.

Park, H., Lee, S., Chen, L., Lee, E. K., Shin, S. Y., Lee, Y. H., Son, S. W., Oh, C. H., Song, J. M., Kang, S. H., *et al.* (2009b). SERS imaging of HER2-overexpressed MCF7 cells using antibody-conjugated gold nanorods. *Phys. Chem. Chem. Phys.* **11**, 7444–7449.

Patel, I. S., Premasiri, W. R., Moir, D. T., and Ziegler, L. D. (2008). Barcoding bacterial cells: a SERS based methodology for pathogen identification. *J. Raman Spectrosc.* **39**, 1660–1672.

Qian, X., Peng, X., Ansari, D., Yin-Goen, Q., Chen, G., Shin, D., Yang, L., Young, A., Wang, M., Nie, S. (2008). In vivo tumor targeting and spectroscopic detection with surface-enhanced Raman nanoparticle tags. *Nat. Biotechnol.* **26**, 83–90.

Rodriguez, L. G., Lockett, S. J., and Holtom, G. R. (2006). Coherent anti-Stokes Raman scattering microscopy: a biological review. *Cytometry A* **69**, 779–791.

Sebba, D. S., Watson, D. A., and Nolan, J. P. (2009). High throughput single nanoparticle spectroscopy. *ACS Nano* **3**, 1477–1484.

Shachaf, C., Elchuri, S., Koh, A., Zhu, J., Nguyen, L., Mitchell, D., Zhang, J., Swartz, K., Sun, L., Chan, S., *et al.* (2009). A novel method for detection of phosphorylation in single cells by surface enhanced Raman scattering (SERS) using composite organic-inorganic nanoparticles (COINs). *PLoS One* **4**, e5206.

Shanmukh, S., Jones, L., Driskell, J., Zhao, Y., Dluhy, R., Tripp, R. A. (2006). Rapid and sensitive detection of respiratory virus molecular signatures using a silver nanorod array SERS substrate. *Nano Lett.* **6**, 2630–2636.

Shanmukh, S., Jones, L., Zhao, Y., Driskell, J. D., Tripp, R. A., Dluhy, R. A. (2008). Identification and classification of respiratory syncytial virus (RSV) strains by surface-enhanced Raman spectroscopy and multivariate statistical techniques. *Anal. Bioanal. Chem.* **390**, 1551–1555.

Short, K., Carpenter, S., Freyer, J., and Mourant, J. (2005). Raman spectroscopy detects biochemical changes due to proliferation in mammalian cell cultures. *Biophys. J.* **88**, 4274–4288.

Stiles, P. L., Dieringer, J. A., Shah, N. C., and Van Duyne, R. P. (2008). Surface-enhanced Raman spectroscopy. *Ann. Rev. Anal. Chem. (Palo Alto Calif.)* **1**, 601–626.

Su, X., Zhang, J., Sun, L., Koo, T., Chan, S., Sundararajan, N., Yamakawa, M., Berlin, A. (2005). Composite organic–inorganic nanoparticles (COINs) with chemically encoded optical signatures. *Nano Lett.* **5**, 49–54.

Sun, L., Sung, K., Dentinger, C., Lutz, B., Nguyen, L., Zhang, J., Qin, H., Yamakawa, M., Cao, M., Lu, Y., *et al.* (2007a). Composite organic-inorganic nanoparticles as Raman labels for tissue analysis. *Nano Lett.* **7**, 351–356.

Sun, L., Sung, K., Dentinger, C., Lutz, B., Nguyen, L., Zhang, J., Qin, H., Yamakawa, M., Cao, M., Lu, Y., *et al.* (2007b). Composite organic-inorganic nanoparticles as Raman labels for tissue analysis. *Nano Lett.* **7**, 351–356.

Sundararajan, N., Mao, D., Chan, S., Koo, T., Su, X., Sun, L., Zhang, J., Sung, K., Yamakawa, M., Gafken, P., *et al.* (2006). Ultrasensitive detection and characterization of posttranslational modifications using surface-enhanced Raman spectroscopy. *Anal. Chem.* **78**, 3543–3550.

Vo-Dinh, T., Houck, K., and Stokes, D. L. (1994). Surface-enhanced Raman gene probes. *Anal. Chem.* **66**, 3379–3383.

Wang, Y., Seebald, J. L., Szeto, D. P., and Irudayaraj, J. (2010). Biocompatibility and biodistribution of surface-enhanced Raman scattering nanoprobes in zebrafish embryos: *in vivo* and multiplex imaging. *ACS Nano* **4**, 4039–4053.

Watson, D., Brown, L., Gaskill, D., Naivar, M., Graves, S., Doorn, S., Nolan, J. (2008). A flow cytometer for the measurement of Raman spectra. *Cytometry A* **73**, 119–128.

Watson, D., Gaskill, D., Brown, L., Doorn, S., and Nolan, J. (2009). Spectral measurements of large particles by flow cytometry. *Cytometry A*. Available at: PM:19199345.

Wiley, B. J., Chen, Y., McLellan, J. M., Xiong, Y., Li, Z., Ginger, D., Xia, Y. (2007). Synthesis and optical properties of silver nanobars and nanorice. *Nano Lett.* **7**, 1032–1036.

Xie, J., Zhang, Q., Lee, J. Y., and Wang, D. I. C. (2008). The Synthesis of SERS-active gold nanoflower tags for *in vivo* applications. *ACS Nano* **2**, 2473–2480.

Yu, K., Lee, S., Han, J., Park, H., Woo, M., Noh, M., Hwang, S., Kwon, J., Jin, H., Kim, Y., *et al.* (2007). Multiplex targeting, tracking, and imaging of apoptosis by fluorescent surface enhanced Raman spectroscopic dots. *Bioconjugate Chem.* **18**, 1155–1162.

Yue, Z., Zhuang, F., Kumar, R., Wang, I., Cronin, S. B., Liu, Y. (2009). Cell kinase activity assay based on surface enhanced Raman spectroscopy. *Spectrochim Acta A Mol. Biomol. Spectrosc.* **73**, 226–230.

Zavaleta, C. L., Smith, B. R., Walton, I., Doering, W., Davis, G., Shojaei, B., Natan, M. J., Gambhir, S. S. (2009). Multiplexed imaging of surface enhanced Raman scattering nanotags in living mice using noninvasive Raman spectroscopy. *Proc. Natl. Acad. Sci. U.S.A* **106**, 13511–13516.

Zhang, X., Young, M. A., Lyandres, O., and Van Duyne, R. P. (2005). Rapid detection of an anthrax biomarker by surface-enhanced Raman spectroscopy. *J. Am. Chem. Soc.* **127**, 4484–4489.

CHAPTER 21

Recent Advances in Flow Cytometric Cell Sorting

Geoffrey W. Osborne

Queensland Brain Institute/Australian Institute for Bioengineering and Nanotechnology, The University of Queensland, Brisbane, Queensland, Australia

Abstract

The classification and separation of one cell type or particle from others is a fundamental task in many areas of science. Numerous techniques are available to perform this task; however, electrostatic cell sorting has gained eminence over others because, when combined with the analysis capabilities of flow cytometry it provides flexible separations based on multiple parameters. Unlike competing technologies, such as gradient or magnetic separations that offer much larger total throughput, flow cytometric cell sorting permits selections based on various levels of fluorescent reporters, rather the complete presence or absence of the reporter. As such, this technology has found application in a huge range of fields. This chapter aims to describe the utility of single-cell sorting with particular emphasis given to index sorting. This is followed by two recently developed novel techniques of sorting cells or particles. The first of these is positional sorting which is useful in cell-based studies where sorting can proceed and produce meaningful results without being inherently dependant on prior knowledge of where gates should be set. Secondly, reflective plate sorting is introduced which positionally links multiwell sample and collection plates in a convenient assay format so that cells in the collection plate "reflect" those in the sample plate.

I. Introduction

Electrostatic cell sorting, often later referred to as fluorescence-activated cell sorting (FACS[TM]: a term introduced by Herzenberg and trademarked by Becton Dickinson), was initially described (Fulwyler, 1965) in 1965 and while the potential was clear, in the next decade there were few publications. However, with the commercialization of technology and increase in the availability of cell sorting instruments, there has been a steady increase in publications utilizing FACS, with a dramatic increase in the last decade (Fig. 1).

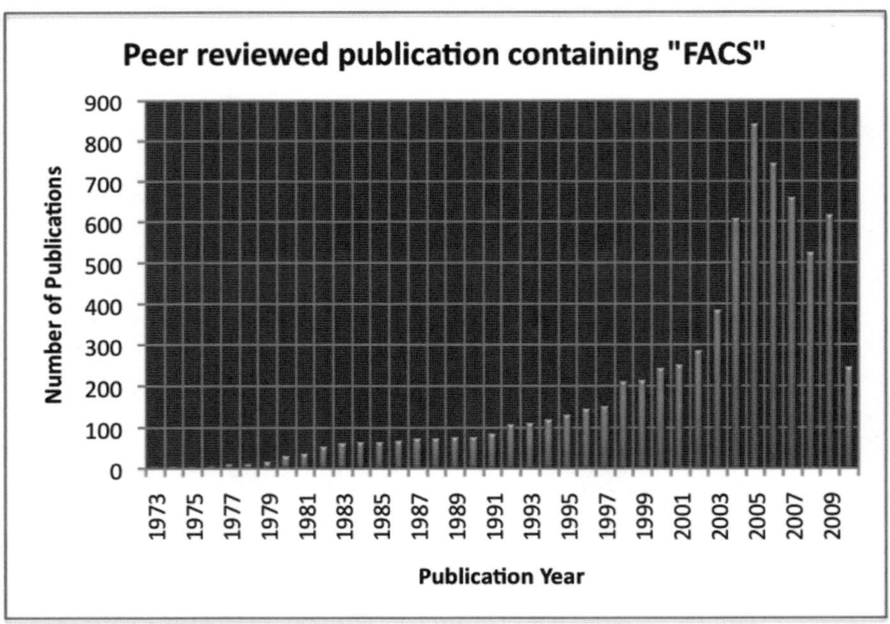

Fig. 1   Summary of the number of publications utilizing Fluorescence Activated Cell Sorting (FACS) as opposed to the term "flow cytometry" in Thomson Reuter ISI Web of Knowledge Databases. Data shown are for the period 1973 to mid 2010.

The reasons for the increasing popularity are many; however, there is little doubt that the rise of stem cell research and the fundamental role of flow cytometry and cell sorting in this field are at least partially linked (Johnson *et al.*, 2007). Indeed investigations into both progenitor (Sugiyama and Kim, 2008) and stem cells would not have made the considerable progress to date without the insight gained using flow cytometry. In stem cell research, cell sorting is the definitive method for the assessment and separation of side population cells (Goodell, 2002; Goodell *et al.*, 1996) that have now been characterized in a range of cell types (Lin and Goodell, 2006; Nadin *et al.*, 2003; Wulf *et al.*, 2003). Stem cells potentially have great clinical applications, and it is likely that sorting will play some role in the preparation of a therapeutic product.

Both patients and physicians hold hopes that purified cells, as a part of cell-based therapy, will aid in the treatment of cancers, and neurodegenerative diseases, such as Parkinson's disease. While much cell sorting involves the enrichment and purification of subpopulations of cells from mixed suspensions, the cancer stem cell hypothesis (Bonnet and Dick, 1997) and other work have shown the dangers than a single contaminating cell in an otherwise pure population may present. With these factors in mind, research experiments that rely on the purification and separation of single cells are of critical importance.

II. Single-Cell Deposition and Index Sorting

The first reports of successful deposition and experimentation using single cells began in the 1970s, and by the end of the decade accurate deposition of cells was reported (Stovel and Sweet, 1979). Since that time, there have been a significant number of publications that have utilized single-cell sorting to answer scientific questions that would have been difficult to answer in any other fashion (Battye *et al.*, 2000). However, while there have been applications that utilize single-cell deposition there have been technical limitations, and lack of technical advances that has seen this area of flow cytometry stagnate until recently. Here we discuss single-cell deposition and in particular focus on the possibilities associated with index sorting. Following that we present two advances in single-cell deposition that cannot be addressed using conventional sorting methods, the first of these is positional cell sorting, and the second is reflective plate cell sorting.

Single-cell deposition has been available for a number of years and yet the power of this approach to cell sorting has, in the opinion of the author, been largely underutilized. This is possibly due to the lack of knowledge in the wider research community as to the availability of this sorting methodology and a thorough understanding of how it may be utilized experimentally.

In the simplest terms, single-cell deposition may be thought of as an elegant limiting dilution assay. As the name implies, single cells meeting selected criteria are sorted to known locations. This technique has been successfully used for the tissue culture (Gasche *et al.*, 2003), single-cell PCR (Bertram *et al.*, 1995) or

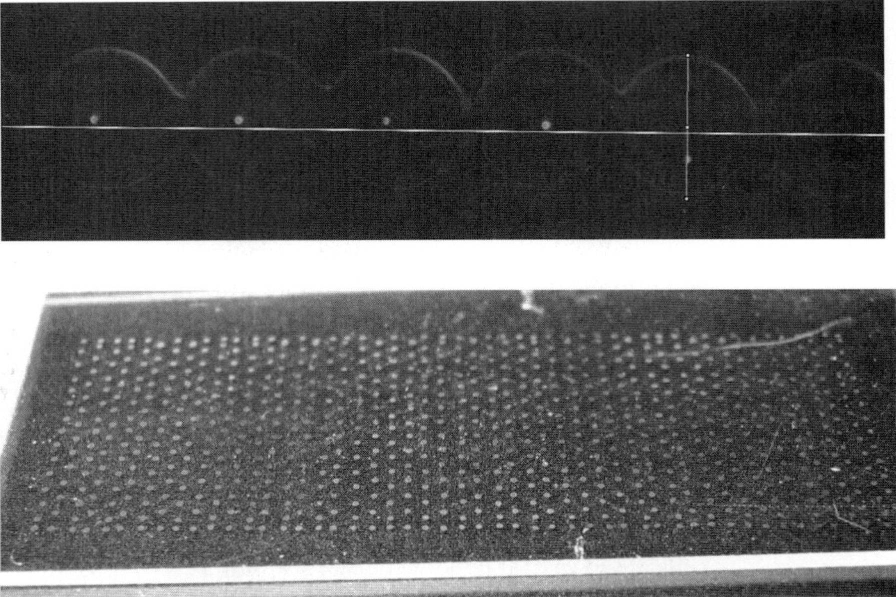

Fig. 2   Upper panel demonstrates the accuracy with which 3 μm beads (Flowcheck, Beckman Coulter) can be deposited on a glass microscope slide. The slide has been dipped in fetal calf serum (FCS) and before the FCS has fully solidified, drops containing beads sorted onto the slide. Each drop forms an impact impression in the semi solid FCS, the larger circular halo, with the bright central spot the bead. In this example 4 of the 5 sorted beads show little deviation from a horizontal plane (line left to right), while the right most drop, the impact puddle of which is bisected by a vertical line, shows some variability in particle position.

microscopic evaluation (Alberti *et al.*, 1984; Kanz *et al.*, 1986). Fig. 2 shows an elegant example of the accuracy with which particles can be deposited, not only with respect to drop positions, but also to the position of particles within the drop. If particles within a drop do not occupy the center position during flight, then they are positioned off center when they are captured on the slide, with respect to the horizontal viewing plane of the photograph.

Dependant on the cell type and application, a variety of cell collection media are suitable, and those applicable to normal cell sorting applications are also suitable to single cell deposition. However, there are a number of technical considerations, related to specimen preparation and instrumentation that are important when undertaking this kind of sorting. Special consideration should be given to the following.

A. Cell Viability

The importance of using viability indicators really cannot be overstated, as all single-cell assays for cell culture require viable cells as a starting point. Thus, it is essential to prepare and maintain the sample for the duration of the sort under conditions that are optimal for cell "health". This can be monitored during cell sorting by comparing the percentage of the total population of cells staining as non-viable, as the sort progresses, with the percentage reported as non-viable when the sort began. Good results can be obtained using traditional viability indicators such as Propidium Iodide or 7-aminoactinomycin-D, used at a final concentration of 3 or 20 $\mu g/\mu L$, respectively. Alternatively, there are now a host of available viability probes, for example Calcein AM, which is particularly bright and pH insensitive. Comprehensive information regarding this fluorochrome or others which may be spectrally more suitable for a given application can be found at: http://www.invitrogen.com/site/us/en/home/References/Molecular-Probes-The-Handbook/Assays-for-Cell-Viability-Proliferation-and-Function/Viability-and-Cytotoxicity-Assay-Reagents.html.

In addition, the time it takes to sort infrequently occurring cells should also be considered in the context of maintenance of cell viability. For cells that are infrequent, it is advisable to minimize the time when the cells sit in the media prior to be sorted. This may be achieved by preparing a number of smaller "aliquots" of cells throughout the sorting period, which may last for a number of hours, rather than providing a more concentrated sample that lasts for the duration of the cell sort. A less concentrated sample also provides additional benefits in decreased likelihood of cells settling and forming clumps that can block the cell sorter nozzle and cause a deflection of the stream into a well containing a single cell, thereby compromising the whole experiment.

Another factor that is worthy of consideration with respect to viability is gas exchange and evaporation in the small volumes of fluid contained in multiwell plates. These conditions are cell type dependant and best optimized prior to the experiment.

B. Pressure and Drop Charge

All electrostatic cell sorters drive fluids and the cells contained within the fluids under pressure. When sorting, this results in cells contained within drops travelling at considerable velocities that, when they impact the collection media, can result in decreased cell viability.

Some common pressures, velocities, and drop volume estimates appear in Table I. Using this simple table as a guide the upper limit of working pressures that in the author's laboratory have found to yield viable cells, work down from these pressure ranges to find system pressures which are not affecting cell viability for a particular experiment.

Table I

Guide to sort pressures and stream velocities in Jet-in-Air cell sorters

Nozzle diameter (μm)	Pressure (PSI)	Typical drop drive frequency (Hz)	Pressure (kPa)	Velocity (m/s)	Estimated drop volume (μL)
70	60	96,180	413	24.90	7.84E−04
80	45	72,882	310	21.56	1.17E−03
90	30	52,896	207	17.61	1.67E−03
100	28	45,992	193	17.01	2.28E−03
120	20	32,392	138	14.38	3.95E−03
150	15	22,442	103	12.45	7.71E−03

Decreases in cell viability can be assessed using fluorochromes previously outlined, and longer term by measuring whether apoptosis increases in the days after sorting, compared with nonsorted, but identically treated experimental controls. There is also anecdotal evidence that cell viability is increased by using larger diameter nozzles that form larger drops. The concept is that larger drops may provide a greater protective capsule around the cell and cushion the impact of the drop and thereby negate some of the detrimental effects of the drop velocity on cell viability.

The effects of pressure are best identified empirically for a given cell type. Certain cells maybe more fragile at a particular stage of development, for example blasting B lymphocytes, or cells derived from solid tissue that are subject to digestion enzymes, compared with cells which may be more hardy. For example, it has been shown (Ashcroft and Lopez, 2000) that some cell populations with low initially viability do not withstand the higher pressures associated with high-speed sorting (upper limits indicated in Table I) hence, in general higher pressure should be avoided especially when accuracy (see the following section) may also be adversely affected.

In the context of single drop/cell charging, charges that actually end up on a single drop are extremely difficult to calculate accurately. The reasons for this are due to the dynamic nature of the drop breakoff, and the possibilities of a drop breaking off prematurely, and the electrical potential of the stream itself. In theory, the charge on an isolate conductive drop will relate to the applied potential, and since the stream is conductive the breaking drop should be equal to the applied voltage. In practice though, because the charge is distributed along the stream it will provide some potential and what actually appears on the drop will be the remaining potential rather than the total applied voltage (Fig. 3). (David Parks, personal communication)

The practical relevance of this is that careful attention needs to be paid to the amounts of charge that are applied to sorted drops when trying to place single cells reproducibly into collection vessels. Because cells contained in drops are heavier than drops without cells, slightly different amounts of charge maybe required for accurate single-cell deposition compared with test deflection settings.

C. Validation

All commercially available cell sorters have the ability to position drops with sufficient accuracy to deposit a single cell into the wells of a multiwell plate. The plate with the smallest target area is a 1536 plate with a surface area of approximately 1 mm^2, and are therefore most difficult to establish the correct settings for. Validate that the drop is correctly positioned by initially sorting on to the lid of the plate sitting on top of the plate. Sort a sufficient number of drops at varying well locations to be able to visualize either the sorted particle, or the dried puddle of saline solution from where a drop dried on the plate.

Validation as to whether a drop containing a cell should be carried out in every instance of single-cell sorting by visual confirmation using a microscope. This task can be extremely frustrating when trying to see small essentially transparent cells in clear media. A partial solution is to let the sort collection vessel sit for some time to

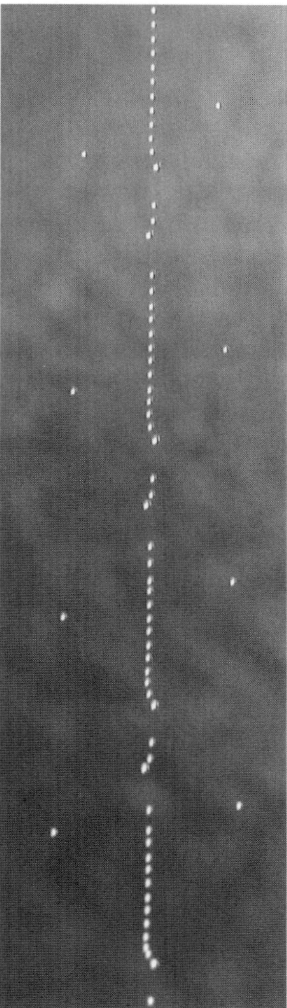

Fig. 3 Demonstration of residual charge on unsorted drops on a single cell "test" sort on a BD Biosciences FACSVantage SE cell sorter where single drops are being deflected both to the left and right. Note that the drops in the main stream following the deflected drop, while still travelling to waste, are holding slight amounts of charge following the grounding of the stream. This causes these drops to deviate slight left or right while still going to waste.

allow the cells to settle at the bottom of the well before viewing. As the cell settling rates are dependent upon the cell size and media viscosity, time periods of up to 1 h post sorting may need to be employed in order to visualize cells. Another suggestion is to perform test sorts using a small sample of the cells of interest, stained with a fluorescent dye, such as DiI or CFSE (see the next section) that can be easily seen under a fluorescence microscope. If a cell type is going to be used frequently for single-cell sorting, then prepare a suitably stained and fixed population of cells that can be used for this purpose.

D. Index Sorting

Index sorting builds on normal single-cell sorting by keeping a list or "index" within a saved flow cytometry file that relates the characteristics of each sorted particle with the well location in the plate into which it was collected. This function adds a parameter to the file to house this index information, or uses an existing "parameter" thus decreasing the number of real parameters that can be saved. However, few experiments require the use of every available detector or parameter and this feature provides a powerful means of relating functional outcome in culture, for example, with the level of fluorescence or scattered light associated with that cell.

E. Outcomes of Single and Index Sorting

A typical graphical representation of an index sorting experiment generated by early CloneCyt® and CELLQuest Pro® software on a BD Biosciences FACSVantage SE flow cytometer is shown in Fig. 4. Sorted events meeting the criteria of both larger regions identified in the plots are deposited into an individual well, with the cell in a particular well identified by the smaller circle within each region. Note that with these early software packages it may be impossible to discern, due to plot resolution and display limitations, which event actually ended up in a well.

It is also important to recognize that instabilities can affect yield (Osborne, 2010) and therefore there may not be a cell in each well, despite what is reported by software.

Recently, more elegant and potentially powerful display options have become available. Fig. 5 shows a typical graphical representation using BD Biosciences

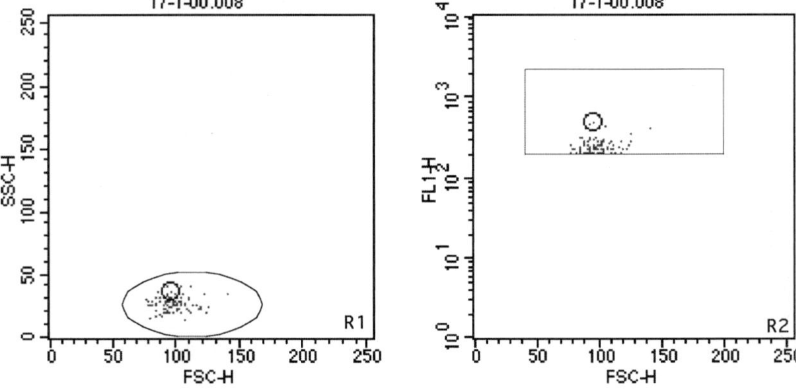

Fig. 4   Data display for single cells sorted in Index Sort mode based on scattered light parameters and fluorescence intensity. Sorted events (coloured red dots) which meet both region 1 (R1) and region 2 (R2) are displayed. When a single well of the plate is selected in software an additional small circular region within both R1 and R2 is draw to highlight measured parameters for the sorted cell. (For interpretation of the references to color in this figure legend, the reader is referred to the Web version of the chapter.)

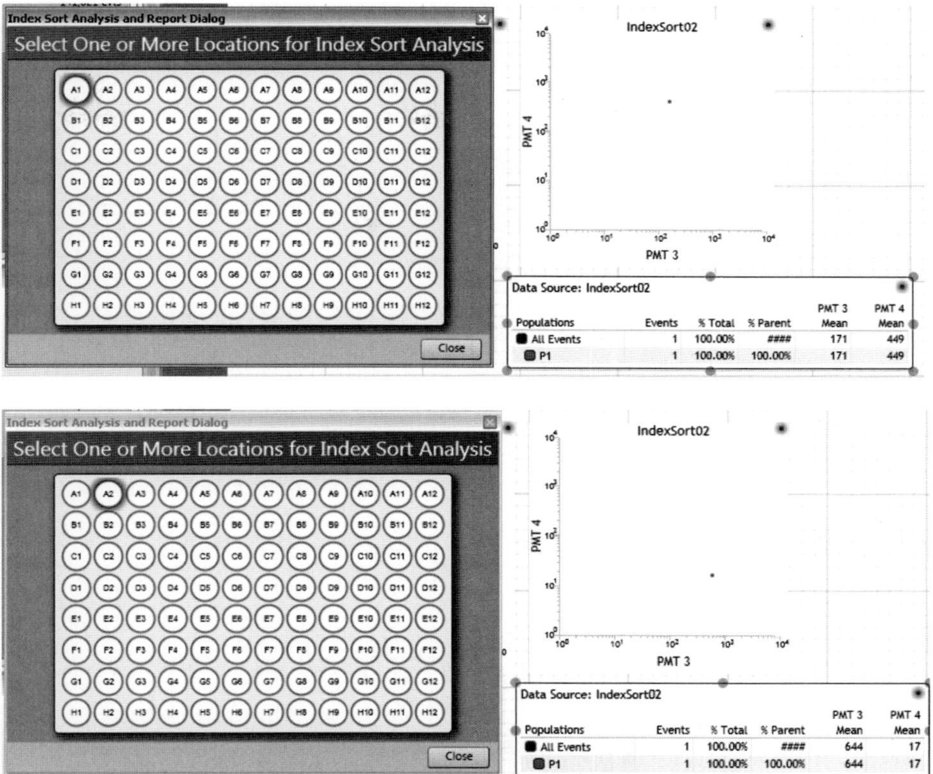

Fig. 5   Data display for single cells sorted and displayed using BD Biosciences Influx software. The upper panel shows the corresponding fluorescence levels (designated PMT3 and PMT4) associated with the cell sorted into well "A1" of a 96 well plate. The lower panel shows the corresponding display for well "A2".

Influx software of a single-cell deposited firstly in well A1 (upper panel), or A2 (lower panel) of a 96 well plate, identifying the fluorescence levels (designated PMT3 and PMT4) associated with each cell.

Alternatively, the range of fluorescence levels on all cells sorted within a column on the plate may be displayed, as in Fig. 6. In the same manner, all cells showing the other region criteria that the cells met, in addition to the primary sort gate criteria, may be displayed as those shown in Fig. 7. This is a powerful means of display that generates a colored map of multiple measured cell characteristics that can easily be taken into consideration when assessing how the cells responded.

Results such as these have been utilized by researchers with a focus on molecular biology in single cells, single cell PCR or DNA screening, and there are many other possibilities. For example, the utility of single-cell index sorting was elegantly demonstrated in a recent paper (Hayashi *et al.*, 2010) that combined index sorting and the single-cell reverse transcription2013polymerase chain reaction (RT2013PCR) to

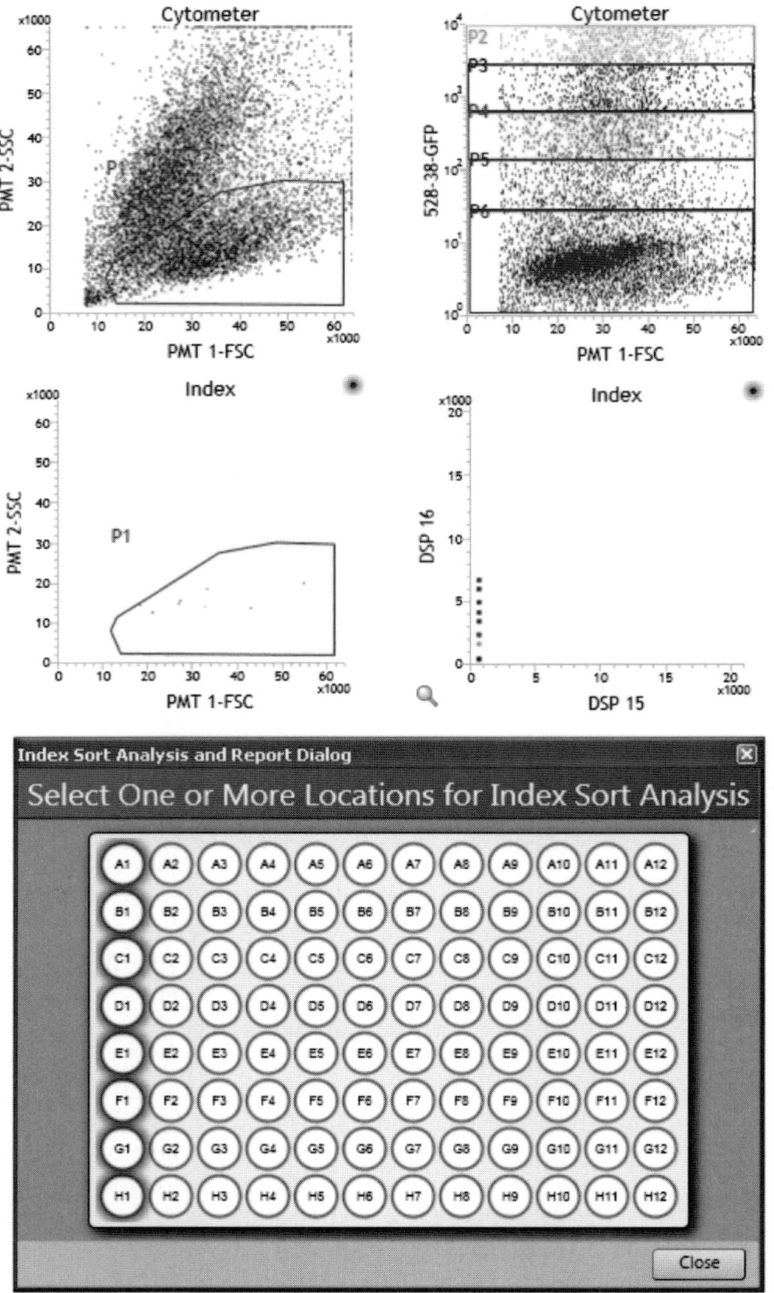
Cytometer
PMT 2-SSC
PMT 1-FSC
Cytometer
528-38-GFP
P2
P3
P4
P5
P6
PMT 1-FSC
Index
PMT 2-SSC
P1
PMT 1-FSC
Index
DSP 16
DSP 15
Index Sort Analysis and Report Dialog
Select One or More Locations for Index Sort Analysis
A1 A2 A3 A4 A5 A6 A7 A8 A9 A10 A11 A12
B1 B2 B3 B4 B5 B6 B7 B8 B9 B10 B11 B12
C1 C2 C3 C4 C5 C6 C7 C8 C9 C10 C11 C12
D1 D2 D3 D4 D5 D6 D7 D8 D9 D10 D11 D12
E1 E2 E3 E4 E5 E6 E7 E8 E9 E10 E11 E12
F1 F2 F3 F4 F5 F6 F7 F8 F9 F10 F11 F12
G1 G2 G3 G4 G5 G6 G7 G8 G9 G10 G11 G12
H1 H2 H3 H4 H5 H6 H7 H8 H9 H10 H11 H12
Close

detect the gene expression and cell cycle state of stem cells. By utilizing the index function and the cDNA information from each cell it was possible to analyze the gene expression and correlate that with components of the cell cycle phase.

These results show that in a routine fashion it is possible to generate single-cell clones. Given the ease with clones can be obtained, and pure population subsequently derived, the lack of impact in research experimentation and publications is surprising. An explanation maybe that the hardware and software required are not widespread, but it is hoped that with improved software that has recently been developed, and with increased publicity, there will be an expansion of experimental use.

III. Positional Sorting

Positional cell sorting deposits cells in two-dimensional arrays that directly reflect the position of two measured parameters on a bivariate dotplot. Thus, a cell that generated a greater signal on the x-axis of a plot is sorted further to the right compared with another cell with a lower signal, while the same principle is applied to the events appearing on the y-axis.

Fig. 8 shows a typical bivariate dotplot (left) from Rainbow 8 peak calibration particles (Spherotech Inc, Chicago, USA, cat # RCP-30-5A) and the resultant positionally sorted populations deposited in a 35 mm diameter culture plate. The positional sort plot plate is shown in the same orientation as the histogram display.

The following outlines the problems that can be addressed using positional cell sorting.

A. Why Positionally Sort?

This newly developed sorting technique has many potential applications. In many experimental situations, the level of protein expressed by a cell, reported by fluorescence, and measured by flow cytometry covers a large biologically useful range. This necessitates the digitization and display of data on a logarithmic scale and the appearance of populations of cells with a log-normal bell-shaped distribution. Fig. 9 shows a single parameter frequency distribution for a relative fluorescence intensity, where typically a number of subpopulations exist but are poorly distinguished in part to due compression of data displayed on a logarithmic scale.

Fig. 6 Column data display for single cells sorted and displayed using BD Biosciences Influx software. Display has three sections: Upper section: upper left shows scattered light dotplot of cells (FSC vs SSC dotplot) with the primary region used for Index Sorting, designated P1. Upper right shows a dotplot of scattered light and the level of green fluorescence associated with each cell (FSC vs GFP) with daughter regions of P1 designated P2 to P6 with a colour associated with each region. Middle section: middle left plot shows the distribution of cells with P1 sorted into column A1 to H1 (lower section), while the middle right plot shows a representation of the sorted events coloured coded based on the daughter regions of P1. Lower section: graphical representation of a 96 well plate, with the event characteristics associated with column A1 to H1 selected. (For interpretation of the references to color in this figure legend, the reader is referred to the Web version of the chapter.)

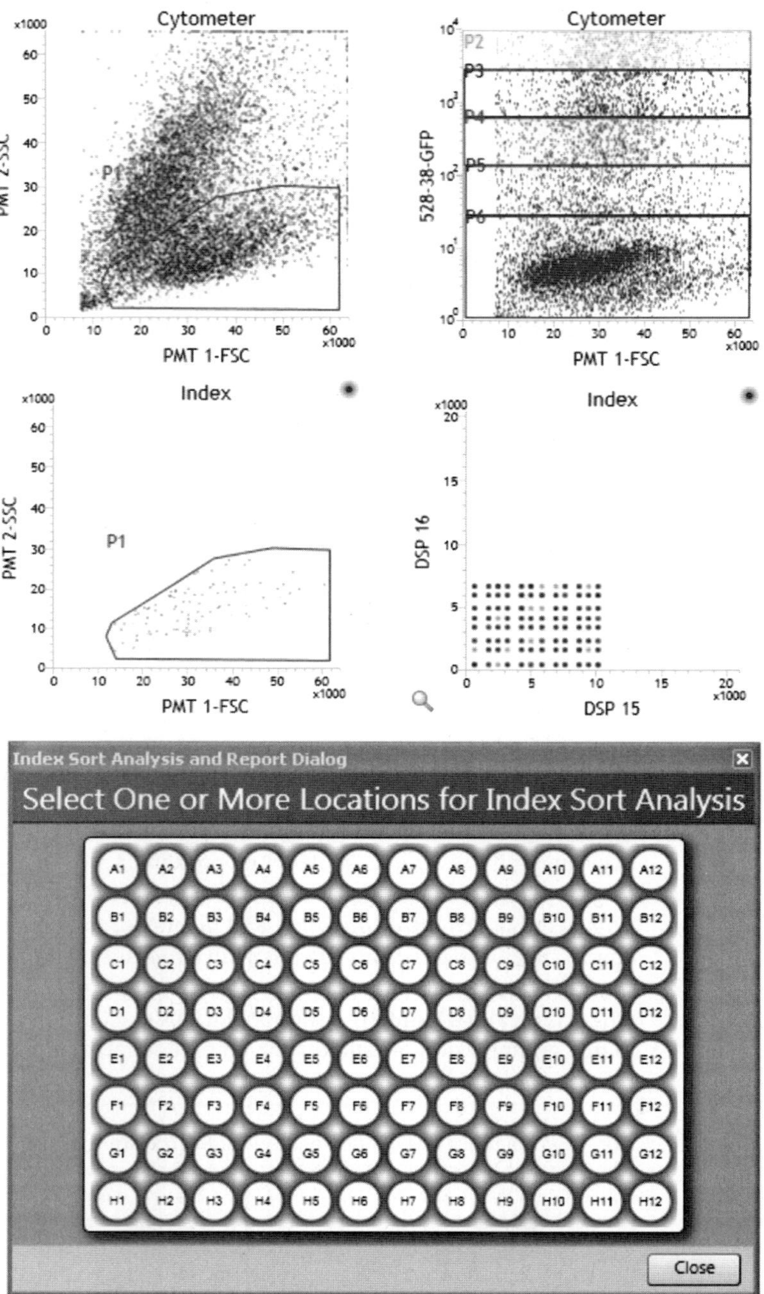
Cytometer
PMT 2-SSC
P1
PMT 1-FSC
x1000
Cytometer
528-38-GFP
P2
P3
P4
P5
P6
PMT 1-FSC
x1000
Index
PMT 2-SSC
P1
PMT 1-FSC
x1000
Index
DSP 16
DSP 15
x1000
Index Sort Analysis and Report Dialog
Select One or More Locations for Index Sort Analysis
A1 A2 A3 A4 A5 A6 A7 A8 A9 A10 A11 A12
B1 B2 B3 B4 B5 B6 B7 B8 B9 B10 B11 B12
C1 C2 C3 C4 C5 C6 C7 C8 C9 C10 C11 C12
D1 D2 D3 D4 D5 D6 D7 D8 D9 D10 D11 D12
E1 E2 E3 E4 E5 E6 E7 E8 E9 E10 E11 E12
F1 F2 F3 F4 F5 F6 F7 F8 F9 F10 F11 F12
G1 G2 G3 G4 G5 G6 G7 G8 G9 G10 G11 G12
H1 H2 H3 H4 H5 H6 H7 H8 H9 H10 H11 H12
Close

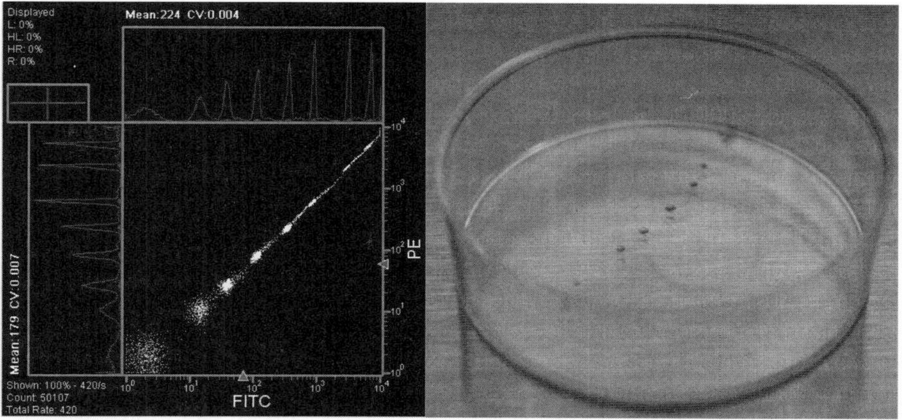

Fig. 8 Positional sorted Rainbow 8 peak calibration particles (Spherotech Inc, Chicago, USA, cat # RCP-30-5A) are shown the in the right panel of the figure, deposited in a 35 mm diameter culture dish in the absence of media for the purposes of alignment and confirmation of the maintenance of positional information. The left panel of the Fig. shows the dotplot display that was used to generate the layout of the positional sort. The positional sort dish is shown in the same orientation as the dotplot display. Note that within the second population of white dots on the left of the dotplot some dots are coloured red indicating at the time of the plot capture the events meeting this criteria were being positionally sorted. (For interpretation of the references to color in this figure legend, the reader is referred to the Web version of the chapter.)

In many instances, for example cell division tracking experiments, regions will be used as the basis for cell sorting, and are created to encompass peaks on the displays. However, as the peaks overlap, as shown by the colored areas under the modeled histogram peaks, the regions used for sorting (boundaries indicated by blue vertical lines) create uncertainty in distinguishing one population from another. It is not known within the heterogeneous population of cells, defined by the region, which cells are the ones that are critical to a successful outcome post cell sorting, and which peak they actually represent.

A conventional approach to answer this problem is to divide the log-normal distribution into smaller segments. The experiment is repeated with these fractions until it is determined whether various levels indicated by the fluorescent reporter are functionally important. Experimentally this procedure would be expensive in terms of time, labor, and reagents, and thus it would be beneficial

Fig. 7 Whole plate data display for single cells sorted and displayed using BD Biosciences Influx software. Display has three sections: Upper section: upper left shows scattered light dotplot of cells (FSC vs SSC dotplot) with the primary region used for Index Sorting, designated P1. Upper right shows a dotplot of scattered light and the level of green fluorescence associated with each cell (FSC vs GFP) with daughter regions of P1 designated P2 to P6 with a colour associated with each region. Middle section: middle left plot shows the distribution of cells with P1 sorted into the entire plate while the middle right plot shows a representation of the sorted events coloured coded based on the daughter regions of P1. Lower section: graphical representation of a 96 well plate, with the event characteristics associated with the entire plate selected. (For interpretation of the references to color in this figure legend, the reader is referred to the Web version of the chapter.)

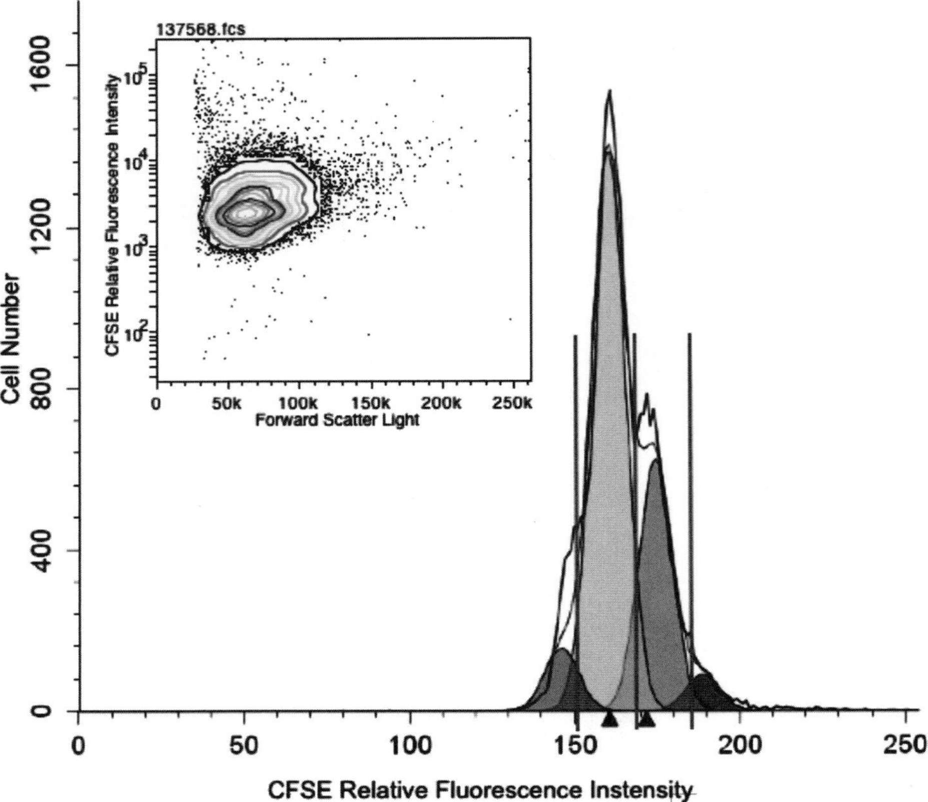

Fig. 9 Frequency histogram display of CFSE Relative Fluorescence Intensity showing logarithmic data displayed on a 0 to 256 channel linear scale. The data has been modelled using Modfit LT® version 3.2 (Verity Software House) to show various populations exist as discrete peaks within the data. However as the peaks overlap, as shown by the coloured areas under the modelled histogram peaks, the regions used for sorting (boundaries indicated by blue vertical lines) create uncertainty in distinguishing one population from another. Inlay plot show the same data displayed in a convention contour plot display emphasising the lack of distinction as to where one peak ends and another begins which makes setting sort regions subjective and often inaccurate. (For interpretation of the references to color in this figure legend, the reader is referred to the Web version of the chapter.)

to have a simpler method to address this problem. This is one area where positional sorting may be of value.

In this context, cell division tracking (Fig. 10) can be analyzed based on the quantitative serial halving of the membrane permeant, stably incorporating fluorescent dye, carboxyfluorescein diacetate, succinimidyl ester (CFSE or CFDA, SE) (Lyons, 1999).

By performing a positional sort of a bivariate display showing scattered light, indicative of the size of a cell, versus CFSE fluorescence intensity, indicative of cell

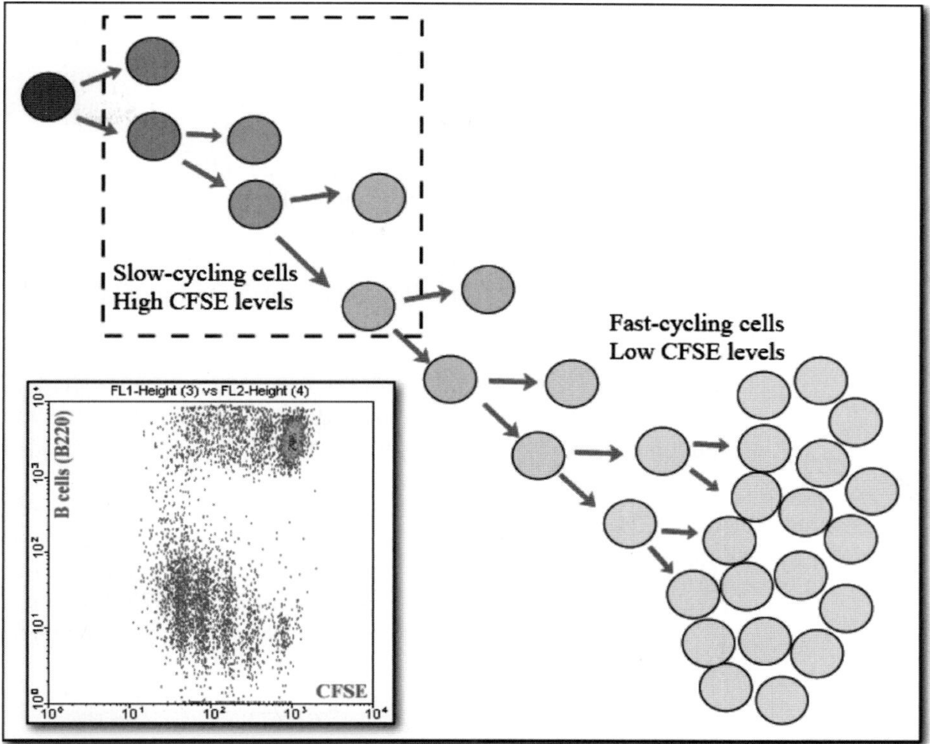

Fig. 10 The principle of cell division tracking using CFSE. Each daughter cell contains half the fluorescence intensity of the parent cell. Brightly stained cells, dark green dots are indicative of cell that are dividing less than more dimly stained cells, yellow dots. Inlay plot demonstrates a combination of CFSE (X axis) vs B220 Antibody staining (Y axis) of mouse B cells, with excellent separation of each cell division population. (For interpretation of the references to color in this figure legend, the reader is referred to the Web version of the chapter.)

division, the question of stem cell division rates and relative size can be addressed without gating, using the Neurosphere assay. This assay is a cell culture survival assay where the ability to survive for extended periods of time and form large colonies of cells (neurospheres) are the key functional readouts. This assay will show whether cells that retain CFSE, and therefore are relatively quiescent, can form large colonies in semi-solid media and how they relate to the flow cytometry plots. This is a hallmark feature of neural stem cells.

In addition, positional sorting can be helpful in applications where fluorescent protein expression levels are used as the sort criteria and the resultant cells used for the tissue culture (Dolnikov *et al.*, 2003). If the sort region encompasses all of the high GFP expressing cells, one does not know whether the very highest cells are critical to culture success, or that GFP level may be associated with toxicity

(Zimmer, 2002) or whether a lower level of protein expression is in any way linked to a better performance in culture. Positional sorting can provide a rapid feedback on where the functionally important cells are in this setting without the need to create regions to sort, which may be incorrectly positioned and exclude the cells of interest.

Other basic research applications utilize cells that are not well characterized, and the cells that will grow and respond in the experimental culture conditions are being tested. Here positional sorting may play a role. A typical scenario is in stem cell research where populations of cells are differentiating and becoming committed to a particular cell lineage. If cells can be sorted into culture plates that allow cell growth and expansion and can be meaningfully linked to the original histograms then time, and reagents, taken to determine the "functional" cells can be greatly reduced compared to sorting many small fractions of the total population. In addition, fewer cells are needed to perform pilot studies by sampling from the whole of the displayed population, so functional results, representative of the total pool, yet derived from individual cells are generated.

Lastly, there are applications for positional sorting combined with direct hybridization of DNA. Cells can be sorted onto nitrocellulose filters and the DNA probed with nucleic acid specific probes for the presence of specific sequences within the DNA (Bianchi *et al.*, 1987). Positional sorting provides the capability of locating the cells expressing the gene of interest without further gating.

B. Positional Sorting Methodology

Two examples of positional sorting using primary tissue derived from the forebrain subventricular zone of adult mice, and a cell line, HEK293T cells follow. Note that these protocols are designed to give results that are representative of a particular histogram layout and will require optimization based on the requirements of each cell type.

Striata from embryonic day 14 C57/BL6 mice and the forebrain subventricular region of adult C57/BL6 mouse brains were microdissected and the tissue enzymatically processed with Trypsin/EDTA (0.05%, Gibco). Cells were plated in neurosphere assay media and supplemented with 10 ng/mL of EGF (10 ng/mL, BD Biosciences). Cells were resuspended in PBS prior to sorting at a concentration of 1×10^6/mL. Cells were labeled with CFSE at a concentration of 2×10^6 cells in Neurosphere Media with 1 µL CFSE (final concentration 5 µM) for 10 min at room temperature, and then the reaction was stopped by washing three times using Neurosphere Media containing 10% fetal calf serum (FCS), then resuspended in 1 mL on PBS. Cells were sorted directly into NeuroCult® Medium (Stem Cell Technologies cat #05702) for expansion of mouse neural stem and progenitor cells.

HEK293T cells were grown in DMEM (Invitrogen) supplemented with 10% FCS and penicillin/streptomycin (Invitrogen), in the CO_2 incubator at 37 °C + 5% CO_2.

Cells were washed once with PBS, trypsinised and resuspended in new media. After centrifugation at 600 g for 5 min, cells were resuspended at a concentration of 1×10^6 cells/mL in PBS.

All the experiments described in this section were performed using an Influx cell sorter (BD Biosciences) that currently provides both index sorting and positional sorting functions.

Perform normal daily instrument alignment and quality control procedure as per the manufacturer's recommendations.

- Ideally system pressures should be kept low to minimize the affect of pressure on cells are they are deflected directly into a semi-solid media. Traditional low speed sorting pressures of 10-12PSI work well and help in achieving good particle deflection.
- Nozzles diameters may also require optimization to ensure the best accuracy for the deflection of drop.

Prepare the instrument for sterile sorting and optimize drop delays for maximum recovery.

- Sterility is important to ensure a good result and cannot be over emphasized, as in functional assays cells may be cultured for up to 21 days.

Follow the manufacturer's recommendations for adjusting and optimizing the plate or well layout for the collection device you will be using.

- In the authors' laboratory 35 mm diameter dishes are used frequently for this application as they provide a suitable compromise between the amount of media that is needed, and the amount of deflection that can be obtained using normal droplet charge voltages and sort deflection plate high voltages.

If using semisolid tissue culture media, set the stage where the tissue culture plate is positioned for the duration of the cell sort, to the appropriate temperature.

- In the case of both media used in this example, 37 °C maintains the media at an appropriate level of solidity. This is a critical step as if the media is too liquid then the location of the embedded cells will be lost as soon as the plate is moved to the incubator. Ensure that media consistency is sufficient to allow droplets and cells to embed well into the media, and yet not so solid as to prohibit cell penetration.

Positional sorting in the current implementation is a very simple procedure involving making the plot you wish to base sorting on as the active plot and selecting "Positional sorting" as the sort mode.

- Place the collection vessel on the sort stage at an appropriate position to receive sorted cells and commence sorting.
- At the conclusion of the sort, remove the plate and place in a 37 °C incubator with 5% CO_2 for two weeks (HEK293T cells) or three weeks (brain cells).

- Remove plates and image using a convention microscopy platform that can combine images to yield a complete picture of the collection vessel.
- Alternatively use a plate based readout system which images the plate, selects clones, and picks and places them in 96 well plates for further assays (e.g., see http://www.genetix.com and the ClonePix FL system).

C. "Functional" Displays of Bivariate Plots

Fig. 11 shows typical results for HEK293T cells after two weeks in culture. Colonies of cells appearing are evident and appear as groupings of "dots" in the two sections of a 90 mm diameter plate that has been used for two sorting experiments, top and bottom of lower left panel.

Following a positional sorting experiment such as this, colonies that are producing antibody, as shown by colorimetric indicators within the media can be selected, and sub-cloned either manually or mechanically. Once the critical area of the flow cytometry plot has been identified on this basis, conventional small regions and gates can be made to accurately select for the cells of interest. Indeed a recommended procedure to follow in this implementation is the deposition of single cells in known locations based on the identified regions using an index sort function previously described so that the most successful clones can once again be tracked back to the source cell and its position on the dotplot.

Fig. 12 shows the resulting large colonies indicative of neural stem cells. These types of results are indicative that pressures within the cell sorter permit cells to be sorted and remain viable. However, it is difficult to calculate the total number of stem cells which may have been present in the starting population as infrequent cells, which stem cell are, may not be sorted during the passage of the sort. Thus, positional sorting essentially provides a representation of cell population frequencies within the sample, which may not be sufficient to allow enough cells to be sorted, survive, and grow in culture.

IV. Reflective Plate Sorting

A. Introduction

The selection and cloning of cells described above result in single cells being deposited and growing in individual wells of multiwell plates. The daughter cells that grow from clones normally still show some diversity in antibody producing ability or protein expression levels. It is convenient to sub-clone these daughter cells into other 96 well plates so that high expressing cells or other desirable cell characteristics can be continually selected. In addition to creating a sub-cloned plate, it is preferable to link the well location of the original clone and plate with the subsequent daughter plates. Thus, a cell sorted into well number A1, as a single

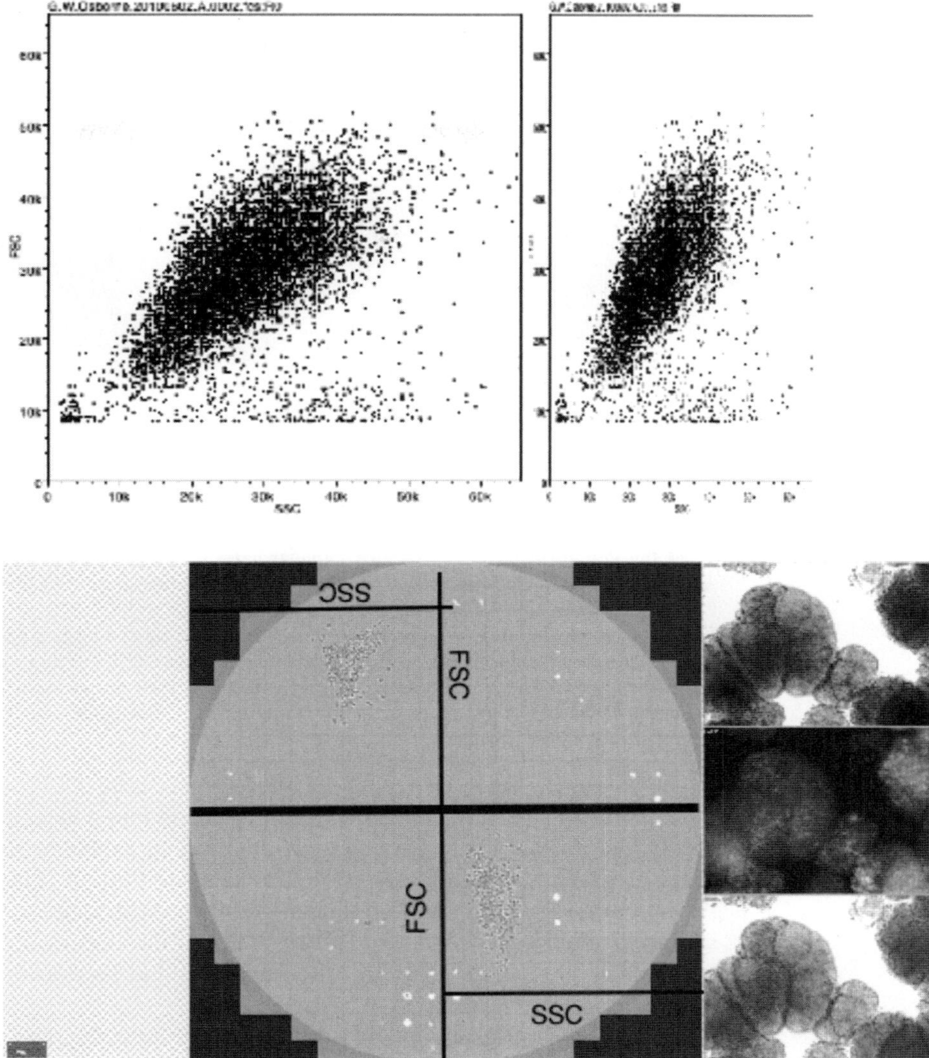

Fig. 11   Typical results for HEK293T cells Positionally sorted at 20 PSI, 120 um nozzle into either 5 or 10 ml of Clonepix™ HEK specific media. Upper panels show in the left dotplot, the original plot layout and the right plot shows the compressed plot that results in the pictures in the lower panel. Spigot® software settings have been modified to increase the spread of data in the "Y" plane, but data spread is limited in the "X" plane by the maximum amount of charge deflection. Note two sorts were done on each plate. Also RH panel of lower figure shows images of embedded cell colonies after two weeks in culture. Far lower left scale bar indicates 5 mm length.

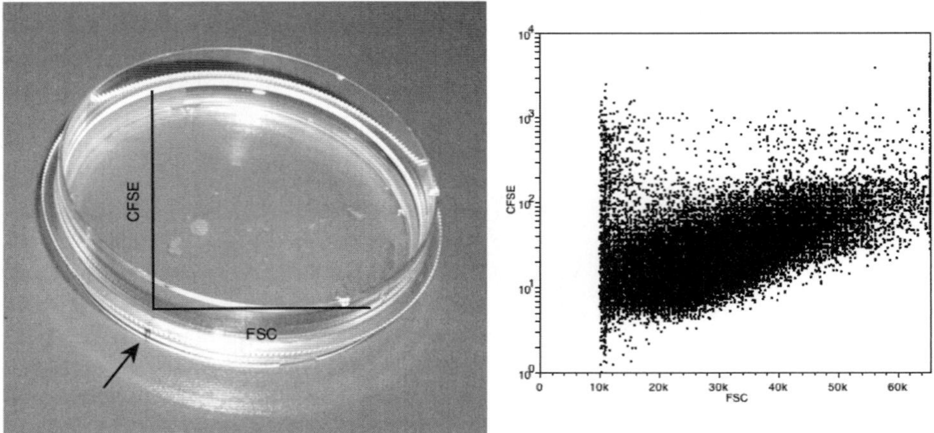

Fig. 12   Positional sorting results of neural directly embedded into semi-solid media to test the CFSE slow cycling cell hypothesis. Left panel of figure shows large colonies (white areas in pink media greater than 2 mm in diameter) indicative of neural stem cells and their origin relative to the X and Y axis of the generating dotplot. Right panel shows the dotplot that was used for the generation of the positional sorting information used to create the left panel. Both panels show the same orientation. This combination of dotplot and resulting positional sort shows that in this instance high CFSE levels, associated with slowly dividing cells was not linked to neural stem cells. This may occur due to the quiescent nature of the stem cells, with lower esterase activity leading to less cleaving of the CFSE to a fluorescent form. Results such as these highlight the benefit of positional sorting as it does not rely on making assumptions as to where to set sorting gates. (For interpretation of the references to color in this figure legend, the reader is referred to the Web version of the chapter.)

clone, can have a single cell sub-cloned by sorting into a daughter plate at location A1 and so on, for as many rounds of sub-cloning as are required. We label this process of linked plate sorting as Reflective Plate Sorting (Fig. 13). This form of sorting is most powerful when used in conjunction with single-cell deposition index sorting, as then a record of the characteristics of the starting cell, which leads to a particularly productive clone, can be identified and used as a putative target for further experimental work.

Reflective plate sorting has possibilities in the following areas:

• Generation of clonal populations derived from a single expanded clone.
• Generation of subplates from a master plate, containing known numbers of cells.

B. An Implementation of Reflective plate Sorting

Reflective plate sorting is described using a BD Biosciences Influx Cell Sorter, fitted with a Cytek Development Automated Microsampler (AMS) control unit and running Spigot 6.1 software.

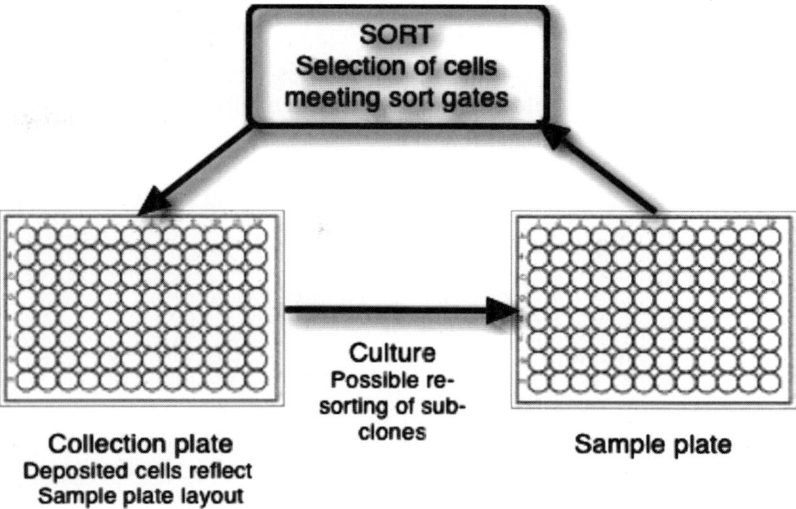

Fig. 13 Schema of Reflective Plate Sorting. The basic principle is to link the movement of a sample multi-well plate, with a collection multi-well plate, so that subsets of sorted cells or particles from the sample plate can be deposited in the collection plate, and maintain the identical well positional information of the sample plate. This can become an iterative process, with the collection plate becoming the sample plate, for example after culturing the cells in the initial collection plate, and further rounds of reflective plate sorting carried out.

- Perform daily alignment and quality control procedures.
- Establish an appropriate instrument drop delay and sort settings for 10 to 12 PSI sheath pressure in Spigot version 6.1 software.
- Run a test sample from a tube placed on the sample injection port on the Influx.
- Establish an appropriate gating protocol to define the populations of interest for subsequent sorting and save configuration.
- Pipette 200 µL of the test sample into well A1 of a 96 well plate and place on the plate holder of the Cytek AMS.
- Remove sample tube and connect the sample injection line from the Influx sample injection port to Cytek AMS sample line.
- Start the Cytek AMS and run a sample with ''test mode'' set on the AMS console to verify that the set differential pressure is sufficient to provide cells to the cytometer from the well A1 test sample.
- Close Spigot software then edit the file ''cytek config.xml'' so that ''sync with sort'' is set to ''1'' enable and the ''timeout threshold'' allows sufficient time for sorting to proceed, with 100 s a good starting point. This time will be affected by differential pressure and the volume in the well. Save the modified file.
- Restart spigot 6.1 and reload the configuration file previously saved.
- Select 96 well plate as the sort configuration.

- Enter the number of cells to be sorted per well. This can be greater than 1 if required.
- Selected the "External Devices" tab, check "Cytek" checkbox as the external device and enter number of events for each file.
- Confirm that override is "ON".
- Confirm Cytek is connected to the cytometer.
- Start Cytek, program settings, etc. and start the sort via the Cytek AMS.

Typical Working Cytek AMS Settings

Mix	1–5
Boost	2–5
Backflush	5–10
Wash	5–10

This procedure takes 30 s for each sample to appear on the cytometer. Usually over 1 min for each well.

C. Expected Results

The procedure detailed above will result in a desired number of cells being sorted from the sample plate well into the corresponding well in the collection plate. As the probe in the sample plate moves from well to well, with a wash step between wells, it is important to recognize that contamination from one sample well to another can occur. Carryover of one sample to another is not acceptable in reflective plate sorting. This might mean that when a colony is grown from a well after sorting, it is possible that the colony of interest came from the previous well rather than the expected deposited/reflected well. Thorough testing of the system is recommended, using labeled cells with different fluorescent labels in alternating wells of the plate. Perform reflective plate sorting and monitor the fluorescence of the cells in the reflected plate (the collection plate) to ensure that each well contains cells with the appropriate fluorescent label based on the parent sample plate. If this is not the case, optimization of the "Backflush" and "Wash" step times on the Cytek AMS unit as required to obtain the expected result. Note that we recommend the use of labeled cells, rather than fluorescent calibration beads in this context as experience has shown that cells have different adherent properties to the sample probe and plastic sample lines, than do beads. If all else fails, and sample carried over is persistent, the current procedure to prevent this occurring is to only seed every second well in the sample plate, with every other well containing buffered saline or a wash solution. This will, however, delay the sorting procedure as there are no events to trigger the instrument on in the wash wells, and thus the "time-out" value previously saved into the configuration file comes into effect.

In the current implementation, reflective plate sorting results in the generation of an individual data file for each well that has been analyzed. The file is closed and the system moves on to the next well when either of two conditions is satisfied:

(1) the total count set for each file has been met and no events meeting the sort criteria were encountered or (2) the required number of events to be sorted, be it one event or more is met, regardless of the file count setting. The system will close a file without sorting and move on to the next well based on the time-out setting in the configuration file. Configured correctly, it will not continue to sample until a sample plate well is empty as this introduces air into the flow cytometers fluidics and disrupts further sorting. Data files are numbered both by well identifier and sequentially in order of acquisition, thus <filename>.<A01>.<0001.fcs>, <filename>.<A02>.<0002.fcs>, and so on. As this system is designed for "walk away" operation during the course of running a whole sample plate, it is recommended that batch analysis be performed on the data files to identify any problems that may have occurred during the experiment. It is important to quantify the number of cells that the system sorted into each well, as even with the sort count criteria set to a single cell, if the cell of interest is infrequent, wells may contain no sorted cells.

V. Summary

This chapter has attempted to illustrate some of the utility and recent developments in electrostatic cell sorting. Single-cell deposition has been shown to be an experimentally powerful procedure and progress in software display of indexed populations should lead to increased usage in the broad range of scientific areas. Two new sorting procedures, positional sorting and reflective plate sorting have also been described. Positional sorting has been demonstrated linked to functional cell-based readout where it provides information about the characteristics of a cell at a particular location on a display histogram without *a priori* gating strategy to determine sorting decisions. The power of this approach in antibody clone generation is currently under development, and has the potential to be extremely useful in this and other areas where cell surface markers are poorly defined or absent. Lastly, reflective plate sorting has been introduced in the context of cloning and sub-cloning of cells, while still maintaining a reference to the original starting population. This form of sorting takes the serial dilution of cells to generate clones to another level. It is hoped that reflective plate sorting will be widely accepted by the flow cytometry community as instrumentation with this capacity becomes more widespread.

Acknowledgments

The author gratefully acknowledges the helpful suggestions provided by John Wilson in the preparation of this chapter. The author also wishes to acknowledge the work of Loic Deleyrolle, Benjamin Hughes, and Michael Song in experiments involving positional sorting, and Barclay Purcell for positional and reflective plate sorting software engineering.

References

Alberti, S., Stovel, R., and Herzenberg, L. (1984). Preservation of cells sorted individually onto microscope slides with a fluorescence-activated cell sorter. *Cytometry* **5**, 644–647.

Ashcroft, R., and Lopez, P. (2000). Commercial high speed machines open new opportunities in high throughput flow cytometry (HTFC). *J. Immunol. Methods* **243**, 13–24.

Battye, F., Light, A., and Tarlinton, D. (2000). Single cell sorting and cloning. *J. Immunol Methods* **243**, 25–32.

Bertram, S., Hufert, F., Neumann-Haefelin, D., and von Laer, D. (1995). Detection of DNA in single cells using an automated cell deposition unit and PCR. *Biotechniques* **19**, 616–620.

Bianchi, D. W., Harris, P., Flint, A., and Latt, S. A. (1987). Direct hybridization to DNA from small numbers of flow-sorted nucleated newborn cells. *Cytometry* **8**, 197–202.

Bonnet, D., and Dick, J. (1997). Human acute myeloid leukemia is organized as a hierarchy that originates from a primitive hematopoietic cell. *Nat. Med.* **3**, 730–737.

Dolnikov, A., Shen, S., Millington, M., Passioura, T., Pedler, M., Rasko, J. E. J., Symonds, G. (2003). A sensitive dual-fluorescence reporter system enables positive selection of ras suppressors by suppression of ras-induced apoptosis. *Cancer Gene Ther.* **10**, 745–754.

Fulwyler, M. J. (1965). An electronic particle separator with potential biological application. *Science* **150**, 371.

Gasche, C., Chang, C. L., Natarajan, L., Goel, A., Rhees, J., Young, D. J., Arnold, C. N., Boland, C. R. (2003). Identification of frame-shift intermediate mutant cells. *PNAS* **100**, 1914–1919.

Goodell, M. A. (2002). Multipotential stem cells and 'side population' cells. *Cytotherapy* **4**, 507–508.

Goodell, M. A., Brose, K., Paradis, G., Conner, A. S., and Mulligan, R. C. (1996). Isolation and functional properties of murine hematopoietic stem cells that are replicating *in vivo*. *J. Exp. Med.* **183**, 1797–1806.

Hayashi, T., Shibata, N., Okumura, R., Kudome, T., Nishimura, O., Tarui, H., Agata, K. (2010). Single-cell gene profiling of planarian stem cells using fluorescent activated cell sorting and its "index sorting" function for stem cell research. *Dev., Growth Differentiation* **52**, 131–144.

Johnson, K., Dooner, M., and Quesenberry, P. (2007). Fluorescence activated cell sorting: a window on the stem cell. *Curr. Pharm. Biotechnol.* **8**, 133–139.

Kanz, L., Bross, K., Mielke, R., Lohr, G., and Fauser, A. (1986). Fluorescence-activated sorting of individual cells onto poly-L-lysine-coated slide areas. *Cytometry* **7**, 491–494.

Lin, K. K., and Goodell, M. A. (2006). Purification of hematopoietic stem cells using the side population. *Methods Enzymol.* **420**, 255–264.

Lyons, A. (1999). Divided we stand: tracking cell proliferation with carboxyfluorescein diacetate succinimidyl ester. *Immunol. Cell Biol.* **77**, 509–515.

Nadin, B. M., Goodell, M. A., and Hirschi, K. K. (2003). Phenotype and hematopoietic potential of side population cells throughout embryonic development. *Blood* **102**, 2436–2443.

Osborne, G. (2010). A method of quantifying cell sorting yield in "real time". *Cytometry A* **77**(10), 983–989.

Stovel, R., and Sweet, R. (1979). Individual cell sorting. *J. Histochem. Cytochem.* **27**, 284–288.

Sugiyama, T., and Kim, S. (2008). Fluorescence-activated cell sorting purification of pancreatic progenitor cells. *Diabetes Obes. Metab.* **10**(Suppl 4), 179–185.

Wulf, G. G., Luo, K. L., Jackson, K. A., Brenner, M. K., and Goodell, M. A. (2003). Cells of the hepatic side population contribute to liver regeneration and can be replenished with bone marrow stem cells. *Haematologica* **88**, 368–378.

Zimmer, M. (2002). Green fluorescent protein (GFP): applications, structure, and related photophysical behavior. *Chem. Rev.* **102**, 759–782.

VOLUMES IN SERIES

Founding Series Editor
DAVID M. PRESCOTT

Volume 1 (1964)
Methods in Cell Physiology
Edited by David M. Prescott

Volume 2 (1966)
Methods in Cell Physiology
Edited by David M. Prescott

Volume 3 (1968)
Methods in Cell Physiology
Edited by David M. Prescott

Volume 4 (1970)
Methods in Cell Physiology
Edited by David M. Prescott

Volume 5 (1972)
Methods in Cell Physiology
Edited by David M. Prescott

Volume 6 (1973)
Methods in Cell Physiology
Edited by David M. Prescott

Volume 7 (1973)
Methods in Cell Biology
Edited by David M. Prescott

Volume 8 (1974)
Methods in Cell Biology
Edited by David M. Prescott

Volume 9 (1975)
Methods in Cell Biology
Edited by David M. Prescott

Volume 10 (1975)
Methods in Cell Biology
Edited by David M. Prescott

Volume 11 (1975)
Yeast Cells
Edited by David M. Prescott

Volume 12 (1975)
Yeast Cells
Edited by David M. Prescott

Volume 13 (1976)
Methods in Cell Biology
Edited by David M. Prescott

Volume 14 (1976)
Methods in Cell Biology
Edited by David M. Prescott

Volume 15 (1977)
Methods in Cell Biology
Edited by David M. Prescott

Volume 16 (1977)
Chromatin and Chromosomal Protein Research I
Edited by Gary Stein, Janet Stein, and Lewis J. Kleinsmith

Volume 17 (1978)
Chromatin and Chromosomal Protein Research II
Edited by Gary Stein, Janet Stein, and Lewis J. Kleinsmith

Volume 18 (1978)
Chromatin and Chromosomal Protein Research III
Edited by Gary Stein, Janet Stein, and Lewis J. Kleinsmith

Volume 19 (1978)
Chromatin and Chromosomal Protein Research IV
Edited by Gary Stein, Janet Stein, and Lewis J. Kleinsmith

Volume 20 (1978)
Methods in Cell Biology
Edited by David M. Prescott

Advisory Board Chairman
KEITH R. PORTER

Volume 21A (1980)
Normal Human Tissue and Cell Culture, Part A:
   Respiratory, Cardiovascular, and Integumentary Systems
Edited by Curtis C. Harris, Benjamin F. Trump, and Gary D. Stoner

Volume 21B (1980)
**Normal Human Tissue and Cell Culture, Part B: Endocrine,
    Urogenital, and Gastrointestinal Systems**
Edited by Curtis C. Harris, Benjamin F. Trump, and Gray D. Stoner

Volume 22 (1981)
Three-Dimensional Ultrastructure in Biology
Edited by James N. Turner

Volume 23 (1981)
Basic Mechanisms of Cellular Secretion
Edited by Arthur R. Hand and Constance Oliver

Volume 24 (1982)
**The Cytoskeleton, Part A: Cytoskeletal Proteins, Isolation and
    Characterization**
Edited by Leslie Wilson

Volume 25 (1982)
The Cytoskeleton, Part B: Biological Systems and In Vitro Models
Edited by Leslie Wilson

Volume 26 (1982)
Prenatal Diagnosis: Cell Biological Approaches
Edited by Samuel A. Latt and Gretchen J. Darlington

Series Editor
LESLIE WILSON

Volume 27 (1986)
Echinoderm Gametes and Embryos
Edited by Thomas E. Schroeder

Volume 28 (1987)
Dictyostelium discoideum: Molecular Approaches to Cell Biology
Edited by James A. Spudich

Volume 29 (1989)
**Fluorescence Microscopy of Living Cells in Culture,
    Part A: Fluorescent Analogs, Labeling Cells, and Basic Microscopy**
Edited by Yu-Li Wang and D. Lansing Taylor

Volume 30 (1989)
**Fluorescence Microscopy of Living Cells in Culture, Part B: Quantitative
    Fluorescence Microscopy—Imaging and Spectroscopy**
Edited by D. Lansing Taylor and Yu-Li Wang

Volume 31 (1989)
Vesicular Transport, Part A
Edited by Alan M. Tartakoff

Volume 32 (1989)
Vesicular Transport, Part B
Edited by Alan M. Tartakoff

Volume 33 (1990)
Flow Cytometry
Edited by Zbigniew Darzynkiewicz and Harry A. Crissman

Volume 34 (1991)
Vectorial Transport of Proteins into and across Membranes
Edited by Alan M. Tartakoff

Selected from Volumes 31, 32, and 34 (1991)
Laboratory Methods for Vesicular and Vectorial Transport
Edited by Alan M. Tartakoff

Volume 35 (1991)
Functional Organization of the Nucleus: A Laboratory Guide
Edited by Barbara A. Hamkalo and Sarah C. R. Elgin

Volume 36 (1991)
***Xenopus laevis:* Practical Uses in Cell and Molecular Biology**
Edited by Brian K. Kay and H. Benjamin Peng

Series Editors
LESLIE WILSON AND PAUL MATSUDAIRA

Volume 37 (1993)
Antibodies in Cell Biology
Edited by David J. Asai

Volume 38 (1993)
Cell Biological Applications of Confocal Microscopy
Edited by Brian Matsumoto

Volume 39 (1993)
Motility Assays for Motor Proteins
Edited by Jonathan M. Scholey

Volume 40 (1994)
A Practical Guide to the Study of Calcium in Living Cells
Edited by Richard Nuccitelli

Volume 41 (1994)
Flow Cytometry, Second Edition, Part A
Edited by Zbigniew Darzynkiewicz, J. Paul Robinson, and Harry A. Crissman

Volume 42 (1994)
Flow Cytometry, Second Edition, Part B
Edited by Zbigniew Darzynkiewicz, J. Paul Robinson, and Harry A. Crissman

Volume 43 (1994)
Protein Expression in Animal Cells
Edited by Michael G. Roth

Volume 44 (1994)
Drosophila melanogaster: Practical Uses in Cell and Molecular Biology
Edited by Lawrence S. B. Goldstein and Eric A. Fyrberg

Volume 45 (1994)
Microbes as Tools for Cell Biology
Edited by David G. Russell

Volume 46 (1995)
Cell Death
Edited by Lawrence M. Schwartz and Barbara A. Osborne

Volume 47 (1995)
Cilia and Flagella
Edited by William Dentler and George Witman

Volume 48 (1995)
Caenorhabditis elegans: Modern Biological Analysis of an Organism
Edited by Henry F. Epstein and Diane C. Shakes

Volume 49 (1995)
Methods in Plant Cell Biology, Part A
Edited by David W. Galbraith, Hans J. Bohnert, and Don P. Bourque

Volume 50 (1995)
Methods in Plant Cell Biology, Part B
Edited by David W. Galbraith, Don P. Bourque, and Hans J. Bohnert

Volume 51 (1996)
Methods in Avian Embryology
Edited by Marianne Bronner-Fraser

Volume 52 (1997)
Methods in Muscle Biology
Edited by Charles P. Emerson, Jr. and H. Lee Sweeney

Volume 53 (1997)
Nuclear Structure and Function
Edited by Miguel Berrios

Volume 54 (1997)
Cumulative Index

Volume 55 (1997)
Laser Tweezers in Cell Biology
Edited by Michael P. Sheetz

Volume 56 (1998)
Video Microscopy
Edited by Greenfield Sluder and David E. Wolf

Volume 57 (1998)
Animal Cell Culture Methods
Edited by Jennie P. Mather and David Barnes

Volume 58 (1998)
Green Fluorescent Protein
Edited by Kevin F. Sullivan and Steve A. Kay

Volume 59 (1998)
The Zebrafish: Biology
Edited by H. William Detrich III, Monte Westerfield, and Leonard I. Zon

Volume 60 (1998)
The Zebrafish: Genetics and Genomics
Edited by H. William Detrich III, Monte Westerfield, and Leonard I. Zon

Volume 61 (1998)
Mitosis and Meiosis
Edited by Conly L. Rieder

Volume 62 (1999)
Tetrahymena thermophila
Edited by David J. Asai and James D. Forney

Volume 63 (2000)
Cytometry, Third Edition, Part A
Edited by Zbigniew Darzynkiewicz, J. Paul Robinson, and Harry Crissman

Volume 64 (2000)
Cytometry, Third Edition, Part B
Edited by Zbigniew Darzynkiewicz, J. Paul Robinson, and Harry Crissman

Volume 65 (2001)
Mitochondria
Edited by Liza A. Pon and Eric A. Schon

Volume 66 (2001)
Apoptosis
Edited by Lawrence M. Schwartz and Jonathan D. Ashwell

Volume 67 (2001)
Centrosomes and Spindle Pole Bodies
Edited by Robert E. Palazzo and Trisha N. Davis

Volume 68 (2002)
Atomic Force Microscopy in Cell Biology
Edited by Bhanu P. Jena and J. K. Heinrich Hörber

Volume 69 (2002)
Methods in Cell–Matrix Adhesion
Edited by Josephine C. Adams

Volume 70 (2002)
Cell Biological Applications of Confocal Microscopy
Edited by Brian Matsumoto

Volume 71 (2003)
Neurons: Methods and Applications for Cell Biologist
Edited by Peter J. Hollenbeck and James R. Bamburg

Volume 72 (2003)
Digital Microscopy: A Second Edition of Video Microscopy
Edited by Greenfield Sluder and David E. Wolf

Volume 73 (2003)
Cumulative Index

Volume 74 (2004)
Development of Sea Urchins, Ascidians, and Other Invertebrate Deuterostomes: Experimental Approaches
Edited by Charles A. Ettensohn, Gary M. Wessel, and Gregory A. Wray

Volume 75 (2004)
Cytometry, 4th Edition: New Developments
Edited by Zbigniew Darzynkiewicz, Mario Roederer, and Hans Tanke

Volume 76 (2004)
The Zebrafish: Cellular and Developmental Biology
Edited by H. William Detrich, III, Monte Westerfield, and Leonard I. Zon

Volume 77 (2004)
The Zebrafish: Genetics, Genomics, and Informatics
Edited by William H. Detrich, III, Monte Westerfield, and Leonard I. Zon

Volume 78 (2004)
Intermediate Filament Cytoskeleton
Edited by M. Bishr Omary and Pierre A. Coulombe

Volume 79 (2007)
Cellular Electron Microscopy
Edited by J. Richard McIntosh

Volume 80 (2007)
Mitochondria, 2nd Edition
Edited by Liza A. Pon and Eric A. Schon

Volume 81 (2007)
Digital Microscopy, 3rd Edition
Edited by Greenfield Sluder and David E. Wolf

Volume 82 (2007)
Laser Manipulation of Cells and Tissues
Edited by Michael W. Berns and Karl Otto Greulich

Volume 83 (2007)
Cell Mechanics
Edited by Yu-Li Wang and Dennis E. Discher

Volume 84 (2007)
Biophysical Tools for Biologists, Volume One: In Vitro Techniques
Edited by John J. Correia and H. William Detrich, III

Volume 85 (2008)
Fluorescent Proteins
Edited by Kevin F. Sullivan

Volume 86 (2008)
Stem Cell Culture
Edited by Dr. Jennie P. Mather

Volume 87 (2008)
Avian Embryology, 2nd Edition
Edited by Dr. Marianne Bronner-Fraser

Volume 88 (2008)
Introduction to Electron Microscopy for Biologists
Edited by Prof. Terence D. Allen

Volume 89 (2008)
Biophysical Tools for Biologists, Volume Two: In Vivo Techniques
Edited by Dr. John J. Correia and Dr. H. William Detrich, III

Volume 90 (2008)
Methods in Nano Cell Biology
Edited by Bhanu P. Jena

Volume 91 (2009)
Cilia: Structure and Motility
Edited by Stephen M. King and Gregory J. Pazour

Volume 92 (2009)
Cilia: Motors and Regulation
Edited by Stephen M. King and Gregory J. Pazour

Volume 93 (2009)
Cilia: Model Organisms and Intraflagellar Transport
Edited by Stephen M. King and Gregory J. Pazour

Volume 94 (2009)
Primary Cilia
Edited by Roger D. Sloboda

Volume 95 (2010)
Microtubules, in vitro
Edited by Leslie Wilson and John J. Correia

Volume 96 (2010)
Electron Microscopy of Model Systems
Edited by Thomas Müeller-Reichert

Volume 97 (2010)
Microtubules: In Vivo
Edited by Lynne Cassimeris and Phong Tran

Volume 98 (2010)
Nuclear Mechanics & Genome Regulation
Edited by G.V. Shivashankar

Volume 99 (2010)
Calcium in Living Cells
Edited by Michael Whitaker

Volume 100 (2010)
The Zebrafish: Cellular and Developmental Biology, Part A
Edited by: H. William Detrich III, Monte Westerfield and Leonard I. Zon

Volume 101 (2011)
The Zebrafish: Cellular and Developmental Biology, Part B
Edited by: H. William Detrich III, Monte Westerfield and Leonard I. Zon

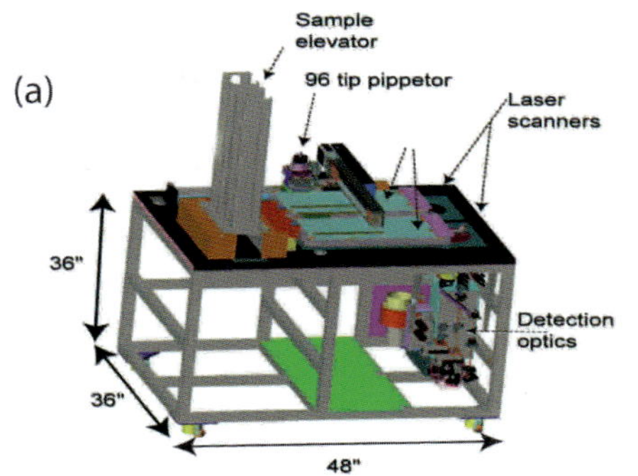

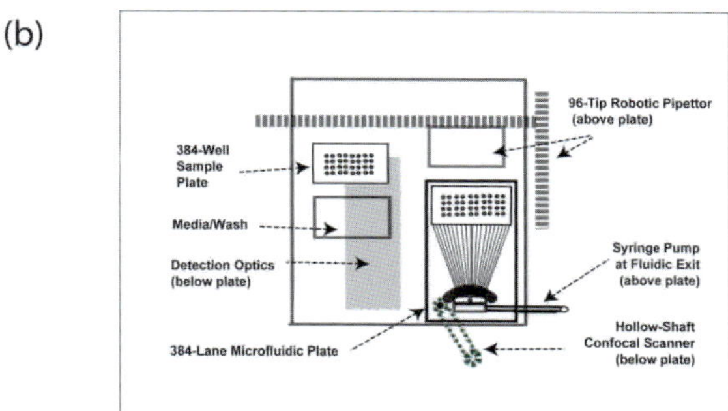

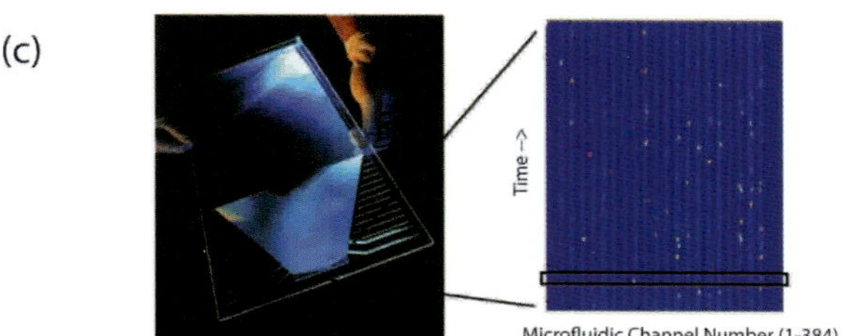

Plate 1   (Figure 3.1 on page 55 of this volume).

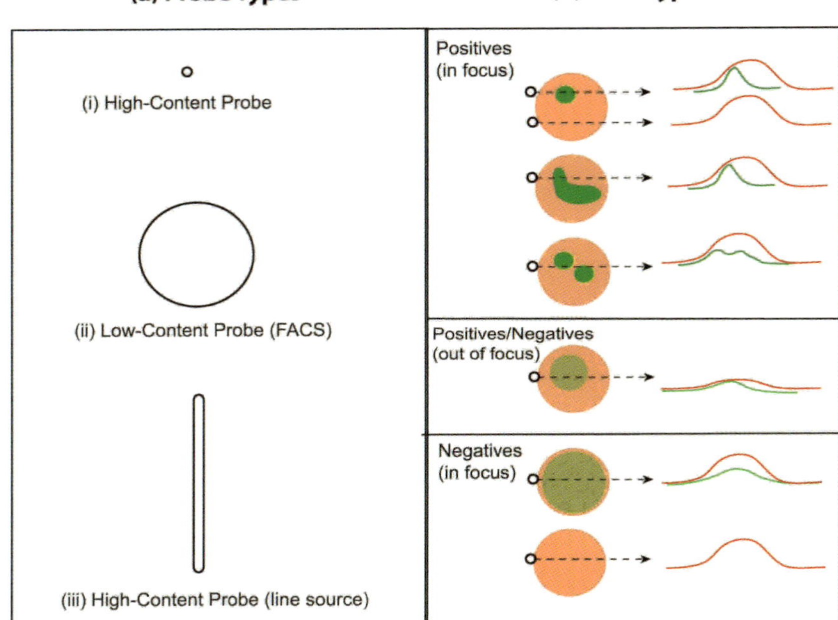

Plate 2   (Figure 3.8 on page 67 of this volume).

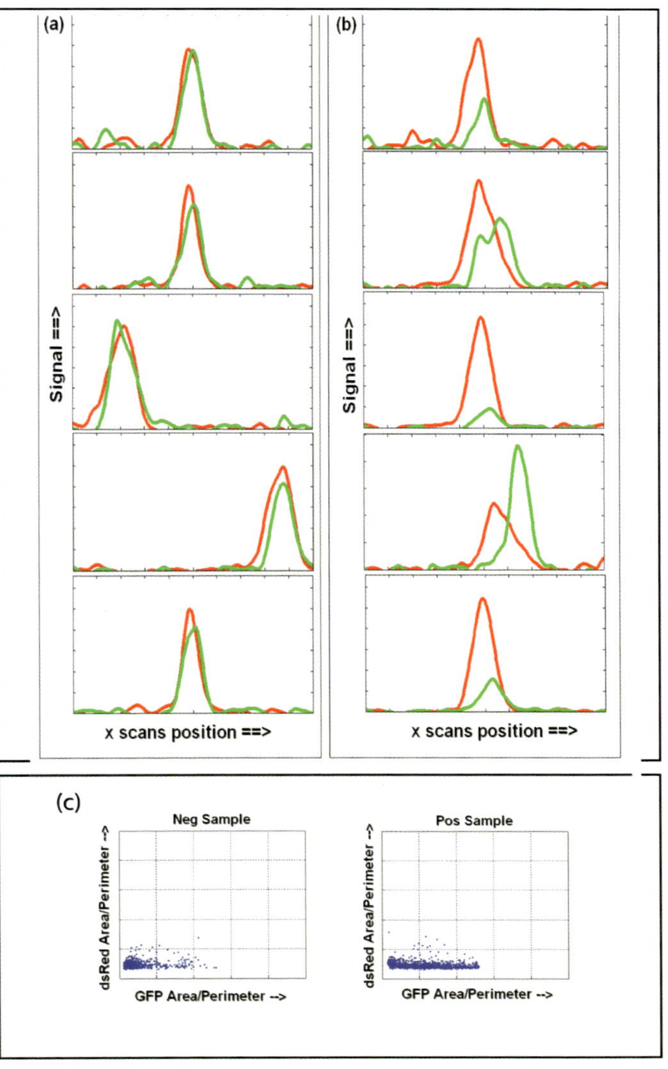

Plate 3   (Figure 3.9 on page 69 of this volume).

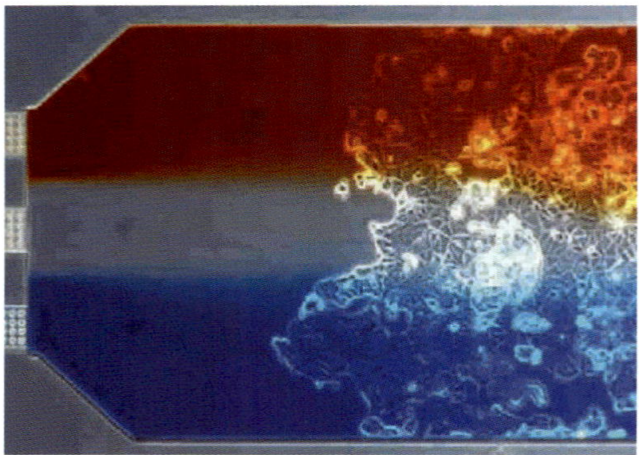

Plate 4   (Figure 4.5 on page 82 of this volume).

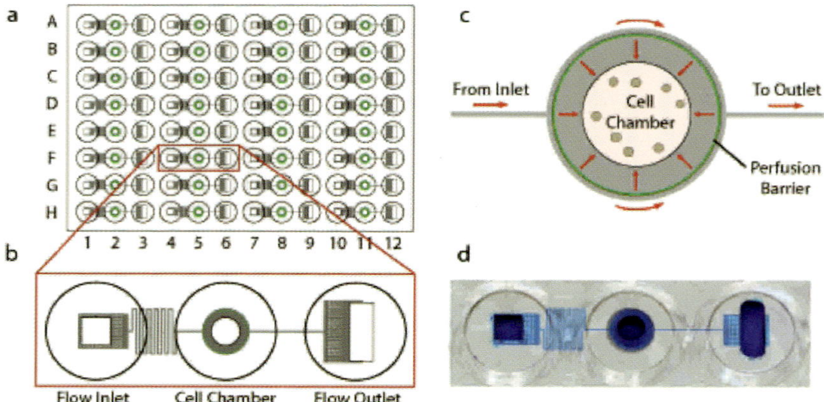

Plate 5   (Figure 4.20 on page 100 of this volume).

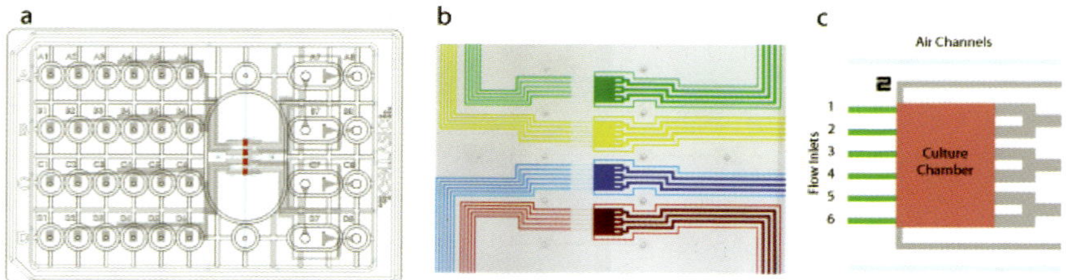

Plate 6   (Figure 4.21 on page 100 of this volume).

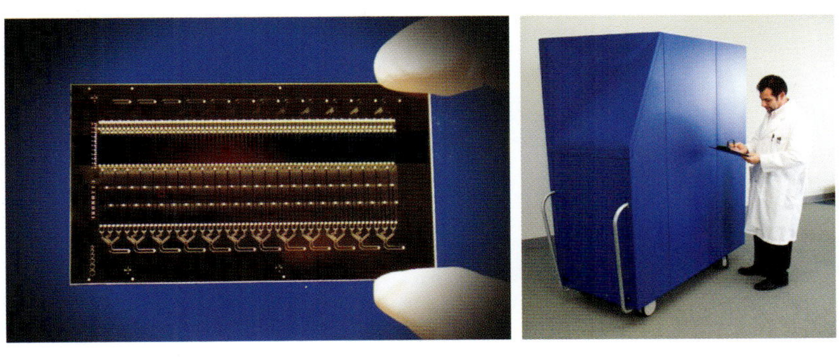

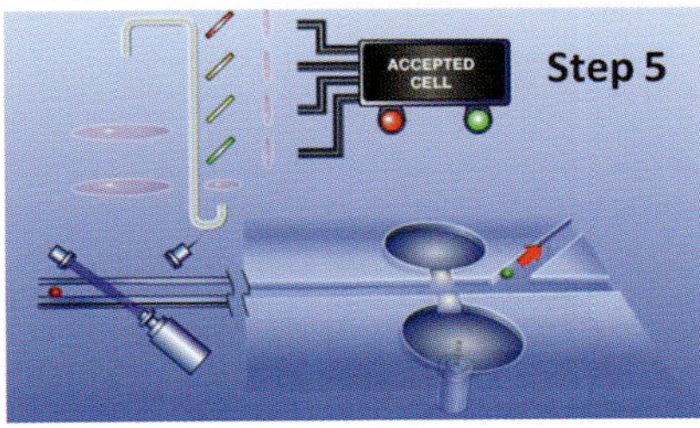

Plate 7   (Figure 5.3 on page 114 of this volume).

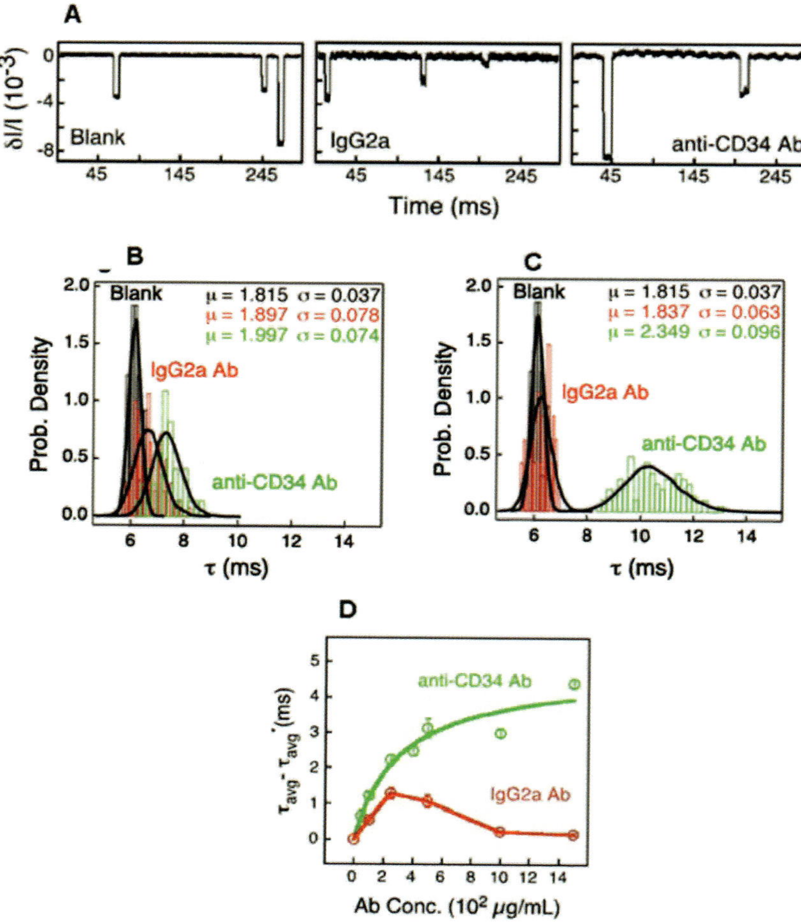

Plate 8   (Figure 6.17 on page 148 of this volume).

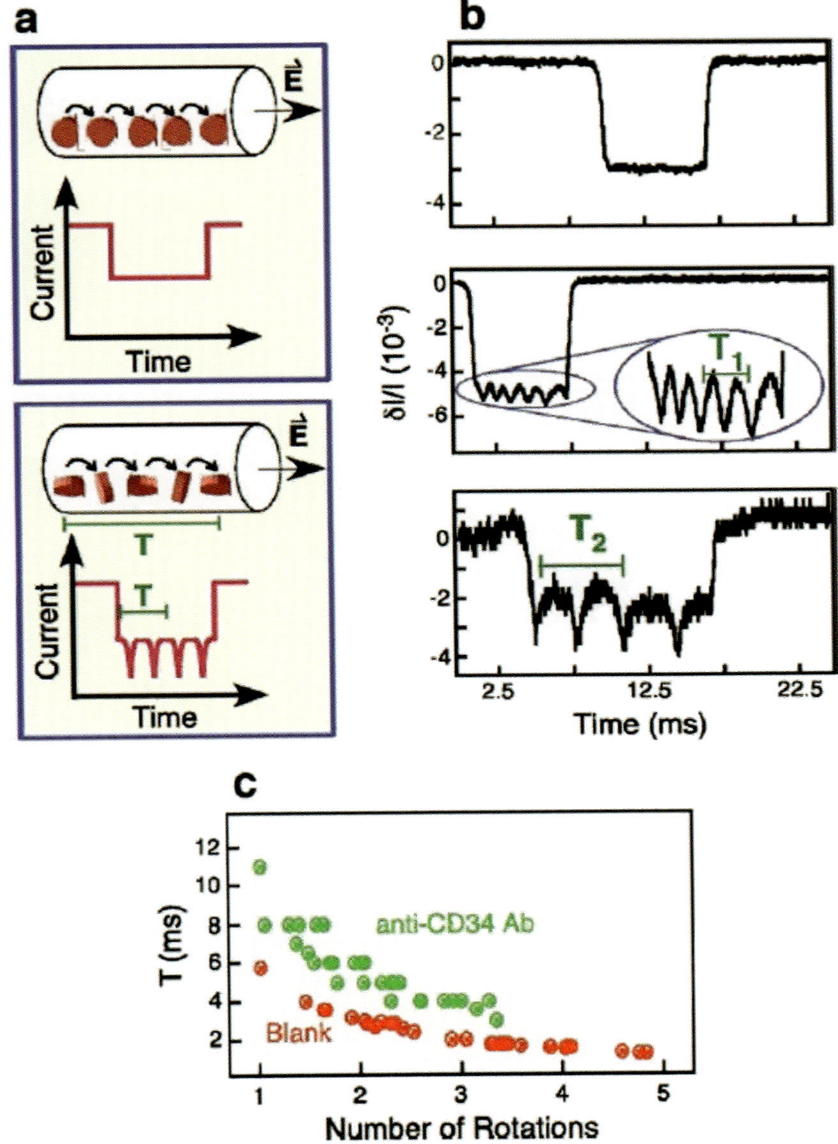

Plate 9   (Figure 6.20 on page 152 of this volume).

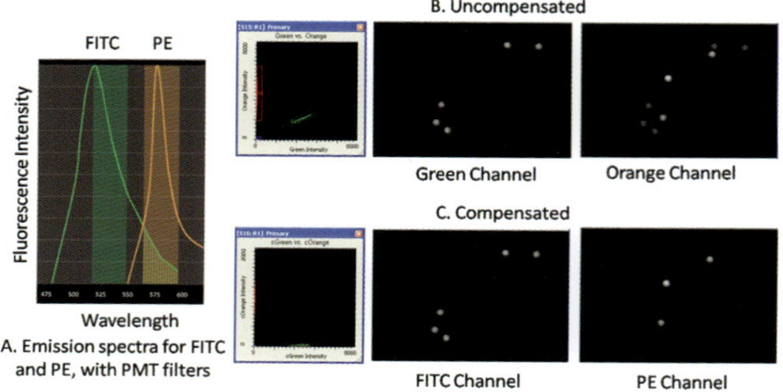

Plate 10   (Figure 7.9 on page 177 of this volume).

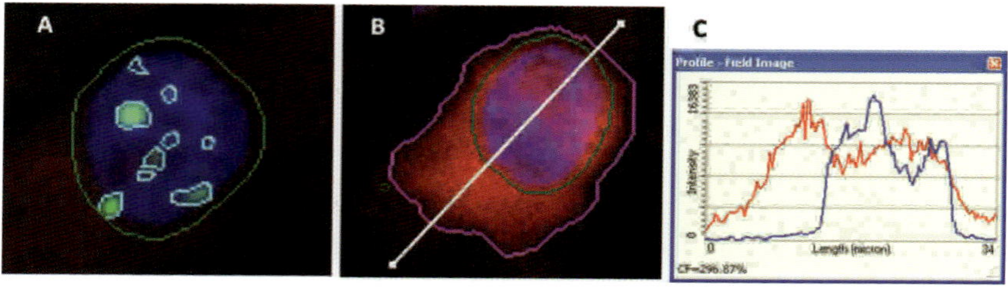

Plate 11   (Figure 7.13 on page 182 of this volume).

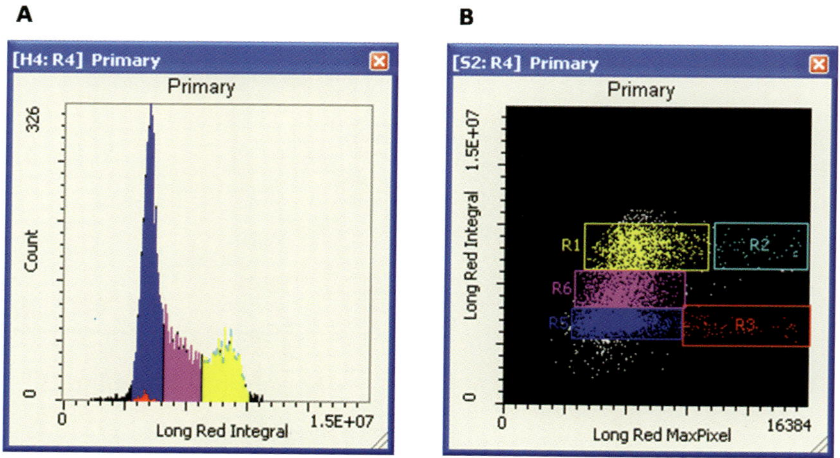

Plate 12   (Figure 7.15 on page 184 of this volume).

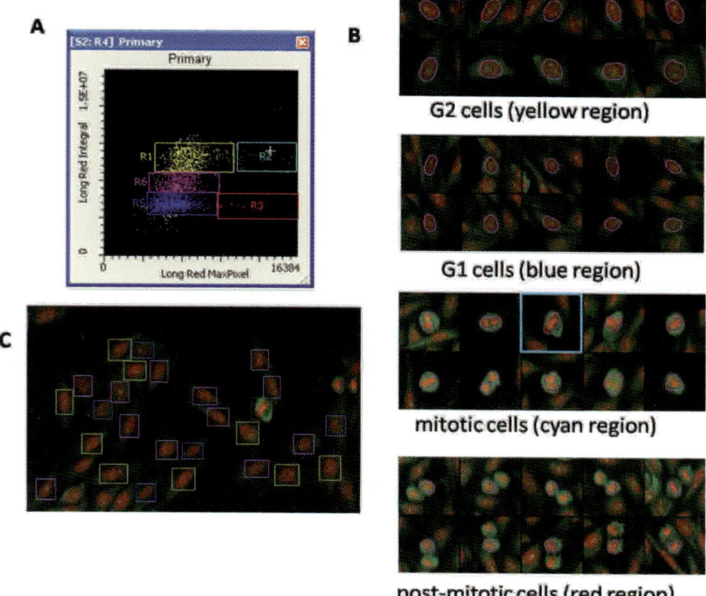

Plate 13   (Figure 7.16 on page 185 of this volume).

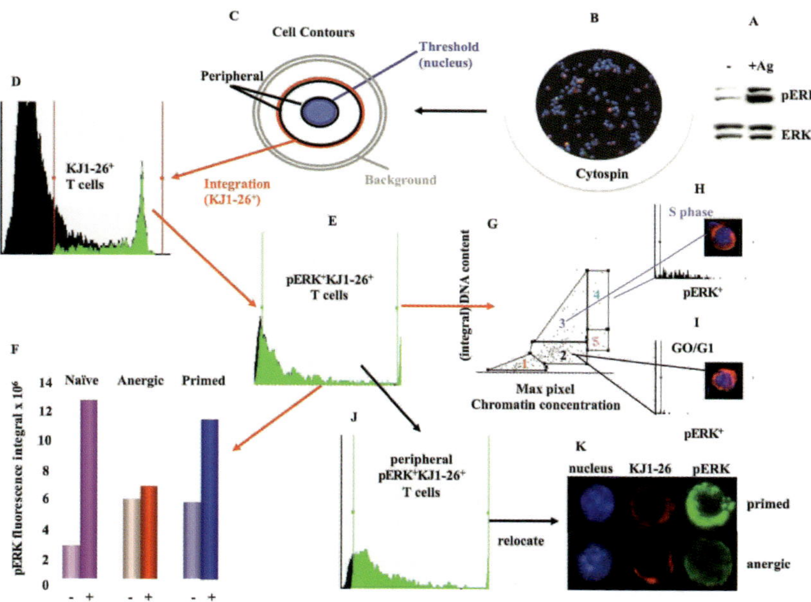

Plate 14   (Figure 9.1 on page 240 of this volume).

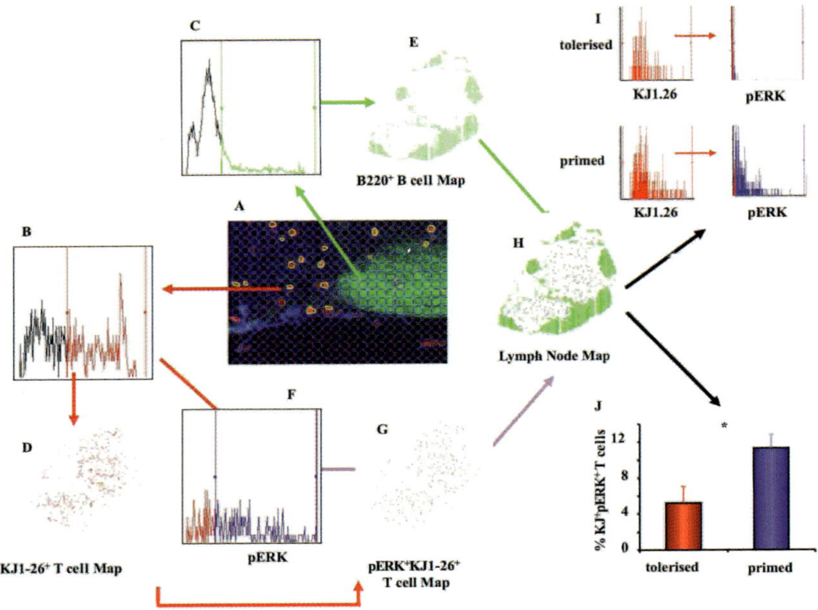

Plate 15   (Figure 9.2 on page 243 of this volume).

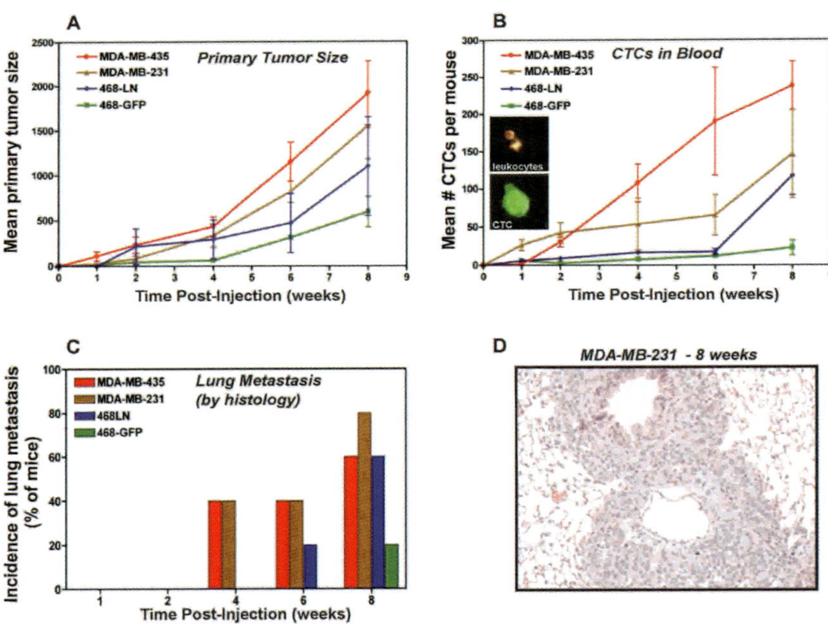

Plate 16   (Figure 10.1 on page 264 of this volume).

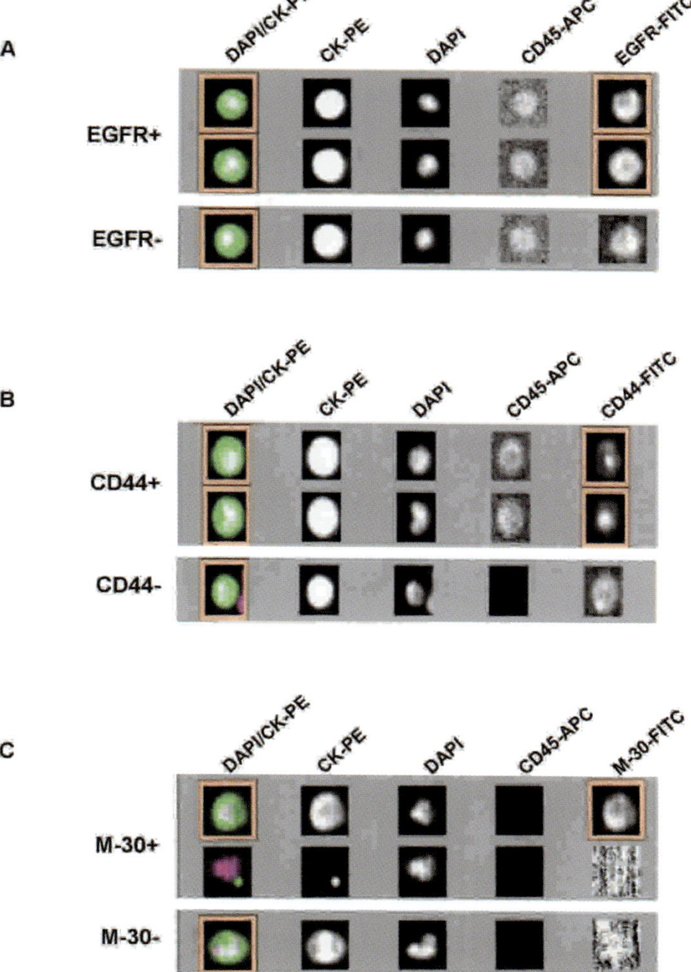

Plate 17 (Figure 10.4 on page 281 of this volume).

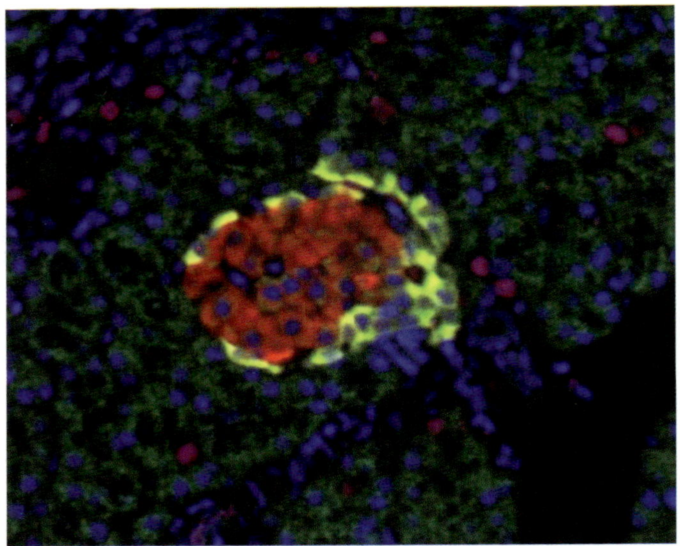

Plate 18   (Figure 11.1 on page 296 of this volume).

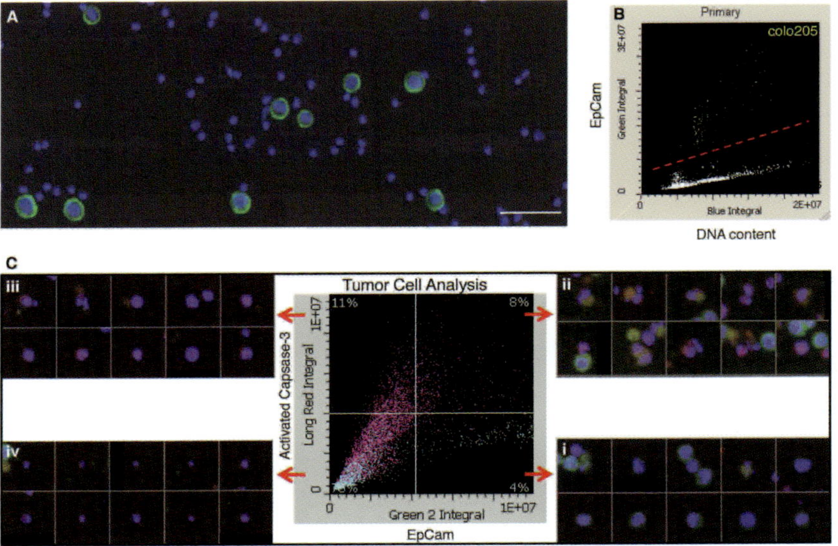

Plate 19   (Figure 12.3 on page 317 of this volume).

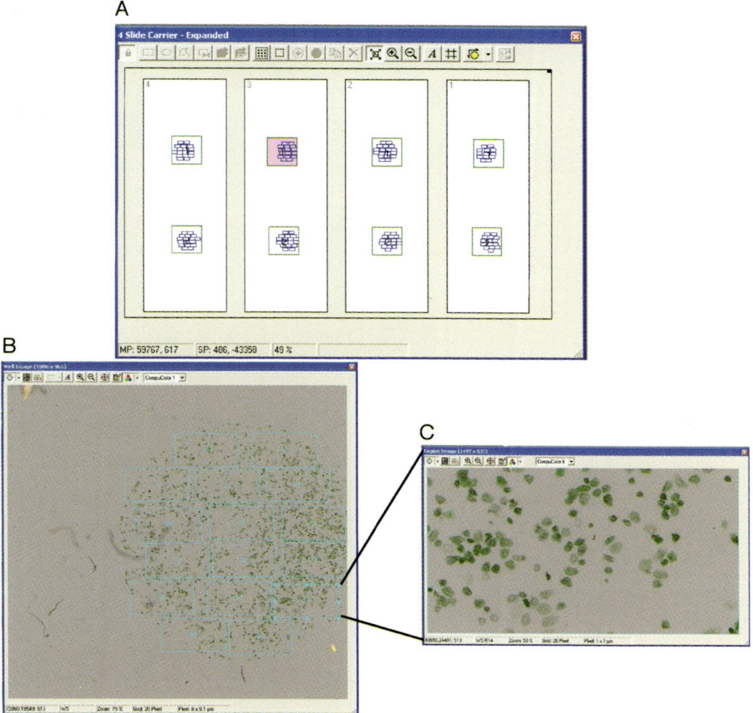

Plate 20   (Figure 13.2 on page 327 of this volume).

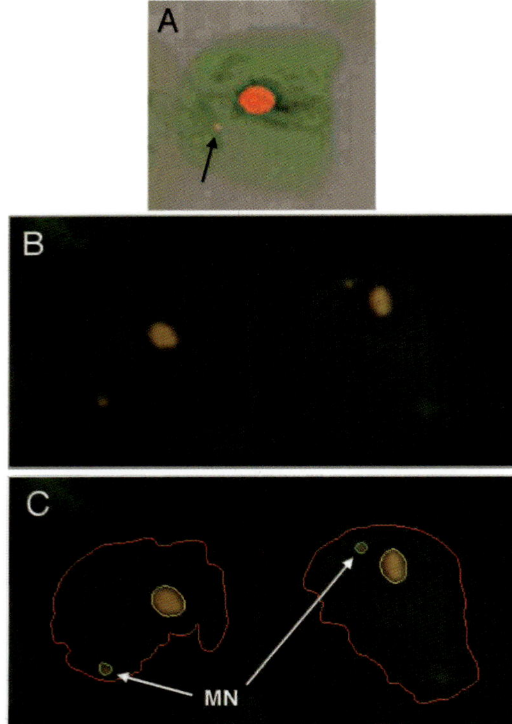

Plate 21   (Figure 13.5 on page 330 of this volume).

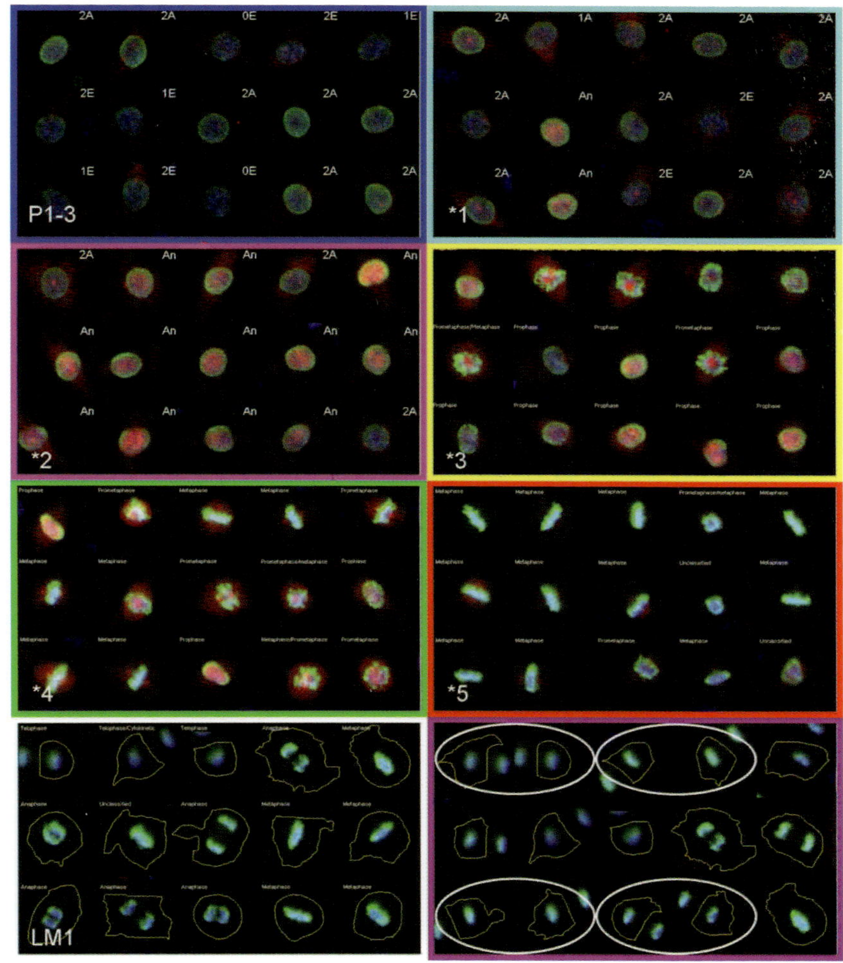

Plate 22   (Figure 14.11 on page 359 of this volume).

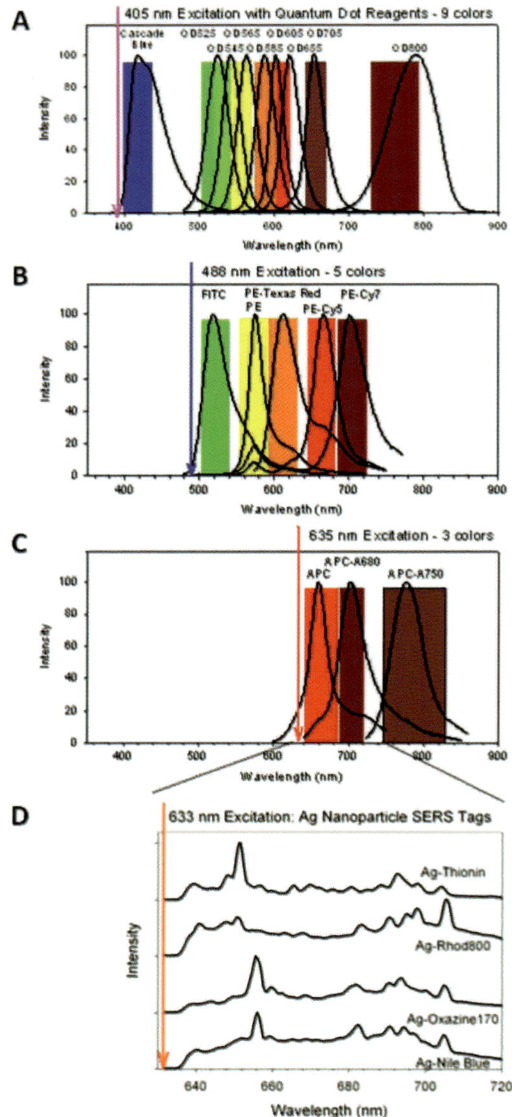

Plate 23   (Figure 20.1 on page 517 of this volume).